ENCYCLOPEDIA OF PHYSICS

EDITOR IN CHIEF
S. FLÜGGE

VOLUME XLIX/7

GEOPHYSICS III

PART VII

BY

G. SCHMIDTKE · K. SUCHY · K. RAWER

EDITOR
K. RAWER

WITH 298 FIGURES

SPRINGER-VERLAG
BERLIN HEIDELBERG NEW YORK TOKYO
1984

HANDBUCH DER PHYSIK

HERAUSGEGEBEN VON

S. FLÜGGE

BAND XLIX/7

GEOPHYSIK III

TEIL VII

VON

G. SCHMIDTKE · K. SUCHY · K. RAWER

BANDHERAUSGEBER

K. RAWER

MIT 298 FIGUREN

SPRINGER-VERLAG
BERLIN HEIDELBERG NEW YORK TOKYO
1984

Professor Dr. SIEGFRIED FLÜGGE
Physikalisches Institut der Universität, D-7800 Freiburg i. Br.

Professor Dr. KARL RAWER
D-7801 March-Hugstetten
Universität Freiburg, D-7800 Freiburg i. Br.

ISBN 3-540-11425-4 Springer-Verlag Berlin Heidelberg New York Tokyo
ISBN 0-387-11425-4 Springer-Verlag New York Heidelberg Berlin Tokyo

Library of Congress Catalog Card Number A 56-2942.
Printed in Germany.

Satz, Druck und Bindearbeiten: Universitätsdruckerei H. Stürtz AG, 8700 Würzburg.
2153/3130-543210

Preface

This Encyclopedia aims, basically, at summarizing the wealth of well-established facts and outlining the relevant theories in the different branches of physics. With this as goal, the writers were asked to present their specific field in such a way that access is possible to any scientist without special *a priori* information in that field; the basic concepts of physics are assumed to be known to the reader. The survey given in each paper was also to be long-lasting, so that even a few years after publication, each volume would be useful, for example as an introduction for newcomers or as a source of information for workers in a neighbouring field.

In the field of geophysics, dealt with in Vols. 47–49 of the Encyclopedia, this task is difficult to achieve because during the last decades there has been a much faster development of basic information and theory than during the decades before. When I came to contribute to this work the famous Julius Bartels, then editor of the geophysical part, told me that Vol. 49 should certainly take into account the results of the "International Geophysical Year" 1957/58 (I.G.Y.), and that we had better wait until these were accessible than produce a kind of information which might be obsolete in a short time. In fact, this international common project has enormously contributed to the basic information in our field, producing data at a rate that otherwise would have needed two decades to gather. Afterwards, this international effort was followed by other projects almost comparable in size, in particular the "Years of the Quiet Sun" 1963/64 (I.Q.S.Y.).

Since then, two events might be noted which have had an unexpected and large impact on the development of geophysics as whole. One of these started with the first SPUTNIK in 1957. Since that date, space research has become a rapidly growing branch of science with particular impact on astronomy, geodesy, meteorology and, of course, geophysics. *In-situ* measurements as well as remote sensing by satellite techniques have produced an enormous thesaurus of data which could not be directly measured before. This is particularly so in the physics of the upper atmosphere, the discussion of which is to be closed with the present volume.

To my feeling, another technical development which also happened during the last two decades has had a large impact on geophysical theory. The growing capacity of computers has lead to a new type of theoretical work. Before that advent, theoretical workers could not de better than simplifying the given problem to such an extent that solutions could be achieved with classical methods. This was a very sensible restriction but it sometimes ended up with misleading results. On the other hand, it was rather easy at that time to summarize theoretical work so that the reader was able to follow it step by step. This was drastically changed with the advent of powerful computers. Nowadays, a theoretical worker finds it quite difficult to describe each detail of his computer program – and periodicals are not eager to publish such programs. So, the reader only gets a rough summary and the numerical results. In contrast to earlier days, he is now unable to check the computation which remains the sole responsibility of the author.

All this implies that it is difficult to reproduce facts and theories in a condensed form. In my own paper, which concludes this volume, I have tried to do so using a large number of figures by which often more information can be transferred than by the same amount of text. I hope that this approach will be accepted well, though I am aware of possible critics.

There are two more papers in this volume dealing with subjects which, to my feeling, needed consideration. K. SUCHY's paper on collisions and transport in aeronomic plasmas could just as easily have appeared in another, non-geophysical volume. However, since it didn't, and since the results of this theory are of particular importance in aeronomy, it was felt that it should appear in this last volume on geophysics.

G. SCHMIDTKE's paper deals with a particular problem which came up after G. NIKOL'SKIJ's contribution to volume 49/6 had been written. Since about a century ago, and particularly after I.G.Y. and I.Q.S.Y., the solar-terrestrial relations have been found of increasing interest. In fact, our Earth is a planet in the solar system, so that geophysics cannot be without a solar-terrestrial link. For the upper atmosphere, the Sun with its wave and particle radiation is the main energy source and the variations of its radiation deeply influence the structure of the whole atmosphere. Therefore, adequate measures of "solar activity" are needed. Only recently could reliable measurements of the energetic radiations of solar origin be made using rockets and satellites. With this in mind, one may now ask whether the classical indices of solar activity might not be replaced by measured intensities of those radiations which do not penetrate the atmosphere, but which are absorbed in it and so are the true cause of the structures observed in the upper atmosphere. Though the discussion is yet under way, this is certainly an important question to be answered definitively in the coming decades.

KARL RAWER

Contents

Modelling of the Solar Extreme Ultraviolet Irradiance for Aeronomic Applications

By

G. Schmidtke

With 34 Figures

I. Temporal variations of the solar irradiance

1. Introduction

α) The "solar constant" S, which is the total solar irradiance over all wavelengths at the top of the Earth's atmosphere (corrected to mean Sun-Earth distance) is not really constant. Since about a century some variability was felt to exist because of the observed climatological changes with the *solar cycle* of eleven years. Long series of high altitude (mountain top) measurements were performed to define the range of possible variation. When reevaluating[1] the data collected from about 1908 through 1952 the variability of the "solar constant" S was found to ly below about $\pm 1\%$. The range of variability could not be identified more accurately in view of missing knowledge about the variability of radiation absorbing atmospheric constituents. Also it is quite difficult to hold the instrumental response to this radiation exactly constant on an almost secular scale.

A more reliable determination of S has been strongly promoted by the modern technological developments of experimental platforms above the atmosphere (balloons, aircraft, rockets and satellites); this measure reduces or avoids atmospheric absorption. The data resulting from such and ground-based measurements apparently did not indicate a variability of S with the solar cycle[2]; $S = 1{,}367 \pm 4\ \mathrm{Wm}^{-2}$ was the most probable value found. On the other hand, from 1966 through 1980 a long-term steady decrease by 0.03% per year seems to be possible to derive from balloon and satellite observations; this apparent change was accompanied by a simultaneous decrease in the global mean temperature.

β) On a *medium time scale* of about one month the total solar electromagnetic radiation is exhibiting an unexpectedly high degree of variability:

---

[1] Aldrich, L.B., Hoover, W.H. (1954): Ann. Ap. Obs. Smithsonian Inst. *7*, 165

[2] Fröhlich, C., Brusa, R.W. (1981): Solar Phys. *74*, 209

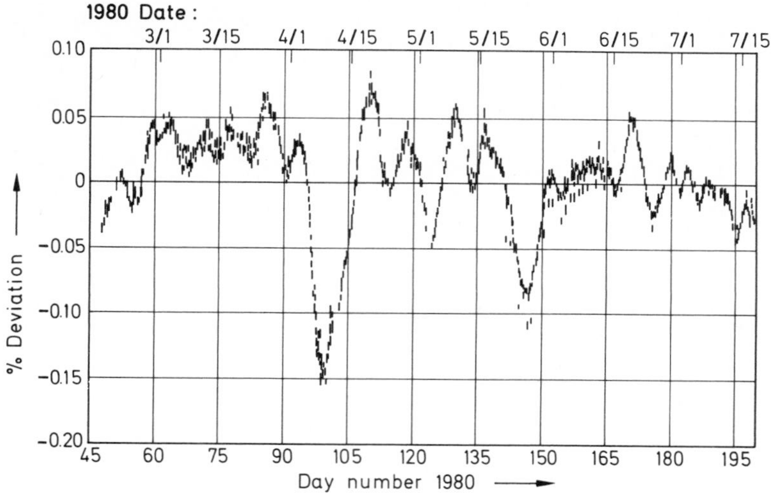

Fig. 1. Solar total irradiance in 1980 (WILLSON, 1981)[3]. Solar Maximum Mission/Active Cavity Radiometer results: traces indicate mean over one satellite orbit and standard deviation. The mean irradiance over the whole period was 1,368.31 W m^{-2}. The ordinates are deviations from this value in %

Decreases up to 0.2% occuring with seven to ten day quasi-periods (see Fig. 1) are reported[3]; these are highly correlated with the development of specific sunspot groups. Satellite data show significant fluctuations even from one orbit to the next one (one orbit$\simeq$96 min). The first high precision space-craft measurements were carried out in 1980 aboard the "Solar Maximum Mission". The "Active Cavity Radiometer Irradiance Monitor" experiment achieved a much better performance than any measurement before: the statistical significance of the observed variations is as high as 99.999%.

γ) In the extreme ultraviolet (EUV) spectral region the solar emissions vary by far more. Of course, the total amount of EUV energy flux is by about five orders of magnitude smaller than *S*, but spectral observations in the EUV range (from about 0.05 to 200 nm) are of special interest for a) solar physics, b) interplanetary physics, c) cometary physics, d) physics of the planetary moons and e) terrestrial and planetary aeronomy. As the physical background of the investigations in this context changes from subject to subject, the observational requirements for the solar radiation measurements strongly differ between these heterogeneous intentions. One requires:

to a) very high spatial and spectral resolutions, moderate temporal resolution and medium absolute accuracy with respect to the radiation flux.

to b) very high spectral resolution of the resonance emissions of hydrogen, helium, oxygen... from the *total* solar *disc* (i.e. without spatial resolution), low temporal resolution and moderate accuracy for the absolute flux.

[3] Willson, R.C. (1981): Solar Phys. *74*, 217

to c) moderate spectral resolution, no spatial resolution, low temporal resolution and moderate accuracy for the absolute flux.

to d) moderate spectral resolution, moderate temporal resolution, moderate accuracy for the absolute flux and no spatial resolution.

to e) moderate to high spectral resolution, high precision and high accuracy for the absolute flux, moderate temporal resolution (except for the investigation of solar flare-associated ionospheric phenomena), daily measurements and no spatial resolution.

These requirements certainly are to be modified for special investigations.

The observations of solar EUV fluxes as needed in the context of solar physics is dealt with in NIKOL'SKIJ's contribution in volume 49/6 of this Encyclopedia. The subject will be reported here as far as it is important in terrestrial and planetary aeronomy (see item e above).

2. Solar EUV measurements of aeronomic interest

α) For aeronomy four *spectral ranges* are of primary interest:

(1) 180 to 125 nm: The solar radiation in the Schumann-Runge continuum produces atomic oxygen due to atmospheric absorption by molecular oxygen and is one of the heating sources in the lower thermosphere.

(2) 121.6 nm: The Lyman-alpha emission of solar hydrogen ionizes nitric oxide in the ionospheric D-region.

(3) 103 to 16 nm: Absorption in this range constitutes the ionization source for the ionospheric E- and F-regions and is the most important heating source for the thermosphere.

(4) 2 to 0.1 nm: This penetrating radiation is important as ionization source for the ionospheric D-region.

These spectral ranges will be reviewed in the following sections. With all of them in determining absolute radiation fluxes one has met typical experimental difficulties, which are mostly due to the particular conditions of measurements in space.

(i) The space instrumentation had to be calibrated in the laboratory; true in-flight calibration is not yet possible, because neither a calibration source nor a stable calibration detector exists for flight applications. In general, the accuracy of laboratory calibrations ranges from 10 to 30%, depending on the wavelength region and the calibration method applied.

(ii) Before launch, rocket and satellite instrumentations need a very long time for qualification, testing and integration into the payload. During this period no thorough recalibration can be performed. In some cases experimental sets were calibrated years before launch (e.g. $2\frac{1}{2}$ years for the Skylab Photoelectric Spectroheliometer[1]). During this whole period the change of the instrumental response to radiation can not be monitored. For example, when turning-on the EUV spectrometer aboard satellite AEROS-A at the beginning of the spacecraft mission, a detector sensitivity degradation by a factor of two as compared to the laboratory calibration was determined by measurements[2].

[1] Reeves, E.M., Huber, M.C.E., Timothy, J.G. (1976): Appl. Opt. *15*, 1976

[2] Schmidtke, G., Knothe, M., Heidinger, F. (1975): Appl. Opt. *14*, 1645

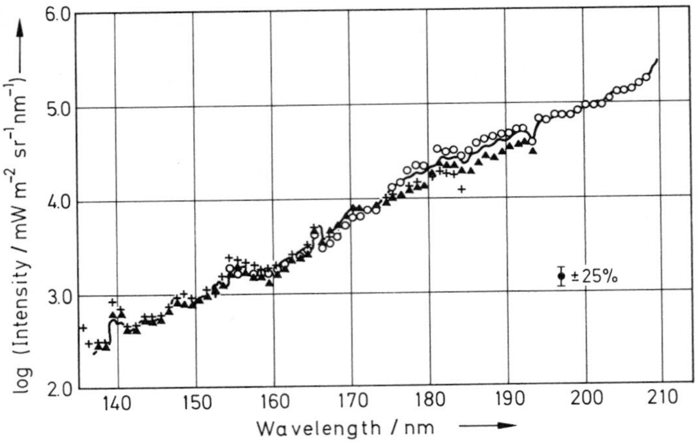

Fig. 2. Average solar radiation intensities, 140 to 210 nm. —— KJELDSETH-MOE et al.[8]; ○ SAMAIN and SIMON[9]; + ROTTMAN[10]; ▲ HEROUX and SWIRBALUS[11]

(iii) Another source of error is the change of sensitivity of the instruments during the mission. While for rocket measurements the problem is that time is often too short to reach stabilisation of the experimental response to the EUV radiation, the mission periods of satellites are too long to keep this response stable. In fact the known sensors are all individually variable under space conditions. For this reason recalibration of satellite EUV spectrometers is sometimes achieved by accompanying rocket programs. In complex payloads, with several independent spectrometer channels of partially overlapping wavelength ranges changes of the instrumental efficiency might be estimated, if the mission lasts long enough. Measurements of highest precision (on a relative scale) were performed during missions exceeding half a year for those wavelength ranges, for which the statistics of the data was adequate.

β) Spectral range from 180 to 125 nm. In this wavelength range (see also Sect. 20 of NIKOL'SKIJ's contribution in volume 49/6) measurements of solar radiation flux from the total disc have been obtained from film-records[3,4], ionization chamber[5] and photoelectric detection[6,7] (see also Fig. 17 in Sect. 5 of THOMAS's contribution, also in volume 49/6, which summarizes data of before 1970). Accuracies of ±15% to ±20% are quoted. In view of these experimental uncertainties and of the more or less sporadic nature of the

[3] Detwiler, C.R., Garret, J.D., Purcell, J.D., Tousey, R. (1961): Ann. Geophys. *17*, 9

[4] Samain, D., Simon, P.C. (1976): Solar Phys. *49*, 33

[5] Carver, J.M., Horton, B.H., Lockey, G.W.A., Rofe, B. (1972): Solar Phys. *27*, 347

[6] Parkinson, W.H., Reeves, E.M. (1969): Solar Phys. *10*, 342

[7] Hinteregger, H.E. (1976): J. Atmos. Terr. Phys. *38*, 791

[8] Kjeldseth-Moe, O., Van Hoosier, M.E., Bartoe, J.-D.F., Brückner, G.E. (1976): Naval Research Laboratory Report Nr. 8057

[9] Samain, D., Simon, P.C. (1976): Solar Phys. *49*, 33

[10] Rottman, C.J. (1974): Trans. Am. Geophys. Union *56*, 1157

[11] Heroux, L., Swirbalus, R.A. (1976): J. Geophys. Res. *81*, 436

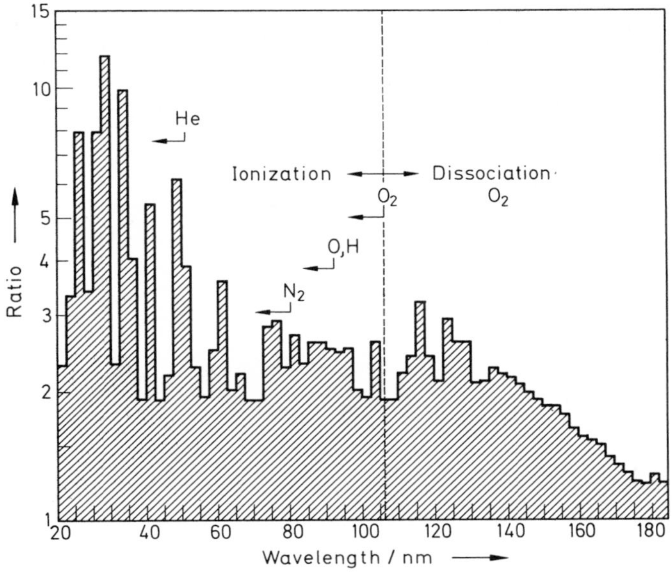

Fig. 3. Overview of results of EUVS experiment on the satellite AE-E[13]. Ordinates give the rate of change (as a ratio) from July 1976 (solar activity minimum) to January 1979 (close to maximum)

measurements, no figures on the solar flux variability could be derived from the results of rocket launches.

The spectra shown in Fig. 2[12] were selected under the condition of a most reliable instrumental calibration taking into account center to limb variation effects as well as the individual spectral resolution. They were aquired from rocket flights during the period from end of 1972 through April 1974. Therefore, average values of this data set represent the solar EUV flux in the Schumann-Runge spectral range for the declining solar cycle 20, roughly 4 to 6 years after its maximum. It would be too speculative to estimate from these measurements the variability of fluxes during the full solar cycle 20, just because measurements prior to that period were too uncertain; in fact at that time the scientific instrumentation was not yet ready to record spectra with an absolute accuracy of $\pm 25\%$.

The variability on a relative scale was monitored by HINTEREGGER[13] from solar minimum to maximum activity conditions during solar cycle 21 (see Fig. 3). Up to now this measurement is the only one that produced reliable data for the Schumann-Runge spectral region from minimum to maximum of a solar cycle recorded with adequate spectral and temporal resolutions.

Proceeding to shorter wavelengths one finds that more strong solar emission lines are superposed to the continuous spectrum. While this latter originates in the lower solar atmospheric layers and shows smaller variability, the emissions from the higher levels of the solar atmosphere are more variable and

[12] Brückner, G.E. (1981): Adv. Space Res. *1*

[13] Hinteregger, H.E. (1981): Adv. Space Res. *1*, 39

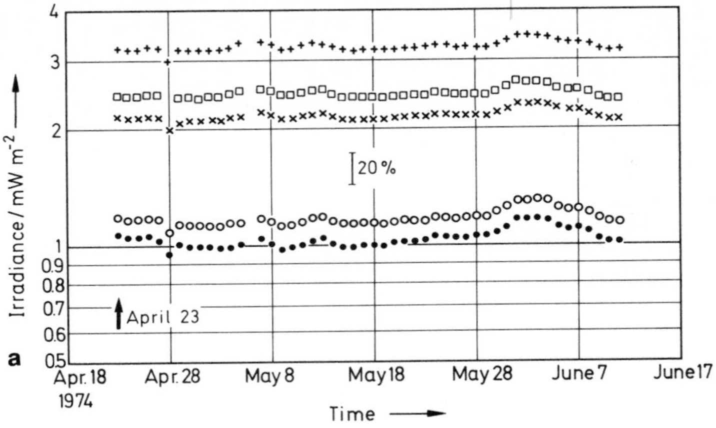

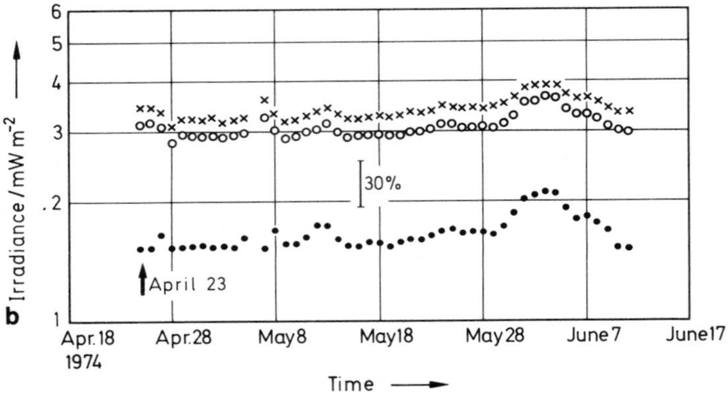

Fig. 4a. Solar irradiation flux variations from 23 April 1974 to 11 June 1974 for the wavelength intervals 158.5 to 165.5 nm (●), 166.1 to 169.7 nm (○), 169.1 to 173.1 nm (×), 172.2 to 176.1 nm (□) and 175.2 to 179.1 nm (+) after HINTEREGGER et al.[14]

Fig. 4b. Solar irradiation flux variations from 23 April 1974 to 11 June 1974 for the wavelength intervals 140.2 to 145.0 nm (●), 147.9 to 153.5 nm (○) and 155.5 to 158.0 nm (×) after HINTEREGGER et al.[14]

occur on shorter wavelengths. In Figs. 4a, b medium-term variations mainly due to the solar rotation of about 27d dominate the spectral features; again the amplitudes of the variation increase with decreasing wavelength. Apart from these variations of a 27d period there are others with timescales of days due to the birth and decay of active regions on the solar disc, and even with periods of minutes due to flares. The spectral signatures of these solar events are largely different from one to the other. They shall not be described here because their aeronomic impact is not important. For example, for a strong (3B) solar flare the increase[15] in the solar flux was less than 1% at wave-

[14] Hinteregger, H.E., Bedo, D.E., Manson, J.E. (1977): Space Research *XVII*, 533
[15] Heath, D.F. (1969): J. Atmos. Sci. *26*, 1157

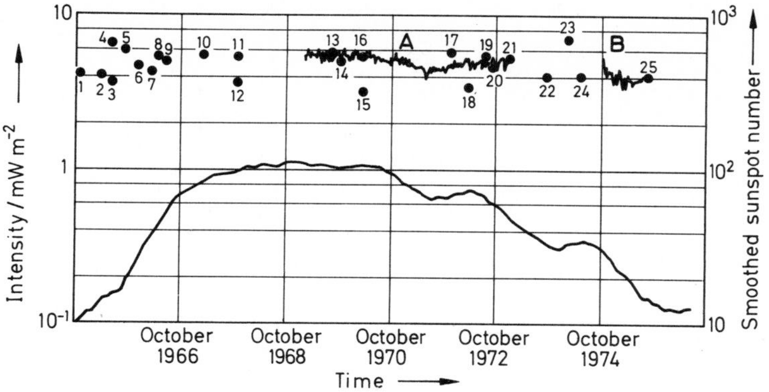

Fig. 5. Comparison of Lyman-alpha (121.6 nm) solar irradiation flux measurements during solar cycle 20 (see original publication for references 1 ... 25)[17]. Upper part: irradiation fluxes (left-hand ordinate); lower part (curve): Smoothed Zürich sunspot number (right-hand ordinate)

lengths longer than 140 nm. Corresponding changes of the dissociation rate of molecular oxygen and of lower thermospheric heating could not be measured, yet.

In summary, in the wavelength range from 180 to 125 nm the investigation of the variability of the solar EUV radiation is to be considered at an early stage, still.

References to publications giving more details on this subject are found in[7, 12, 13, 15–17] and in NIKOL'SKIJ's contribution in volume 49/6 of this Encyclopedia.

γ) Solar emission at 121.6 nm. The Lyman-alpha line of hydrogen at 121.6 nm is a very bright chromospheric emission with a flux intensity of the same order of magnitude as the integrated intensities from all solar emissions below 120 nm. Beside of its important role in solar physics, in the interplanetary medium, in the upper planetary and terrestrial atmospheres etc., it is controlling the (quiet) D-region ionization.

This spectral line can easily be generated in the laboratory. It can be detected by ionization chambers, photodiodes, Schumann-Runge film and by open and closed secondary electron multipliers. For these reasons reliable measurements exist since many years so that for this emission our knowledge is by far more advanced than for any other spectral range reviewed here[17, 18].

Before solar cycle 20 the Lyman-alpha measurements were at an exploratory state with less emphasis on high absolute accuracy with respect to the photometric calibration. Strong technological progress was made in this field from about 1960 on. The measurements performed during solar cycle 20 with rockets (circles) and satellites (curves) are compiled in Fig. 5[17]. Though the absolute accuracy is quoted with $\pm 30\%$[18] in general, it scarcely exceeds the

[16] Simon, P.C. (1978): Planet. Space Sci. *26*, 355

[17] Delaboudinière, J.P., Donnelly, R.F., Hinteregger, H.E., Schmidtke, G., Simon, P.C. (1978): COSPAR Technique Manual No. 7

[18] Vidal-Madjar, A. (1977): The solar output and its variation. Colorado. Boulder, CO: Associated University Press, p. 213

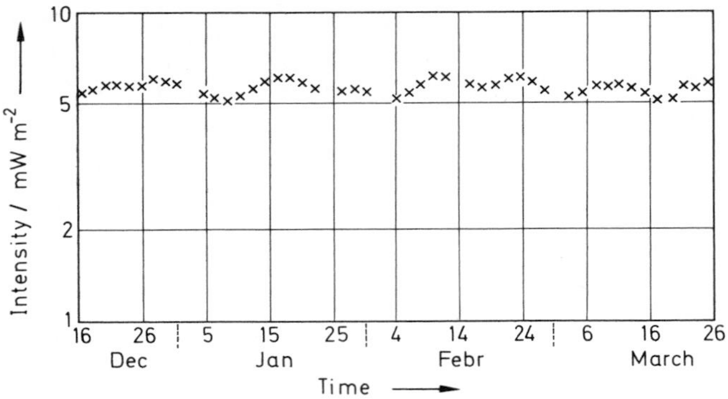

Fig. 6. Lyman-alpha solar irradiation flux variations from 16 Dec. 1969 to 26 March 1970 after VIDAL-MADJAR[18]

variability of this line during solar cycle 20; the decrease from solar maximum to minimum is about 40% (considered from solar minimum to maximum conditions this is an increase by 70%). By the way, the solar "EUV minimum"[19] i.e. the lowest emissivity of radiation in the 1 to 200 nm range appeared on 20 April 1975 – more than one year ahead of the classical solar minimum e.g. with respect to the sunspot number. Figure 3 shows a flux increase during solar cycle 21 at 121.6 nm exceeding 100% from July 1976 to January 1979. Since the level of EUV radiation of July 1976 is higher than the one at the "EUV minimum", and since in January 1979 the solar maximum was not yet passed, the total flux increase from solar maximum of cycle 21 should be about 200%. Thus at this spectral line solar cycles 20 and 21 showed a quite different change of activity. One must, however, bear in mind, that the data evaluation of these satellite observations are based on certain assumptions concerning the changes of the instrumental response[20]. These assumptions are primarily based on experience.

The variability of Lyman-alpha with solar rotation is apparently different in different periods (see Fig. 6) with amplitudes up to 30%[18]. Flare-associated enhancements range from 2% for a class 1 flare and 8% for a class 2 flare to 18% for a (3B) solar flare[21,22]. Various measurements performed by a great number of laboratories all over the world give us a good picture on the variability of this interesting solar emission.

δ) EUV range from 103 to 16 nm. The solar flux in this wavelength range is totally absorbed in the thermosphere, in the 103 to 91 nm range mainly by molecular oxygen, from 91 to 80 nm both by molecular and atomic oxygen and below 80 nm by molecular oxygen and nitrogen and atomic oxygen; resulting are ionization and heating of the thermosphere. Since the EUV irradiance in the whole range from 103 to 16 nm is largely variable with time,

---

[19] Hinteregger, H.E. (1977): Geophys. Res. Lett. *4*, 231
[20] Vidal-Madjar, A., Phissamay, B. (1980): Solar Phys. *66*, 259
[21] Hall, L.A. (1971): Solar Phys. *21*, 167
[22] Heath, D.F. (1973): ibid *78*, 2779

it represents one of the fundamental parameters for ionospheric and thermospheric physics.

Systematic rocket measurements for aeronomic purposes were first performed by HINTEREGGER[23]. From subsequent measurements a tabulation of solar EUV fluxes was presented[24]. Though the interpretation of the measurements on the OSO-1 satellite[25] showed some variability of the radiation, no numerical estimate of this latter could be presented at that time. For this reason the tabulation was used as an "EUV standard flux" for many years (see also Table 5 in Sect. 5 of THOMAS's contribution in volume 49/6 of this Encyclopedia).

With the operation of the EUV spectrometer aboard the OSO-3 satellite in 1967[26] and of the 30.4 nm monochromator aboard the OSO-4 satellite in 1967 and 1968[27] short-term (for periods of a few to ten min) and medium-term (periods of the order of a month) variations have been observed. The latter are mainly due to the solar rotation of 27d. On a relative intensity scale the short-term variations comprise up to about 25% of the total EUV but variations are different for the different emission lines and continua. A total flux increase of up to 40% has been observed resulting from the solar rotation[26]. Examples are presented in Fig. 7. Similar levels of variation for short events are reported from later missions[28] (see Fig. 8); unfortunately, the measurements performed covered only a very small fraction of the total time. The actual maximum variations that really occurred may be somewhat greater than shown in the figures.

Short-term and medium-term variations have been observed at different levels of solar activity, but the determination of long-term variations over years, e.g. a solar cycle is still difficult. Technological problems related with the calibration of the instruments and changes of the photometric efficiency have been encountered. With an uncertainty of the laboratory calibration of about 10 to 15% and more severe errors due to efficiency changes during instrumental spacecraft integration and launch as well as during the mission, the best actually achievable accuracy of absolute flux determination is probably about ±30%. An analysis of rocket flights from 1960 through 1968 ended-up with the result that the long-term variability of most of the solar EUV emissions lies within this experimental uncertainty, with the exception of coronal emission lines of Fe XV and Fe XVI[29]. Since these emissions do contribute only a small fraction to the total EUV energy in this wavelength range, no estimate could be presented for the total EUV variability during solar cycle 19.

This situation improved when better EUV-instrumentation was flown in a few satellite missions during solar cycle 20. Aboard the OSO-4 satellite the (He II Lyman-alpha) line at 30.4 nm could be recorded from October to De-

---

[23] Hinteregger, H.E. (1961): J. Geophys. Res. *66*, 2367
[24] Hinteregger, H.E., Hall, L.A., Schmidtke, G. (1965): Space Research *V*, 1175
[25] Neupert, W.M., Behring, W.E., Lindsay, W.C. (1964): Space Research *IV*, 719
[26] Hall, L.A., Hinteregger, H.E. (1970): J. Geophys. Res. *75*, 6959
[27] Timothy, A.F., Timothy, J.G. (1970): J. Geophys. Res. *75*, 6950
[28] Schmidtke, G. (1978): Planet. Space Sci. *26*, 347
[29] Hall, L.A., Higgins, J.E., Chagnon, C.W., Hinteregger, H.E. (1969): J. Geophys. Res. *74*, 4181

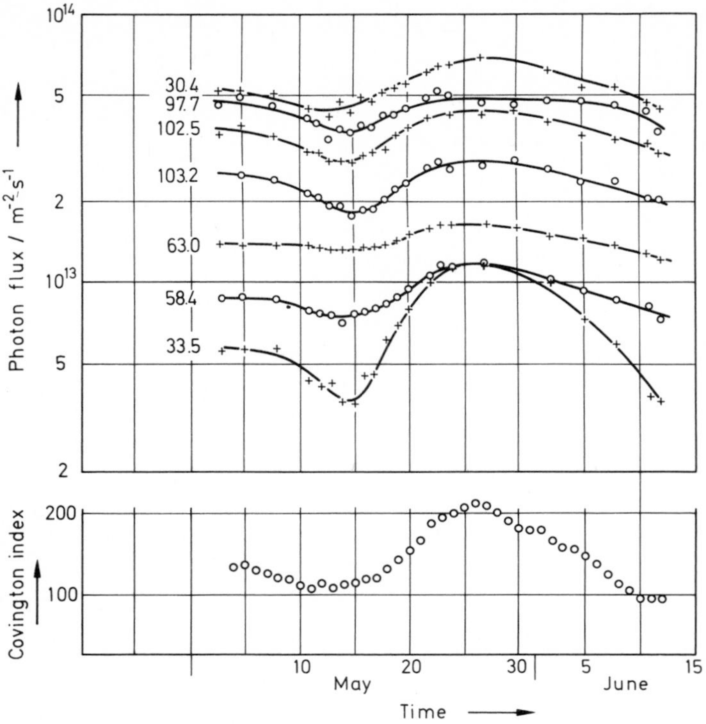

Fig. 7. Variability with solar rotation of a few solar emission lines (wavelength/nm = parameter of the curves) after HALL and HINTEREGGER[26] and Covington's (solar noise) index for comparison (May/June 1969)

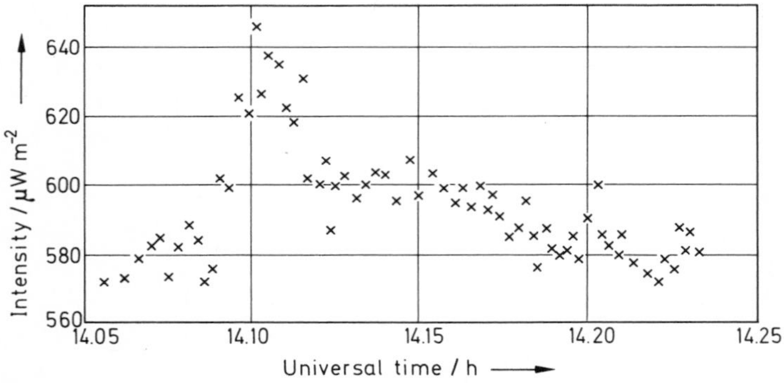

Fig. 8. Increased intensity of the 30.4 nm emission line (Lyman-alpha of He^+) during a (2B) solar flare (recorded on 15 June 1973 aboard AEROS-A by SCHMIDTKE[28])

cember 1967 and from June 1968 to December 1969. A change by 20% was derived[27]. The EUV spectrometers aboard the AEROS satellites[30,31] traced

---

[30] Schmidtke, G. (1976): Geophys. Res. Lett. *3*, 573

[31] Schmidtke, G., Rawer, K., Botzek, H., Norbert, D., Holzer, K. (1977): J. Geophys. Res. *82*, 2423

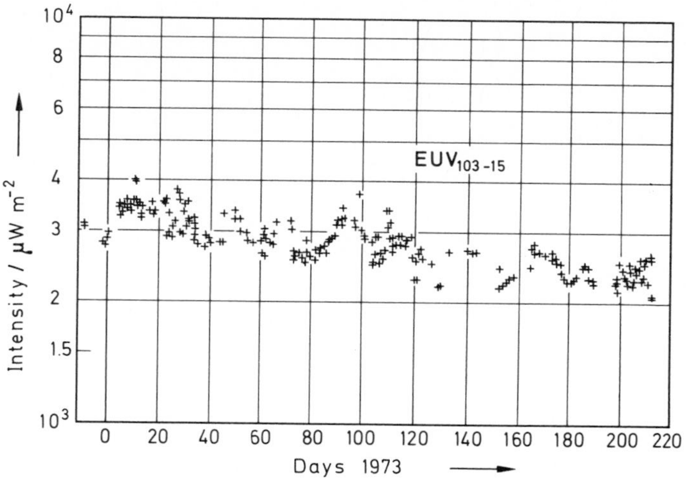

Fig. 9. Solar flux index EUV_{103-15}, i.e. the flux intensity in the main EUV range

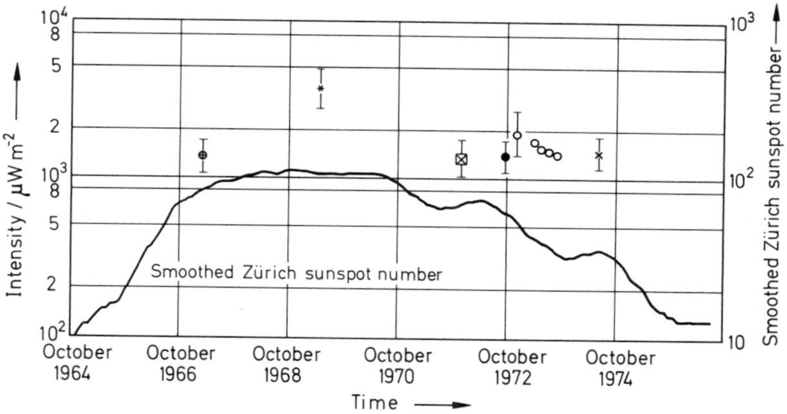

Fig. 10. Intercomparison of solar flux index EUV_{103-28} during the solar cycle 20[17], namely the flux intensity in this wavelength range (left-hand scale). For comparison smoothed Zürich sunspot numbers are shown (curve, right-hand scale). Symbols: ⊕ Hall and Hinteregger[26]; ● Heroux et al.[33]; ∗ Timothy[34]; ○ Schmidtke[30]; ⊠ Higgins[35]; × Hinteregger[7]

an average decrease of the *total* EUV flux in the 103 to 16 nm range by about 30% during the first 200 days of 1973, while the smoothed sunspot numbers decreased from 55 to 36 (see Fig. 9)[30]. This was the first set of measurements covering the complete wavelength range from 103 to 16 nm and a period of about eight months. In-flight calibration of the multipliers was applied for the first time. Usually they are the most sensitive parts of an EUV spectrometer. Very important changes of their efficiency to the detected radiation were found[32,2]. The mission time, however, is too short for these measurements to

[32] Schmidtke, G., Schweizer, W., Knothe, M. (1974): J. Geophys. *40*, 577
[33] Heroux, L., Cohen, M., Higgins, J.E. (1972): J. Geophys. Res. *79*, 5237
[34] Timothy, J.G. (1976): Center for Astrophysics Report Nr. 525
[35] Higgins, J.E. (1976): J. Geophys. Res. *81*, 1301

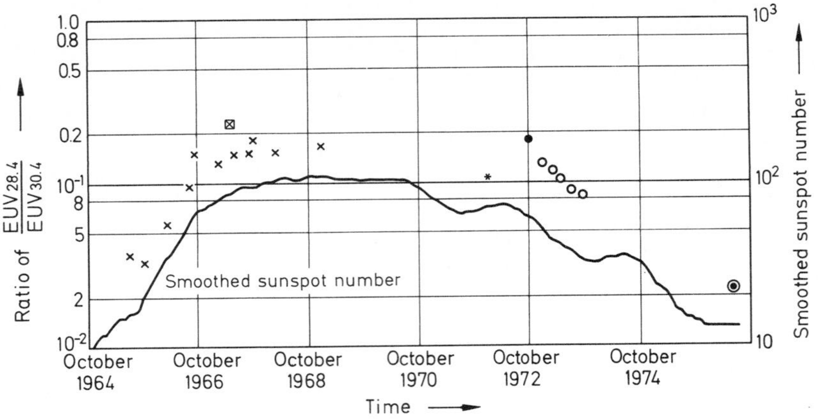

Fig. 11. Variation of the ratio of emissions from FeXV at 28.4 nm and from HeII at 30.4 nm (left-hand scale) during solar cycle 20 after SCHMIDTKE[28]. Symbols: × HALL et al.[29]; ● HEROUX et al.[33]; ⊠ HALL and HINTEREGGER[26]; ○ SCHMIDTKE et al.[31]; ∗ HIGGINS[35]; ⊙ HINTEREGGER[38]. Smoothed sunspot numbers (curve) for comparison (right-hand scale)

be conclusive for estimating the flux level at the preceded solar maximum (about five years earlier). Unfortunately, the changes during the whole solar cycle 20 can even not be derived quantitatively, if the data available from other experiments are taken into account, too (see Fig. 10)[17].

From ionospheric model calculations[36] using ground-based incoherent scatter measurements (see Sect. 5 of RAWER's contribution, p. 256) a solar EUV flux increase by about 100% from solar minimum to maximum conditions has been concluded; this is in contradiction to the (sporadic) rocket measurements[37] of solar cycle 20 indicating a 30% increase only for the total wavelength range from 103 to 16 nm. However, considering the measurements at 121.6 nm[19] and those aboard the satellite AEROS-A[30] during eight months, which is a short period as compared with the solar cycle, a higher variability than 30% during cycle 20 cannot be ruled out. Another indicator, namely the ratio of simultaneously measured fluxes of the emission lines of FeXV at 28.4 nm and of HeII at 30.4 nm (see Fig. 11)[28] has been found to characterize solar activity quite well; such data support the view that the variability over

→

Fig. 12a. Irradiance variations of H Ly-α, H Ly-β and HeI 58.4 nm after AE-E satellite observations during solar cycle 21 after HINTEREGGER[39] (monthly mean values in arbitrary units; absolute values apply only to $\langle F \rangle$ and $\langle R_Z + 100 \rangle$)

Fig. 12b. Irradiance variations of some solar coronal emissions after AE-E satellite observations during solar cycle 21 after HINTEREGGER[39] (monthly mean values in arbitrary units; absolute values apply only to $\langle F \rangle$ and $\langle R_Z + 100 \rangle$)

---

[36] Roble, R.G. (1976): J. Geophys. Res. *81*, 265

[37] Heroux, L., Higgins, J.E. (1977): J. Geophys. Res. *82*, 3307

[38] Hinteregger, H.E. (1977): Private communication

[39] Hinteregger, H.E. (1981): NATO AGARD Conference Proceedings, No. 295, I-1

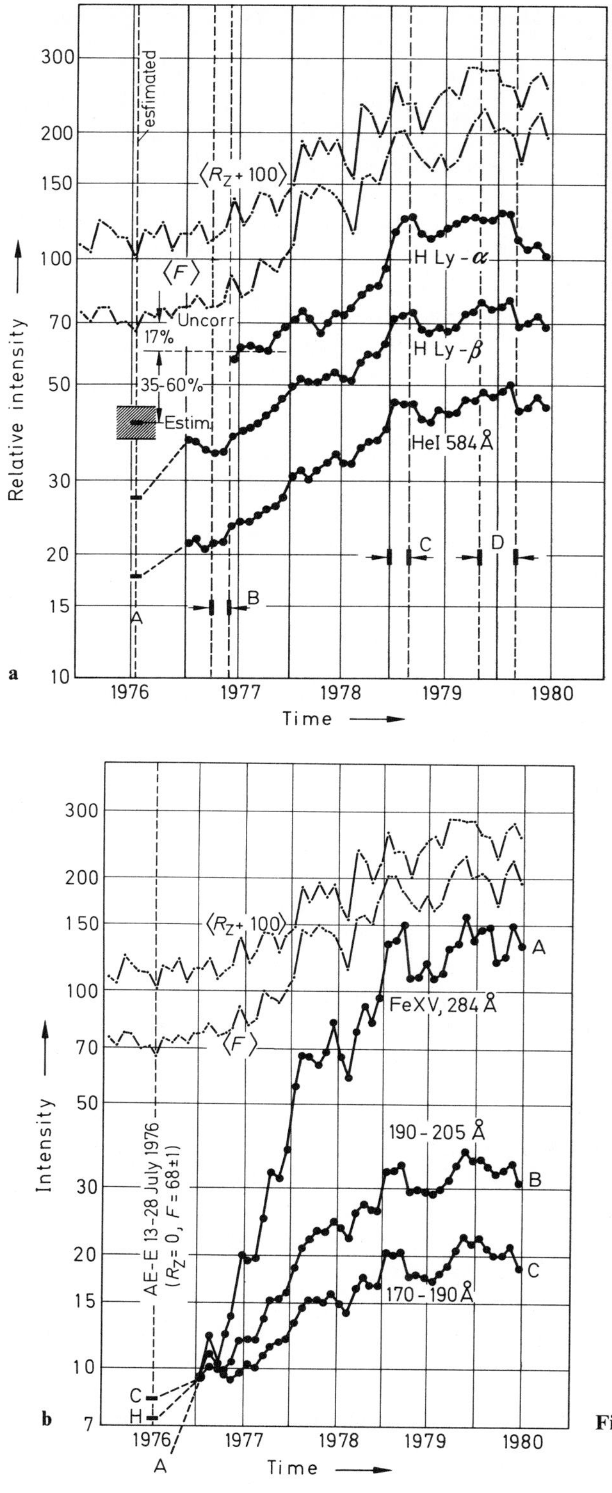

a
estimated
⟨R_Z+ 100⟩
H Ly - α
⟨F⟩
Uncorr
17%
35-60%
Estim
H Ly - β
HeI 584 Å
A
B
C
D
Relative intensity
Time
b
⟨R_Z+ 100⟩
⟨F⟩
FeXV, 284 Å
190 - 205 Å
170 - 190 Å
AE-E 13-28 July 1976 (R_Z= 0, F = 68±1)
Intensity
Time

Fig. 12a, b

cycle 20 was larger. In addition, an analysis of the unique set of data[19] covering times of low and high activity during solar cycle 21 leads to a flux increase of about 200% for the same wavelength range. Figures 12a and b show some details for selected EUV emission lines.

As a general rule coronal lines emitted from hot solar regions (with temperature greater than 10^6 K) increase significantly more with solar activity than do the other emissions. The only exception from this rule are resonance lines of Be-like ions, e.g. Mg IX and Si XI.

Similarly to the Lyman-alpha emission of hydrogen at 121.6 nm, the emissivity in the wavelength range from 103 to 16 nm was apparently much smaller in solar cycle 20 than in cycle 21.

ε) Soft X-ray region from 2 to 0.1 nm. The wavelength range from 16 to 2 nm is not considered here, since it contains not enough energy to control any effect of geophysical interest. Also, the technical problems involved in continuous measurements are rather difficult to solve. For these reasons this field of research is almost abandoned since 1972; the earlier observations are described in Sect. 5 of NIKOL'SKIJ's contribution in volume 49/6 of this Encyclopedia; see also Table 6 in Sect. 5 of THOMAS's contribution in the same volume. Hence, little is known concerning the variability in this part of the soft X-ray spectral region.

One of the best known Sun-related terrestrial effects is the ionization enhancement in the ionospheric D-region during solar flares and during high levels of solar activity caused by increased X-ray emission from active regions of the Sun. This radiation in the range from 2 to 0.1 nm can be measured with relatively simple devices such as Geiger or proportional counters or ionization chambers; such instruments are used since long in the laboratory[40,41]. Since the beginning of the space research era this type of instrumentation was integrated into rocket and satellite payloads. The measurements indicate a very high degree of variability ranging from two orders of magnitude during flares up to five orders of magnitude and more over a solar cycle (see Fig. 13)[40], at the shortest wavelengths around 0.1 nm. The SOLRAD satellite measurements monitored the fluxes integrated over the 0.8 to 0.1 nm range at an average level of 10^{-3} mW m^{-2} during solar maximum conditions of cycle 20 (1969 to 1970) and smaller by a factor of 500 at solar minimum. As an example, daily averages from SOLRAD 9 and OSO-5 satellites are presented in Fig. 14. – Aboard the SOLRAD 11 satellite similar broadband sensors were operated from solar minimum to solar maximum conditions during the ascending part of solar cycle 21[41]. Later the geostationary satellites GOES-2, 3 and 4 monitored the solar X-rays within the spectral bands of 0.4 to 0.05 nm and 0.8 to 0.1 nm with the extremely high sampling rate of $\frac{1}{3}$ s.

As a major conclusion from this condensed presentation, we may say from an aeronomic point of view that the solar EUV radiation and its variability has become better known during the past decade of space research. The

---

[40] Kreplin, R.W. (1977): The solar output and its variation. Boulder, CO: Colorado Associated University Press, p. 287

[41] Horan, D.M., Kreplin, R.W. (1980): J. Geophys. Res. *85*, 4257

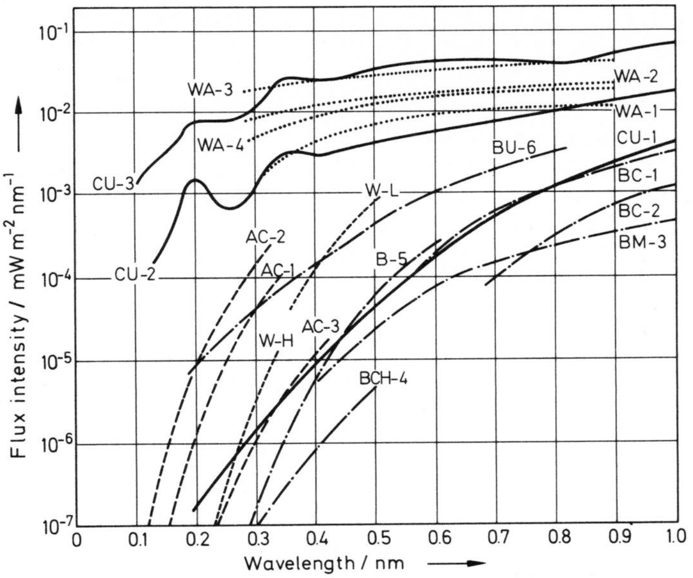

Fig. 13. Solar continuum spectra obtained with proportional counter and Bragg crystal spectrometers by various authors demonstrating strong variability in this wavelength region[40]

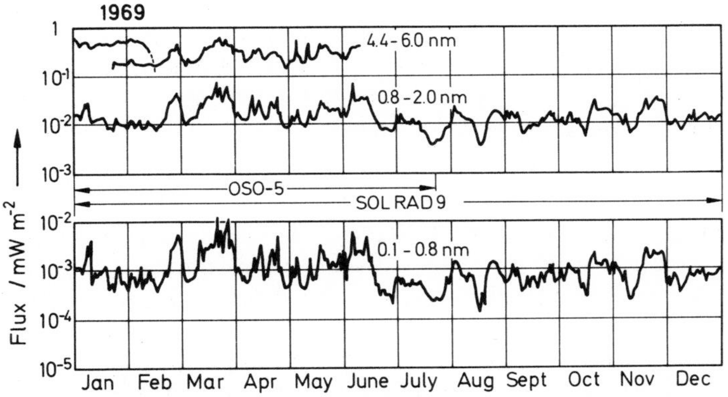

Fig. 14. Daily averages for 0.1 to 0.8 nm and 0.8 to 2.0 nm flux levels for 1969 derived from photometers aboard SOLRAD 9 and OSO-5. Flare X-ray flux enhancements have been removed from the averages as far as possible[40]

variability of energy fluxes could, for the first time, be identified on different time scales, from short-term to long-term. We know now that in spite of certain regular features the degree of variability differs strongly for emissions of different origin and with time.

We do not expect this work to be continued as intensely during the next decade, since it is very expensive. What is still needed in this field of research is: a statistical analysis of all now available data, a descriptive representation of them and further elaboration of empirical relations between classical solar indices and EUV fluxes. The inherent limitations of such relations need to be classified. A few more satellite missions can be expected but unfortunately, an international coordinated program for specific EUV measurements is not in view though additional observational data are still needed.

As an approach to one of these goals the next chapters deal with the descriptive representation of solar EUV data and with empirical relations between classical activity indices and measured EUV intensities; a few examples will also be given of model calculations using EUV fluxes as an input. In this context, see for comparison Chapt. E of RAWER's contribution (p. 430) and also Figs. 20 to 22 in Sect. 7 of THOMAS's contribution in volume 49/6 of this Encyclopedia.

II. Representation and estimation of solar EUV fluxes

The variations of solar activity are regularly observed on Earth since a few centuries now. It is, of course, easier to achieve ground-based measurements than space experiments. Therefore, from an early stage of solar EUV measurements on efforts have been undertaken to correlate these latter with classical solar indices obtained from ground-based observations such as the Zürich sunspot number R_Z, the calcium plage index PI or the Covington index F (radiowave flux at 2,800 MHz). This is an important task in view of the lack of continuous recording of solar EUV fluxes. For periods without space measurement such empirical relations as derived from correlation analysis may give a rough estimate of missing data. In this chapter schemes for representing the measured solar EUV fluxes, empirical relations between classical solar indices and EUV fluxes and proposals for indices based on EUV data will be reviewed.

3. Representation of solar EUV fluxes. The very early predictions of the solar EUV radiation were given by astronomers assuming a solar black body model of 5,000 to 6,000 K; first quantitative determinations of the flux were undertaken in order to understand the results of ionospheric observations introducing an "ultraviolet excess-factor"[1,2] as large as 10^6. This factor was required to get a production term which, in the limits of contemporary aeronomy might explain the important plasma density in the ionosphere as revealed by ionosondes. One numerical value only was derived namely the total EUV energy of about 0.1 $\mathrm{mW\,m^{-2}}$ as the solar flux in the spectral region of interest. With the use of rockets for space experiments more detailed information became available[3] (compare Table 1 in NIKOL'SKIJ's contribution in volume 49/6 of this Encyclopedia). With more advanced photoelectric measurements HINTEREGGER et al.[4] established a tabulation that was widely considered as "EUV standard flux" for many years. This table contains about 80 emission

[1] Saha, N. (1937): Proc. Roy. Soc. *A160*, 155
[2] Kiepenheuer, K.O. (1945): Ann. Astrophys. *8*, 210
[3] Rawer, K. (1981): Adv. Space Res. *1*, 87
[4] Hinteregger, H.E., Hall, L.A., Schmidtke, G. (1965): Space Research *V*, 1175

lines or subranges of continua or unresolved line groups. A later edition is found as Table 2 in NIKOL'SKIJ's contribution (l.c., p. 322). Next the data from the two AEROS satellites were tabulated representing the variable solar EUV fluxes from 106 through 16 nm as daily average values in 42 spectral lines or ranges for the periods from December 1972 to August 1973 (AEROS-A)[5] and July 1974 to September 1975 (AEROS-B)[6]. From this set examples were chosen[7] to identify typical conditions (one example is reproduced as Table 9 in NIKOL'SKIJ's contribution, l.c., p. 366):

Of course, full information about the spectrum with the resolution as suitable for aeronomic purposes would be ideal. However, since in recent satellite missions instead of a full-scale spectrum smaller scale parts were more often recorded (but with better resolution, down to 0.1 nm), another kind of description was felt to be desirable in which one could obtain an estimate of the full spectrum from measurements in narrow spectral ranges or lines only. It is the aim of this approach[8,9] to provide as much detailed spectral information as possible from a minimum set of daily numbers. As an input two tabulations are proposed:

a) The reference spectrum of 1972 identifying different solar wavelengths or wavelength intervals between 200 and 1.8 nm with fluxes ($\Phi_{\lambda\text{Ref}}$) assigned to each of them. As such reference spectrum the average of an especially selected period of two weeks was chosen for solar quiet conditions, namely the period 13 through 28 July, 1976 (satellite AE-E data).

b) With statistical analysis, for each of these lines or ranges a variability index C_λ is derived in order to fit the following equation

$$\Phi_\lambda = \Phi_{\lambda\text{Ref}}[1 + (R_K - 1)\, C_\lambda] \tag{3.1}$$

where R_K is the ratio of the intensities on the "key" wavelength K_λ observed on the given day and in the reference spectrum. For $K_\lambda = 1$ the "key" wavelength is $\lambda_{\text{Ref}} = 102.7$ nm of hydrogen and for $K_\lambda = 2$ it is $\lambda_{\text{Ref}} = 33.5$ nm of Fe XVI.

This is, of course, an approximation because the relation with C_λ was established by *statistical* analysis. Therefore, it is not strictly valid for each individual day. It is, however, better than older relations based on just one activity index, which stems from measurements being alien of solar EUV data, e.g. sunspots, plages or solar microwave-noise.

From the established coefficient C_λ in the relevant class K_λ of the desired wavelength λ, the solar flux Φ_λ for this λ is obtained by equation (3.1) using the reference flux $\Phi_{\lambda\text{Ref}}$. By summing up over all ranges (λ) we get approximately the total solar EUV flux

$$\Phi_{\text{EUV}} = \sum_\lambda \Phi_{\lambda\text{Ref}}[1 + (R_K - 1)\, C_\lambda]. \tag{3.2}$$

---

[5] Schmidtke, G. (1977): Report IPW-W.B.3. Freiburg i. Br.: Institut für Physikalische Weltraumforschung

[6] Schmidtke, G. (1979): Report IPW-W.B.11. Freiburg i. Br.: Institut für Physikalische Weltraumforschung

[7] Roble, R.G., Schmidtke, G. (1979): J. Atmos. Terr. Phys. *41*, 153

[8] Hinteregger, H.E. (1981): Adv. Space Res. *1*, 39

[9] Hinteregger, H.E., Fukui, K., Gilson, B.R. (1981): Geophys. Res. Lett. *8*, 1147

The values required for representing of the solar EUV fluxes as measured during the mission period of satellite AE-E (1976 to 1980) are given in Tables 4 and 5 (see annex). While Table 4 presents the reference wavelengths λ in mm, the reference fluxes $\Phi_{\lambda\text{Ref}}$ (unit $m^{-2}s^{-1}$), the identification of the emissions, the variability class index K_λ ($=1$ or $=2$) and the variability index C_λ, Table 5 is giving the values R_K for the two "key" wavelengths for a selected period (Julian date).

If one consideres the greatly diversified nature and history of solar activity in the different atmospheric layers, it is not surprising that the data-adjusted values of C_λ may differ for the listed ones (see Table 4) and this depending on the wavelength. Deviations of the order of 5 to 35% are encountered[9]. This has to be taken into account for studies based on this representation of solar EUV fluxes.

4. Estimation of the solar EUV fluxes from ground-based measurements. Also by correlation and regression analysis empirical relations between measured EUV fluxes and classical solar activity indices were derived in order to represent the radiation, but also for forecasting or estimating these fluxes. This may either be done for periods without direct measurements or for the future. (Since very long data series exist for example for the sunspot number R_Z more elaborate prediction methods can be applied to these.)

α) For the range of longer wavelengths the EUV variability was also determined with a more involved two component model[1] using as indicator the ratio of plage to quiet Sun contrast values, since the area of enhanced EUV emission is indentical with the Ca II plage area. The fraction f of the solar disc covered by plages apparently varies with the solar cycle as function of the sunspot number R_Z[2] by

$$f = 6.25 \cdot 10^{-4} \cdot R_Z .$$

It is supposed that

(i) the variability is caused by the appearance and disappearance of plages only, while the background quiet solar radiation remains constant;

(ii) the contrast value plage/quiet Sun for a particular wavelength remains constant over the solar cycle; and

(iii) the area of enhanced EUV emission is identical with the Ca II plage area.

Of course, these assumptions will often not be correct for individual days. It is also clear that use of such simplifying relations "by-passes the entire problem of spectral detail in both solar EUV emission and terrestrial atmospheric absorption"[3]. Only if the above simplifications are admitted, the solar flux from the total solar disc may be estimated by[1]

$$F_\lambda = \pi \cdot \langle I_Q \rangle_{\text{disc}} \cdot [f \cdot C_\lambda^* + (1-f)]. \tag{4.1}$$

[1] Cook, J.W., Brückner, G.E., VanHoosier, M.E. (1980): J. Geophys. Res. *85*, 2257

[2] Sheeley, N.R. (1967): Astrophys. J. *147*, 1106

[3] Hinteregger, H.E. (1981): AGARD Conference Proc. No. 295, I-1

Table 1. Two component model[1]; full disk flux variability

A. Continuum			*B. Lines total energy flux of the lines*			
λ/nm	C^*_λ	$\frac{\langle I_Q\rangle_{\text{disk}}}{\text{mW m}^{-2}\,\text{nm}^{-1}\,\text{sr}^{-1}}$		λ/nm	C^*_λ	$\frac{\langle I_Q\rangle_{\text{disk}}}{\text{mW m}^{-2}\,\text{sr}^{-1}}$
140	9.0	3.13 (0)	C III	117.6	14	2.8 (2)
142	7.8	3.98 (0)	Si III	120.6	14	8.6 (2)
144	7.2	4.72 (0)	HL_α	121.5	4.6	7.1 (4) t
146	6.7	5.57 (0)	NV	123.8	11	1.8 (2)
148	6.1	6.07 (0)		124.2	10	9.8 (1)
150	5.4	7.15 (0)	OI	130.2	10	2.8 (2) t
152	4.8	8.76 (0)		130.4	7.7	3.0 (2) t
154	4.3	9.77 (0)		130.6	8.9	2.8 (2) t
156	4.0	1.08 (1)	Si III	129.4	11	1.3 (1)
158	3.8	1.22 (1)		129.6	11	1.5 (1)
160	3.6	1.36 (1)	(blend)	129.8	13	4.0 (1)
162	3.4	1.68 (1)	C II	133.4	13	4.1 (2) t
164	3.3	2.06 (1)		133.5	10	6.5 (2) t
166	3.1	2.56 (1)	O I	135.5	2.9	1.0 (2)
168	2.9	3.34 (1)		135.8	3.2	4.0 (1)
170	2.8	5.68 (1)	Si IV	139.3	18	3.4 (2)
172	2.7	6.22 (1)		140.2	16	1.9 (2)
174	2.6	8.20 (1)	Si II	152.6	13	1.8 (2)
176	2.5	1.39 (2)		153.3	15	2.0 (2)
178	2.3	2.11 (2)	C IV	154.8	5.4	9.2 (2)
180	2.1	2.65 (2)		155.0	5.6	5.0 (2)
182	2.0	3.26 (2)	Si II	180.8	13	1.7 (3)
184	2.0	3.84 (2)		181.6	9.2	2.6 (3)
186	2.0	4.15 (2)		181.7	5.5	1.2 (3)
188	2.0	5.11 (2)				
190	2.0	5.60 (2)				
192	2.0	6.15 (2)				
194	2.0	6.80 (2)				
196	2.0	8.40 (2)				
198	2.0	8.79 (2)				
200	2.0	9.77 (2)				
202	2.0	1.20 (3)				
204	1.8	1.54 (3)				
206	1.7	1.71 (3)				
208	1.4	2.11 (3)				
210	1.0	5.37 (3)				

Numerical indication $a(b)$ means $a \cdot 10^b$
t = Optically thick

Values for the average quiet Sun intensity $\langle I_Q\rangle_{\text{disc}}$ and the contrast factors C^*_λ are given in Table 1. – The Zürich sunspot number R_Z is the index with the longest series of observations (more than 250a).

In Fig. 15 predictions obtained from this model are compared with the measurements of Fig. 3. The agreement is quite satisfying except for the wavelength region from 170 to 150 nm. Hopefully, an analysis based on data from the more recent satellite AE-E might improve the accuracy of the prediction.

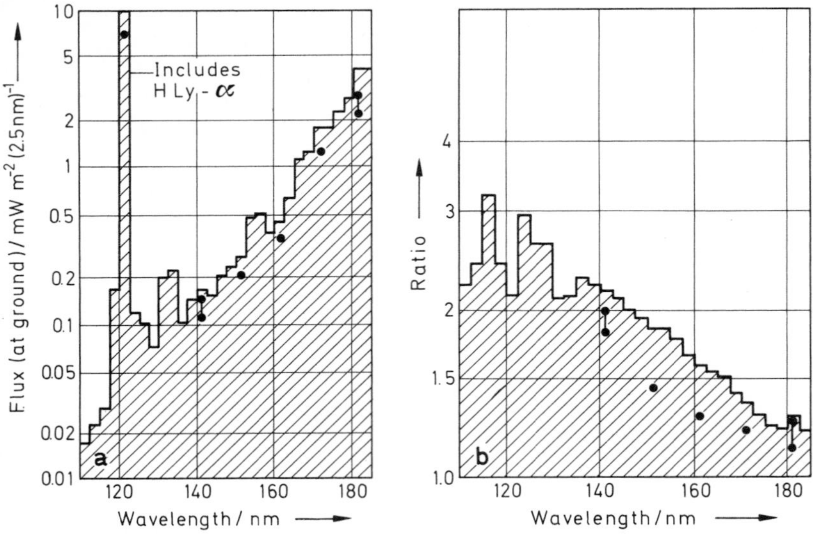

Fig. 15 a, b. Preliminary results from the EUVS experiment on satellite AE-E[3] compared with results from the two component model (●). This figure is adapted from an original provided by HINTEREGGER. (a) Fluxes in 2.5 nm intervals for 22 Jan. 1979 ($R_Z = 178$). (b) Ratio of fluxes from same date to an average of 24 sets of observations obtained during July 1976 on days with $R_Z = 0$ and $F \leqq 70$ (July monthly average $\langle R_Z \rangle = 1.9$). The 140 to 142.5 nm and 180 to 182.5 nm results in (a) and (b) are shown both for the continuum alone (lower values in all cases) and for the 140 to 142.5 nm continuum plus the SiIV 140.2 nm line and the 180 to 182.5 nm continuum plus the SiII 180.8 nm, 181.6 nm and 181.7 nm lines

In Fig. 16 an estimate of the variability of the continuum radiation during solar cycle 19 is given which was obtained from terrestrial observations of plage/quiet Sun contrast values by the two component method. This cycle had rather high activity during this solar maximum period.

Figure 16 includes the hydrogen Lyman-alpha radiation. This is a prediction which cannot be compared with direct radiation measurements, since such measurements are not available for the total cycle 19. However, a comparison of predicted and observed values could be made for the four years from 1969 to 1972 of solar cycle 20, see Fig. 17.

β) One empirical representation for the most important EUV line, Lyman-alpha of hydrogen (121.6 nm), gives the quantum flux for high solar activity as[4]:

$$\frac{F_{L\alpha}}{10^{15}\,\mathrm{m}^{-2}\,\mathrm{s}^{-1}} = 0.77 \cdot 10^{-2} \cdot \langle R_Z \rangle + 0.38 \cdot 10^{-2} \cdot R_Z + 2.10 \tag{4.2a}$$

where $\langle R_Z \rangle$ is the sliding monthly average of the Zürich sunspot number R_Z. This parameter was introduced in order to identify the slowly varying (quiet) flux. For $R_Z \geqq 70$ Eqs. (4.1) and (4.2a) are not too largely different, either may

[4] Vidal-Madjar, A. (1975): Solar Phys. *40*, 69

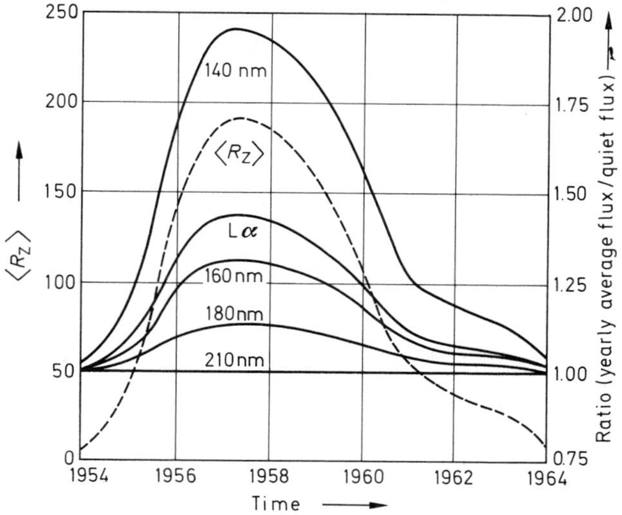

Fig. 16. Variability of the Lyman continuum and Lyman-alpha in solar cycle 19 as estimated from the two component model[1]

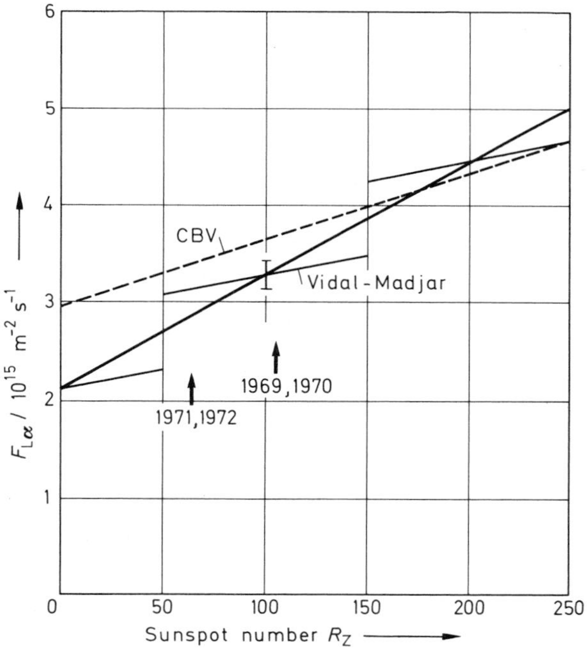

Fig. 17. Lyman-alpha flux at the Earth after the representation of VIDAL-MADJAR[6] and flux predicted from the two component model (broken line)[1]. We illustrate VIDAL-MADJAR's results by a straight line representing days with daily R_Z equal to the yearly average and by straight line segments illustrating flux versus daily R_Z for years with $\langle R_Z \rangle_{year} = 0$, 100 and 200. Values of $\langle R_Z \rangle_{year}$ for 1969 to 1972 are indicated

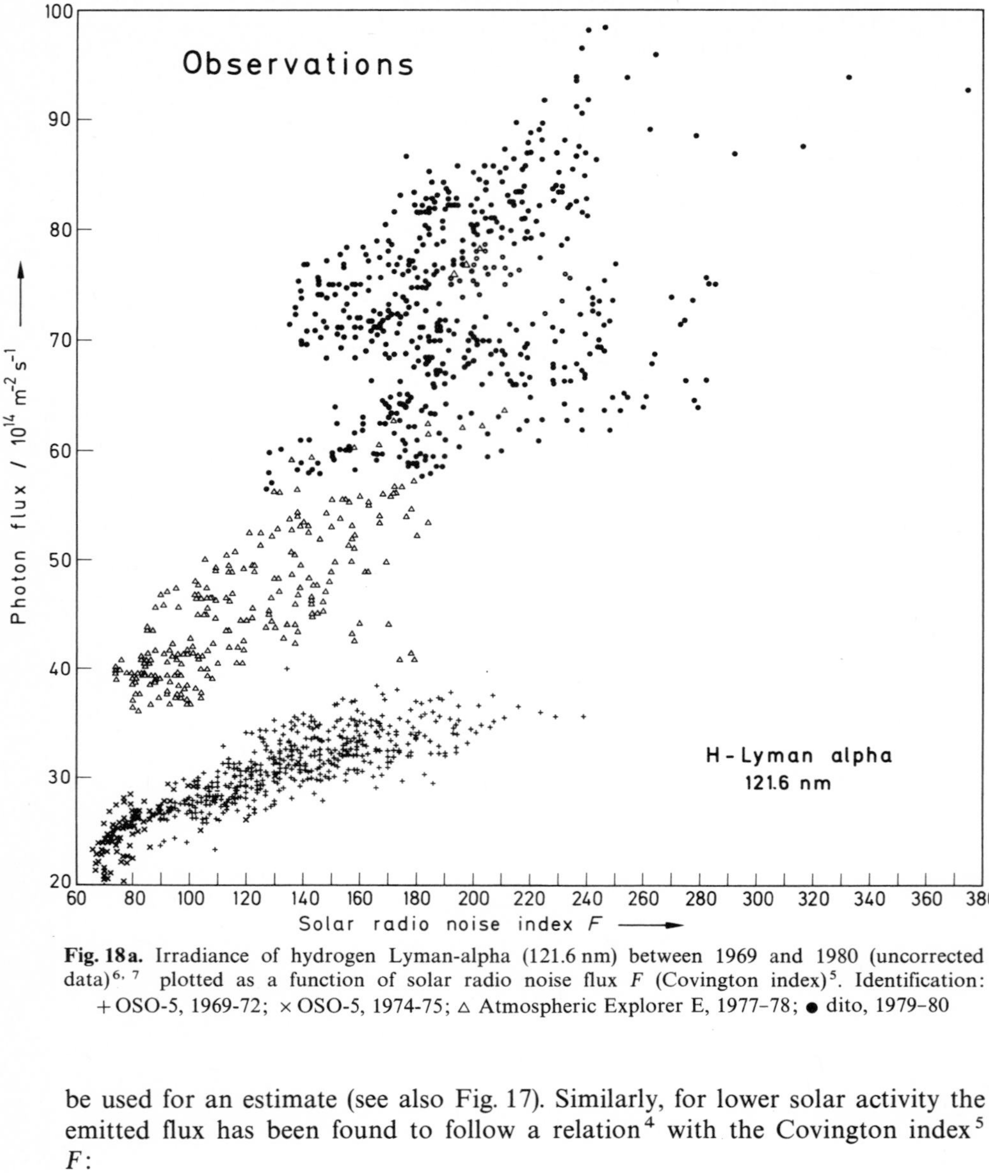

Fig. 18a. Irradiance of hydrogen Lyman-alpha (121.6 nm) between 1969 and 1980 (uncorrected data)[6, 7] plotted as a function of solar radio noise flux F (Covington index)[5]. Identification: + OSO-5, 1969-72; × OSO-5, 1974-75; △ Atmospheric Explorer E, 1977–78; ● dito, 1979–80

be used for an estimate (see also Fig. 17). Similarly, for lower solar activity the emitted flux has been found to follow a relation[4] with the Covington index[5] F:

$$\frac{F_{L\alpha}}{10^{15}\,\mathrm{m}^{-2}\,\mathrm{s}^{-1}} = 0.63 \cdot 10^{-2} \cdot \langle F \rangle + 0.54 \cdot 10^{-2} \cdot F + 1.49. \qquad (4.2\mathrm{b})$$

[5] The Covington index F is equal to the numerical value (in units of $10^{-22}\,\mathrm{W\,m^{-2}\,Hz^{-1}}$) of the solar radio noise at 2800 MHz, i.e. at a wavelength of 10.7 cm

[6] Vidal-Madjar, A. (1975): Solar Phys. *40*, 69; Vidal-Madjar, A., Phissamay, B. (1980): Solar Phys. *66*, 259

[7] Hinteregger, H.E. (1980): Proc. Workshop on Solar UV irradiance. July 31–August 1, 1980, Boulder, Colorado; (1981): Adv. Space Res. *1*, 39

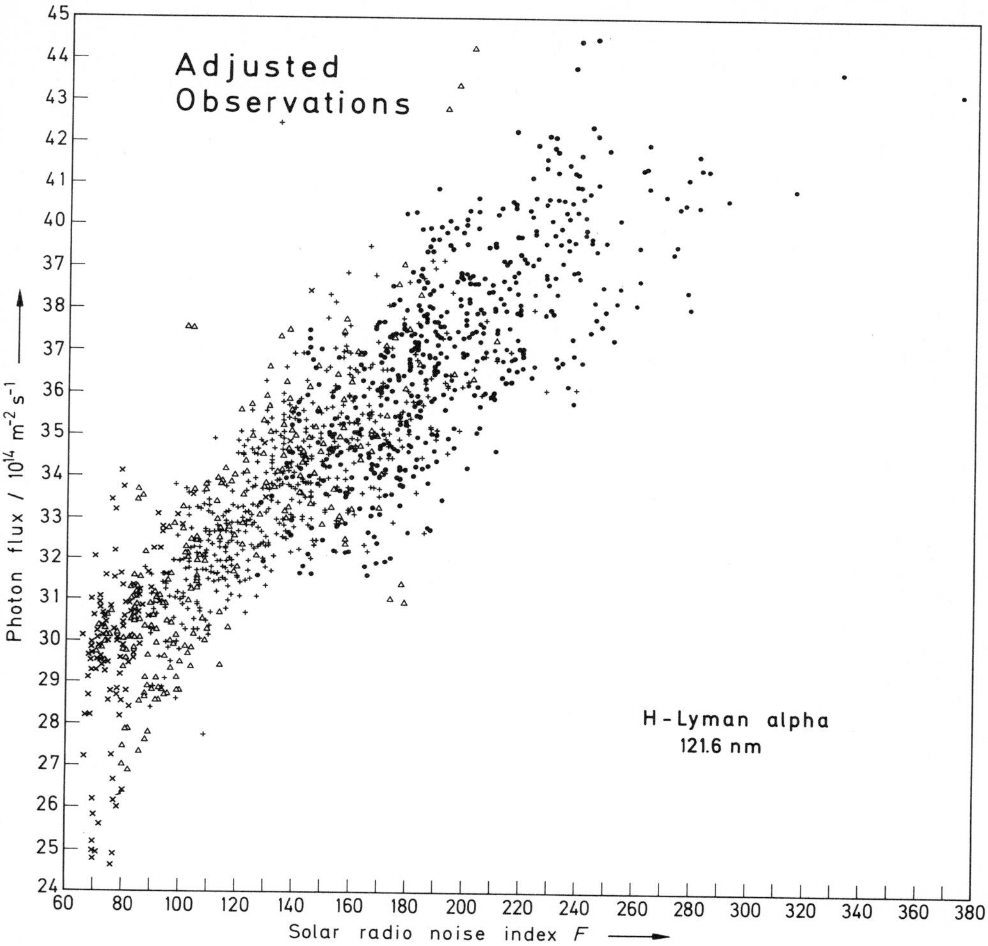

Fig. 18b. Irradiance of Lyman-alpha plotted as a function of solar radio noise flux at 10.7 cm as in Fig. 18a after adjusting with the Bossy-Nicolet procedure [8]

In this latter formula the sliding average is performed over 81 days, centered on the 41th day of the period; this corresponds to three solar rotations.

Recently, with quite a few sets of Lyman-alpha data of different sources Bossy and Nicolet[8] have plotted these as function of the Covington index (Fig. 18a). Linearizing this relation these authors then undertook a correction of the measured data *feeling* that the sets of different authors were inconsistent; they attributed this to erroneous calibrations or to changes of the sensitivity with time. Giving equal weight to all inputs they obtained corrections by a

---

[8] Bossy, L., Nicolet, M. (1981): Planet. Space Sci. *29*, 907; Bossy, L. (1983): Planet. Space Sci. *31*, 977

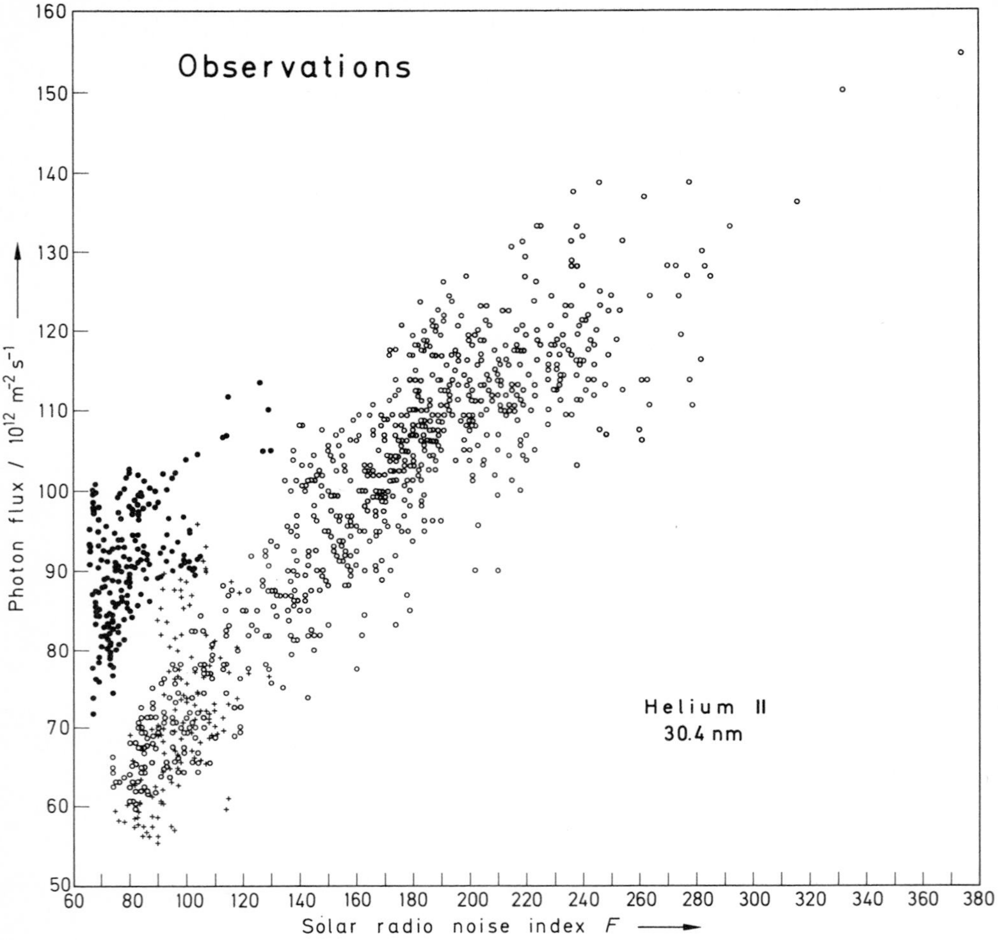

Fig. 18c. Same as Fig. 18a for ionized helium Lyman-alpha (30.4 nm) [uncorrected data from satellites AEROS-A (crosses), -B (dots) and Atmospheric Explorer -E (circles)][8]

best fit procedure forcing all data into an essentially linear relation with F. The result of this manipulation is shown as Fig. 18b. From this latter plot the relation

$$\frac{F'_{L\alpha}}{10^{15}\,\mathrm{m}^{-2}\,\mathrm{s}^{-1}} = 2.55 + 0.011 \cdot (F - 65) \tag{4.3}$$

was derived. The authors feel that their relation should be used for aeronomic purposes in the future. It is, however, not yet clear whether the experimenters agree with their conclusions.

Another correction method is proposed by OSTER[9]. He also uses an assumed linear relation admitting linear errors only of the measurements; using

[9] Oster, L. (1983): J. Geophys. Res. *88*, 1953 and 9037

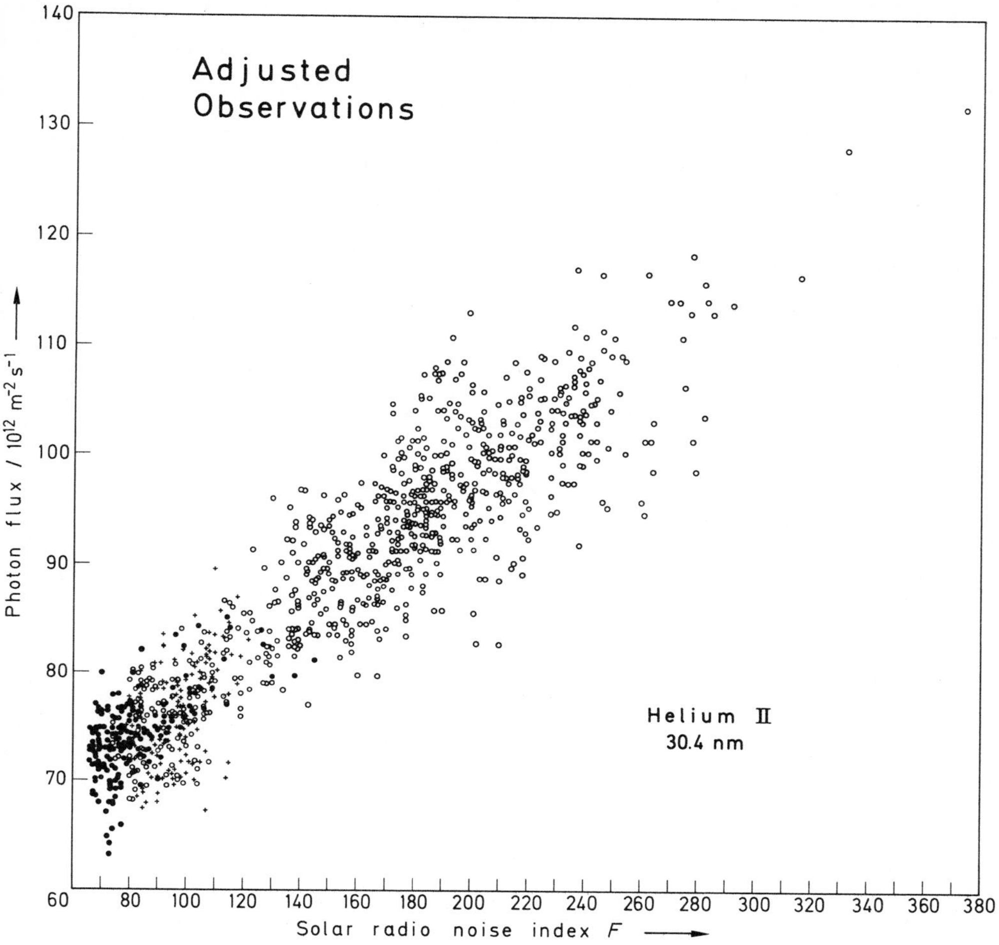

Fig. 18d. Data as in Fig. 18c after adjusting with the BOSSY-NICOLET procedure[8]

data of satellite AE-E corrections of up to 30% are quoted. Compare also recent investigations with simulating the Lα-variations by a three index scheme based on detailed solar astronomical observations[10].

Figures 18c and d have been obtained by BOSSY applying his method[8] to the L_α-emission of He^+ at 30.4 nm.

γ) For wavelengths below 100 nm a correlation analysis has been performed with the EUV spectral fluxes measured by the AEROS-A satellite[11] on one

[10] Cook, J.W., Brueckner, G.E., van Hoosier, M.E. (1980): J. Geophys. Res. *85*, 2257; Lean, J.L., White, O.R., Livingston, W.C., Heath, D.F., Donnely, R.F., Skumanich, A. (1982): J. Geophys. Res. *87*, 10307

[11] Rawer, K., Emmenegger, G., Schmidtke, G. (1979): Space Research *XIX*, 199

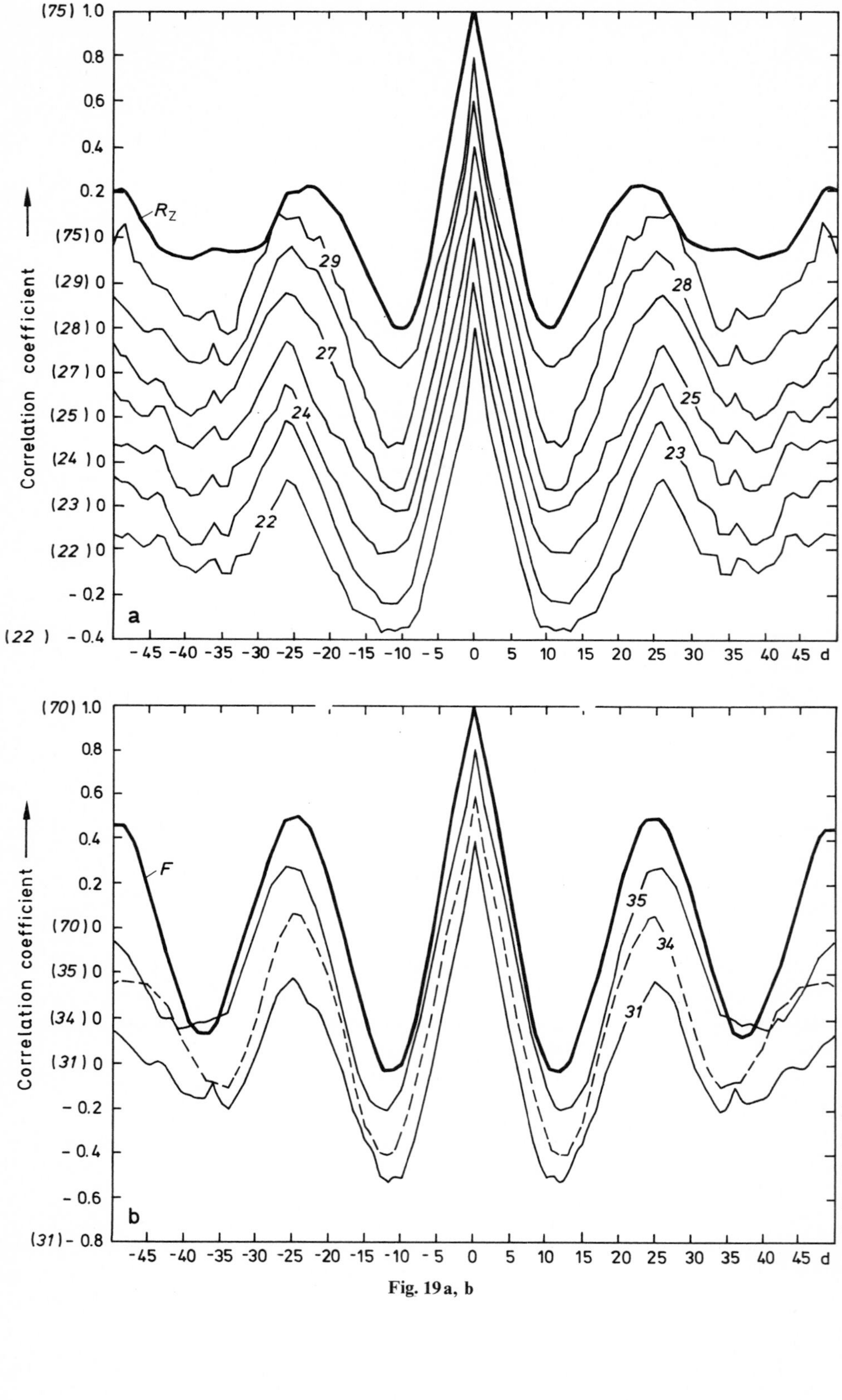
Correlation coefficient
(75) 1.0
0.8
0.6
0.4
0.2
(75) 0
(29) 0
(28) 0
(27) 0
(25) 0
(24) 0
(23) 0
(22) 0
- 0.2
(22) - 0.4
R_Z
29
28
27
25
24
23
22
a
-45 -40 -35 -30 -25 -20 -15 -10 -5 0 5 10 15 20 25 30 35 40 45 d
(70) 1.0
0.8
0.6
0.4
0.2
(70) 0
(35) 0
(34) 0
(31) 0
- 0.2
- 0.4
- 0.6
(31) - 0.8
F
35
34
31
b
29
27
25
23
75
-45 -40 -35 -30 -25 -20 -15 -10 -5 0 5 10 15 20 25 30 35 40 45 d

Fig. 19 a, b

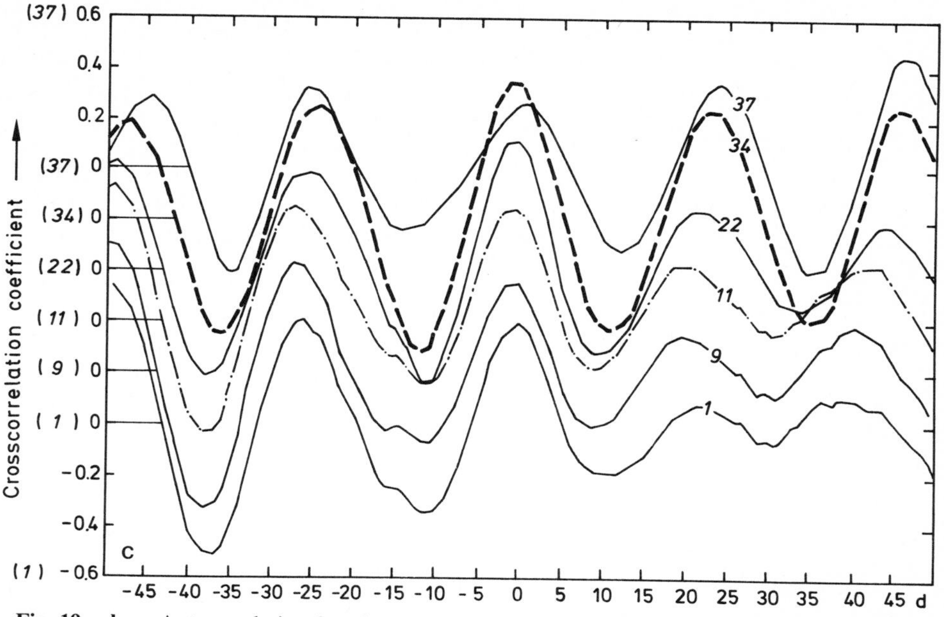

Fig. 19a, b, c. Autocorrelation functions of solar EUV fluxes in some of the 42 ranges with lags ranging from −50 to +50 days, and with displaced zeros. Bold curve R_Z in upper (**a**), F in lower diagram (**b**). (**a**) Ranges 22 (=OV 63 nm) to 29 (=SiXII 50 nm). (**b**) Identification: 31 = background 63...46 nm; 35 = FeXVI (36 and 33 nm). (Range numeration is that used in Table 9 of NIKOL'SKIJ's contribution in volume 49/6, p. 222). (**c**) Crosscorrelation functions with Covington-index F as leader. Note strong coherence of 34 = FeXVI

Table 2. Medium-term negative slope of activity indices (year 1973)[11]

Index	I	II	III	IV	V	VI	VII	VIII	R_Z	F
Slope ($\log_{10}$)	−0.13	−0.09	−0.10	−0.14	−0.14	−0.31	−0.48	−0.36	−0.59	−0.11
Ratio to R_Z	0.22	0.15	0.17	0.24	0.24	0.53	0.81	0.61	1.00	0.19
Ratio to F	1.18	0.82	0.91	1.27	1.27	2.82	4.36	3.27	5.36	1.00
Range of λ	103 ... 91 ... 80 ... 63 ... 46 ... 37 ... 28 ... 20 ... 15 nm									

side and the two most used classical solar indices R_Z and F on the other. The influence of different absolute levels of fluxes and indices was removed by using logarithms everywhere. The long-term slopes over all eight months of the mission (during decreasing solar activity) was negative for all inputs, however, with very different absolute values according to the wavelength range (see Table 2). Comparing with the indices one may roughly say that the long-term (logarithmic) variation of R_Z was considerably larger than for any of the eight EUV ranges in Table 2, while that of the Covington index F falls between the less important range variations. At least during that period the short wavelength ranges showed much stronger long-term variation than F. Autocorrelation analysis (Figs. 19 a, b) clearly shows a repetition tendency after about 26d

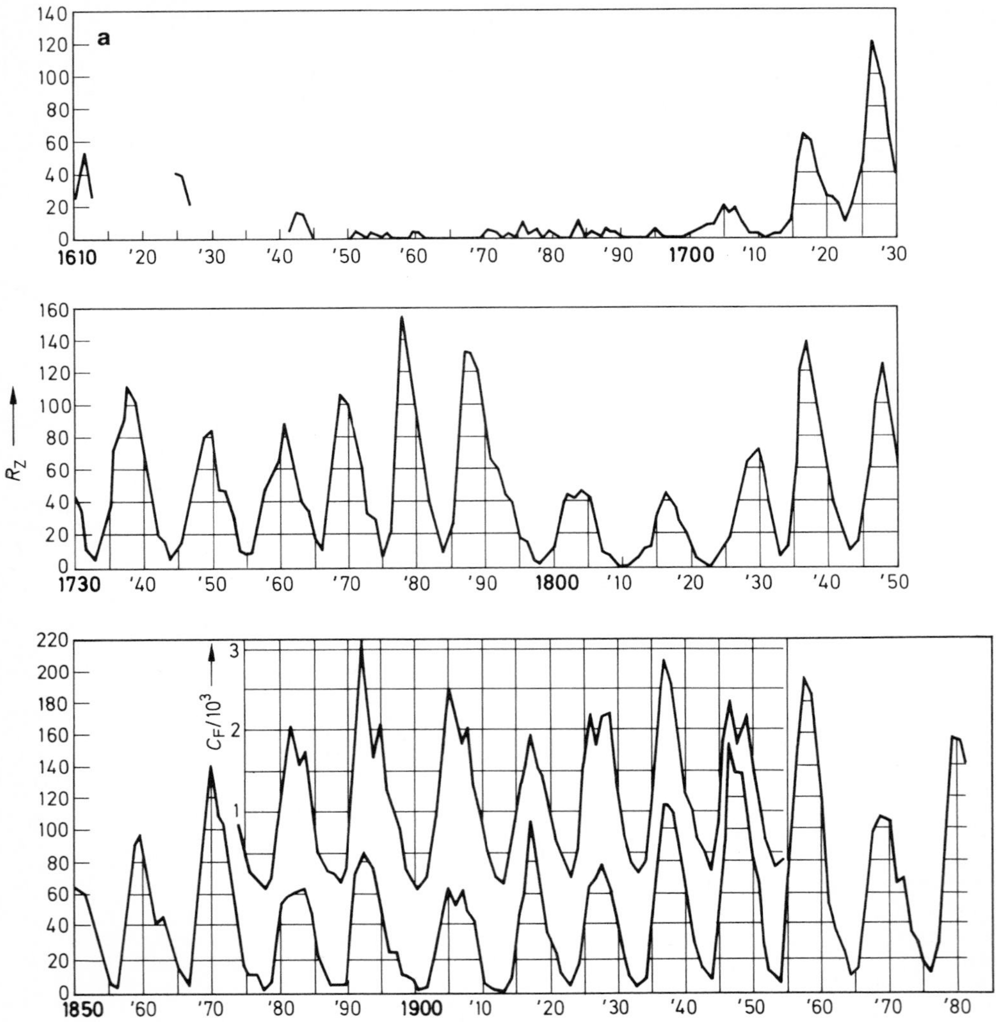

Fig. 20a. Smoothed sunspot number (Zürich sunspot index $\bar{R}_Z$) from 18th to 20th century. The upper curve shows the Greenwhich 'faculae area index', C_F, which is available since 1875 and was, unfortunately, discontinued in 1955

(certainly due to the solar rotation) but, again there are large differences after the wavelength. *F* (top curve in Fig. 19b) has highest repetition tendency, even over two solar rotations; it is followed by Fe XVI emission lines. Less ionized species, however, show much smaller repetition rate (Fig. 19a). This shows the narrow limitations of EUV-predictions made by the Covington index *F*. This is most clearly shown in Fig. 19c where this index is used as leader for cross-correlation analysis. Strong coherence appears with the emissions of a few, strongly ionized species only. For others correlation values of about 0.25 after one solar rotation cannot provide too good predictions.

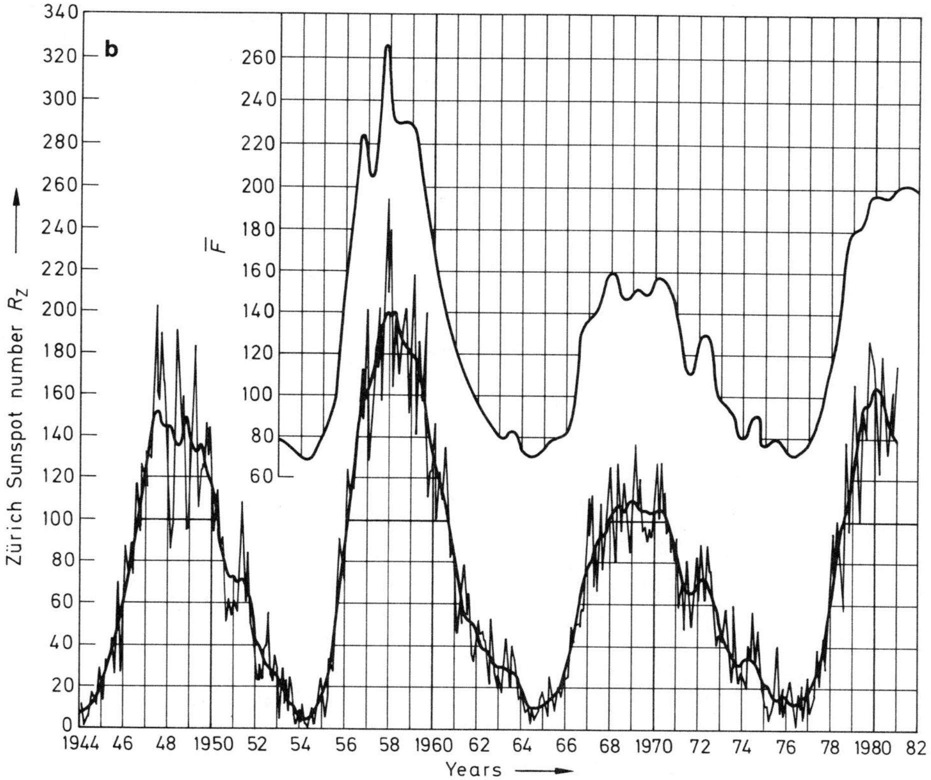

Fig. 20b. Monthly and smoothed sunspot numbers, R_Z and $\bar{R}_Z$, since 1944; smoothed Covington index F for comparison on top

The drawbacks of the classical solar activity indices, in particular the Zürich sunspot number R_Z and the Covington index F, are often overlooked. Their background is in fact purely empirical. As for R_Z this appears from the rather artificial (but quite successful) combination of 'spot' and 'group' counts in its very definition. On the other hand, the solar decimetric radiation from which F is derived[5] has no effects in the terrestrial atmosphere. It is, of course, linked with the solar plasma conditions in just one layer of the solar atmosphere (the lower corona) but this is also so for the spots (the photosphere). The main merit of these indices lies in the fact that they are observed since long time: some 25 solar cycles for R_Z, just three for F (see Fig. 20) while consistent EUV observations do not yet really cover one cycle (see Figs. 10, 11 above). On behalf of these two, mainly used indices it might also be noted that in the long term relations with the ionospheric peak electron densities there appear a few queer deviations[12] which are not existing with another empirical index[13].

BOSSY's analysis[8] which is based on an assumed *linear* dependence of the solar EUV emission on the Covington index F, and by which observed data are corrected, was also applied to the He^+ Lα-line at 30.4 nm, see Figs. 18c, d above.

The good correlation between the Covington index F and higher ionized Fe-lines (Fe X...XIII) appears also from Fig. 21 where the variation of their

---

[12] Smith, P.A., King, J.W. (1981): J. Atmos. Terr. Phys. *43*, 1057

[13] the 'faculae area index' which was during many decades determined at the Royal Greenwhich Observatory (U.K.) [see Fig. 20a]

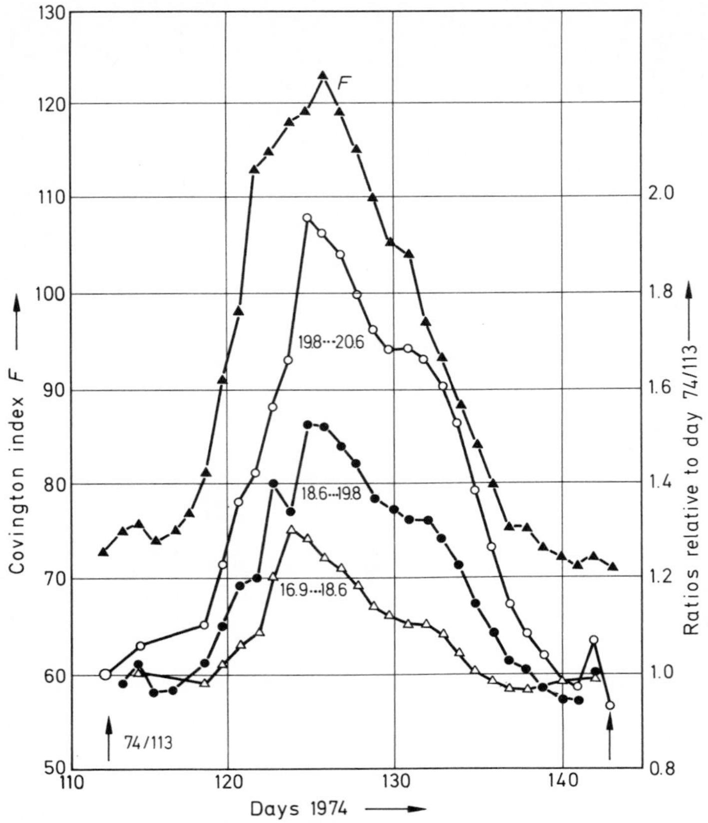

Fig. 21. EUV fluxes during one solar rotation for different Fe X ... XIII emissions, illustrating good correlation with *F*. The EUV data were acquired by the AFGL experiment on the AE-C satellite[14]

intensity, and that of *F*, is shown during just one solar rotation[14]. On the other hand Fig. 22 shows an example of rather poor correlation during solar activity minimum (supposed to have happened in April 1975) with data from satellites AE-C[15,16] and AEROS-B[17]. For certain periods *F* and the He I emission at 58.4 nm seem to be almost anticorrelated. The correlation might be better during periods of high solar activity.

The fact that, in general, solar EUV radiation below 120 nm and solar radio noise flux are not so well correlated is not surprising from the viewpoint of solar physics: The bulk of the emission in this spectral region is generated in the solar chromosphere and chromosphere-corona transition region, whereas the radio flux at 10.7 cm wavelength is of coronal origin. The physics of these

---

[14] Hinteregger, H.E. (1976): J. Atmos. Terr. Phys. *38*, 791
[15] Hinteregger, H.E. (1977): Geophys. Res. Lett. *4*, 231
[16] Hinteregger, H.E., Fukui, K., Gilson, B.R. (1981): Geophys. Res. Lett. *8*, 1147
[17] Schmidtke, G., Börsken, N., Sünder, G. (1981): J. Geophys. *49*, 146

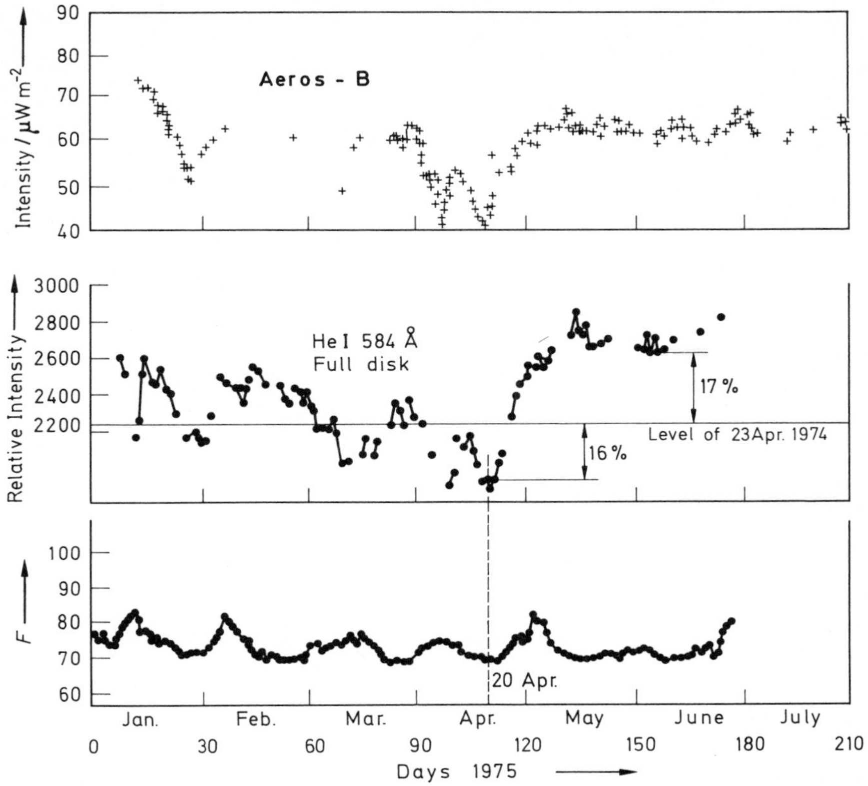

Fig. 22. The solar "EUV Minimum" between solar cycles 20 and 21 in April 1975 observed at 58.4 nm (He I) on satellites AE-C[15] and AEROS-B[17]

different atmospheric layers is dominated by different mechanisms. Emissions from highly ionized atoms such as Fe XV or Fe XVI stem almost exclusively from active regions. For these reasons one would not expect good correlation for all of those emissions of very different origin, from H I through Fe XVI, with one specific solar index such as F.

Still, accurate solar EUV fluxes are requested also for periods outside of the missions of AEROS and Atmospheric Explorer (AE) satellites. Estimates based on empirical relations with indices obtainable on Earth as described before, in spite of their draw-backs (see above) may be the best possible way-out, at present. In view of this fact and of the easy availability of the Covington index, the following relation is proposed[16] for estimating the solar EUV fluxes below 125 nm, Φ_λ,

$$\frac{\Phi_\lambda}{\Phi_{\lambda\mathrm{Ref}}} = B_0 + B_1 \cdot (\langle F \rangle - 71.5) + B_2 \cdot (F - \langle F \rangle + 3.9) \tag{4.4}$$

using the reference fluxes $\Phi_{\lambda\mathrm{Ref}}$ as given on Table 4. Examples for coefficients B_0, B_1 and B_2 obtained by fitting are shown on Table 3. The analysis of this

Table 3. EUV data regression analysis for a two-variable association with F^{16} (see Eq. (4.4) in text)

Wavelength or wavelength range Solar emission	$\Phi_{\lambda\text{Ref}}$ / $10^{10}\,\text{m}^{-2}\,\text{s}^{-1}$	B_0	B_1	B_2
16.8–19.0 nm dominated by Fe VIII–XII	1,380	1.110	0.01026	0.00251
19.0–20.6 nm dominated by Fe XII, XIII	490	1.509	0.02678	0.00567
20.6–25.5 nm various lines	850	1.184	0.02228	0.00708
25.5–30.0 nm various lines	1,160	1.196	0.03656	0.01254
"30.4" nm blend of He II and Si XI	6,240	0.952	0.00647	0.00326
51.0–58.0 nm various lines	1,440	1.286	0.00714	0.00242
58.4 nm resonance line He I	1,580	1.290	0.00876	0.00425
59.0–60.6 nm Mg X, O V, ...	2,360	1.311	0.00830	0.00230
102.6 nm H Ly-β "key" for $K=1$	4,410	1.310	0.01106	0.00492
33.5 nm Fe XVI "key" for $K=2$	34	−6.618	0.66159	0.38319
121.6 nm H Ly-α	300,000	1.046	0.01136	0.00305
28.4 nm Fe XV	114	0.731	0.23050	0.11643
20.0–20.4 nm dominated by Fe XIII	122	1.726	0.03928	0.00885
17.8–18.3 nm dominated by Fe XI	265	1.169	0.01225	0.00282
16.9–17.3 nm dominated by Fe IX	342	1.018	0.00600	0.00177

presentation is not yet finished. Another, more direct method uses a statistically established dependence of the different emissions on three indices, namely the actual intensities of the H Lβ emission at 102.5 nm and of Fe XVI at 33.5 nm, and the longterm average of Lβ[18].

More data from broad EUV spectral ranges have been obtained with the more recent SOLRAD satellites of the Naval Research Laboratories (NRL); final data evaluation is still under way[19]. From preliminary data sets rather good correlation is claimed between ranges 17 to 50 nm, 72.5 to 105 nm and F.

δ) In the *soft X-ray range* at wavelengths below 6 nm there exists the long series of observations due to the SOLRAD satellites of the american NRL[19,20]. The data were obtained with broadband detectors i.e. "unresolved"

[18] Tai, H., Rawer, K. (1985): Ann. Geophysicae *3* (in press)
[19] Horan, D.M., Kreplin, R.W. (1980): J. Geophys. Res. *85*, 4257
[20] Dere, K.P., Horan, D.M., Kreplin, R.W. (1974): J. Atmos. Terr. Phys. *36*, 989

(see Sect. 5 of NIKOL'SKIJ's contribution in volume 49/6)[21]. Since on these wavelengths solar flare effects with strong changes in a few minutes are characteristic and most prominent, it makes not much sense to establish relations with daily activity indices.

5. EUV fluxes as solar indices. There are good reasons to assume that variations of the exospheric temperature of the main neutral constituents are predominantly controlled by the solar EUV radiation. The long-term variability of these fluxes was, unfortunately, not known until the AEROS-A mission (end of 1972); even worse, continuous monitoring is not envisaged after 1981. Thus, another preliminary measure had to be chosen for representing the solar energy input into planetary and the terrestrial atmospheres when making model computations. JACCHIA[1] has proposed a measure derived from the Covington index F. For geomagnetically quiet conditions the terrestrial exospheric temperature T_{ex} was shown to follow the empirical relation

$$\frac{T_{ex}}{\mathrm{K}} = 379 + 3.25 \cdot \langle F \rangle + 1.3 \cdot (F - \langle F \rangle). \tag{5.1}$$

Since the coefficients were obtained by fitting tracking data from many satellites the relation is well founded statistically. Of course, the special dependence on "two variables" from solar noise was assumed and should not be taken as describing the relevant solar-terrestrial physics. The same "solar activity variables", F and $\langle F \rangle$, are used in most recent atmospheric models, see Sect. 11 of RAWER's contribution in this volume. However, the coefficients vary from one to the other models[2–5], because they were obtained with data from different techniques and periods or missions. The application of the models outside of the period for which they were established is often of limited accuracy. One of the reasons for discrepancies is the lack of full correlation between radio noise of coronal origin F and the solar EUV flux, which constitutes the main heating source of the thermosphere. Bearing this in mind we have proposed activity indices EUV_X and/or EUV_{Y-Z}[6] which are directly derived from the solar radiation intensity at suitably chosen EUV lines or ranges. Table 11 (p. 222) in NIKOL'SKIJ's contribution in volume 49/6 of this Encyclopedia gives an example. The absolute radiation fluxes are directly given by the indices because they are equal to the numerical value of the relevant solar radiation in units of $\mu\mathrm{W\,m^{-2}}$ ($=10^{-3}\,\mathrm{erg\,cm^{-2}\,s^{-1}}$). X indicating a wavelength (in nm) and $Y-Z$ defining a wavelength range from Y to Z.

Comparing with R_Z or F, EUV_{103-15} (for a fixed time) and correspondingly $\langle \mathrm{EUV}_{103-15} \rangle$ (for the average of 81 days) might be a helpful index for

[21] Printing error in heading of NIKOL'SKIJ's Table 5 (p. 349 in volume 49/6): read unit $\mathrm{mW\,m^{-2}}$ instead of $\mathrm{nW\,m^{-2}}$

[1] Jacchia, L.G. (1971): Smithonian Astrophys. Obs. Report No. 332, 21

[2] Jung, M., Krankowsky, D., Lake, L.R., Römer, M. (1976): AEROS Symposium, Bonn

[3] Thuillier, G., Falin, J.L., Wachtel, C. (1977): J. Atmos. Terr. Phys. *39*, 399

[4] v. Zahn, U., Köhnlein, W., Fricke, K.H., Laux, U., Trinks, H., Volland, H. (1977): Geophys. Res. Lett. *4*, 33

[5] Hedin, A.E., Salah, J.E., Evans, J.V., Reber, C.A., Newton, G.P., Spencer, N.W., Kayser, D.C., Alcaydè, D., Bauer, P., Cogger, L., McClure, J.P. (1977): J. Geophys. Res. *82*, 2139

[6] Schmidtke, G. (1976): Geophys. Res. Lett. *3*, 573

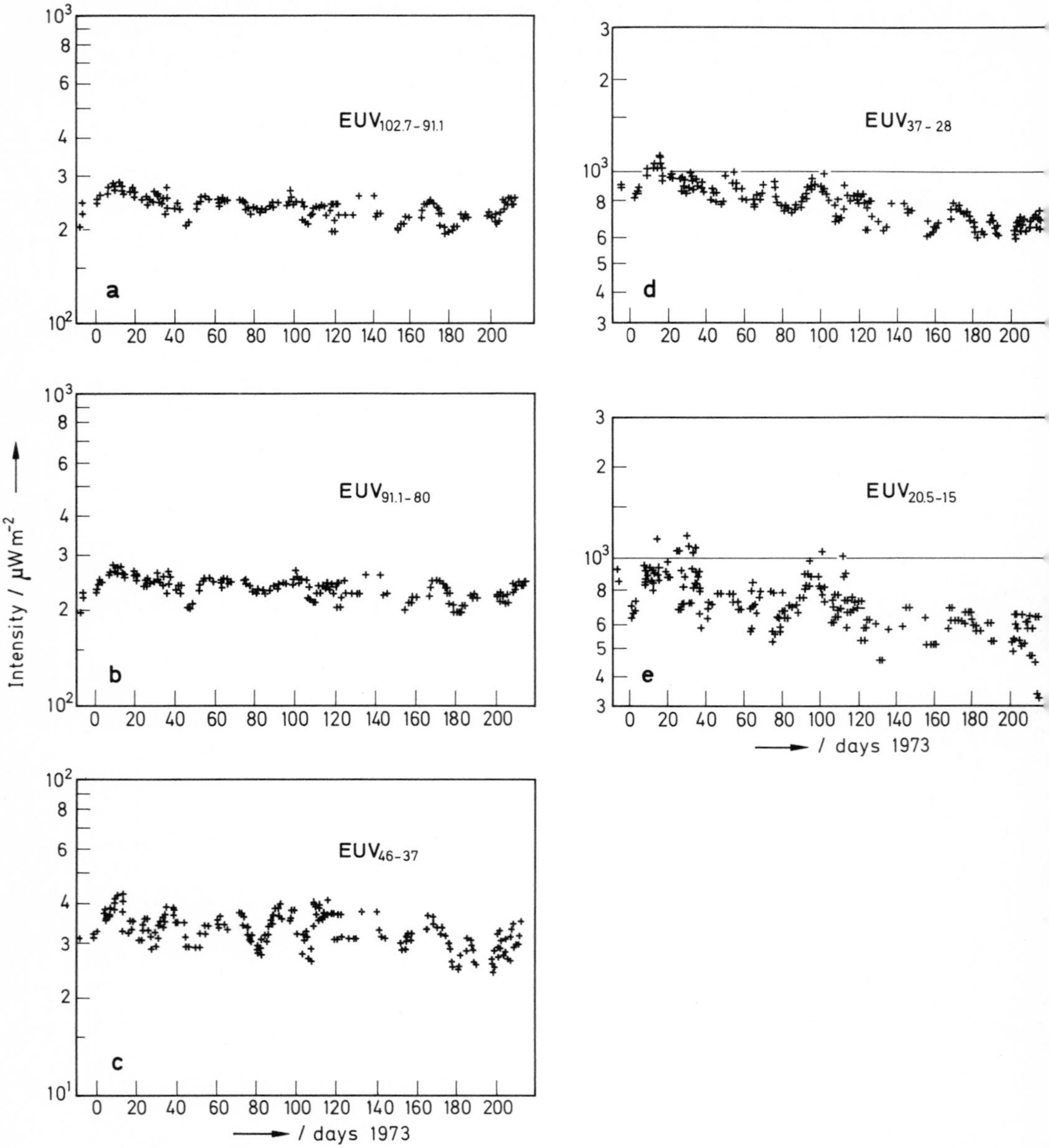

Fig. 23a–e. Variation of five of SCHMIDTKE's EUV-indices during the AEROS-A mission[6]

application in quite different solar-terrestrial fields or relations. Similarly, EUV$_{185-125}$ and/or EUV$_{185-125}$ might be meaningful for meso- and lower thermosphere as well as EUV$_{121.6}$ and EUV$_{2-0.1}$ etc.

If one wishes to express the solar fluxes in terms of no more than two variables, e.g. for simplified global atmospheric modelling, the use of

EUV_{103-80} and EUV_{80-15} is proposed[6]. The first index characterizes the variability of the less energetic solar radiation including the spectral interval from 103 to 91 nm, which is absorbed primarily by molecular oxygen in the lower thermosphere. The energy within the range from 80 to 15 nm is converted at greater heights thus controlling the essential parameters of the upper atmosphere up to the exosphere.

We shall report in the next chapter about an ionospheric study[7] that is based on eight EUV indices covering all-together the EUV range from 102.7 to 15.5 nm (see Table 11, p. 369, in NIKOL'SKIJ's contribution in volume 49/6 of this Encyclopedia). Range limits were especially chosen so as to reflect threshold wavelengths of aeronomical importance or characteristic features of the solar EUV emission spectrum. Figure 23 shows day-by-day data for five of these range indices, it may give an impression of the largely different variability. It can be concluded that a procedure with more than two variables is certainly advantageous as compared with procedures taking into account one variable only.

For more detailed aeronomical studies the index tabulations are available for the two AEROS missions[8,9]. As for the Atmospheric Explorer satellites the relative flux values R_1 and R_2, including the reference spectrum and the tabulation of C_λ might be used as EUV indices, too (see the preceding Chap. II.3). They are available on tape[10]. It will be an important task for the future to establish by thorough statistical analysis better empirical relations between the solar classical and EUV indices and improve the accuracy. Further measurements in space are highly desirable to the same end.

III. Aeronomical implications of solar EUV radiation

The absorption of the solar EUV radiation in the Earth's upper atmosphere initiates a great number of physical processes: dissociation of molecular oxygen, predissociation and dissociation of molecular nitrogen, ionization of molecular nitrogen and oxygen and atomic oxygen, fluorescent and resonant scattering by atmospheric species, excitation of neutral particles etc. These processes control secondary ones the most important of which is probably the chain caused by supra-thermal photo-electrons.

Out of this complex response of the upper atmosphere to irradiation by the Sun in the EUV spectral region, two aspects will be discussed shortly in the following two sections, namely the ionospheric and the thermal impacts of the energy deposit due to absorbed solar radiation.

6. Computation of ionospheric parameters. One of the schemes used at the theoretical computation of ionospheric and atmospheric heating is outlined in

---

[7] Roble, R.G., Schmidtke, G. (1979): J. Atmos. Terr. Phys. *41*, 153

[8] Schmidtke, G. (1977): Report IPW-W.B.3. Freiburg i. Br.: Institut für Physikalische Weltraumforschung

[9] Schmidtke, G. (1979): Report IPW-W.B.11. Freiburg i. Br.: Institut für Physikalische Weltraumforschung

[10] Hinteregger, H.E. (1981): AFGL-LKO, Hanscom Air Force Base, Bedford, Mass. 01730, USA

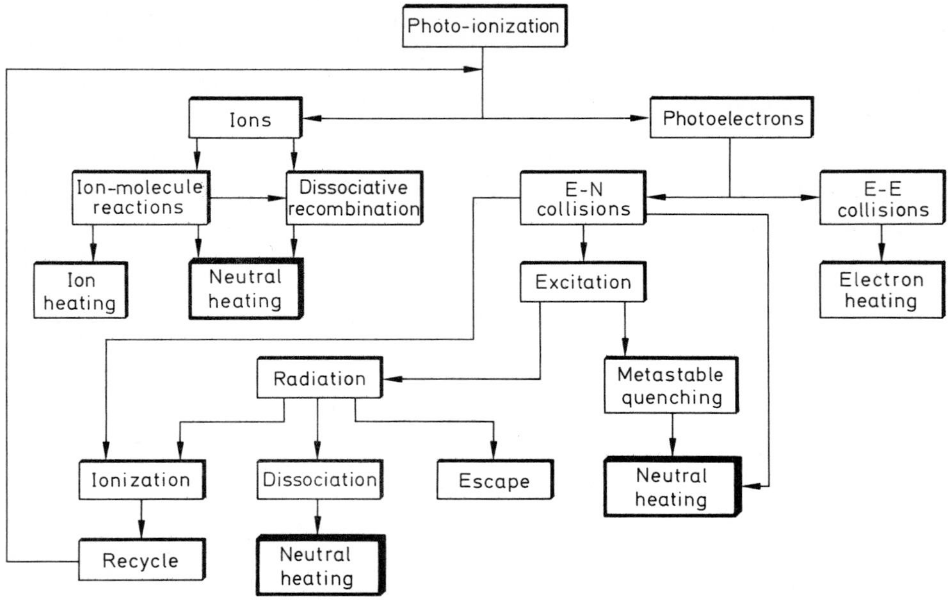

Fig. 24. Flow graph of the most important physical processes leading to neutral gas heating[1]

Fig. 24[1]. The solar energy deposit from wavelengths below 103 nm is split between ion-electron production becoming stored as chemical energy (Fig. 25) and kinetic energy of photo-electrons producing secondary ion-electron pairs. These supra-thermal electron fluxes have been measured aboard the AEROS satellites[2] allowing computations of this kind to be checked. In these latter electron fluxes moving from lower heights where they are produced along the Earth's magnetic field up to the satellite's orbital height; collisional loss undergone on that move was calculated using the MSIS atmospheric model[3,4]. The absorption cross sections needed in this computation were weighted from 35 publications[5] taking into account the spectral ranges used when establishing the AEROS EUV flux tabulation. Agreement between the measured and calculated supra-thermal electron fluxes is good (see Fig. 26)[6].

A more detailed computational model was elaborated[7,8] and tested[9] in order to compare measured and calculated electron density as well as tempera-

---

[1] Stolarski, R.S., Hays, P.B., Roble, R.G. (1975): J. Geophys. Res. *80*, 2266

[2] Spenner, K., Dumbs, A., Lotze, W., Wolf, H. (1974): Space Research *XIV*, 259

[3] Hedin, A.E., Salah, J.E., Evans, J.V., Reber, C.A., Newton, G.P., Spencer, N.W., Kayser, D.C., Alcaydè, D., Bauer, P., Cogger, L., McClue, J.P. (1977): J. Geophys. Res. *82*, 2139

[4] Hedin, A.E., Reber, C.A., Newton, G.P., Spencer, N.W., Brinton, H.C., Mayr, H.G., Potter, W.E. (1977): J. Geophys. Res. *82*, 2156

[5] Schmidtke, G. (1979): Ann. Geophys. *35*, 141

[6] Schmidtke, G. (1979): J. Geomag. Geoelectr. *31*, 581

[7] Roble, R.G. (1975): Planet. Space Sci *23*, 1017

[8] Roble, R.G., Stewart, A.I., Torr, M.R., Rusch, D.W., Wand, R.H. (1978): J. Atmos. Terr. Phys. *40*, 21

[9] Roble, R.G., Schmidtke, G. (1979): J. Atmos. Terr. Phys. *41*, 153

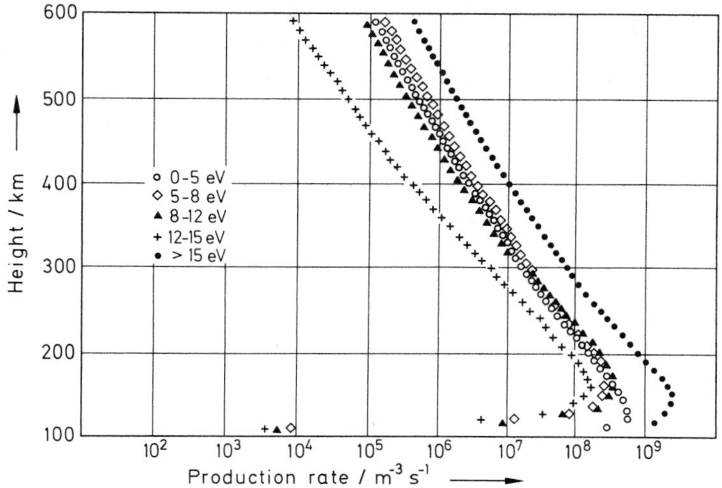

Fig. 25. Electron production rate profiles computed with the solar EUV radiation observed aboard AEROS-B for 1 Aug. 1974 (position 45° N, 11° E)

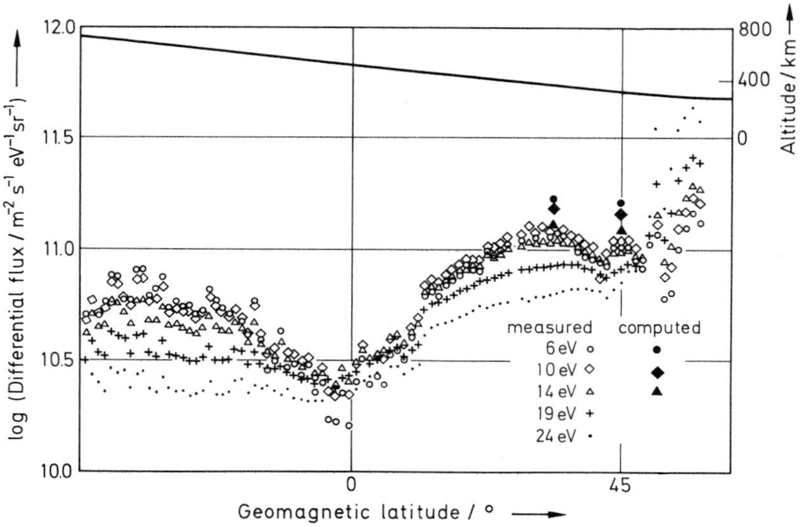

Fig. 26. Comparison of suprathermal electron fluxes measured by the retarding potential analyzer, with the photoelectron flux computed for the observed solar radiation (AEROS-B, 1 Aug. 1974). Line on top: satellite altitude (right-hand scale)

ture changes caused by the variability of the solar EUV fluxes. The latter were monitored by the AEROS-A EUV experiment (see Figs. 27 and 28). An important simplification was, however, made in these calculations since the same structure of the upper atmosphere was assumed for all of the six typical cases of spectra which were considered: two of them may be taken as to characterize the average long-term variation during the mission, two others allow to see the effect of medium-term variations due to the solar rotation,

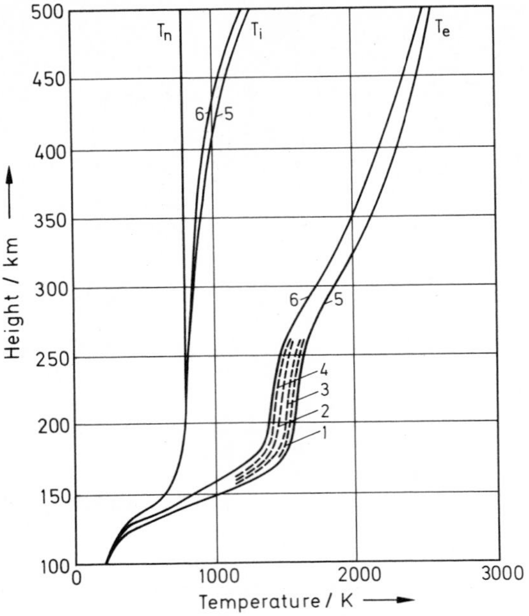

Fig. 27. Electron and ion temperature profiles calculated for the various measured solar flux values identified as cases 1 to 6 (see text)

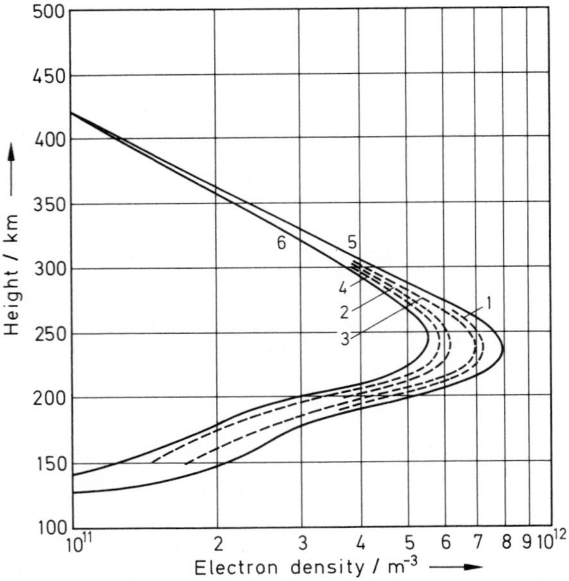

Fig. 28. The calculated electron density profiles for the various measured solar flux values given as cases 1 to 6 (see text)

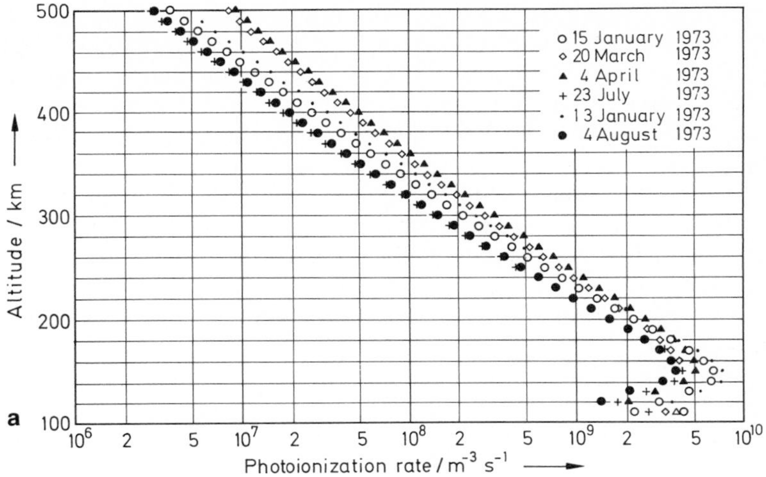

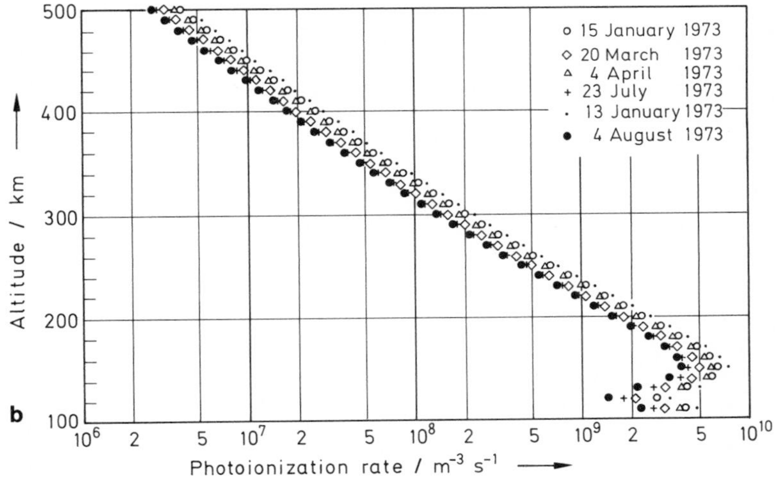

Fig. 29a. Primary electron production rates computed with a constant atmosphere

Fig. 29b. Primary electron production rates computed with a variable atmosphere

and, finally, two more cases were chosen as those days where the solar EUV emissivity was minimum and maximum during the AEROS-A mission period (the latter two cases must be interpreted with some care due to the data statistics and non-reproducible instrumental artifacts of EUV technology). In a later study[5] the computation was improved by taking account of the variability of the upper atmosphere with the variable solar (F) and magnetic (Kp) activities for the six cases. Figure 29 shows the profiles of primary production rate in the case of a constant (Fig. 29a) and a variable[3,4] atmospheric model (Fig. 29b). The difference is rather small such that we reach similar conclusions in both cases[5,9]. Comparing these semi-theoretical results with those of ground-based[10–12] and satellite[2] ionospheric observations we find that changes of the solar EUV fluxes cannot directly explain the large fluctuations in space and time as observed in the ionospheric electron density and electron temperature which have been measured by numerous experiments[2,10–12]. Ionospheric and atmospheric dynamics seem to have stronger impacts on the actual parameters than flux changes. Unfortunately, similar studies on a global scale were not yet carried out because such investigation would involve the availability of a global atmospheric model taking correctly account of effects of variable activity, further the introduction of numerous primary and secondary processes (see the contribution by THOMAS in volume 49/6 of this Encyclopedia), in addition one should take into account upper atmospheric dynamics and the interaction with the neutral atmosphere (see STUBBE's contribution in the same volume), further the night-time ionization and compare the results with the International Reference Ionosphere (see Sect. 13 in RAWER's preceding contribution). Finally at least at higher latitudes the contributions due to corpuscular heating and other disturbance effects by the solar wind including Joule heating must be taken account of. The scheme introduced by ROBLE[1,7] includes a few of these effects but not all of them (see Sects. 17, 18 of RAWER's preceding contribution). In full generality this problem is of great complexity.

7. Atmospheric heating by solar EUV radiation. In the preceding, ionospheric aspects of the interaction between the solar EUV radiation and the upper atmosphere were discussed. Atmospheric heating is another topic of great aeronomic importance.

In Fig. 24[1] different channels are indicated in which the ionizing radiation below 103 nm can be converted into heat. In addition, absorption at longer wavelengths, in the Schumann-Runge spectral region (see Sect. 4 of THOMAS's contribution in volume 49/6 of this Encyclopedia), strongly contributes to the thermal energy of the neutral particles in the lower thermosphere, say above 100 km; Fig. 30 shows the scheme of actions and reactions. One of the parameters that represents particularly well the atmospheric energy balance is the temperature of the exosphere. In the past two decades it has been derived by various methods. Among them are atmospheric drag analysis, mass spectrometric measurements, solar EUV radiation extinction analysis and Fabry-Perot interferometer measurements from ground or satellites. For details see Chap. A of RAWER's contribution, p. 233. Based on experimental data derived from such measurements, semi-empirical atmospheric models were derived (see ibidem Chap. C). Similarly, the thermopause temperature can be computed from these models as is presented in Fig. 31[2]. Because routine solar EUV measurements are lacking, the Covington index F was used to characterize the EUV

---

[10] Evans, J.V. (1965): Planet. Space Sci. *13*, 1031
[11] Evans, J.V. (1973): Planet. Space Sci. *21*, 763
[12] Evans, J.V. (1973): J. Geophys. Res. *78*, 2344
[1] Stolarski, R.S., Hays, P.B., Roble, R.G. (1975): J. Geophys. Res. *80*, 2266
[2] Kockarts, G. (1981): Solar Phys. *74*, 295

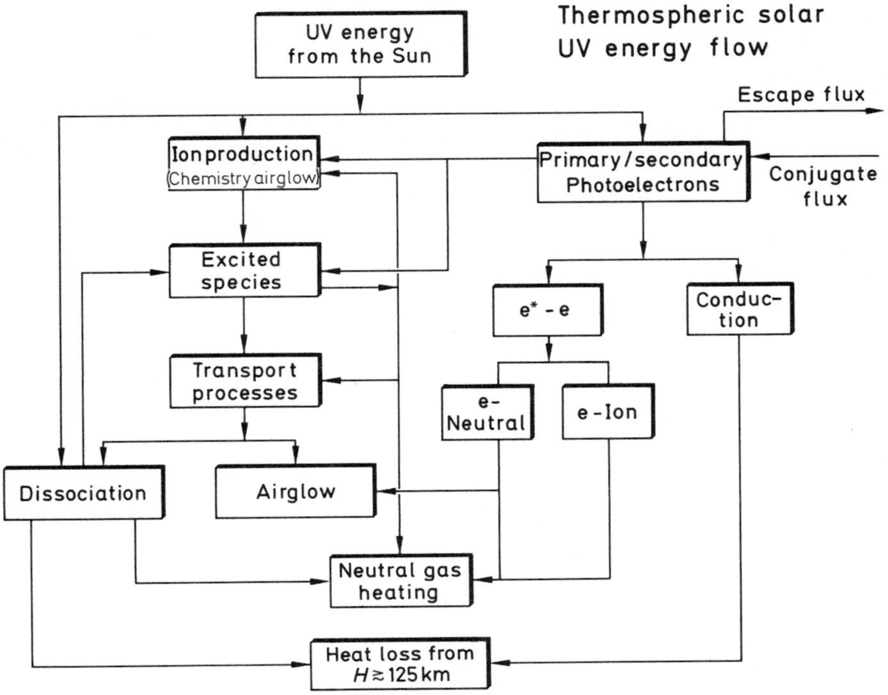

Fig. 30. Graph of the major solar ultraviolet energy flow channels[5]

energy input. Also the energy input into atmospheric heating due to the solar wind, more precisely to the solar particle flow via the magnetosphere into the atmosphere, is indirectly taken into account; it is parametrized by the *Ap* index characterizing the degree of geomagnetic disturbance[3].

One of the interesting features in Fig. 31 is the variation of the thermopause temperature with the eleven years solar cycle being different from one cycle to the next one. The annual variation of the temperature is not correlated with solar activity characterized by the Covington index *F*, thus probably resulting more indirectly from seasonal atmospheric circulation effects than directly from irradiation changes. Peaks in the daily index *F* do not necessarily lead to peaks in the thermopause temperature. For geomagnetically disturbed periods the influence of particle dissipation in the auroral regions may be quite strong, even in midlatitudinal regions as applies to Scheveningen.

Incoherent scatter observations allowed also to determine the diurnal variations of exospheric temperature. The amplitudes of the diurnal (24 h) and semidiurnal (12 h) components were found to increase strongly with increasing Covington index $\langle F \rangle$, see Fig. 32[4].

---

[3] Siebert, M. (1971): This Encyclopedia, volume 49/3, p. 225

[4] Forbes, J.M., Garrett, H.B. (1979): J. Geophys. Res. *84*, 1947

[5] Torr, M.R., Richards, P.G., Torr, D.G. (1981): Adv. Space Res. *1*, 53

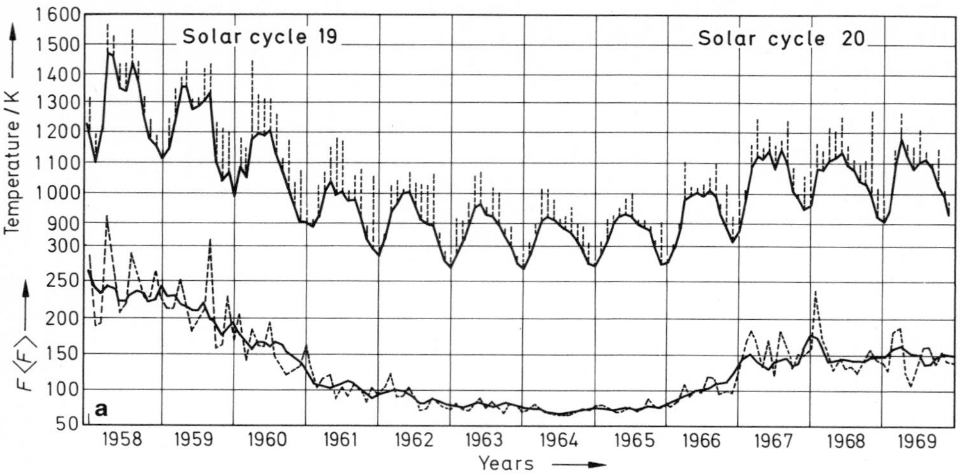

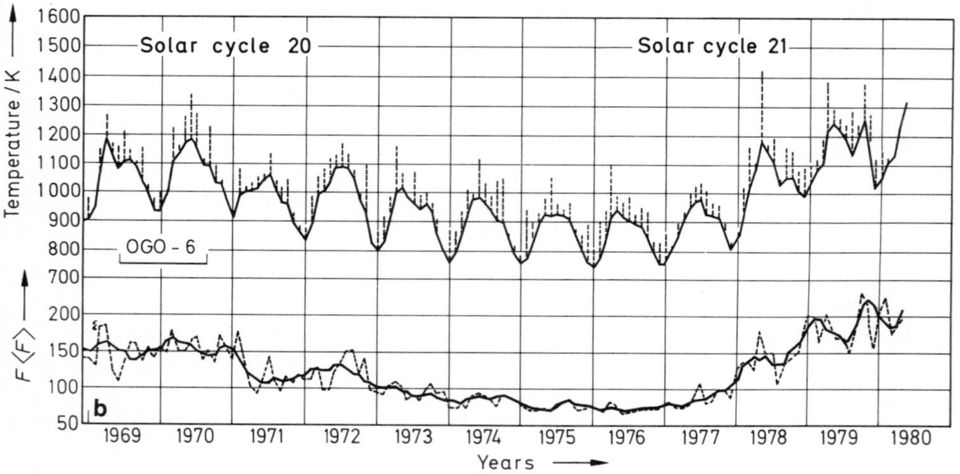

Fig. 31a. Upper part: Thermopause temperature (at 17 LT) above Scheveningen (52.08° N) obtained by applying the "OGO-6" model. The temperature was computed for the first of each month from 1958 to 1969. Vertical dashed lines indicate temperature increases resulting from geomagnetic activity. Lower part: Dashed curve gives the daily 10.7 cm solar flux F; full curve represents the value $\langle F \rangle$ obtained by averaging over three solar rotations (i.e. 81 d, centered on day no. 40)

Fig. 31b. Continuation of Fig. 31a from 1969 to 1980. Horizontal line labeled OGO-6 indicates the period over which temperature data were really measured with a Fabry-Perot interferometer aboard that satellite (see RAWER's contribution, p. 273, his Sect. 6β) from which data the computation schedule was derived[2]

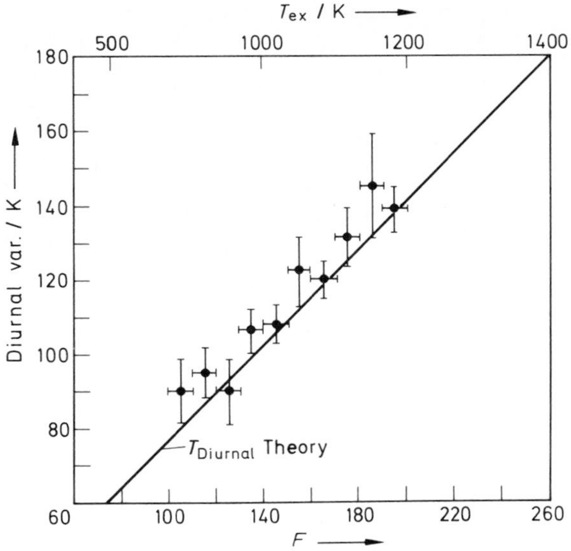

Fig. 32. Diurnal amplitude of exospheric temperature from 1969 through 1972 mid-latitude incoherent scatter observations[4] as function of Covington-index F (corresponding average exospheric temperature on top)

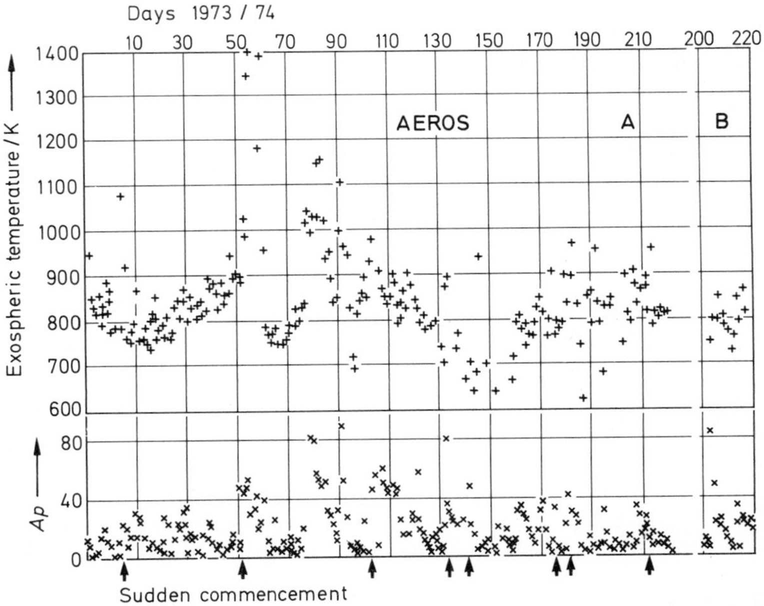

Fig. 33. Exospheric temperatures derived from EUV extinction measurements[6] and Ap Index. Full triangles identify the starting time of geomagnetic disturbances

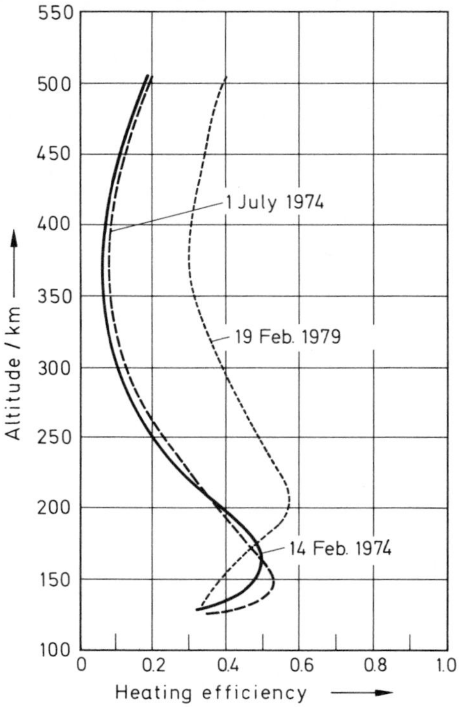

Fig. 34. Heating efficiency profiles for solar minimum and maximum conditions. Results of involved aeronomic model computations[5]

The exospheric temperature T_{ex} was experimentally derived from AEROS-A measurements by EUV extinction analysis[6]; the total solar EUV flux measured by the same experiment was taken as indicator of solar activity. With data of 1973 it was found that exospheric temperatures at midlatitudes neither reflect the long-term decrease of the solar EUV irradiance nor the medium-term changes with solar rotations (see Fig. 33). Geomagnetic activity is, apparently, controlling the exospheric temperature stronger since the starts of geomagnetic disturbances reappear in increased values of T_{ex}. Similar observations were made by the Atmospheric Explorer satellites[7]. The decrease of the ionizing EUV irradiance during the mission of AEROS-A was by about 30%, or of the order of 10^{11} W for the whole Earth; this is just about the magnitude of the energy involved in a geomagnetic storm. Thus during disturbances Joule heating and dissipation of particles of solar wind origin are competing with solar EUV radiation as upper atmospheric heat sources, at least during solar minimum conditions.

It is rather difficult to identify the influence of the different heat sources in the upper atmosphere[8] separately in view of the dynamics of the Earth-

[6] Schmidtke, G., Börsken, N., Sünder, G. (1981): J. Geophys. *49*, 146

[7] Hinteregger, H.E. (1979): Private communication

[8] Schmidtke, G. (editor) (1981): Adv. Space Res. *1*, Sects. 1 and 2, p. 3

atmosphere system. In addition, most of the parameters involved have different time variation. The result is that the heating efficiency of the solar EUV radiation is not constant[5], neither with atmospheric height nor with time (Fig. 34). Therefore, this efficiency is a very important parameter which must be taken into account in a quantitative analysis of the energy budget of the upper atmosphere. This, however, is not yet possible, because too little is known about the atmospheric heating efficiency and variability of the solar wind and the role of the atmospheric airglow. These fields are yet to be investigated in more detail.

Caption to Tables 4 and 5 (Tables see pages 46–55)

Table 4. Computation of EUV fluxes after Eq. (3.1). To each wavelength λ and identification 'Ident.' the table gives:
in column 2 the reference flux ('Flux')/10^{10} [photons] $m^{-2} s^{-1}$,
in column 4 the variability class index K_λ ('Type'),
in column 5 the variability index C_λ ('Intens.')

Table 5. Variability of key emissions: ratio R_K of observed to reference intensity [Eqs. (3.1) and (3.2)] as 'I_1' for H Lyman-beta and 'I_2' for Fe XVI (33.5 nm). First column gives Julian date. [Reference period was 13–28 July 1976]

Table 4 (Caption see page 45)

λ/nm	Flux Ident.	Type	Intens.	λ/nm	Flux Ident.	Type	Intens.	λ/nm	Flux Ident.	Type	Intens.
1.86	.1 O VII	2	.20	7.73	2.9*	1	.83	12.27	4.1NE VI	1	.48
1.90	.1 O VIII	2	.70	7.77	3.9MG IX	2	.01	12.35	2.6*	1	.83
2.16	.3 O VII	2	.20	7.86	3.0*	1	.83	12.76	4.5*	1	.83
2.18	.1 O VII	2	.20	7.87	2.8NI XI	2	.02	12.99	3.4 O VI	1	1.04
2.21	.3 O VII	2	.20	7.91	1.9*	1	.83	13.03	.1NE VI	1	.48
2.85	.5 C VI	2	.18	7.95	1.8FE XII	2	.05	13.10	5.2$	1	.83
2.88	2.5 N VI	2	.10	7.98	2.3*	1	.83	13.12	4.8$	1	.83
2.95	2.2 N VI	2	.10	8.00	1.4FE XII	2	.05	13.62	.1NE VI	1	.48
3.00	.9*	1	.83	8.05	2.3FE XII	2	.05	13.63	.1NE VI	1	.49
3.04	.6 S XIV	2	.22	8.24	4.9FE IX	2	.01	13.63	.1NE VI	1	.48
3.37	1.1 C VI	2	.18	8.27	2.6FE XII	2	.05	13.64	.1NE VI	1	.48
4.09	.6SI XII	2	.15	8.28	2.6FE XII	2	.05	13.65	.1NE VI	1	.48
4.38	2.1SI XI	2	.07	8.34	4.5MG VII	2	.01	14.12	11.3*	1	.83
4.40	.8SI XII	2	.15	8.37	3.8MG VII	2	.01	14.43	1.1NI X	2	.01
4.42	.9SI XII	2	.15	8.40	5.1MG VII	2	.01	14.50	14.1NI X	2	.01
4.57	.5SI XII	2	.15	8.68	4.6FE XI	2	.03	14.84	29.4NI XI	2	.02
4.64	2.7SI XI	2	.07	8.69	1.9NI X	2	.01	15.01	7.6 O VI	1	1.04
4.67	4.0*	1	.83	8.70	3.1FE XI	2	.03	15.21	13.1NI XII	2	.07
4.79	4.5*	1	.83	8.73	2.4*	1	.83	15.42	6.9NI XII	2	.07
4.92	4.3SI XI	2	.07	8.76	2.0*	1	.83	15.77	6.2NI XIII	2	.12
5.05	5.6SI X	2	.04	8.81	4.5NE VIII	2	.02	15.84	14.2NI X	2	.01
5.07	5.6SI X	2	.04	8.81	6.1$	1	.83	16.00	13.0NI X	2	.01
5.23	3.5SI XI	2	.07	8.81	.1NE VIII	2	.02	16.04	11.0*	1	.83
5.29	.1FE XV	2	.41	8.84	1.9*	1	.83	16.20	5.5*	1	.83
5.41	8.2$	1	.83	8.86	2.4*	1	.83	16.41	3.7NI XIV	2	.20
5.44	3.6*	1	.83	8.89	3.8FE XI	2	.03	16.75	19.5FE VIII	1	.42
5.51	4.0MG IX	2	.01	8.91	2.7FE XI	2	.03	16.82	36.0FE VIII	1	.42
5.53	10.6SI IX	2	.02	8.97	3.0FE XI	2	.03	16.85	20.5FE VIII	1	.42
5.61	2.4 S IX	2	.02	9.01	3.1FE XI	2	.03	16.89	12.6FE VIII	1	.42
5.69	7.4*	1	.83	9.04	2.0FE XI	2	.03	16.97	23.4*	1	.83
5.74	6.2*	1	.83	9.07	3.0*	1	.83	17.11	307.5FE IX	2	.01
5.76	5.0*	1	.83	9.10	3.9*	1	.83	17.22	11.2 O V	1	.47
5.79	4.3MG X	2	.03	9.15	1.9*	1	.83	17.31	21.4$	1	.83
5.90	.7FE XIV	2	.15	9.17	5.4NI X	2	.01	17.46	268.8FE X	2	.02
5.96	.7FE XIV	2	.15	9.18	4.8*	1	.83	17.53	33.1$	1	.83
6.03	2.5*	1	.83	9.21	3.8*	1	.83	17.72	162.5FE X	2	.02
6.08	3.6*	1	.83	9.28	3.8*	1	.83	17.80	20.5FE XI	2	.03
6.11	5.8SI VIII	2	.02	9.36	5.6*	1	.83	17.93	.4NI XV	2	.30
6.16	2.9 S VIII	2	.02	9.41	8.2FE X	2	.02	17.97	20.6FE XI	2	.03
6.19	5.0$	1	.83	9.42	.1NE VII	2	.01	18.04	179.0FE XI	2	.03
6.23	.1NE VIII	2	.02	9.44	1.7*	1	.83	18.11	21.2FE XI	2	.03
6.23	1.1FE XII	2	.05	9.49	.1NE VII	2	.01	18.22	23.1FE XI	2	.03
6.28	3.4MG IX	2	.01	9.54	4.8FE X	2	.02	18.34	1.5CA XIV	2	.15
6.32	3.3MG X	2	.03	9.55	2.9*	1	.83	18.45	43.8FE X	2	.02
6.33	5.4MG X	2	.03	9.58	2.9*	1	.83	18.48	2.3FE XI	2	.03
6.36	4.1*	1	.83	9.60	8.2FE X	2	.02	18.52	18.2FE VIII	B1	.42
6.41	1.1FE XIII	2	.09	9.65	1.9*	1	.83	18.66	3.2CA XIV	2	.15
6.46	2.5*	1	.83	9.68	2.7FE X	2	.02	18.69	23.1 S XI	B2	.03
6.52	3.0*	1	.83	9.71	5.2FE X	2	.02	18.79	.7AR XIV	2	.40
6.57	4.1*	1	.83	9.75	3.0$	1	.83	18.82	9.7FE XII	2	.05
6.58	3.0MG X	2	.03	9.79	2.3FE X	2	.02	18.83	120.0FE XI	2	.03
6.63	.3FE XVI	2	1.00	9.81	2.7NE VIII	2	.02	19.00	42.3FE X	2	.02
6.63	3.8FE XII	2	.05	9.83	2.7NE VIII	2	.02	19.10	10.5FE XII	2	.05
6.64	.4FE XVI	2	1.00	9.85	2.6*	1	.83	19.13	9.0FE XIII	B2	.09
6.71	3.0MG IX	2	.01	9.97	2.0*	1	.83	19.24	40.0FE XII	2	.05
6.73	2.0FE XII	2	.05	10.00	2.7*	1	.83	19.28	58.8FE XI	2	.03
6.83	2.3*	1	.83	10.05	8.4*	1	.83	19.35	65.5FE XII	2	.05
6.96	13.8*	1	.83	10.30	1.5NE VIII	2	.02	19.51	107.0FE XII	2	.05
7.00	.1FE XV	2	.41	10.31	.1NE VII	2	.01	19.65	9.7FE XIII	2	.09
7.05	3.3*	1	.83	10.36	6.1FE IX	2	.01	19.66	2.9FE XII	2	.05
7.07	3.0*	1	.83	10.39	6.1*	1	.83	19.74	4.0FE XIII	2	.09
7.10	4.4*	1	.83	10.52	5.1FE IX	2	.01	19.86	7.1FE XII	B2	.05
7.19	1.2FE XIV	2	.15	10.62	2.0NE VII	2	.01	20.00	15.0FE XIII	2	.09
7.23	6.4$	1	.83	10.80	1.5FE VIII	1	.42	20.11	25.5FE XIII	B2	.09
7.26	1.6FE XI	2	.03	11.00	.1NE VII	2	.01	20.20	39.5FE XIII	2	.09
7.28	2.1*	1	.83	11.06	.1NE VII	2	.01	20.26	22.1$	1	.83
7.29	3.4*	1	.83	11.06	.1NE VII	2	.01	20.38	19.5FE XIII	2	.09
7.35	.5FE XV	2	.41	11.08	1.5*	1	.83	20.42	7.2FE XIII	2	.09
7.35	1.9NE VIII	2	.02	11.12	.1NE VI	1	.48	20.49	4.9FE XIII	2	.09
7.42	2.5*	1	.83	11.12	4.8*	1	.83	20.63	1.5*	1	.83
7.44	1.3*	1	.83	11.38	3.0*	1	.83	20.64	1.5*	1	.83
7.48	4.0$	1	.83	11.41	2.6*	1	.83	20.75	1.5*	1	.83
7.50	4.6MG VIII	2	.01	11.42	.1NE VI	1	.48	20.83	1.8 S X	B2	.02
7.53	2.5*	1	.83	11.54	.1NE VII	2	.01	20.96	.9FE XIII	2	.09
7.55	3.8*	1	.83	11.58	2.4*	1	.83	20.98	1.1*	1	.83
7.57	2.5*	1	.83	11.67	3.9*	1	.83	21.13	27.4FE XIV	2	.15
7.60	3.2FE XIII	2	.09	11.72	2.6*	1	.83	21.21	8.0 S XII	2	.07
7.65	1.0FE XIII	2	.09	12.04	.1NE VII	2	.01	21.38	2.5FE XIII	2	.09
7.68	4.1$	1	.83	12.11	.1NE VI	1	.48	21.47	4.3SI VIII	2	.02
7.69	3.3*	1	.83	12.18	1.1*	1	.83	21.52	14.5 S XII	2	.07

Table 4 (continued)

λ/nm	Flux Ident.	Type	Intens.	λ/nm	Flux Ident.	Type	Intens.	λ/nm	Flux Ident.	Type	Intens.
21.69	10.0SI VIII	B2	.02	36.08	17.5FE XVI	2	1.00	59.96	190.0 O III	1	.42
21.82	18.5 S XII	2	.07	36.45	75.6FE XII	2	.05	60.98	450.0MG X	2	.03
21.91	7.2FE XIV	2	.15	36.81	739.4MG IX	2	.01	61.66	16.7 O II	1	.42
22.01	10.3FE XIV	2	.15	39.98	15.6NE VI	1	.48	62.49	100.0MG X	2	.03
22.14	17.8 S XII	2	.07	40.11	34.5NE VI	1	.48	62.97	1500.0 O V	1	.47
22.18	.9FE XIII	2	.09	40.19	91.2NE VI	1	.48	63.85	24.9*	1	.83
22.47	41.7 S IX	B2	.02	40.33	53.4NE VI	1	.48	64.04	5.9 S II	1	.42
22.51	76.5$	1	.83	41.72	8.8FE XV	2	.41	64.09	7.0 S II	1	.42
22.70	41.5SI IX	2	.02	43.05	82.3MG VIII	2	.01	64.18	9.2 S II	1	.42
22.72	.2FE XV	2	.41	43.67	122.4MG VIII	2	.01	64.41	10.9 O II	1	.42
22.75	29.5 S XII	2	.05	45.30	1.3HE I [C]	1	.83	65.03	15.6*	1	.83
22.87	20.4 S X	2	.02	45.40	1.3HE I [C]	1	.83	65.73	6.2 S IV	1	.42
23.06	13.0HE II	1	.42	45.50	1.3HE I [C]	1	.83	66.14	6.2 S IV	1	.42
23.15	15.8HE II	1	.42	45.60	1.3HE I [C]	1	.83	67.15	10.6 N II	1	.50
23.26	23.0HE II	1	.42	45.70	1.3HE I [C]	1	.83	68.17	37.9NA IX	2	.01
23.38	3.6FE XV	2	.41	45.80	1.3HE I [C]	1	.83	68.57	101.3 N III	G1	.52
23.44	33.1$	1	.83	45.90	1.3HE I [C]	1	.83	69.08	20.9*	1	.83
23.71	3.9FE XII	2	.05	46.00	1.6HE I [C]	1	.83	69.43	22.6NA IX	2	.01
23.72	.2SI XIV	2	1.50	46.10	1.6HE I [C]	1	.83	70.00	1.0 H LY[C]	1	1.00
23.73	19.3HE II	1	.37	46.20	1.6HE I [C]	1	.83	70.10	1.0 H LY[C]	1	1.00
23.99	20.0 S XI	2	.03	46.30	1.6HE I [C]	1	.83	70.20	1.0 H LY[C]	1	1.00
24.07	16.7FE XIII	2	.09	46.40	1.6HE I [C]	1	.83	70.30	1.1 H LY[C]	1	1.00
24.17	89.4FE IX	2	.01	46.50	2.0HE I [C]	1	.83	70.34	391.5 O III	G1	.42
24.30	87.0HE II	1	.37	46.52	180.0NE VII	2	.01	70.40	1.1 H LY[C]	1	1.00
24.38	5.7FE XV	2	.41	46.60	2.0HE I [C]	1	.83	70.50	1.1 H LY[C]	1	1.00
24.49	58.0FE IX	2	.01	46.70	2.0HE I [C]	1	.83	70.60	1.1 H LY[C]	1	1.00
24.59	1.0$	1	.83	46.80	2.5HE I [C]	1	.83	70.70	1.1 H LY[C]	1	1.00
24.62	22.8FE XIII	2	.09	46.90	2.5HE I [C]	1	.83	70.80	1.1 H LY[C]	1	1.00
24.69	11.7$	1	.83	47.00	2.5HE I [C]	1	.83	70.90	1.1 H LY[C]	1	1.00
24.72	24.9$	1	.83	47.10	2.5HE I [C]	1	.83	71.00	1.3 H LY[C]	1	1.00
24.92	1.2NI XVII	2	.90	47.20	2.9HE I [C]	1	.83	71.10	1.3 H LY[C]	1	1.00
25.11	.8FE XVI	2	1.00	47.30	2.9HE I [C]	1	.83	71.20	1.3 H LY[C]	1	1.00
25.19	22.8FE XIII	2	.09	47.40	3.3HE I [C]	1	.83	71.27	12.4 S VI	1	.50
25.22	16.0FE XIV	2	.15	47.50	3.3HE I [C]	1	.83	71.30	1.4 H LY[C]	1	1.00
25.38	16.5SI X	2	.04	47.60	3.3HE I [C]	1	.83	71.40	1.4 H LY[C]	1	1.00
25.63	235.0HE II	1	.37	47.70	3.8HE I [C]	1	.83	71.50	1.5 H LY[C]	1	1.00
25.64	36.3SI X	2	.04	47.80	3.8HE I [C]	1	.83	71.60	1.5 H LY[C]	1	1.00
25.66	15.0 S XIII	2	.12	47.90	4.1HE I [C]	1	.83	71.70	1.5 H LY[C]	1	1.00
25.69	2.5FE XV	2	.41	48.00	4.1HE I [C]	1	.83	71.80	1.7 H LY[C]	1	1.00
25.72	65.7 S X	B2	.02	48.10	4.5HE I [C]	1	.83	71.85	53.1 O II	1	.42
25.74	17.9FE XIV	B2	.15	48.20	4.5HE I [C]	1	.83	71.90	1.7 H LY[C]	1	1.00
25.84	85.0SI X	2	.04	48.30	5.0HE I [C]	1	.83	72.00	1.7 H LY[C]	1	1.00
25.95	36.2 S X	2	.02	48.40	5.4HE I [C]	1	.83	72.10	1.7 H LY[C]	1	1.00
26.10	42.0SI X	2	.04	48.50	5.4HE I [C]	1	.83	72.20	1.8 H LY[C]	1	1.00
26.30	2.0FE XVI	2	1.00	48.60	5.8HE I [C]	1	.83	72.30	1.8 H LY[C]	1	1.00
26.42	31.2 S X	2	.02	48.70	6.1HE I [C]	1	.83	72.40	1.9 H LY[C]	1	1.00
26.48	30.0FE XIV	2	.15	48.80	6.6HE I [C]	1	.83	72.50	1.9 H LY[C]	1	1.00
27.05	25.0FE XIV	2	.15	48.90	6.6HE I [C]	1	.83	72.60	1.9 H LY[C]	1	1.00
27.20	45.6SI X	2	.04	48.95	11.0NE III	1	.33	72.70	1.9 H LY[C]	1	1.00
27.26	8.8SI VII	2	.01	49.00	7.0HE I [C]	1	.83	72.80	2.0 H LY[C]	1	1.00
27.42	50.0FE XIV	2	.15	49.10	7.4HE I [C]	1	.83	72.90	2.0 H LY[C]	1	1.00
27.53	21.5SI VII	2	.01	49.20	7.9HE I [C]	1	.83	73.00	2.2 H LY[C]	1	1.00
27.57	17.9SI VII	2	.01	49.30	8.3HE I [C]	1	.83	73.10	2.2 H LY[C]	1	1.00
27.62	6.1MG VII	2	.01	49.40	8.6HE I [C]	1	.83	73.20	2.2 H LY[C]	1	1.00
27.68	7.5$	1	.83	49.50	9.5HE I [C]	1	.83	73.30	2.4 H LY[C]	1	1.00
27.70	3.0$	1	.83	49.60	9.9HE I [C]	1	.83	73.40	2.4 H LY[C]	1	1.00
27.73	63.0$	1	.83	49.70	10.4HE I [C]	1	.83	73.50	2.4 H LY[C]	1	1.00
27.84	24.1MG VII	B2	.01	49.80	10.8HE I [C]	1	.83	73.60	2.5 H LY[C]	1	1.00
28.14	10.5 S XI	2	.03	49.90	11.5HE I [C]	1	.83	73.70	2.5 H LY[C]	1	1.00
28.41	100.0FE XV	2	.41	49.94	77.5SI XII	2	.15	73.80	2.7 H LY[C]	1	1.00
28.57	14.4 S XI	G2	.03	50.00	12.0HE I [C]	1	.83	73.90	2.7 H LY[C]	1	1.00
28.92	11.8FE XIV	2	.15	50.10	12.7HE I [C]	1	.83	74.00	2.7 H LY[C]	1	1.00
29.07	30.0SI IX	2	.02	50.20	13.3HE I [C]	1	.83	74.10	2.8 H LY[C]	1	1.00
29.17	15.0 S XI	G2	.03	50.30	14.0HE I [C]	1	.83	74.20	2.9 H LY[C]	1	1.00
29.28	44.0SI IX	2	.02	50.40	14.9HE I [C]	1	.83	74.30	2.9 H LY[C]	1	1.00
29.62	50.0SI IX	2	.02	50.79	76.5 O III	G1	.42	74.40	3.0 H LY[C]	1	1.00
29.95	8.7 S XII	2	.07	51.56	33.5HE I-3	1	.73	74.50	3.0 H LY[C]	1	1.00
30.33	235.0SI XI	2	.12	52.07	35.6SI XII	2	.15	74.60	3.2 H LY[C]	1	1.00
30.38	6000.0HE II	1	.37	52.58	75.6 O III	1	.42	74.70	3.3 H LY[C]	1	1.00
31.50	133.5MG VIII	2	.01	53.70	186.0HE I-2	1	.73	74.80	3.4 H LY[C]	1	1.00
31.62	112.5SI VIII	2	.02	54.28	23.2NE IV	1	.33	74.90	3.4 H LY[C]	1	1.00
31.90	17.5NI XV	2	.30	55.00	23.2AL XI	1	.58	75.00	3.5 H LY[C]	1	1.00
31.98	146.3SI VIII	2	.02	55.44	799.2 O IV	G1	.46	75.00	40.5*	1	.83
32.06	4.7NI XVIII	B2	1.10	55.86	61.5NE VI	1	.48	75.10	3.7 H LY[C]	1	1.00
33.54	34.0FE XVI	2	1.00	56.28	81.9NE VI	1	.48	75.20	3.7 H LY[C]	1	1.00
34.51	96.3SI IX	2	.02	56.85	51.7$	1	.83	75.30	3.8 H LY[C]	1	1.00
34.57	84.0FE X	2	.02	57.23	68.9$	1	.83	75.40	3.9 H LY[C]	1	1.00
34.74	141.0SI X	2	.04	58.04	11.1 O II	1	.42	75.50	3.9 H LY[C]	1	1.00
34.98	96.3SI IX	2	.02	58.43	1580.0HE I	1	.99	75.60	4.1 H LY[C]	1	1.00
35.60	103.4SI X	2	.04	59.24	19.2*	1	.83	75.70	4.2 H LY[C]	1	1.00

Table 4 (continued)

λ/nm	Flux	Ident.	Type	Intens.
75.80	4.3	H LY[C]	1	1.00
75.87	34.9	O V	1	.47
75.90	4.6	H LY[C]	1	1.00
75.94	26.7	O V	1	.47
76.00	4.6	H LY[C]	1	1.00
76.03	93.0	O V	1	.47
76.10	4.7	H LY[C]	1	1.00
76.11	23.2	O V	1	.47
76.20	4.8	H LY[C]	1	1.00
76.20	34.9	O V	1	.47
76.30	4.9	H LY[C]	1	1.00
76.40	5.1	H LY[C]	1	1.00
76.50	5.2	H LY[C]	1	1.00
76.51	199.7	N IV	1	.58
76.60	5.3	H LY[C]	1	1.00
76.70	5.4	H LY[C]	1	1.00
76.80	5.6	H LY[C]	1	1.00
76.90	5.7	H LY[C]	1	1.00
77.00	5.8	H LY[C]	1	1.00
77.04	242.5	NE VIII	2	.02
77.10	6.1	H LY[C]	1	1.00
77.20	6.2	H LY[C]	1	1.00
77.30	6.3	H LY[C]	1	1.00
77.40	6.6	H LY[C]	1	1.00
77.50	6.6	H LY[C]	1	1.00
77.60	11.8	N II	1	.50
77.60	6.8	H LY[C]	1	1.00
77.70	7.1	H LY[C]	1	1.00
77.80	7.2	H LY[C]	1	1.00
77.90	7.3	H LY[C]	1	1.00
78.00	7.6	H LY[C]	1	1.00
78.03	130.6	NE VIII	2	.02
78.10	7.7	H LY[C]	1	1.00
78.20	8.0	H LY[C]	1	1.00
78.30	8.1	H LY[C]	1	1.00
78.40	8.3	H LY[C]	1	1.00
78.50	8.6	H LY[C]	1	1.00
78.60	8.7	H LY[C]	1	1.00
78.65	146.3	S V	1	.42
78.70	9.1	H LY[C]	1	1.00
78.77	277.5	O IV	1	.46
78.80	9.2	H LY[C]	1	1.00
78.90	9.5	H LY[C]	1	1.00
79.00	9.6	H LY[C]	1	1.00
79.01	477.3	O IV	G1	.46
79.10	10.0	H LY[C]	1	1.00
79.20	10.2	H LY[C]	1	1.00
79.30	10.3	H LY[C]	1	1.00
79.40	10.7	H LY[C]	1	1.00
79.50	11.0	H LY[C]	1	1.00
79.60	11.2	H LY[C]	1	1.00
79.70	11.5	H LY[C]	1	1.00
79.80	11.9	H LY[C]	1	1.00
79.90	12.1	H LY[C]	1	1.00
80.00	12.4	H LY[C]	1	1.00
80.10	12.9	H LY[C]	1	1.00
80.20	13.1	H LY[C]	1	1.00
80.30	13.5	H LY[C]	1	1.00
80.40	13.9	H LY[C]	1	1.00
80.50	14.1	H LY[C]	1	1.00
80.60	14.5	H LY[C]	1	1.00
80.70	14.9	H LY[C]	1	1.00
80.80	15.3	H LY[C]	1	1.00
80.90	15.6	H LY[C]	1	1.00
81.00	16.0	H LY[C]	1	1.00
81.10	16.5	H LY[C]	1	1.00
81.20	16.9	H LY[C]	1	1.00
81.30	17.3	H LY[C]	1	1.00
81.40	17.8	H LY[C]	1	1.00
81.50	18.3	H LY[C]	1	1.00
81.60	18.6	H LY[C]	1	1.00
81.70	19.1	H LY[C]	1	1.00
81.80	19.5	H LY[C]	1	1.00
81.90	20.2	H LY[C]	1	1.00
82.00	20.5	H LY[C]	1	1.00
82.10	21.2	H LY[C]	1	1.00
82.20	21.7	H LY[C]	1	1.00
82.30	22.3	H LY[C]	1	1.00
82.40	22.9	H LY[C]	1	1.00
82.50	23.4	H LY[C]	1	1.00

λ/nm	Flux	Ident.	Type	Intens.
82.60	24.0	H LY[C]	1	1.00
82.70	24.7	H LY[C]	1	1.00
82.80	25.2	H LY[C]	1	1.00
82.90	25.9	H LY[C]	1	1.00
83.00	26.6	H LY[C]	1	1.00
83.10	27.2	H LY[C]	1	1.00
83.20	27.9	H LY[C]	1	1.00
83.30	28.7	H LY[C]	1	1.00
83.40	29.3	H LY[C]	1	1.00
83.42	666.5	OII, III	G1	.42
83.50	30.2	H LY[C]	1	1.00
83.60	30.8	H LY[C]	1	1.00
83.70	31.6	H LY[C]	1	1.00
83.80	32.4	H LY[C]	1	1.00
83.90	33.3	H LY[C]	1	1.00
84.00	34.1	H LY[C]	1	1.00
84.10	35.1	H LY[C]	1	1.00
84.20	35.8	H LY[C]	1	1.00
84.30	36.8	H LY[C]	1	1.00
84.40	37.7	H LY[C]	1	1.00
84.50	38.7	H LY[C]	1	1.00
84.60	39.7	H LY[C]	1	1.00
84.70	40.7	H LY[C]	1	1.00
84.80	41.7	H LY[C]	1	1.00
84.90	42.7	H LY[C]	1	1.00
85.00	43.8	H LY[C]	1	1.00
85.10	45.0	H LY[C]	1	1.00
85.20	46.1	H LY[C]	1	1.00
85.30	47.4	H LY[C]	1	1.00
85.40	48.5	H LY[C]	1	1.00
85.50	49.7	H LY[C]	1	1.00
85.60	51.0	H LY[C]	1	1.00
85.70	52.4	H LY[C]	1	1.00
85.80	53.6	H LY[C]	1	1.00
85.90	55.0	H LY[C]	1	1.00
86.00	56.4	H LY[C]	1	1.00
86.10	57.9	H LY[C]	1	1.00
86.20	59.2	H LY[C]	1	1.00
86.30	60.7	H LY[C]	1	1.00
86.40	62.2	H LY[C]	1	1.00
86.50	63.9	H LY[C]	1	1.00
86.60	65.5	H LY[C]	1	1.00
86.70	67.1	H LY[C]	1	1.00
86.80	68.9	H LY[C]	1	1.00
86.90	70.6	H LY[C]	1	1.00
87.00	72.4	H LY[C]	1	1.00
87.10	74.2	H LY[C]	1	1.00
87.20	76.1	H LY[C]	1	1.00
87.30	78.1	H LY[C]	1	1.00
87.40	80.1	H LY[C]	1	1.00
87.50	82.1	H LY[C]	1	1.00
87.60	84.1	H LY[C]	1	1.00
87.70	86.2	H LY[C]	1	1.00
87.80	88.5	H LY[C]	1	1.00
87.90	90.7	H LY[C]	1	1.00
88.00	93.0	H LY[C]	1	1.00
88.10	95.4	H LY[C]	1	1.00
88.20	97.9	H LY[C]	1	1.00
88.30	100.4	H LY[C]	1	1.00
88.40	102.9	H LY[C]	1	1.00
88.50	105.5	H LY[C]	1	1.00
88.60	108.1	H LY[C]	1	1.00
88.70	110.8	H LY[C]	1	1.00
88.80	113.7	H LY[C]	1	1.00
88.90	116.6	H LY[C]	1	1.00
89.00	119.6	H LY[C]	1	1.00
89.10	122.6	H LY[C]	1	1.00
89.20	125.7	H LY[C]	1	1.00
89.30	128.8	H LY[C]	1	1.00
89.40	132.1	H LY[C]	1	1.00
89.50	135.4	H LY[C]	1	1.00
89.60	138.9	H LY[C]	1	1.00
89.70	142.4	H LY[C]	1	1.00
89.80	145.9	H LY[C]	1	1.00
89.90	149.7	H LY[C]	1	1.00
90.00	153.5	H LY[C]	1	1.00
90.10	157.4	H LY[C]	1	1.00
90.20	161.4	H LY[C]	1	1.00
90.30	165.5	H LY[C]	1	1.00
90.40	169.7	H LY[C]	1	1.00

λ/nm	Flux	Ident.	Type	Intens.
90.41	116.9	C II	G1	.50
90.50	174.0	H LY[C]	1	1.00
90.60	178.3	H LY[C]	1	1.00
90.70	182.9	H LY[C]	1	1.00
90.80	187.6	H LY[C]	1	1.00
90.90	192.3	H LY[C]	1	1.00
91.00	197.2	H LY[C]	1	1.00
91.10	202.1	H LY[C]	1	1.00
91.20	207.3	H LY[C]	1	1.00
91.30	2.4	C I [C]	1	.50
91.40	155.0$		1	.83
91.40	2.4	C I [C]	1	.50
91.50	130.0$		1	.83
91.50	2.4	C I [C]	1	.50
91.60	115.0$		1	.83
91.60	2.5	C I [C]	1	.50
91.70	100.0$		1	.83
91.70	2.5	C I [C]	1	.50
91.80	2.7	C I [C]	1	.50
91.90	100.0$		1	.83
91.90	2.7	C I [C]	1	.50
92.00	2.8	C I [C]	1	.50
92.10	115.0	H LY-9	1	1.00
92.10	2.8	C I [C]	1	.50
92.20	2.8	C I [C]	1	.50
92.30	3.0	C I [C]	1	.50
92.31	120.0	H LY-8	B1	1.00
92.40	3.0	C I [C]	1	.50
92.50	3.1	C I [C]	1	.50
92.60	3.1	C I [C]	1	.50
92.62	125.0	H LY-7	1	1.00
92.70	3.2	C I [C]	1	.50
92.80	3.2	C I [C]	1	.50
92.90	3.3	C I [C]	1	.50
93.00	3.3	C I [C]	1	.50
93.07	130.0	H LY-6	B1	1.00
93.10	3.4	C I [C]	1	.50
93.20	3.4	C I [C]	1	.50
93.30	3.5	C I [C]	1	.50
93.34	98.0	S VI	1	.50
93.40	3.5	C I [C]	1	.50
93.50	3.6	C I [C]	1	.50
93.60	3.6	C I [C]	1	.50
93.70	3.7	C I [C]	1	.50
93.78	190.0	H LY-5	1	1.00
93.80	3.8	C I [C]	1	.50
93.90	3.8	C I [C]	1	.50
94.00	3.9	C I [C]	1	.50
94.10	3.9	C I [C]	1	.50
94.20	4.1	C I [C]	1	.50
94.30	4.3	C I [C]	1	.50
94.40	4.3	C I [C]	1	.50
94.45	65.0	S VI	1	.50
94.50	4.4	C I [C]	1	.50
94.60	4.5	C I [C]	1	.50
94.70	4.5	C I [C]	1	.50
94.80	4.6	C I [C]	1	.50
94.90	4.7	C I [C]	1	.50
94.97	350.0	H LY-4	1	1.00
95.00	4.7	C I [C]	1	.50
95.10	4.8	C I [C]	1	.50
95.20	4.9	C I [C]	1	.50
95.30	5.0	C I [C]	1	.50
95.40	5.0	C I [C]	1	.50
95.50	5.1	C I [C]	1	.50
95.60	5.3	C I [C]	1	.50
95.70	5.4	C I [C]	1	.50
95.80	5.5	C I [C]	1	.50
95.90	5.6	C I [C]	1	.50
96.00	5.6	C I [C]	1	.50
96.10	5.7	C I [C]	1	.50
96.20	5.8	C I [C]	1	.50
96.30	5.9	C I [C]	1	.50
96.40	6.1	C I [C]	1	.50
96.50	6.2	C I [C]	1	.50
96.60	6.3	C I [C]	1	.50
96.70	6.5	C I [C]	1	.50
96.80	6.6	C I [C]	1	.50
96.90	6.7	C I [C]	1	.50
97.00	6.8	C I [C]	1	.50

Table 4 (continued)

λ/nm	Flux	Ident.	Type	Intens.	λ/nm	Flux	Ident.	Type	Intens.	λ/nm	Flux	Ident.	Type	Intens.
97.10	6.9	C I [C]	1	.50	104.10	24.3	C I [C]	1	.50	125.00	125.1	QCON[01]	1	.82
97.20	7.0	C I [C]	1	.50	104.20	24.8	C I [C]	1	.50	125.10	121.2	QCON[01]	1	.82
97.25	824.0	H LY-3	1	1.00	104.30	25.3	C I [C]	1	.50	125.20	117.3	QCON[01]	1	.82
97.30	7.1	C I [C]	1	.50	104.40	25.7	C I [C]	1	.50	125.30	113.7	QCON[01]	1	.82
97.40	7.2	C I [C]	1	.50	104.50	26.1	C I [C]	1	.50	125.40	109.6	QCON[01]	1	.82
97.50	7.4	C I [C]	1	.50	104.60	26.7	C I [C]	1	.50	125.50	106.1	QCON[01]	1	.82
97.60	7.6	C I [C]	1	.50	104.70	27.1	C I [C]	1	.50	125.60	102.2	QCON[01]	1	.82
97.70	7.8	C I [C]	1	.50	104.80	27.6	C I [C]	1	.50	125.70	98.5	QCON[01]	1	.82
97.70	5957.0	C III	1	.61	104.90	28.0	C I [C]	1	.50	125.80	94.9	QCON[01]	1	.82
97.80	7.9	C I [C]	1	.50	105.00	28.7	C I [C]	1	.50	125.90	91.2	QCON[01]	1	.82
97.90	8.0	C I [C]	1	.50	105.10	29.1	C I [C]	1	.50	126.00	62.1	QCON[01]	1	.82
98.00	8.2	C I [C]	1	.50	105.20	29.6	C I [C]	1	.50	126.04	154.3	SI II	1	.42
98.10	8.3	C I [C]	1	.50	105.30	30.2	C I [C]	1	.50	126.10	59.5	QCON[01]	1	.82
98.20	8.4	C I [C]	1	.50	105.40	30.7	C I [C]	1	.50	126.20	57.0	QCON[01]	1	.82
98.30	8.5	C I [C]	1	.50	105.50	31.3	C I [C]	1	.50	126.30	54.5	QCON[01]	1	.82
98.40	8.8	C I [C]	1	.50	105.60	31.9	C I [C]	1	.50	126.40	52.1	QCON[01]	1	.82
98.50	8.9	C I [C]	1	.50	105.70	32.5	C I [C]	1	.50	126.47	286.6	SI II	1	.42
98.60	9.0	C I [C]	1	.50	105.80	33.0	C I [C]	1	.50	126.50	49.6	QCON[01]	1	.82
98.70	9.2	C I [C]	1	.50	105.90	33.7	C I [C]	1	.50	126.60	47.3	QCON[01]	1	.82
98.80	9.3	C I [C]	1	.50	106.00	34.2	C I [C]	1	.50	126.70	45.0	QCON[01]	1	.82
98.90	9.6	C I [C]	1	.50	106.10	34.8	C I [C]	1	.50	126.80	42.8	QCON[01]	1	.82
98.98	191.3	N III	1	.52	106.20	35.5	C I [C]	1	.50	126.90	40.3	QCON[01]	1	.82
99.00	9.7	C I [C]	1	.50	106.27	60.5	S IV	1	.42	127.00	84.4	QCON[01]	1	.82
99.10	10.0	C I [C]	1	.50	106.30	36.1	C I [C]	1	.50	127.10	82.3	QCON[01]	1	.82
99.15	382.5	N III	1	.52	106.40	36.9	C I [C]	1	.50	127.20	80.1	QCON[01]	1	.82
99.20	10.1	C I [C]	1	.50	106.50	37.4	C I [C]	1	.50	127.30	78.1	QCON[01]	1	.82
99.30	10.3	C I [C]	1	.50	106.60	38.1	C I [C]	1	.50	127.40	75.9	QCON[01]	1	.82
99.40	10.4	C I [C]	1	.50	106.70	38.8	C I [C]	1	.50	127.50	73.8	QCON[01]	1	.82
99.50	10.6	C I [C]	1	.50	106.80	39.5	C I [C]	1	.50	127.60	71.9	QCON[01]	1	.82
99.60	10.8	C I [C]	1	.50	106.90	40.2	C I [C]	1	.50	127.70	69.7	QCON[01]	1	.82
99.70	10.9	C I [C]	1	.50	107.00	40.9	C I [C]	1	.50	127.80	67.7	QCON[01]	1	.82
99.80	11.3	C I [C]	1	.50	107.10	41.8	C I [C]	1	.50	127.90	65.8	QCON[01]	1	.82
99.90	11.5	C I [C]	1	.50	107.20	42.4	C I [C]	1	.50	128.00	48.0	QCON[01]	1	.82
100.00	11.7	C I [C]	1	.50	107.30	43.2	C I [C]	1	.50	128.10	48.5	QCON[01]	1	.82
100.10	11.8	C I [C]	1	.50	107.40	44.1	C I [C]	1	.50	128.20	48.9	QCON[01]	1	.82
100.20	12.0	C I [C]	1	.50	107.50	44.7	C I [C]	1	.50	128.30	49.3	QCON[01]	1	.82
100.30	12.3	C I [C]	1	.50	107.60	45.6	C I [C]	1	.50	128.40	49.8	QCON[01]	1	.82
100.40	12.5	C I [C]	1	.50	107.70	46.4	C I [C]	1	.50	128.50	50.2	QCON[01]	1	.82
100.50	12.8	C I [C]	1	.50	107.71	118.2	S III	B1	.42	128.60	50.6	QCON[01]	1	.82
100.60	13.0	C I [C]	1	.50	107.80	47.4	C I [C]	1	.50	128.70	51.1	QCON[01]	1	.82
100.70	13.2	C I [C]	1	.50	107.90	48.1	C I [C]	1	.50	128.80	51.5	QCON[01]	1	.82
100.80	13.5	C I [C]	1	.50	108.00	49.0	C I [C]	1	.50	128.90	51.9	QCON[01]	1	.82
100.90	13.7	C I [C]	1	.50	108.10	49.9	C I [C]	1	.50	129.00	82.6	QCON[01]	1	.82
101.00	13.9	C I [C]	1	.50	108.20	50.9	C I [C]	1	.50	129.10	83.4	QCON[01]	1	.82
101.02	85.0	C II	1	.50	108.30	51.6	C I [C]	1	.50	129.20	84.1	QCON[01]	1	.82
101.10	14.1	C I [C]	1	.50	108.40	52.6	C I [C]	1	.50	129.30	84.8	QCON[01]	1	.82
101.20	14.4	C I [C]	1	.50	108.50	53.7	C I [C]	1	.50	129.40	85.7	QCON[01]	1	.82
101.30	14.8	C I [C]	1	.50	108.51	559.0	N II	G1	.50	129.50	86.4	QCON[01]	1	.82
101.40	15.0	C I [C]	1	.50	108.60	54.6	C I [C]	1	.50	129.60	87.2	QCON[01]	1	.82
101.50	15.3	C I [C]	1	.50	108.70	55.6	C I [C]	1	.50	129.70	87.9	QCON[01]	1	.82
101.60	15.5	C I [C]	1	.50	108.80	56.5	C I [C]	1	.50	129.80	88.7	QCON[01]	1	.82
101.70	15.8	C I [C]	1	.50	108.90	57.6	C I [C]	1	.50	129.90	89.3	QCON[01]	1	.82
101.80	16.1	C I [C]	1	.50	109.00	58.6	C I [C]	1	.50	130.00	88.1	QCON[01]	1	.82
101.90	16.4	C I [C]	1	.50	109.10	59.7	C I [C]	1	.50	130.10	89.1	QCON[01]	1	.82
102.00	16.7	C I [C]	1	.50	109.20	60.8	C I [C]	1	.50	130.20	89.9	QCON[01]	1	.82
102.10	17.0	C I [C]	1	.50	109.30	61.8	C I [C]	1	.50	130.22	1155.0	O I	1	.42
102.20	17.3	C I [C]	1	.50	109.40	63.0	C I [C]	1	.50	130.30	90.5	QCON[01]	1	.82
102.30	17.6	C I [C]	1	.50	109.50	64.2	C I [C]	1	.50	130.40	91.2	QCON[01]	1	.82
102.40	17.8	C I [C]	1	.50	109.60	65.3	C I [C]	1	.50	130.49	1186.5	O I	1	.42
102.50	18.3	C I [C]	1	.50	109.70	66.5	C I [C]	1	.50	130.50	92.2	QCON[01]	1	.82
102.57	4375.0	H LY-2	1	1.00	109.80	67.7	C I [C]	1	.50	130.60	93.0	QCON[01]	1	.82
102.60	18.6	C I [C]	1	.50	109.90	68.8	C I [C]	1	.50	130.60	1291.5	O I	1	.42
102.70	18.9	C I [C]	1	.50	110.00	70.1	C I [C]	1	.50	130.70	94.0	QCON[01]	1	.82
102.80	19.3	C I [C]	1	.50	112.25	311.7	SI IV	B1	.58	130.80	94.8	QCON[01]	1	.82
102.90	19.7	C I [C]	1	.50	112.83	387.0	SI IV	1	.58	130.90	95.5	QCON[01]	1	.82
103.00	20.0	C I [C]	1	.50	117.57	2783.0	C III	1	.61	130.93	219.9	SI III	1	.60
103.10	20.3	C I [C]	1	.50	120.65	4250.0	SI III	1	.50	131.00	65.2	QCON[01]	1	.82
103.19	3184.0	O VI	1	1.04	121.57	300000.0	H LY-1	1	.83	131.10	65.6	QCON[01]	1	.82
103.20	20.7	C I [C]	1	.50	123.88	191.2	N V	1	1.25	131.20	66.0	QCON[01]	1	.82
103.30	21.0	C I [C]	1	.50	124.00	59.7	QCON[01]	1	.82	131.30	66.6	QCON[01]	1	.82
103.40	21.5	C I [C]	1	.50	124.10	61.5	QCON[01]	1	.82	131.40	67.1	QCON[01]	1	.82
103.50	21.9	C I [C]	1	.50	124.20	59.7	QCON[01]	1	.82	131.50	67.6	QCON[01]	1	.82
103.60	22.2	C I [C]	1	.50	124.28	83.9	N V	1	1.50	131.60	68.1	QCON[01]	1	.82
103.63	418.0	C II	1	.50	124.30	57.9	QCON[01]	1	.82	131.70	68.6	QCON[01]	1	.82
103.70	22.6	C I [C]	1	.50	124.40	56.1	QCON[01]	1	.82	131.80	69.3	QCON[01]	1	.82
103.70	470.0	C II	1	.50	124.50	54.4	QCON[01]	1	.82	131.89	132.5*		1	.83
103.76	1703.0	O VI	1	1.04	124.60	52.7	QCON[01]	1	.82	131.90	69.8	QCON[01]	1	.82
103.80	23.1	C I [C]	1	.50	124.70	50.9	QCON[01]	1	.82	132.00	75.0	QCON[01]	1	.82
103.90	23.5	C I [C]	1	.50	124.80	49.2	QCON[01]	1	.82	132.10	75.7	QCON[01]	1	.82
104.00	24.0	C I [C]	1	.50	124.90	47.6	QCON[01]	1	.82	132.20	76.4	QCON[01]	1	.82

Table 4 (continued)

λ/nm	Flux Ident.	Type	Intens.	λ/nm	Flux Ident.	Type	Intens.	λ/nm	Flux Ident.	Type	Intens.
132.30	77.0QCON[01]	1	.82	139.80	144.1QCON[02]	1	.81	147.60	337.2QCON[04]	1	.44
132.40	77.6QCON[01]	1	.82	139.90	145.3QCON[02]	1	.81	147.70	340.0QCON[04]	1	.44
132.50	78.3QCON[01]	1	.82	140.00	172.1QCON[03]	1	.63	147.80	342.9QCON[04]	1	.44
132.60	79.0QCON[01]	1	.82	140.10	173.7QCON[03]	1	.63	147.90	345.7QCON[04]	1	.44
132.70	79.6QCON[01]	1	.82	140.20	174.9QCON[03]	1	.63	148.00	356.2QCON[04]	1	.44
132.80	80.3QCON[01]	1	.82	140.28	978.2SI IV	1	.70	148.10	359.1QCON[04]	1	.44
132.90	80.9QCON[01]	1	.82	140.30	176.5QCON[03]	1	.63	148.20	362.0QCON[04]	1	.44
133.00	90.7QCON[01]	1	.82	140.40	178.1QCON[03]	1	.63	148.30	365.2QCON[04]	1	.44
133.10	91.3QCON[01]	1	.82	140.50	179.7QCON[03]	1	.63	148.40	368.3QCON[04]	1	.44
133.20	92.0QCON[01]	1	.82	140.60	181.3QCON[03]	1	.63	148.50	371.4QCON[04]	1	.44
133.30	92.8QCON[01]	1	.82	140.70	182.9QCON[03]	1	.63	148.60	374.6QCON[04]	1	.44
133.40	93.7QCON[01]	1	.82	140.80	184.5QCON[03]	1	.63	148.70	377.6QCON[04]	1	.44
133.45	1955.0 C II	1	.60	140.90	186.1QCON[03]	1	.63	148.80	380.9QCON[04]	1	.44
133.50	94.4QCON[01]	1	.82	141.00	153.0QCON[03]	1	.63	148.90	384.1QCON[04]	1	.44
133.57	2677.5 C II	1	.60	141.10	154.3QCON[03]	1	.63	149.00	327.0QCON[04]	1	.44
133.60	95.1QCON[01]	1	.82	141.20	155.5QCON[03]	1	.63	149.10	329.9QCON[04]	1	.44
133.70	96.0QCON[01]	1	.82	141.30	156.9QCON[03]	1	.63	149.20	332.7QCON[04]	1	.44
133.80	96.7QCON[01]	1	.82	141.40	158.3QCON[03]	1	.63	149.30	335.4QCON[04]	1	.44
133.90	97.6QCON[01]	1	.82	141.50	159.6QCON[03]	1	.63	149.40	338.3QCON[04]	1	.44
134.00	88.4QCON[01]	1	.82	141.60	160.9QCON[03]	1	.63	149.50	341.2QCON[04]	1	.44
134.10	89.2QCON[01]	1	.82	141.70	162.3QCON[03]	1	.63	149.60	344.0QCON[04]	1	.44
134.20	90.0QCON[01]	1	.82	141.80	163.6QCON[03]	1	.63	149.70	347.0QCON[04]	1	.44
134.30	90.8QCON[01]	1	.82	141.90	165.1QCON[03]	1	.63	149.80	349.9QCON[04]	1	.44
134.40	91.6QCON[01]	1	.82	142.00	182.9QCON[03]	1	.63	149.90	353.0QCON[04]	1	.44
134.50	92.4QCON[01]	1	.82	142.10	184.4QCON[03]	1	.63	150.00	375.2QCON[05]	1	.37
134.60	93.1QCON[01]	1	.82	142.20	185.9QCON[03]	1	.63	150.10	378.5QCON[05]	1	.37
134.70	93.9QCON[01]	1	.82	142.30	187.5QCON[03]	1	.63	150.20	381.8QCON[05]	1	.37
134.80	94.8QCON[01]	1	.82	142.40	189.1QCON[03]	1	.63	150.30	384.9QCON[05]	1	.37
134.90	95.4QCON[01]	1	.82	142.50	190.6QCON[03]	1	.63	150.40	388.2QCON[05]	1	.37
135.00	102.1QCON[02]	1	.81	142.60	192.4QCON[03]	1	.63	150.50	391.6QCON[05]	1	.37
135.10	102.7QCON[02]	1	.81	142.70	194.0QCON[03]	1	.63	150.60	394.9QCON[05]	1	.37
135.20	103.4QCON[02]	1	.81	142.80	195.6QCON[03]	1	.63	150.70	398.2QCON[05]	1	.37
135.30	104.0QCON[02]	1	.81	142.90	197.3QCON[03]	1	.63	150.80	401.5QCON[05]	1	.37
135.40	104.8QCON[02]	1	.81	143.00	202.4QCON[03]	1	.63	150.90	405.0QCON[05]	1	.37
135.50	105.4QCON[02]	1	.81	143.10	204.0QCON[03]	1	.63	151.00	433.0QCON[05]	1	.37
135.56	346.5 O I	1	.42	143.20	205.5QCON[03]	1	.63	151.10	436.6QCON[05]	1	.37
135.60	106.2QCON[02]	1	.81	143.30	207.4QCON[03]	1	.63	151.20	440.3QCON[05]	1	.37
135.70	106.7QCON[02]	1	.81	143.40	209.2QCON[03]	1	.63	151.30	444.1QCON[05]	1	.37
135.80	107.5QCON[02]	1	.81	143.50	210.8QCON[03]	1	.63	151.40	447.8QCON[05]	1	.37
135.85	93.4 O I	1	.42	143.60	212.7QCON[03]	1	.63	151.50	451.5QCON[05]	1	.37
135.90	108.2QCON[02]	1	.81	143.70	214.5QCON[03]	1	.63	151.60	455.4QCON[05]	1	.37
136.00	101.5QCON[02]	1	.81	143.80	216.3QCON[03]	1	.63	151.70	459.1QCON[05]	1	.37
136.10	102.1QCON[02]	1	.81	143.90	218.1QCON[03]	1	.63	151.80	463.0QCON[05]	1	.37
136.20	102.8QCON[02]	1	.81	144.00	202.2QCON[03]	1	.63	151.90	467.0QCON[05]	1	.37
136.30	103.7QCON[02]	1	.81	144.10	203.9QCON[03]	1	.63	152.00	447.7QCON[05]	1	.37
136.40	104.5QCON[02]	1	.81	144.20	205.6QCON[03]	1	.63	152.10	451.9QCON[05]	1	.37
136.50	105.4QCON[02]	1	.81	144.30	207.3QCON[03]	1	.63	152.20	456.2QCON[05]	1	.37
136.60	106.3QCON[02]	1	.81	144.40	209.1QCON[03]	1	.63	152.30	460.3QCON[05]	1	.37
136.70	107.0QCON[02]	1	.81	144.50	211.0QCON[03]	1	.63	152.40	464.6QCON[05]	1	.37
136.80	107.9QCON[02]	1	.81	144.60	212.7QCON[03]	1	.63	152.50	468.8QCON[05]	1	.37
136.90	108.8QCON[02]	1	.81	144.70	214.5QCON[03]	1	.63	152.60	473.2QCON[05]	1	.37
137.00	108.8QCON[02]	1	.81	144.80	216.3QCON[03]	1	.63	152.67	871.5SI II	1	.42
137.10	109.7QCON[02]	1	.81	144.90	218.1QCON[03]	1	.63	152.70	477.5QCON[05]	1	.37
137.20	110.6QCON[02]	1	.81	145.00	221.5QCON[04]	1	.44	152.80	482.0QCON[05]	1	.37
137.30	111.7QCON[02]	1	.81	145.10	223.3QCON[04]	1	.44	152.90	486.3QCON[05]	1	.37
137.40	112.5QCON[02]	1	.81	145.20	225.2QCON[04]	1	.44	153.00	502.1QCON[05]	1	.37
137.50	113.5QCON[02]	1	.81	145.30	227.1QCON[04]	1	.44	153.10	506.8QCON[05]	1	.37
137.60	114.4QCON[02]	1	.81	145.40	229.0QCON[04]	1	.44	153.20	511.3QCON[05]	1	.37
137.70	115.5QCON[02]	1	.81	145.50	230.9QCON[04]	1	.44	153.30	515.9QCON[05]	1	.37
137.80	116.4QCON[02]	1	.81	145.60	232.9QCON[04]	1	.44	153.34	703.5SI II	1	.42
137.90	117.4QCON[02]	1	.81	145.70	234.9QCON[04]	1	.44	153.40	520.7QCON[05]	1	.37
138.00	100.1QCON[02]	1	.81	145.80	236.9QCON[04]	1	.44	153.50	525.4QCON[05]	1	.37
138.10	101.0QCON[02]	1	.81	145.90	238.9QCON[04]	1	.44	153.60	530.0QCON[05]	1	.37
138.20	101.8QCON[02]	1	.81	146.00	288.8QCON[04]	1	.44	153.70	534.8QCON[05]	1	.37
138.30	102.7QCON[02]	1	.81	146.10	291.3QCON[04]	1	.44	153.80	539.7QCON[05]	1	.37
138.40	103.6QCON[02]	1	.81	146.20	293.8QCON[04]	1	.44	153.90	544.4QCON[05]	1	.37
138.50	104.4QCON[02]	1	.81	146.30	296.3QCON[04]	1	.44	154.00	550.9QCON[05]	1	.37
138.60	105.4QCON[02]	1	.81	146.40	298.8QCON[04]	1	.44	154.10	555.8QCON[05]	1	.37
138.70	106.2QCON[02]	1	.81	146.50	301.3QCON[04]	1	.44	154.20	560.5QCON[05]	1	.37
138.80	107.1QCON[02]	1	.81	146.60	303.8QCON[04]	1	.44	154.30	565.5QCON[05]	1	.37
138.90	108.1QCON[02]	1	.81	146.70	306.4QCON[04]	1	.44	154.40	570.3QCON[05]	1	.37
139.00	134.7QCON[02]	1	.81	146.80	309.1QCON[04]	1	.44	154.50	575.3QCON[05]	1	.37
139.10	135.8QCON[02]	1	.81	146.90	311.7QCON[04]	1	.44	154.60	580.3QCON[05]	1	.37
139.20	136.9QCON[02]	1	.81	147.00	320.5QCON[04]	1	.44	154.70	585.4QCON[05]	1	.37
139.30	138.2QCON[02]	1	.81	147.10	323.5QCON[04]	1	.44	154.80	590.5QCON[05]	1	.37
139.38	1397.5SI IV	1	.70	147.20	326.2QCON[04]	1	.44	154.82	4241.3 C IV	1	.70
139.40	139.5QCON[02]	1	.81	147.30	493.5 S I	1	.42	154.90	595.7QCON[05]	1	.37
139.50	140.5QCON[02]	1	.81	147.30	329.0QCON[04]	1	.44	155.00	611.4QCON[06]	1	.31
139.60	141.7QCON[02]	1	.81	147.40	331.7QCON[04]	1	.44	155.08	2193.8 C IV	1	.70
139.70	143.0QCON[02]	1	.81	147.50	334.5QCON[04]	1	.44	155.10	617.0QCON[06]	1	.31

Table 4 (continued)

λ/nm	Flux Ident.	Type	Intens.	λ/nm	Flux Ident.	Type	Intens.	λ/nm	Flux Ident.	Type	Intens.
155.20	622.3QCON[06]	1	.31	162.70	1152.4QCON[07]	1	.31	170.10	4069.1QCON[09]	1	.16
155.30	627.7QCON[06]	1	.31	162.80	1170.1QCON[07]	1	.31	170.20	4132.8QCON[09]	1	.16
155.40	633.5QCON[06]	1	.31	162.90	1188.3QCON[07]	1	.31	170.30	4197.3QCON[09]	1	.16
155.50	638.8QCON[06]	1	.31	163.00	1120.2QCON[07]	1	.31	170.40	4263.1QCON[09]	1	.16
155.60	644.5QCON[06]	1	.31	163.10	1137.4QCON[07]	1	.31	170.50	4329.7QCON[09]	1	.16
155.70	650.3QCON[06]	1	.31	163.20	1154.7QCON[07]	1	.31	170.60	4397.6QCON[09]	1	.16
155.80	655.9QCON[06]	1	.31	163.30	1172.1QCON[07]	1	.31	170.70	4466.5QCON[09]	1	.16
155.90	661.7QCON[06]	1	.31	163.40	1189.8QCON[07]	1	.31	170.80	4536.6QCON[09]	1	.16
155.91	526.7FE II	1	.42	163.50	1207.8QCON[07]	1	.31	170.90	4607.7QCON[09]	1	.16
156.00	720.4QCON[06]	1	.31	163.60	1226.1QCON[07]	1	.31	171.00	3920.0QCON[09]	1	.16
156.10	725.8QCON[06]	1	.31	163.70	1244.6QCON[07]	1	.31	171.10	3985.7QCON[09]	1	.16
156.10	1099.7 C I	G1	.60	163.80	1263.5QCON[07]	1	.31	171.20	4051.3QCON[09]	1	.16
156.20	731.0QCON[06]	1	.31	163.90	1282.8QCON[07]	1	.31	171.30	4118.1QCON[09]	1	.16
156.30	736.4QCON[06]	1	.31	164.00	1235.7QCON[07]	1	.31	171.30	1870.5FE II	1	.42
156.38	526.7FE II	1	.42	164.03	1612.5FE II	B1	.42	171.40	4186.0QCON[09]	1	.16
156.40	741.7QCON[06]	1	.31	164.04	587.5*	1	.83	171.50	4254.8QCON[09]	1	.16
156.50	747.2QCON[06]	1	.31	164.10	1256.6QCON[07]	1	.31	171.60	4324.8QCON[09]	1	.16
156.60	752.6QCON[06]	1	.31	164.20	1277.5QCON[07]	1	.31	171.70	4396.0QCON[09]	1	.16
156.70	758.2QCON[06]	1	.31	164.30	1298.7QCON[07]	1	.31	171.80	4468.3QCON[09]	1	.16
156.80	763.8QCON[06]	1	.31	164.40	1320.1QCON[07]	1	.31	171.90	4541.8QCON[09]	1	.16
156.90	769.5QCON[06]	1	.31	164.50	1342.1QCON[07]	1	.31	172.00	4440.4QCON[09]	1	.16
157.00	714.7QCON[06]	1	.31	164.60	1364.4QCON[07]	1	.31	172.10	4475.5QCON[09]	1	.16
157.02	311.7FE II	1	.42	164.70	1386.9QCON[07]	1	.31	172.20	4509.8QCON[09]	1	.16
157.10	720.5QCON[06]	1	.31	164.80	1409.8QCON[07]	1	.31	172.30	4544.5QCON[09]	1	.16
157.20	726.4QCON[06]	1	.31	164.90	1433.3QCON[07]	1	.31	172.40	4579.4QCON[09]	1	.16
157.30	732.2QCON[06]	1	.31	165.00	1968.6QCON[08]	1	.24	172.50	4614.4QCON[09]	1	.16
157.40	738.0QCON[06]	1	.31	165.10	1996.2QCON[08]	1	.24	172.60	4649.9QCON[09]	1	.16
157.50	743.9QCON[06]	1	.31	165.20	2023.5QCON[08]	1	.24	172.70	4685.7QCON[09]	1	.16
157.60	749.8QCON[06]	1	.31	165.30	2051.3QCON[08]	1	.24	172.80	4721.7QCON[09]	1	.16
157.70	756.0QCON[06]	1	.31	165.40	2079.7QCON[08]	1	.24	172.90	4758.1QCON[09]	1	.16
157.80	761.9QCON[06]	1	.31	165.50	2108.3QCON[08]	1	.24	173.00	4052.4QCON[09]	1	.16
157.90	768.0QCON[06]	1	.31	165.60	2137.7QCON[08]	1	.24	173.10	4084.8QCON[09]	1	.16
158.00	684.9QCON[06]	1	.31	165.63	1857.1 C I	G1	.60	173.20	4116.8QCON[09]	1	.16
158.10	690.6QCON[06]	1	.31	165.70	2167.3QCON[08]	G1	.24	173.30	4148.9QCON[09]	1	.16
158.20	696.0QCON[06]	1	.31	165.74	2230.6 C I	G1	.60	173.40	4181.5QCON[09]	1	.16
158.30	701.5QCON[06]	1	.31	165.80	2197.5QCON[08]	1	.24	173.50	4214.3QCON[09]	1	.16
158.40	707.1QCON[06]	1	.31	165.90	2329.0QCON[08]	1	.24	173.60	4247.3QCON[09]	1	.16
158.50	712.6QCON[06]	1	.31	166.00	1553.4QCON[08]	1	.24	173.70	4280.6QCON[09]	1	.16
158.60	718.5QCON[06]	1	.31	166.10	1755.1QCON[08]	1	.24	173.80	4314.1QCON[09]	1	.16
158.70	724.1QCON[06]	1	.31	166.20	1783.2QCON[08]	1	.24	173.90	4347.9QCON[09]	1	.16
158.80	729.9QCON[06]	1	.31	166.30	1811.9QCON[08]	1	.24	174.00	4823.2QCON[09]	1	.16
158.90	735.6QCON[06]	1	.31	166.40	1840.8QCON[08]	1	.24	174.10	4866.0QCON[09]	1	.16
159.00	679.1QCON[06]	1	.31	166.50	1870.2QCON[08]	1	.24	174.20	4904.1QCON[09]	1	.16
159.10	685.0QCON[06]	1	.31	166.60	1900.2QCON[08]	1	.24	174.30	4942.6QCON[09]	1	.16
159.20	690.9QCON[06]	1	.31	166.70	1930.6QCON[08]	1	.24	174.40	4981.2QCON[09]	1	.16
159.30	696.7QCON[06]	1	.31	166.80	1961.6QCON[08]	1	.24	174.50	5020.2QCON[09]	1	.16
159.40	702.7QCON[06]	1	.31	166.90	1993.0QCON[08]	1	.24	174.60	5059.6QCON[09]	1	.16
159.46	177.5*	1	.83	167.00	1850.7QCON[08]	1	.24	174.70	5099.2QCON[09]	1	.16
159.50	708.5QCON[06]	1	.31	167.08	2945.5FE II	B1	.42	174.80	5139.2QCON[09]	1	.16
159.60	714.5QCON[06]	1	.31	167.10	1881.9QCON[08]	1	.24	174.90	5179.4QCON[09]	1	.16
159.70	720.5QCON[06]	1	.31	167.20	1913.3QCON[08]	1	.24	175.00	5500.4QCON[10]	1	.11
159.80	726.7QCON[06]	1	.31	167.30	1945.2QCON[08]	1	.24	175.10	5543.9QCON[10]	1	.11
159.90	732.8QCON[06]	1	.31	167.40	1977.6QCON[08]	1	.24	175.20	5587.4QCON[10]	1	.11
160.00	762.0QCON[07]	1	.31	167.50	2010.4QCON[08]	1	.24	175.30	5631.1QCON[10]	1	.11
160.10	768.1QCON[07]	1	.31	167.60	2043.9QCON[08]	1	.24	175.40	5675.2QCON[10]	1	.11
160.20	774.4QCON[07]	1	.31	167.68	322.5FE II	1	.42	175.50	5719.6QCON[10]	1	.11
160.30	780.6QCON[07]	1	.31	167.70	2077.8QCON[08]	1	.24	175.60	5764.5QCON[10]	1	.11
160.40	786.8QCON[07]	1	.31	167.80	2112.4QCON[08]	1	.24	175.70	5809.7QCON[10]	1	.11
160.50	793.1QCON[07]	1	.31	167.90	2147.3QCON[08]	1	.24	175.80	5855.2QCON[10]	1	.11
160.60	799.6QCON[07]	1	.31	168.00	2514.9QCON[08]	1	.24	175.90	5901.1QCON[10]	1	.11
160.70	805.9QCON[07]	1	.31	168.10	2554.7QCON[08]	1	.24	176.00	5981.2QCON[10]	1	.11
160.80	812.5QCON[07]	1	.31	168.20	2594.4QCON[08]	1	.24	176.10	6028.5QCON[10]	1	.11
160.90	819.0QCON[07]	1	.31	168.30	2634.9QCON[08]	1	.24	176.20	6075.9QCON[10]	1	.11
161.00	905.9QCON[07]	1	.31	168.40	2675.8QCON[08]	1	.24	176.30	6123.5QCON[10]	1	.11
161.10	913.3QCON[07]	1	.31	168.50	2717.4QCON[08]	1	.24	176.40	6171.5QCON[10]	1	.11
161.20	920.7QCON[07]	1	.31	168.60	2760.0QCON[08]	1	.24	176.50	6219.7QCON[10]	1	.11
161.30	928.2QCON[07]	1	.31	168.70	2803.1QCON[08]	1	.24	176.60	6268.5QCON[10]	1	.11
161.40	935.8QCON[07]	1	.31	168.80	2846.8QCON[08]	1	.24	176.70	6317.6QCON[10]	1	.11
161.50	943.4QCON[07]	1	.31	168.90	2891.3QCON[08]	1	.24	176.80	6367.1QCON[10]	1	.11
161.60	951.0QCON[07]	1	.31	169.00	3459.9QCON[08]	1	.24	176.90	6417.1QCON[10]	1	.11
161.70	958.9QCON[07]	1	.31	169.10	3512.1QCON[08]	1	.24	177.00	7143.8QCON[10]	1	.11
161.80	966.8QCON[07]	1	.31	169.20	3564.3QCON[08]	1	.24	177.10	7200.5QCON[10]	1	.11
161.90	974.6QCON[07]	1	.31	169.30	3617.2QCON[08]	1	.24	177.20	7256.9QCON[10]	1	.11
162.00	1035.0QCON[07]	1	.31	169.40	3671.0QCON[08]	1	.24	177.30	7313.8QCON[10]	1	.11
162.10	1051.1QCON[07]	1	.31	169.50	3725.6QCON[08]	1	.24	177.40	7371.1QCON[10]	1	.11
162.20	1067.4QCON[07]	1	.31	169.60	3781.0QCON[08]	1	.24	177.50	7428.9QCON[10]	1	.11
162.30	1083.9QCON[07]	1	.31	169.70	3837.5QCON[08]	1	.24	177.60	7487.0QCON[10]	1	.11
162.40	1100.5QCON[07]	1	.31	169.80	3894.8QCON[08]	1	.24	177.70	7545.7QCON[10]	1	.11
162.50	1117.5QCON[07]	1	.31	169.90	3953.1QCON[08]	1	.24	177.80	7604.9QCON[10]	1	.11
162.60	1134.7QCON[07]	1	.31	170.00	4005.4QCON[09]	1	.16	177.90	7664.3QCON[10]	1	.11

Table 4 (continued)

λ/nm	Flux Ident.	Type	Intens.	λ/nm	Flux Ident.	Type	Intens.	λ/nm	Flux Ident.	Type	Intens.
178.00	7817.3QCON[10]	1	.11	185.40	11657.5QCON[12]	1	.08	193.10	19919.9QCON[12]	1	.08
178.10	7879.1QCON[10]	1	.11	185.50	11748.6QCON[12]	1	.08	193.20	20076.0QCON[12]	1	.08
178.20	7940.8QCON[10]	1	.11	185.60	11840.7QCON[12]	1	.08	193.30	20233.3QCON[12]	1	.08
178.30	8003.1QCON[10]	1	.11	185.70	11933.6QCON[12]	1	.08	193.40	20391.9QCON[12]	1	.08
178.40	8065.7QCON[10]	1	.11	185.80	12027.0QCON[12]	1	.08	193.50	20551.6QCON[12]	1	.08
178.50	8129.0QCON[10]	1	.11	185.90	12121.2QCON[12]	1	.08	193.60	20712.7QCON[12]	1	.08
178.60	8192.6QCON[10]	1	.11	186.00	13611.6QCON[12]	1	.08	193.70	20874.9QCON[12]	1	.08
178.70	8256.8QCON[10]	1	.11	186.10	13719.1QCON[12]	1	.08	193.80	21038.4QCON[12]	1	.08
178.80	8321.5QCON[10]	1	.11	186.20	13826.6QCON[12]	1	.08	193.90	21203.2QCON[12]	1	.08
178.90	8386.7QCON[10]	1	.11	186.30	13935.0QCON[12]	1	.08	194.00	21284.9QCON[12]	1	.08
179.00	8007.6QCON[10]	1	.11	186.40	14044.1QCON[12]	1	.08	194.10	25236.9QCON[12]	1	.08
179.10	8070.8QCON[10]	1	.11	186.50	14154.1QCON[12]	1	.08	194.20	25434.7QCON[12]	1	.08
179.20	8133.9QCON[10]	1	.11	186.60	14265.1QCON[12]	1	.08	194.30	25633.9QCON[12]	1	.08
179.30	8197.7QCON[10]	1	.11	186.70	14376.8QCON[12]	1	.08	194.40	25834.8QCON[12]	1	.08
179.40	8261.9QCON[10]	1	.11	186.80	14489.5QCON[12]	1	.08	194.50	26037.2QCON[12]	1	.08
179.50	8326.7QCON[10]	1	.11	186.90	14602.9QCON[12]	1	.08	194.60	26241.2QCON[12]	1	.08
179.60	8391.9QCON[10]	1	.11	187.00	14964.1QCON[12]	1	.08	194.70	26446.8QCON[12]	1	.08
179.70	8457.6QCON[10]	1	.11	187.10	15082.3QCON[12]	1	.08	194.80	26654.0QCON[12]	1	.08
179.80	8523.9QCON[10]	1	.11	187.20	15200.5QCON[12]	1	.08	194.90	26862.8QCON[12]	1	.08
179.90	8590.6QCON[10]	1	.11	187.30	15319.5QCON[12]	1	.08	195.00	27073.2QCON[12]	1	.08
180.00	9060.2QCON[11]	1	.09	187.40	15439.6QCON[12]	1	.08	195.10	27285.3QCON[12]	1	.08
180.10	9130.3QCON[11]	1	.09	187.50	15560.5QCON[12]	1	.08	195.20	27499.1QCON[12]	1	.08
180.20	9200.3QCON[11]	1	.09	187.60	15682.4QCON[12]	1	.08	195.30	27714.5QCON[12]	1	.08
180.30	9271.1QCON[11]	1	.09	187.70	15805.3QCON[12]	1	.08	195.40	27931.7QCON[12]	1	.08
180.40	9342.2QCON[11]	1	.09	187.80	15929.1QCON[12]	1	.08	195.50	28150.5QCON[12]	1	.08
180.50	9414.0QCON[11]	1	.09	187.90	16053.9QCON[12]	1	.08	195.60	28371.1QCON[12]	1	.08
180.60	9486.4QCON[11]	1	.09	188.00	16601.1QCON[12]	1	.08	195.70	28593.3QCON[12]	1	.08
180.70	9559.3QCON[11]	1	.09	188.10	16732.2QCON[12]	1	.08	195.80	28817.4QCON[12]	1	.08
180.80	9632.7QCON[11]	1	.09	188.20	16863.6QCON[12]	1	.08	195.90	29043.1QCON[12]	1	.08
180.80	9660.0SI II	1	.42	188.30	16995.3QCON[12]	1	.08	196.00	29270.7QCON[12]	1	.08
180.90	9706.8QCON[11]	1	.09	188.40	17128.4QCON[12]	1	.08	196.10	29500.0QCON[12]	1	.08
181.00	10271.0QCON[11]	1	.09	188.50	17262.6QCON[12]	1	.08	196.20	29731.1QCON[12]	1	.08
181.10	10349.7QCON[11]	1	.09	188.60	17397.9QCON[12]	1	.08	196.30	29964.0QCON[12]	1	.08
181.20	10428.5QCON[11]	1	.09	188.70	17534.2QCON[12]	1	.08	196.40	30198.8QCON[12]	1	.08
181.30	10507.8QCON[11]	1	.09	188.80	17671.6QCON[12]	1	.08	196.50	30435.4QCON[12]	1	.08
181.40	10587.5QCON[11]	1	.09	188.90	17810.1QCON[12]	1	.08	196.60	30673.8QCON[12]	1	.08
181.50	10668.1QCON[11]	1	.09	189.00	18531.8QCON[12]	1	.08	196.70	30914.2QCON[12]	1	.08
181.60	10749.2QCON[11]	1	.09	189.10	18678.8QCON[12]	1	.08	196.80	31156.4QCON[12]	1	.08
181.69	14910.0SI II	1	.42	189.20	18825.2QCON[12]	1	.08	196.90	31400.5QCON[12]	1	.08
181.70	10831.0QCON[11]	1	.09	189.30	18972.7QCON[12]	1	.08	197.00	31646.5QCON[12]	1	.08
181.74	5775.0SI II	1	.42	189.40	19121.4QCON[12]	1	.08	197.10	31894.4QCON[12]	1	.08
181.80	10913.4QCON[11]	1	.09	189.50	19271.1QCON[12]	1	.08	197.20	32144.3QCON[12]	1	.08
181.90	10996.5QCON[11]	1	.09	189.60	19422.1QCON[12]	1	.08	197.30	32396.1QCON[12]	1	.08
182.00	12742.0QCON[11]	1	.09	189.70	19574.3QCON[12]	1	.08	197.40	32649.9QCON[12]	1	.08
182.10	12842.6QCON[11]	1	.09	189.80	19727.6QCON[12]	1	.08	197.50	32905.7QCON[12]	1	.08
182.20	12943.2QCON[11]	1	.09	189.90	19882.2QCON[12]	1	.08	197.60	33163.5QCON[12]	1	.08
182.30	13044.6QCON[11]	1	.09	190.00	18291.8QCON[12]	1	.08	197.70	33423.4QCON[12]	1	.08
182.40	13146.9QCON[11]	1	.09	190.10	18273.4QCON[12]	1	.08	197.80	33685.2QCON[12]	1	.08
182.50	13249.8QCON[11]	1	.09	190.20	18416.6QCON[12]	1	.08	197.90	33949.2QCON[12]	1	.08
182.60	13353.6QCON[11]	1	.09	190.30	18560.9QCON[12]	1	.08	198.00	34215.1QCON[12]	1	.08
182.70	13458.4QCON[11]	1	.09	190.40	18706.3QCON[12]	1	.08	198.10	34483.2QCON[12]	1	.08
182.80	13563.7QCON[11]	1	.09	190.50	18852.9QCON[12]	1	.08	198.20	34753.3QCON[12]	1	.08
182.90	13670.1QCON[11]	1	.09	190.60	19000.6QCON[12]	1	.08	198.30	35025.5QCON[12]	1	.08
183.00	12735.4QCON[11]	1	.09	190.70	19149.3QCON[12]	1	.08	198.40	35300.0QCON[12]	1	.08
183.10	12837.2QCON[11]	1	.09	190.80	19299.5QCON[12]	1	.08	198.50	35576.5QCON[12]	1	.08
183.20	12937.8QCON[11]	1	.09	190.90	19450.6QCON[12]	1	.08	198.60	35855.2QCON[12]	1	.08
183.30	13039.3QCON[11]	1	.09	191.00	21963.1QCON[12]	1	.08	198.70	36136.2QCON[12]	1	.08
183.40	13141.3QCON[11]	1	.09	191.10	21743.4QCON[12]	1	.08	198.80	36419.3QCON[12]	1	.08
183.50	13244.3QCON[11]	1	.09	191.20	21913.7QCON[12]	1	.08	198.90	36704.6QCON[12]	1	.08
183.60	13348.0QCON[11]	1	.09	191.30	22085.4QCON[12]	1	.08	199.00	36992.2QCON[12]	1	.08
183.70	13452.7QCON[11]	1	.09	191.40	22258.5QCON[12]	1	.08	199.10	37282.0QCON[12]	1	.08
183.80	13558.1QCON[11]	1	.09	191.50	22432.9QCON[12]	1	.08	199.20	37574.1QCON[12]	1	.08
183.90	13664.3QCON[11]	1	.09	191.60	22608.5QCON[12]	1	.08	199.30	37868.5QCON[12]	1	.08
184.00	10709.2QCON[11]	1	.09	191.70	22785.7QCON[12]	1	.08	199.40	38165.1QCON[12]	1	.08
184.10	10793.8QCON[11]	1	.09	191.80	22964.2QCON[12]	1	.08	199.50	38464.1QCON[12]	1	.08
184.20	10878.3QCON[11]	1	.09	191.90	23144.1QCON[12]	1	.08	199.60	38765.5QCON[12]	1	.08
184.30	10963.5QCON[11]	1	.09	192.00	19005.2QCON[12]	1	.08	199.70	39069.2QCON[12]	1	.08
184.40	11049.4QCON[11]	1	.09	192.10	20016.8QCON[12]	1	.08	199.80	39375.3QCON[12]	1	.08
184.50	11136.0QCON[11]	1	.09	192.20	20173.7QCON[12]	1	.08	199.90	39683.8QCON[12]	1	.08
184.60	11223.2QCON[11]	1	.09	192.30	20331.6QCON[12]	1	.08	200.00	39994.7QCON[12]	1	.08
184.70	11311.1QCON[11]	1	.09	192.40	20491.1QCON[12]	1	.08				
184.80	11399.8QCON[11]	1	.09	192.50	20651.5QCON[12]	1	.08				
184.90	11489.1QCON[11]	1	.09	192.60	20813.4QCON[12]	1	.08				
185.00	11298.3QCON[12]	1	.08	192.70	20976.3QCON[12]	1	.08				
185.10	11387.6QCON[12]	1	.08	192.80	21140.7QCON[12]	1	.08				
185.20	11476.8QCON[12]	1	.08	192.90	21306.4QCON[12]	1	.08				
185.30	11566.8QCON[12]	1	.08	193.00	26973.5QCON[12]	1	.08				

@RESUME RZUB04

Table 5 (Caption see page 45)

day count	I_1	I_2	day count	I_1	I_2	day count	I_1	I_2	day count	I_1	I_2
77182	1.52	12.59	77334	1.79	9.78	78193	2.21	51.01	79019	2.69	88.02
77183	1.51	11.73	77335	1.82	9.18	78199	2.26	31.25	79020	2.63	84.62
77184	1.48	11.43	77337	1.90	11.61	78202	2.16	42.55	79022	2.85	90.58
77185	1.48	9.06	77340	1.84	11.49	78203	2.08	35.07	79023	2.82	78.78
77186	1.44	6.26	77341	1.84	14.35	78205	1.97	27.36	79025	2.81	78.54
77187	1.45	5.65	77347	1.71	17.39	78207	1.83	21.10	79026	2.78	70.70
77188	1.40	6.87	77348	1.72	13.80	78209	1.82	22.95	79028	2.66	89.61
77194	1.46	5.48	77356	1.60	5.23	78211	1.82	21.15	79030	2.68	90.27
77195	1.45	5.53	77359	1.62	8.57	78212	1.77	18.48	79031	2.72	95.45
77203	1.45	8.45	77360	1.65	12.59	78213	1.81	20.61	79040	2.90	89.67
77204	1.52	8.70	77362	1.75	16.96	78215	1.82	20.85	79041	2.71	90.15
77206	1.58	9.24	77363	1.83	18.54	78216	1.83	23.64	79042	2.68	85.35
77207	1.52	9.30	77364	1.86	25.35	78217	1.93	22.74	79043	2.68	90.34
77209	1.50	9.00	78001	2.03	31.18	78219	1.94	26.81	79044	2.64	78.84
77210	1.50	8.39	78002	2.06	28.94	78221	2.06	29.12	79046	2.64	72.28
77212	1.45	7.97	78003	2.03	29.06	78224	2.08	28.69	79047	2.81	72.71
77213	1.41	7.18	78006	2.00	28.88	78225	2.00	30.58	79048	2.97	73.01
77214	1.42	6.02	78008	1.85	13.89	78227	2.07	33.25	79050	3.00	81.27
77215	1.42	6.02	78012	1.67	12.53	78229	2.08	32.16	79051	3.01	84.56
77217	1.46	6.93	78014	1.70	9.18	78230	2.03	25.90	79052	3.10	83.29
77218	1.38	7.35	78018	1.63	6.69	78232	1.89	13.37	79053	2.91	96.84
77219	1.44	8.02	78022	1.69	20.18	78235	1.95	18.54	79054	2.78	83.10
77220	1.45	6.99	78027	1.83	25.08	78236	1.86	21.04	79055	2.61	85.89
77221	1.45	5.05	78029	1.92	28.15	78241	1.96	28.63	79056	2.57	73.01
77222	1.46	7.72	78030	2.00	32.53	78242	2.03	30.64	79057	2.67	78.00
77223	1.47	8.21	78031	2.06	32.10	78244	2.17	46.08	79058	2.69	67.36
77224	1.45	5.05	78033	2.11	40.85	78247	2.33	47.54	79060	2.88	74.71
77232	1.55	6.81	78035	2.05	39.33	78250	2.22	42.01	79061	2.82	76.78
77234	1.61	10.88	78049	1.76	22.50	78252	2.15	41.52	79062	2.88	75.44
77236	1.55	6.26	78050	1.87	28.21	78253	2.17	45.71	79063	2.94	80.37
77237	1.63	6.57	78053	1.96	32.34	78254	2.04	42.98	79065	2.93	90.51
77239	1.53	5.65	78072	2.09	44.93	78256	2.00	40.73	79066	2.86	87.18
77242	1.55	5.65	78073	2.07	50.09	78257	2.05	37.87	79067	2.90	83.77
77243	1.54	5.72	78075	2.00	42.13	78260	2.19	37.39	79068	2.81	87.72
77245	1.52	7.54	78076	1.86	32.47	78262	2.15	39.03	79069	2.75	83.22
77246	1.52	7.78	78078	1.81	29.12	78263	2.13	34.53	79070	2.81	90.70
77248	1.51	8.39	78081	1.88	26.39	78267	2.18	44.74	79071	2.80	88.02
77249	1.54	9.42	78087	1.87	23.53	78268	2.16	43.58	79072	2.90	85.78
77251	1.53	13.32	78092	1.83	26.26	78270	2.12	41.52	79073	2.95	87.24
77252	1.57	11.49	78093	1.80	22.31	78273	2.08	37.33	79074	2.87	86.75
77254	1.56	11.07	78094	1.84	32.77	78274	2.07	34.77	79075	2.91	79.40
77255	1.63	12.83	78096	1.91	31.91	78275	2.11	31.67	79076	2.94	77.02
77257	1.56	12.59	78097	1.92	30.21	78279	2.19	35.87	79077	2.94	74.78
77258	1.59	12.28	78098	1.96	38.72	78281	2.11	43.04	79078	2.95	78.60
77259	1.60	15.56	78106	1.87	33.61	78282	2.14	42.55	79079	2.85	81.03
77260	1.53	10.70	78108	1.97	34.10	78283	2.08	48.60	79081	2.75	84.92
77261	1.54	13.26	78109	2.05	36.17	78286	2.35	53.87	79082	2.71	86.62
77262	1.54	12.28	78114	2.01	40.18	78288	2.40	52.77	79083	2.68	88.27
77263	1.54	10.83	78115	2.03	33.44	78292	2.37	45.23	79084	2.67	95.20
77265	1.50	9.85	78118	1.96	34.47	78294	2.37	48.45	79085	2.63	94.47
77266	1.50	11.61	78120	1.89	38.23	78295	2.32	48.52	79086	2.76	88.57
77269	1.59	9.30	78121	1.97	32.34	78298	2.28	47.79	79087	2.72	75.87
77275	1.50	7.42	78126	1.96	39.09	78300	2.29	42.98	79088	2.80	76.60
77276	1.50	9.37	78133	1.88	31.74	78301	2.29	42.07	79089	2.91	78.00
77277	1.50	9.24	78135	1.93	33.37	78307	2.42	48.63	79090	3.03	75.44
77278	1.45	13.07	78136	1.97	34.90	78312	2.27	54.71	79091	3.09	80.73
77279	1.54	15.02	78138	1.96	35.56	78316	2.26	40.49	79092	2.93	72.22
77280	1.57	17.75	78139	2.03	36.90	78322	2.15	29.72	79093	2.87	73.62
77281	1.58	14.89	78140	1.92	38.55	78323	2.14	28.94	79095	2.53	57.33
77282	1.60	15.99	78147	2.04	36.23	78336	2.60	55.87	79096	2.49	62.01
77284	1.66	15.10	78148	2.06	33.68	78341	2.48	66.27	79098	2.39	58.00
77285	1.64	16.84	78150	2.06	37.57	78342	2.51	64.98	79099	2.26	53.01
77286	1.68	11.97	78153	2.00	37.14	78344	2.57	63.95	79100	2.32	54.90
77289	1.75	13.86	78154	1.94	35.50	78348	2.39	58.00	79101	2.33	50.52
77291	1.71	8.27	78156	1.85	27.36	78351	2.35	55.74	79102	2.43	53.98
77292	1.72	12.64	78159	1.79	20.67	78352	2.24	48.33	79103	2.45	52.65
77295	1.58	11.91	78160	1.77	18.61	78353	2.21	51.37	79104	2.46	47.60
77301	1.64	10.83	78166	1.89	34.59	78354	2.14	40.49	79105	2.54	56.17
77304	1.61	13.26	78168	2.03	47.17	78355	2.14	39.58	79106	2.54	58.49
77305	1.62	11.67	78169	2.10	44.20	78365	2.51	51.61	79107	2.48	58.54
77307	1.57	12.28	78170	2.17	58.12	79003	2.69	65.17	79109	2.52	65.90
77309	1.56	12.04	78172	2.26	60.12	79005	2.78	68.94	79110	2.43	68.76
77313	1.55	9.78	78173	2.27	60.19	79006	2.73	71.86	79111	2.36	62.38
77314	1.61	12.04	78176	2.29	64.08	79007	2.72	73.98	79112	2.31	70.40
77315	1.53	9.72	78177	2.29	58.12	79008	2.74	78.78	79114	2.44	66.27
77322	1.83	11.97	78182	1.95	36.42	79010	2.66	68.57	79115	2.51	62.43
77323	1.74	7.11	78185	1.89	25.53	79011	2.70	62.49	79116	2.63	67.41
77325	1.82	8.64	78187	1.91	26.45	79013	2.85	75.20	79117	2.74	71.06
77326	1.78	6.81	78190	1.98	39.69	79016	2.73	78.67	79118	2.85	64.38
77327	1.73	8.27	78191	2.13	46.44	79017	2.61	68.63	79119	2.80	61.70

Table 5 (continued)

day count	I_1	I_2	day count	I_1	I_2	day count	I_1	I_2	day count	I_1	I_2
79120	2.77	63.77	79228	2.46	49.18	79352	2.91	85.53	80116	2.66	60.43
79122	2.55	68.45	79229	2.51	57.33	79354	2.73	82.86	80118	2.59	59.33
79125	2.50	68.33	79231	2.53	66.81	79357	2.54	61.33	80119	2.68	69.18
79126	2.55	66.20	79232	2.70	80.37	79358	2.54	56.90	80120	2.56	66.14
79129	2.54	61.70	79234	2.84	78.42	79360	2.54	59.76	80121	2.48	68.51
79130	2.70	59.82	79235	2.91	77.38	79361	2.54	62.92	80122	2.48	67.84
79131	2.73	64.44	79237	3.02	86.45	79363	2.67	70.40	80123	2.51	73.38
79132	2.74	64.44	79238	2.85	92.77	79364	2.79	82.19	80124	2.58	74.41
79133	2.75	58.12	79240	2.83	76.17	80001	2.84	83.89	80125	2.54	84.68
79134	2.74	58.79	79241	2.80	69.00	80002	2.90	93.07	80126	2.64	81.16
79135	2.67	55.87	79242	2.69	72.83	80004	3.09	97.63	80127	2.73	79.82
79136	2.53	59.03	79244	2.61	65.05	80005	3.24	104.74	80128	2.71	82.19
79137	2.38	65.17	79245	2.68	71.98	80007	3.26	104.74	80130	2.78	79.46
79138	2.34	59.82	79247	2.60	65.23	80008	3.27	99.45	80131	2.77	77.75
79139	2.20	63.59	79248	2.62	61.40	80011	3.38	110.76	80132	2.76	77.94
79140	2.23	60.92	79251	2.71	70.12	80012	3.26	115.14	80133	2.65	75.02
79141	2.32	64.14	79253	2.77	68.63	80014	2.98	98.24	80134	2.64	74.05
79142	2.40	60.61	79254	2.83	71.06	80015	2.94	104.56	80137	2.73	82.80
79143	2.42	58.12	79256	2.72	77.94	80017	2.77	86.02	80138	2.79	89.48
79144	2.54	58.90	79257	2.80	91.67	80018	2.78	73.07	80139	2.74	88.81
79145	2.55	57.33	79259	2.78	76.35	80019	2.65	51.25	80140	2.80	92.83
79146	2.49	52.16	79260	2.89	77.21	80021	2.55	63.04	80141	2.87	93.50
79147	2.52	52.89	79262	2.98	81.59	80022	2.60	64.92	80142	2.93	91.97
79148	2.50	55.44	79263	2.97	82.80	80024	2.55	62.19	80143	3.02	98.05
79149	2.46	56.47	79264	3.00	84.44	80025	2.64	66.38	80144	3.15	108.09
79150	2.43	59.21	79267	3.12	88.57	80027	2.70	74.84	80145	3.04	98.30
79151	2.40	70.64	79269	3.12	82.49	80028	2.77	70.58	80146	3.09	104.01
79152	2.52	71.00	79271	2.97	90.10	80030	2.91	70.16	80147	3.06	104.37
79153	2.66	77.33	79272	2.95	83.40	80031	3.03	72.03	80148	2.93	104.13
79154	2.67	75.68	79273	2.80	87.97	80032	3.12	71.49	80149	2.96	101.64
79155	2.78	80.79	79275	2.72	78.84	80033	3.23	85.72	80150	2.78	92.16
79156	2.84	83.16	79276	2.73	84.62	80035	3.29	87.60	80151	2.62	80.18
79157	2.76	88.81	79278	2.75	69.79	80036	3.26	88.51	80153	2.48	69.73
79158	2.94	79.15	79279	2.82	75.81	80039	3.16	88.64	80154	2.37	68.57
79159	2.93	82.37	79281	2.84	75.20	80040	3.04	87.72	80155	2.43	61.33
79160	2.87	80.67	79282	2.80	71.55	80042	2.87	88.88	80156	2.35	68.03
79161	2.76	77.21	79284	2.89	75.75	80043	2.91	87.60	80157	2.37	72.59
79162	2.65	69.49	79285	2.94	75.56	80045	2.86	79.15	80158	2.41	57.76
79163	2.59	68.63	79287	2.94	82.59	80046	2.95	77.94	80159	2.36	59.82
79164	2.50	66.63	79289	2.88	83.29	80047	2.79	71.31	80160	2.36	64.44
79165	2.42	60.30	79291	2.96	86.51	80075	2.45	50.03	80161	2.40	61.09
79166	2.30	57.76	79293	3.09	92.89	80076	2.49	52.22	80162	2.51	68.39
79167	2.44	55.74	79294	2.91	83.40	80077	2.48	56.84	80163	2.46	67.41
79168	2.45	55.81	79296	2.89	85.48	80078	2.44	56.71	80164	2.62	74.29
79169	2.40	57.39	79297	2.85	87.29	80079	2.53	62.55	80165	2.57	73.49
79170	2.47	56.30	79299	2.59	87.24	80080	2.48	65.60	80166	2.51	73.80
79171	2.55	55.99	79300	2.74	85.23	80081	2.48	62.55	80168	2.59	78.84
79172	2.53	50.46	79302	2.73	95.32	80082	2.43	59.39	80169	2.59	85.59
79174	2.51	46.33	79303	2.79	92.77	80083	2.48	68.27	80170	2.73	90.21
79175	2.51	49.30	79305	2.93	94.96	80084	2.63	72.76	80171	2.78	96.84
79184	2.75	85.78	79306	2.94	88.88	80085	2.80	73.13	80173	2.77	87.91
79185	2.80	93.68	79308	2.96	86.36	80086	2.85	73.25	80174	3.00	101.04
79187	2.78	79.03	79309	3.03	93.68	80087	2.90	82.19	80175	2.87	99.88
79188	2.82	77.27	79311	3.22	124.87	80088	2.96	87.11	80176	2.93	86.93
79190	2.70	76.41	79312	3.32	124.87	80089	2.94	78.36	80177	2.86	85.59
79191	2.57	70.33	79314	3.64	147.00	80090	2.92	71.79	80178	2.80	91.73
79193	2.37	54.28	79315	3.59	144.74	80091	2.85	67.97	80179	2.77	84.38
79194	2.26	47.96	79317	3.34	125.11	80092	2.80	67.66	80180	2.61	78.30
79196	2.31	46.14	79318	3.37	112.22	80093	2.70	67.17	80181	2.49	79.46
79197	2.24	45.60	79320	3.26	100.31	80094	2.56	71.43	80182	2.39	65.29
79199	2.38	42.92	79321	3.25	95.14	80095	2.57	71.06	80183	2.37	57.81
79200	2.47	43.17	79323	2.95	89.97	80096	2.63	75.56	80184	2.39	50.52
79202	2.40	41.09	79324	2.77	83.29	80097	2.65	79.88	80185	2.30	50.52
79203	2.44	53.50	79326	2.65	71.06	80098	2.62	77.69	80186	2.25	56.66
79205	2.53	60.79	79327	2.55	60.85	80099	2.71	81.70	80187	2.28	51.68
79207	2.58	64.50	79329	2.51	62.98	80100	2.72	81.34	80188	2.34	55.44
79208	2.62	53.92	79330	2.42	53.25	80101	2.87	88.75	80193	2.34	61.52
79209	2.68	53.50	79332	2.45	41.77	80102	2.98	91.00	80194	2.41	78.73
79211	2.63	58.12	79333	2.36	45.77	80103	2.93	88.94	80195	2.46	83.53
79212	2.62	55.26	79335	2.43	51.31	80104	2.95	84.32	80196	2.55	97.81
79214	2.57	52.52	79336	2.48	50.58	80105	2.83	85.96	80197	2.71	102.86
79215	2.55	49.18	79337	2.62	62.31	80106	2.67	73.98	80198	2.76	112.58
79217	2.54	55.68	79340	2.94	94.53	80107	2.71	68.57	80199	2.88	104.93
79218	2.40	54.11	79342	3.12	114.96	80108	2.57	69.73	80200	3.01	105.90
79220	2.46	49.60	79343	3.19	107.12	80110	2.65	72.46	80202	3.04	108.26
79223	2.41	44.26	79345	3.35	105.72	80111	2.59	70.58	80203	3.04	102.80
79225	2.39	49.06	79346	3.34	108.02	80112	2.59	68.21	80204	2.94	105.47
79226	2.47	57.09	79348	3.50	108.33	80113	2.64	69.79	80206	2.75	93.37
79226	2.38	53.68	79350	3.15	95.62	80114	2.55	57.20	80207	2.75	92.22
79227	2.50	55.57	79351	2.95	90.88	80115	2.54	56.60	80209	2.58	76.60

Table 5 (continued)

day count	I_1	I_2	day count	I_1	I_2	day count	I_1	I_2	day count	I_1	I_2
80210	2.49	70.94	80242	2.43	63.83	80277	2.65	73.13	80326	2.62	65.95
80211	2.42	57.81	80243	2.37	73.86	80278	2.75	77.81	80327	2.63	68.33
80212	2.27	49.85	80244	2.55	76.24	80279	2.64	89.00	80328	2.81	80.79
80213	2.26	47.23	80245	2.47	67.88	80280	2.72	85.96	80329	2.64	74.41
80214	2.14	45.04	80246	2.31	77.75	80281	2.77	88.15	80330	2.66	81.40
80215	2.22	44.14	80247	2.54	77.33	80283	2.77	93.56	80331	2.62	78.91
80217	2.18	40.31	80249	2.65	79.88	80284	2.78	82.62	80332	2.71	80.18
80218	2.19	42.37	80250	2.62	81.59	80286	2.92	92.16	80337	2.97	82.13
80219	2.21	52.34	80251	2.62	79.76	80287	2.81	87.11	80339	2.72	76.24
80220	2.23	48.15	80252	2.62	74.22	80288	2.89	97.75	80341	2.52	66.14
80221	2.39	52.95	80253	2.55	72.46	80289	2.87	93.56	80343	2.43	73.19
80222	2.45	53.19	80254	2.59	67.05	80290	2.70	91.80	80344	2.42	76.17
80223	2.57	62.01	80255	2.61	64.62	80299	2.57	72.71	80345	2.44	73.13
80224	2.69	76.24	80256	2.66	64.92	80300	2.49	70.46	80347	2.69	90.46
80225	2.75	72.83	80257	2.62	60.49	80301	2.45	68.57	80349	2.78	98.24
80226	2.88	81.40	80258	2.54	57.69	80302	2.59	70.52	80350	2.88	106.45
80227	2.80	80.97	80259	2.48	57.20	80303	2.49	68.39	80351	2.90	106.99
80228	2.86	77.69	80260	2.56	57.14	80304	2.50	71.00	80352	2.83	103.29
80229	2.84	80.54	80261	2.45	53.50	80308	2.67	77.87	80353	2.93	102.43
80230	2.88	85.96	80262	2.51	48.33	80309	2.69	86.32	80354	2.94	98.30
80231	2.94	95.93	80263	2.40	59.52	80311	2.78	80.86	80355	2.89	99.88
80232	2.78	91.37	80264	2.36	52.52	80312	2.90	97.75	80356	2.79	105.90
80233	2.70	79.21	80265	2.31	54.04	80314	2.66	84.02	80357	2.79	108.26
80234	2.70	79.57	80266	2.32	58.90	80316	2.60	89.48	80358	2.71	106.50
80235	2.61	78.91	80267	2.32	59.21	80317	2.52	90.51	80359	2.79	97.45
80236	2.49	74.41	80268	2.33	58.00	80318	2.79	99.21	80360	2.83	101.40
80237	2.45	65.17	80269	2.39	59.52	80319	2.56	99.39	80361	2.77	87.66
80238	2.38	62.68	80273	2.54	71.67	80320	2.47	84.80	80362	2.82	94.65
80239	2.30	63.04	80274	2.67	65.11	80323	2.59	80.91	80363	2.82	94.59
80240	2.40	63.35	80275	2.57	67.11	80324	2.57	73.92	80365	2.89	88.02
80241	2.28	59.88	80276	2.65	70.82	80325	2.50	67.17			

Transport Coefficients and Collision Frequencies for Aeronomic Plasmas

By

K. Suchy

With 40 Figures

Introduction

For an understanding of many aeronomic processes, diffusion coefficients D, heat conductivities κ, and viscosities η are needed. To study the propagation of electromagnetic waves through a planetary atmosphere, the electrical conductivity σ of the atmosphere must be known. Since the motion of charged particles is strongly influenced by a magnetic field $\boldsymbol{B}$ pervading the plasma, e.g., the Earth's magnetic field $\boldsymbol{B}_{♁}$, the atmospheric plasma is an *anisotropic medium* with $\hat{\boldsymbol{B}} := \boldsymbol{B}/B$ as distinguished direction. Therefore the *transport coefficients* $\mathbf{D}$, $\boldsymbol{\kappa}$, and $\boldsymbol{\sigma}$ for charged particles are tensors of second rank, the viscosity is a fourth-rank tensor.

If the rotation of a planetary atmosphere has to be taken into account, the vector $\boldsymbol{\omega}$ of its circular frequency determines a distinguished direction for all particles. Then their transport coefficients $\mathbf{D}$, $\boldsymbol{\kappa}$ are also tensors of second rank, and the viscosity is a tensor of fourth rank.

Before a rigorous derivation of the transport coefficients by means of the kinetic theory of gases is given in Chaps. B, C, D, a crude derivation of $\mathbf{D}$, $\boldsymbol{\sigma}$, and $\boldsymbol{\kappa}$ will be given in Chap. A. A *mean-collision-frequency method* will there be used which is slightly more general than the "mean-free-path method" often used in elementary treatises on transport coefficients. Although the limits of nonrigorous kinetic methods will appear, some features of the tensorial character of $\mathbf{D}$, $\boldsymbol{\sigma}$, $\boldsymbol{\kappa}$ can already be demonstrated.

A. Transport coefficients derived with a mean-collision-frequency method

1. Momentum and energy flux balance. To establish balance equations for an ensemble of particles, one needs the *equation of motion*

$$m\dot{\boldsymbol{c}}=\boldsymbol{F}+\boldsymbol{\Omega}\times m\boldsymbol{c} \tag{1.1}$$

for one particle with mass m and velocity $\boldsymbol{c}$. The velocity-independent force

$$\boldsymbol{F}=m\boldsymbol{g}_0+q\boldsymbol{E} \tag{1.2}$$

has a contribution $m\boldsymbol{g}_0$ from the gravitational field $\boldsymbol{g}_0$ and, if the particle has a charge q, from an electric (wave) field $\boldsymbol{E}$. In the axial vector

$$\boldsymbol{\Omega}:=-2\boldsymbol{\omega}+\boldsymbol{\omega}_B \quad \text{with } \boldsymbol{\omega}_B:=-\frac{q\boldsymbol{B}}{m c_0\sqrt{\varepsilon_0\mu_0}} \tag{1.3}$$

of the velocity-dependent force $\boldsymbol{\Omega}\times m\boldsymbol{c}$, the first term $-2\boldsymbol{\omega}$ stems from the *Coriolis force*; the second term $\boldsymbol{\omega}_B$ is the gyro-pulsation, thus 2π times the *gyro-frequency*, i.e., the circular velocity of the gyration of a charged particle in a homogeneous magnetic field $\boldsymbol{B}$.

To derive the heat conductivity κ, the equation of motion, Eq. (1.1) is multiplied by $c^2/2$:

$$\frac{m}{2}c^2\dot{\boldsymbol{c}}=\frac{c^2}{2}\boldsymbol{F}+\boldsymbol{\Omega}\times\frac{m}{2}c^2\boldsymbol{c}. \tag{1.4}$$

The influence of collisions is expressed by an averaging process over a small macroscopic volume containing many particles[1]. If $\boldsymbol{\varphi}(\boldsymbol{c})$ is a vectorial quantity assigned to a particle with velocity $\boldsymbol{c}$, e.g., $\boldsymbol{\varphi}=\boldsymbol{c}$ or $\boldsymbol{\varphi}=(m/2)\,c^2\boldsymbol{c}$, then the mean value of $\boldsymbol{\varphi}$ for a group of particles having had their last collisions at a time $t=\tau$ is denoted by $\tilde{\boldsymbol{\varphi}}(t,\tau)$, and

$$\tilde{\boldsymbol{\varphi}}(\tau,\tau)=0, \qquad \tilde{\dot{\boldsymbol{\varphi}}}(t,\tau)=\dot{\tilde{\boldsymbol{\varphi}}}(t,\tau), \tag{1.5}$$

where the dot designates $\mathrm{d}/\mathrm{d}t$. The number of particles of that group at a time t is proportional to $\exp[-\nu(t-\tau)]$, where

$$\nu:=\text{mean collision frequency of a particle} \\ \approx\frac{\text{mean random (thermal) speed}}{\text{mean free path}}. \tag{1.6}$$

The spatial mean value over all particles in a small macroscopic volume is denoted by a bar and reads

$$\bar{\boldsymbol{\varphi}}(t)=\nu\int_{-\infty}^{t}\mathrm{d}\tau\,\mathrm{e}^{-\nu(t-\tau)}\,\tilde{\boldsymbol{\varphi}}(t,\tau). \tag{1.7}$$

---

[1] Burckhardt, G. (1950): Ann. Phys. (Leipzig) 5, 373–380

With partial integration, using both Eqs. (1.5), one obtains for the time derivative of Eq. (1.7)

$$\dot{\bar{\varphi}}(t) = \overline{\dot{\varphi}}(t) - \nu\bar{\varphi}(t). \tag{1.8}$$

For $\boldsymbol{\varphi} = \boldsymbol{c}$ this means in particular

$$\overline{\dot{\boldsymbol{c}}} = \dot{\bar{\boldsymbol{c}}} + \nu\bar{\boldsymbol{c}}, \tag{1.9}$$

whereas for $\boldsymbol{\varphi} = (m/2)\,c^2\boldsymbol{c} = (m/2)\,\boldsymbol{c}\cdot\boldsymbol{c}\boldsymbol{c}$ one obtains

$$\overline{\frac{m}{2}c^2\dot{\boldsymbol{c}}} = \dot{\overline{\frac{m}{2}c^2\boldsymbol{c}}} + \nu\overline{\frac{m}{2}c^2\boldsymbol{c}}\,; \tag{1.10}$$

on the left-hand side the vanishing of

$$\overline{\boldsymbol{c}\frac{\mathrm{d}}{\mathrm{d}t}\frac{m}{2}\boldsymbol{c}\cdot\boldsymbol{c}} = m\overline{\boldsymbol{c}\dot{\boldsymbol{c}}\cdot\boldsymbol{c}}$$

has been used. Because the angle between $\dot{\boldsymbol{c}}$ and $\boldsymbol{c}$ may take any value, the average over this angle vanishes.

Averaging the equation of motion, Eq. (1.1), using Eq. (1.9) with the *mean velocity*

$$\boldsymbol{v} := \bar{\boldsymbol{c}} \tag{1.11}$$

and adding heuristically as force density the negative gradient of the scalar pressure

$$p = N\,k\,T = N\frac{m}{2}\overline{c^2}, \tag{1.12}$$

with N:= number density and k:= Boltzmann's constant, one obtains the *momentum balance*

$$N\,m(\dot{\boldsymbol{v}} + \nu\,\boldsymbol{v} - \boldsymbol{\Omega}\times\boldsymbol{v}) = N\boldsymbol{F} - k\,T\frac{\partial N}{\partial \boldsymbol{r}} - N\,k\frac{\partial T}{\partial \boldsymbol{r}}. \tag{1.13}$$

The pressure gradient expresses the existence of a diffusion velocity $\boldsymbol{v} = -(k\,T/N\,m\,\nu)\,\partial N/\partial\boldsymbol{r}$ in a steady-state case ($\dot{\boldsymbol{v}} = 0$) without outer forces $\boldsymbol{F} + \boldsymbol{\Omega}\times m\boldsymbol{v}$ and for a constant temperature T.

If we take the average of Eq. (1.4), thereby using Eq. (1.10) with the *energy-flux vector*

$$\boldsymbol{q} := N\overline{\frac{m}{2}c^2\boldsymbol{c}}, \tag{1.14}$$

and adding heuristically the force density $-\partial p/\partial\boldsymbol{r}$, we obtain the *energy-flux balance*

$$\dot{\boldsymbol{q}} + \nu\,\boldsymbol{q} - \boldsymbol{\Omega}\times\boldsymbol{q} = \frac{3kT}{2m}\left(N\boldsymbol{F} - k\,T\frac{\partial N}{\partial\boldsymbol{r}} - N\,k\frac{\partial T}{\partial\boldsymbol{r}}\right). \tag{1.15}$$

The temperature gradient causes an energy flux $\boldsymbol{q} = -(3Nk^2T/2m\nu)\,\partial T/\partial \boldsymbol{r}$ in a steady-state case with $\boldsymbol{F}=0=\boldsymbol{\Omega}=\partial N/\partial \boldsymbol{r}$.

To derive an expression for the viscosity, a balance for the stress tensor must be established. After dyadic multiplication of the equation of motion (1.1) with $\boldsymbol{c}$, the averaging process has to be applied. Since the first of Eqs. (1.5) does not hold for a tensorial quantity $\boldsymbol{\varphi}=m\boldsymbol{c}\boldsymbol{c}$ nor does Eq. (1.8) the procedure becomes quite involved[2]. Therefore the viscosity will not be dealt with in this contribution.

2. Transport equations

α) The mobility tensor. For time-harmonic motions $\boldsymbol{v} \sim \exp - i\omega t \sim \boldsymbol{q}$ and constant $\boldsymbol{\Omega}$ one can solve the balance equations (1.13) (1.15) most conveniently by introducing the *mobility tensor*

$$\boldsymbol{\beta} := [(\nu - i\omega)\,\mathbf{U} - \boldsymbol{\Omega}\times\mathbf{U}]^{-1}. \tag{2.1}$$

Here

$$\mathbf{U} := [\hat{\boldsymbol{x}} \quad \hat{\boldsymbol{y}} \quad \hat{\boldsymbol{z}}] \begin{bmatrix} 1 & 0 & 0 \\ 0 & 1 & 0 \\ 0 & 0 & 1 \end{bmatrix} \begin{bmatrix} \hat{\boldsymbol{x}} \\ \hat{\boldsymbol{y}} \\ \hat{\boldsymbol{z}} \end{bmatrix} \tag{2.2}$$

is the unit tensor (in a Cartesian frame $\hat{\boldsymbol{x}}, \hat{\boldsymbol{y}}, \hat{\boldsymbol{z}}$) and

$$\boldsymbol{\Omega}\times\mathbf{U} = [\hat{\boldsymbol{x}} \quad \hat{\boldsymbol{y}} \quad \hat{\boldsymbol{z}}] \begin{bmatrix} 0 & -\Omega_z & +\Omega_y \\ +\Omega_z & 0 & -\Omega_x \\ -\Omega_y & +\Omega_x & 0 \end{bmatrix} \begin{bmatrix} \hat{\boldsymbol{x}} \\ \hat{\boldsymbol{y}} \\ \hat{\boldsymbol{z}} \end{bmatrix} = \mathbf{U}\times\boldsymbol{\Omega} \tag{2.3}$$

is an antisymmetric tensor. Both tensors $\boldsymbol{\beta}$ and $\mathbf{U}$ are special cases of the class of *axial tensors.* Their properties are discussed in Appendix A, primarily the calculation of the eigenvalues and of the reciprocal.

The eigenvalues of the reciprocal mobility tensor $\boldsymbol{\beta}^{-1}$, Eq. (2.1), are [see Eqs. (A.15) (A.16) in Appendix A]

$$\beta_0^{-1} = \nu - i\omega \qquad\qquad \beta_{\pm 1}^{-1} = \nu - i\omega \pm i\Omega \tag{2.4}$$

with the linear combinations

$$\frac{\beta_{+1}^{-1} + \beta_{-1}^{-1}}{2} = \nu - i\omega \qquad\qquad \frac{\beta_{+1}^{-1} - \beta_{-1}^{-1}}{2} = i\Omega \tag{2.5}$$

$$\frac{\beta_{+1} + \beta_{-1}}{2} = \frac{\nu - i\omega}{(\nu - i\omega)^2 + \Omega^2} \qquad \frac{\beta_{+1} - \beta_{-1}}{2} = \frac{-i\Omega}{(\nu - i\omega)^2 + \Omega^2}. \tag{2.6}$$

---

[2] Ecker, G., Kaschuba, K., Riemann, K.-U., Schumacher, A. (1979): Phys. Fluids *22*, 1203–1207, sect. III.B

Thus the mobility tensor $\boldsymbol{\beta}$ can be written according to Eq. (A.11c) as

$$\boldsymbol{\beta}=\frac{(\nu-i\omega)\,\mathbf{U}+\boldsymbol{\Omega}\times\mathbf{U}+(\nu-i\omega)^{-1}\boldsymbol{\Omega}\boldsymbol{\Omega}}{(\nu-i\omega)^2+\Omega^2} \tag{2.7a}$$

$$=\left[\frac{\mathbf{U}}{\nu-i\omega}+\frac{\boldsymbol{\Omega}\times\mathbf{U}}{(\nu-i\omega)^2}\right]\left(1-\frac{\Omega^2}{(\nu-i\omega)^2}\right)+\frac{\boldsymbol{\Omega}\boldsymbol{\Omega}}{(\nu-i\omega)^3}+O\left(\frac{\Omega^4}{(\nu-i\omega)^5}\right) \tag{2.7b}$$

$$=\left[\frac{\hat{\boldsymbol{\Omega}}\hat{\boldsymbol{\Omega}}}{\nu-i\omega}+\frac{\hat{\boldsymbol{\Omega}}\times\mathbf{U}}{\Omega}\right]\left(1-\frac{(\nu-i\omega)^2}{\Omega^2}\right)+\frac{\nu-i\omega}{\Omega^2}\mathbf{U}+O\left(\frac{(\nu-i\omega)^3}{\Omega^4}\right) \tag{2.7c}$$

For $\Omega\ll|\nu-i\omega|$ the plasma is *collision dominated* ($\Omega\ll\nu\gg\omega$) or in a *high-frequency limit* ($\Omega\ll\omega\gg\nu$) and is therefore almost isotropic. For $\Omega\gg|\nu-i\omega|$ the plasma is *field dominated* and therefore *strongly anisotropic*. In this case the absolute value Ω of the vector $\boldsymbol{\Omega}$, Eq. (1.3), does not appear in the first term of the expression Eq. (2.7c) for $\boldsymbol{\beta}$ but only its direction $\hat{\boldsymbol{\Omega}}$.

β) Fick's diffusion law and Ohm's law. For time-harmonic motions the solution

$$N\,m\,\boldsymbol{v}=\boldsymbol{\beta}\cdot\left(N\boldsymbol{F}-kT\frac{\partial N}{\partial\boldsymbol{r}}-Nk\frac{\partial T}{\partial\boldsymbol{r}}\right) \tag{2.8}$$

of the momentum balance, Eq. (1.13), can be written as

$$N\,\boldsymbol{v}=\frac{\boldsymbol{\sigma}}{q}\cdot\frac{\mathbf{F}}{q}-\mathbf{D}\cdot\frac{\partial N}{\partial\boldsymbol{r}}-\mathbf{D}\cdot\frac{N}{T}\frac{\partial T}{\partial\boldsymbol{r}} \tag{2.9}$$

with

$$\boldsymbol{\sigma}:=\frac{q^2N}{m}\boldsymbol{\beta}\qquad \mathbf{D}:=\frac{kT}{m}\boldsymbol{\beta}=\frac{kT}{q^2N}\boldsymbol{\sigma}. \tag{2.10}$$

For $\partial N/\partial\boldsymbol{r}=0=\partial T/\partial\boldsymbol{r}=\boldsymbol{g}_0$ this is *Ohm's law* $N\,q\,\boldsymbol{v}=\boldsymbol{\sigma}\cdot\boldsymbol{E}$ and for $\boldsymbol{F}=0=\partial T/\partial\boldsymbol{r}$ *Fick's diffusion law* $N\,\boldsymbol{v}=-\mathbf{D}\cdot\partial N/\partial\boldsymbol{r}$.

In the context of the mean-collision-frequency method the *Einstein relation* $\mathbf{D}=(kT/m)\boldsymbol{\beta}$, Eq. (2.10), is merely the statement that $-\partial p/\partial\boldsymbol{r}$ was the correct form for the heuristically added force in the momentum balance (1.13). The scalar dc mobility for $\boldsymbol{\Omega}=0$ is $\beta=\nu^{-1}$, Eq. (2.1). With the definition of ν, Eq. (1.6), and $\sqrt{kT/m}$ proportional to the mean random speed, the Einstein relation expresses the well-known proportionality of D to the product of the mean free path and the mean random speed.

γ) Fourier's law. The solution of the energy-flux balance, Eq. (1.15), for time harmonic motions can be written as

$$\boldsymbol{q}=\check{\mathbf{D}}_T\cdot\left(N\boldsymbol{F}-kT\frac{\partial N}{\partial\boldsymbol{r}}\right)-\boldsymbol{\kappa}\cdot\frac{\partial T}{\partial\boldsymbol{r}} \tag{2.11}$$

with

$$\check{\mathbf{D}}_T:=\frac{3kT}{2m}\boldsymbol{\beta}=\tfrac{3}{2}\mathbf{D}\qquad \boldsymbol{\kappa}:=\frac{3pk}{2m}\boldsymbol{\beta}=\tfrac{3}{2}Nk\,\mathbf{D}. \tag{2.12}$$

In an isotropic medium the relation $\kappa = \frac{3}{2} N k D$ is a well-known result from the mean-free-path theory[1]. Thus the heuristically added force density, $-Nk\,\partial T/\partial \boldsymbol{r}$, in the energy-flux balance, Eq. (1.15), yields the same results as the mean-free-path method in an isotropic medium. But with rigorous kinetic theory we shall show that the relation $\kappa = \frac{3}{2} N k D$ holds only in the direct current (dc) case for a particular kind of collisions as well as the relation $\breve{D}_T = \frac{3}{2} D$ between the Dufour coefficient $\breve{D}_T$ for diffusive heat transfer and the diffusion coefficient D (see Sect. 28 β).

Putting the representation of $\boldsymbol{\beta}$, Eq. (2.7), into the relations between $\boldsymbol{\beta}$ and $\mathbf{D}$, $\boldsymbol{\sigma}$ and $\boldsymbol{\kappa}$, Eqs. (2.9) (2.10) (2.12) respectively, one obtains representations for these tensorial transport coefficients. The approximation

$$\boldsymbol{\beta} \approx \hat{\boldsymbol{\Omega}}\hat{\boldsymbol{\Omega}}/(\nu - i\omega), \qquad [2.7\text{c}]$$

valid for strong anisotropy, has been used for $\boldsymbol{\kappa}$, Eq. (2.12), by HOLLWEG[2].

δ) Onsager-Casimir relations. The above mean-collision-frequency derivation of $\mathbf{D}$, $\boldsymbol{\sigma}$, and $\boldsymbol{\kappa}$ gave a generalization of some standard results of the mean-free-path method including the effect of anisotropy which is expressed by the tensorial character of the transport coefficients. But the *Onsager-Casimir relations* of the thermodynamics of irreversible processes require a certain symmetry in the combination of Eqs. (2.9) (2.11) for $N\boldsymbol{v}$ and $\boldsymbol{q}$ resulting in the replacement of $\mathbf{D}\cdot\partial T/\partial\boldsymbol{r}$ in Eq. (2.9) by $\mathbf{D}_T\cdot\partial T/\partial\boldsymbol{r}$. Thus the coupling terms on the right-hand sides of Eqs. (2.9) (2.11) are not correctly derived by the mean-collision-frequency method.

Furthermore, the value of the mean collision frequency ν, Eq. (1.6), dependent on the kind of interaction during the collisions, is outside the scope of the mean-collision-frequency theory. To calculate these values, rigorous kinetic gas theory must be employed.

B. Gas kinetic cross sections and collision frequencies

3. General methods of the kinetic theory. In a rigorous kinetic theory the transport coefficients must be calculated by averaging over suitable quantities of the particles, thereby using their properties, e.g., interaction laws, etc. The starting point is usually the Liouville equation, a conservation equation for the phase points of an ensemble of systems.

There exist two different methods to calculate transport coefficients starting from the Liouville equation:

1) The older method establishes a hierarchy for many-particle distribution functions, ending up with a kinetic equation for the one-particle distribution function. According to various assumptions, different *kinetic equations* can be derived; the oldest and most famous is the *Boltzmann equation*. In a second

[1] Present, R.D.: See [3], eqs. 3–22 and 3–26

[2] Hollweg, J.V. (1974): J. Geophys. Res. *79*, 3845–3850

step the kinetic equation must be solved in a way that allows expressions for transport coefficients to be established. (For a survey see SUCHY[1].)

2) The modern *correlation-function method* (other names are "linear response method" and "Kubo method") represents transport coefficients via correlation functions. With this method expressions of the *DC* electrical conductivity ($\omega=0$) for very dense plasmas with high temperatures were derived[2,3]. Because aeronomic plasmas have low densities and temperatures this contribution will deal only with the first method.

Out of several approaches to solve the Boltzmann equation, three somewhat different methods can be used to give results needed for the calculation of transport coefficients.

α) The moment method[4–6] must be used in the general case, where the motions of more than one species of particles contribute to the current density or to the heat flux. This method must also be used if, though the contribution of one of such species (e.g., electrons) is dominant, different interactions with other particle species (e.g., neutrals, ions, electrons) must be taken into account. The moment method is used in Chap. B.

β) For the application of the *Lorentz method* the following conditions must be satisfied: (i) The motion of electrons only contributes to the fluxes, (ii) Coulomb collisions can be neglected. (This holds for example for electromagnetic waves with frequencies high above the ion gyrofrequency, Eq. (1.3), when propagating through the *D* and $\mathscr{E}$ regions of the terrestrial ionosphere.) With these assumptions the mutual collisions between electrons can be neglected as compared with electron-neutral collisions. The plasma can then be treated as a *Lorentz plasma* using the Lorentz method when dealing with kinetic gas theory[6–8], see Chap. C.

γ) The *Chapman-Enskog method* has some features in common with the moment method, but it is restricted to the so-called normal solutions of the Boltzmann equation. These solutions vary with space and time merely via the number densities N_j of the different particle species, the mass velocitiy $\boldsymbol{v}$ of the plasma as a whole, and the common temperature T. Hence they are not suited for the description of the temporal variation of electrical current densities in a plasma, which is needed for the calculation of the ac conductivity. Therefore the Chapman-Enskog method is not used in this contribution.

I. Transfer cross sections and transfer collision frequencies

4. Collision cross sections and collision frequencies. It is not intended to derive transport coefficients completely from the kinetic theory in this contri-

[1] Suchy, K. (1972): Radio Sci. 7, 871–884, Fig. 1
[2] Boercker, D.B. (1981): Phys. Rev. A*23*, 1969–1981
[3] Boercker, D.B., Rogers, F.J., De Witt, H.E. (1982): Phys. Rev. A*25*, 1623–1631
[4] Maxwell, J.C. (1879): Sci. Pap. *2*, 681–712
[5] Grad, H. (1949): Comm. Pure Appl. Math. *2*, 331-407
[6] Suchy, K.: See [*8*]
[7] Lorentz, H.A.: See [*1*], note 29
[8] Allis, W.P.: this Encyclopedia, vol. 21, pp. 383–444

bution, but we must explain the quantities which appear in the final expressions for these coefficients. It is convenient to use the concept of collision frequencies and collision cross sections, but their definitions need some careful explanations.

A *collision frequency* for a particle of species j with particles of species k is the product of the *relative speed* g_{jk} between the colliding particles, the *collision cross section* Q_{jk}, and the number density N_k of the collision partners. It is only for the collisions between rigid spheres that Q_{jk} is independent of g_{jk}. This well-known billard-ball model seldom describes the real interaction during a collision; in general a speed-dependent $Q_{jk}(g_{jk})$ has to be considered. Instead of the relative speed g_{jk}, the *collision energy*,

$$\mathscr{E}_{jk} := \tfrac{1}{2}\mu_{jk}\, g_{jk}^2 \quad \text{with} \quad \mu_{jk} := \frac{m_j m_k}{m_j + m_k} \tag{4.1}$$

as *reduced mass*, is often used as variable, mostly measured in electron volts:

$$1\,\text{eV} = 1.6022 \cdot 10^{-19}\,\text{joule} = 1.6022 \cdot 10^{-12}\,\text{erg}.$$

The collision energy $\mathscr{E}_{jk}$ is the sum of the kinetic energies of both collision partners in the center-of-mass system.

α) Differential cross sections. In a rigorous kinetic theory the collisions have to be weighted according to the *deflection angle* χ between the directions of the relative velocities $\boldsymbol{g}$ before and after the collision. In order to do this, *differential cross sections* $q(\Omega, \mathscr{E})$ are introduced with $q\,\mathrm{d}\Omega$ as cross section for collisions deflecting incoming particles into the differential solid angle $\mathrm{d}\Omega = \mathrm{d}\varepsilon\,\mathrm{d}(\cos\chi)$. For spherically symmetric particles, $q(\chi, \varepsilon; \mathscr{E})$ does not depend on the azimuthal angle ε. For *isotropic scattering*, $q(\Omega, \mathscr{E})$ does not depend at all on the solid angle Ω. An example are rigid spheres, where q does not even depend on the energy, see Eq. (6.18). For nonspherical collisions the influence of the azimuth-dependent part of $q(\chi, \varepsilon; \mathscr{E})$ is treated as "averaged out" in the following. This is not always correct but the influence of the nonspherical part is often small (see Sects. 9 and 10).

For spherically symmetric particles the classical limit is[1]

$$\lim_{h \to \infty} q(\chi, \mathscr{E})\,\mathrm{d}(\cos\chi) = b(\chi, \mathscr{E})\,\mathrm{d}b \tag{4.2}$$

with b as *impact parameter*. The dependence on the deflection angle χ can then be calculated by means of the integral of the trajectory, if the interaction potential is known (see Sect. 6).

For quantum-mechanical interactions *scattering amplitudes* $f(\chi)$ must be calculated with the Schrödinger equation. The expressions for differential cross sections are different for collisions between particles of different species (distinguishable particles) and particles of the same species (indistinguishable par-

---

[1] Massey, H.S.W. (1968): Endeavour *27*, 114–119, eq. 8

ticles). One has [2]

$$q(\chi)=|f(\chi)|^2 \quad \text{for distinguishable particles} \tag{4.3a}$$

and [3, 4]

$$[q(\chi)]_{\substack{n\\n/2}}=\frac{s+1}{2s+1}\,\frac{|f(\chi)\pm f(\pi-\chi)|^2}{2}+\frac{s}{2s+1}\,\frac{|f(\chi)\mp f(\pi-\chi)|^2}{2} \tag{4.3b}$$

for indistinguishable particles with

$$\left\{\begin{matrix}\text{integer}\\ \text{half-integer}\end{matrix}\right\}\ \text{spin}\ s=\left\{\begin{matrix}n\\ n/2\end{matrix}\right\}\ \text{obeying}\ \left\{\begin{matrix}\text{Bose-Einstein}\\ \text{Fermi-Dirac}\end{matrix}\right\}\ \text{statistics.}$$

The scattering amplitudes [5]

$$f(\chi)=\frac{1}{2i|\boldsymbol{k}|}\sum_{L=0}^{\infty}(2L+1)\,[\exp(2i\eta_L)-1]\ \mathrm{P}_L(\cos\chi) \tag{4.4a}$$

with

$$|\boldsymbol{k}|:=\frac{\mu g}{\hbar}=\frac{\sqrt{2\mu\mathscr{E}}}{\hbar} \tag{4.4b}$$

have the dimension of a length and are expressed by the *phase shifts* $\eta_L(|\boldsymbol{k}|)$ which can only be determined by resolving the Schrödinger equation. In this way the phase shifts η_L (and hence the scattering amplitude f and the differential cross section q) depend on the *interaction potential* $\varphi(r)$. The $\mathrm{P}_L(\cos\chi)$ are *Legendre polynomials* with the *angular momentum quantum number* L as subscript. Because of $\mathrm{P}_L(\cos(\pi-\chi))=\mathrm{P}_L(-\cos\chi)=(-1)^L\,\mathrm{P}_L(\cos\chi)$ one obtains

$$f(\chi)\pm f(\pi-\chi)=\frac{1}{i|\boldsymbol{k}|}\sum_{\substack{L\,\text{even}\\ L\,\text{odd}}}(2L+1)\,(\mathrm{e}^{2i\eta_L}-1)\ \mathrm{P}_L(\cos\chi) \tag{4.5}$$

for the linear combinations of f occurring in the expression (4.3b) of the differential cross section $q(\chi)$ for indistinguishable particles [6].

Since the quantum-mechanical expression for the differential cross section $q(\chi)$ is quite complicated, one should try to use its classical limit whenever possible. Its dependence on the interaction potential $\varphi(r)$ via the integral for the trajectory

$$\chi=\pi-\int_{-\infty}^{+\infty}\mathrm{d}t\,\dot\theta=\pi-2\int_{r_\mathrm{P}}^{\infty}\frac{\mathrm{d}r}{\dot r}\,\dot\theta=\pi-2\int_{r_\mathrm{P}}^{\infty}\frac{\mathrm{d}r}{g\sqrt{1-\dfrac{b^2}{r^2}-\dfrac{\varphi(r)}{\mathscr{E}}}}\,\frac{bg}{r^2}, \tag{4.6a}$$

with the pericenter r_P determined by

$$1-\frac{b^2}{r_\mathrm{P}^2}-\frac{\varphi(r_\mathrm{P})}{\mathscr{E}}=0, \tag{4.6b}$$

---

2 Massey, H.S.W.: this Encyclopedia, vol. 36, pp. 232–306, eq. 1.10

3 Dalgarno, A. (1958): Philos. Trans. R. Soc. London Ser. A *250*, 426–439, eqs. 1 and 2

4 Hahn, H., Mason, E.A., Smith, F.J. (1971): Phys. Fluids *14*, 278–287, eq. 28

5 Massey, H.S.W.: this Encyclopedia, vol. 36, p. 232–306, eq. 2.17

6 Hirschfelder, J.O., Curtiss, C.F., Bird, R.B.: See [5], eqs. 1.7–27 and 1.7–28

is much less involved than via the phase shifts η_L. However, the transition to the classical limit must be carefully investigated so as to clarify the conditions for this limit to hold. If necessary, correction terms can be found and used. It appears then that the classical limit of the differential cross section $q(\chi, \mathscr{E})$ is more difficult to obtain than the classical limit of weighted spherical means of $q(\chi, \mathscr{E})$. The most obvious mean is the *integrated cross section*

$$Q^{\mathrm{int}}(\mathscr{E}) := \oint \mathrm{d}\Omega\, q(\Omega, \mathscr{E}), \tag{4.7}$$

with a constant weighting factor (equal to unity). This integrated cross section Q^{int} does not appear in a rigorous kinetic theory since it does not properly take account of the persistence of velocities after the collision.

β) Transfer cross sections. In order to express the results of the rigorous kinetic theory, various weighting factors (depending on χ) must be used. They must have some common features, namely, that grazing collisions with $\chi=0$ have the weight zero while collisions with $\chi>0$ have positive weighting factors. These conditions are fulfilled for *transfer cross sections*

$$Q^{(l)}(\mathscr{E}) := \oint \mathrm{d}\Omega\, q(\Omega, \mathscr{E})\,[1-\mathrm{P}_l(\cos\chi)] \tag{4.8}$$

and

$$\phi^{(l)}(\mathscr{E}) := \oint \mathrm{d}\Omega\, q(\Omega, \mathscr{E})\,[1-\cos^l\chi], \tag{4.9}$$

using Legendre polynomials [7, 8] $\mathrm{P}_l(\cos\chi)$ with $\mathrm{P}_l(1)=1$ or powers [9] of $\cos\chi$.

Both types of transfer cross sections are related by [10]

$$Q^{(l)} = \sum_{\lambda=0}^{\mathrm{ent}\frac{l-1}{2}} (-1)^{\lambda} \frac{(2l-2\lambda-1)!!}{(2\lambda)!!\,(l-2\lambda)!}\,\phi^{(l-2\lambda)} \tag{4.10a}$$

and

$$\phi^{(l)} = l! \sum_{\lambda=0}^{\mathrm{ent}\frac{l-1}{2}} \frac{2l-4\lambda+1}{(2\lambda)!!\,(2l-2\lambda+1)!!}\,Q^{(l-2\lambda)}, \tag{4.10b}$$

in particular

$$Q^{(1)} = \phi^{(1)} \qquad Q^{(2)} = \tfrac{3}{2}\phi^{(2)}. \tag{4.10c}$$

The cross section $Q^{(1)}(\mathscr{E})$ is often named *momentum transfer cross section.*

The coefficients in Eqs. (4.10a, b) are the same as in the expansions of the Legendre polynomials $\mathrm{P}_l(x)$ in powers $x^{l-2\lambda}$, and of x^l in terms of Legendre polynomials $\mathrm{P}_{l-2\lambda}(x)$, respectively [11].

If the differential cross sections does not depend on the solid angle Ω, one obtains (because of the orthogonality of the Legendre polynomials)

$$Q^{(l)}(\mathscr{E}) = Q^{\mathrm{int}}(\mathscr{E}) \qquad \text{for isotropic scattering.} \tag{4.11}$$

---

[7] Maxwell, J.C. (1879): Sci. Pap. *2*, 681–712 page 688

[8] Allis, W.P.: this Encyclopedia, vol. 21, pp. 383–444, eq. 27.4

[9] Chapman, S., Cowling, T.G.: See [*12*], eq. 9.33, 4

[10] Kumar, K. (1967): Aust. J. Phys. *20*, 205–252, eq. A.23

[11] Lense, J. (1947): Reihenentwicklungen in der mathematischen Physik, 2nd edn., sects. V.11 and V.18. Berlin: de Gruyter

This holds for rigid-sphere interaction. More realistic interaction potentials, however, do not lead to isotropic scattering.

Inserting the quantum-mechanical expressions Eqs. (4.3) (4.4) (4.5) for the differential cross section $q(\chi)$ into the definitions Eqs. (4.7) (4.8) for the integrated cross section and the transfer cross sections and using the orthogonality of the Legendre polynomials[12, 13], one finds

$$Q^{\mathrm{int}}(\mathscr{E})=\frac{4\pi}{|\boldsymbol{k}|^2}\sum_{L=0}^{\infty}(2L+1)\sin^2\eta_L=\frac{4\pi}{|\boldsymbol{k}|}\operatorname{Im} f(0) \tag{4.12a}$$

$$Q^{(1)}(\mathscr{E})=\frac{4\pi}{|\boldsymbol{k}|^2}\sum_{L=0}^{\infty}(L+1)\sin^2(\eta_{L+1}-\eta_L) \tag{4.12b}$$

$$Q^{(2)}(\mathscr{E})=\frac{6\pi}{|\boldsymbol{k}|^2}\sum_{L=0}^{\infty}\frac{(L+1)(L+2)}{2L+3}\sin^2(\eta_{L+2}-\eta_L) \tag{4.12c}$$

for distinguishable particles and

$$\begin{aligned}[Q^{\mathrm{int}}(\mathscr{E})]_{\substack{n\\n/2}}=[Q^{(1)}(\mathscr{E})]_{\substack{n\\n/2}}&=\frac{s+1}{2s+1}\frac{8\pi}{|\boldsymbol{k}|^2}\sum_{\substack{L\,\mathrm{even}\\L\,\mathrm{odd}}}(2L+1)\sin^2\eta_L\\&+\frac{s}{2s+1}\frac{8\pi}{|\boldsymbol{k}|^2}\sum_{\substack{L\,\mathrm{odd}\\L\,\mathrm{even}}}(2L+1)\sin^2\eta_L\end{aligned} \tag{4.13a}$$

$$\begin{aligned}[Q^{(2)}(\mathscr{E})]_{\substack{n\\n/2}}&=\frac{s+1}{2s+1}\frac{6\pi}{|\boldsymbol{k}|^2}\sum_{\substack{L\,\mathrm{even}\\L\,\mathrm{odd}}}\frac{(L+1)(L+2)}{2L+3}\sin^2(\eta_{L+2}-\eta_L)\\&+\frac{s}{2s+1}\frac{6\pi}{|\boldsymbol{k}|^2}\sum_{\substack{L\,\mathrm{odd}\\L\,\mathrm{even}}}\frac{(L+1)(L+2)}{2L+3}\sin^2(\eta_{L+2}-\eta_L)\end{aligned} \tag{4.13b}$$

for indistinguishable particles with integer/half-integer spin s.

Transfer cross sections $Q^{(l)}(\mathscr{E})$ with $l\geqq 3$ will not be used in this contribution. With the expressions of Eqs. (4.12) for $Q^{(l)}(\mathscr{E})$ the transition to the classical limit will be discussed (see Sect. 5).

γ) Transfer collision frequencies. In the kinetic gas theory the transfer cross sections $Q^{(l)}(\mathscr{E})$ appear always multiplied with the relative speed

$$g=\sqrt{\frac{2\mathscr{E}}{\mu}}. \qquad [4.1]$$

The *transfer collision rate* $g_{jk}\,Q_{jk}^{(l)}(\mathscr{E}_{jk})\,N_j\,N_k$ is the number of collisions of j particles with k particles per unit volume and unit time, with the relative speed g and the weight $1-\mathrm{P}_l(\cos\chi)$. The role of j and k may be interchanged without

[12] Massey, H.S.W.: this Encyclopedia, vol. 36, pp. 232–306, eq. 2.18
[13] Ferziger, J.H., Kaper, H.G.: See [*18*], eqs. 9.3–16 and 9.3–17

altering the transfer collision rate as well as the *transfer collision coefficient* $g_{jk}Q_{jk}^{(l)}(\mathscr{E}_{jk})$.

The quantity

$$\nu_{jk}^{(l)}(\mathscr{E}_{jk}) := g_{jk}\,Q_{jk}^{(l)}(\mathscr{E}_{jk})\,N_k = \sqrt{2\mathscr{E}_{jk}/\mu_{jk}}\;Q_{jk}^{(l)}(\mathscr{E}_{jk})\,N_k \tag{4.14}$$

describes the number of collisions of a j particle with k particles per unit time, when the relative speed is g and the weight $1-P_l(\cos\chi)$. It is therefore called a *transfer collision frequency*. Obviously

$$N_j\,\nu_{jk}^{(l)} = N_k\,\nu_{kj}^{(l)}. \tag{4.15}$$

The particular collision frequency $\nu^{(1)}(\mathscr{E})$ is often called *momentum transfer collision frequency*.

For the calculation of transport coefficients the transfer collision frequencies $\nu^{(l)}(\mathscr{E})$, Eq. (4.14), must be averaged over the collision energies $\mathscr{E}$, Eq. (4.1). This has to be done with the velocity distribution functions for the two species j and k of the collision partners; these contain the temperatures T_j and T_k as parameters. Different averaging procedures must be used for the moment method and the Lorentz method (see Sect. 3). These will be discussed in Chaps. C and D, respectively. In Chap. B we shall show how the transfer cross sections $Q^{(l)}(\mathscr{E})$ can be calculated for the different types of interactions between the various species of particles in an aeronomic plasma.

5. Quantum corrections to classical transfer cross sections. The transition from Eqs. (4.12), i.e., the quantum-mechanical expressions for the transfer cross sections $Q^{(l)}(\mathscr{E})$, to their classical limits is provided by a WKB solution of the radial wave equation. This yields WKB approximations[1] for the phase shifts $\eta_L(|\boldsymbol{k}|)$. A quantity $b(L) := \sqrt{L(L+1)}/|\boldsymbol{k}|$ is introduced formally and another quantity $\chi(b(L),\mathscr{E})$ defined by the right-hand side of the expression (4.6a) for the deflection angle χ with $b(L)$ instead of b. With these quantities the phase shift differences occuring in the expressions (4.12) for $Q^{(l)}(\mathscr{E})$[2,3] can be represented as

$$\eta_{L+l} - \eta_L = \frac{l}{2}\,\chi(b,\mathscr{E};\hbar^0) + O\left(\frac{1}{|\boldsymbol{k}|\,b}\right). \tag{5.1}$$

The summation over L is replaced by an integration over $\mathrm{d}L$ and with $b := \sqrt{L(L+1)}/|\boldsymbol{k}|$ by an integration over

$$\mathrm{d}b = \mathrm{d}L\,\frac{1}{|\boldsymbol{k}|}\sqrt{1+\frac{1}{4(|\boldsymbol{k}|\,b)^2}} = \mathrm{d}L\,\frac{1}{|\boldsymbol{k}|}\left[1-\frac{1}{8(|\boldsymbol{k}|\,b)^2}+O\left(\frac{1}{|\boldsymbol{k}|^4\,b^4}\right)\right]^{-1}.$$

---

[1] de Boer, J., Bird, R.B. (1954): Physica *20*, 185–198

[2] Hirschfelder, J.O., Curtiss, C.F., Bird, R.B.: See [*5*], eqs. 10.3-12 and 10.3-13

[3] McDaniel, E.W., Mason, E.A.: See [*23*], eq. 5-3-40

The result is a *de Boer-Bird expansion*[4, 5]

$$Q^{(l)}(\mathscr{E})=Q^{(l)}_{\text{class}}(\mathscr{E})+\frac{1}{|\boldsymbol{k}|^2}R^{(l)}(\mathscr{E})+\frac{1}{|\boldsymbol{k}|^4}S^{(l)}(\mathscr{E})+O\left(\frac{1}{|\boldsymbol{k}|^6}\right) \tag{5.2}$$

with

$$Q^{(l)}_{\text{class}}(\mathscr{E})=2\pi\int_0^\infty \mathrm{d}b\, b(\chi,\mathscr{E})\,[1-\mathrm{P}_l(\cos\chi)]. \tag{5.3}$$

Because of $|\boldsymbol{k}|=\mu g/\hbar$ this is an expansion after increasing powers of $\hbar^2$, the initial term $Q^{(l)}_{\text{class}}(\mathscr{E})$ being of the order $\hbar^0$. On the other hand it is an expansion after decreasing powers of $(\mu g)^2=2\mu\mathscr{E}$. Since the coefficients Q, R, S depend also on $\mathscr{E}$ one cannot denote the de Boer-Bird expansion, Eq. (5.2), as a "high-energy" expansion. But the reduced mass μ appears only in $|\boldsymbol{k}|^2=2\mu\mathscr{E}/\hbar^2$, not in the coefficients. Hence the de Boer-Bird expansion can be regarded as a *high-mass expansion.*

From this property of the de Boer-Bird expansion one can conclude the following. We compare two collisions with different μ but with the same (or similar) energy dependence, i.e., the same (or similar) interaction potential, at the same given collision energy $\mathscr{E}$. Then the collision with the higher μ will show less quantum influence than that with lower μ. The energy range, where the influence of quantum effects becomes remarkable, increases[6] with $1/\mu$. Since the collisions of ions and electrons with neutrals have similar interaction potentials (see Sect. 9), the energy limit for quantum effects is $\mu_{\text{in}}/\mu_{\text{en}}$ times higher for electrons than for ions (cf. Fig. 1, Sect. 6β).

In Eq. (5.2) the energy range for quantum effects is limited by

$$|\boldsymbol{k}|^2 Q^{(l)}_{\text{class}}(\mathscr{E})=\frac{2\mu\mathscr{E}}{\hbar^2}Q^{(l)}_{\text{class}}(\mathscr{E})=\frac{\mu}{m_{\mathrm{e}}}\frac{\mathscr{E}}{\mathrm{Ry}}\frac{Q^{(l)}_{\text{class}}(\mathscr{E})}{a_0^2}\lesssim 1, \tag{5.4}$$

where the *Bohr radius* a_0 and the atomic energy unit, *rydberg*, are defined by

$$a_0:=\frac{4\pi\varepsilon_0}{\mathrm{u}\,q_{\mathrm{e}}^2}\frac{\hbar^2}{m_{\mathrm{e}}}=5.292\cdot 10^{-11}\,\mathrm{m} \tag{5.5a}$$

$$1\,\mathrm{Ry}:=h\,c\,R_\infty=\frac{\hbar^2}{2m_{\mathrm{e}}a_0^2}=\frac{\mathrm{u}\,q_{\mathrm{e}}^2}{4\pi\varepsilon_0}\frac{1}{2a_0}$$

$$=\left(\frac{\mathrm{u}\,q_{\mathrm{e}}^2}{4\pi\varepsilon_0}\right)^2\frac{m_{\mathrm{e}}}{2\hbar^2}=13.606\,\mathrm{eV}. \tag{5.5b}$$

The letter u denotes 1 or 4π for rationalized or nonrationalized systems of units, respectively. Examples are the Système International or the Gauss System, respectively[7].

For ion-neutral collisions with $\mu_{\text{in}}/m_{\mathrm{e}}\approx 6\cdot 10^4$ and $Q^{(l)}\approx 20a_0^2$ the range is about $\mathscr{E}_{\text{in}}\lesssim 10^{-5}$ eV, whereas for electron-neutral collisions with $\mu_{\text{en}}/m_{\mathrm{e}}\approx 1$ the range is $\mathscr{E}_{\text{en}}\lesssim 0.7$ eV. Thus for aeronomic temperatures collisions between

---

[4] Hirschfelder, J.O., Curtiss, C.F., Bird, R.B.: See [5], eq. 10.3-21

[5] Kihara, T. (1964): J. Phys. Soc. Jpn. *19*, 108–116, eqs. 20 and 21

[6] Lin, S.L., Gatland, I.R., Mason, E.A. (1979): J. Phys. B*12*, 4179–4188, end of sect. 3.1

[7] Rawer, K., Suchy, K.: this Encyclopedia, vol. 49/2, pp. 1–546, appendix

heavy particles can be treated classically, but electron-neutral collisions must be treated quantum mechanically. In the latter case the de Boer-Bird expansion, Eq. (5.2), converges so slowly that it is practically useless for calculations. Quantum-mechanical scattering theory has to be applied instead. A low-energy expansion after increasing powers of $|\boldsymbol{k}|$ has been used, e.g., for this purpose (see Sect. 8 ζ).

The above estimates were made for slowly varying values of $Q^{(l)}_{\text{class}}(\mathscr{E})$. For Coulomb collisions the transfer cross sections vary roughly with $(q_j q_k \mathrm{u}/4\pi\varepsilon_0 \mathscr{E}_{jk})^2$, Eq. (8.5). Hence the energy range is given by

$$\frac{2\mu\mathscr{E}}{\hbar^2} Q^{(l)}_{\text{class}}(\mathscr{E}) \approx \frac{2\mu}{\hbar^2}\left(\frac{\mathrm{u}\, q_j q_k}{4\pi\varepsilon_0}\right)^2 \frac{1}{\mathscr{E}_{jk}} = 4\frac{\mu}{m_\mathrm{e}} Z_j^2 Z_k^2 \frac{\mathrm{Ry}}{\mathscr{E}_{jk}} \lesssim 1. \tag{5.6}$$

This means $\mathscr{E}_{jk} \gtrsim 3\cdot 10^6$ eV for $\mu_{jk}/m_\mathrm{e} \approx 6\cdot 10^4$ and $\mathscr{E}_{\mathrm{ei}} \gtrsim 54$ eV for $\mu_{\mathrm{ei}}/m_\mathrm{e} \approx 1$. Thus for aeronomic temperatures quantum effects can be neglected for Coulomb collisions.

It should be noted that the quantum influence on the differential cross sections $q(\chi, \mathscr{E})$ is much more pronounced than on the transfer cross sections

$$Q^{(l)}(\mathscr{E}) = \oint \mathrm{d}\Omega\, q(\Omega, \mathscr{E})\,[1 - \mathrm{P}_l(\cos\chi)]. \qquad [4.8]$$

To explain this difference we recall

$$q(\chi, \mathscr{E}) = |f(\chi)|^2 \qquad [4.3\mathrm{a}]$$

and the series expansion of $f(\chi)$, Eq. (4.4), in Legendre polynomials $\mathrm{P}_l(\cos\chi)$. The χ averaging for the calculation of $Q^{(l)}(\mathscr{E})$, which uses the orthogonality of the Legendre polynomials, smears out much of the oscillations of $q(\chi)$ which are due to quantum-mechanical interference phenomena [8].

The integral cross section $Q^{\text{int}}(\mathscr{E})$, Eq. (4.12a), does not depend on phase shift differences $\eta_{L+l} - \eta_L$, but on the phase shifts η_L themselves. Therefore in this case the introduction of a quantity $\chi(b, \mathscr{E}; \hbar^0)$ equivalent to the classical deflection angle is more complicated than in the case of the transfer cross sections $Q^{(l)}(\mathscr{E})$, Eq. (4.12b, c), via Eq. (5.1) [9].

6. Transfer cross sections. In this section some general features of classical cross sections and their dependence on the interaction potential will be discussed. The subscript "class" will in general be omitted and only written out where really needed.

Comparing the general definition Eq. (4.8) for the transfer cross sections with the classical limit Eq. (5.3), one confirms Eq. (4.2) as the differential cross section q, which can be written in the form

$$q(\chi, \mathscr{E}) = \frac{1}{2}\frac{\partial b^2(\chi, \mathscr{E})}{\partial(\cos\chi)}. \tag{6.1}$$

The connection between the impact parameter $b(\chi, \mathscr{E})$ and the deflection angle χ is given by the integral in Eq. (4.6a) for the trajectory, which is written in the

[8] Toennies, J.P. (1974): Proc. 9th Int. Symp. Rarefied Gas Dynamics, Göttingen, A.1-1–A.1-28, fig. 7. Porz-Wahn: DFVLR-Press

[9] McDaniel, E.W., Mason, E.A.: See [*23*], eq. 5-3-41

form

$$\chi = \pi - 2\int_0^{\rho_P} \frac{d\rho}{\sqrt{1-\rho^2-\frac{\varphi(b/\rho)}{\mathscr{E}}}} \qquad \text{with } \rho := \frac{b}{r}. \tag{6.2a}$$

The upper limit is given by the value ρ_P at the pericenter determined by

$$1-\rho_P^2-\frac{\varphi(b/\rho_P)}{\mathscr{E}}=0. \tag{6.2b}$$

α) Multiparameter interaction potentials. If the interaction potential $\varphi(r)$ depends on at least two parameters, it can be written unambigously in a dimensionless form

$$\varphi(r)=\varphi_R\,\varphi^*\left(\frac{r}{r_R}\right) \qquad \text{with } \varphi^*(1)=1 \quad \text{and } \varphi^*(\infty)=0. \tag{6.3}$$

Any length or energy may then be normalized by the *reference length* r_R and the corresponding *reference energy* φ_R. With the *normalized energy*

$$\mathscr{E}^* := \frac{\mathscr{E}}{\varphi_R} \tag{6.4}$$

Eqs. (6.2) can be written as

$$\chi = \pi - 2\int_0^{\rho_P} \frac{d\rho}{\sqrt{1-\rho^2-\frac{1}{\mathscr{E}^*}\varphi^*\left(\frac{b/r_R}{\rho}\right)}}, \tag{6.5a}$$

$$1-\rho_P^2-\frac{1}{\mathscr{E}^*}\varphi^*\left(\frac{b/r_R}{\rho_P}\right)=0. \tag{6.5b}$$

The pericenter value ρ_P is a function of $\mathscr{E}^*$ and b/r_R, hence $\chi=\chi(\mathscr{E}^*, b/r_R)$, too. From the expression

$$q(\chi,\mathscr{E})=\frac{1}{2}\frac{\partial b^2(\chi,\mathscr{E})}{\partial(\cos\chi)} \tag*{[6.1]}$$

for the differential cross section it follows that this quantity can be normalized with r_R^2 as

$$q(\chi,\mathscr{E})=: r_R^2\, q^*(\cos\chi, \mathscr{E}^*). \tag{6.6}$$

The same reference length r_R can then be used to normalize the transfer cross sections

$$Q^{(l)}(\mathscr{E})=\oint d\Omega\, q(\chi,\mathscr{E})\,[1-P_l(\cos\chi)] \tag*{[4.8]}$$

yielding

$$Q^{(l)}(\mathscr{E})=: \pi\, r_R^2\, Q^{(l)*}(\mathscr{E}^*). \tag{6.7}$$

Note the simple relationship between $Q^{(l)}$ and r_R and that the collision energy $\mathscr{E}$ can be normalized with the reference value $\varphi_R = \varphi(r_R)$ of the interaction potential.

Inspection of the expressions for the quantum corrections $R^{(l)}(\mathscr{E})$, $S^{(l)}(\mathscr{E})$, etc. of the de Boer-Bird expansion (5.2) shows that these quantities can also be normalized with πr_R^2 and φ_R:

$$\frac{Q^{(l)}(\mathscr{E})}{\pi r_R^2} = Q_{\text{class}}^{(l)*}(\mathscr{E}^*) + \frac{1}{\pi(|\boldsymbol{k}|\, r_R)^2} R^{(l)}(\mathscr{E}^*) + \frac{1}{\pi(|\boldsymbol{k}|\, r_R)^4} S^{(l)*}(\mathscr{E}^*) + O\left(\frac{1}{|\boldsymbol{k}|^6 r_R^6}\right). \tag{6.8}$$

Introducing the energy independent *de Boer-Bird parameter* for multiparameter potentials[1]

$$\frac{\mathscr{E}^*}{|\boldsymbol{k}|^2 r_R^2} = \frac{\mathscr{E}}{|\boldsymbol{k}|^2 r_R^2 \varphi_R} = \frac{\hbar^2}{2\mu \varphi_R r_R^2} = \frac{m_e}{\mu} \frac{\text{Ry}}{\varphi_R} \frac{a_0^2}{r_R^2} \tag{6.9}$$

the expansion can be regarded as a high-mass expansion with energy dependent coefficients $Q_{\text{class}}^{(l)*}(\mathscr{E}^*)$, $(1/\mathscr{E}^*) R^{(l)}(\mathscr{E}^*)$ etc.

The influence of quantum corrections for $|\mathscr{E}^*| \lesssim 10$ has been illustrated by *Toennies* and *Winkelmann*[2] for He + He collisions with a de Boer-Bird parameter of 0.122.

Interaction potentials with a repulsive core and an attractive tail have a minimum at r_m which may serve as reference length. It is given uniquely by

$$d\varphi(r)/dr = 0 \qquad \text{for } r = r_m. \tag{6.10}$$

The *potential minimum* φ_m is often denoted by $-\varepsilon$ in the literature, consequently $-\varphi_m = \varepsilon$ as the *well depth*; its absolute value $|\varphi_m| = -\varphi_m = \varepsilon$ often serves as reference energy φ_R.

For monotonic potentials a characteristic length describing the strength of the descent or ascent can be chosen as reference length r_R (see Sect. 8).

β) Inverse power potentials. For inverse power potentials

$$\varphi(r) = \frac{C_n}{r^n} \tag{6.11}$$

with only one parameter C_n, factorization according to Eq. (6.3) would have to split the parameter C_n into two factors depending on the choice of the reference length r_R. In order to avoid this, an energy-dependent *collision length* $r_n(\mathscr{E})$ is introduced by

$$r_n(\mathscr{E}) := \left(\frac{C_n}{\mathscr{E}}\right)^{\frac{1}{n}}. \tag{6.12}$$

For repulsive potentials ($C_n > 0$) this is the *distance of closest approach* in a head-on collision. With this physical interpretation the interaction of rigid spheres with the radius sum r_d may be described by assuming the limit $n \to \infty$ as

$$\lim_{n\to\infty} r_n(\mathscr{E}) = \lim_{n\to\infty} \left(\frac{C_n}{\mathscr{E}}\right)^{\frac{1}{n}} = r_d. \tag{6.13}$$

---

[1] Hirschfelder, J.O., Curtiss, C.F., Bird, R.B.: See [5], eq. 10.4–2

[2] Toennies, J.P., Winkelmann, K. (1977): J. Chem. Phys. *66*, 3965–3979, fig. 3

With the collision length r_n, Eq. (6.12), the integral in Eq. (6.2) for the trajectory is written as

$$\chi = \pi - 2\int_0^{\rho_\mathrm{P}} \frac{\mathrm{d}\rho}{\sqrt{1-\rho^2-\left(\frac{r_n}{b}\right)^n \rho^n}} \tag{6.14a}$$

with

$$1-\rho_\mathrm{P}^2-\left(\frac{r_n}{b}\right)^n \rho_\mathrm{P}^n = 0. \tag{6.14b}$$

This can be written as a $\varphi(b)/\mathscr{E}$ expansion[3]:

$$\chi = -\sum_{\kappa=1}^{\infty} \frac{\frac{n\kappa-1}{2}!\,(-\frac{1}{2})!}{\kappa!\,\frac{\kappa(n-2)}{2}!}\left(\frac{-r_n^n(\mathscr{E})}{b^n}\right)^{\kappa} \tag{6.15a}$$

for $n>2$ and

$$\frac{-r_n^n}{b^n} < \frac{2}{n-2}\left(\frac{n-2}{n}\right)^{\frac{n}{2}}. \tag{6.15b}$$

The root ρ_P of Eq. (6.14b) is a function of $\rho_n = b/r_n$ and likewise χ, Eq. (6.14a). The differential cross section $q(\chi, \mathscr{E})$ can now be normalized with $|r_n|^2$ as[4]

$$q(\chi, \mathscr{E}) =: |r_n(\mathscr{E})|^2\, q^*(\cos\chi, n, \operatorname{sgn} C_n) = \frac{|C_n|^{2/n}}{\mathscr{E}^{2/n}}\, q^*. \tag{6.16}$$

In contrast to the normalization, Eq. (6.6), for multiparameter potentials the energy dependence of $q(\chi, \mathscr{E})$ is now contained in the first factor $|r_n(\mathscr{E})|^2$. The dimensionless factor $q^*(\cos\chi, n, \operatorname{sgn} C_n)$ does not depend on $\mathscr{E}$. The functional form of its variation with $\cos\chi$ depends on the power n of the interaction potential C_n/r^n, Eq. (6.11), and on the sign of the constant C_n, which is positive (negative) for repulsive (attractive) potentials.

Examples are:

$$q_\mathrm{class}(\chi, \mathscr{E}) = r_1^2(\mathscr{E}) \frac{1}{4(1-\cos\chi)^2} \quad \text{for Coulomb interaction } (n=1) \tag{6.17}$$

$$q_\mathrm{class}(\chi, \mathscr{E}) = r_\mathrm{d}^2 \cdot \tfrac{1}{4} \quad \text{for rigid spheres } (n\to\infty). \tag{6.18}$$

The latter expression is isotropic and independent of $\mathscr{E}$. From the Coulomb expression (6.17) with $r_1 = C_1/\mathscr{E}$, Eq. (6.12), and $1-\cos\chi = 2\sin^2(\chi/2)$ one obtains the Rutherford form of the differential cross section. It does not depend on $\operatorname{sgn} C_1$, i.e., it holds for repulsion and attraction as well, and it is valid also in quantum mechanics.

The energy-dependent collision length $r_n(\mathscr{E}) := (C_n/\mathscr{E})^{1/n}$, Eq. (6.12), normalizes not only the differential cross section Eq. (6.16), but also the transfer cross

[3] Mott-Smith, H.M. (1960): Phys. Fluids 3, 721–724, eq. 1

[4] McDaniel, E.W.: See [6], sect. 3-9

sections

$$Q^{(l)}(\mathscr{E}) =: \pi |r_n(\mathscr{E})|^2 Q^{(l)*}(n, \operatorname{sgn} C_n) = \pi \frac{|C_n|^{2/n}}{\mathscr{E}^{2/n}} Q^{(l)*}. \tag{6.19}$$

The dimensionless factors $Q^{(l)*}$ depend only on the power n and the sign of the potential constant C_n of Eq. (6.11). They are listed in Table 1. The limits $n=1$ and $n \to \infty$, giving Coulomb interaction and rigid-sphere collisions are dealt with in Subsect. γ below.

Table 1. Normalized transfer cross sections $Q^{(l)*}(n, \operatorname{sgn} C_n) = Q^{(l)}(\mathscr{E})/\pi |r_n(\mathscr{E})|^2$, Eq. (6.19), for inverse power potentials, Eq. (6.11). For attractive potentials with $C_n<0$ a thin repulsive rigid core is assumed (after HIRSCHFELDER et al.[5]; KIHARA et al.[6]; HIGGINS and SMITH[7])

n	Repulsive		Attractive	
	$Q^{(1)*}(n, +)$	$Q^{(2)*}(n, +)$	$Q^{(1)*}(n, -)$	$Q^{(2)*}(n, -)$
2	1.5904	3.1668	3.2276	4.2660
3	1.2951	2.2046	2.6671	2.8924
4	1.1936	1.8510	2.2092	2.3076
5	1.1426	1.6619	1.8355	1.9857
6	1.1117	1.5438	1.5784	1.7859
8	1.0770	1.4016	1.3004	1.5444
10	1.0581	1.3194	1.1738	1.4107
12	1.0471	1.2665		
14	1.0380	1.2246		
15	1.0348	1.2113	1.0609	1.2440
20		1.1571		
24		1.1299		
25	1.0187	1.1259	1.0153	1.1290
50	1.0080	1.0623	1.0026	1.0574
∞	1.0000	1.0000	1.0000	1.0000

Inspection of the expressions $R^{(l)}(\mathscr{E})$, $S^{(l)}(\mathscr{E})$, etc. for the quantum corrections of the de Boer-Bird expansion, Eq. (5.2), reveals that these quantities can likewise be normalized with $\pi |r_n|^2$ and that the remaining dimensionless terms are also independent[8] of the energy $\mathscr{E}$:

$$\frac{Q^{(l)}(\mathscr{E})}{\pi |r_n(\mathscr{E})|^2} = Q^{(l)*}_{\text{class}}(n, \operatorname{sgn} C_n) + \frac{1}{\pi(|\boldsymbol{k}|\,|r_n|)^2} R^{(l)}(n, \operatorname{sgn} C_n)$$
$$+ \frac{1}{\pi(|\boldsymbol{k}|\,|r_n|)^4} S^{(l)*}(n, \operatorname{sgn} C_n) + O\left(\frac{1}{|\boldsymbol{k}|^6 |r_n|^6}\right). \tag{6.20}$$

Since the coefficients $Q^{(l)*}_{\text{class}}$, $R^{(l)}$, $S^{(l)*}$ in the de Boer-Bird expansion, Eq. (6.20), of $Q^{(l)}(\mathscr{E})/\pi |r_n(\mathscr{E})|^2$ do not depend on the collision energy $\mathscr{E}$, this expan-

[5] Hirschfelder, J.O., Curtiss, C.F., Bird, R.B.: See [5], table 8.3-3
[6] Kihara, T., Taylor, M.H., Hirschfelder, J.O. (1960): Phys. Fluids *3*, 715–720, table I
[7] Higgins, L.D., Smith, F.J. (1968): Mol. Phys. *14*, 399–400, table 2
[8] Hahn, H., Mason, E.A., Smith, F.J. (1971): Phys. Fluids *14*, 278–287, eq. 9

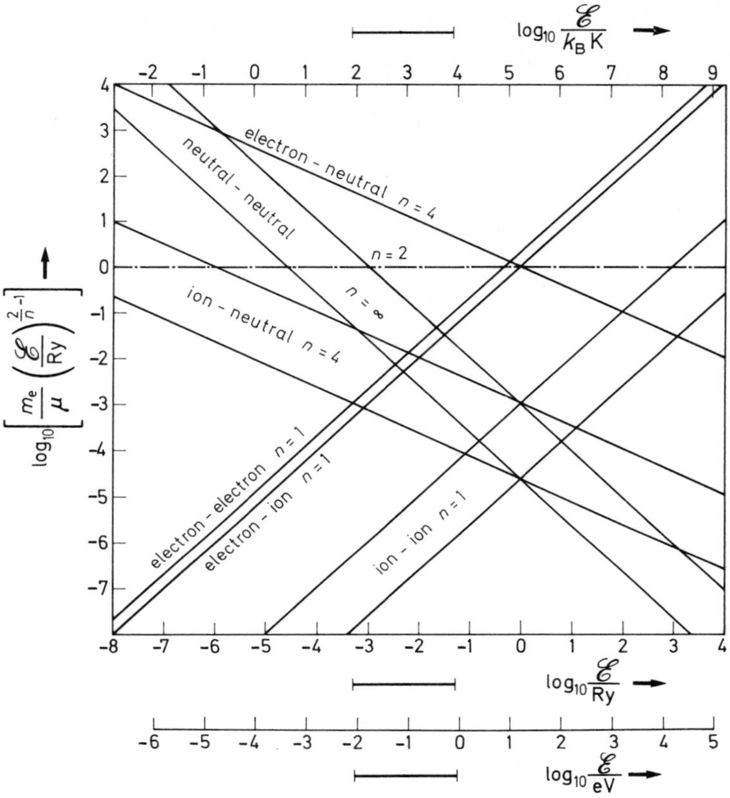

Fig. 1. Energy dependence of the normalized Hahn-Mason-Smith parameter, Eq. (6.21c), for inverse power potentials with $n=1, 2, 4, \infty$. The aeronomic energy range is marked on the abscissa by ⊢⊣, masses between m_H and m_{CO_2} being considered

sion can now be taken as an energy expansion after increasing powers of the energy dependent *Hahn-Mason-Smith parameter* for inverse power potentials

$$\frac{1}{|\boldsymbol{k}|^2 r_n^2} = \frac{m_e}{\mu}\left[\frac{C_e/a_0}{2C_n/a_0^n}\right]^{\frac{2}{n}}\left(\frac{\mathscr{E}}{\mathrm{Ry}}\right)^{\frac{2}{n}-1} \tag{6.21a}$$

with

$$C_e := \frac{\mathrm{u}q_e^2}{4\pi\varepsilon_0} = 2.3071\cdot 10^{-28}\,\mathrm{Jm} = 1.4300\cdot 10^{-9}\,\mathrm{eVm}. \tag{6.21b}$$

For $n<2 (n>2)$ this parameter increases (decreases) with increasing energy. The three classes of collisions in plasmas – Coulomb collisions, charged-neutral collisions, and neutral-neutral collisions – can be characterized by $n=1$, $n=4$ (see Sect. 9), and $n=\infty$ (rigid spheres), respectively. The factor in brackets is 1/2 for Coulomb interaction, but for charged-neutral interactions it is $-\alpha_n^* Z_c^2 \approx 10^0 \ldots 10^1$, see Eqs. (9.2) (9.6) and Table 4, Sect. 9. Figure 1 shows a plot of the energy dependence of the normalized Hahn-Mason-Smith parameter

$$\frac{1}{|\boldsymbol{k}|^2 r_n^2}\left[\frac{2C_n/a_0^n}{C_e/a_0}\right]^{\frac{2}{n}}=\frac{m_e}{\mu}\left(\frac{\mathscr{E}}{\mathrm{Ry}}\right)^{\frac{2}{n}-1} \tag{6.21c}$$

for $n=1$, 2, 4, ∞. It demonstrates the dominance of quantum effects for electron-neutral collisions in the temperature range of aeronomic interest and also the slight influence of quantum effects on collisions between light neutrals for low energies[2] and on Coulomb collisions for high energies[9].

γ) Special cases. For *rigid spheres* the definition, Eq. (4.8), of the transfer cross sections $Q^{(l)}(\mathscr{E})$ with the expression (6.18) for $q_{\text{class}}(\chi, \mathscr{E})$ yields the result

$$Q^{(l)}_{\text{class}}(\mathscr{E})=\pi r_d^2[1-\delta_{l0}], \tag{6.22}$$

which does not depend on the collision energy $\mathscr{E}$.

In contrast to this very simple result for the transfer cross sections $Q^{(l)}(\mathscr{E})$, Eq. (4.8), the transfer cross sections $\phi^{(l)}(\mathscr{E})$, Eq. (4.9), become[10]

$$\phi^{(l)}_{\text{class}}(\mathscr{E})=\pi r_d^2\left[1-\frac{1}{2}\frac{1+(-1)^l}{l+1}\right]. \tag{6.23}$$

Since $r_\infty(\mathscr{E})=r_d$, Eq. (6.13), was found as limiting value of the collision length $r_n(\mathscr{E})$, Eq. (6.12), comparison of Eq. (6.19) with Eq. (6.22) yields for $n\to\infty$ the limiting value

$$Q^{(l)*}(\infty, +)=1-\delta_{l0}. \tag{6.24}$$

Table 1 (Subsect. β above) shows that $Q^{(1)*}(\infty, \pm)=1=Q^{(2)*}(\infty, \pm)$ is indeed the limiting value of $Q^{(1)*}(n, \pm)$ and $Q^{(2)*}(n, \pm)$ for $n\to\infty$.

For *Coulomb interaction* ($n=1$) the angular integral in Eq. (4.8) for $Q^{(l)}(\mathscr{E})$ diverges at low deflection angles χ for Rutherford's differential cross section Eq. (6.17). To secure convergence the Coulomb potential must be modified for large distances r, e.g., by screening (see Sect. 8γ). The result is a logarithmic energy dependence of $Q^{(l)*}(1, \pm)$ where $r_1=C_1/\mathscr{E}$, Eq. (6.12), in Eq. (6.19) is defined for the pure Coulomb potential, cf. Eq. (8.6). The quantum corrections in the de Boer-Bird expansion Eq. (6.20) can be calculated for a pure Coulomb potential without divergencies, Eq. (8.8).

7. Transfer cross sections for nonmonotonic interaction potentials. Transfer collision frequencies of the different particle species in an aeronomic plasma shall be discussed in more detail in Sects. 9 through 17. In Sects. 7 and 8 we shall consider various types of interaction potentials and the corresponding transfer cross sections; the latter are currently used for describing aeronomic collisions. Usually, transfer cross sections $Q^{(l)}(\mathscr{E})$ are numerically computed from a given potential $\varphi(r)$. Only for inverse power potentials C_n/r^n can

---

[9] Biolsi, L. (1978): J. Geophys. Res. *83*, 1125–1131

[10] Hirschfelder, J.O., Curtiss, C.F., Bird, R.B.: See [5], eq. 8.2–4

analytic expressions be calculated. Sect. 7 deals with interaction potentials with a repulsive core and an attractive tail. Monotonic interaction potentials will be discussed in Sect. 8.

α) The *Morse potential* may be described by any one of the three following expressions:

$$\frac{\varphi(r)}{-\varphi_m}=\exp\left[2\alpha\left(1-\frac{r}{r_m}\right)\right]-2\exp\left[\alpha\left(1-\frac{r}{r_m}\right)\right] \tag{7.1a}$$

$$=4\exp\left[2(\alpha-\ln 2)\left(1-\frac{r}{r_0}\right)\right]-4\exp\left[(\alpha-\ln 2)\left(1-\frac{r}{r_0}\right)\right] \tag{7.1b}$$

$$=\exp\left[2(\alpha-\ln 2)\frac{r_m-r}{r_0}\right]-2\exp\left[(\alpha-\ln 2)\frac{r_m-r}{r_0}\right]. \tag{7.1c}$$

They differ in the normalization for the interparticle distance r. Eq. (7.1a) applies the distance r_m of the potential minimum φ_m as reference length, according to Eq. (6.3). The second form, Eq. (7.1b), uses the zero point r_0 of the potential as normalization length which is related to r_m by

$$\frac{r_0}{r_m}=1-\frac{\ln 2}{\alpha}. \tag{7.2}$$

The third form, Eq. (7.1c), is often used in the literature with $\alpha-\ln 2=:C$; it introduces a fourth, apparently independent parameter r_0 which is in fact not independent of the remaining three parameters but related with r_m and α by Eq. (7.2).

The repulsive core should be "harder" than the attractive tail; therefore the first exponential must exceed the second one for $r=0$. This yields the condition

$$\alpha>\ln 2=0.6931 \quad \text{or} \quad \alpha-\ln 2=:C>0. \tag{7.3}$$

The parameter α (or $C:=\alpha-\ln 2$) is a direct measure of the *reduced curvature* at the minimum[1]:

$$\frac{\varphi''(r_m)}{-\varphi_m/r_m^2}=2\alpha^2 \qquad \frac{\varphi''(r_m)}{-\varphi_m/r_0^2}=2(\alpha-\ln 2)^2. \tag{7.4}$$

Normalized transfer cross sections

$$Q^{(l)*}(\mathscr{E}^*):=\frac{Q^{(l)}(\mathscr{E})}{\pi r_m^2} \qquad [6.7]$$

have been computed by LOVELL and HIRSCHFELDER[2] and are plotted in Fig. 2.

The small oscillations near $\mathscr{E}^*=1$ are due to the phenomenon of "orbiting". In this energy range the two colliding particles spend a considerable amount of time rotating around one

[1] Bernstein, R.B., Muckermann, J.T. (1967): Adv. Chem. Phys. *12*, 389–486, eq. 19

[2] Lovell, S.E., Hirschfelder, J.O. (1961): Transport properties of gases obeying the Morse potential. Report WIS-AF-19, Ser. 5, 20 Dec. 1961, table IV. Madison: University of Wisconsin

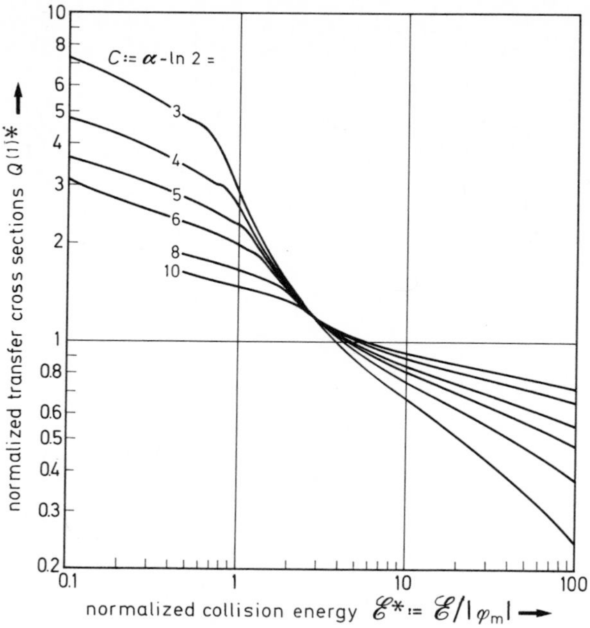

Fig. 2a. Normalized transfer cross sections $Q^{(1)*}(\mathscr{E}^*) := Q^{(1)}(\mathscr{E})/\pi r_m^2$, Eq. (6.7), for Morse potentials, Eq. (7.1), with $C := \alpha - \ln 2$ as parameter (after LOVELL and HIRSCHFELDER [2])

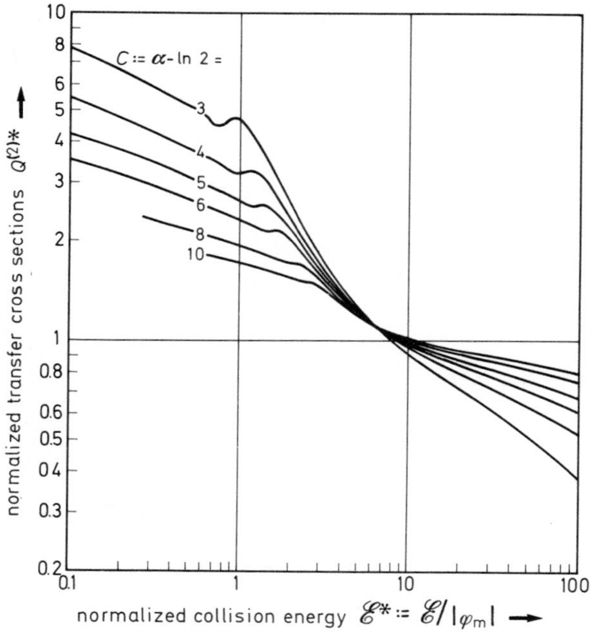

Fig. 2b. Normalized transfer cross sections $Q^{(2)*}(\mathscr{E}^*) := Q^{(2)}(\mathscr{E})/\pi r_m^2$, Eq. (6.7), for Morse potentials, Eq. (7.1), with $C := \alpha - \ln 2$ as parameter (after LOVELL and HIRSCHFELDER [2])

another[3]. With decreasing parameter $C := \alpha - \ln 2$ the potential well becomes more and more shallow, thus increasing the prerequisites for orbiting.

β) The Buckingham (exp β, n) *potential.* The expression

$$\frac{\varphi(r)}{-\varphi_m} = \frac{n}{\beta - n} \exp\left[\beta\left(1 - \frac{r}{r_m}\right)\right] - \frac{\beta}{\beta - n}\left(\frac{r_m}{r}\right)^n \tag{7.5}$$

has a "soft" exponentially repulsive core and an inverse power attractive tail for $\beta > n$. Near the origin, for $r \to 0$, $\varphi(r)$ decreases as $-r^{-n}$ to minus infinity which is not acceptable for physical reasons. There appears a zero at $r_0/r_m \approx (\beta/n)^{1/n} \exp(-\beta/n)$ which should not occur. Therefore it must be very small. This can be achieved either with a small β/n or a large β/n. Because of $\beta > n$ the latter choice must be taken.

Table 2 gives the ratio r_0/r_m of an (exp β, 6) potential for several β values.

Table 2. Ratio of the distance r_0 of the unwanted zero of a Buckingham (exp β, 6) potential, Eq. (7.5), to the distance r_m of the minimum (after MASON[4])

β	12	13	14	15
r_0/r_m	0.8761	0.8832	0.8891	0.8942

The reduced curvature is

$$\frac{\varphi''(r_m)}{-\varphi_m/r_m^2} = \frac{\beta n(\beta - n - 1)}{\beta - n}. \tag{7.6}$$

Figure 3 shows computated values of the normalized transfer cross sections $Q^{(1)*}$, $Q^{(2)*}$ for $n=6$. For high energies $\mathscr{E}/|\varphi_m|$ the exponential term dominates the interaction. In this case the transfer cross sections, Eq. (8.3), for a repulsive exponential potential, Eq. (8.2), can be used with the parameters

$$\varphi_R = \frac{n|\varphi_m|}{\beta - n} e^{\beta - 1} \qquad r_R = \frac{r_m}{\beta} \qquad \text{for } \mathscr{E} \gg |\varphi_m|. \tag{7.7a}$$

For low energies ($\mathscr{E}/|\varphi_m|$) the attractive r^{-n} term dominates the interaction. Therefore the transfer cross sections vary for low energies, according to Eq. (6.19), as

$$\frac{Q^{(l)}(\mathscr{E})}{\pi r_m^2} \approx \left(\frac{|\varphi_m|}{\mathscr{E}}\right)^{\frac{2}{n}} \left(\frac{\beta}{\beta - n}\right)^{\frac{2}{n}} Q^{(l)*}(n, -) \qquad \text{for } \mathscr{E} \ll |\varphi_m| \tag{7.7b}$$

with the numerical values of $Q^{(l)*}(n, -)$ from Table 1, Sect. 6β.

γ) The Mason-Schamp-Kihara (n, n_1, n_0) *potential.* The expression

$$\frac{\varphi(r)}{\varphi_m} = \frac{\dfrac{1+\gamma}{n}\left(\dfrac{r_m - r_K}{r - r_K}\right)^n - 2\dfrac{\gamma}{n_1}\left(\dfrac{r_m - r_K}{r - r_K}\right)^{n_1} - \dfrac{1-\gamma}{n_0}\left(\dfrac{r_m - r_K}{r - r_K}\right)^{n_0}}{\dfrac{1+\gamma}{n} - 2\dfrac{\gamma}{n_1} - \dfrac{1-\gamma}{n_0}} \tag{7.8a}$$

[3] Hirschfelder, J.O., Curtiss, C.F., Bird, R.B.: See [5], sect. 8.4a

[4] Mason, E.A. (1954): J. Chem. Phys. *22*, 169–186, tables I and III

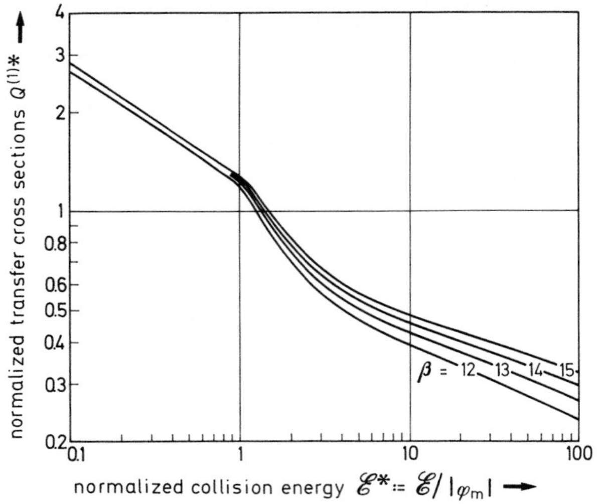

Fig. 3a. Normalized transfer cross sections $Q^{(1)*}(\mathscr{E}^*)=Q^{(1)}(\mathscr{E})/\pi r_m^2$, Eq. (6.7), for a Buckingham (exp β, 6) potential, Eq. (7.5) (after MASON[4])

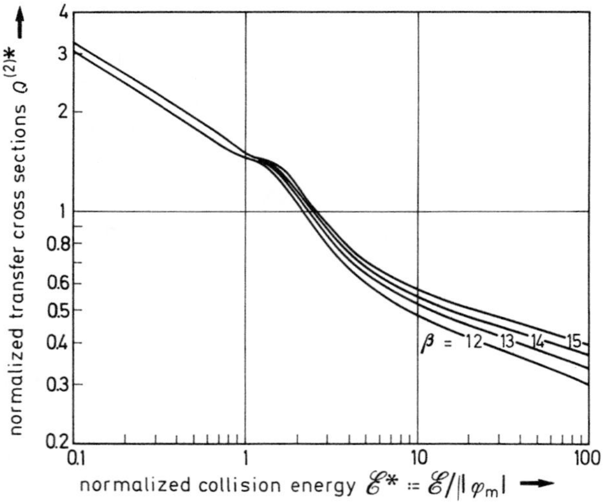

Fig. 3b. Normalized transfer cross sections $Q^{(2)*}(\mathscr{E}^*)=Q^{(2)}(\mathscr{E})/\pi r_m^2$, Eq. (6.7), for a Buckingham (exp β, 6) potential, Eq. (7.5) (after MASON[4])

combines three inverse power expressions: a repulsive core $\sim r^{-n}$ requiring $\gamma > -1$, a tail $\sim r^{-n_0}$, and an intermediate expression $\sim r^{-n_1}$. If the latter two terms should represent attractive potentials, the *Mason-Schamp parameter* γ may only vary in the range $0 \leqq \gamma \leqq 1$. But if a repulsive tail and an attractive intermediate potential should be represented, then γ must exceed unity. In Fig. 4 (Subsect. δ below) curves of *Mason-Schamp* (12, 6, 4) *potentials* (with $r_K = 0$) are plotted for $0 \leqq \gamma \leqq 6$.

Another notation

$$\frac{\varphi(r)}{-\varphi_m}=\frac{n_0+(n_1-n_0)\gamma_{KH}}{n-n_0}\left(\frac{r_m-r_K}{r-r_K}\right)^n-\gamma_{KH}\left(\frac{r_m-r_K}{r-r_K}\right)^{n_1}-\frac{n-(n-n_1)\gamma_{KH}}{n-n_0}\left(\frac{r_m-r_K}{r-r_K}\right)^{n_0} \tag{7.8b}$$

with

$$\gamma_{KH}:=\frac{2nn_0\gamma}{n_1(n-n_0)-(nn_1+n_0n_1-2nn_0)\gamma}=:2\gamma_{RK} \tag{7.8c}$$

was introduced by KLEIN and HANLEY[5] and with $\gamma_{RK}:=\frac{1}{2}\gamma_{KH}$ by ROBSON and KUMAR[6]. It has a simpler denominator than (7.8a) but somewhat more complicated numerators. For $n_1=n/2$ and $n_0=n/3$ two notations coalesce, viz. $\gamma=\gamma_{RK}$. For $n_1=2n/3$ and $n_0=n/2$ the Klein-Hanley notation with $C_8^*:=\gamma_{KH}$ was used by TOENNIES and WINKELMANN[7].

The denominators of Eqs. (7.8a) and (7.8c) become independent of γ for $nn_1+n_1n_0-2nn_0=0$. This is satisfied for $n_1=n/2$ and $n_0=n/3$ which is therefore often used in the literature (with $n=12$).

The *Kihara distance* r_K shifts the pole of the potential from the origin ($r=0$), for $r_K=0$, to the finite value r_K. Therefore r_K is limited to $r_K<r_m$. For positive r_K the pole is situated at the positive r axis. Therefore the part of a potential curve with $r\leqq r_K$ is not physically acceptable for $r_K>0$.

The reduced curvature is

$$\frac{\varphi''(r_m)}{-\varphi_m/r_m^2}=\frac{nn_1n_0}{\left(1-\frac{r_K}{r_m}\right)^2}\,\frac{n-n_0+(n+n_0-2n_1)\gamma}{n_1(n-n_0)-(nn_1+n_1n_0-2nn_0)\gamma}. \tag{7.9a}$$

The denominator is proportional to the denominator of the Mason-Schamp-Kihara potential, Eq. (7.8a), and therefore becomes independent of γ under the same condition:

$$\frac{\varphi''(r_m)}{-\varphi_m/r_m^2}=\frac{n^2(2+\gamma)}{6\left(1-\frac{r_K}{r_m}\right)^2}\qquad \text{for } n_1=\frac{n}{2} \text{ and } n_0=\frac{n}{3}. \tag{7.9b}$$

For special values of the Mason-Schamp parameter γ and the Kihara distance r_K, the Mason-Schamp-Kihara potential, Eq. (7.8a), provides simpler shapes which will be dealt with in the following Subsects. δ, ε and ζ.

δ) The Mason-Schamp (n, n_1, n_0) *potential*[8] is the special case of Eq. (7.8a) for $r_K=0$:

$$\frac{\varphi(r)}{\varphi_m}=\frac{\frac{1+\gamma}{n}\left(\frac{r_m}{r}\right)^n-2\frac{\gamma}{n_1}\left(\frac{r_m}{r}\right)^{n_1}-\frac{1-\gamma}{n_0}\left(\frac{r_m}{r}\right)^{n_0}}{\frac{1+\gamma}{n}-2\frac{\gamma}{n_1}-\frac{1-\gamma}{n_0}}. \tag{7.10}$$

Plots for a (12, 6, 4) potential with $0\leqq\gamma\leqq 6$ are given in Fig. 4.

---

[5] Klein, M., Hanley, H.J.M. (1970): J. Chem. Phys. *53*, 4722–4723
[6] Robson, R.E., Kumar, K. (1973): Aust. J. Phys. *26*, 187–201, eqs. 19 to 22
[7] Toennies, J.P., Winkelmann, K. (1977): J. Chem. Phys. *66*, 3965–3979, eq. 36
[8] Mason, E.A., Schamp, H.W. (1958): Ann. Phys. N.Y. *4*, 233–270, eq. 1, table II

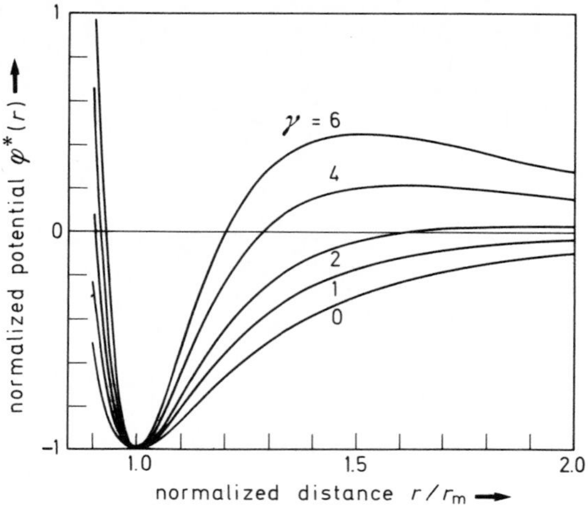

Fig. 4. Normalized MASON-SCHAMP (12, 6, 4) potential $\varphi^*(r) := \varphi(r)/|\varphi_{\rm m}|$, Eq. (7.10)

The Mason-Schamp potential, Eq. (7.10), with $\gamma > 1$ may be a good representation for some potential curves describing the interaction of neutral particles. The potential curves for the $^1\Sigma_g^+$ and $^1\Sigma_u^+$ states of He+He collisions seem to have repulsive cores and tails and an intermediate attractive contribution[9]. They could therefore be fitted with a Mason-Schamp potential with $\gamma > 1$ (Fig. 4). Computations of transfer cross sections for $\gamma > 1$ are still not available.

For the same potential with $0 \leqq \gamma \leqq 1$ computed values of the transfer cross sections $Q^{(1)*}$ and $Q^{(2)*}$ are given in Fig. 5. For high (normalized) energies ($\mathscr{E}^* = \mathscr{E}/|\varphi_{\rm m}|$) the repulsive r^{-n} term in Eq. (7.10) dominates the interaction. Therefore the transfer cross sections of a $(n, n/2, n/3)$ potential vary in the high-energy range, according to Eq. (6.19), as

$$\frac{Q^{(l)}(\mathscr{E})}{\pi r_{\rm m}^2} = \left(\frac{|\varphi_{\rm m}|}{\mathscr{E}}\right)^{\frac{2}{n}} \left(\frac{1+\gamma}{2}\right)^{\frac{2}{n}} Q^{(l)*}(n, +) \quad \text{for } \mathscr{E} \gg |\varphi_{\rm m}|. \tag{7.11a}$$

For low energies they take the limiting values

$$\frac{Q^{(l)}(\mathscr{E})}{\pi r_{\rm m}^2} = \left(\frac{|\varphi_{\rm m}|}{\mathscr{E}}\right)^{\frac{6}{n}} \left(\frac{3}{2}\right)^{\frac{6}{n}} Q^{(l)*}\left(\frac{n}{3}, -\right) \quad \text{for } \mathscr{E} \ll |\varphi_{\rm m}| \quad \text{and } \gamma = 0 \tag{7.11b}$$

or

$$\frac{Q^{(l)}(\mathscr{E})}{\pi r_{\rm m}^2} = \left(\frac{|\varphi_{\rm m}|}{\mathscr{E}}\right)^{\frac{4}{n}} 2^{\frac{4}{n}} Q^{(l)*}\left(\frac{n}{2}, -\right) \quad \text{for } \mathscr{E} \ll |\varphi_{\rm m}| \quad \text{and } \gamma = 1 \tag{7.11c}$$

with the numerical values for $Q^{(l)*}(n, \pm)$ from Table 1, Sect. 6 β.

[9] Andresen, B., Kuppermann, A. (1975): Mol. Phys. *30*, 997–1004, fig. 2

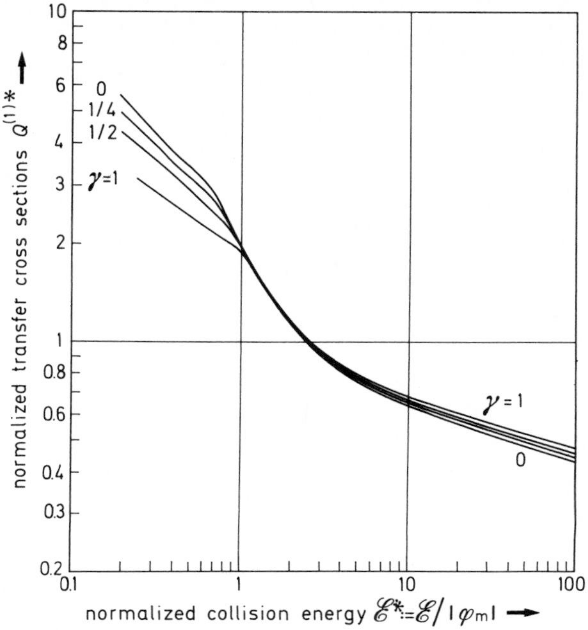

Fig. 5a. Normalized transfer cross sections $Q^{(1)*}(\mathscr{E}^*)=Q^{(1)}(\mathscr{E})/\pi r_m^2$, Eq. (6.7), for a MASON-SCHAMP (12, 6, 4) potential, Eq. (7.10) (after KIHARA and KOTANI[10]; MASON and SCHAMP[8]; MCDANIEL and MASON[11]; ST.-MAURICE and SCHUNK[12])

ε) The *Kihara* (n, n_γ) *potential* is a special case of Eq. (7.8a) for either $\gamma=0$ or $\gamma=1$:

$$\frac{\varphi(r)}{-\varphi_m}=\frac{n_\gamma}{n-n_\gamma}\left(\frac{r_m-r_K}{r-r_K}\right)^n-\frac{n}{n-n_\gamma}\left(\frac{r_m-r_K}{r-r_K}\right)^{n_\gamma}. \tag{7.12}$$

Its zero r_0 is related to the distance r_m of the potential minimum φ_m by

$$\frac{r_0}{r_m}=\frac{r_K}{r_m}+\left(1-\frac{r_K}{r_m}\right)\left(\frac{n_\gamma}{n}\right)^{\frac{1}{n-n_\gamma}}. \tag{7.13a}$$

This yields

$$\frac{r_m-r_K}{r-r_K}=\frac{r_0-r_K}{r-r_K}\left(\frac{n}{n_\gamma}\right)^{\frac{1}{n-n_\gamma}}, \tag{7.13b}$$

and the insertion of this expression into Eq. (7.12) gives a representation of the Kihara potential with the distance r_0 instead of r_m as parameter.

For $n_\gamma=n/2$ the expression (7.12) for the Kihara (n, $n/2$) potential becomes particularly simple:

$$\frac{\varphi(r)}{-\varphi_m}=\left(\frac{r_m-r_K}{r-r_K}\right)^n-2\left(\frac{r_m-r_K}{r-r_K}\right)^{\frac{n}{2}}=4\left(\frac{r_0-r_K}{r-r_K}\right)^n-4\left(\frac{r_0-r_K}{r-r_K}\right)^{\frac{n}{2}}. \tag{7.13c}$$

[10] Kihara, T., Kotani, M. (1943): Proc. Phys. Math. Soc. Jpn. *25*, 602–614, table I
[11] McDaniel, E.W., Mason, E.A.: See [*23*], tables I
[12] St.-Maurice, J.-P., Schunk, R.W. (1977): Planet. Space Sci. *25*, 243–260, fig. 2

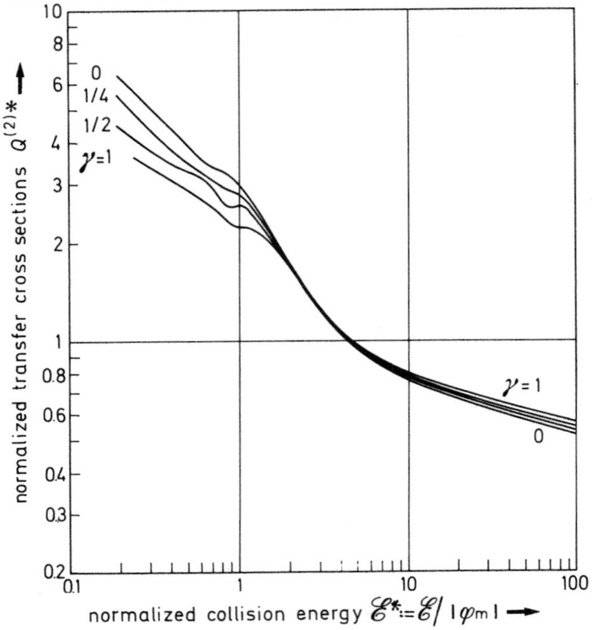

Fig. 5b. Normalized transfer cross sections $Q^{(2)*}(\mathscr{E}^*) = Q^{(2)}(\mathscr{E})/\pi r_m^2$, Eq. (6.7), for a MASON-SCHAMP (12, 6, 4) potential, Eq. (7.10) (after KIHARA and KOTANI[10]; MASON and SCHAMP[8]; MCDANIEL and MASON[11])

Normalized transfer cross sections $Q^{(1)*}$, $Q^{(2)*}$ were computed by MASON et al.[13] for a Kihara (12, 4) potential, but are not published in their paper.

ζ) The Lennard-Jones (n, n_γ) *potential* is a special case of the Kihara (n, n_γ) potential, Eq. (7.12), for $r_K = 0$, or of the Mason-Schamp (n, n_1, n_0) potential, Eq. (7.10), for either $\gamma = 0$ or $\gamma = 1$ (see Fig. 4, Subsect. δ above):

$$\frac{\varphi(r)}{-\varphi_m} = \frac{n_\gamma}{n-n_\gamma}\left(\frac{r_m}{r}\right)^n - \frac{n}{n-n_\gamma}\left(\frac{r_m}{r}\right)^{n_\gamma} \tag{7.14a}$$

$$= \frac{n_\gamma}{n-n_\gamma}\left(\frac{n}{n_\gamma}\right)^{\frac{n}{n-n_\gamma}}\left(\frac{r_0}{r}\right)^n - \frac{n}{n-n_\gamma}\left(\frac{n}{n_\gamma}\right)^{\frac{n_\gamma}{n-n_\gamma}}\left(\frac{r_0}{r}\right)^{n_\gamma}. \tag{7.14b}$$

This expression takes a simple form for $n_\gamma = n/2$, which is therefore often used in the literature (with $n = 12$):

$$\frac{\varphi(r)}{-\varphi_m} = \left(\frac{r_m}{r}\right)^n - 2\left(\frac{r_m}{r}\right)^{\frac{n}{2}} \tag{7.14c}$$

$$= 4\left(\frac{r_0}{r}\right)^n - 4\left(\frac{r_0}{r}\right)^{\frac{n}{2}} \quad \text{with } \frac{r_0}{r_m} = \left(\frac{1}{2}\right)^{\frac{2}{n}}. \tag{7.14d}$$

[13] Mason, E.A., O'Hara, H., Smith, F.J. (1972): J. Phys. B5, 169–176

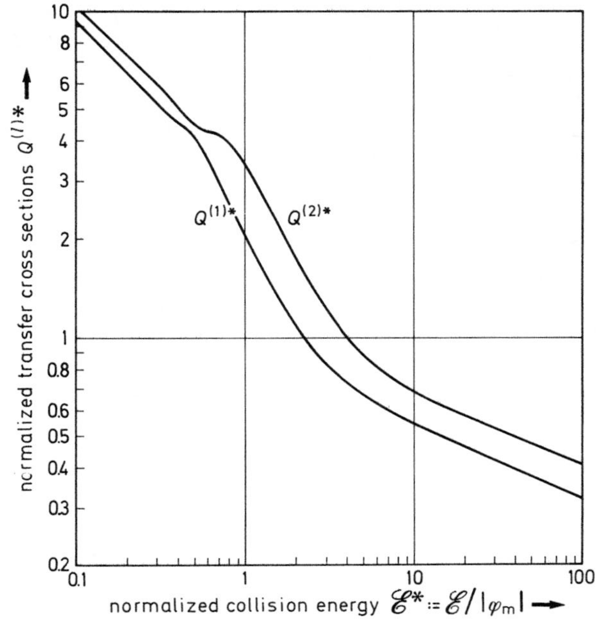

Fig. 6. Normalized transfer cross sections $Q^{(l)*}(\mathscr{E}^*)=Q^{(l)}(\mathscr{E})/\pi r_m^2$, Eq. (6.7), for Lennard-Jones (8, 4) potentials. Eq. (7.14) (after HASSÉ and COOK[14]; MCDANIEL and MASON[11]; ST.-MAURICE and SCHUNK[12])

Figure 6 shows computed values of the normalized transfer cross sections $Q^{(1)*}$ and $Q^{(2)*}$ for an (8, 4) potential. According to Eq. (6.19), they vary like

$$\frac{Q^{(l)}(\mathscr{E})}{\pi r_m^2}=\left(\frac{|\varphi_m|}{\mathscr{E}}\right)^{\frac{2}{n}} Q^{(l)*}(n,+) \qquad \text{for } \mathscr{E}\gg|\varphi_m| \tag{7.15a}$$

and

$$\frac{Q^{(l)}(\mathscr{E})}{\pi r_m^2}=\left(\frac{|\varphi_m|}{\mathscr{E}}\right)^{\frac{4}{n}} 2^{\frac{4}{n}} Q^{(l)*}\left(\frac{n}{2},-\right) \qquad \text{for } \mathscr{E}\ll|\varphi_m|, \tag{7.15b}$$

with $Q^{(l)*}(n,+)$ and $Q^{(l)*}(n/2,-)$ from Table 1, Sect. 6β. For (12, 6) and (12, 4) potentials the corresponding curves are contained for $\gamma=1$ and $\gamma=0$, respectively, in Fig. 5 (Subsect. δ above).

η) The Sutherland (∞, n) *potential*

$$\begin{aligned}\varphi(r)&=\varphi_m\left(\frac{r_m}{r}\right)^n && \text{for } r\geqq r_m\\ &=\infty && \text{for } r<r_m\end{aligned} \tag{7.16}$$

[14] Hassé, H.R., Cook, W.R. (1929): Proc. R. Soc. London Ser. A *125*, 196–221

has the hard core of rigid spheres with a radius sum r_m and an attractive inverse power tail for $\varphi_m < 0$. It can therefore be regarded as the limit of a Lennard-Jones potential, Eq. (7.14), with $n \to \infty$ and $n_\gamma \to n$. The zero r_0 equals the position r_m of the minimum. Of course, a curvature at r_m cannot be determined.

The transfer cross sections have been calculated by HASSÉ[15], HASSÉ and COOK[16], and CHAPMAN and COWLING[17]. The results (for $\varphi_m < 0$) contain the high-energy limit

$$Q^{(l)}(\mathscr{E}) = Q^{(l)}_{\text{rigid spheres}} + r_m^2 \frac{|\varphi_m|}{\mathscr{E}} i^{(l)}(n) \qquad \text{for } \mathscr{E} \gg |\varphi_m|. \tag{7.17a}$$

The rigid-sphere value is simply πr_m^2, Eq. (6.22). The collision factors $i^{(l)}(n)$ are given in Table 3. For low energies the attractive part $\sim r^{-n}$ dominates and the

Table 3. Collision factors $i^{(l)}(n)$, Eq. (7.17a), for Sutherland (∞, n) potentials, Eq. (7.16) [after CHAPMAN and COWLING[17], Table 5 with $i^{(1)}(n) \equiv i_1(\infty, n+1)$ and $i^{(2)}(n) \equiv \frac{3}{2} i_2(\infty, n+1)$]

n	$i^{(1)}(n)$	$i^{(2)}(n)$
2	$\frac{1}{8}(12-\pi^2) = 0.2663$	$\frac{3}{16}(\pi^2-8) = 0.3506$
3	$3-4\ln 2 = 0.2274$	$4(3\ln 2-2) = 0.3178$
4	$\frac{1}{8}(3\pi^2-28) = 0.2011$	$\frac{9}{4}(10-\pi^2) = 0.2934$
6	$\frac{1}{6} = 0.1667$	$\frac{5}{16}(9\pi^2-88) = 0.2585$
8	$\frac{13}{90} = 0.1444$	$\frac{7}{30} = 0.2333$

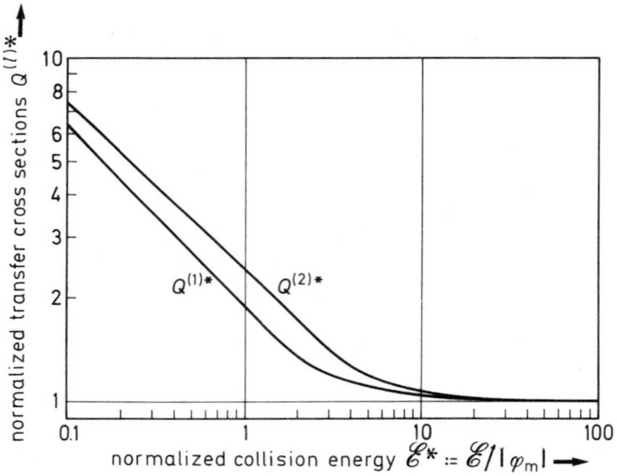

Fig. 7. Normalized transfer cross sections $Q^{(l)*}(\mathscr{E}^*) = Q^{(l)}(\mathscr{E})/\pi r_m^2$, Eq. (6.7), for a Sutherland $(\infty, 4)$ potential, Eq. (7.16) (after HASSÉ[15]; HASSÉ and COOK[16]; MCDANIEL and MASON[11]; ST.-MAURICE and SCHUNK[12])

[15] Hassé, H.R. (1926): Philos. Mag. *1*, 139–160, tables I and II
[16] Hassé, H.R., Cook, W.R. (1927): Philos. Mag. *3*, 977–990, tables I–III
[17] Chapman, S., Cowling, T.G.: See [*12*], sect. 10.41

$Q^{(l)}(\mathscr{E})$ vary, according to Eq. (6.19), as

$$\frac{Q^{(l)}(\mathscr{E})}{\pi r_m^2} = \left(\frac{|\varphi_m|}{\mathscr{E}}\right)^{\frac{2}{n}} Q^{(l)*}(n, -) \qquad \text{for } \mathscr{E} \ll |\varphi_m|, \tag{7.17b}$$

with the numerical values for $Q^{(l)*}(n, -)$ from Table 1, Sect. 6β. For $n=4$ and $\varphi_m<0$ the normalized transfer cross sections $Q^{(1)*}$, $Q^{(2)*}$ are given in Fig. 7.

8. Transfer cross sections for monotonic interaction potentials

α) The *Sato potential*

$$\varphi(r) = \frac{1}{3}\varphi_R\left\{\exp\left[2\alpha\left(1-\frac{r}{r_R}\right)\right] + 2\exp\left[\alpha\left(1-\frac{r}{r_R}\right)\right]\right\} \tag{8.1}$$

is a repulsive potential (for $\varphi_R>0<\alpha$). It has physical connections with the Morse potential, Eq. (7.1), as will be shown in Sect. 15γ. No values for transfer cross sections seem to be published.

β) The *Exponential* (*Born-Mayer*) *potential*

$$\varphi(r) = \varphi_R \exp\left(1-\frac{r}{r_R}\right) \tag{8.2}$$

is repulsive (attractive) for positive (negative) φ_R. Values for normalized transfer cross sections $Q^{(l)*}(\mathscr{E}/|\varphi_R|)$ are given in Fig. 8. The repulsive potential

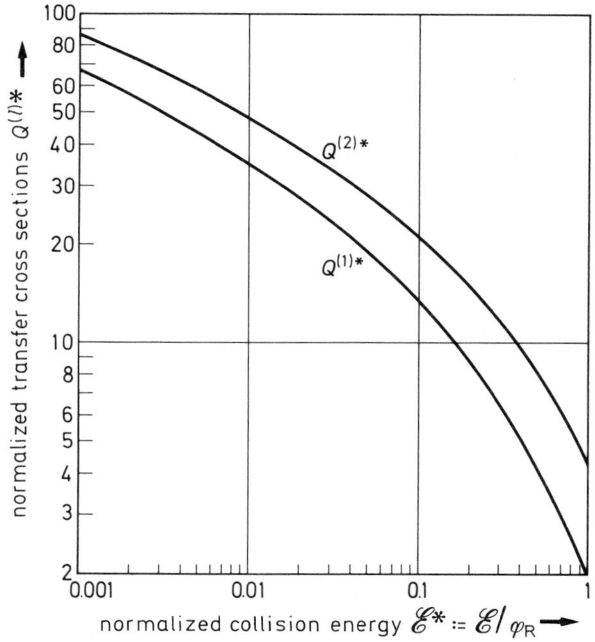

Fig. 8. Normalized transfer cross sections $Q^{(l)*}(\mathscr{E}^*) = Q^{(l)}(\mathscr{E})/\pi r_R^2$, Eq. (6.7), for a repulsive exponential (Born-Mayer) potential Eq. (8.2) (after Monchick[1])

[1] Monchick, L. (1959): Phys. Fluids 2, 695–700, table I

calculations of MASON and VANDERSLICE[2] indicate that the square root of $Q^{(l)*}(\mathscr{E}^*)$ is accurately represented by a linear polynomial of $\log_{10} E^*$. If we plot the values for $\sqrt{Q^{(l)*}(\mathscr{E}^*)}$ from Fig. 8 versus $\log_{10} E^*$ we obtain straight lines:

$$\sqrt{Q^{(l)*}(\mathscr{E}^*)} = A^{(l)*} - B^{(l)*} \log_{10} \mathscr{E}^* \qquad \text{for } 10^{-3} \lesssim \mathscr{E}^* \lesssim 1 \tag{8.3a}$$

with

$$A^{(1)*} = 1.50, \quad A^{(2)*} = 2.15, \quad B^{(1)*} = 2.25, \quad B^{(2)*} = 2.45. \tag{8.3b}$$

For later calculations (Sect. 18γ) it is convenient to write the transfer cross section, Eq. (8.3a), in the form

$$\frac{Q^{(l)}(\mathscr{E})}{\pi r_{(l)}^2} = \left(\ln \frac{\mathscr{E}_{(l)}}{\mathscr{E}}\right)^2 \qquad \text{for } 2 \cdot 10^{-4} \lesssim \frac{\mathscr{E}}{\mathscr{E}_{(l)}} \lesssim 2 \cdot 10^{-1} \tag{8.3c}$$

with the normalization quantities

$$r_{(l)} := r_{\mathrm{R}} \frac{B^{(l)*}}{\ln 10} \qquad \mathscr{E}_{(l)} := \varphi_{\mathrm{R}}\, 10^{A^{(l)*}/B^{(l)*}}. \tag{8.3d}$$

Their ratios to the parameters r_{R} and φ_{R} of expression (8.2) are

$$\frac{r_{(1)}}{r_{\mathrm{R}}} = 0.977, \quad \frac{r_{(2)}}{r_{\mathrm{R}}} = 1.064, \quad \frac{\mathscr{E}_{(1)}}{\varphi_{\mathrm{R}}} = 4.64, \quad \frac{\mathscr{E}_{(2)}}{\varphi_{\mathrm{R}}} = 7.54. \tag{8.3e}$$

γ) The *Screened Coulomb potential*

$$\varphi(r) = \varphi_{\mathrm{R}} \frac{r_{\mathrm{R}}}{r} \exp\left(1 - \frac{r}{r_{\mathrm{R}}}\right) \tag{8.4}$$

has its main region of interest in collision energies of $\mathscr{E}^* = \mathscr{E}/|\varphi_{\mathrm{R}}| \gg 1$ (see Sect. 12); this inequality holds for aeronomic and for most other plasmas. For collisions with small impact parameters $b \ll r_{\mathrm{R}}$ the main part of the interaction will take place at distances $r \ll r_{\mathrm{R}}$, where the screening by the exponential function is negligible. These collisions can therefore approximately be dealt with as pure Coulomb collisions. Collisions with large impact parameters $b \gg r_{\mathrm{R}}$ are influenced mainly by the screened part where the potential is already weak and so is the deflection angle χ. A "small χ approximation" is therefore sufficient for these collisions. Because of the smallness of $\mathscr{E}^{*-1}$, the two regions can be linked at some distance b_0 with $r_{\mathrm{R}}/\mathscr{E}^* \ll b_0 \ll r_{\mathrm{R}}$. Adding the results for both regions, the linking distance b_0 drops out and the result is[3]

$$Q^{(l)}(\mathscr{E}) = \frac{l(l+1)}{2} \pi r_{\mathrm{R}}^2 \left(\frac{|\varphi_{\mathrm{R}}|\, \mathrm{e}^1}{\mathscr{E}}\right)^2 \left[\ln\left(\frac{4\mathscr{E}}{\mathrm{e}^{\gamma+1} |\varphi_{\mathrm{R}}|}\right) - \frac{l}{2}\right] \tag{8.5}$$

for $\mathscr{E} \gg |\varphi_{\mathrm{R}}|$ and $l = 1, 2$ with $\gamma := 0.5772 = \ln 1.781$.

[2] Mason, E.A., Vanderslice, J.T. (1957): J. Chem. Phys. *27*, 917–927, eq. 34 and appendix D
[3] Liboff, R. (1959): Phys. Fluids *2*, 40–46, eqs. 4.14 and 4.21

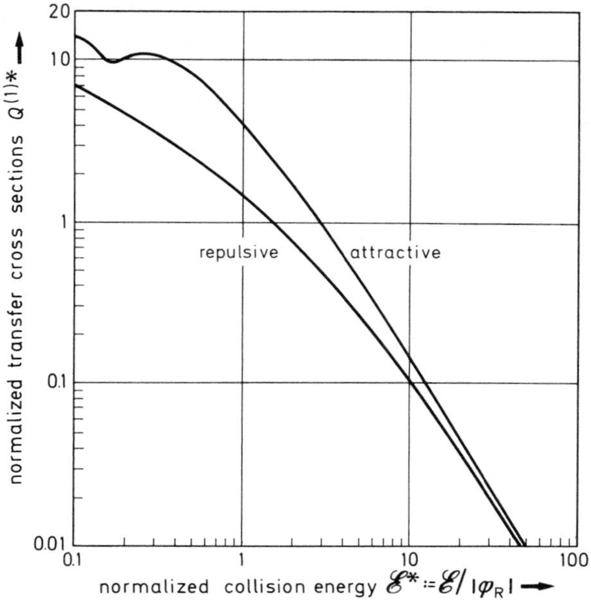

Fig. 9a. Normalized transfer cross sections $Q^{(1)*}(\mathscr{E}^*) = Q^{(1)}(\mathscr{E})/\pi r_R^2$, Eq. (6.7) for a screened Coulomb potential, Eq. (8.4) (after HAHN et al.[4])

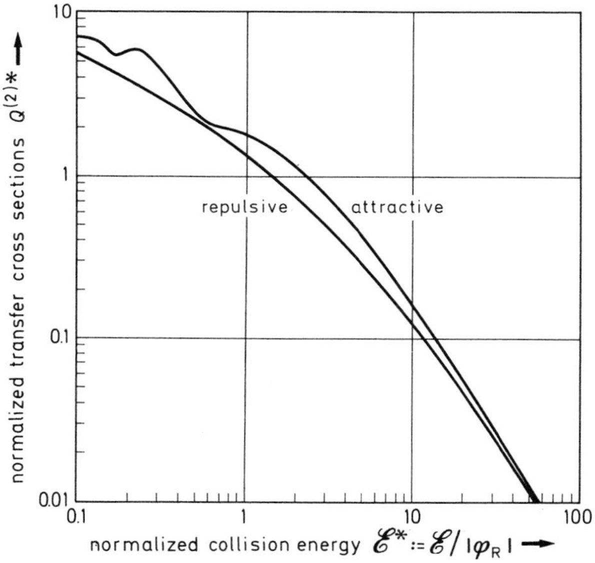

Fig. 9b. Normalized transfer cross sections $Q^{(2)*}(\mathscr{E}^*) = Q^{(2)}(\mathscr{E})/\pi r_R^2$, Eq. (6.7) for a screened Coulomb potential, Eq. (8.4) (after HAHN et al.[4])

[4] Hahn, H., Mason, E.A., Smith, F.J. (1971): Phys. Fluids *14*, 278–287, table I

Although the screening length r_{R} appears as a natural normalization length in Eq. (8.5), it is physically sometimes more instructive to use the collision length $r_1 = C_1/\mathscr{E}$, Eq. (6.12), of the pure Coulomb potential $\varphi = C_1/r$ as a normalization length. Because of $C_1 = \varphi_{\mathrm{R}}\,\mathrm{e}^1\, r_{\mathrm{R}}$, we obtain

$$\begin{aligned} Q^{(l)}(\mathscr{E}) &= \frac{l(l+1)}{2}\,\pi \left(\frac{C_1}{\mathscr{E}}\right)^2 \left[\ln\left(\frac{4 r_{\mathrm{R}}\,\mathscr{E}}{\mathrm{e}^{\gamma}\,|C_1|}\right) - \frac{l}{2}\right] \\ &= \frac{l(l+1)}{2}\,\pi\, r_1^2(\mathscr{E}) \left[\ln\left(\frac{4 r_{\mathrm{R}}}{\mathrm{e}^{\gamma}\,|r_1(\mathscr{E})|}\right) - \frac{l}{2}\right] \\ &\text{for } \mathscr{E} \gg |C_1|/r_{\mathrm{R}} \text{ and } l = 1, 2. \end{aligned} \tag{8.6}$$

If the high-energy condition $\mathscr{E}^* \gg 1$ becomes invalid, the transfer cross sections must be computed numerically. The result is shown in Fig. 9. Note the difference between repulsive and attractive interaction in the intermediate- and low-energy region. For high energies this difference vanishes.

δ) The *inverse power potential*

$$\varphi(r) = \frac{C_n}{r^n} = \left(\frac{r_n}{r}\right)^n \mathscr{E} \quad \text{with} \quad r_n(\mathscr{E}) := \left(\frac{C_n}{\mathscr{E}}\right)^{\frac{1}{n}} \qquad [6.11]\ [6.12]$$

yields transfer cross sections of the form

$$Q^{(l)}(\mathscr{E}) = \pi\, |r_n(\mathscr{E})|^2\, Q^{(l)*}(n, \operatorname{sgn} C_n) \qquad [6.19]$$

as already discussed in Sect. 6β. The energy dependence is contained in the collision length r_n which serves as normalization length here. The dimensionless factors $Q^{(l)*}(n, \operatorname{sgn} C_n)$ are listed in Table 1 (Sect. 6β).

The quantities $R^{(1)}$ and $R^{(2)}$ in the quantum corrections of the de Boer-Bird expansion, Eq. (6.20), have been calculated for $n = 12$, $\operatorname{sgn} C_{12} = +1$. They are[5]

$$R^{(1)}(12, +) = 0.410, \qquad R^{(2)}(12, +) = 4.949. \tag{8.7}$$

ε) For a *pure Coulomb potential*

$$\varphi(r) = \frac{C_1}{r} \qquad [6.11]$$

the collision length $r_1 = C_1/\mathscr{E}$, Eq. (6.12), is twice the major axis of the trajectory (for attractive as well as for repulsive interaction) and is sometimes named *Coulomb length*. The classical expression for the transfer cross section

$$Q^{(l)}_{\mathrm{class}}(\mathscr{E}) = 2\pi \int \mathrm{d}(\cos\chi)\, q_{\mathrm{class}}(\chi, \mathscr{E})\,[1 - \mathrm{P}_l(\cos\chi)] \qquad [4.8]$$

diverges for $\chi \to 0$ upon insertion of Rutherford's differential cross section $q_{\mathrm{class}}(\chi, \mathscr{E}) \sim (1 - \cos\chi)^{-2}$, Eq. (6.17). Therefore the classical limit for the transfer

[5] Hirschfelder, J.O., Curtiss, C.F., Bird, R.B.: See [5], eq. 10.3-32

cross sections $Q^{(l)}(\mathscr{E})$, Eq. (8.5), is usually calculated for a screened Coulomb potential, Eq. (8.4), which can be physically justified (Sect. 12).

The quantum corrections, however, in the de Boer-Bird expansion, Eq. (6.20), can be calculated without divergencies. The results are[6-8]

$$R^{(l)}(1) = -\frac{l(l+1)}{2}\frac{\pi}{3} \qquad S^{(l)*}(1) = -\frac{l(l+1)}{2}\frac{2\pi}{15}. \tag{8.8}$$

For Coulomb interactions the de Boer-Bird expansion converges very slowly[9]. But it can be replaced by the closed expressions[10]

$$Q^{(l)}(\mathscr{E}) = Q^{(l)}_{\text{class}}(\mathscr{E}) - \frac{l(l+1)}{2}\pi r_1^2(\mathscr{E})\left[\ln\frac{2}{|\boldsymbol{k} r_1|} + \operatorname{Re}\psi\left(i\frac{|\boldsymbol{k} r_1|}{2}\right)\right] \tag{8.9a}$$

for distinguishable particles with

$$|\boldsymbol{k}| := \frac{\mu g}{\hbar}, \qquad \frac{|\boldsymbol{k} r_1|}{2} = \frac{|C_1|}{\hbar g} = \left|\frac{C_1}{C_e}\right|\sqrt{\frac{\mu\,\text{Ry}}{m_e \mathscr{E}}} \qquad [4.4\text{b}]\ [6.12]\ [6.21\text{a}]$$

and

$$Q^{(l)}(\mathscr{E}) = Q^{(l)}(\text{distinguishable}) \pm \frac{l(l+1)}{2}\frac{\pi r_1^2(\mathscr{E})}{2s+1}\frac{\frac{\pi}{4}|\boldsymbol{k} r_1|}{\sinh\frac{\pi}{2}|\boldsymbol{k} r_1|} \tag{8.9b}$$

for indistinguishable particles with integer/half-integer spin s; here

$$\operatorname{Re}\psi(iy) := \operatorname{Re}\frac{\Gamma'(iy)}{\Gamma(iy)} = \ln y + \frac{1}{12y^2} + \frac{1}{120y^4} + O\left(\frac{1}{y^6}\right)$$
$$= -\gamma + y^2\zeta(3) + O(y^4) \tag{8.10}$$

is the real part of the digamma function (Appendix B γ). For small energies $\mathscr{E} \ll (\mu/m_e)$ Ry Eq. (8.9a) yields the de Boer-Bird expansion, Eq. (6.20), with the coefficients of Eq. (8.8). For high energies $\mathscr{E} \gg (\mu/m_e)$ Ry the dominant terms of Eqs. (8.9) are the same as the dominant terms of the quantum-mechanical Born approximation[11,12].

ζ) The *Maxwell potential*

$$\varphi(r) = \frac{C_4}{r^4} \tag{8.11}$$

plays an important role (with $C_4 < 0$) for the description of collisions between charged and neutral particles (Sect. 9). As was shown in Sect. 5 electron-neutral collisions can only be dealt with by full quantum mechanics, at least for energies of aeronomic interest.

In the quantum-mechanical calculations one uses a "low-energy approximation" for the determination of the phase shifts η_L in the series for the transfer cross sections $Q^{(l)}(\mathscr{E})$, Eq. (4.12). The *effective-range approximation*[13]

---

6 Kihara, T. (1964): J. Phys. Soc. Jpn. *19*, 108–116, eq. 4.7
7 Honda, N. (1964): J. Phys. Soc. Jpn. *19*, 1201–1206
8 Hahn, H., Mason, E.A., Smith, F.J. (1971): Phys. Fluids *14*, 278–287, eq. 15
9 Hahn, H., Mason, E.A., Smith, F.J. (1971): Phys. Fluids *14*, 278–287, fig. 1
10 Hahn, H., Mason, E.A., Smith, F.J. (1971): Phys. Fluids *14*, 278–287, eq. 35
11 Massey, H.S.W.: this Encyclopedia, vol. 36, pp. 232–306, sect. 6
12 Hahn, H., Mason, E.A., Smith, F.J. (1971): Phys. Fluids *14*, 278–287, sect. IV
13 Joachain, C.J.: See [*28*], sect. 11.6

yields for the phase shifts[14–16] in the attractive case $C_4<0$:

$$\tan\eta_L = -a|\boldsymbol{k}|\,\delta_{L0} + \frac{\pi}{(2L+3)(2L+1)(2L-1)}\,\frac{|\boldsymbol{k}|^2}{k_4^2} - \frac{4}{3}\,\frac{a|\boldsymbol{k}|^3}{k_4^4}\ln\left(\frac{|\boldsymbol{k}|}{4k_4}\right)\delta_{L0} + O(|\boldsymbol{k}|^{4-\delta_{L0}-\delta_{L1}}) \tag{8.12}$$

with

$$k_4^2 := \frac{\hbar^2}{2\mu|C_4|}. \tag{8.13a}$$

Here

$$a := -\lim_{|\boldsymbol{k}|\to 0} f(\chi) \tag{8.13b}$$

is the *scattering length.* It contains all the information from the inner part of the colliding particles, which has not much influence on the shaping of the attractive Maxwell potential. The calculation is very involved. Therefore the scattering length is mostly used as an adjustable parameter. It is positive for collisions of electrons with all aeronomic neutrals save CO_2 (Table 5, Sect. 10). A formula for $\tan\eta_0$ different from (8.12) has been proposed by PAUL[17].

For low energies the phase shifts η_L become very small. Hence $\sin^2(\eta_{L+n}-\eta_L)$ in the series (4.12) for $Q^{(l)}(\mathscr{E})$ can be replaced by $(\eta_{L+n}-\eta_L)^2$ and $\tan\eta_L$ in Eq. (8.12) by η_L. Inserting the phase shift values of Eq. (8.12) with $|\boldsymbol{k}|^2=2\mu\mathscr{E}/\hbar^2$, Eq. (4.4b), into the series of Eq. (4.12) for $Q^{(l)}(\mathscr{E})$, one obtains O'MALLEY's *modified effective-range expansion*[14]

$$\frac{Q^{(1)}(\mathscr{E})}{4\pi a^2} = 1 + \frac{4\pi}{5}\,\frac{1}{a k_4}\left(\frac{\mathscr{E}}{\mathscr{E}_4}\right)^{\frac{1}{2}} + \frac{4}{3}\,\frac{\mathscr{E}}{\mathscr{E}_4}\ln\left(\frac{1}{16}\,\frac{\mathscr{E}}{\mathscr{E}_4}\right) + \sum_{2\lambda=2}^{\infty} Q_\lambda^{(1)*}\left(\frac{\mathscr{E}}{\mathscr{E}_4}\right)^{\lambda} \tag{8.14a}$$

$$\frac{Q^{(2)}(\mathscr{E})}{4\pi a^2} = 1 + \frac{24\pi}{35}\,\frac{1}{a k_4}\left(\frac{\mathscr{E}}{\mathscr{E}_4}\right)^{\frac{1}{2}} + \frac{4}{3}\,\frac{\mathscr{E}}{\mathscr{E}_4}\ln\left(\frac{1}{16}\,\frac{\mathscr{E}}{\mathscr{E}_4}\right) + \sum_{2\lambda=2}^{\infty} Q_\lambda^{(2)*}\left(\frac{\mathscr{E}}{\mathscr{E}_4}\right)^{\lambda} \tag{8.14b}$$

with

$$\mathscr{E}_4 := \frac{\hbar^2}{2\mu}k_4^2 = \left(\frac{\hbar^2}{2\mu}\right)^2\frac{1}{|C_4|}. \tag{8.15}$$

(The summation subscript λ takes integer and half integer values 1, 3/2, 2, 5/2, etc.)

The corresponding expansion for the differential cross section $q(\chi,\mathscr{E})$ is[14]

$$\frac{q(\chi,\mathscr{E})}{a^2} = 1 + \frac{\pi}{a k_4}\left(\frac{\mathscr{E}}{\mathscr{E}_4}\right)^{\frac{1}{2}}\sin\frac{\chi}{2} + \frac{4}{3}\,\frac{\mathscr{E}}{\mathscr{E}_4}\ln\left(\frac{1}{16}\,\frac{\mathscr{E}}{\mathscr{E}_4}\right) + O(\mathscr{E}). \tag{8.16}$$

---

[14] O'Malley, T.F. (1963): Phys. Rev. *130*, 1020–1029, eqs. 2.3, 2.5, 2.6
[15] O'Malley, T.F., Spruch, L., Rosenberg, L. (1961): J. Math. Phys. N.Y. *2*, 491–498, eq. 5.7
[16] Fabrikant, I.I. (1979): J. Phys. B*12*, 3599–3610, eqs. 46 and 47
[17] Paul, D. (1980): Can. J. Phys. *58*, 134–137

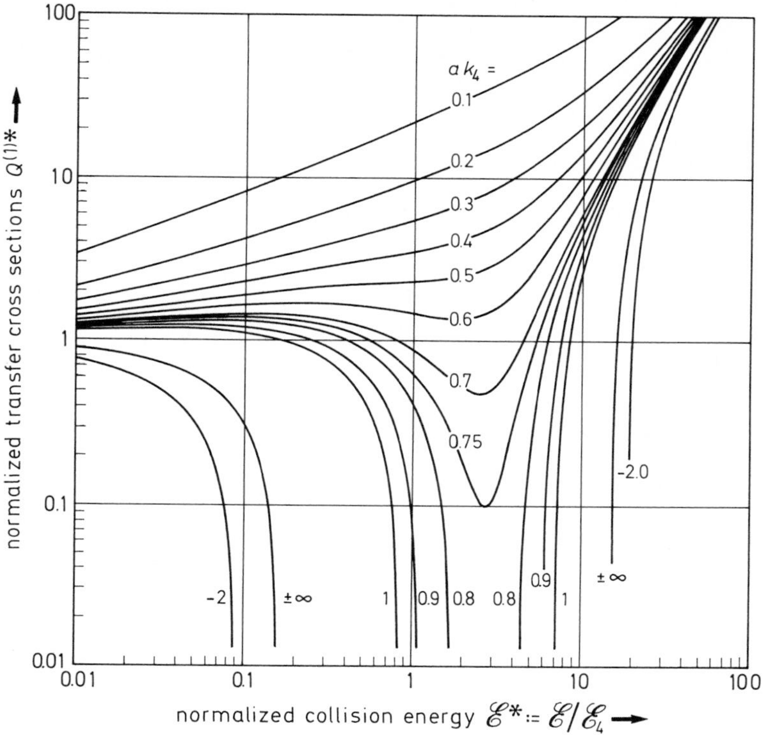

Fig. 10a. First three terms of the modified effective-range expansion (8.14a) for the normalized quantum-mechanical momentum transfer cross section $Q^{(1)*}(\mathscr{E}^*) := Q^{(1)}(\mathscr{E})/4\pi a^2$ for an attractive Maxwell potential C_4/r^4, Eq. (8.11). The parameter ak_4 is the product of the scattering length a of Eq. (8.13b) with $k_4 := \hbar^2/2\mu|C_4| = \sqrt{2\mu\,\mathscr{E}_4}/\hbar$, Eq. (8.15)

For the first three terms of $Q^{(1)}(\mathscr{E})$, contributions stem from the phase shifts η_0 and η_1 only, i.e., from s and p waves. For the first three terms of $Q^{(2)}(\mathscr{E})$ only η_0 and η_2 contribute, i.e., s and d waves.

Fig. 10 shows the first three terms of the transfer cross sections $Q^{(l)}(\mathscr{E})$, Eq. (8.14). The minimum of some curves near $\mathscr{E} \approx 3\mathscr{E}_4$ is known experimentally as *Ramsauer effect*.

For the first three terms of $Q^{(1)}$ and $Q^{(2)}$ in Eq. (8.14) the minimum becomes negative for $ak_4 = 0.76547$ and $ak_4 = 0.6561$, respectively. For a fit of experimental values the higher-order terms with $\lambda \geqq 1$ can be used with their coefficients $Q_\lambda^{(l)*}$ as additional adjustable parameters.

With the further approximation $\tan\eta_L \approx \sin\eta_L$ the sum of contributions $\sim |\boldsymbol{k}|^2$ to the scattering amplitude $f(\chi)$, Eq. (4.4), for all phase shifts η_L with $L > 2$ can be given in closed form according to Thompson[18]. With this result Nesbet[19] obtained the corresponding expression for the momentum transfer

[18] Thompson, D.G. (1966): Proc. R. Soc. London Ser. A *294*, 160–174
[19] Nesbet, R.K. (1979): Phys. Rev. A*20*, 58–70, eq. 11

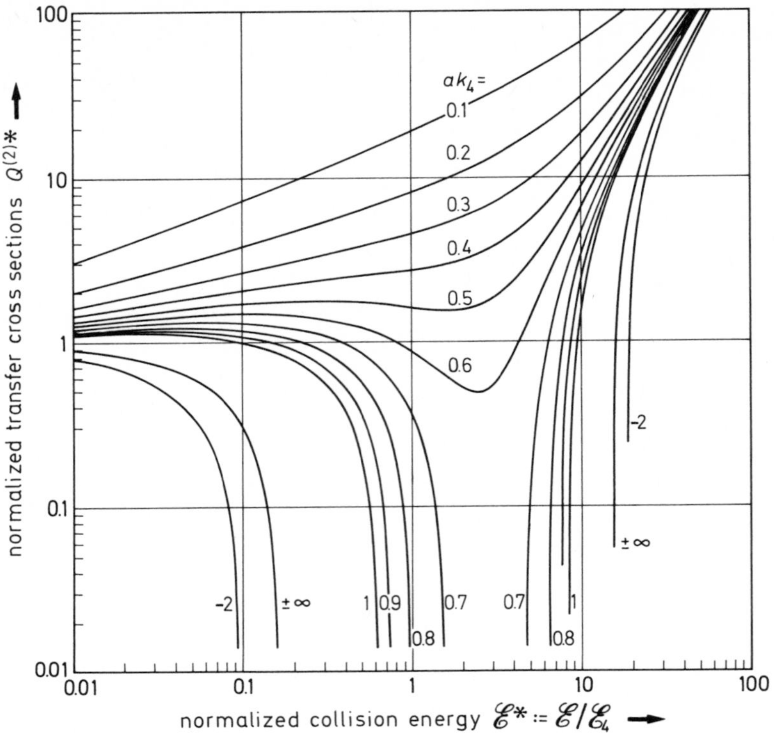

Fig. 10b. First three terms of the modified effective-range expansion (8.14b) for the normalized quantum-mechanical transfer cross section $Q^{(2)*}(\mathscr{E}^*) := Q^{(2)}(\mathscr{E})/4\pi a^2$ for an attractive Maxwell potential C_4/r^4, Eq. (8.11). The parameter ak_4 is the product of the scattering length a of Eq. (8.13b) with $k_4 := \hbar^2/2\mu|C_4| = \sqrt{2\mu\mathscr{E}_4}/\hbar$, Eq. (8.15)

cross section, Eq. (4.12b), as

$$Q^{(1)}(\mathscr{E}) = \frac{4\pi}{|\boldsymbol{k}|^2} \sum_{L=0}^{2} (L+1)\sin^2(\eta_{L+1} - \eta_L) + \frac{2\pi}{33075 k_4^2} \frac{\mathscr{E}}{\mathscr{E}_4} + O(\mathscr{E}^2). \tag{8.17}$$

II. Transfer cross sections for particular interactions

9. Collisions between charged and neutral particles. If a charged particle c encounters a neutral particle n, its charge q_c induces a dipole in the neutral particle. The interaction energy of the dipole moment $\boldsymbol{\alpha}_n \cdot \boldsymbol{E}$ with the field $\boldsymbol{E} = \mathrm{u}\, q_c \hat{\boldsymbol{r}}/4\pi\varepsilon_0 r^2$ of the charge q_c is

$$U(r) = -\int_0^{E} \mathrm{d}\boldsymbol{E}\cdot\boldsymbol{\alpha}_\mathrm{n}\cdot\boldsymbol{E} = -\tfrac{1}{2}\boldsymbol{E}\cdot\boldsymbol{\alpha}_\mathrm{n}\cdot\boldsymbol{E}$$
$$= -\frac{1}{2}\frac{\mathrm{u}\,q_\mathrm{c}^2}{4\pi\,\varepsilon_0}\frac{\hat{\boldsymbol{r}}}{r^2}\cdot\frac{\mathrm{u}\,\boldsymbol{\alpha}_\mathrm{n}}{4\pi\,\varepsilon_0}\cdot\frac{\hat{\boldsymbol{r}}}{r^2}. \tag{9.1}$$

Here $\boldsymbol{\alpha}_\mathrm{n}$ is the *polarizability* of the neutral particle[1]. The spherical average over all directions $\hat{\boldsymbol{r}}$ between the charged and neutral particle yields the attractive Maxwell potential, Eq. (8.11)

$$\varphi(r) = \frac{C_4}{r^4} \qquad [8.11]$$

with

$$C_4 = -\tfrac{1}{2}Z_\mathrm{c}^2\,C_\mathrm{e}\frac{\mathrm{u}\,\alpha_\mathrm{n}}{4\pi\,\varepsilon_0} = -1.0595\cdot 10^{-40}\,Z_\mathrm{c}^2\frac{\mathrm{u}\,\alpha_\mathrm{n}}{4\pi\,\varepsilon_0\,a_0^3}\,\mathrm{eVm}^4, \tag{9.2}$$

where Z_c is the charge excess, $C_\mathrm{e} = \mathrm{u}\,q_\mathrm{e}^2/4\pi\,\varepsilon_0$, Eq. (6.21b), is the constant of the Coulomb potential for electron-electron collisions, and

$$\alpha_\mathrm{n} := \frac{1}{4\pi}\oint \mathrm{d}\Omega\,\hat{\boldsymbol{r}}\cdot\boldsymbol{\alpha}_\mathrm{n}\cdot\hat{\boldsymbol{r}} \tag{9.3}$$

is the spherical mean of the tensor $\boldsymbol{\alpha}_\mathrm{n}$, i.e., one third of its trace.

If the neutral particle is a (spherical symmetric) atom the polarizability $\boldsymbol{\alpha}_\mathrm{n}$ is the product of a scalar α_n and the unit tensor $\mathbf{U}$ whose trace is 3. For a diatomic molecule with $\boldsymbol{s}$ as internuclear separation the polarizability is

$$\boldsymbol{\alpha}_\mathrm{n} = \alpha_\mathrm{n}^{\parallel}\,\hat{\boldsymbol{s}}\hat{\boldsymbol{s}} + \alpha_\mathrm{n}^{\perp}(\mathbf{U} - \hat{\boldsymbol{s}}\hat{\boldsymbol{s}})$$
$$= \frac{\alpha_\mathrm{n}^{\parallel} + 2\alpha_\mathrm{n}^{\perp}}{3}\,\mathbf{U} + (\alpha_\mathrm{n}^{\parallel} - \alpha_\mathrm{n}^{\perp})\,[\hat{\boldsymbol{s}}\hat{\boldsymbol{s}} - \tfrac{1}{3}\mathbf{U}], \tag{9.4}$$

yielding[2]

$$\alpha_n = \frac{\alpha_\mathrm{n}^{\parallel} + 2\alpha_\mathrm{n}^{\perp}}{3}. \tag{9.5}$$

Values of $\mathrm{u}\,\alpha_\mathrm{n}/4\pi\,\varepsilon_0$ for aeronomic neutral particles are given in Table 4. Since those values for atoms are of the vector of a_0^3, a dimensionless *reduced polarizability*

$$\boldsymbol{\alpha}_\mathrm{n}^* := \frac{\mathrm{u}\,\boldsymbol{\alpha}_\mathrm{n}}{4\pi\,\varepsilon_0\,a_0^3} \tag{9.6}$$

is introduced, which will facilitate the writing of the results.

---

[1] Hirschfelder, J.O., Curtiss, C.F., Bird, R.B.: See [5], eq. 12.2-2b

[2] Arthurs, A.M., Dalgarno, A. (1960): Proc. R. Soc. London Ser. A*256*, 552–558, eqs. 23 and 31

Table 4. Polarizabilities $\alpha_n^{\parallel}$, $\alpha_n^{\perp}$, Eq. (9.4), and $\alpha_n=(\alpha_n^{\parallel}+2\alpha_n^{\perp})/3$, Eq. (9.5), of aeronomic neutral particles (after TAKAYANAGI and ITIKAWA[3]; MCDANIEL and MASON[4]; MULDER et al.[5] for N_2)

	$\frac{\alpha_n^{\parallel}\,u/4\pi\varepsilon_0}{10^{-30}\,m^3}$	$\frac{\alpha_n^{\perp}\,u/4\pi\varepsilon_0}{10^{-30}\,m^3}$	$\frac{\alpha_n\,u/4\pi\varepsilon_0}{10^{-30}\,m^3}$	$\alpha_n^{\parallel *}:=\frac{\alpha_n^{\parallel}\,u/4\pi\varepsilon_0}{a_0^3}$	$\alpha_n^{\perp *}:=\frac{\alpha_n^{\perp}\,u/4\pi\varepsilon_0}{a_0^3}$	$\alpha_n^{*}:=\frac{\alpha_n\,u/4\pi\varepsilon_0}{a_0^3}$
H			0.667			4.500
H_2	1.03	0.715	0.82	6.94	4.82	5.53
He			0.205			1.383
O			0.77			5.19
N_2	2.16	1.515	1.735	14.7	10.3	11.8
O_2	2.32	1.23	1.60	15.75	8.33	10.80
CO_2	4.00	1.90	2.59	27.15	12.98	17.49

Because of $\hat{\boldsymbol{r}}\cdot(\hat{\boldsymbol{s}}\hat{\boldsymbol{s}}-\frac{1}{3}\mathbf{U})\cdot\hat{\boldsymbol{r}}=(\hat{\boldsymbol{r}}\cdot\hat{\boldsymbol{s}})^2-\frac{1}{3}=\frac{2}{3}P_2(\hat{\boldsymbol{r}}\cdot\hat{\boldsymbol{s}})$, the anisotropic part of the polarizability $\boldsymbol{\alpha}_n$, Eq. (9.4), adds an anisotropic term[2]

$$-\tfrac{1}{2}Z_c^2\,C_1\,a_0^3\,\tfrac{2}{3}(\alpha_n^{\parallel *}-\alpha_n^{\perp *})\frac{P_2(\hat{\boldsymbol{r}}\cdot\hat{\boldsymbol{s}})}{r^4} \tag{9.7}$$

to the Maxwell potential, Eq. (9.2). For the rotational ground state it has no influence on the transfer cross sections[6]. These are given by Eq. (6.19) with $n=4$ and C_4 from Eq. (9.2).

For the description of collisions between electrons and neutrals, quantum effects play a dominant role for aeronomic energies (see Fig. 1, Sect. 6β). For low energies the modified effective-range approximation (Sect. 8ζ) is sometimes sufficient. Its parameters

$$k_4^2:=\frac{\hbar^2}{2\mu|C_4|} \qquad \mathscr{E}_4:=\frac{\hbar^2}{2\mu}k_4^2 \qquad [8.13a]\ [8.15]$$

are expressed with the reduced polarizability α_n^*, Eq. (9.6), the charge excess $|Z_e|=1$, the Bohr radius a_0, Eq. (5.5a), and the rydberg, Eq. (5.5b) as

$$a_0^2\,k_4^2=\frac{m_e}{\mu}\frac{1}{\alpha_n^*}\approx\frac{1}{\alpha_n^*}, \qquad \frac{\mathscr{E}_4}{\mathrm{Ry}}=\left(\frac{m_e}{\mu}\right)^2\frac{1}{\alpha_n^*}\approx\frac{1}{\alpha_n^*}. \tag{9.8}$$

10. Collisions between electrons and neutral particles. General features. As was shown at the end of Sect. 6β the collisions of electrons with neutral particles are governed by quantum mechanics under conditions which are far from a classical limit. In Sect. 9 it was shown that the main characteristic of charge-neutral interaction is an attractive Maxwell potential

$$\varphi(r)=\frac{C_4}{r^4} \quad \text{with } C_4=-\tfrac{1}{2}Z_c^2\,C_e\,\alpha_n^*\,a_0^3. \qquad [9.2]$$

[3] Takayanagi, K., Itikawa, Y.: See [*13*], table II
[4] McDaniel, E.W., Mason, E.A.: See [*23*], tables II-1 and II-2
[5] Mulder, F., van Dijk, G., van der Avoird, A. (1980): Mol. Phys. *39*, 407–425, table 2
[6] Chang, E.S. (1974): Phys. Rev. A*9*, 1644–1655, eq. 7

Therefore the low-energy approximation, leading to O'Malley's effective range expansion, Eq. (8.14), should give reasonable results in its range of validity. With the expressions of Eq. (9.8) for k_4 and $\mathscr{E}_4$ the formulas (8.14) for the transfer cross sections now read

$$\frac{Q^{(1)}(\mathscr{E})}{4\pi a^2} = 1 + \frac{4\pi}{5}\frac{a_0}{a}\alpha_n^*\left(\frac{\mathscr{E}}{\mathrm{Ry}}\right)^{\frac{1}{2}} + \frac{4}{3}\alpha_n^*\frac{\mathscr{E}}{\mathrm{Ry}}\ln\left(\frac{1}{16}\alpha_n^*\frac{\mathscr{E}}{\mathrm{Ry}}\right) + \sum_{2\lambda=2}^{\infty} Q_\lambda^{(1)*}\alpha_n^{*\lambda}\left(\frac{\mathscr{E}}{\mathrm{Ry}}\right)^\lambda, \tag{10.1a}$$

$$\frac{Q^{(2)}(\mathscr{E})}{4\pi a^2} = 1 + \frac{24\pi}{25}\frac{a_0}{a}\alpha_n^*\left(\frac{\mathscr{E}}{\mathrm{Ry}}\right)^{\frac{1}{2}} + \frac{4}{3}\alpha_n^*\frac{\mathscr{E}}{\mathrm{Ry}}\ln\left(\frac{1}{16}\alpha_n^*\frac{\mathscr{E}}{\mathrm{Ry}}\right) + \sum_{2\lambda=2}^{\infty} Q_\lambda^{(2)*}\alpha_n^{*\lambda}\left(\frac{\mathscr{E}}{\mathrm{Ry}}\right)^\lambda. \tag{10.1b}$$

The sign of the term $\sim\sqrt{\mathscr{E}}$ in Eq. (10.1) depends on the sign of the scattering length a, which is positive for all aeronomic neutrals save CO_2 (Table 5). The sign of the $\mathscr{E}\ln\mathscr{E}$ term is negative for energy ranges of aeronomic interest, $\mathscr{E} \lesssim 10^{-2}$ Ry (see Fig. 1, Sect. 6β). The strength of the influence of the $\mathscr{E}\ln\mathscr{E}$ term depends, of course, on the values of the higher-order terms.

Table 5. Scattering lengths a, Eq. (8.13b), for collisions between electrons and neutrals

Neutral particle	$\frac{a}{10^{-10}\,\mathrm{m}}$	$\frac{a}{a_0}$	Authors
H (singlet)	3.926	7.419	Callaway et al.
H (triplet)	0.8869	1.676	Callaway et al.
H_2	0.635	1.20	Morrison and Lane
He	0.624	1.18	Nesbet
O	0.257	0.485	Kutcher et al.
N_2	0.23	0.44	Engelhardt et al., Chang
O_2	0.24	0.46	Hake and Phelps, Chang
CO_2	−3.8	−7.2	Singh

Callaway, J., La Bahn, R.W., Pu, R.T., Duxler, W.M. (1968): Phys. Rev. *168*, 12–21, sect. V
Chang, E.S. (1974): Phys. Rev. A *9*, 1644–1655, sect. IV. B, IV. C; J. Phys. B *14*, 893–901 (1981), sect. 3.2
Engelhardt, A.G., Phelps, A.V., Risk, C.G. (1964): Phys. Rev. A *135*, 1556–1574
Hake, R.D., Phelps, A.V. (1967): Phys. Rev. *158*, 70–84
Kutcher, G.J., Szydlik, P.P., Green, A.E.S. (1974): Phys. Rev. A *10*, 842–850, end of sect. III. B
Morrison, M.M., Lane, N.F. (1975): Phys. Rev. A *12*, 2361–2368, sect. III. B
Nesbet, R.K. (1979): J. Phys. B *12*, L243–248, eq. 3
Singh, Y. (1970): J. Phys. B *3*, 1222–1231, eq. 20

The modified effective range expansion, Eq. (10.1), is strictly valid only for neutral particles with an isotropic polarizability $\boldsymbol{\alpha}_n = \alpha_n \mathbf{U}$ and without a permanent dipole and quadrupole moment. This holds for atoms in their ground states. For excited atoms and for molecules the calculation approach of the phase shifts η_L must be generalized so as to include the anisotropic part of the polarizability and the dipole and quadrupole moments. This was done by SINGH[1] and CHANG[2] as well as ONDA and TEMKIN[3].

The dipole moment $\boldsymbol{p}_n = p_n \hat{\boldsymbol{s}}$ and the quadrupole moment $\mathbf{Q}_n = Q_n(\hat{\boldsymbol{s}}\hat{\boldsymbol{s}} - \frac{1}{3}\mathbf{U})$ contribute the following anisotropic terms

$$-\frac{p_n}{4\pi\varepsilon_0}\frac{\mathrm{P}_1(\hat{\boldsymbol{r}}\cdot\hat{\boldsymbol{s}})}{r^2} - \frac{Q_n}{4\pi\varepsilon_0}\frac{\mathrm{P}_2(\hat{\boldsymbol{r}}\cdot\hat{\boldsymbol{s}})}{r^3} \tag{10.2}$$

to the interaction potential. This is now composed of the isotropic attractive Maxwell potential, Eq. (9.2), the anisotropic parts, Eqs. (9.7) and (10.2), and another part taking into account the other contributions of the molecule. Permanent dipole moments and quadrupole moments are listed in Table 6.

Table 6. Permanent dipole moments p_n and permanent quadrupole moments Q_n, Eq. (10.2), of molecules (after TAKAYANAGI and ITIKAWA[4] for H_2 and O_2; MULDER et al.[5] for N_2; VUCELIĆ et al.[6], MORRISON et al.[7] for CO_2)

Molecule	p_n	$\frac{Q_n}{10^{-40}\,\mathrm{C\,m^2}}$	$\frac{Q_n}{\lvert q_e\rvert a_0^2}$
H_2	0	+ 2.20	+0.490
N_2	0	− 4.68	−1.04
O_2	0	− 1.30	−0.29
CO_2	0	−17.5	−3.9

For nonpolar molecules ($\boldsymbol{p}_n = 0$) the *extended effective range expansion* yields the addition of[2]

$$\frac{\Delta Q^{(1)}(\mathscr{E})}{4\pi a_0^2} = \frac{Q_n^2}{q_e^2 a_0^4}\frac{j(j+1)}{(2j-1)(2j+3)}\left(\frac{481}{2700} + \frac{2}{45}\ln 2\right) + \frac{2\pi}{25}\frac{\alpha_n^{\|*} - \alpha_n^{\perp *}}{3}\frac{Q_n}{\lvert q_e\rvert a_0^2}\frac{j(j+1)}{(2j-1)(2j+3)}\left(\frac{\mathscr{E}}{\mathrm{Ry}}\right)^{\frac{1}{2}} + O(\sqrt{\mathscr{E}}\ln\mathscr{E}) \tag{10.3}$$

to O'Malley's expression (10.1a) of the momentum transfer cross section $Q^{(1)}(\mathscr{E})$. It vanishes for the molecule in the rotational ground state ($j = 0$).

For polar molecules ($\boldsymbol{p}_n \neq 0$) in the *high moment of inertia approximation* the same theory yields[1]

$$\frac{\Delta Q^{(1)}(\mathscr{E})}{4\pi a_0^2} = \frac{2}{3}\frac{p_n^2}{q_e^2 a_0^2}\left(\frac{\mathscr{E}}{\mathrm{Ry}}\right)^{-1} + \frac{4}{45}\frac{Q_n^2}{q_e^2 a_0^4} + \frac{4\pi}{45}\frac{\alpha_n^{\|*} - \alpha_n^{\perp *}}{3}\frac{Q_n}{\lvert q_e\rvert a_0^2}\left(\frac{\mathscr{E}}{\mathrm{Ry}}\right)^{\frac{1}{2}} + \frac{\pi^2}{120}\left(\frac{\alpha_n^{\|*} - \alpha_n^{\perp *}}{3}\right)^2\frac{\mathscr{E}}{\mathrm{Ry}}. \tag{10.4}$$

[1] Singh, Y. (1970): J. Phys. B*3*, 1222–1231, eq. 14
[2] Chang, E.S. (1974): Phys. Rev. A*9*, 1644–1655, eq. 45
[3] Onda, K., Temkin, A. (1983): Phys. Rev. A*28*, 621–631
[4] Takayanagi, K., Itikawa, Y.: See [*13*], table II
[5] Mulder, F., van Dijk, G., van der Avoird, A. (1980): Mol. Phys. *39*, 407–425
[6] Vucelić, M., Öhrn, Y., Sabin, J.R. (1973): J. Chem. Phys. *59*, 3003–3007
[7] Morrison, M.A., Lane, N.F., Collins, L.A. (1977): Phys. Rev. A*15*, 2186–2201

Comparing the two results, Eqs. (10.3) and (10.4), for nonpolar molecules ($p_n = 0$), it seems obvious that Eq. (10.4) contains an averaging procedure over the rotational states j. This is probably due to the high-moment-of-inertia approximation and restricts[8] the applicability of Eq. (10.4) to energies above 0.01 eV.

In Sect. 11 we intend to discuss electron collision with aeronomic neutrals in particular. All available experimental data refer to the momentum transfer cross section $Q^{(1)}(\mathscr{E})$, there are none for the higher transfer cross sections $Q^{(l)}(\mathscr{E})$ with $l \geqq 2$. Thus values obtained theoretically from the effective range expansion, Eqs. (10.1) (10.3) (10.4), could only be compared for $Q^{(1)}(\mathscr{E})$. It would certainly be desirable to calculate the extensions of Eqs. (10.3) (10.4) to this expansion also for $Q^{(2)}(\mathscr{E})$. With the parameters obtained from the comparison of experimental and theoretical data for $Q^{(1)}(\mathscr{E})$ one could then establish theoretical values for $Q^{(2)}(\mathscr{E})$.

11. Collisions between electrons and aeronomic neutrals. In this section calculations and measurements of momentum transfer cross sections $Q^{(1)}(\mathscr{E})$ for e + n collisions will be discussed in the order of increasing mass of the neutral particles.

Until now higher-order transfer cross sections $Q^{(2)}$, $Q^{(3)}$ etc. have mostly been neglected. But some calculated and measured values of differential cross sections $q(\chi, \mathscr{E})$ are available. Angular averaging with weight functions $1 - P_l(\cos\chi)$ yields the transfer cross sections $Q^{(l)}(\mathscr{E})$, Eq. (4.8). For even values of l it is not important that the experiments be extended up to $\chi = 180°$ since in these cases $P_l(-1) = 1$ such that collisions with high deflection χ have the same small weights as those with $\chi \approx 0$. Maximum weight occurs at $\chi \approx 90°$ for l even. Recent reviews of theoretical methods are given by RESCIGNO et al.[1] as well as LANE[2].

α) For collisions of electrons with *hydrogen atoms* the phase shifts η_L were calculated up to $L = 7$ by HUNTER and KURIYAN[3]. Figure 11 shows the mo-

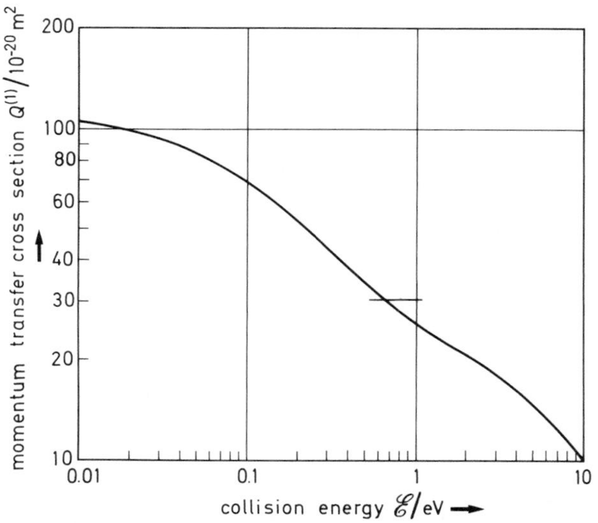

Fig. 11. Momentum transfer cross section $Q^{(1)}(\mathscr{E})$ for collisions of electrons with hydrogen atoms. Theoretical curve after HUNTER and KURIYAN[3]. Experimental value (−) from KÜHN and MOTSCHMANN[4]

[8] Singh. Y. (1970): J. Phys. B*3*. 1222–1231. end of sect. 3
[1] Rescigno. T.. McKoy. V.. Schneider. B. (eds): See [*33*]
[2] Lane. N.F. (1980): Rev. Mod. Phys. *52*. 29–119
[3] Hunter, G., Kuriyan, M. (1975): Proc. R. Soc. London Ser. A*341*, 491–515, table 5
[4] Kühn. V.. Motschmann. H. (1964): Z. Naturforsch. Teil A *19*. 658–659

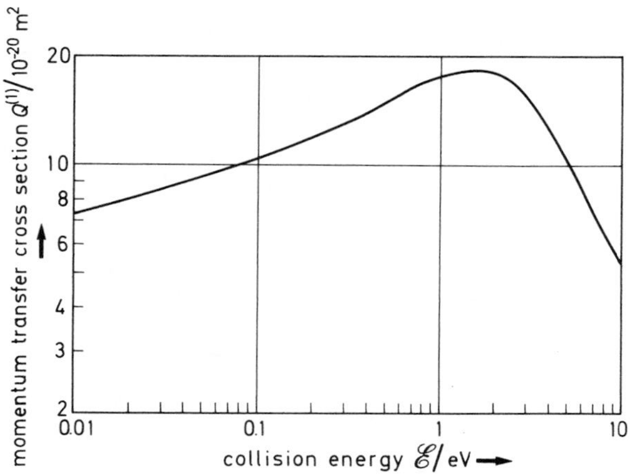

Fig. 12. Momentum transfer cross section $Q^{(1)}(\mathscr{E})$ for collisions of electrons with hydrogen molecules (after ITIKAWA[9])

mentum transfer cross section $Q^{(1)}(\mathscr{E})$ calculated with these phase shifts. The only existing experimental value, $3\cdot 10^{-19}\,\text{m}^2$ for $0.5 \leqq \mathscr{E}/\text{eV} \leqq 1$, is also indicated. Measured values[5-7] for $\mathscr{E} > 1\,\text{eV}$ of the integrated collision cross section $Q^{\text{int}}(\mathscr{E})$, Eq. (4.7), support the theoretical curve of Fig. 11.

β) For collisions of electrons with *hydrogen molecules* the theoretical values of the momentum transfer cross section $Q^{(1)}(\mathscr{E})$ of CROMPTON et al.[8] for $0.01\,\text{eV} \leqq \mathscr{E} \leqq 2.0\,\text{eV}$ have been connected by ITIKAWA[9] with experimental data of SRIVASTAVA and CHUTJIAN[10] for $\mathscr{E} = 5$, 7 and 10 eV (Fig. 12). The theoretical values are consistent with experimental data.

γ) For collisions of electrons with *helium atoms* the momentum transfer cross section $Q^{(1)}(\mathscr{E})$ was expressed by NESBET[11] with the Born approximation, Eq. (8.17), without the term $O(\mathscr{E}^2)$. The phase shifts for the s and p waves ($L=0$ and 1) were calculated with a variational method. The result is in agreement with the measured values of CROMPTON et al.[12] for $0.01 \leqq \mathscr{E}/\text{eV} \leqq 6$ and MILLOY and CROMPTON[13] for $6 \leqq \mathscr{E}/\text{eV} \leqq 10$ within less than 1.4% and is shown in Fig. 13.

δ) For collisions of electrons with *nitrogen atoms* integrated cross sections $Q^{\text{int}}(\mathscr{E})$, Eq. (4.7), have been calculated by THOMAS and NESBET[14] assuming the ^{3}P state of the negative nitrogen ion

[5] Smith, S.J. (1969) in: Physics of the one- and two-electron atoms. Bopp, F., Kleinpoppen, H. (eds.), fig. 15. Amsterdam: North-Holland

[6] Bederson, B., Kieffer, L.J.: See [*15*], fig. 9

[7] Alton, G.D., Garrett, W.R., Reeves, M., Turner, J.E. (1972): Phys. Rev. A*6*, 2138–2146, figs. 3 and 6

[8] Crompton, R.W., Gibson, D.K., McIntosh, A.I. (1969): Aust. J. Phys. *22*, 715–731, table 4

[9] Itikawa, Y. (1978): At. Data Nucl. Data Tables *21*, 69–75

[10] Srivastava, S.K., Chutjian, A. (1975): J. Chem. Phys. *63*, 2659–2665

[11] Nesbet, R.K. (1979): Phys. Rev. A*20*, 58–70, eq. 11 and table XI

[12] Crompton, R.W., Elford, M.T., Robertson, A.G. (1970): Aust. J. Phys. *23*, 667–681

[13] Milloy, H.B., Crompton, R.W. (1977): Phys. Rev. A*15*, 1847–1850

[14] Thomas, L.D., Nesbet, R.K. (1975): Phys. Rev. A*12*, 2369–2377, fig. 2

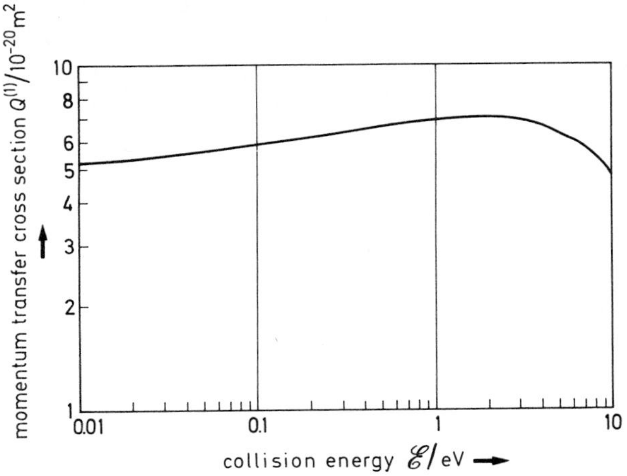

Fig. 13. Momentum transfer cross section $Q^{(1)}(\mathscr{E})$ for collisions of electrons with helium atoms after NESBET[11]

N^- to be a free resonance state and not a bound state. For the resonance energy 0.105 eV the peak value of $Q^{\text{int}}(\mathscr{E})$ was calculated as $583\pi a_0^2 = 513 \cdot 10^{-20}\,\text{m}^2$. This exceeds the cross section for atomic oxygen by about two orders of magnitude (see Fig. 14, Subsect. ε below). Although there are experimental indications[15] for an increase of $Q^{\text{int}}(\mathscr{E})$ until $66 \cdot 10^{-20}\,\text{m}^2$ for a decreasing collision energy $\mathscr{E}$ down to 0.13 eV, the uncertainty about the character of the ^3P state of N^- prevents using the theoretical values of THOMAS and NESBET[14] for the integrated cross section. Moreover, for the Earth's atmosphere the ratio of the densities of atomic nitrogen to atomic oxygen[16] is generally smaller than 0.04. Therefore the cross section of atomic nitrogen would have to exceed that of atomic oxygen by more than one order of magnitude to have an influence in the Earth's atmosphere comparable to that of atomic oxygen. Assuming the ^3P state of N^- to be a resonance state and not a bound state, and the peak value of $Q^{\text{int}}(\mathscr{E})$ to be $513 \cdot 10^{-20}\,\text{m}^3$, AGGARWAL et al.[17] have found that the largest influence of e + N collisions in the Earth's atmosphere is smaller by a factor 0.4 than the dominant values of e + O and e + N_2 collisions.

ε) For collisions of electrons with *oxygen atoms* the transfer cross sections $Q^{(l)}(\mathscr{E})$ with $l = 1, 2$ were obtained for $\mathscr{E} \geqq 0.136$ eV from theoretical values of differential cross sections $q(\chi, \mathscr{E})$, Eq. (4.2) by angular avaraging[18], Eq. (4.8) (Fig. 14). The curve for the momentum transfer cross section $Q^{(1)}(\mathscr{E})$ has been extrapolated down to 0.01 eV by ITIKAWA[9] using the theoretical values of HENRY and McELROY[20]. The measured value of $2 \cdot 10^{-20}\,\text{m}^2$ at 0.5 eV is identified as a momentum transfer cross section $Q^{(1)}$ by SHKAROFSKY et al.[21]. All other measured values are integrated cross sections $Q^{\text{int}}(\mathscr{E})$, Eq. (4.7).

---

15 Miller, T.M., Aubrey, B.B., Eisner, P.N., Bederson, B. (1970): Bull. Am. Phys. Soc. *15*, 416

16 Mauersberger, K., Engebretson, M.J., Kayser, D.C., Potter, W.E. (1976): J. Geophys. Res. *81*, 2413–2416, fig. 1 and table 2

17 Aggarwal, K.M., Nath, N., Setty, C.S.G.K. (1979): Planet. Space Sci. *27*, 753–768, fig. 2

18 Thomas, L.D., Nesbet, R.K. (1975): Phys. Rev. A*11*, 170–173

19 Lin, S.C., Kivel, B. (1959): Phys. Rev. *114*, 1026–1027

20 Henry, R.J.W., McElroy, M.B. (1968) in: The atmospheres of venus and mars. Brandt, J.C., McElroy, M.B. (eds.), pp. 251–285, fig. 9. New York: Gordon & Breach

21 Shkarofsky, I.P., Johnston, T.W., Bachynski, M.P.: See [9], footnote on page 198

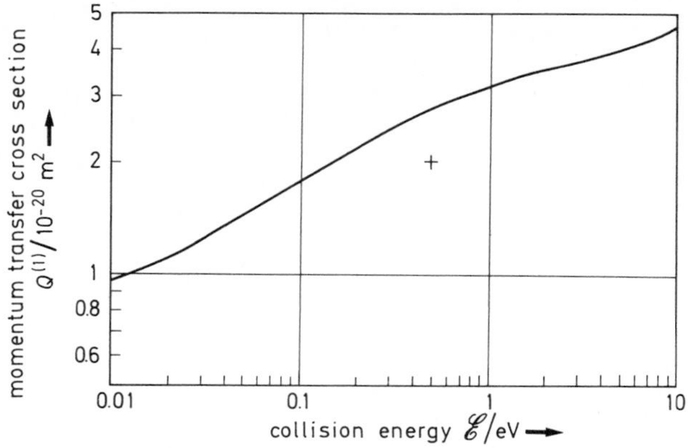

Fig. 14. Momentum transfer cross section $Q^{(1)}(\mathscr{E})$ for collisions of electrons with oxygen atoms. Theoretical curve from ITIKAWA[9]. Experimental value (+) from LIN and KIVEL[19]

BALUJA et al.[22] adapted the first four terms of the modified effective range expansion, Eq. (10.1a) – using the reduced polarizability $\alpha_0^* = 5.19$, Eq. (9.6), from Table 4 (Sect. 9) – to two theoretical values of $Q^{(1)}(\mathscr{E})$ of THOMAS and NESBET[18] at $\mathscr{E} = 0.136$ eV and $\mathscr{E} = 0.544$ eV with the parameter values $a = 0.39046\, a_0 = 2.0662 \cdot 10^{-11}$ m and $Q_1^{(1)*} = -7.35359$.

An extrapolation of THOMAS and NESBET's theoretical results for $\mathscr{E} \geqq 0.136$ eV (Fig. 14) to lower energies by means of the modified effective range expansion, Eq. (10.1a) causes difficulties. Between 0.001 eV and 1.0 eV the theoretical values for the s-wave phase shift η_0, Eq. (8.12), of KUTCHER et al.[23] are in good agreement with those obtained with the first three terms of the modified effective range expansion, Eq. (8.12), of η_0 if the latter is adapted to the former with $a = 0.485 a_0 = 2.57 \cdot 10^{-11}$ m, using $\alpha_0^* = 5.19$, Eq. (9.6), from Table 4, Sect. 9. Since the scattering length a, Eq. (8.13b), is the zero-energy limit of $\sqrt{Q^{(l)}(\mathscr{E})/4\pi}$, Eq. (10.1), the value $2.57 \cdot 10^{-11}$ m, obtained from energy values down to 0.001 eV, seems to be more justified than $2.0662 \cdot 10^{-11}$ m, obtained from two energy values 0.136 eV and 0.544 eV. But with only the first three terms of the modified effective range expansion, Eq. (10.1), for $Q^{(1)}(\mathscr{E})$ and $Q^{(2)}(\mathscr{E})$ one gets values at 0.2 eV which exceed the results of THOMAS and NESBET[18] in Fig. 14 by about 40%. Therefore it seems necessary to include the fourth term[22] in the modified effective range expansion, Eq. (10.1), with the preliminary value of -7.354 for its coefficient $Q_1^{(1)*}$.

The difficulty in applying the modified effective range method may be due to the low electron affinity of the oxygen atom which gives rise to a joint e + O system to form a bound state O^- at low energies[24].

ζ) For collisions of electrons with *nitrogen molecules* the most reliable experimental data for the momentum transfer cross section $Q^{(1)}(E)$ are today those of ENGELHARDT et al.[25], which are shown in Fig. 15. They include inelastic collisions according to SRIVASTAVA et al.[26], CHANDRA and TEMKIN[27], POTTER et al.[28].

---

[22] Baluja, K.L., Aggarwal, K.M., Setty, C.S.G. (1977): Indian J. Radio Space Phys. *6*, 296–298, sect. 4

[23] Kutcher, G.J., Szydlik, P.P., Green, A.E.S. (1974): Phys. Rev. A*10*, 842–850, table IV

[24] Chang, E.S. (1974): Phys. Rev. A*9*, 1644–1655, sect. IV.C

[25] Engelhardt, A.G., Phelps, A.V., Risk, C.G. (1964): Phys. Rev. A*135*, 1556–1574

[26] Srivastava, S.K., Chutjian, A., Trajmar, S. (1976): J. Chem. Phys. *64*, 1340–1344, fig. 4

[27] Chandra, N., Temkin, A. (1976): Phys. Rev. A*13*, 188–203, eq. 5.8b and fig. 16

[28] Potter, J.E., Steph, N.C., Dwivedi, P.H., Golden, D.E. (1977): J. Chem. Phys. *66*, 5557–5563

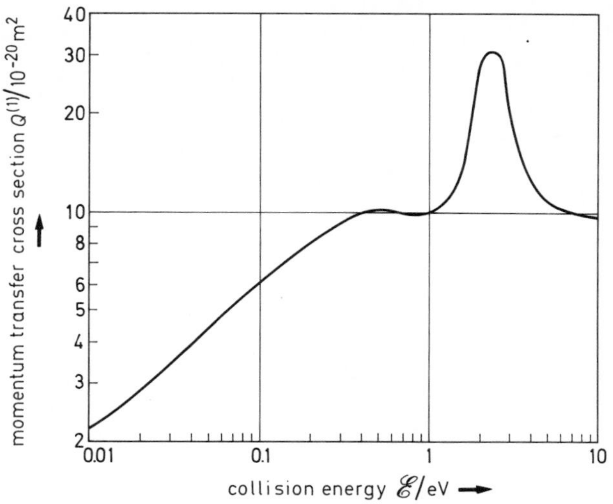

Fig. 15. Momentum transfer cross section $Q^{(1)}(\mathscr{E})$ for collisions of electrons with nitrogen molecules. Experimental values from ENGELHARDT et al.[25]

Fitting the low-energy values with the modified effective range expansion, Eq. (10.1 a), the additional terms, Eq. (10.3) of CHANG[29] must be taken into account, which include the effect of the quadrupole moment of the nitrogen molecule[30].

The low-energy values in Fig. 15 vary approximately as

$$\frac{Q^{(1)}(\mathscr{E})}{10^{-20}\,\mathrm{m}^2} = 20\left(\frac{\mathscr{E}}{\mathrm{eV}}\right)^{\frac{1}{2}} \quad \text{for } 0.01 < \frac{\mathscr{E}}{\mathrm{eV}} < 0.1. \tag{11.1}$$

η) For collisions of electrons with *oxygen molecules* the most quoted experimental data for the momentum transfer cross section $Q^{(1)}(\mathscr{E})$ are those of HAKE and PHELPS[31]. Some doubts on their reliability for low energies have been raised by CROMPTON and ELFORD[32], by ITIKAWA[33], and by CHANG[24]. Therefore LAWTON and PHELPS[34] have combined the results of HAKE and PHELPS[31] for energies above 0.1 eV with the experimental results of MENTZONI[35] for energies below 0.1 eV. Their values are given in Fig. 16. For low energies they vary approximately as

$$\frac{Q^{(1)}(\mathscr{E})}{10^{-20}\,\mathrm{m}^2} = 10\left(\frac{\mathscr{E}}{\mathrm{eV}}\right)^{\frac{4}{7}} \quad \text{for } 0.01 < \frac{\mathscr{E}}{\mathrm{eV}} < 0.05. \tag{11.2}$$

---

[29] Chang, E.S. (1974): Phys. Rev. A*9*, 1644, eq. 45
[30] Chang, E.S. (1974): Phys. Rev. A*9*, 1644, sect. IV.B, C; (1981): J. Phys. B*14*, 893–901
[31] Hake, R.D., Phelps, A.V. (1967): Phys. Rev. *158*, 70–84
[32] Crompton, R.W., Elford, M.T. (1973): Aust. J. Phys. *26*, 771–782, end of sect. V
[33] Itikawa, Y. (1972): Argonne Natl. Lab. Rep. ANL-7939, sect. 5
[34] Lawton, S.A., Phelps, A.V. (1978): J. Chem. Phys. *69*, 1055–1068, fig. 7
[35] Mentzoni, M.H. (1965): J. Res. Natl. Bur. Stand. Sect. D*69*, 213–217, sect. 3

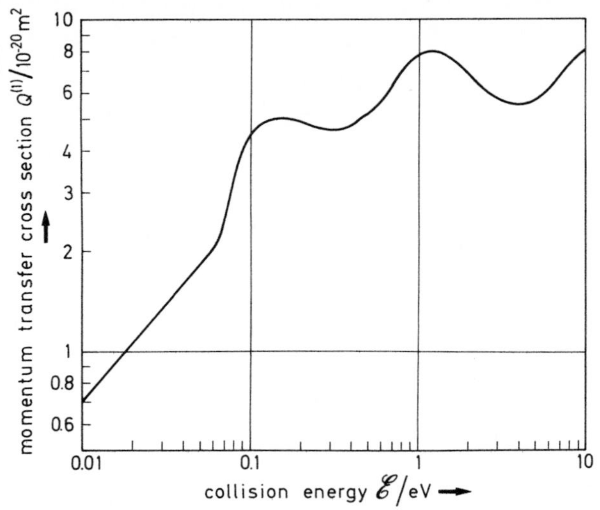

Fig. 16. Momentum transfer cross section $Q^{(1)}(\mathscr{E})$ for collisions of electrons with oxygen molecules. Experimental values according to LAWTON and PHELPS[34]

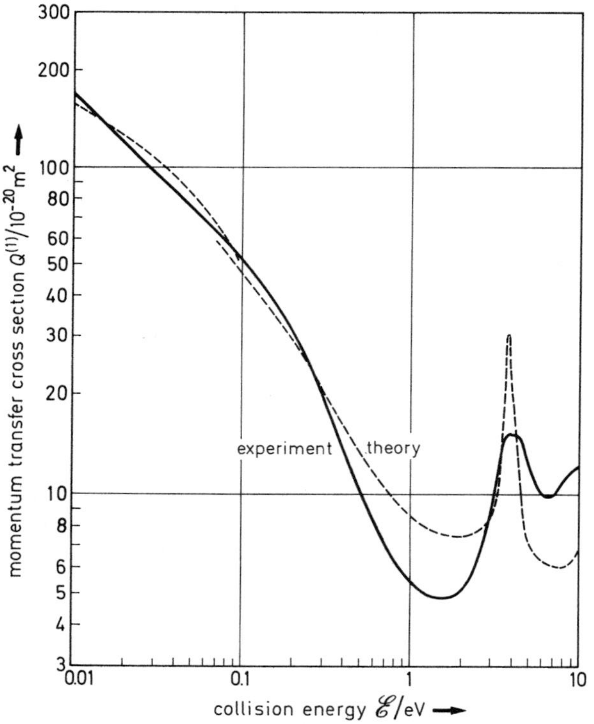

Fig. 17. Momentum transfer cross section $Q^{(1)}(\mathscr{E})$ for collisions of electrons with carbon dioxide molecules. Experimental values from HAKE and PHELPS[31], theoretical curves from SINGH[36] and MORRISON et al.[37]

[36] Singh, Y. (1970): J. Phys. B*3*, 1222–1231, fig. 1

[37] Morrison, M.A., Lane, N.F., Collins, L.A. (1977): Phys. Rev. A*15*, 2186–2201, fig. 3 and table IV

ϑ) For collisions of electrons with *carbon dioxide molecules* the most quoted experimental values for the momentum transfer cross section $Q^{(1)}(\mathscr{E})$ are those of HAKE and PHELPS[31]. Theoretical values for $0.01\,\text{eV} < \mathscr{E} < 0.1\,\text{eV}$ are given by SINGH[36] using his extended effective range expansion, Eq. (10.4), and for $0.07\,\text{eV} < \mathscr{E} < 10\,\text{eV}$ by MORRISON et al.[37]. All curves are shown in Fig. 17.

The low-energy part can be approximated as a classical Maxwell interaction with

$$\frac{Q^{(1)}(\mathscr{E})}{10^{-20}\,\text{m}^2} = 17\left(\frac{\mathscr{E}}{\text{eV}}\right)^{-\frac{1}{2}} \qquad \text{for } 0.01 < \frac{\mathscr{E}}{\text{eV}} < 0.1. \tag{11.3}$$

12. Collisions between charged particles. For collisions between charged particles the Coulomb potential leads to divergent expressions for the transfer cross sections (see Sect. 6γ). To remedy this inconvenience one introduces modifications either to the potential shape or to the integration range of the collision parameter b. Cutting off the range of the potential or b yields very similar expressions for $Q^{(l)}(\mathscr{E})$; these, of course, depend on the cutoff length. Another way is "screening" of the original COULOMB potential C_1/r, Eq. (6.11), by multiplying it with a factor $\exp(-r/r_s)$, viz.,

$$\varphi(r) = \frac{C_1}{r} \exp - \frac{r}{r_s}. \tag{12.1}$$

Rather similar expressions are so obtained for $Q^{(l)}(\mathscr{E})$ depending in this case on the *screening length* r_s.

The difficulties appearing with infinite integrals express a physical phenomenon, namely, that considering Coulomb interaction as an effect of binary collisions only is not adequate at larger distances. An adequate treatment has to take into account the *collective interaction* of many particles simultaneously. This is done, for instance, in the DEBYE-HÜCKEL[1] theory of electrolytes and in the modern theory of "hot" plasmas[2]. However, in these theories it is very difficult to compare the influence of Coulomb interactions with short-range interactions if both are of similar order of magnitude. Therefore one tries to adhere to the concept of binary collisions, because by comparing collision frequencies for Coulomb interactions and for short-range interactions it is most easy to estimate their influence on the transport coefficients.

To avoid the aforementioned divergencies, one takes advantage in "borrowing" from the theories of collective interaction the *Debye-Hückel length*

$$\lambda_D := [4\pi \sum_c l_{cc} N_c]^{-\frac{1}{2}} \tag{12.2}$$

as the screening length r_s for an exponential screening $\exp(-r/r_s)$ of the Coulomb potential. Here

$$l_{jk} := r_1(k\hat{T}_{jk}) = \frac{C_1}{k\hat{T}_{jk}} = \frac{u q_j q_k / 4\pi\varepsilon_0}{k T_{jk}} = 1.684 \cdot 10^{-5} \frac{Z_j Z_k}{\hat{T}_{jk}/\text{K}}\, m \tag{12.3}$$

---

[1] Debye, P., Hückel, E. (1923): Phys. Z. *24*, 185–206, 305–325

[2] Montgomery, D.C., Tidman, D.A.: See [7]

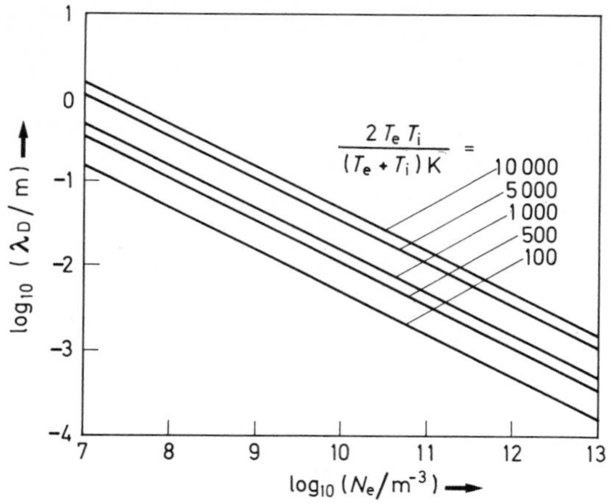

Fig. 18. DEBYE-HÜCKEL length λ_D, Eq. (12.5) for a quasi-neutral plasma with one singly charged species of ions. The parameter is the harmonic mean of the electron temperature T_e and the ion temperature T_i

is the *Landau length*, i.e., the distance of closest approach for a repulsive Coulomb collision between charged particles of species j and k having the *thermal collision energy*

$$k\hat{T}_{jk} := \mu_{jk}\left(\frac{kT_j}{m_j} + \frac{kT_k}{m_k}\right), \tag{12.4}$$

which is expressed by the *combined temperature* $\hat{T}_{jk}$ of the two particle species (Sect. 18 ζ).

If there is only one species of (positive) ions with $q_i = -q_e$ and $N_i = N_e$ then the Debye-Hückel length becomes (Fig. 18)

$$\lambda_D(\mathrm{e}, \mathrm{i}) := \left[\frac{k}{4\pi C_e N_e} \frac{T_e T_i}{T_e + T_i}\right]^{\frac{1}{2}} = 48.61 \left[\frac{2 T_e T_i}{(T_e + T_i)\,\mathrm{K}} \frac{\mathrm{m}^{-3}}{N_e}\right]^{\frac{1}{2}} \mathrm{m}. \tag{12.5}$$

For the classical transfer cross sections $Q^{(l)}(\mathscr{E})$ the values from Fig. 9, Sect. 8γ, and the high-energy expressions (8.5) can be taken with the Debye-Hückel length λ_D, Eq. (12.2), as reference length r_R. The reference potential φ_R of Eq. (8.4) becomes $C_1/\lambda_D \exp 1$ according to Eq. (12.1).

But the Debye-Hückel length λ_D, Eq. (12.2), depends on two different kinds of parameters: densities N_c and temperatures T_c of charged particle species c. For later use (Sect. 20 ϑ) it is desirable to separate the dependence on these two parameter families. Therefore we introduce for the representation of transfer cross sections for collisions between charged particles of species j and k the *normalization length*

$$r_N := (|l_{jk}|\,\lambda_D^2)^{\frac{1}{3}} = \left[4\pi \sum_c \frac{l_{cc}}{|l_{jk}|} N_c\right]^{-\frac{1}{3}}. \tag{12.6a}$$

This length describes a mean distance between charged particles and depends essentially on the number densities N_c of all charged particle species. Moreover, because of

$$\frac{l_{cc}}{l_{jk}}=\frac{q_c^2}{q_j q_k}\frac{\hat{T}_{jk}}{T_c}, \tag{12.6b}$$

see Eq. (12.3), it depends on charge ratios and temperature ratios, serving as weights for the summation over the number densities N_c in the definition Eq. (12.6a) of the normalization length r_N.

Consequently, the *normalization energy*

$$\mathscr{E}_N := \frac{C_1}{r_N} = \frac{C_1}{(|l_{jk}|\lambda_D^2)^{\frac{1}{3}}} = \frac{u q_j q_k}{4\pi\varepsilon_0}\left[4\pi\sum_c \frac{l_{cc}}{|l_{jk}|}N_c\right]^{\frac{1}{3}} \tag{12.7}$$

is introduced. For densities N_c of charged particles between 10^7 and $10^{13}\,\mathrm{m}^{-3}$ (Fig. 18) the normalization energy $\mathscr{E}_N$ varies between 10^{-6} and 10^{-4} eV, corresponding to a temperature range between 0.01 and 1 K.

With the reference quantities r_N, Eq. (12.6a), and $\mathscr{E}_N$, Eq. (12.7), the Debye-Hückel length λ_D, Eq. (12.2), can now be factorized as

$$\lambda_D = r_N\left(\frac{C_1}{|l_{jk}|\mathscr{E}_N}\right)^{\frac{1}{2}} = r_N\left(\frac{k\hat{T}_{jk}}{|\mathscr{E}_N|}\right)^{\frac{1}{2}}. \tag{12.8}$$

Besides the temperature ratios $\hat{T}_{jk}/T_c$, Eq. (12.6b), in the definition (12.6a) of the normalization length r_N, the temperature dependence of λ_D is contained in the factor $(k\hat{T}_{jk}/|\mathscr{E}_N|)^{1/2}$. Here the normalization energy $\mathscr{E}_N$ normalizes the thermal collision energy $k\hat{T}_{jk}$, Eq. (12.4), of the collision partners.

The transfer cross sections $Q^{(l)}(\mathscr{E})$, Eq. (8.6), can now be written in the normalized form

$$\frac{Q^{(l)}(\mathscr{E})}{\pi r_N^2} = \frac{l(l+1)}{2}\left(\frac{\mathscr{E}_N}{\mathscr{E}}\right)^2\left[\ln\left(\frac{4}{e^\gamma}\sqrt{\frac{k\hat{T}_{jk}}{|\mathscr{E}_N|}}\frac{\mathscr{E}}{|\mathscr{E}_N|}\right)-\frac{l}{2}\right]$$
$$\text{for } l=1,2 \text{ and } \mathscr{E} \gg \frac{|C_1|}{\lambda_D} = \frac{|\mathscr{E}_N|^{3/2}}{\sqrt{k\hat{T}_{jk}}} \tag{12.9}$$

with $\gamma=0.5772=\ln 1.781$. For Debye-Hückel lengths λ_D between 10^{-4} and 10^0 m (Fig. 18) the Coulomb energy C_e/λ_D varies between 10^{-5} and 10^{-9} eV.

The quantum corrections can be taken from those for the pure Coulomb potential, Eq. (8.9).

13. Collisions between ions and neutrals. Polarization part. It was shown in Sect. 6β that collisions of ions with neutral particles are in general not influenced by quantum effects, at least as far as the attractive Maxwell part C_4/r^4 of the interaction potential is concerned, which is caused by the induced dipole field, Eq. (9.1). Therefore it is often sufficient to calculate cross sections by classical procedures as long as the attractive Maxwell part plays the dominant rôle in the interaction potential.

It should be stressed that for plasmas with temperatures exceeding 500 K the influence of the chosen form of the potential curve on the transport coefficients is by no means small[1]. Therefore care should be taken to collect as much information as possible from all available sources in order to secure a reasonable choice of the potential curve and its parameters. For lower temperatures the transport coefficients become less sensitive to the chosen potential curve[2].

α) Attractive Maxwell potential. If no information about the inner part of the interaction potential is available, e.g., for the oxygen atom, then a pure attractive Maxwell potential must be used. The transfer cross sections

$$Q^{(l)}(\mathscr{E}) = \pi\, |r_4(\mathscr{E})|^2\, Q^{(l)*}(4, -) \qquad [6.19]$$

are simple functions of the energy because of $r_4^2(\mathscr{E}) = \sqrt{C_4/\mathscr{E}}$, Eq. (6.12). The numerical values

$$Q^{(1)*}(4, -) = 2.2092 \qquad Q^{(2)*}(4, -) = 2.3076$$

are taken from Table 1, Sect. 6*β*.

β) Sutherland $(\infty, 4)$ *potential.* Because of the long-range attractive character of the Maxwell part, an expression for the core of the interaction potential must be added. The first attempt was made by LANGEVIN[3] who used a Sutherland $(\infty, 4)$ potential, Eq. (7.16), i.e., an attractive Maxwell potential C_4/r^4 with a hard core. This means physically that the colliding particles are regarded as rigid spheres as long as the charge of the ion is not taken into account. Thus with the given polarizability $\alpha_n \sim -C_4$ of the neutral particle and the given diameters $r_0(A^+)$ and $r_0(B)$ of both the collision partners A^+ and B, the constants of the Sutherland potential, Eq. (7.16), can be determined by taking

$$\varphi_m r_m^4 = C_4 = -\tfrac{1}{2} Z_c^2\, C_e\, a_0^3\, \alpha_n^* \qquad [9.2]\ [9.6]$$

and[4]

$$r_m(A^+ + B) = \tfrac{1}{2}[r_0(A) + r_0(B)] \qquad (13.1a)$$

$$r_m(A^+ + A) = \sqrt{2}\, r_0(A). \qquad (13.1b)$$

The values for the polarizabilities α_n are listed in Table 4, Sect. 9, the values for the rigid sphere diameters r_0 in Table 7. Then the transfer cross section can be taken from Fig. 7, Sect. 7*η*.

---

[1] Mason, E.A. (1970): Planet. Space Sci. *18*, 137–144, fig. 1

[2] Milloy, H.B., Watts, R.O., Robson, R.E., Elford, M.T. (1974): Aust. J. Phys. *27*, 787–794, figs. 1 and 3

[3] Langevin, P. (1905): Ann. Chim. Phys. *5*, 245–288

[4] McDaniel, E.W.: See [6], table 9-2-2

Table 7. Well center r_m, zero r_0, and well depth $|\varphi_m|$ of the Lennard-Jones (12, 6) potential, Eq. (7.14) for neutral particles (after HIRSCHFELDER et al.[5])

Neutral particle	$\frac{r_m}{10^{-10}\,m}$	$\frac{r_0}{10^{-10}\,m}$	$\frac{\lvert\varphi_m\rvert}{10^{-3}\,eV}$	Neutral particle	$\frac{r_m}{10^{-10}\,m}$	$\frac{r_0}{10^{-10}\,m}$	$\frac{\lvert\varphi_m\rvert}{10^{-3}\,eV}$
H_2	3.148	2.915	3.27	NO	3.748	3.470	10.25
He	2,782	2.576	0.881	O_2	3.824	3.541	7.58
N_2	4.049	3.749	6.88	CO_2	4.189	3.897	18.35

γ) Mason-Schamp (12, 6, 4) potential. Of course, the rigid-sphere simulation of the inner part of the colliding particles is a rather crude approximation. An improvement is the use of a Lennard-Jones potential, Eq. (7.14) for this purpose. Combining a Lennard-Jones (n, n_1) potential with an attractive inverse power potential C_{n_0}/r^{n_0}, Eq. (6.11), leads to a Mason-Schamp (n, n_1, n_0) potential, Eq. (7.10). Its parameters $\varphi_m(\text{Mason-Schamp}) =: \varphi_{MS}$, $r_m(\text{Mason-Schamp}) =: r_{MS}$ and γ must be calculated from the parameters $\varphi_m(\text{Lennard-Jones}) =: \varphi_{LJ}$, $r_m(\text{Lennard-Jones}) =: r_{LJ}$ and C_{n_0}.

Comparison of terms with equal powers of r leads immediately to

$$\frac{r_{MS}}{r_{LJ}} = \left(\frac{2\gamma}{1+\gamma}\right)^{\frac{1}{n-n_1}} \tag{13.2a}$$

and

$$\frac{\varphi_{MS}}{-C_{n_0}/r_{LJ}^{n_0}} = n_0\left[\frac{1}{n}\frac{1+\gamma}{1-\gamma} - \frac{1}{n_1}\frac{2\gamma}{1-\gamma} - \frac{1}{n_0}\right]\left(\frac{1+\gamma}{2\gamma}\right)^{\frac{n_0}{n-n_1}}. \tag{13.2b}$$

The determination of γ is more involved. Comparing equal powers of r one deduces the algebraic equation

$$\left(\frac{2\gamma}{1+\gamma}\right)^{\frac{n+n_0}{n-n_1}} - \frac{1}{2}\left(\frac{2\gamma}{1+\gamma}\right)^{\frac{n_1+n_0}{n-n_1}} - \frac{n_0(n-n_1)}{2nn_1}\,\frac{C_{n_0}/r_{LJ}^{n_0}}{\varphi_{LJ}}\,\frac{2}{1-\gamma} = 0. \tag{13.3}$$

Its solution yields γ as a function of given $C_{n_0}/r_{LJ}^{n_0}\varphi_{LJ}$. The functional form depends on the powers n, n_1, n_0. With this solution the other Mason-Schamp parameters r_{MS} and φ_{MS} can be expressed by r_{LJ}, $C_{n_0}/r_{LJ}^{n_0}$, φ_{LJ}, Eqs. (13.2).

For $n_1 = n/2$ and $n_0 = n/3$ the solution $\gamma = \gamma(C_{n/3}/r_{LJ}^{n/3}\varphi_{LJ})$ is plotted in Fig. 19. For the Maxwell case $n/3 = 4$ the other powers are then to be taken as $n = 12$, $n_1 = 6$.

As for the Sutherland potential the parameter

$$C_4 = -\tfrac{1}{2} Z_c^2 C_e a_0^3 \alpha_n^* \qquad [9.2]\ [9.6]$$

can be determined by the value for the reduced polarizability α_n^* from Table 4, Sect. 9. The Lennard-Jones value

$$r_m(A^+ + B) = \tfrac{1}{2}[r_m(A) + r_m(B)] \tag{13.4}$$

for the collision of an ion A^+ with a neutral particle B has to be taken as the arithmetic mean of the Lennard-Jones values $r_m(A)$ and $r_m(B)$ (Table 7) if no other information is at hand[6]. For the

[5] Hirschfelder. J.O.. Curtiss. C.F.. Bird. R.B.: See [5]. appendix. table I-A
[6] Hirschfelder, J.O., Curtiss, C.F., Bird, R.B.: See [5], eqs. 3.6–8 and 3.6–9

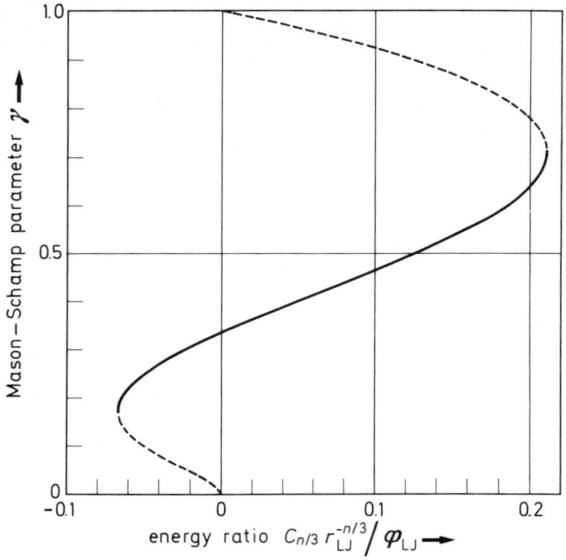

Fig. 19. Parameter γ of a MASON-SCHAMP $(n, n/2, n/3)$ potential, Eq. (7.10), obtained by combining – solving Eq. (13.3) – a LENNARD-JONES $(n, n/2)$ potential, Eq. (7.14) (with parameters φ_{LJ}, r_{LJ}) with an attractive inverse power potential $C_{n/3}/r^{n/3}$, Eq. (6.11)

well depths $-\varphi_m$ of the Lennard-Jones potential the geometric mean

$$-\varphi_m(A^+ + B) = \sqrt{\varphi_m(A)\,\varphi_m(B)} \tag{13.5}$$

may be taken if no better information is at hand[6]. The values of φ_m are also listed in Table 7. All the parameters of the Mason-Schamp (12, 6, 4) potential so being determined, the values for the transfer cross sections $Q^{(l)}(\mathscr{E})$ may then be taken from Fig. 5, Sect. 7δ.

As compared with the Sutherland potential (Subsect. β above) the Mason-Schamp (12, 6, 4) potential is a better approximation to real conditions. The price to pay for this improvement is the determination of the Mason-Schamp parameters which is quite involved, in particular that of γ.

14. Collisions between atomic ions and atoms. The description of the interaction potential for a $i+n$ collision by an (attractive) Maxwell part and another part (rigid spheres or Lennard-Jones type) for the inner range of the potential is too crude when collisions between lighter atoms, like H and He, are considered. Such particles, during their collisional encounter, form an intermediate molecule which must be described by its quantum-mechanical states.

The ground state of an intermediate molecule is usually a Σ^+ state. Its multiplicity $2S+1$, written as left superscript on the term symbol, indicates the degeneracy[1] due to the different directions of the spin S.

[1] Landau, L.D., Lifšic, E.M.: Kvantovaja mehanika. Moskva 1962. English transl.: Quantum mechanics. London: Pergamon 1977. German transl.: Quantenmechanik. Berlin: Akademie-Verlag 1965, §78

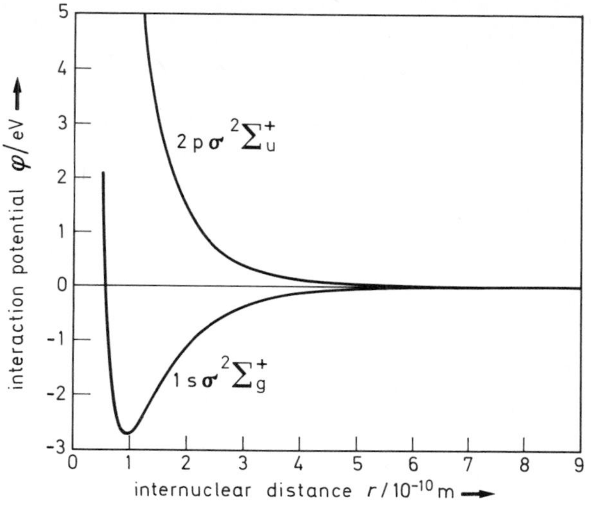

Fig. 20. Potential curves of the ground state $^2\Sigma^+$ of H_2^+ (after HERZBERG[6])

Another two-fold degeneracy occurs for all terms with $\Lambda \neq 0$ e.g., for the Π and Δ terms, due to the change of the sign of Λ upon reflection on a plane perpendicular to the symmetry axis of the (intermediate) molecule[2].

The cross sections corresponding to different states must be combined using a weighted mean[3]. The weights are proportional to the degeneracies, i.e., to the multiplicities for the Σ states and to twice the multiplicities for the other states[4].

If the intermediate molecule has two identical nuclei, which is the case for an atomic ion colliding with its corresponding "parent" atom (e.g., $H^+ + H$), a further degeneracy has to be taken into account, since the system "ion + atom" has the same energy as the system "atom + ion". This "resonance degeneracy" splits the potential curve into two branches when the separation distance r runs from $+\infty$ to lower values. The wave function of one of these branches has even *(gerade)* parity with respect to the coordinates of all electrons, and that for the other branch has odd *(ungerade)* parity. This parity (g, u) is indicated as a right subscript on the term symbol[5]. Usually one of the branches is mainly attractive, the other one repulsive (Fig. 20).

Both branches contain the same contribution caused by the polarizability of the colliding atom (Sect. 9). This contribution is described by an attractive

[2] Landau, L.D., Lifšic, E.M.: Kvantovaja mehanika. Moskva 1962. English transl.: Quantum mechanics. London: Pergamon 1977. German transl.: Quantenmechanik. Berlin: Akademie-Verlag 1965, §§79 and 88

[3] Mason, E.A., Vanderslice, J.T., Yos, J.M. (1959): Phys. Fluids *2*, 688–694

[4] McDaniel, E.W., Mason, E.A.: See [*23*], end of sect. 5–7B

[5] Landau, L.D., Lifšic, E.M.: Kvantovaja mehanika. Moskva 1962. English transl.: Quantum mechanics. London: Pergamon 1977. German transl.: Quantenmechanik. Berlin: Akademie-Verlag 1965, §§78 and 86

[6] Herzberg, G. (1950): Spectra of diatomic molecules, fig. 166. Princeton: van Nostrand

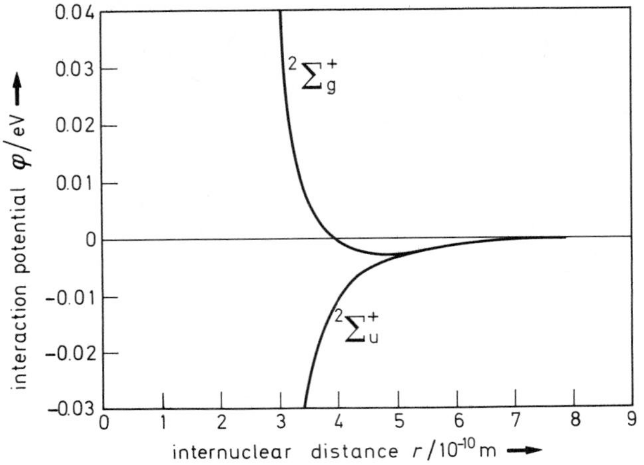

Fig. 21. The two branches of the interaction potential of $HeHe^+$ for large internuclear distances r (after HOLSTEIN[7]; denotation after MASON and VANDERSLICE[8])

Maxwell potential $C_4/r^4<0$, Eq. (9.2). It is important only for internuclear distances r large compared with the Bohr radius $a_0=5.292\cdot 10^{-11}$ m, Eq. (5.5a) (Fig. 21). There the potential curves of the two branches can be approximated by the functions

$$\varphi_{\text{attr}}^{\text{rep}}(r)=\pm\frac{1}{2}\varphi_{\Delta}\left(\frac{r}{r_{\Delta}}\right)^{\delta}\exp\left(1-\frac{r}{r_{\Delta}}\right)+\frac{C_4}{r^4}\qquad\text{for } r\gg a_0 \tag{14.1}$$

with $C_4<0$ from Eq. (9.2) and $\delta=0$ or 1. For $H^+ + H$ the value $\delta=1$ was obtained from quantum-mechanical calculations by LANDAU and LIFŠIC[9]; for $O^+ + O$ the potential curves were fitted with $\delta=0$ by YOS[10].

For all internuclear distances r a potential curve for an attractive branch can mostly be fitted by a Morse potential, Eq. (7.1) with parameters φ_m, r_m, α. In this case the corresponding repulsive branch can be represented by a Sato potential, Eq. (8.1) with parameters[11]

$$\varphi_R=-\tfrac{3}{2}\varphi_m \qquad r_R=r_m \qquad \alpha(\text{SATO})=\alpha(\text{MORSE}). \tag{14.2}$$

While the transfer cross sections for a Morse potential can be taken from Figs. 2 (Sect. 7α) no cross sections are available for Sato potentials. Therefore a repulsive exponential potential, Eq. (8.2), was used by MASON and VANDER-

[7] Holstein, T. (1952): J. Phys. Chem. *56*, 832–836, fig. 4

[8] Mason, E.A., Vanderslice, J.T. (1957): Phys. Rev. *180*, 293–294, fig. 1

[9] Landau, L.D., Lifšic, E.M.: Kvantovaja mehanika. Moskva 1962. English transl.: Quantum mechanics. London: Pergamon 1977. German transl.: Quantenmechanik. Berlin: Akademie-Verlag 1965, § 81, eq. 4

[10] Yos, J.M. (1965): AVCO Corp. Res. Rep. RAD-TR-65-7, eq. 115b

[11] Sato, S. (1955): J. Chem. Phys. *23*, 592–593, eq. 4

SLICE[12] to represent the repulsive branch with

$$\varphi_R = -2\varphi_m \exp(\alpha - 1) \qquad r_R = r_m/\alpha. \tag{14.3}$$

Instead of using these formulas, the parameters φ_R and r_R of the repulsive exponential potential are usually obtained directly by curve fitting of the repulsive branch. Only those parameters are listed in Table 8. Sometimes a screened Coulomb potential, Eq. (8.4), is used to approximate the repulsive branch.

Table 8. Approximation for states of (intermediate) systems ion + atom

	State	Potential	$\varphi_m, \varphi_R e^1, \varphi_\Delta e^1$ / eV	r_m, r_R, r_Δ / 10^{-10} m	$\alpha - \ln 2$, (n, n_γ), δ	$Q^{(l)}$	Authors
$H^+ + H$	$^2\Sigma_g^+$	Morse, Eq. (7.1)	−2.8	1.06	1.0	Fig. 2, Sect. 7α	Devoto
			−3.52	0.939	0.843		Ulmschneider
			−2.800	1.100	1.230		Biolsi
	$^2\Sigma_u^+$	expon., Eq. (8.2)	56.38	0.582		Fig. 8, Sect. 8β	Biolsi
$H^+ + He$	$X\,^1\Sigma^+$	Morse, Eq. (7.1)	−1.944	0.763	1.460	Fig. 2, Sect. 7α	Ulmschneider
			−1.905	0.762	1.230		Biolsi
		Lennard-Jones,	−2.04	0.83	(12, 4)		Whealton, Mason, Vu
		Eq. (7.14)	−2.040	0.774			Kolos and Peek
$He^+ + H$	$^1\Sigma$	expon., Eq. (8.2)	149.2	0.331		Fig. 8, Sect. 8β	Biolsi
	$^3\Sigma$		157.75	0.269			Biolsi
$He^+ + He$	$X\,^2\Sigma_u^+$	Morse, Eq. (7.1)	−2.16	1.080	1.637	Fig. 2, Sect. 7α	Mason and Vanderslice
			−2.7	1.06	1.52		Devoto and Lee
			−2.48				Sinha
	$^2\Sigma_g^+$	expon., Eq. (8.2)	179.7	0.344		Fig. 8, Sect. 8β	Devoto
$O^+ + O$	$^2\Sigma_{g,u}^+$	Eq. (14.1)	± 66.4	0.5522	0	Fig. 8, Sect. 8β	Yos
	$^4\Sigma_{g,u}^+$		±133.0	0.5522	0		Yos
	$^6\Sigma_{g,u}^+$		±199.4	0.5522	0		Yos
	$^2\Pi_{g,u}$		± 24.4	0.4866	0		Yos
	$^4\Pi_{g,u}$		± 49.0	0.4866	0		Yos
	$^6\Pi_{g,u}$		± 73.4	0.4866	0		Yos

Biolsi, L. (1978): J. Geophys. Res. *83*, 1125–1131
Devoto, R.S. (1968): J. Plasma Phys. *2*, 617–631, table 2
Devoto, R.S., Li, C.P. (1968): J. Plasma Phys. *2*, 17–32
Kolos, W., Peek, J.M. (1976): Chem. Phys. *12*, 381–386, table I
Mason, E.A., Vanderslice, J.T. (1958): J. Chem. Phys. *29*, 361–365, sect. IV
Sinha, S., Lin, S.L., Bardsley, J.N. (1979): J. Phys. B *12*, 1613–1622
Ulmschneider, P. (1970): Astron. Astrophys. *4*, 144–151
Whealton, J.H., Mason, E.A., Vu, T.H. (1974): Chem. Phys. Lett. *28*, 125–129
Yos, J.M. (1965): AVCO Corp. Res. Rep. RAD-TR-65-7, eq. 115b

[12] Mason, E.A., Vanderslice, J.T. (1958): J. Chem. Phys. *29*, 361–365, eq. 2

In Sect. 6, Fig. 1, it was shown that quantum effects should be of minor importance for ion-neutral collisions. DALGARNO and DICKINSON[13] have shown that quantum-mechanical values for $H^+ + He$ collisions differ not more than 20% from classical values with an attractive Maxwell potential. This result is illustrated by WHEALTON et al.[14].

15. Collisions of ions in their parent gas. Charge exchange. As was already pointed out in Sect. 14 there exists the phenomenon of "resonance degeneracy" for the collision of a neutral with an ion of the same species as the colliding neutral. This resonance degeneracy is the quantum-mechanical expression for the exchange of the charge between the particles during a collision. Such a charge exchange is, of course, also possible if an ion collides with a neutral of another species. But then additional energy is necessary to exchange the charges and thus the collision becomes inelastic. Only for ions in their parent gas is no additional energy necessary for charge exchange, the collisions remain elastic, and the phenomenon is called *resonant charge exchange.* A review was given by HASTED[1].

α) The quantum-mechanical *impact parameter method* that deals with the phenomenon of resonant charge exchange can only be sketched here. In Sect. 5 we described the transition from the quantum-mechanical expressions, Eq. (4.10), of the transfer cross sections for distinguishable particles to their classical limit, Eq. (5.3). The connection between the function $\chi(b, \mathscr{E})$, occurring in Eqs. (5.1) (5.3), was given by the integral in Eq. (4.6) for the trajectory of the classical path, where the integrand contains the interaction potential $\varphi(r)$.

Due to the resonance degeneracy the interaction potential is split into two branches $\varphi^+(r)$ and $\varphi^-(r)$ (Sect. 14), $\varphi^+(r)$ being symmetric in the nuclei and $\varphi^-(r)$ being antisymmetric. The corresponding phase shifts (Sect. 4α) are $\eta_L^+(|\boldsymbol{k}|)$ and $\eta_L^-(|\boldsymbol{k}|)$, respectively. The momentum transfer cross section is expressed by[2]

$$[Q^{(1)}(\mathscr{E})]_{\substack{n\\ n/2}} = \frac{s+1}{2s+1} Q^{\pm} + \frac{s}{2s+1} Q^{\mp} \tag{15.1a}$$

with

$$Q^{\pm} := \frac{4\pi}{|\boldsymbol{k}|^2} \Big[\sum_{\substack{L\,\text{odd}\\ L\,\text{even}}} (L+1) \sin^2(\eta_{L+1}^+ - \eta_L^-) + \sum_{\substack{L\,\text{even}\\ L\,\text{odd}}} (L+1) \sin^2(\eta_{L+1}^- - \eta_L^+) \Big], \tag{15.1b}$$

valid for particles with integer (half integer) spin $s=n$ ($s=n/2$). The "impact parameter method" uses a WKB approximation for the phase shifts $\eta_L^{\pm}(|\boldsymbol{k}|)$.

This leads to the expression[3]

$$\eta_{L+1}^{\pm} - \eta_L^{\mp} = \zeta^{\pm}(b, \mathscr{E}; \hbar^{-1}) + O(\hbar^0) \tag{15.2a}$$

---

[13] Dalgarno, A., Dickinson, A.S. (1968): Planet. Space Sci. *16*, 911–914

[14] Whealton, J.H., Mason, E.A., Vu, T.H. (1974): Chem. Phys. Lett. *28*, 125–129, fig. 1

[1] Hasted, J.B.: See [*32*]

[2] Dalgarno, A. (1958): Philos. Trans. R. Soc. London Ser. A *250*, 426–439, eqs. 1–3

[3] Dalgarno, A. (1958): Philos. Trans. R. Soc. London Ser. A *250*, 426–439, eq. 6

with

$$\zeta^{\pm}(b,\mathscr{E}) := \frac{1}{2\hbar} \int_{-\infty}^{+\infty} \mathrm{d}t^{\pm}(r)\,[\varphi^{+}(r)-\varphi^{-}(r)]. \tag{15.2b}$$

The difference between $\zeta^{+}(\mathscr{E})$ and $\zeta^{-}(\mathscr{E})$ stems from the difference in the trajectories determined by[4]

$$\mathrm{d}t^{\pm} = \frac{\mathrm{d}r}{\dot{r}^{\pm}} = \frac{\mathrm{d}r}{g\sqrt{1-\frac{b^2}{r^2}-\frac{\varphi^{\pm}(r)}{\mathscr{E}}}}.$$

In the high-energy limit $\mathscr{E} \gg \varphi^{\pm}(r)$ one has straight line trajectories and this difference vanishes.

The impact parameter method introduces a quantity $b(L) := (L+\frac{1}{2})/|\boldsymbol{k}|$ and replaces the summation over L in Eq. (15.1b) by an integration over $\mathrm{d}L \sim \mathrm{d}b$. One obtains[5]

$$Q^{(1)}(\mathscr{E}) \approx \pi \int_{0}^{\infty} \mathrm{d}b^2 [\sin^2 \zeta^{+}(b,\mathscr{E}) + \sin^2 \zeta^{-}(b,\mathscr{E})]. \tag{15.3}$$

Because of $\zeta \sim \hbar^{-1}$, Eq. (15.2b), expression (15.3) is not a classical limit, in contrast to the initial term $Q^{(1)}_{\text{class}}(\mathscr{E})$, Eq. (5.3), of the de Boer-Bird expansion, Eq. (5.2). This clearly shows that the resonance degeneracy, i.e., resonance charge exchange, is a typical quantum effect.

If instead of the momentum transfer cross section

$$Q^{(1)}(\mathscr{E}) = \oint \mathrm{d}\Omega\, q(\Omega,\mathscr{E})\,[1-\mathrm{P}_1(\cos\chi)] \tag*{[4.8]}$$

the integrated cross section

$$Q^{\mathrm{ex}}(\mathscr{E}) = \oint \mathrm{d}\Omega\, q(\Omega,\mathscr{E}) \tag*{[4.7]}$$

is calculated, the result is[6]

$$Q^{\mathrm{ex}}(\mathscr{E}) \approx \pi \int_{0}^{\infty} \mathrm{d}b^2\, \tfrac{1}{2} [\sin^2 \zeta^{+}(b,\mathscr{E}) + \sin^2 \zeta^{-}(b,\mathscr{E})] = \tfrac{1}{2} Q^{(1)}(\mathscr{E}). \tag{15.4}$$

It can be shown that $\sin^2 \zeta^{\pm}(b,\mathscr{E})$ are the probabilities for charge exchange, if the trajectories are governed by the branches $\varphi^{\pm}(r)$ of the potential curve[7].

β) Up to a certain *critical impact parameter* $b_c^{\pm}(\mathscr{E})$ the probabilities $\sin^2 \zeta^{\pm}(b,\mathscr{E})$ oscillate rapidly for all values of b. The mean values are approximately 1/2 for both $\sin^2 \zeta^{+}$ and $\sin^2 \zeta^{-}$ in the range $0 < b < b_c^{\pm}$. For $b > b_c^{\pm}$ both functions decrease rapidly to zero for increasing b. This indicates that only collisions with impact parameters b below $b_c^{\pm}(\mathscr{E})$ contribute to the charge

[4] Mason, E.A., Vanderslice, J.T., Yos, J.M. (1959): Phys. Fluids *2*, 688–694, eq. 23
[5] Dalgarno, A. (1958): Philos. Trans. R. Soc. London Ser. A *250*, 426–439, eq. 4
[6] Dalgarno, A. (1958): Philos. Trans. R. Soc. London Ser. A *250*, 426–439, eq. 10
[7] Holstein, T. (1952): J. Phys. Chem. *56*, 832–836, eq. 10

exchange process, with an average probability of about one. For collisions with $b > b_c^{\pm}(\mathscr{E})$ charge exchange is very ineffective:

$$Q^{ex}(\mathscr{E}) \approx \frac{\pi}{4}[(b_c^+)^2 + (b_c^-)^2] + \frac{\pi}{2} \int_{(b_c^+)^2}^{\infty} db^2 \sin^2 \zeta^+ + \frac{\pi}{2} \int_{(b_c^-)^2}^{\infty} db^2 \sin^2 \zeta^- . \qquad (15.5a)$$

According to the considerations described above, the critical impact parameter $b_c^{\pm}(\mathscr{E})$ is defined implicitly by the largest root of $\sin^2 \zeta^{\pm}(b, \mathscr{E}) - \frac{1}{2} = 0$, i.e., of

$$|\zeta^{\pm}(b_c^{\pm}, \mathscr{E})| - \frac{\pi}{4} = 0. \qquad (15.5b)$$

To avoid the integrations in Eq. (15.5a), an *equivalent impact parameter* $b_e^{\pm}(\mathscr{E})$ is chosen somewhat higher than $b_c^{\pm}(\mathscr{E})$. Instead of the expression (15.5a) for $Q^{ex}(\mathscr{E})$, the approximation

$$Q^{ex}(\mathscr{E}) \approx \frac{\pi}{4}[(b_e^+)^2 + (b_e^-)^2] \qquad (15.6a)$$

is then used and $b_e^{\pm}$ is determined by the largest root of[8]

$$|\zeta^{\pm}(b_e^{\pm}, \mathscr{E})| - \frac{\pi}{6} = 0. \qquad (15.6b)$$

With this procedure one puts[8]

$$\int_{(b_c^{\pm})^2}^{\infty} db^2 \sin^2 \zeta^{\pm} \approx (b_e^{\pm})^2 - (b_c^{\pm})^2 .$$

To find $b_c^{\pm}(\mathscr{E})$ or $b_e^{\pm}(\mathscr{E})$, one has to evaluate the integral $\int dt^{\pm}(r)[\varphi^+(r) - \varphi^-(r)]$, Eq. (15.2b), for $\zeta^{\pm}(b, \mathscr{E})$ and then to solve Eqs. (15.5b) or (15.6b) for $b_c^{\pm}$ or $b_e^{\pm}$, respectively.

In many cases the critical impact parameters $b_c^{\pm}(\mathscr{E})$ are so large that for $b > b_c^{\pm}(\mathscr{E})$ the deflection is very small, i.e., the trajectories are almost straight lines. In this *high-energy approximation* one has

$$dt^+ = dt^- = \frac{dr}{\dot{r}} = \frac{dr}{g\sqrt{1 - \frac{b^2}{r^2}}} . \qquad [4.6a]$$

Thus the difference between ζ^+ and ζ^- vanishes and therefore $b_c^+ - b_c^-$ too[9]:

$$\zeta(b, \mathscr{E}) \approx \frac{1}{\hbar} \int_b^{\infty} dr \frac{\varphi^+(r) - \varphi^-(r)}{g\sqrt{1 - \frac{b^2}{r^2}}} . \qquad (15.7)$$

STUBBE[10] has shown by a semiclassical procedure that for low energies $\mathscr{E} \lesssim 0.1$ eV the curvature of the trajectories enhances $Q^{ex}(\mathscr{E})$ more and more, leading to an increase of about a factor of two for $\mathscr{E} \approx 0.01$ eV; see also SINHA and BARDSLEY[11].

[8] Rapp. D., Francis. W.E. (1962): J. Chem. Phys. *37*. 2631–2645. eq. 13 and fig. 1

[9] Bates. D.R., Massey. H.S.W., Stewart. A.L. (1953): Proc. R. Soc. London Ser. A *216*, 437–458, eq. 134

[10] Stubbe. P. (1968): J. Atmos. Terr. Phys. *30*. 1965–1985

[11] Sinha. S., Bardsley. J.N. (1976): Phys. Rev. *14*. 104–113. sects. III and VII

γ) Cross sections and equivalent impact parameter. The potential curves $\varphi^{\pm}(r)$ represent an attractive branch and a repulsive branch (Fig. 21, Sect. 14). The attractive branch can often be fitted by a Morse potential, Eq. (7.1) with parameters φ_m, r_m, α. The corresponding repulsive branch can then be represented by a Sato potential[12], Eq. (8.1) with parameters $\varphi_R = -\frac{3}{2}\varphi_m$, $r_R = r_m$, $\alpha(\text{Sato}) = \alpha(\text{Morse})$. The difference of the two potential branches so approximated is[13,14]

$$\Delta\varphi(r) = \varphi(\text{Sato}) - \varphi(\text{Morse})$$
$$= \tfrac{1}{2}\varphi_m \exp\left[2\alpha\left(1-\frac{r}{r_m}\right)\right] - 3\varphi_m \exp\left[\alpha\left(1-\frac{r}{r_m}\right)\right]. \tag{15.8}$$

For large impact parameters b the difference $\Delta\varphi(r)$ is often approximated by[15]

$$\Delta\varphi(r) = \varphi_\Delta \left(\frac{r}{r_\Delta}\right)^{\delta} \exp\left(1-\frac{r}{r_\Delta}\right) \tag{15.9}$$

[see also Eq. (14.1)]. For $\delta = 0$ this leads in the high-energy approximation, Eq. (15.7) to[16]

$$\zeta(b, \mathscr{E}) \approx \frac{\varphi_\Delta e^1 b}{\hbar g} K_1\left(\frac{b}{r_\Delta}\right) \tag{15.10}$$

with $K_1(b/r_\Delta)$ as a modified Bessel function[17]. For $\delta \neq 0$ the high-energy result can be obtained by applying the operator[18]

$$\left(-\frac{1}{r_\Delta}\right)^{\delta} \left(\frac{d}{dr_\Delta^{-1}}\right)^{\delta} = \left(-\frac{b}{r_\Delta}\right)^{\delta} \left(\frac{d}{d(b r_\Delta^{-1})}\right)^{\delta} \tag{15.11}$$

to Eq. (15.10) for $\delta = 0$. This leads to the more general expression[19]

$$\zeta(b, \mathscr{E}) \approx \frac{\varphi_\Delta e^1 b}{\hbar g} \left(\frac{b}{2r_\Delta}\right)^{\delta} \sum_{l=0}^{\delta} \binom{\delta}{l} K_{1-\delta-2l}\left(\frac{b}{r_\Delta}\right). \tag{15.12}$$

Since we are dealing in the high-energy approximations with impact parameters b above the critical impact parameters $b_c^{\pm}(\mathscr{E})$ which are usually rather large, the b values are very large compared with r_Δ. For large arguments of the

---

[12] Sato. S. (1955): J. Chem. Phys. *23*. 592–593. eq. 4

[13] Flannery. M.R.. Cosby. P.C.. Moran. T.F. (1973): J. Chem. Phys. *59*. 5494–5510. eq. 13

[14] Moran. T.F.. Flannery. M.R.. Cosby. P.C. (1974): J. Chem. Phys. *61*. 1261–1273. eq. 9

[15] Bardsley. J.N.. Holstein. T.. Junker. B.R.. Sinha. S. (1975): Phys. Rev. A*11*. 1911–1920. eq. 37

[16] Dalgarno. A. (1958): Philos. Trans. R. Soc. London Ser. A *250*. 426–439. eq. 18

[17] Abramowitz. M.. Stegun. I. (eds.) (1964): Handbook of mathematical functions. sect. 9.6. New York: Dover

[18] Rapp. D.. Francis. W.E. (1962): J. Chem. Phys. *37*. 2631–2645. eq. 8

[19] Abramowitz, M., Stegun, J. (eds.) (1964): Handbook. of mathematical functions, eq. 9.6.29. New York: Dover

modified Bessel functions $K_\nu(z)$ the leading term $\sqrt{\pi/2z}\exp -z$ of the asymptotic expansion is independent[20] of the order ν. Therefore we obtain[21]

$$\zeta(b,\mathscr{E})\approx\frac{\varphi_\Delta e^1 b}{\hbar g}\left(\frac{b}{r_\Delta}\right)^\delta\left(\frac{\pi r_\Delta}{2b}\right)^{\frac{1}{2}}\exp\left(-\frac{b}{r_\Delta}\right)=\frac{\sqrt{\pi b r_\Delta/2}}{\hbar g}\Delta\varphi(b). \tag{15.13}$$

The function $\zeta(b,\mathscr{E})$ increases exponentially with decreasing b, thus $\sin\zeta$ oscillates rapidly for small b. The equivalent impact parameter $b_e(\mathscr{E})$ is determined with Eqs. (15.6b) and (15.13) by the largest root of

$$\frac{\varphi_\Delta e^1 r_\Delta}{\hbar g}\left(\frac{b_e}{r_\Delta}\right)^{\delta+\frac{1}{2}}\left(\frac{\pi}{2}\right)^{\frac{1}{2}}\exp\left(-\frac{b_e}{r_\Delta}\right)-\frac{\pi}{6}=0. \tag{15.14}$$

To solve Eq. (15.14) for b_e in a limited energy range, one replaces b_e in the pre-exponential factors by an average value $\overline{b_e}$ and obtains

$$\frac{b_e(\mathscr{E})}{r_\Delta}=\ln\left[3\left(\frac{2}{\pi}\right)^{\frac{1}{2}}\left(\frac{\overline{b_e}}{r_\Delta}\right)^{\delta+\frac{1}{2}}\frac{\varphi_\Delta e^1 r_\Delta}{\hbar g}\right]=\ln\sqrt{\frac{\mathscr{E}_\Delta}{\mathscr{E}}} \tag{15.15a}$$

with the reference energy

$$\mathscr{E}_\Delta:=\frac{9}{\pi}\left(\frac{\overline{b_e}}{r_\Delta}\right)^{2\delta+1}\mu\left(\frac{\varphi_\Delta e^1 r_\Delta}{\hbar}\right)^2. \tag{15.15b}$$

The range of validity was found by YOS[22] to be

$$5.6\leqq b_e/r_\Delta\leqq 11.2,\qquad \text{i.e.}\quad 0.0037\leqq\frac{\mathscr{E}}{\mathscr{E}_\Delta}\leqq 0.061. \tag{15.15c}$$

An equivalent for $b_e(\mathscr{E})$ can be obtained with a Morse potential, Eq. (7.1) with parameters φ_m, r_m, α, for the attractive branch and an repulsive exponential potential Eq. (8.2) with parameters

$$\varphi_R=-2\varphi_m\exp(\alpha-1)\qquad r_R=r_m/\alpha \tag{14.3}$$

for the repulsive branch:[23]

$$\frac{b_e(\mathscr{E})}{r_m}=\frac{\sqrt{2}}{\alpha}\left[C(\alpha)+1.7204\log_{10}\frac{|\varphi_m|\,r_m}{\hbar g}\right]=\frac{\sqrt{2}}{\alpha}0.7472\left[\frac{C(\alpha)}{0.7472}+\ln\frac{|\varphi_m|\,r_m}{\hbar g}\right] \tag{15.16a}$$

for

$$\frac{\hbar g}{|\varphi_m|\,r_m}<5000. \tag{15.16b}$$

The function $C(\alpha)$ varies[23] between 3.03 and 4.80 for α between 1 and 5.

The expression (15.16a) can be brought into a form analogous to Eq. (15.15a), viz.,

$$b_e(\mathscr{E})=r_\alpha\ln\sqrt{\frac{\mathscr{E}_\alpha}{\mathscr{E}}}\qquad\text{for}\quad\frac{\mathscr{E}}{\mathscr{E}_\alpha}<1000 \tag{15.16c}$$

[20] Abramowitz, M., Stegun, J. (eds.) (1964): Handbook of mathematical functions, eq. 9.7.2. New York: Dover

[21] Hodgkinson, D.P., Briggs, J.S. (1976): J. Phys. B 9, 225–267, eqs. 14 and 17

[22] Yos, J.M. (1965): AVCO Corp. Res. Rep. RAD-TR-65-7, eq. 118a

[23] Mason, E.A., Vanderslice, J.T. (1958): J. Chem. Phys. *29*, 361–365, eq. 8 and table 1

with the normalization length r_α and normalization energy $\mathscr{E}_\alpha$ defined by

$$r_\alpha := 0.7472 \frac{\sqrt{2}}{\alpha} r_m \tag{15.16d}$$

$$\mathscr{E}_\alpha := \frac{\mu}{2} \left(\frac{\varphi_m r_m}{\hbar}\right)^2 \exp\left(\frac{2C(\alpha)}{0.7472}\right). \tag{15.16e}$$

If the potential difference $\varphi^+(r) - \varphi^-(r)$, Eq. (15.2b), is expressed by the difference φ(Sato) $-\varphi$(Morse), Eq. (15.8), the logarithmic representation for the equivalent impact parameter $b_e(\mathscr{E})$ is also obtained[24].

δ) Logarithmic energy dependence of the cross sections. If we put the logarithmic expression (15.15a) for $b_e(\mathscr{E})$ into Eq. (15.6a) for the charge exchange cross section $Q^{ex}(\mathscr{E})$, we obtain

$$2Q^{ex} = \pi b_e^2(\mathscr{E}) = \pi r_\Delta^2 (\ln \sqrt{\mathscr{E}_\Delta/\mathscr{E}})^2 \tag{15.17a}$$

$$= 2\left(A - B \log_{10} \frac{\mathscr{E}}{\text{eV}}\right)^2 \tag{15.17b}$$

with

$$A = \sqrt{\frac{\pi}{2}}\, r_\Delta \ln \frac{\mathscr{E}_\Delta}{\text{eV}} \qquad B = \sqrt{\frac{\pi}{8}}\, r_\Delta \ln 10. \tag{15.17c}$$

The results of measurements of $Q^{ex}(\mathscr{E})$ are usually expressed in the form of Eq. (15.17b). Table 9 shows measured values of A, B and the corresponding values of

$$r_\Delta = \sqrt{\frac{8}{\pi}} \frac{B}{\ln 10} \qquad \frac{\mathscr{E}_\Delta}{\text{eV}} = 10^{A/B}. \tag{15.17d}$$

Table 9. Parameters A and B, normalization length r_Δ, and normalization energy $\mathscr{E}_\Delta$ in the expressions $\sqrt{Q^{ex}} = A - B \log_{10}(\mathscr{E}/\text{eV}) = \sqrt{\pi/2}\, r_\Delta \ln \sqrt{\mathscr{E}_\Delta/\mathscr{E}}$, Eq. (15.17), for the charge exchange cross section $Q^{ex}(\mathscr{E})$

	$\frac{A}{10^{-10}\text{m}}$	$\frac{B}{10^{-10}\text{m}}$	$\frac{r_\Delta}{10^{-10}\text{m}}$	$\frac{\mathscr{E}_\Delta}{\text{eV}}$	Authors
H	7.60	1.06	0.735	$1.472 \cdot 10^7$	Stubbe
H_2	3.47	0.563	0.390	$1.457 \cdot 10^6$	Mason and Vanderslice
He	5.09	0.69	0.48	$2.381 \cdot 10^7$	Rundel et al.
O	5.95	0.63	0.44	$2.782 \cdot 10^9$	Stubbe
N_2	7.36	0.68	0.47	$6.666 \cdot 10^{10}$	Stubbe, Kobayashi
O_2	4.85	1.1	0.76	$2.565 \cdot 10^4$	Baer et al.

Baer, T., Murray, P.T., Squires, L. (1978): J. Chem. Phys. *68*, 4901–4906, fig. 4
Kobayashi, N. (1975): J. Phys. Soc. Japan *38*, 519–523, fig. 2
Mason, E.A., Vanderslice J.T. (1959): Phys. Rev. *114*, 497–502, eq. 18
Rundel, R.D., Nitz, D.E., Smith, K.A., Geis, M.W., Stebbings, R.F. (1979): Phys. Rev. A*19*, 33–42, sect. VI
Stubbe, P. (1968): J. Atmos. Terr. Phys. *30*, 1965–1985, table 1

[24] Moran, T.F., Flannery, M.R., Cosby, P.C. (1974): J. Chem. Phys. *61*, 1261–1273, fig. 7

ε) Transfer cross sections $Q^{(l)}(\mathscr{E})$ with l odd are the only ones to be influenced by charge exchange, as was shown by MASON et al.[25]. The authors used a semiclassical argument. It should be confirmed by a derivation of $Q^{(l)}(\mathscr{E})$ for $l \geqq 2$ from the rigorous quantum-mechanical expression in the same manner as the derivation of $Q^{(1)}(\mathscr{E})$, Eq. (15.3), from $[Q^{(1)}(\mathscr{E})]_n$, Eq. (15.1), was made.

16. Collisions of ions on their parent gas. Polarization. In Sect. 15 the influence of the resonant charge exchange on the momentum transfér cross section was discussed. To calculate this phenomenon, the ion + atom system was taken as an intermediate molecule. Due to resonance degeneracy each of its potential curves had to be splitted into an attractive branch and a repulsive branch.

It turned out that only collisions with impact parameters b up to a certain critical impact parameter $b_c(\mathscr{E})$ contributed essentially to the resonant charge exchange process. The contributions of highimpact parameters $b > b_c$ decreased exponentially. In this range of large values of the impact parameter the contribution of the polarization, i.e., the induced dipole force (attractive Maxwell potential) must now be considered. Figure 21, Sect. 14, shows the influence on both branches of the interaction potential.

The fact that the influence of charge exchange is very strong for $b < b_c$ and very weak for $b > b_c$ suggests that charge exchange can be completely neglected for $b > b_c$. For the influence of the polarization force the reverse assumption is made: neglect of $b < b_c$ in favour of charge exchange, but full consideration for $b > b_c$. The result would be[1]

$$Q^{(1)}(\mathscr{E}) = \pi b_c^2(\mathscr{E}) + \pi \int_{b_c^2}^{\infty} db^2 [1 - P_1(\cos\chi)]. \tag{16.1}$$

A somewhat refined and generalized expression is given by MASON et al.[2]:

$$Q^{(l)}(\mathscr{E}) = \pi b_c^2(\mathscr{E}) + \pi \int_{b_c^2}^{\infty} db^2 (1 - 2\sin^2\zeta)\,[1 - P_l(\cos\chi)] \quad \text{for } l \text{ odd.} \tag{16.2}$$

(A "unified" expression has been proposed by HODGKINSON[3].)

The integral $\int_{b_c^2}^{\infty} db^2 [1 - P_l(\cos\chi)]$ can be calculated classically with both of the two branches of the split potential curve. The arithmetic mean of the results should be taken. If not enough information is available on these two branches an attractive Maxwell potential, Eq. (9.2), may be supplemented with a suitable core[4]. For a Sutherland $(\infty, 4)$ potential, Eq. (7.16), STUBBE[1] made the calculations and evaluated them numerically.

The appearance, instead of zero, of the finite lower integration limit $b_c^2(\mathscr{E})$ in the integral of Eq. (16.1) complicates the calculations. Its replacement by zero[5,6] means a simple addition of the classical momentum transfer cross

---

[25] Mason, E.A., Vanderslice, J.T., Yos, J.M. (1959): Phys. Fluids *2*, 688–694, eq. 13

[1] Stubbe, P. (1968): J. Atmos. Terr. Phys. *30*, 1965–1985, eq. 16

[2] Mason, E.A., Vanderslice, J.T., Yos, J.M. (1959): Phys. Fluids *2*, 688–694, eq. 15

[3] Hodgkinson, D.P. (1975): Report T.P. 640, A.E.R.E., Harwell Didcot, October 1975

[4] Dalgarno, A. (1961): Planet. Space Sci. *3*, 217–220, sect. 4

[5] Devoto, R.S., Li, C.P. (1968): J. Plasma Phys. *2*, 17–32

section

$$Q^{(1)}_{\text{class}}(\mathscr{E}) = \pi \int_0^\infty db^2 [1 - P_1(\cos\chi)] = \pi \left(\frac{|C_4|}{\mathscr{E}}\right)^{\frac{1}{2}} Q^{(1)*}(4, -)$$

$$= 7.1439 \cdot 10^{-20} Z_c \left(\frac{\alpha_n^*}{\mathscr{E}/\text{eV}}\right)^{\frac{1}{2}} \text{m}^2 \qquad [6.19]$$

for the induced dipole potential, Eq. (9.1), with

$$C_4 = -\tfrac{1}{2} Z_c^2 C_e \alpha_n^* a_0^3 = -1.0595 \cdot 10^{-40} Z_c^2 \alpha_n^* \text{eV m}^4 \qquad [9.2]$$

and the momentum transfer cross section for charge exchange

$$2Q^{\text{ex}}(\mathscr{E}) = \pi b_e^2(\mathscr{E}) = \pi r_\Delta^2 \left(\ln\sqrt{\frac{\mathscr{E}_\Delta}{\mathscr{E}}}\right)^2 = 2\left(A - B \log_{10} \frac{\mathscr{E}}{\text{eV}}\right)^2, \qquad [15.17]$$

yielding

$$Q^{(1)} = Q^{(1)}_{\text{class}} + 2Q^{\text{ex}}. \qquad (16.3)$$

The values for $Q^{(1)*}(4, -)$, α_n^*, r_Δ, $\mathscr{E}_\Delta$, A, B can be taken from Tables 1 [Sect. 6β], 4 [Sect. 9], 9 [Sect. 15].

A modification of this simple addition has been proposed by Banks[7]:

$$Q^{(1)} = Q^{(1)}_{\text{class}}\, \theta(Q^{(1)}_{\text{class}} - 2Q^{\text{ex}}) + 2Q^{\text{ex}}\, \theta(2Q^{\text{ex}} - Q^{(1)}_{\text{class}}) \qquad (16.4\text{a})$$

with the Heaviside step function

$$\theta(x) = \begin{cases} 1 & \text{for } x > 0 \\ 0 & \text{for } x < 0. \end{cases} \qquad (16.4\text{b})$$

This expression for $Q^{(1)}(\mathscr{E})$ is noncontinuous at $Q^{(1)}_{\text{class}} = 2Q^{\text{ex}}$ and has the limit values $Q^{(1)}_{\text{class}}$ for low energies and $2Q^{\text{ex}}$ for high energies.

Wolf and Turner[8] write

$$Q^{(1)} = f Q^{(1)}_{\text{class}} + \frac{1}{Q^{\text{ex}}}\left(Q^{\text{ex}} - \frac{Q^{(1)}_{\text{class}}}{4}\right)^2 \theta\left(Q^{\text{ex}} - \frac{Q^{(1)}_{\text{class}}}{4}\right) \qquad (16.5)$$

and determine f from measurements.

For an analytic averaging of the transfer cross sections $Q^{(1)}(\mathscr{E})$ needed for the calculation of transport collision frequencies (Sect. 19α), the Heaviside step functions in Eqs. (16.4) and (16.5) lead to considerable complications. On the other side, the simple addition expression (16.3) may be too crude for certain requirements. In this case it is convenient to make the analytic energy averaging of $Q^{(1)}_{\text{class}}(\mathscr{E})$ and $2Q^{\text{ex}}(\mathscr{E})$ separately and afterwards combine the results similar to Eq. (16.4a). To avoid a noncontinuity in the final result, one can apply a smoothing procedure as proposed by Banks[7].

Banks[7] as well as Wolf and Turner[8] use an attractive Maxwell potential, Eq. (9.2) for $Q^{(1)}_{\text{class}}(\mathscr{E})$. For collisions of light ions in their parent gas other potential functions have been adapted, see Table 8, Sect. 14.

---

6 Ulmschneider, P. (1970): Astron. Astrophys. *4*, 144–151, sect. 3.B4

7 Banks, P. (1966): Planet. Space Sci. *14*, 1105–1122, sect. III.2

8 Wolf, F.A., Turner, B.R. (1968): J. Chem. Phys. *48*, 4226–4233, eqs. 11 and 12

17. Collisions between neutrals. As in the case of ion-neutral collisions, during the collision time the two colliding neutral particles should be taken as an intermediate molecule. This is particularly true for atom-atom collisions including the case of collisions between atoms of the same species.

In recent years potential curves of the different states of diatomic molecules were calculated by quantum-mechanicals methods (e.g., NANBU[1]). These values were fitted with analytic potentials as discussed in Sects. 7 and 8. For attractive states, Morse potentials, Eq. (7.1), and Buckingham (exp β, 6) potentials, Eq. (7.5), were usually taken, while for repulsive states, exponential potentials after Eq. (8.2) are adequate. They are listed in Table 10. The ground state for He+He collisions may have – besides a repulsive core – a repulsive tail, too, with an intermediate attractive part[2]. Curves of this shape could be approximated by a Mason-Schamp potential, Eq. (7.10) with $\gamma > 1$ (Fig. 4, Sect. 7δ).

The parameters for collisions between unlike neutrals not listed in Table 10 can be calculated from those for collisions between neutrals of the same species using *empirical combination laws.* For Lennard-Jones (12, 6) potentials, Eq. (7.14), these laws are[3]

$$r_{\mathrm{m}}(\mathrm{A+B}) = \tfrac{1}{2}[r_{\mathrm{m}}(\mathrm{A+A}) + r_{\mathrm{m}}(\mathrm{B+B})] \tag{17.1}$$

$$-\varphi_{\mathrm{m}}(\mathrm{A+B}) = [\varphi_{\mathrm{m}}(\mathrm{A+A})\,\varphi_{\mathrm{m}}(\mathrm{B+B})]^{\frac{1}{2}}. \tag{17.2}$$

They relate the parameters between unlike particles A and B to those between particles of same species A+A and B+B. Of course, the using of parameters which were calculated with these empirical laws must be regarded as provisional; it is much less justified than when the parameters are obtained by curve fitting. Another combination law is described by KIHARA[4].

Table 10. Approximations for states of intermediate molecules

	State	Potential	$\varphi_{\mathrm{m}}, \varphi_{\mathrm{R}}\mathrm{e}^1$ / eV	$r_{\mathrm{m}}, r_{\mathrm{R}}$ / 10^{-10} m	$\alpha - \ln 2$, β, (n, n_γ)	$Q^{(l)}$	Authors
H+H	$\mathrm{X}\,{}^1\Sigma_g^+$	Morse,	−4.747	0.740	1.0	Fig. 2,	Devoto
		Eq. (7.1)	−4.750	0.7412	0.7282	Sect. 7α	Kafri
	$\mathrm{b}\,{}^3\Sigma_u^+$	expon.,	60.42	0.3319		Fig. 8,	Vanderslice
$H+H_2$		Eq. (8.2)	61.5	0.3388		Sect. 8β	et al.
H+He	$\mathrm{X}\,{}^2\Sigma^+$	Buckingham	$-3.9\cdot10^{-4}$	3.59	12.145	Fig. 3,	Gengenbach
		(exp β, 6),				Sect. 7β	et al.
		Eq. (7.5)					
$H+N_2$		expon.,	540	0.216		Fig. 8,	Leonas
$H+O_2$		Eq. (8.2)	330	0.211		Sect. 8β	Leonas
H_2+H_2		(exp β, 6)	$-31.37\cdot10^{-4}$	3.43	11.1	Fig. 3	Ree
		(exp β, 8)	$-1.79\cdot10^{-4}$	4.26	15.7		Ree and Bender

[1] Nanbu. K. (1974): J. Chem. Phys. *61*. 2189–2192
[2] Andresen. B.. Kuppermann. A. (1975): Mol. Phys. *30*. 997–1004. fig. 2
[3] Hirschfelder, J.O., Curtiss, C.F., Bird, R.B.: See [5], eqs. 3.6–8 and 3.6–9
[4] Kihara, T. (1978): Intermolecular forces, eqs. 6.28 and 6.29. New York: Wiley & Sons

Table 10 (continued)

	State	Potential	$\varphi_m, \varphi_R e^1$ / eV	r_m, r_R / 10^{-10}m	$\alpha - \ln 2$, $\beta, (n, n_\gamma)$	$Q^{(l)}$	Authors
H_2+He		(exp β, 6)	$-13.36 \cdot 10^{-4}$	3.37	12.7	Fig. 3,	Ree
He+He	$X\,^1\Sigma_g^+$	(exp β, 6)	$-9.11 \cdot 10^{-4}$	2.97	13.6	Sect. 7 β	
		Morse	$-9.45 \cdot 10^{-4}$	2.963	5.497	Fig. 2	Farrar and Lee
He+O		Le.-Jones	$-2.48 \cdot 10^{-3}$	3.08	(12, 6)	Fig. 5	Aquilanti et al.
$He+N_2$		(exp β, 6)	$-1.62 \cdot 10^{-3}$	3.83	15.15	Fig. 3	Habitz et al.
$He+O_2$		Morse	$-2.52 \cdot 10^{-3}$	3.52	5.447	Fig. 2	Faubel et al.
$He+CO_2$		Le.-Jones	$-3.55 \cdot 10^{-3}$	3.5	(12, 6)	Fig. 5	Buck
O+O	$X\,^3\Sigma_g^-$ I	Morse,	-3.81	1.30	3.346	Fig. 2,	Capitelli
	$a\,^1\Delta_g$ I	Eq. (7.1)	-2.81	1.33	3.582	Sect. 7 α	and
	$b\,^1\Sigma_g^+$ I		-2.44	1.34	3.636		Ficocelli
	$c\,^1\Sigma_u^-$ I		-0.90	1.56	5.105		
	$C\,^3\Delta_u$ I		-0.67	1.55	7.798		
	$A\,^3\Sigma_u^+$ I		-0.61	1.56	6.23		
	$^1\Sigma_g^+$ II	expon.	+1040	0.2998		Fig. 8,	Capitelli
	$^3\Sigma_u^+$ II	Eq. (8.2)	+ 803	0.3118		Sect. 8 β	and
	$^5\Sigma_g^+$ I		+3414	0.2349			Ficocelli
	$^5\Sigma_g^+$ II		+4856	0.2431			
	$^5\Sigma_u^-$		+1649	0.2508			
	$^1\Pi_g$ I		+ 587	0.2963			
	$^1\Pi_u$ I		+ 821	0.2731			
	$^3\Pi_g$ I		+ 158	0.3652			
	$^3\Pi_u$ I		+ 427	0.2893			
	$^5\Pi_g$		+5367	0.1968			
	$^5\Pi_u$		+4551	0.2261			
	$^5\Delta_g$		+3527	0.2333			
$O+N_2$		expon.,	$+2.88 \cdot 10^{+3}$	0.1912		Fig. 8,	Leonas
$O+O_2$		Eq. (8.2)	$2.56 \cdot 10^{+3}$	0.200		Sect. 8 β	Leonas
$O+CO_2$			$1.00 \cdot 10^{+3}$	0.246			Leonas
N_2+N_2		(exp β, 6),	$-8.72 \cdot 10^{-3}$	4.011	17	Fig. 3,	Yun and Mason
N_2+O_2		Eq. (7.5)	$-1.00 \cdot 10^{-2}$	3.861	17	Sect. 7 β	Yun and Mason
N_2+CO_2		expon.	$1.17 \cdot 10^{+3}$	0.246		Fig. 8	Leonas
O_2+O_2		(exp β, 6)	$-1.14 \cdot 10^{-2}$	3.726	17	Fig. 3	Yun and Mason
O_2+CO_2		expon.	490	0.279		Fig. 8	Leonas
CO_2+CO_2		(exp β, 6)	$-1.605 \cdot 10^{-2}$	5.48	15	Fig. 3	Pandey et al.

Aquilanti, V., Liuti, G., Pirani, F., Vecchiocattivi, F., Volpi, G.G. (1976): J. Chem. Phys. *65*, 4751–4755, table II

Buck, U. (1975): Adv. Chem. Phys. *30*, 313–388, table VIII

Capitelli, M., Ficocelli, E. (1972): J. Phys. B*5*, 2066–2073, table 1

Devoto, R.S. (1968): J. Plasma Phys. *2*, 617–631, table 2

Farrar, J.M., Lee, Y.T. (1972): J. Chem. Phys. *56*, 5801–5807, p. 5805

Faubel, M., Kohl, K.H., Toennies, J.P., Gianturco, F.A. (1983): J. Chem. Phys. *78*, 5629–5636, table I

Gengenbach, R., Hahn, C., Toennies, J.P. (1973): Phys. Rev. A*7*, 98–103, table II

Habitz, P., Tang, K.T., Toennies, J.P. (1982): Chem. Phys. Lett. *85*, 461–466

Kafri, O. (1979): Chem. Phys. Lett. *61*, 538–541

Leonas, V.B. (1973): Sov. Phys. Usp. *15*, 266–281, tables IV and V

Pandey, L., Reddy, C.P.K., Sarkar, K.L. (1983): Canad. J. Phys. *61*, 664–670, table 1

Ree, F.H. (1983): J. Chem. Phys. *78*, 409–415

Ree, F.H., Bender, C.F. (1979): J. Chem. Phys. *71*, 5362–5375, eq. 15

Vanderslice, J.T., Weissman, S., Mason, E.A., Fallon, R.J. (1962): Phys. Fluids *5*, 155–164, table I

Yun, K.S., Mason, E.A. (1962): Phys. Fluids *5*, 380–386, tables I and III

A combination law for Buckingham (exp β, 6) potentials, Eq. (7.5) is given by MASON and RICE[5]. Combination laws for various interaction potentials are discussed by MASON and MARRERO[6] and also by DIAZ, PENA, PANDO, and RENUNCIO[7] as well as MAITLAND, RIGBY, SMITH, and WAKEHAM[8].

For H+H and He+He collisions the ground state is split into two branches, for O+O collisions into eighteen. The transfer cross sections $Q^{(l)}(\mathscr{E})$ have to be calculated separately for each potential curve (Table 10). A weighted mean must then be taken. The statistical weight of a molecular state is proportional to its degeneracy. This latter equals the spin multiplicity for Σ states (given as superscripts on the Σ symbol) but is twice the multiplicity for other states (Sect. 14). For H + H collisions the transfer cross sections for $^1\Sigma$ and $^3\Sigma$ states are therefore weighted with 1/4 and 3/4, respectively. For O + O collisions the transfer cross sections for the $^3\Sigma$ and the $^1\Delta$ states are weighted with 3/84 and 2/84, respectively.

18. Analytic expressions and normalizations for transfer cross sections and transfer collision frequencies

α) General considerations. With the exception of inverse powers potentials $\varphi(r) = C_n/r^n$, Eq. (6.11), we were able to normalize the collision energy values E of all transfer cross sections $Q^{(l)}(\mathscr{E})$ with a reference energy $\mathscr{E}_R$, thus working with a dimensionless energy variable $\mathscr{E}^* := \mathscr{E}/\mathscr{E}_R$, Eq. (6.4). With the same exception we could normalize all transfer cross sections $Q^{(l)}(\mathscr{E})$ themselves with a reference length r_R, which is independent of $\mathscr{E}$, thus introducing dimensionless transfer cross sections

$$Q^{(l)*}(\mathscr{E}^*) := \frac{Q^{(l)}(\mathscr{E})}{\pi r_R^2}. \qquad [6.7]$$

For the transfer collision frequencies

$$\nu_{jk}^{(l)}(\mathscr{E}_{jk}) := g_{jk} Q_{jk}^{(l)}(\mathscr{E}_{jk}) N_k = \sqrt{2\mathscr{E}_{jk}/\mu_{jk}}\, Q_{jk}^{(l)}(\mathscr{E}_{jk}) N_k \qquad [4.14]$$

the same possibilities for normalizations hold. With a reference energy $\mathscr{E}_R$ we can further introduce a reference relative speed

$$g_R := \sqrt{2\mathscr{E}_R/\mu} \qquad (18.1\text{a})$$

for the normalization

$$g^* := \frac{g}{g_R} = \sqrt{\frac{\mathscr{E}}{\mathscr{E}_R}} = \sqrt{\mathscr{E}^*} \qquad (18.1\text{b})$$

of the relative speed $g_{jk} = \sqrt{2\mathscr{E}_{jk}/\mu_{jk}}$, Eq. (4.1).

---

[5] Mason. E.A.. Rice. W.E. (1954): J. Chem. Phys. *22*. 522–535. sect. C
[6] Mason, E.A., Marrero, T.R.: See [*14*], sect. K. 3
[7] Diaz Pena, M., Pando, C., Renuncio, J.A.R. (1982): J. Chem. Phys. *76*, 325–332, 333–339
[8] Maitland, G.C., Rigby, M., Smith, E.B., Wakeham, W.A.: See [*35*], sect. 9.9.2

If there exists a reference length r_{R} independent of $\mathscr{E}$ one can introduce a dimensionless transfer collision frequency

$$\nu_{jk}^{(l)*}(\mathscr{E}^*) := \frac{\nu_{jk}^{(l)}(\mathscr{E})}{\sqrt{2\mathscr{E}_{\mathrm{R}}/\mu_{jk}}\,\pi r_{\mathrm{R}}^2 N_k} = \sqrt{\mathscr{E}^*}\,Q^{(l)*}(\mathscr{E}^*). \tag{18.2}$$

Except those for inverse power potentials, all analytic expressions for transfer cross sections $Q^{(l)}(\mathscr{E})$ and transfer collision frequencies $\nu^{(l)}(\mathscr{E})$ in this contribution can be written in the forms

$$Q^{(l)}(\mathscr{E}) = \sum_p Q_p^{(l)} \mathscr{E}^{*p} + Q_{p_l,\ln}^{(l)} \mathscr{E}^{*p_l} \ln \mathscr{E}^* + Q_{\ln^2}^{(l)} (\ln \mathscr{E}^*)^2 \tag{18.3}$$

and

$$\nu^{(l)}(\mathscr{E}) = \sum_p \nu_p^{(l)} \mathscr{E}^{*p+\frac{1}{2}} + \nu_{p_l,\ln}^{(l)} \mathscr{E}^{*p_l+\frac{1}{2}} \ln \mathscr{E}^* + \nu_{\ln^2}^{(l)} \mathscr{E}^{*\frac{1}{2}} (\ln \mathscr{E}^*)^2 \tag{18.4a}$$

$$\nu^{(l)}(g) = \sum_p \nu_p^{(l)} g^{*2p+1} + 2\nu_{p_l,\ln}^{(l)} g^{*2p_l+1} \ln g^* + 4\nu_{\ln^2}^{(l)} g^* (\ln g^*)^2 \tag{18.4b}$$

with the following relation between the coefficients $Q_p^{(l)}$ etc. of Eq. (18.3) and $\nu_p^{(l)}$ etc. of Eq. (18.4):

$$\nu_p^{(l)} = g_{\mathrm{R}} Q_p^{(l)} N_k = \sqrt{2\mathscr{E}_{\mathrm{R}}/\mu}\, Q_p^{(l)} N_k \qquad \text{etc.} \tag{18.5}$$

The powers p and p_l can be integers, half-integers, or rational numbers, positive or negative.

If there exists a reference length r_{R} independent of $\mathscr{E}$, the expression (18.3) for $Q^{(l)}(\mathscr{E})$ divided by πr_{R}^2 reads

$$Q^{(l)*}(\mathscr{E}^*) = \sum_p Q_p^{(l)*} \mathscr{E}^{*p} + Q_{p_l,\ln}^{(l)*} \mathscr{E}^{*p_l} \ln \mathscr{E}^* + Q_{\ln^2}^{(l)*} (\ln \mathscr{E}^*)^2 \tag{18.6}$$

with dimensionless coefficients $Q_p^{(l)*}$ etc. The expressions (18.4) for $\nu^{(l)}(\mathscr{E})$ are divided by $\sqrt{2\mathscr{E}_{\mathrm{R}}/\mu}\,\pi r_{\mathrm{R}}^2 N_k = g_{\mathrm{R}} \pi r_{\mathrm{R}}^2 N_k$ [see Eq. (18.2)] and become

$$\nu^{(l)*}(\mathscr{E}^*) = \sum_p Q_p^{(l)*} \mathscr{E}^{*p+\frac{1}{2}} + Q_{p_l,\ln}^{(l)*} \mathscr{E}^{*p_l+\frac{1}{2}} \ln \mathscr{E}^* + Q_{\ln^2}^{(l)*} \mathscr{E}^{*\frac{1}{2}} (\ln \mathscr{E}^*)^2 \tag{18.7a}$$

$$\nu^{(l)*}(g^*) = \sum_p Q_p^{(l)*} g^{*2p+1} + 2Q_{p_l,\ln}^{(l)*} g^{*2p_l+1} \ln g^* + 4Q_{\ln^2}^{(l)*} g^* (\ln g^*)^2. \tag{18.7b}$$

Since the factor $\sqrt{\mathscr{E}^*}$ between $\nu^{(l)*}(\mathscr{E}^*)$ and $Q^{(l)*}(\mathscr{E}^*)$ in Eq. (18.2) is contained in the powers of $\mathscr{E}^*$ in Eq. (18.7), the coefficients in Eqs. (18.6) and (18.7) are identical.

In the following subsections the expressions (18.3) and (18.6) for $Q^{(l)}(\mathscr{E})$ and $Q^{(l)*}(\mathscr{E}^*)$ will be discussed for particular interactions.

β) For inverse power potentials $\varphi(r) = C_n/r^n$, Eq. (6.11), there exists neither a reference energy $\mathscr{E}_{\mathrm{R}}$ nor a reference length r_{R} independent of $\mathscr{E}$. However, for the averaging of the transfer collision frequencies $\nu^{(l)}(\mathscr{E})$ over $\mathscr{E}$ necessary for the moment method of kinetic gas theory (Sect. 3α), it is useful to normalize at least the energy scale. To do this we introduce for collisions with inverse power potentials the thermal collision energy

$$k\hat{T}_{jk} := \mu_{jk} k \left(\frac{T_j}{m_j} + \frac{T_k}{m_k}\right) \qquad [12.4]$$

as reference energy and write instead of Eq. (6.19)

$$Q^{(l)}(\mathscr{E}) = Q^{(l)}(k\hat{T}) \left(\frac{k\hat{T}}{\mathscr{E}}\right)^{\frac{2}{n}} = Q^{(l)}(k\hat{T})\, \mathscr{E}^{*-\frac{2}{n}}. \qquad (18.8\text{a})$$

The factor

$$Q^{(l)}(k\hat{T}) = \pi \left(\frac{|C_n|}{k\hat{T}}\right)^{\frac{2}{n}} Q^{(l)*}(n, \operatorname{sgn} C_n) \qquad (18.8\text{b})$$

is called *monoenergetic transfer cross section.* It is the transfer cross section for particles whose collision energy $\mathscr{E}$ is $k\hat{T}$ and is the only nonvanishing coefficient $Q^{(l)}_{-2/n}$ in expression (18.3) in the case of inverse power potentials. The dimensionless quantities $Q^{(l)*}(n, \operatorname{sgn} C_n)$ are listed in Table 1, Sect. 6β.

Large differences between the electron temperature T_e, the ion temperature T_i, and the neutral gas temperature T_n in the planetary ionospheres are well established[1-3] and must therefore be taken into account[4-6].

If the species j and k of the collision partners have the same temperature $T_j = T_k$, the combined temperature $\hat{T}_{jk}$, Eq. (12.4), coincides with $T_j = T_k$. If one of the collision partners is an electron, the other a heavy particle h (i.e., an atom or ion), then $\mu_{eh} \approx m_e$ and $\hat{T}_{eh} \approx T_e$.

The reference relative speed according to Eq. (18.1 a), viz.,

$$\hat{g}_{jk} := \left(\frac{2k\hat{T}_{jk}}{\mu_{jk}}\right)^{\frac{1}{2}} = \left(\frac{2kT_j}{m_j} + \frac{2kT_k}{m_k}\right)^{\frac{1}{2}}, \qquad (18.9)$$

is the *most probable relative speed* if the distribution functions of the two colliding particle species are Maxwell distributions. Since we shall deal only with small deviations from (local) thermodynamic equilibrium, the distribution functions are almost Maxwellian and therefore $\hat{g}$, Eq. (18.9), may carry the label "most probable" without serious error.

The transfer collision frequency $\nu^{(l)}(\mathscr{E})$ for inverse power potentials is, according to Eqs. (4.14), (18.5), (18.8a)

$$\nu^{(l)}(\mathscr{E}) = \nu^{(l)}(k\hat{T})\, \mathscr{E}^{*\frac{1}{2}-\frac{2}{n}} = \nu^{(l)}(k\hat{T})\, g^{*1-\frac{4}{n}} \qquad (18.10\text{a})$$

---

[1] Bauer, S.J.: See [21], chapt. III

[2] Knudsen, W.C., Spenner, K., Whitten, R.C., Spreiter, J.R., Miller, K.L., Novak, V. (1979): Science *203*, 757–763

[3] Oyama, K.I., Hirao, K., Banks, P.M., Williamson, P.R. (1980): Planet. Space Sci. *28*, 207–211

[4] St. Maurice, J.-P., Schunk, R.W. (1977): Planet. Space Sci. *25*, 907–920

[5] Conrad, J.R., Schunk, R.W. (1979): J. Geophys. Res. *84*, 811–822

[6] Nakada, M.P., Sullivan, E.C. (1980): J. Geophys. Res. *85*, 171–176

with the *monoenergetic transfer collision frequency*

$$\nu^{(l)}(k\hat{T}) = \sqrt{2k\hat{T}/\mu}\, Q^{(l)}(k\hat{T})\, N_k$$
$$= \pi \sqrt{\frac{2}{\mu}}\, |C_n|^{\frac{2}{n}} (k\hat{T})^{\frac{1}{2}-\frac{2}{n}}\, Q^{(l)*}(n, \operatorname{sgn} C_n)\, N_k \tag{18.10b}$$

as the only nonvanishing coefficient $\nu^{(l)}_{-2/n}$ in expression (18.4) for $\nu^{(l)}(\mathscr{E})$. The monoenergetic momentum transfer collision frequency $\nu^{(1)}(k\hat{T})$ is often written as ν_{m}.

With the *monoenergetic collision length*

$$r_n(k\hat{T}) = \left(\frac{C_n}{k\hat{T}}\right)^{\frac{1}{n}} \tag{18.11}$$

the transfer cross sections $Q^{(l)}(\mathscr{E})$ can also be normalized, viz.,

$$Q^{(l)*}(\mathscr{E}^*) = \frac{Q^{(l)}(\mathscr{E})}{\pi |r_n(k\hat{T})|^2} = Q^{(l)*}(n, \operatorname{sgn} C_n)\, \mathscr{E}^{*\,-\frac{2}{n}}. \tag{18.12}$$

The dimensionless quantity $Q^{(l)*}(n, \operatorname{sgn} C_n)$ listed in Table 1, Sect. 6β is the only nonvanishing coefficient $Q^{(l)*}_{-2/n}$ in expression (18.6) for $Q^{(l)*}(\mathscr{E}^*)$ as well as in expression (18.7) for $\nu^{(l)*}(\mathscr{E}^*)$.

γ) For *repulsive exponential potentials* $\varphi(r) = \varphi_{\mathrm{R}} \exp(1 - r/r_{\mathrm{R}})$, Eq. (8.2) with reference quantities φ_{R}, r_{R}, and for the *charge exchange* part of collisions of ions in their parent gas the transfer cross sections are of the form

$$Q^{(l)*}(\mathscr{E}^*) = (\ln \mathscr{E}^*)^2. \qquad [8.3\mathrm{c}]\ [15.17\mathrm{a}]$$

In the first case the normalization quantities $r_{(l)}$ and $\mathscr{E}_{(l)}$, Eq. (8.3d), are related to the reference quantities r_{R} and φ_{R} of the potential, Eq. (8.2), by Eqs. (8.3e). In the second case the normalization quantities r_{Δ} and $\mathscr{E}_{\Delta}$ are listed in Table 9, Sect. 15.

The normalized transfer cross sections can be represented by the general expression (18.6) for $Q^{(l)*}(\mathscr{E}^*)$ with the only nonvanishing coefficient

$$Q^{(l)*}_{\ln^2} = 1. \tag{18.13}$$

δ) For *screened Coulomb potentials* $\varphi(r) = \varphi_{\mathrm{R}}(r_{\mathrm{R}}/r) \exp(1 - r/r_{\mathrm{R}})$, Eq. (8.4) with reference energy $|\varphi_{\mathrm{R}}|$ and reference length r_{R}, the high-energy approximation was

$$Q^{(l)*}(\mathscr{E}^*) = \frac{l(l+1)}{2} \left(\frac{\mathrm{e}^1}{\mathscr{E}^*}\right)^2 \left[\ln \frac{4\mathscr{E}^*}{\mathrm{e}^{1+\gamma}} - \frac{l}{2}\right] \quad \text{for } \mathscr{E}^* \gg 1 \text{ and } l = 1, 2. \qquad [8.5]$$

Comparison with the general expression (18.6) yields the nonvanishing coefficients

$$Q^{(l)*}_{-2} = \frac{l(l+1)}{2} \mathrm{e}^2 \left(\ln 4 - 1 - \gamma - \frac{l}{2}\right), \tag{18.14a}$$

$$Q^{(l)*}_{-2,\ln} = \frac{l(l+1)}{2} e^2. \tag{18.14b}$$

If the screening length r_R is the Debye-Hückel length

$$\lambda_D := [4\pi \sum_c l_{cc} N_c]^{-\frac{1}{2}} \qquad \text{with } l_{jk} := C_1/k\hat{T}_{jk} \qquad [12.2, 3]$$

as Landau length, it is useful for later energy averaging with $\hat{T}$ as parameter (Sect. 19α) not to take the temperature dependent Debye-HKckel length λ_D as reference length, but the (density dependent) normalization length

$$r_N := (|l_{jk}|\, \lambda_D^2)^{\frac{1}{3}} \qquad \text{with } |\mathscr{E}_N| := |C_1|/r_N \qquad [12.6, 7]$$

as reference energy. The high-energy approximation is then written as

$$Q^{(l)*}(\mathscr{E}^*) = \frac{l(l+1)}{2} \mathscr{E}^{*-2} \left[\ln \frac{4\lambda_D \mathscr{E}^*}{e^{\gamma} r_N} - \frac{l}{2}\right]$$
$$\text{for } \mathscr{E}^* \gg \frac{r_N}{\lambda_D} = \sqrt{\frac{|\mathscr{E}_N|}{k\hat{T}_{jk}}} \quad \text{and } l = 1, 2. \qquad [12.8, 9]$$

Comparing this result with the general expression (18.6) for $Q^{(l)*}(\mathscr{E}^*)$, we obtain the nonvanishing coefficients

$$Q^{(l)*}_{-2} = \frac{l(l+1)}{2} \left[\ln \frac{4\lambda_D}{r_N} - \gamma - \frac{l}{2}\right] \qquad Q^{(l)*}_{-2,\ln} = \frac{l(l+1)}{2}. \tag{18.15}$$

Because of the very small normalization energy $\mathscr{E}_N$ (Sect. 12) the condition $\mathscr{E}\sqrt{k\hat{T}} \gg |\mathscr{E}_N|^{3/2}$ for the high-energy approximation, Eq. (12.9), is satisfied for all aeronomic plasmas.

ε) For *the modified effective range expansion*, Eq. (8.14), the reference energy is $\mathscr{E}_4 := (\hbar^2/2\mu)^2/|C_4| = \hbar^2 k_4^2/2\mu$, Eq. (8.15), and the reference length is twice the scattering length a, Eq. (8.13b). The powers of $\mathscr{E}^*$ are $p = 0, \frac{1}{2}, 1, \frac{3}{2}$ etc. and the coefficients of the general expression (18.6) for $Q^{(l)*}(\mathscr{E}^*)$ are

$$Q_0^{(l)*} = 1, \qquad Q_{1/2}^{(1)*} = \frac{4\pi}{5} \frac{1}{a k_4}, \qquad Q_{1/2}^{(2)*} = \frac{6}{7} Q_{1/2}^{(1)*},$$
$$Q_1^{(l)*} = \frac{4}{3} \ln \frac{1}{16}, \qquad Q_{1,\ln}^{(l)*} = \frac{4}{3}, \qquad Q_{\ln 2}^{(l)*} = 0. \tag{18.16}$$

If the attractive Maxwell potential $\varphi(r) = C_4/r^4 = -|C_4|/r^4$ is caused by an induced dipole during the collision between an electron and a neutral particle with reduced polarizability α_n^*, Eq. (9.6), the reference energy $\mathscr{E}_4$ and the parameter k_4 are expressed by α_n^* as

$$\mathscr{E}_4 = \frac{\mathrm{Ry}}{\alpha_n^*} \qquad k_4^2 = \frac{1}{a_0^2\,\alpha_n^*}. \qquad [9.8]$$

ζ) Approximations of transfer cross sections and transfer collision frequencies. For most of the interaction potentials of Sects. 7 and 8 the transfer cross sections $Q^{(l)}(\mathscr{E})$ are not given as analytic expressions but as numerical results listed in tables and/or figures. The same holds of the transfer cross sections for collisions between electrons and neutrals in Sect. 11. In these cases it is sometimes sufficient to approximate the numerical results by an analytic expression of the type Eq. (18.3), valid for a certain energy range. This energy range is determined by the thermal collision energy

$$k\hat{T}_{jk} := \mu_{jk} k \left(\frac{T_j}{m_j} + \frac{T_k}{m_k}\right). \qquad [12.4]$$

Usually it is sufficient to approximate $Q^{(l)}(\mathscr{E})$ in an interval of about two decades around the thermal energy $k\hat{T}$ (Sect. 19).

The most frequently used approximations are polynomials[7, 8] in $\mathscr{E}$ or in $\sqrt{\mathscr{E}}$. But rational functions yield much more accurate results[9] for the energy averaging in Sect. 19.

Sometimes a *power law approximation*

$$Q^{(l)}(\mathscr{E}) = Q_p^{(l)} \mathscr{E}^{*p} = Q_p^{(l)} (\mathscr{E}/\mathscr{E}_\mathrm{R})^p \qquad (18.17\mathrm{a})$$

$$\nu^{(l)}(\mathscr{E}) = \nu_p^{(l)} \mathscr{E}^{*p+\frac{1}{2}} = \nu_p^{(l)} g^{*2p+1} \qquad (18.17\mathrm{b})$$

is sufficient. Fitting with a straight line the numerical values for $Q^{(l)}(\mathscr{E})$ on a doubly logarithmic plot, the power p and the factor $Q_p^{(l)}$ can be obtained. The latter is proportional to the power p of the reference energy $\mathscr{E}_\mathrm{R}$, usually 1 eV. The fitting should be done in one or two energy decades around the *normalized* combined *temperature*

$$\hat{T}_{jk}^* := \frac{k\hat{T}_{jk}}{\mathscr{E}_\mathrm{R}} = \mu_{jk} \frac{k}{\mathscr{E}_\mathrm{R}} \left(\frac{T_j}{m_j} + \frac{T_k}{m_k}\right). \qquad (18.18)$$

Then we can write Eqs. (18.17) analogous to Eqs. (18.8) as

$$Q^{(l)}(\mathscr{E}) = Q^{(l)}(k\hat{T})\,(\mathscr{E}/k\hat{T})^p \qquad (18.19\mathrm{a})$$

with

$$Q^{(l)}(k\hat{T}) = Q_p^{(l)}\,\hat{T}^{*p} \qquad (18.19\mathrm{b})$$

---

[7] Itikawa. Y. (1971): Planet. Space Sci. *19*. 993–1007. eq. 23

[8] Mantas. G.P. (1974): J. Atoms. Terr. Phys. *36*. 1587–1600. table 3

[9] Weinert. U.. Kratzsch. K.-A.. Oberhage. H.-R. (1978): Z. Naturforsch. Teil A *33*. 1423–1427

as monoenergetic transfer cross sections, furthermore

$$\nu^{(l)}(\mathcal{E})=\nu^{(l)}(k\hat{T})\,(\mathcal{E}/k\hat{T})^{p+\frac{1}{2}} \tag{18.20a}$$

with

$$\begin{aligned}\nu^{(l)}(k\hat{T})=\nu_p^{(l)}\,\hat{T}^{*p+\frac{1}{2}}&=\sqrt{2\mathcal{E}_{\mathrm{R}}/\mu}\,Q_p^{(l)}N_k\,\hat{T}^{*p+\frac{1}{2}}\\&=\sqrt{2/\mu}\,\mathcal{E}_{\mathrm{R}}^{-p}\,Q_p^{(l)}N_k(k\hat{T})^{p+\frac{1}{2}}\end{aligned} \tag{18.20b}$$

as monoenergetic transfer collision frequencies. The monoenergetic momentum transfer collision frequency $\nu^{(1)}(k\hat{T})$ is often written ν_{m} in the literature.

The fitting on a doubly logarithmic plot can be done immediately for the low-energy values of the transfer cross sections $Q^{(l)}(\mathcal{E})$ for collisions between electrons and N_2, O_2, CO_2, respectively in Sect. 11 ζ, η, ϑ. With 1 eV as reference energy we obtain for collisions

$$\mathrm{e}+\mathrm{N}_2:\quad Q_{1/2}^{(1)}=20\cdot10^{-20}\,\mathrm{m}^2\quad\text{for } 0.01<\frac{\mathcal{E}}{\mathrm{eV}}<0.1 \tag{[11.1]}$$

$$\mathrm{e}+\mathrm{O}_2:\quad Q_{4/7}^{(1)}=10\cdot10^{-20}\,\mathrm{m}^2\quad\text{for } 0.01<\frac{\mathcal{E}}{\mathrm{eV}}<0.05 \tag{[11.2]}$$

$$\mathrm{e}+\mathrm{CO}_2:\quad Q_{-1/2}^{(1)}=17\cdot10^{-20}\,\mathrm{m}^2\quad\text{for } 0.01<\frac{\mathcal{E}}{\mathrm{eV}}<0.1\,. \tag{[11.3]}$$

In the same energy ranges the following monoenergetic momentum transfer collision frequencies $\nu^{(1)}(k\hat{T})\equiv\nu_{\mathrm{m}}$ are obtained with Eq. (18.20b):

$$\mathrm{e}+\mathrm{N}_2:\quad \frac{\nu_{\mathrm{m}}}{\mathrm{Hz}}=7.4\cdot10^{5}\,\frac{kT_{\mathrm{e}}}{\mathrm{J}}\,\frac{N_{\mathrm{N}_2}}{\mathrm{m}^{-3}} \tag{18.21a}$$

$$\mathrm{e}+\mathrm{O}_2:\quad \frac{\nu_{\mathrm{m}}}{\mathrm{Hz}}=8.15\cdot10^{6}\left(\frac{kT_{\mathrm{e}}}{\mathrm{J}}\right)^{\frac{15}{14}}\frac{N_{\mathrm{O}_2}}{\mathrm{m}^{-3}} \tag{18.21b}$$

$$\mathrm{e}+\mathrm{CO}_2:\quad \frac{\nu_{\mathrm{m}}}{\mathrm{Hz}}=1.01\cdot10^{-13}\,\frac{N_{\mathrm{CO}_2}}{\mathrm{m}^{-3}}\,. \tag{18.21c}$$

The power law approximation, Eq. (18.17) with $k\hat{T}$ as reference energy and $p=-2/n$, yields an exact expression for inverse power potentials $\varphi(r)=C_n/r^n$, Eq. (6.11) [see Eq. (18.8)].

III. Transport collision frequencies and their calculation

19. Transport collision frequencies

α) Definitions of transport collision coefficients and transport collision frequencies. In the moment method of kinetic gas theory (see Sect. 3α) the transfer collision frequencies

$$\nu_{jk}^{(l)}(\mathscr{E}_{jk}) := g_{jk} Q_{jk}^{(l)}(\mathscr{E}_{jk}) N_k = \sqrt{2\mathscr{E}_{jk}/\mu_{jk}}\, Q_{jk}^{(l)}(\mathscr{E}_{jk}) N_k \qquad [4.14]$$

and the transfer cross sections

$$\phi^{(l)}(\mathscr{E}) := \oint \mathrm{d}\Omega\, q(\Omega, \mathscr{E}) [1 - \cos^l \chi] \qquad [4.9]$$

must be averaged over $\mathscr{E}$, weighted with the velocity distribution functions of the two species j and k of the collision partners. These distribution functions contain the temperatures T_j and T_k as parameters.

For weak deviations from (local) thermodynamic equilibrium the weight functions for the energy averaging must contain a factor

$$\exp\left(-\frac{\mathscr{E}}{k\hat{T}}\right) = \exp\left(-\frac{g^2}{\hat{g}^2}\right)$$

because the Maxwell contributions to the distribution functions are dominant. Small modifications of the Maxwell contributions may then be expressed with powers of $\mathscr{E}$ as often done in the literature, a procedure leading to the so-called *omega integrals*[1,2]

$$\Omega^{(l,s)}(\hat{T}) := \frac{\hat{g}}{4\sqrt{\pi}} \int_0^\infty \mathrm{d}w\, \mathrm{e}^{-w} w^{s+1} \phi^{(l)}(k\hat{T}w) \qquad (19.1\,\mathrm{a})$$

where

$$w := \frac{\mathscr{E}}{k\hat{T}} = \frac{g^2}{\hat{g}^2}. \qquad (19.1\,\mathrm{b})$$

But it can be shown[3] that, instead of powers of $\mathscr{E}$, suitable orthogonal polynomials are more useful when representating gas kinetic results. Since the weight function is $\exp - w$, in the interval $0 \leqq w \leqq \infty$ the corresponding orthogonal polynomials are generalized Laguerre polynomials with half-integer superscripts[4]

$$\mathrm{L}_s^{(l+\frac{1}{2})}(w) := \sum_{r=0}^{s} \binom{l+\frac{1}{2}+s}{s-r} \frac{(-w)^r}{r!}. \qquad (19.2\,\mathrm{a})$$

These are also named *Sonine polynomials.*

The first three Sonine polynomials are

$$\mathrm{L}_0^{(l+\frac{1}{2})}(w) = 1, \qquad \mathrm{L}_1^{(l+\frac{1}{2})}(w) = l + \tfrac{3}{2} - w, \qquad (19.2\,\mathrm{b})$$

$$\mathrm{L}_2^{(l+\frac{1}{2})}(w) = \tfrac{1}{2}(l+\tfrac{3}{2})(l+\tfrac{5}{2}) - (l+\tfrac{5}{2})\, w + \tfrac{1}{2} w^2. \qquad (19.2\,\mathrm{c})$$

---

[1] Hirschfelder. J.O.. Curtiss. C.F.. Bird. R.B.: See [*5*]. eq. 8.2-2

[2] Chapman. S.. Cowling. T.G.: See [*12*]. eq. 9.33, 5

[3] Suchy. K. (1969): Proc. Ninth Internat. Conf. on Phenomena in Ionized Gases. pp. 397. Bucharest: Editura Academiei Republicii Socialiste Romania

[4] Abramowitz, M., Stegun, I.A. (eds.) (1964): Handbook of mathematical functions, eq. 22.3.9. New York: Dover

The *transport collision frequencies* are defined by

$$\nu^{(ls)}(\hat{T}) := \frac{(-1)^s s!}{(l+\frac{1}{2}+s)!} \int_0^\infty dw\, e^{-w} L_s^{(l+\frac{1}{2})}(w)\, w^{l+\frac{1}{2}} \nu^{(l)}(k\hat{T}w) \tag{19.3}$$

as the coefficients of the orthogonal expansion

$$\nu^{(l)}(\mathscr{E}) = \sum_{s=0}^{\infty} (-1)^s \nu^{(ls)}(\hat{T})\, L_s^{(l+\frac{1}{2})}(w) \tag{19.4}$$

of the transfer collision frequencies $\nu^{(l)}(\mathscr{E})$, Eq. (4.14).

The factorial $\alpha!$ of a noninteger argument α is given by the gamma function $\Gamma(\alpha+1)$.

Thus (for l fixed) the absolute values $|\nu^{(ls)}|$ decrease with increasing s while the signs of the $\nu^{(ls)}$ alternate. This property makes the transport collision frequencies $\nu^{(ls)}$, Eq. (19.3), suitable for estimating approximations which may be labelled by the superscript s.

This can only be achieved with orthogonal polynomials $L_s^{(l+\frac{1}{2})}(w)$ and not with powers $w^{s+\frac{1}{2}}$ in the definition of the omega integrals $\Omega^{(l,s)}$, Eq. (19.1). The factor $(-1)^s$ in the definition, Eq. (19.3), for $\nu^{(ls)}$ is introduced to make $\nu^{(l1)}$ positive for most of the short-range interactions (see Fig. 24, Subsect. β below).

The ratios

$$\frac{\nu_{jk}^{(ls)}}{N_k} = \frac{\nu_{kj}^{(ls)}}{N_j} \qquad [4.15]$$

have the same physical dimension (volume per unit time) as the transfer collision coefficients $gQ^{(l)}(\mathscr{E})$ in Sect. 4γ and as a recombination coefficient. Therefore they are called *transport collision coefficients.* They have a linear relation with the omega integrals $\Omega^{(l,s)}$, Eq. (19.1 a).

With the relations (4.10) among the $Q^{(l)}$ and the $\phi^{(l)}$, the representations Eqs. (19.2) of the generalized Laguerre polynomials $L_s^{(l+\frac{1}{2})}(w)$, and the definitions Eq. (19.1) of $\Omega^{(l,s)}$ and Eq. (19.3) of $\nu^{(ls)}$, one obtains[5]

$$\frac{\nu_{jk}^{(ls)}}{N_k} = \sum_{\lambda=0}^{\operatorname{ent}\frac{l-1}{2}} \sum_{r=0}^{s} a_{\lambda r}^{ls}\, \Omega^{(l-2\lambda,\, l+r)} \tag{19.5a}$$

with

$$a_{\lambda r}^{ls} = (-1)^{s+\lambda+r} \frac{(2l-2\lambda-1)!!}{(l-2\lambda)!\,(2\lambda)!!} \binom{s}{r} \frac{2^{l+r+3}}{(2l+2r+1)!!}. \tag{19.5b}$$

In particular

$$\frac{\nu_{jk}^{(10)}}{N_k} = \frac{16}{3}\Omega^{(1,1)}, \qquad \frac{\nu_{jk}^{(11)}}{N_k} = \frac{32}{15}\Omega^{(1,2)} - \frac{16}{3}\Omega^{(1,1)}, \tag{19.5c, d}$$

---

[5] Weinert, U., Kratzsch, K.-A., Oberhage, H.-R. (1978): Z. Naturforsch. Teil A *33*, 1423–1427, eqs. A.19 and A.16

$$\frac{\nu_{jk}^{(12)}}{N_k}=\frac{16}{3}\Omega^{(1,1)}-\frac{64}{15}\Omega^{(1,2)}+\frac{64}{105}\Omega^{(1,3)}, \tag{19.5e}$$

$$\frac{\nu_{jk}^{(20)}}{N_k}=\frac{16}{5}\Omega^{(2,2)}, \qquad \frac{\nu_{jk}^{(21)}}{N_k}=\frac{32}{35}\Omega^{(2,3)}-\frac{16}{5}\Omega^{(2,2)}. \tag{19.5f, g}$$

The double factorial $(2\lambda)!!$ of an even integer means $2\lambda(2\lambda-2)(2\lambda-4)\cdots 2$ with $0!!:=1$; the double factorial $(2r+1)!!$ of an odd integer means $(2r+1)(2r-1)(2r-3)\cdots 1$ with $(-1)!!:=1$.

The inverse formula reads[5]

$$\Omega^{(l,r)}=\sum_{\lambda=0}^{\mathrm{ent}\frac{l-1}{2}}\sum_{s=0}^{r-l} b_{\lambda s}^{lr}\frac{\nu_{jk}^{(l-2\lambda,s)}}{N_k} \tag{19.6a}$$

with

$$b_{\lambda s}^{lr}=\frac{2l-4\lambda+1}{2^{r-2\lambda+3}}\,\frac{(2l+2s-4\lambda+1)!!}{(2l-2\lambda+1)!!\,(2\lambda)!!}\,\frac{l!(r-l)!}{(r-l-s)!}\,\frac{(2r+1)!!}{(2l+2s+1)!!}. \tag{19.6b}$$

In particular

$$\Omega^{(1,1)}=\frac{3}{16}\frac{\nu_{jk}^{(10)}}{N_k}, \qquad \Omega^{(1,2)}=\frac{15}{32}\frac{\nu_{jk}^{(10)}+\nu_{jk}^{(11)}}{N_k}, \tag{19.6c, d}$$

$$\Omega^{(1,3)}=\frac{105}{64}\frac{\nu_{jk}^{(10)}+2\nu_{jk}^{(11)}+\nu_{jk}^{(12)}}{N_k}, \tag{19.6e}$$

$$\Omega^{(2,2)}=\frac{5}{16}\frac{\nu_{jk}^{(20)}}{N_k}, \qquad \Omega^{(2,3)}=\frac{35}{32}\frac{\nu_{jk}^{(20)}+\nu_{jk}^{(21)}}{N_k}. \tag{19.6f, g}$$

The definition Eq. (19.3) of the transport collision frequencies $\nu^{(ls)}(\hat{T})$ allows us to specify the energy range for approximations of the transfer cross section $Q^{(l)}(\mathscr{E})$, mentioned in Sect. 18ζ. The weight function in the integral of Eq. (19.3) for the energy averaging of the transfer cross section $Q^{(l)}(k\hat{T}w)\sim w^{-1/2}\nu^{(l)}(k\hat{T}w)$ is

$$\begin{aligned} W^{(ls)}(w):&=\frac{s!}{(l+\frac{1}{2}+s)!}\,\mathrm{e}^{-w}\,\mathrm{L}_s^{(l+\frac{1}{2})}(w)\,w^{l+1}\sqrt{k\hat{T}} \\ &=\frac{s!}{(l+\frac{1}{2}+s)!}\,\mathrm{e}^{-\frac{\mathscr{E}^*}{\hat{T}^*}}\,\mathrm{L}_s^{(l+\frac{1}{2})}\left(\frac{\mathscr{E}^*}{\hat{T}^*}\right)\left(\frac{\mathscr{E}^*}{\hat{T}^*}\right)^{l+1}\sqrt{\mathscr{E}_{\mathrm{R}}\hat{T}^*}. \end{aligned} \tag{19.7a}$$

Fig. 22 shows curves of the normalized weight function

$$\frac{W^{(l0)}(\mathscr{E}^*/\hat{T}^*)}{W_{\mathrm{Max}}^{(l0)}}=\left(\frac{\mathscr{E}^*/\hat{T}^*}{l+1}\right)^{l+1}\exp\left(-\frac{\mathscr{E}^*}{\hat{T}^*}+l+1\right) \tag{19.7b}$$

for $l=1$. The main contribution stems from one or two decades around $\mathscr{E}^*=(l+1)\hat{T}^*$; thus only the $Q^{(l)}(\mathscr{E})$ values in this energy range contribute essentially to the transport collision frequency $\nu^{(l0)}(\hat{T})$. It is therefore often sufficient to approximate $Q^{(l)}(\mathscr{E})$ with analytical expressions in one or two decades around $w=l+1$.

β) Transport collision frequencies of transfer cross sections proportional to a power of the kinetic energy. For *inverse power potentials* the thermal energy $k\hat{T}$,

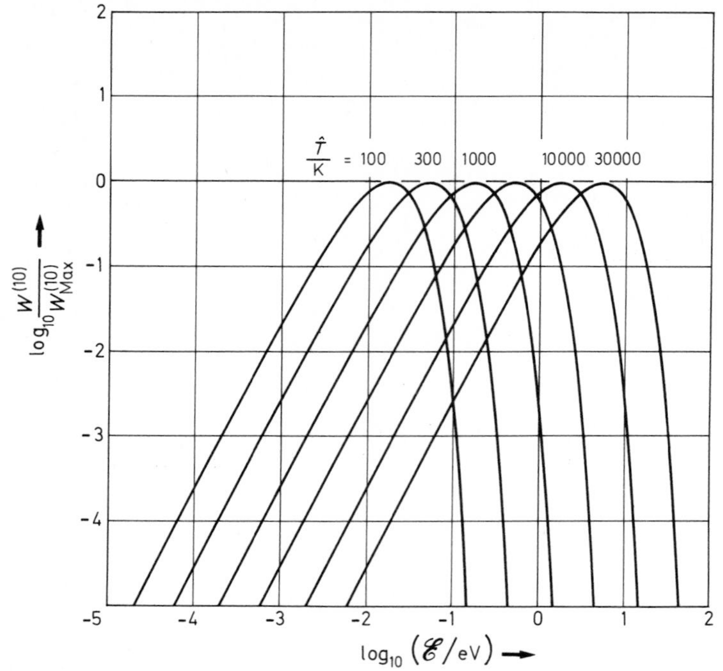

Fig. 22. Normalized weight function $W^{(10)}(\mathscr{E}/\mathscr{E}_R\,\hat{T}^*)/W^{(10)}_{\text{Max}}$ of Eq. (19.7b) to be used for the calculation of the momentum transport collision frequency $\nu^{(10)}(\hat{T})$, Eq. (19.3), from the momentum transfer cross section $Q^{(1)}(\mathscr{E})$, Eq. (4.8). The reference energy is $\mathscr{E}_R = 1$ eV, thus $\hat{T}^* = k\hat{T}/\text{eV} = \hat{T}/11\,604.5$ K. The parameter of the curves is $\hat{T}/\text{K} = 11\,604.5\,\hat{T}^*$

Eq. (12.4), was already introduced in Sect. 18β as reference energy. Thus in this case the normalized energy $\mathscr{E}^*$ is equal to $w := \mathscr{E}/k\hat{T}$, Eq. (19.1b).

The transfer cross sections $Q^{(l)}(\mathscr{E})$ and $\phi^{(l)}(\mathscr{E})$ can therefore be written as

$$Q^{(l)}(\mathscr{E}) = Q^{(l)}(k\hat{T})\,w^{-\frac{2}{n}} \qquad \phi^{(l)}(\mathscr{E}) = \phi^{(l)}(k\hat{T})\,w^{-\frac{2}{n}} \tag{[18.8a]}$$

with $Q^{(l)}(k\hat{T}) \sim \hat{T}^{-2/n}$, Eq. (18.8b), as the monoenergetic transfer cross section. The transfer collision frequency is

$$\nu^{(l)}(\mathscr{E}) = \nu^{(l)}(k\hat{T})\,w^{\frac{1}{2}-\frac{2}{n}} \tag{[18.10a]}$$

with $\nu^{(l)}(k\hat{T}) \sim \hat{T}^{\frac{1}{2}-\frac{2}{n}}$, Eq. (18.10b), as the monoenergetic transfer collision frequency.

For transfer cross sections $Q^{(l)}(\mathscr{E})$ approximated by a *power law*

$$Q^{(l)}(\mathscr{E}) = Q^{(l)}_p\,\mathscr{E}^{*p}, \tag{[18.19a]}$$

the transfer collision frequency is $\nu^{(l)}(\mathscr{E}) = \nu^{(l)}_p\,\mathscr{E}^{*p+\frac{1}{2}}$, Eq. (18.20a). Here the kinetic energy $\mathscr{E} = \mathscr{E}^*\,\mathscr{E}_R$ is normalized with a given reference energy $\mathscr{E}_R$.

With the normalized combined temperature $\hat{T}^* := k\hat{T}/\mathscr{E}_R$, Eq. (18.18), we can write

$$Q^{(l)}(\mathscr{E}) = Q^{(l)}(k\hat{T})\,w^p \tag{[18.19a]}$$

$$\nu^{(l)}(\mathscr{E}) = \nu^{(l)}(k\hat{T})\,w^{\frac{1}{2}+p} \tag{[18.20a]}$$

with

$$Q^{(l)}(k\hat{T}) = Q_p^{(l)} \hat{T}^{*p}$$ [18.19b]

as the monoenergetic transfer cross section and

$$\nu^{(l)}(k\hat{T}) = \nu_p^{(l)} \hat{T}^{*p+\frac{1}{2}} = \sqrt{2\mathscr{E}_R/\mu}\, Q_p^{(l)} N_k \hat{T}^{*p+\frac{1}{2}}$$ [18.20b]

as the monoenergetic transfer collision frequency.

The energy integrations in the definitions (19.1a) for $\Omega^{(l,s)}$ and (19.3) for $\nu^{(ls)}$ can now be performed (Appendix B, Eqs. B 1a, b, c) yielding

$$\Omega^{(l,s)}(\hat{T}) = \hat{g}\phi^{(l)}(k\hat{T}) \frac{\left(s+1-\frac{2}{n}\right)!}{4\sqrt{\pi}} \tag{19.8a}$$

$$\nu^{(ls)}(\hat{T}) = \nu^{(l)}(k\hat{T}) \frac{\left(l+1-\frac{2}{n}\right)! \left(\frac{1}{2}-\frac{2}{n}\right)!}{(l+\frac{1}{2}+s)! \left(\frac{1}{2}-\frac{2}{n}-s\right)!} \tag{19.8b}$$

$$= \nu^{(l0)}(\hat{T}) \frac{(l+\frac{1}{2})! \left(\frac{1}{2}-\frac{2}{n}\right)!}{(l+\frac{1}{2}+s)! \left(\frac{1}{2}-\frac{2}{n}-s\right)!} \tag{19.8c}$$

for inverse power potentials and

$$\nu^{(ls)}(\hat{T}) = \nu^{(l)}(k\hat{T}) \frac{(l+p+1)!\,(p+\frac{1}{2})!}{(l+\frac{1}{2}+s)!\,(p+\frac{1}{2}-s)!} \tag{19.8d}$$

$$= \nu^{(l0)}(\hat{T}) \frac{(l+\frac{1}{2})!\,(p+\frac{1}{2})!}{(l+\frac{1}{2}+s)!\,(p+\frac{1}{2}-s)!}. \tag{19.8e}$$

for power low transfer cross sections.

For fixed values l and n the omega integrals $\Omega^{(l,s)}$, Eq. (19.1a), increase with increasing s, whereas the absolute values $|\nu^{(ls)}|$ of the transport collision frequencies, Eq. (19.3), decrease with increasing s (Figs. 23 and 24). This different behavior is due to the different weight functions in the definitions Eq. (19.1a) of $\Omega^{(l,s)}$ and Eq. (19.3) for $\nu^{(ls)}$. The powers $w^{s+\frac{1}{2}}$ in the definition Eq. (19.1a) of $\Omega^{(l,s)}$ cause an increase with s, whereas the orthogonal polynomials $L_s^{(l+\frac{1}{2})}(w)$ in the definition Eq. (19.3) of $\nu^{(ls)}$ cause the decrease of $|\nu^{(ls)}|$ with increasing s.

For (classical) Maxwell interaction ($n=4$) and energy independent transfer collision frequencies $\nu^{(l)}$ ($p=-1/2$) one obtains

$$\nu^{(ls)} = \nu^{(l)} \delta_{s0} \quad \text{for } n=4 \quad \text{or } p=-\tfrac{1}{2}, \tag{19.8f}$$

i.e., all transport collision frequencies $\nu^{(ls)}$, Eqs. (19.8b, d), vanish except $\nu^{(l0)}$, but the omega integrals $\Omega^{(l,s)}$, Eqs. (19.8a), do not [see Eq. (19.6a)].

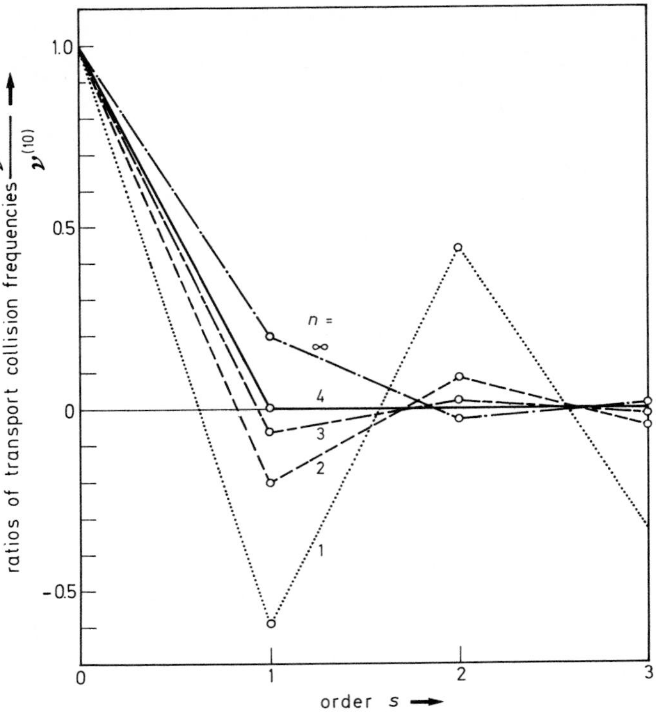

Fig. 23. Ratios $\nu^{(1s)}/\nu^{(10)}$, Eq. (19.8c), of classical transport collision frequencies $\nu^{(1s)}$, Eq. (19.3), for inverse power potentials C_n/r^n. Values for integer s; the connecting straight lines are drawn for clarity only (after SUCHY[6])

The ratio

$$\frac{\nu^{(l1)}}{\nu^{(l0)}}=\frac{p+\frac{1}{2}}{l+\frac{3}{2}}=\frac{2p+1}{2l+3} \tag{19.9}$$

reproduces the power $p+\frac{1}{2}$ of the energy $w^{p+\frac{1}{2}}$ (up to the denominator $l+\frac{3}{2}$), or the power $2p+1$ of the relative speed $g\sim w^{1/2}$ in Eq. (18.20a) for $\nu^{(l)}(\mathscr{E})$. Therefore the expressions

$$(2l+3)\frac{\nu^{(l1)}}{\nu^{(l0)}}=:r_l \tag{19.10a}$$

have been called "effective velocity powers"[8]. Of course, the definition Eq. (19.10a) for r_l is not restricted to power law transfer collision frequencies $\nu^{(l)}(\mathscr{E})\sim\mathscr{E}^{p+\frac{1}{2}}$, but can be used for any form of energy dependence of $\nu^{(l)}(\mathscr{E})$.

Calculations of

$$r_1=5\frac{\nu^{(11)}}{\nu^{(10)}}=6C^*-5 \tag{19.10b}$$

---

[6] Suchy, K.: See [8], fig. 5

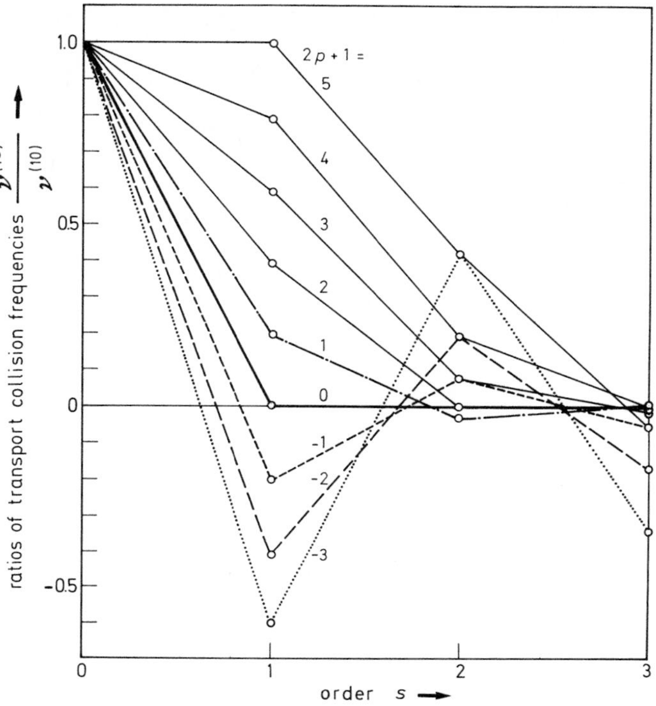

Fig. 24. Ratios $\nu^{(1s)}/\nu^{(10)}$, Eq. (19.8e), of transport collision frequencies $\nu^{(1s)}$ for transfer collision frequencies $\nu^{(1)}(E) \sim \mathscr{E}^{p+\frac{1}{2}}$. Values for integer s, the connecting straight lines are drawn for clarity only (after SUCHY and RAWER[7])

[see Eq. (19.16a) below] for polynomial approximations of $\nu^{(l)}(\mathscr{E})$ (Sect. 18ζ) have been done by COMFORT[8]. With the designation $6C^*-5$ tables of r_1 for various interaction potentials can be found in the literature (see Sect. 20). Curves for eight potentials are shown by DHAM and GUPTA[9].

γ) Normalizations. The omega integrals $\Omega^{(l,s)}(\hat{T})$, Eq. (19.1a), for rigid spheres ($n=\infty$)

$$\Omega^{(l,s)}_{\text{rig sph}}(\hat{T}) = \sqrt{\frac{2k\hat{T}}{\mu}}\, \phi^{(l)}_{\text{rig sph}} \frac{(s+1)!}{4\sqrt{\pi}} \tag{19.11}$$

with

$$\phi^{(l)}_{\text{rig sph}} = \pi r_{\text{d}}^2 \left[1 - \frac{1}{2}\frac{1+(-1)^l}{l+1}\right] \qquad [6.23]$$

[7] Suchy, K., Rawer, K. (1971): J. Atmos. Terr. Phys. *33*, 1853–1868, fig. 1
[8] Comfort, R.H. (1975): Planet. Space Sci. *23*, 533–539, eqs. 10, 12, 16
[9] Dham, A.K., Gupta, S.C. (1976): Indian J. Pure Appl. Phys. *14*, 765–766, fig. 1

are often used to normalize the transport collision coefficients by introducing[10]

$$\Omega^{(l,s)\star}(\hat{T}^*):=\left(\frac{r_d}{r_R}\right)^2\frac{\Omega^{(l,s)}(\hat{T})}{\Omega^{(l,s)}_{\text{rig sph}}(\hat{T})}=\frac{1}{1-\frac{1}{2}\frac{1+(-1)^l}{l+1}}\,\frac{1}{(s+1)!}\int_0^\infty dw\,e^{-w}w^{s+1}\frac{\phi^{(l)}(k\hat{T}w)}{\pi r_R^2} \tag{19.12}$$

with the normalized combined temperature $\hat{T}^*:=k\hat{T}/\mathscr{E}_R$, Eq. (18.18). Note that $\Omega^{(l,s)\star}(\hat{T}^*)$ describes the temperature dependence of $\Omega^{(l,s)}(\hat{T})$ only up to a factor $\sim\hat{T}^{1/2}\sim\Omega_{\text{rig sph}}(\hat{T})$.

For the transport collision frequencies $\nu^{(ls)}(\hat{T})$, Eq. (19.3), we introduce the same normalization as for the transfer collision frequencies $\nu^{(l)}(\mathscr{E})$, Eq. (18.2), viz.,

$$\nu_{jk}^{(ls)*}(\hat{T}^*):=\frac{\nu_{jk}^{(ls)}(\hat{T})}{\sqrt{2\mathscr{E}_R/\mu_{jk}}\,\pi r_R^2 N_k}. \tag{19.13}$$

The normalized transport collision frequency $\nu^{(ls)*}(\hat{T}^*)$ describes completely the temperature dependence of $\nu^{(ls)}(\hat{T})$.

From the connection Eq. (19.5) between the transport collision frequencies $\nu^{(ls)}$ and the omega integrals $\Omega^{(l,s)}$ we deduce the following relations between the corresponding normalized quantities $\nu^{(ls)*}$, Eq. (19.13), and $\Omega^{(l,s)\star}$, Eq. (19.12):

$$\nu^{(ls)*}=\sqrt{\hat{T}^*}\sum_{\lambda=0}^{\text{ent}\frac{l-1}{2}}\sum_{r=0}^{s}c_{\lambda r}^{ls}\,\Omega^{(l-2\lambda,\,l+r)\star} \tag{19.14a}$$

with

$$c_{\lambda r}^{ls}=a_{\lambda r}^{ls}\left[1-\frac{1}{2}\frac{1+(-1)^{l-2\lambda}}{l-2\lambda+1}\right]\frac{(l+r+1)!}{4\sqrt{\pi}} \tag{19.14b}$$

and $a_{\lambda r}^{ls}$ from Eq. (19.5b).

In particular

$$\frac{\nu^{(10)*}}{\sqrt{\hat{T}^*}}=\frac{8}{3\sqrt{\pi}}\Omega^{(1,1)\star}, \tag{19.14c}$$

$$\frac{\nu^{(11)*}}{\sqrt{\hat{T}^*}}=\frac{16}{15\sqrt{\pi}}\Omega^{(1,2)\star}-\frac{8}{\sqrt{\pi}}\Omega^{(1,1)\star}, \tag{19.14d}$$

$$\frac{\nu^{(12)*}}{\sqrt{\hat{T}^*}}=\frac{8}{3\sqrt{\pi}}\Omega^{(1,1)\star}-\frac{32}{5\sqrt{\pi}}\Omega^{(1,2)\star}+\frac{128}{35\sqrt{\pi}}\Omega^{(1,3)\star}, \tag{19.14e}$$

$$\frac{\nu^{(20)*}}{\sqrt{\hat{T}^*}}=\frac{16}{5\sqrt{\pi}}\Omega^{(2,2)\star}, \tag{19.14f}$$

$$\frac{\nu^{(21)*}}{\sqrt{\hat{T}^*}}=\frac{128}{35\sqrt{\pi}}\Omega^{(2,3)\star}-\frac{16}{5\sqrt{\pi}}\Omega^{(2,2)\star}. \tag{19.14g}$$

These relations allow conversion of the $\Omega^{(l,s)\star}$ tables into tables for the $\nu^{(ls)*}$.

[10] Hirschfelder. J.O.. Curtiss. C.F.. Bird. R.B.: See [5]. eq. 8.2–8

Often the following ratios of different $\Omega^{(l,s)\star}$ are tabulated:

$$A^* := \frac{\Omega^{(2,2)\star}}{\Omega^{(1,1)\star}} = \frac{1}{2}\frac{\Omega^{(2,2)}}{\Omega^{(1,1)}} = \frac{5}{6}\frac{\nu^{(20)}}{\nu^{(10)}} = \frac{5}{6}\frac{\nu^{(20)*}}{\nu^{(10)*}}, \tag{19.15a}$$

$$B^* := \frac{5\Omega^{(1,2)\star} - 4\Omega^{(1,3)\star}}{\Omega^{(1,1)\star}} = \frac{5\Omega^{(1,2)} - \Omega^{(1,3)}}{3\Omega^{(1,1)}}$$

$$= \frac{5}{6}\left(\frac{5}{2} - \frac{\nu^{(10)} + 2\nu^{(11)} + \frac{7}{2}\nu^{(12)}}{\nu^{(10)}}\right), \tag{19.15b}$$

$$C^* := \frac{\Omega^{(1,2)\star}}{\Omega^{(1,1)\star}} = \frac{1}{3}\frac{\Omega^{(1,2)}}{\Omega^{(1,1)}} = \frac{5}{6}\left(1 + \frac{\nu^{(11)}}{\nu^{(10)}}\right), \tag{19.15c}$$

$$E^* := \frac{\Omega^{(2,3)\star}}{\Omega^{(2,2)\star}} = \frac{1}{4}\frac{\Omega^{(2,3)}}{\Omega^{(2,2)}} = \frac{7}{8}\left(1 + \frac{\nu^{(21)}}{\nu^{(20)}}\right). \tag{19.15d}$$

With given $\nu^{(10)}$ or $\nu^{(10)*}$ the other $\nu^{(ls)}$ can be calculated using the ratios Eq. (19.15):

$$\frac{\nu^{(11)}}{\nu^{(10)}} = \frac{6}{5}C^* - 1, \tag{19.16a}$$

$$\frac{\nu^{(12)}}{\nu^{(10)}} = 1 - \frac{12}{35}B^* - \frac{24}{35}C^*, \tag{19.16b}$$

$$\frac{\nu^{(20)}}{\nu^{(10)}} = \frac{6}{5}A^* \qquad \frac{\nu^{(21)}}{\nu^{(10)}} = \left(\frac{8}{7}E^* - 1\right)\frac{6}{5}A^*, \tag{19.16c, d}$$

$$\frac{\nu^{(10)} + 2\nu^{(11)} + \frac{7}{2}\nu^{(12)}}{\nu^{(10)}} = \frac{5}{2} - \frac{6}{5}B^*. \tag{19.16e}$$

The last linear combination of $\nu^{(10)}$, $\nu^{(11)}$, $\nu^{(12)}$ will play an important rôle for the calculation of transport coefficients [see Eq. (23.10f)].

In Sect. 20 normalized transport collision frequencies $\nu^{(ls)*}$ $(\hat{T}^*)$, Eq. (19.13), are shown which were computed from tabulated values of the normalized transport coefficients $\Omega^{(l,s)\star}$ $(\hat{T}^*)$, Eq. (19.12).

In Sect. 21α curves of transport collision frequencies $\nu^{(ls)}$ $(\hat{T})$, Eq. (19.3), will be presented which were computed by energy averaging of those transfer collision frequencies $\nu^{(l)}(\mathscr{E}) = \sqrt{2\mathscr{E}/\mu}\, Q^{(l)}(\mathscr{E})\, N$, Eq. (4.14), for collisions between electrons and neutrals, for which the transfer cross sections $Q^{(l)}(\mathscr{E})$ are given in Sect. 11 as the result of measurements or quantum theoretical calculations. Finally in Sect. 21β analytic expressions of transport collision frequencies $\nu^{(ls)}(\hat{T})$ will be given for those cases where the transfer collision cross sections $Q^{(l)}(\mathscr{E})$ can be expressed as analytic functions of the kinetic energy $\mathscr{E}$.

20. Transport collision frequencies for multiparameter interaction potentials. In this section curves of the (temperature dependent) normalized transport collision frequencies $\nu^{(ls)*}(\hat{T}^*)$, Eq. (19.13), will be presented for most of those interaction potentials for which curves of (energy dependent) normalized transfer collision cross sections $Q^{(l)*}(\mathscr{E}^*)$, Eq. (6.7), were given in Sects. 7 and 8. They are calculated from published tables of reduced omega integrals $\Omega^{(l,s)\star}(\hat{T}^*)$, Eq. (19.12), using the representation Eq. (19.14) of the $\nu^{(ls)*}$ by linear combinations of $\Omega^{(l-2\lambda, l+r)\star}$. The published tables of $\Omega^{(l,s)\star}(\hat{T}^*)$ were computed by numerical energy averaging of transfer cross sections $\phi^{(l)}(\mathscr{E})$ according to the definition of the omega integrals $\Omega^{(l,s)}(\hat{T})$, Eq. (19.1a).

α) For the Morse potential

$$\frac{\varphi(r)}{-\varphi_{\mathrm{m}}}=\exp\left[2\alpha\left(1-\frac{r}{r_{\mathrm{m}}}\right)\right]-2\exp\left[\alpha\left(1-\frac{r}{r_{\mathrm{m}}}\right)\right] \qquad [7.1]$$

the normalized transfer cross sections $Q^{(l)*}(\mathscr{E}^*)=Q^{(l)}(\mathscr{E})/\pi r_{\mathrm{m}}^2$ with $\mathscr{E}^*=\mathscr{E}/|\varphi_{\mathrm{m}}|$ and $l=1,2$ are shown in Fig. 2, Sect. 7α, for the parameter values $C:=\alpha-\ln 2=3, 4, 5, 6, 8, 10$. Corresponding curves for normalized transport collision frequencies

$$\nu_{jk}^{(ls)*}(\hat{T}^*)=\frac{\nu^{(ls)}(\hat{T})}{\sqrt{2|\varphi_{\mathrm{m}}|/\mu_{jk}}\,\pi r_{\mathrm{m}}^2 N_k} \qquad [19.13]$$

with $\hat{T}^*=k\hat{T}/|\varphi_{\mathrm{m}}|$, Eq. (18.18), are shown in Fig. 25.

They were computed from tables[1] for

$$\Omega^{(l,l)\star}(\hat{T}^*)=\left(\frac{r_{\mathrm{d}}}{r_0}\right)^2\frac{\Omega^{(l,l)}(\hat{T})}{\Omega^{(l,l)}_{\mathrm{rig\,sph}}(\hat{T})} \qquad l=1,2 \qquad [19.12]$$

and B^*, C^*, E^*, Eqs. (19.15). The distance r_0, defined by $\varphi(r_0)=0$, was converted to r_{m}, the distance of the potential minimum φ_{m}, by means of

$$\frac{r_0}{r_{\mathrm{m}}}=\frac{\alpha-\ln 2}{\alpha}=\frac{C}{C+\ln 2}. \qquad [7.2]$$

The extrema of the curves for the higher-order transport collision frequencies $\nu^{(ls)*}$ with $s=1,2$ in the range $0.3<\hat{T}^*<1$ are caused by the oscillations of the transfer collision cross sections $Q^{(l)*}(\mathscr{E}^*)$ near $\mathscr{E}^*=1$ (Fig. 2, Sect. 7α), which are due to "orbiting" of the colliding particles. These oscillations decrease with increasing values of the parameter $C:=\alpha-\ln 2$, and therefore the extrema of the $\nu^{(ls)*}$ values become less pronounced as the value of the parameter increases.

For all values of the parameter α in Figs. 25 the values of $\nu^{(l1)}$ and $\nu^{(l2)}$ are smaller by about one order of magnitude than $\nu^{(l0)}$. Comparison with Fig. 24 (Sect. 19β) leads then to the conclusion that the transfer cross sections $Q^{(l)}(\mathscr{E})$ vary in sections approximately like $\mathscr{E}^p$ with negative slopes p in the range $-1\lesssim 2p+1\lesssim +1$, i.e., $-1\lesssim p\lesssim 0$. This is confirmed by Fig. 2 (Sect. 7α). For increasing temperature $\hat{T}$ the weight function $W^{(ls)}$, Eq. (19.7a), selects increasing energy decades from the $Q^{(l)}(\mathscr{E})$ values as main parts in the energy integrals, Eq. (19.3), for the calculation of the transport collision frequencies $\nu^{(ls)}$.

For low energies the slopes of the $Q^{(l)}(\mathscr{E})$ curves indicate very small negative p values with $-\frac{1}{2}\lesssim p$, i.e., $0\lesssim 2p+1$. This yields positive values for $\nu^{(l1)}$ and negative ones for $\nu^{(l2)}$, according to

[1] Smith, F.J., Munn, R.J. (1974): J. Chem. Phys. *41*, 3560–3568, table IV

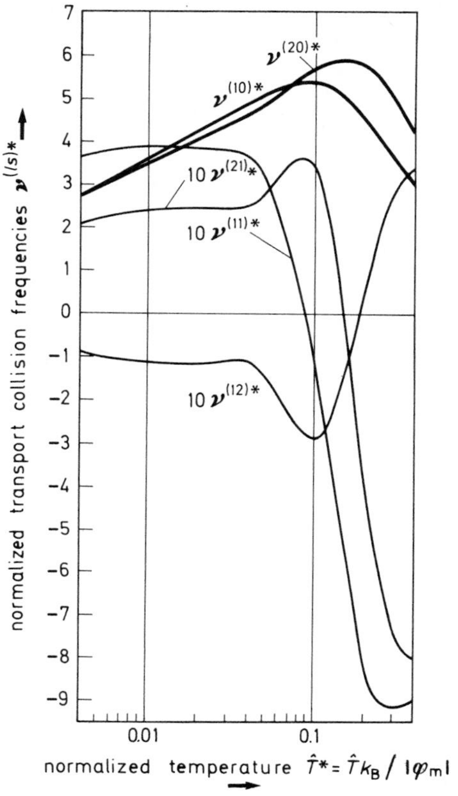

Fig. 25a. Normalized transport collision frequencies $\nu^{(ls)*}(\hat{T}^*) = \nu^{(ls)}(\hat{T})/\sqrt{2|\varphi_m|/\mu}\,\pi r_m^2 N_k$, Eq. (19.13), for a MORSE potential, Eq. (7.1) with parameter $C := \alpha - \ln 2 = 1$

Fig. 24 (Sect. 19β). With increasing energy the negative slopes p steepen and decrease under $-\frac{1}{2}$, i.e., $2p+1$ becomes negative. Therefore, according to Fig. 2 (Sect. 7α), the $\nu^{(l1)}$ become negative, the $\nu^{(l2)}$ positive, whereas for high energies the same signs as for low energies must hold. With increasing temperature these changes in the signs of $\nu^{(l1)}$ and $\nu^{(l2)}$ are reproduced in Figs. 25 for all values of the parameter $C := \alpha - \ln 2$.

As will be shown in Sect. 26α, the transport collision frequency $\nu^{(11)} \sim D_T$ is proportional to the thermal diffusion current (or Soret current), Eq. (26.4). The sign change of $\nu^{(11)}$ with varying temperature is observed experimentally as inversion of the thermal diffusion current at the "inversion temperatures"[2].

For high energies the negative slope p of the $Q^{(l)}(\mathscr{E})$ curves in Fig. 2 (Sect. 7α) increases from about $-\frac{1}{2}$ for $\alpha - \ln 2 = 3$ to very small negative p values for $\alpha - \ln 2 = 10$, i.e., $2p+1$ increases from about zero to nearly unity. The corresponding $\nu^{(l1)}$ values for high temperatures increase by about one order of magnitude according to Figs. 25, which corresponds to a similar increase of $\nu^{(11)}/\nu^{(10)}$ in Fig. 24 (Sect. 19β) for $0 \leqq 2p+1 \leqq 1$.

[2] Dham, A.K., Gupta, S.C. (1976): Indian J. Pure Appl. Phys. *14*, 765–766

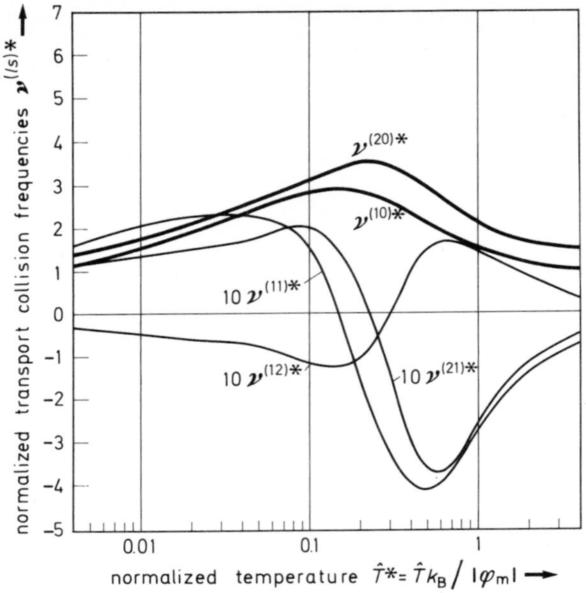

Fig. 25b. Normalized transport collision frequencies $\nu^{(ls)*}(\hat{T}^*) = \nu^{(ls)}(\hat{T})/\sqrt{2|\varphi_m|/\mu}\,\pi r_m^2 N_k$, Eq. (19.13), for a Morse potential, Eq. (7.1) with parameter $C := \alpha - \ln 2 = 2$

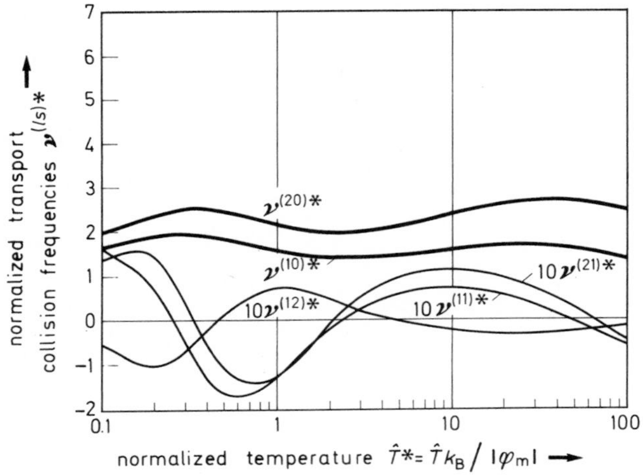

Fig. 25c. Normalized transport collision frequencies $\nu^{(ls)*}(\hat{T}^*) = \nu^{(ls)}(\hat{T})/\sqrt{2|\varphi_m|/\mu}\,\pi r_m^2 N_k$, Eq. (19.13), for a Morse potential, Eq. (7.1) with parameter $C := \alpha - \ln 2 = 3$

Morse potentials have been used to fit several attractive states of intermediate molecules formed during collisions between ions and neutrals (Table 8, Sect. 14) and also between neutrals of the same species (Table 10, Sect. 17).

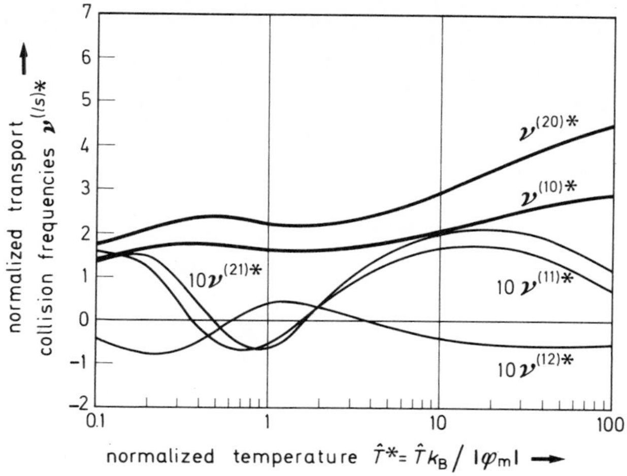

Fig. 25d. Normalized transport collision frequencies $\nu^{(ls)*}(\hat{T}^*) = \nu^{(ls)}(\hat{T})/\sqrt{2|\varphi_m|/\mu}\,\pi r_m^2 N_k$, Eq. (19.13), for a Morse potential, Eq. (7.1) with parameter $C := \alpha - \ln 2 = 4$

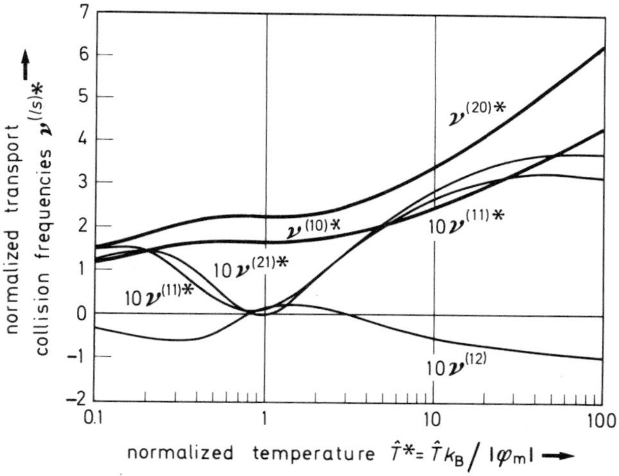

Fig. 25e. Normalized transport collision frequencies $\nu^{(ls)*}(\hat{T}^*) = \nu^{(ls)}(\hat{T})/\sqrt{2|\varphi_m|/\mu}\,\pi r_m^2 N_k$, Eq. (19.13), for a Morse potential, Eq. (7.1) with parameter $C := \alpha - \ln 2 = 5$

β) For a *Buckingham* (exp *β*, *n*) *potential*

$$\frac{\varphi(r)}{-\varphi_m} = \frac{n}{\beta - n} \exp\left[\beta\left(1 - \frac{r}{r_m}\right)\right] - \frac{\beta}{\beta - n}\left(\frac{r_m}{r}\right)^n \qquad [7.5]$$

Fig. 3 (Sect. 7β) presents normalized transfer cross sections $Q^{(1)*}$, $Q^{(2)*}$ for $n=6$ and $\beta = 12, 13, 14, 15$, varying for low energies as

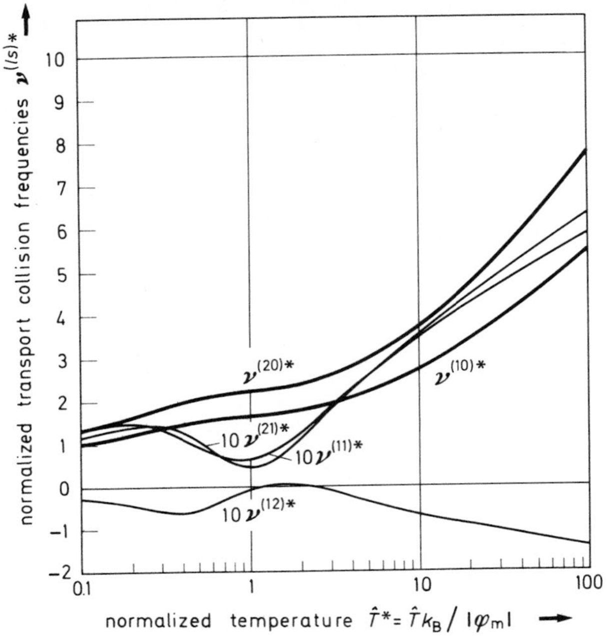

Fig. 25f. Normalized transport collision frequencies $\nu^{(ls)*}(\hat{T}^*) = \nu^{(ls)}(\hat{T})/\sqrt{2|\varphi_m|/\mu}\,\pi r_m^2 N_k$, Eq. (19.13), for a MORSE potential, Eq. (7.1) with parameter $C := \alpha - \ln 2 = 6$

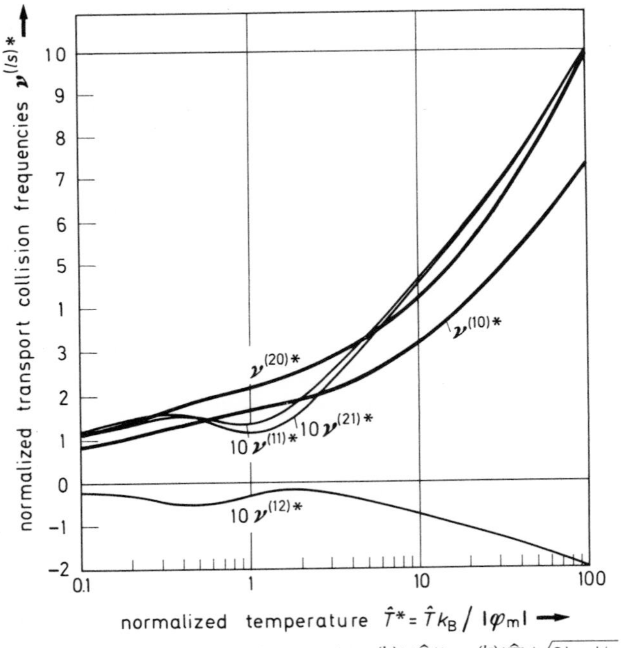

Fig. 25g. Normalized transport collision frequencies $\nu^{(ls)*}(\hat{T}^*) = \nu^{(ls)}(\hat{T})/\sqrt{2|\varphi_m|/\mu}\,\pi r_m^2 N_k$, Eq. (19.13), for a MORSE potential, Eq. (7.1) with parameter $C := \alpha - \ln 2 = 8$

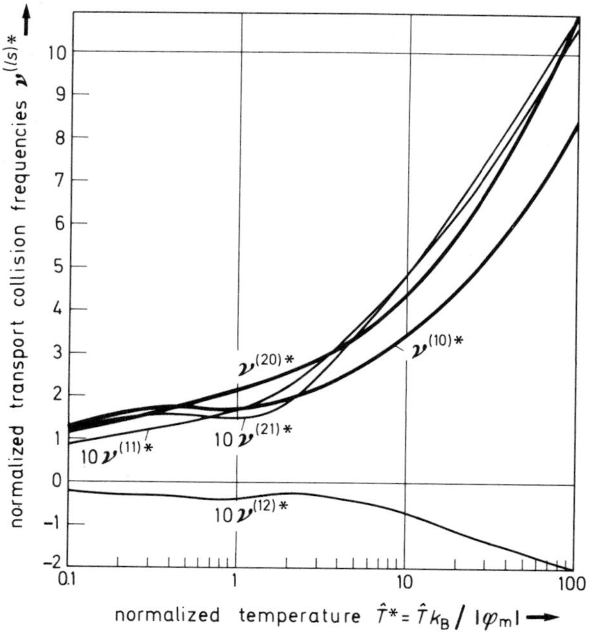

Fig. 25h. Normalized transport collision frequencies $\nu^{(ls)*}(\hat{T}^*)=\nu^{(ls)}(\hat{T})/\sqrt{2|\varphi_m|/\mu}\,\pi r_m^2 N_k$, Eq. (19.13), for a MORSE potential, Eq. (7.1) with parameter $C:=\alpha-\ln 2=10$

$$Q^{(l)*}(\mathscr{E}^*)=\mathscr{E}^{*-\frac{2}{n}}\left(\frac{\beta}{\beta-n}\right)^{\frac{2}{n}} Q^{(l)*}(n,-)\qquad \text{for } \mathscr{E}^*\ll 1 \qquad [7.7]$$

with $n=6$.

The corresponding curves for normalized transport collision frequencies $\nu^{(ls)*}$ in Figs. 26 are calculated from tables [3,4] for

$$Z^{(l,l)}:=\left[\hat{T}^*\left(\frac{\beta-6}{\beta}\right)\right]^{\frac{1}{3}}\left(\frac{r_0}{r_m}\right)^2 \Omega^{(l,l)\star}\qquad (l=1,2)$$

with

$$\Omega^{(l,l)\star}:=\left(\frac{r_d}{r_0}\right)^2\frac{\Omega^{(l,l)}}{\Omega^{(l,l)}_{\text{rig sph}}} \qquad [19.12]$$

and for B^*, C^*, E^*, Eq. (19.15).

The low-temperature behavior can be taken from Eqs. (7.7) and (19.8b) as

$$\nu^{(ls)*}(\hat{T}^*)=\hat{T}^{*\frac{1}{2}-\frac{2}{n}}\left(\frac{\beta}{\beta-n}\right)^{\frac{2}{n}} Q^{(l)*}(n,-)\frac{\left(l+1-\frac{2}{n}\right)!\left(\frac{1}{2}-\frac{2}{n}\right)!}{\left(l+\frac{1}{2}+s\right)!\left(\frac{1}{2}-\frac{2}{n}-s\right)!} \qquad (20.1)$$

$$\text{for } \hat{T}\ll|\varphi_m|$$

with $Q^{(l)*}(n,-)$ from Table 1 (Sect. 6β).

---

[3] Mason, E.A. (1954): J. Chem. Phys. *22*, 169–186, table IV

[4] Hirschfelder, J.O., Curtiss, C.F., Bird, R.B.: See [5], table VII-B

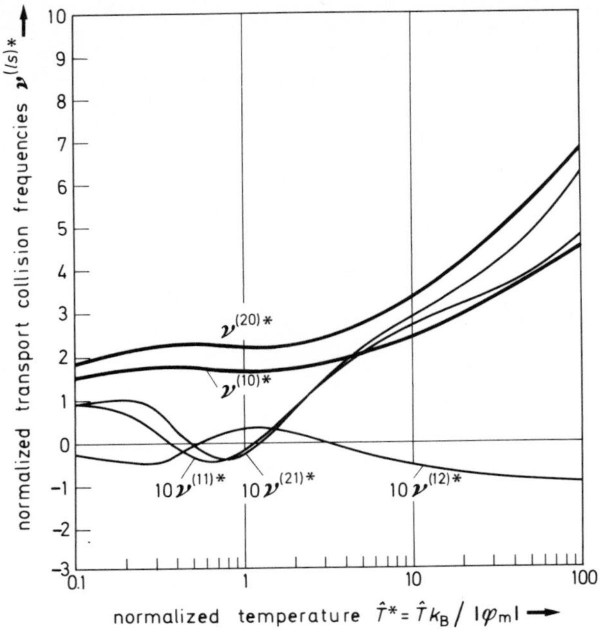

Fig. 26a. Normalized transport collision frequencies $\nu^{(ls)*}(\hat{T}^*) = \nu^{(ls)}(\hat{T})/\sqrt{2|\varphi_m|/\mu}\,\pi r_m^2 N_k$, Eq. (19.13), for a BUCKINGHAM (exp 12, 6) potential, Eq. (7.5)

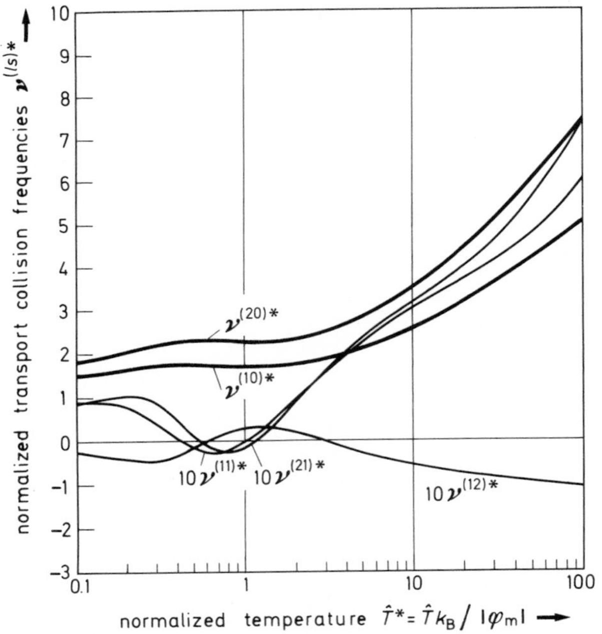

Fig. 26b. Normalized transport collision frequencies $\nu^{(ls)*}(\hat{T}^*) = \nu^{(ls)}(\hat{T})/\sqrt{2|\varphi_m|/\mu}\,\pi r_m^2 N_k$, Eq. (19.13), for a BUCKINGHAM (exp 13, 6) potential, Eq. (7.5)

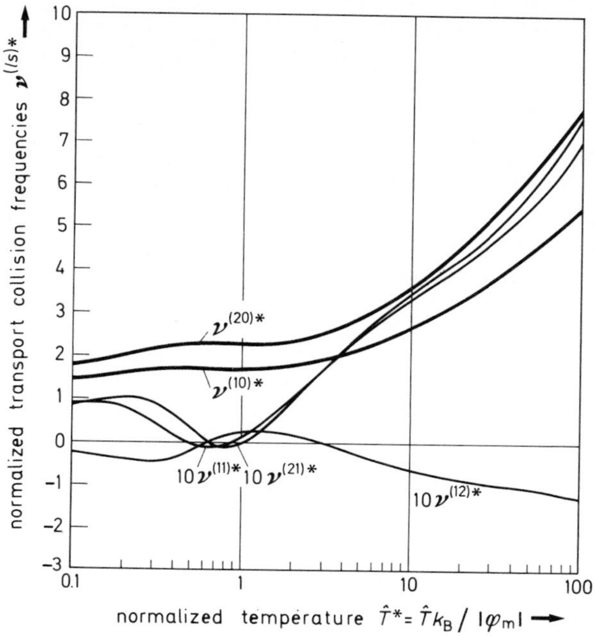

Fig. 26c. Normalized transport collision frequencies $\nu^{(ls)*}(\hat{T}^*) = \nu^{(ls)}(\hat{T})/\sqrt{2|\varphi_m|/\mu}\,\pi r_m^2 N_k$, Eq. (19.13), for a BUCKINGHAM (exp 14, 6) potential, Eq. (7.5)

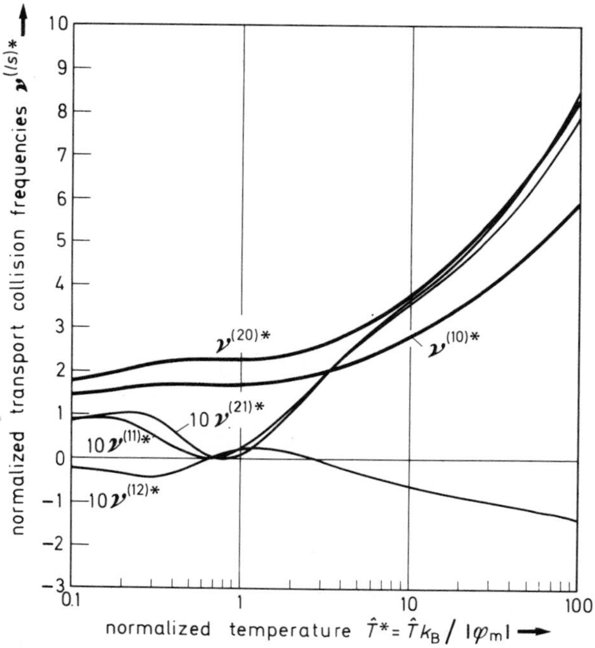

Fig. 26d. Normalized transport collision frequencies $\nu^{(ls)*}(\hat{T}^*) = \nu^{(ls)}(\hat{T})/\sqrt{2|\varphi_m|/\mu}\,\pi r_m^2 N_k$, Eq. (19.13), for a BUCKINGHAM (exp 15, 6) potential, Eq. (7.5)

At high temperatures the transport collision frequencies $\nu^{(ls)}$ show a similar behaviour as those for a Morse potential in Figs. 25, since the high-energy variation of the corresponding transfer cross sections $Q^{(l)}(\mathscr{E})$ in Figs. 2 and 3 (Sect. 7α, β) are also similar. The same holds for the sign variations of $\nu^{(11)}$, $\nu^{(12)}$, $\nu^{(21)}$, because of similar slope variations of the $Q^{(l)}(\mathscr{E})$ in Figs. 2 and 3 (Sect. 7α, β).

Buckingham potentials have been used to fit some of the interaction potentials for collisions between neutrals (Table 10, Sect. 17). It can be expected that they will be used more extensively in the future, since experimental results show that potential cores are best fitted by exponential functions, whereas potential tails vary with inverse powers of the distance (Sect. 36ε).

γ) A *Mason-Schamp* (n, n_1, n_0) *potential*

$$\frac{\varphi(r)}{\varphi_m}=\frac{\frac{1+\gamma}{n}\left(\frac{r_m}{r}\right)^n-2\frac{\gamma}{n_1}\left(\frac{r_m}{r}\right)^{n_1}-\frac{1-\gamma}{n_0}\left(\frac{r_m}{r}\right)^{n_0}}{\frac{1+\gamma}{n}-2\frac{\gamma}{n_1}-\frac{1-\gamma}{n_0}} \qquad [7.10]$$

yields normalized transport collision frequencies $\nu^{(ls)*}(\hat{T}^*)$ presented in Figs. 27 for $(n, n_1, n_0)=(12, 6, 4)$ with $\gamma=0, \frac{1}{4}, \frac{1}{2}, 1$. The corresponding transfer cross sections $Q^{(l)}(\mathscr{E})$ are shown in Fig. 5 (Sect. 7δ).

The $\nu^{(ls)*}$ were calculated from tables[5] of $\Omega^{(1,1)*}$, A^*, B^*, C^*, E^* for $\gamma=0, 0.2, 0.4, 0.6, 0.8, 1.0$. The plotted values for $\gamma=0.25$ and 0.5 were interpolated. The tables[5] give also values for (8, 6, 4) and (16, 6, 4) potentials, others[6] for (n, 8, 6) potentials.

For low and high temperatures the limit values of the transport collision frequencies in Fig. 27 are for a $\left(n, \frac{n}{2}, \frac{n}{3}\right)$ potential [see Eq. (7.11b)]

$$\nu^{(ls)*}(\hat{T}^*)=\hat{T}^{*\frac{1}{2}-\frac{6}{n}}\left(\frac{3}{2}\right)^{\frac{6}{n}}Q^{(l)*}\left(\frac{n}{3}, -\right)\frac{\left(l+1-\frac{6}{n}\right)!\left(\frac{1}{2}-\frac{6}{n}\right)!}{\left(l+\frac{1}{2}+s\right)!\left(\frac{1}{2}-\frac{6}{n}-s\right)!} \qquad (20.2a)$$

$$\text{for } \hat{T}\ll|\varphi_m|, \quad \gamma=0$$

and [see Eq. (7.11c)]

$$\nu^{(ls)*}(\hat{T}^*)=\hat{T}^{*\frac{1}{2}-\frac{4}{n}}2^{\frac{4}{n}}Q^{(l)*}\left(\frac{n}{2}, -\right)\frac{\left(l+1-\frac{4}{n}\right)!\left(\frac{1}{2}-\frac{4}{n}\right)!}{\left(l+\frac{1}{2}+s\right)!\left(\frac{1}{2}-\frac{4}{n}-s\right)!} \qquad (20.2b)$$

$$\text{for } \hat{T}\ll|\varphi_m|, \quad \gamma=1.$$

[5] Viehland, L.A., Mason, E.A., Morrison, W.F., Flannery, M.R. (1975): At. Data Nucl. Data Tables *16*, 495–514, tables I and II

[6] Klein, M., Hanley, H.J.M., Smith, F.J., Holland, P. (1974): Tables of collision integrals and second virial coefficients for the (m, 6, 8) intermolecular potential function, NSRDS-NBS 47. Washington: U.S. Government Printing Office

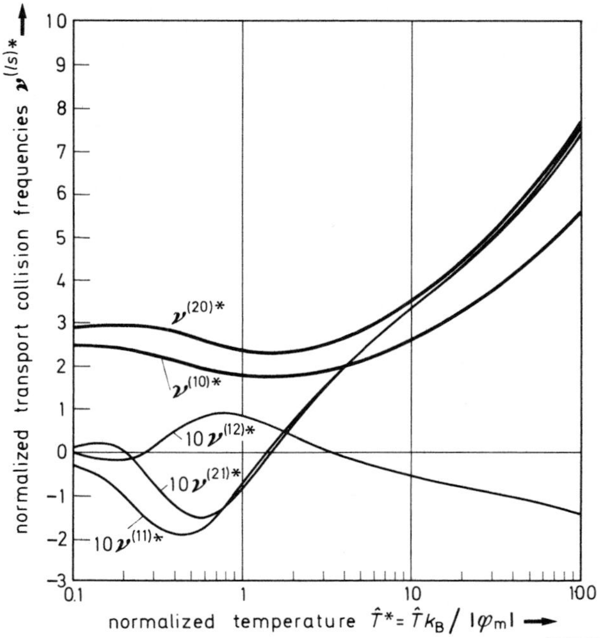

Fig. 27a. Normalized transport collision frequencies $\nu^{(ls)*}(\hat{T}^*) = \nu^{(ls)}(\hat{T})/\sqrt{2|\varphi_m|/\mu}\,\pi r_m^2 N_k$, Eq. (19.13), for a MASON–SCHAMP (12, 6, 4) potential, Eq. (7.10), with parameter $\gamma = 0$, i.e. a LENNARD–JONES (12, 4) potential, Eq. (7.14)

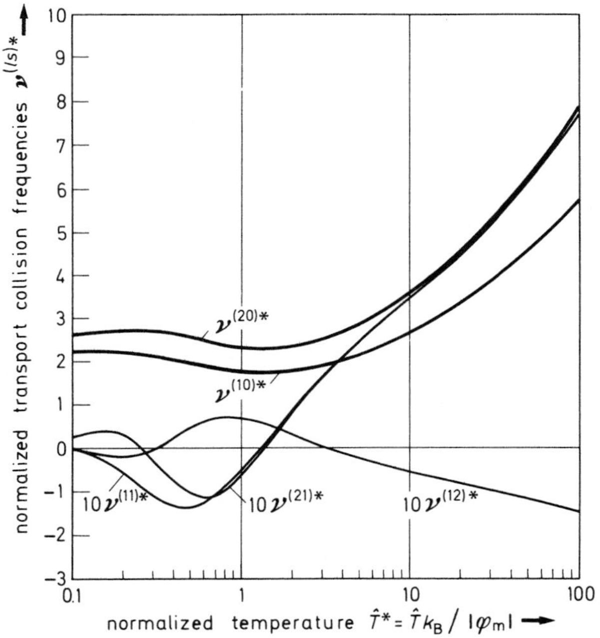

Fig. 27b. Normalized transport collision frequencies $\nu^{(ls)*}(\hat{T}^*) = \nu^{(ls)}(\hat{T})/\sqrt{2|\varphi_m|/\mu}\,\pi r_m^2 N_k$, Eq. (19.13) for a MASON–SCHAMP (12, 6, 4) potential, Eq. (7.10), with parameter $\gamma = 1/4$

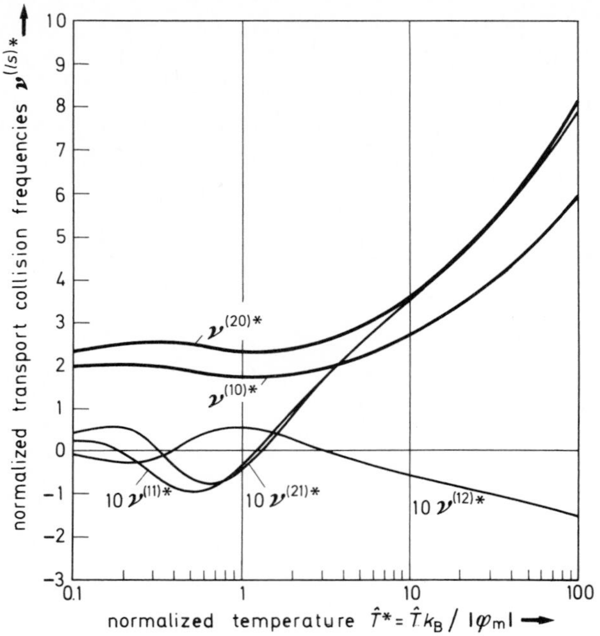

Fig. 27c. Normalized transport collision frequencies $\nu^{(ls)*}(\hat{T}^*) = \nu^{(ls)}(\hat{T})/\sqrt{2|\varphi_m|/\mu}\,\pi r_m^2 N_k$, Eq. (19.13), for a MASON-SCHAMP (12, 6, 4) potential, Eq. (7.10), with parameter $\gamma = 1/2$

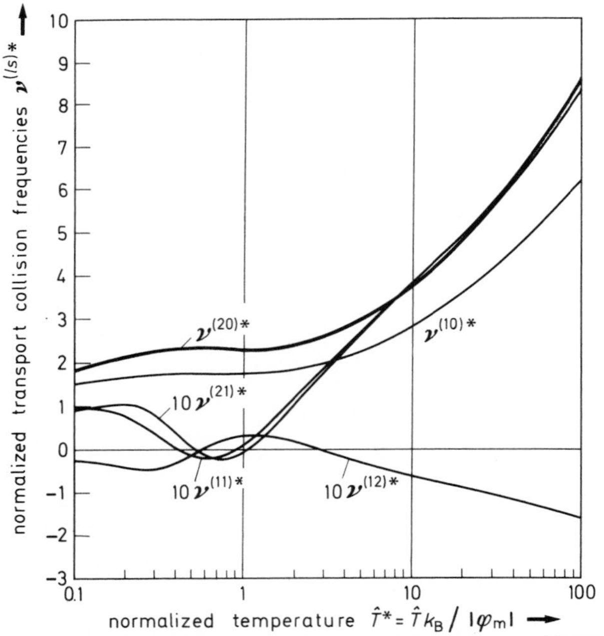

Fig. 27d. Normalized transport collision frequencies $\nu^{(ls)*}(\hat{T}^*) = \nu^{(ls)}(\hat{T})/\sqrt{2|\varphi_m|/\mu}\,\pi r_m^2 N_k$, Eq. (19.13), for a MASON-SCHAMP (12, 6, 4) potential, Eq. (7.10) with parameter $\gamma = 1$, i.e. a LENNARD-JONES (12, 6) potential, Eq. (7.14)

The high-temperature behavior is given by Eq. (7.11a) as

$$\nu^{(ls)*}(\hat{T}^*) = T^{*\frac{1}{2}-\frac{2}{n}} \left(\frac{1+\gamma}{2}\right)^{\frac{2}{n}} Q^{(l)*}(n,+) \frac{\left(l+1-\frac{2}{n}\right)!\left(\frac{1}{2}-\frac{2}{n}\right)!}{\left(l+\frac{1}{2}+s\right)!\left(\frac{1}{2}-\frac{2}{n}-s\right)!} \tag{20.2c}$$

$$\text{for } \hat{T} \gg |\varphi_m|$$

The reason for the sign changes of $\nu^{(11)*}$, $\nu^{(12)*}$, $\nu^{(21)*}$ with varying temperature are the varying slopes of the $Q^{(l)*}$ curves in Fig. 5 (Sect. 7δ), as explained for the Morse potential in Subsect. α.

Mason-Schamp (12, 6, 4) potentials are mainly used to fit interaction potentials for ion-neutral collisions (see Sect. 13γ).

δ) For *Kihara* (n, n_γ) *potentials*

$$\frac{\varphi(r)}{-\varphi_m} = \frac{n_\gamma}{n-n_\gamma}\left(\frac{r_m-r_K}{r-r_K}\right)^n - \frac{n}{n-n_\gamma}\left(\frac{r_m-r_K}{r-r_K}\right)^{n_\gamma} \qquad [7.12]$$

tables[7] of $\Omega^{(1,1)\star}$, A^* and $6C^*-5$ for $n=12$, $n_\gamma=4$, and $r_K/r_m=0$, 0.1, 0.2, ... 0.8 can be used to compute normalized transport collision frequencies $\nu^{(10)*}$, Eq. (19.14c), $\nu^{(11)*}$, Eq. (19.14d) and $\nu^{(20)*}$, Eq. (19.14f).

A Kihara (12, 4) potential was used[8] to describe collisions between SF_5^- and SF_6^- ions and SF_6 neutrals.

For collisions between helium atoms, nitrogen molecules, oxygen molecules and carbon dioxide molecules Kihara (6, 3) and (12, 6) potentials were used[9], but not for transport processes.

ε) *Lennard-Jones* (n, n_γ) *potentials*

$$\frac{\varphi(r)}{-\varphi_m} = \frac{n_\gamma}{n-n_\gamma}\left(\frac{r_m}{r}\right)^n - \frac{n}{n-n_\gamma}\left(\frac{r_m}{r}\right)^{n_\gamma} \qquad [7.14]$$

are special cases of Mason-Schamp potentials, Eq. (7.10), Subsect. γ, for $\gamma=0$ or $\gamma=1$. Curves for normalized transport collision frequencies $\nu^{(ls)*}$ of (12, 4) and (12, 6) potentials are therefore given in Figs. 27a and 27d, respectively. For (8, 4) potentials tables[10] for $\Omega^{(1,1)\star}$, A^*, and $6C^*-5$ were used to compute $\nu^{(10)*}$, $\nu^{(11)*}$ and $\nu^{(20)*}$ in Fig. 28.

Limit expressions for low and high temperatures are special cases of Eqs. (20.2b) and (20.2c) with $n=8$, $\gamma=1$.

For other $(n, 4)$ potentials with $n=6$, 10, 14, 18, 300 tables[11] for $\Omega^{(1,1)\star}$, A^*, B^*, C^*, E^* allow computation of $\nu^{(10)*}$, $\nu^{(11)*}$, $\nu^{(12)*}$, $\nu^{(20)*}$, $\nu^{(21)*}$, using the connections Eq. (19.16).

A Lennard-Jones (12, 4) potential was used to fit the interaction potential for $H^+ + He$ collisions (Table 8, Sect. 14), and (12, 6) potentials are used for collisions between neutrals (Table 10; Sect. 17).

---

[7] McDaniel, E.W., Mason, E.A.: See [*23*], tables I-8, I-9, I-10

[8] Mason, E.A., O'Hara, H., Smith, F.J. (1972): J. Phys. B*5*, 169–176

[9] Thakkar, A.J., Smith, V.H. (1975): Mol. Phys. *29*, 731–744; Diaz Pena, M., Pando, C., Renuncio, J.A.R. (1982): J. Chem. Phys. *76*, 325–332

[10] McDaniel, E.W., Mason, E.A.: See [*23*], table I-4

[11] Viehland, L.A., Mason, E.A., Morrison, W.F., Flannery, M.R. (1975): At. Data Nucl. Data Tables *16*, 495–514, table IV

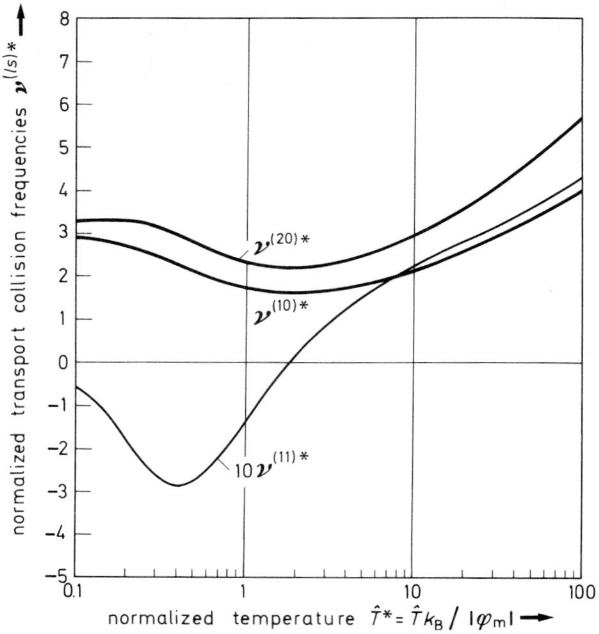

Fig. 28. Normalized transport collision frequencies $\nu^{(ls)*}(\hat{T}^*)=\nu^{(ls)}(\hat{T})/\sqrt{2|\varphi_m|/\mu}\,\pi r_m^2 N_k$, Eq. (19.3), for a LENNARD–JONES (8, 4) potential, Eq. (7.12)

ζ) The *Sutherland* (∞, n) *potential*

$$\varphi(r)=\varphi_m\left(\frac{r_m}{r}\right)^n \quad \text{for } r\geqq r_m, \qquad \varphi(r)=\infty \quad \text{for } r<r_m \qquad [7.16]$$

is a combination of a rigid-sphere potential for the hard core ($r<r_m$) with an attractive inverse power potential (for $\varphi_m<0$) for the tail. Normalized transfer cross sections $Q^{(l)*}(\mathscr{E})$ are plotted in Fig. 7 (Sect. 7η), limit values for high and low energies are given by Eqs. (7.17).

Values for the normalized transport collison frequencies $\nu^{(10)*}$, $\nu^{(11)*}$, $\nu^{(20)*}$ were computed from tables[10] for $\Omega^{(1,1)\star}$, A^*, and $6C^*-5$ and are shown in Fig. 29.

For low temperatures we obtain from Eqs. (7.17b) and (19.8b)

$$\nu^{(ls)*}(\hat{T}^*)=\hat{T}^{*\frac{1}{2}-\frac{2}{n}}\,Q^{(l)*}(n,-)\,\frac{\left(l+1-\frac{2}{n}\right)!\,\left(\frac{1}{2}-\frac{2}{n}\right)!}{\left(l+\frac{1}{2}+s\right)!\,\left(\frac{1}{2}-\frac{2}{n}-s\right)!} \qquad (20.3\,\mathrm{a})$$

for $\hat{T}\ll|\varphi_m|$.

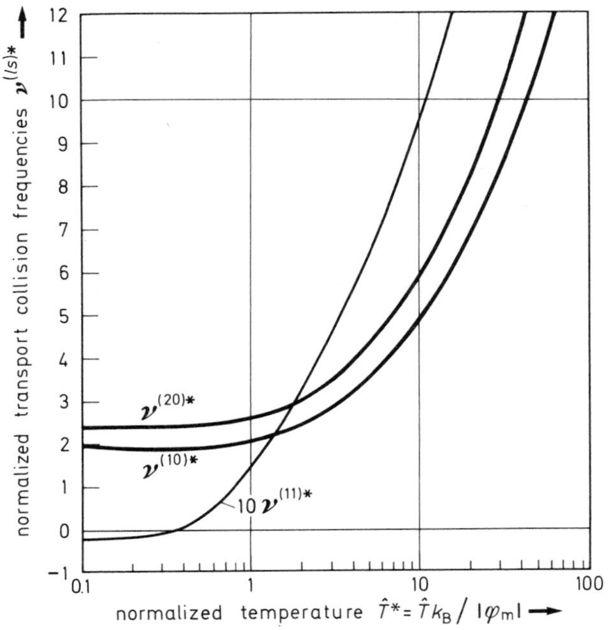

Fig. 29. Normalized transport collision frequencies $\nu^{(ls)*}(\hat{T}^*) = \nu^{(ls)}(\hat{T})/\sqrt{2|\varphi_m|/\mu}\,\pi r_m^2 N_k$, Eq. (19.3), for a SUTHERLAND $(\infty, 4)$ potential, Eq. (7.16)

The limit for high temperatures is calculated from Eqs. (7.17a), (18.3), (18.4), (18.5), (19.8d) as

$$\nu^{(ls)*}(\hat{T}^*) = \sqrt{\hat{T}^*}\,\frac{(l+1)!\,\frac{1}{2}!}{(l+\frac{1}{2}+s)!\,(\frac{1}{2}-s)!} + \frac{i^{(l)}(n)}{\pi\sqrt{\hat{T}^*}}\,\frac{l!\,(-\frac{1}{2})!}{(l+\frac{1}{2}+s)!\,(-\frac{1}{2}-s)!} \qquad (20.3\,\text{b})$$

for $\hat{T} \gg |\varphi_m|$.

The numerical values for $Q^{(l)*}(n, -)$ and $i^{(l)}(n)$ are found in Tables 1 (Sect. 6β) and 3 (Sect. 7η), respectively.

Like the Mason-Schamp (12, 6, 4) potential (Subsect. γ) and the Lennard-Jones (12, 4) potential (Subsect. ε), the Sutherland $(\infty, 4)$ potential has been used to fit the interaction potentials for ion-neutral collisions (Sect. 13β).

η) For a *repulsive exponential (Born-Mayer) potential*

$$\varphi(r) = \varphi_R \exp\left(1 - \frac{r}{r_R}\right) \quad \text{with } \varphi_R > 0 \qquad [8.2]$$

the normalized transfer cross sections $Q^{(l)*}(\mathscr{E}^*) = Q^{(l)}(\mathscr{E})/\pi r_R^2$ with $\mathscr{E}^* = \mathscr{E}/\varphi_R$ and $l = 1, 2$ are shown in Fig. 8, Sect. 8β. They can be approximated very accurately

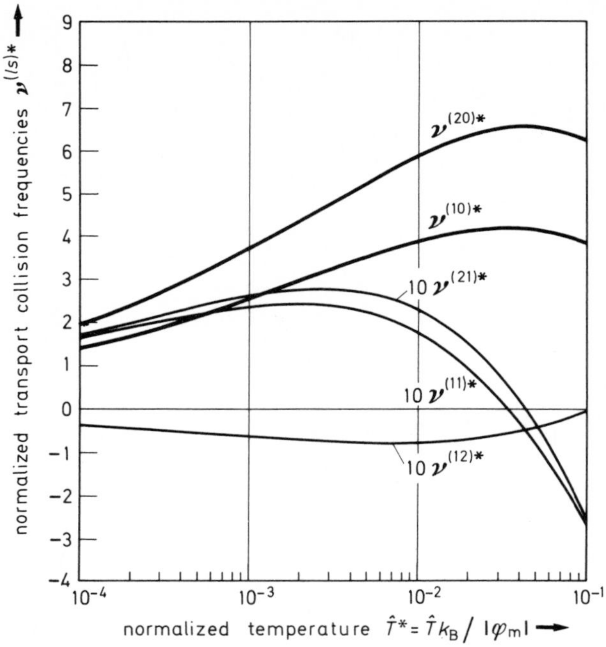

Fig. 30. Normalized transport collision frequencies $\nu^{(ls)*}(\hat{T}^*) = \nu^{(ls)}(\hat{T})/\sqrt{2\varphi_R/\mu}\,\pi r_R^2 N_k$, Eq. (19.3), for a repulsive exponential (Born-Mayer) potential, Eq. (8.2)

by

$$\frac{Q^{(l)}(\mathscr{E})}{\pi r_{(l)}^2} = \left(\ln\frac{\mathscr{E}_{(l)}}{\mathscr{E}}\right)^2 \qquad \text{for } 2\cdot 10^{-4} \lesssim \frac{\mathscr{E}}{\mathscr{E}_{(l)}} \lesssim 2\cdot 10^{-1}. \tag{[8.3c]}$$

The ratios of the normalization quantities $r_{(l)}$ and $\mathscr{E}_{(l)}$ to the reference quantities r_R and φ_R, respectively, of the potential Eq. (8.2) are given in Eqs. (8.3e). (They are of order one.) The transport collision frequencies $\nu^{(ls)*}(\hat{T}^*)$, normalized with r_R and φ_R, are shown in Fig. 30.

They were computed from tables[12] for

$$I_{(l,l)} := (l+1)!\left[1-\frac{1}{2}\frac{1+(-1)^l}{1+l}\right]\frac{1}{8(1-\ln\hat{T}^*)^2}\,\Omega^{(l,l)\star}(\hat{T}^*) \qquad (l=1,2)$$

with

$$\Omega^{(l,l)\star} := \left(\frac{r_d}{r_R}\right)^2 \frac{\Omega^{(l,l)}}{\Omega^{(l,l)}_{\text{rig sph}}} \tag{[19.12]}$$

and for B^*, C^*, E^*, Eqs. (19.15).

The changes of signs of the $\nu^{(11)}$ and $\nu^{(12)}$ with increasing temperature can be explained analogously to the case of the Morse potential (see Subsect. α).

With the approximation Eq. (8.3c) for $Q^{(l)}(\mathscr{E})$ the transport collision frequencies $\nu^{(ls)}$ can be calculated analytically. The result is given in Eq. (22.3).

[12] Monchick, L. (1959): Phys. Fluids 2, 695–700, table II

There the ratios of the normalized transport collision frequencies $\nu^{(ls)*}(\hat{T}^*)$ to the square root of the normalized temperature $\hat{T}^* := \hat{T}k/\mathscr{E}_{ij}$ are represented as polynomials of second degree in $\ln \hat{T}^*$.

The repulsive exponential potential is often used to represent the repulsive branch of a ground state of an (intermediate) diatomic molecule. This in turn is used to describe a collisional encounter between light ions and light atoms (see Table 8, Sect. 14) or between atoms (see Table 10, Sect. 17).

ϑ) The *screened Coulomb potential*

$$\varphi(r) = \frac{C_1}{r} \exp\left(-\frac{r}{r_s}\right) \qquad [12.1]$$

with the Debye-Hückel length

$$\lambda_D := [4\pi \sum_c l_{cc} N_c]^{-\frac{1}{2}} \qquad [12.2]$$

as screening length r_s is mostly used for the description of collisions between charged particles. For a reference length $r_R = r_s$ and $C_1/e^1 r_R$ as reference energy φ_R the normalized transfer cross sections $Q^{(l)}(\mathscr{E})/\pi r_R^2 = Q^{(l)*}(\mathscr{E}/|\varphi_R|)$ are shown in Figs. 9 (Sect. 8γ). Their high-energy limits are given in Eqs. (8.5).

The Landau length

$$l_{jk} := \frac{\mathrm{u}\, q_j q_k}{4\pi \varepsilon_0} \frac{1}{k \hat{T}_{jk}} \qquad [12.3]$$

was used to introduce the normalization length

$$r_N := \left[4\pi \sum_c \frac{l_{cc}}{|l_{jk}|} N_c\right]^{-\frac{1}{2}} = \frac{\lambda_D}{\sqrt{\hat{T}^*}} \qquad [12.6a, 8]$$

and the normalization energy

$$\mathscr{E}_N := \frac{C_1}{r_N} = \frac{\mathrm{u}\, q_j q_k}{4\pi \varepsilon_0} \frac{1}{r_N}, \qquad [12.7]$$

both varying with temperatures merely via the ratios $\hat{T}_{jk}/T_c$. The high-energy expressions for the transfer cross sections $Q^{(l)}(\mathscr{E})$ normalized with r_N and $\mathscr{E}_N$ are given by Eqs. (12.9).

From tables[13] of

$$T^{*2} \Omega^{(l,l)*} := \left(\frac{k\hat{T}\lambda_D}{C_1}\right)^2 \left(\frac{r_d}{\lambda_D}\right)^2 \frac{\Omega^{(l,l)}}{\Omega^{(l,l)}_{\text{rig sph}}}$$

$$= \left(\frac{k\hat{T}}{\mathscr{E}_N/\sqrt{\hat{T}^*}}\right)^2 \left(\frac{r_d}{\sqrt{\hat{T}^*}\, r_N}\right)^2 \frac{\Omega^{(l,l)}}{\Omega^{(l,l)}_{\text{rig sph}}} = \hat{T}^{*2} \Omega^{(l,l)\star}$$

for $l = 1, 2$ and B^*, C^*, E^*, curves of normalized transport collision frequencies

$$\nu_{jk}^{(ls)*} = \frac{\nu_{jk}^{(ls)}}{\sqrt{2|\mathscr{E}_N|/\mu}\, \pi r_N^2 N_k} \qquad [19.13]$$

are computed with the relations (19.14) between the $\Omega^{(l,l)\star}$ and $\nu^{(l0)*}$ for $l = 1, 2$ and Eqs. (19.16) for the connections between B^*, C^*, E^* and $\nu^{(11)*}$, $\nu^{(12)*}$, $\nu^{(21)*}$. The results are shown in Fig. 31.

[13] Mason, E.A., Munn, R.J., Smith, F.J. (1967): Phys. Fluids *10*, 1827–1832

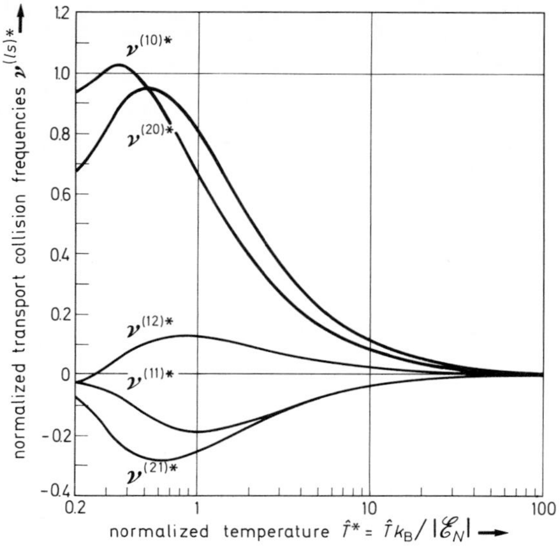

Fig. 31a. Normalized transport collision frequencies $\nu^{(ls)*}(\hat{T}^*) = \nu^{(ls)}(\hat{T})/\sqrt{2|\mathscr{E}_N|/\mu}\,\pi r_N^2 N_k$, Eq. (19.13), for an *attractive* Debye-Hückel screened Coulomb potential, Eq. (12.1). Normalization length r_N and normalization energy $\mathscr{E}_N$ after Eqs. (12.6) and (12.7), respectively

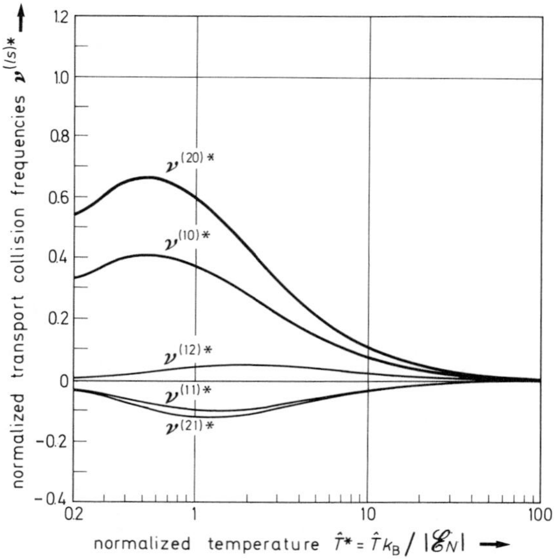

Fig. 31b. Normalized transport collicion frequencies $\nu^{(ls)*}(\hat{T}^*) = \nu^{(ls)}(\hat{T})/\sqrt{2\,\mathscr{E}_N/\mu}\,\pi r_N^2 N_k$, Eq. (19.13) for a *repulsive* Debye-Hückel screened Coulomb potential, Eq. (12.1). Normalization length r_N and normalization energy $\mathscr{E}_N$ after Eqs. (12.6) and (12.7), respectively

The high-temperature approximation, Eq. (22.1b) with Eq. (B.7c) from Appendix B δ, reads

$$\nu^{(ls)*}\left(\frac{k\hat{T}}{|\mathscr{E}_N|}\right)=\frac{l(l+1)}{2}\left(\frac{k\hat{T}}{|\mathscr{E}_N|}\right)^{-\frac{3}{2}}\frac{(l-1)!\,(-\frac{3}{2})!}{(l+\frac{1}{2}+s)!\,(-\frac{3}{2}-s)!}$$
$$\times\left\{\ln\left[\frac{4}{e^{2\gamma}}\left(\frac{k\hat{T}}{|\mathscr{E}_N|}\right)^{\frac{3}{2}}\right]+O(1)\right\} \tag{20.4}$$

and shows the alternation of the sign with varying s as $\nu^{(ls)}\sim(-1)^s$ because of $(-\frac{3}{2}-s)!\equiv\Gamma(-\frac{1}{2}-s)$ in the denominator. The temperature independent ratios $\nu^{(1s)}/\nu^{(10)}$ are shown in Figs. 23 and 24 (Sect. 19 β) with the inverse power $n=1$ for the Coulomb potential and the power $p=-3$ for the (dominant) energy dependence of the transfer collision frequencies, Eqs. (4.14) and (8.5). The ratios $\nu^{(1s)}/\nu^{(10)}$ in these figures converge very slowly to zero for increasing s, in contrast to ratios for greater powers $p>-3$ in Fig. 24 (Sect. 19 β). This is the reason why the $\nu^{(ls)*}$ in Fig. 31 for high temperatures are of about the same order of magnitude in contrast to other interaction potentials in Figs. 25 to 30, where the $\nu^{(l0)}$ exceed the $\nu^{(ls)}$ with $s\geqq 1$ by about one order of magnitude.

It is interesting to note that – according to Eqs. (12.6a) and (12.8) – the dimensionless quantity

$$\left(\frac{k\hat{T}_{jk}}{|\mathscr{E}_N|}\right)^{\frac{3}{2}}=:\hat{T}_{jk}^{*\frac{3}{2}}=\left(\frac{r_N}{|l_{jk}|}\right)^{\frac{3}{2}}=\left(\frac{\lambda_D}{r_N}\right)^3=4\pi\sum_c\frac{l_{cc}}{|l_{jk}|}N_c\,\lambda_D^3=\frac{\lambda_D}{|l_{jk}|} \tag{20.5}$$

is three times the number N_D of charged particles in a Debye-Hückel sphere (of radius λ_D). The reciprocal of this quantity, the *plasma* or *discreteness parameter*, must be small for almost all kinetic theories of *hot plasmas*, corresponding to high values of $\hat{T}^{*3/2}$. For *non-ideal* (or *strongly coupled*) *plasmas* with high densities N_c and low temperatures T_c this prerequisite is no longer satisfied. But our model of Debye-Hückel screened binary Coulomb collisions yields results even for values of $\hat{T}^*$ of order unity or smaller, see Figs. 31. For very small values of $\hat{T}^*$ the exponential screening dominates the pure Coulomb interaction and the transport collision frequencies $\nu^{(ls)}(\hat{T})$ in Fig. 31b vary similar to those for a repulsive exponential (Born-Mayer) potential in Fig. 30.

The normalization frequency

$$\nu_N:=\sqrt{2\,|\mathscr{E}_N|/\mu_{jk}}\,\pi\,r_N^2\,N_k, \tag{20.6}$$

which was used to normalize the transport collision frequencies $\nu_{jk}^{(ls)}$ according to Eq. (19.13) as

$$\nu_{jk}^{(ls)*}=\nu_{jk}^{(ls)}/\nu_N, \tag{20.7}$$

is proportional to the *plasma circular frequency*

$$\omega_{jk}:=\left(\frac{\mathrm{u}\,|q_j\,q_k|\,N_k}{\varepsilon_0\,\mu_{jk}}\right)^{\frac{1}{2}}. \tag{20.8}$$

With the expressions Eqs. (12.7) for $\mathscr{E}_N$ and (12.6a) for r_N we find

$$\nu_N^2 = \frac{\omega_{jk}^2}{8 \sum\limits_{c} \frac{l_{cc}}{|l_{jk}|} \frac{N_c}{N_k}} = \frac{\pi r_N^3 N_k}{2} \omega_{jk}^2. \tag{20.9}$$

Because of $l_{jk} \sim q_j q_k / \hat{T}_{jk}$, Eq. (12.3), the proportionality factor between ν_N and ω_{jk} depends on ratios of charges, temperatures and number densities.

21. Numerical and analytical results for transport collision frequencies

α) Transport collision frequencies for collisions between electrons and neutrals. In Sect. 20 the transport collision frequencies $\nu^{(ls)}(\hat{T})$ for some interaction potentials were taken from published tables of omega integrals $\Omega^{(l,s)}(\hat{T})$, which in turn were obtained by numerical energy averaging of transfer cross sections $\phi^{(l)}(\mathscr{E})$ as described in the corresponding literature. To obtain *momentum transport collision frequencies* $\nu^{(1s)}(\hat{T})$ for collisions between electrons and neutrals, the curves of momentum transfer cross sections $Q^{(1)}(\mathscr{E})$ in Sect. 11 must be averaged.

The computational procedure for this averaging is described by WEINERT et al.[1]. Due to the orthogonality properties of the generalized Laguerre (Sonine) polynomials $L_s^{(l+\frac{1}{2})}(w)$, used as weight functions for the averaging, Eq. (19.3), it was sufficient[2] to do the numerical integration only for one or two values of the (combined) temperature $\hat{T} \approx T_e$, Eq. (12.4). The results[3] are represented in Figs. 32 through 38. Shown are the ratios $\nu_{en}^{(1s)}(T_e)/N_n$ of momentum transport collision frequencies $\nu_{en}^{(1s)}(T_e)$ to number densities N_n of the neutrals. These ratios are called *momentum transport collision coefficients* analogous to the (energy dependent) transfer collision coefficients $gQ^{(l)}(\mathscr{E})$ in Sect. 4γ (see Sect. 19α).

For collisions between electrons and the heavier aeronomic molecules N_2, O_2, CO_2, the low-energy behavior of the momentum transfer cross sections $Q^{(1)}(\mathscr{E})$ was roughly represented by a power law approximation $Q_{en}^{(1)}(\mathscr{E}) = Q_p^{(1)}(\mathscr{E}/\text{eV})^p$, Eq. (18.17a) with the coefficients $Q_p^{(1)}$ collected at the end of Sect. 18ζ. The corresponding momentum transfer collision frequencies $\nu_{en}^{(1)}(\mathscr{E}) = \nu_{en}^{(1)}(kT_e)\, w^{p+\frac{1}{2}}$, Eq. (18.20a), are proportional to the monoenergetic momentum transfer collision frequencies [see Eq. (18.20b)],

$$\nu_m := \nu_{en}^{(1)}(kT_e) = \sqrt{2\,\text{eV}/m_e}\; Q_p^{(1)} (kT_e/\text{eV})^{p+\frac{1}{2}} N_n, \tag{21.1a}$$

which can be written as

$$\frac{\nu_m}{\text{Hz}} = 5.931 \cdot 10^5 \frac{Q_p^{(1)}}{\text{m}^2} (8.617 \cdot 10^{-5})^{p+\frac{1}{2}} \left(\frac{T_e}{\text{K}}\right)^{p+\frac{1}{2}} \frac{N_n}{\text{m}^{-3}}. \tag{21.1b}$$

[1] Weinert, U., Kratzsch, K.-A., Oberhage, H.-R. (1978): Z. Naturforsch. Teil A *33*, 1423–1427

[2] Weinert, U. (1978): Z. Naturforsch. Teil A *33*, 1428–1431

[3] Weinert, U., Kratzsch, K.-A., Oberhage, H.-R. (1980): private communication

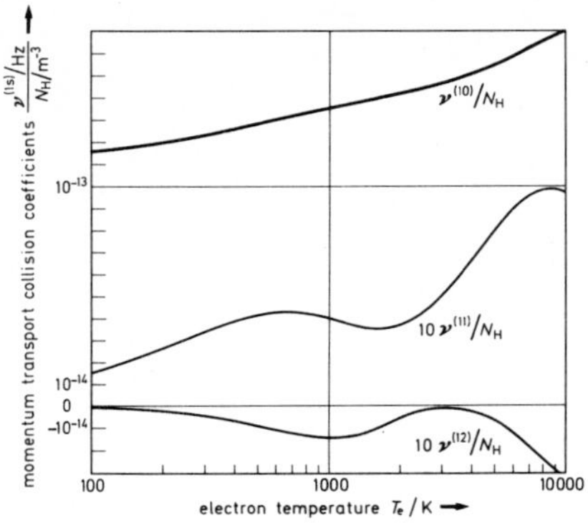

Fig. 32

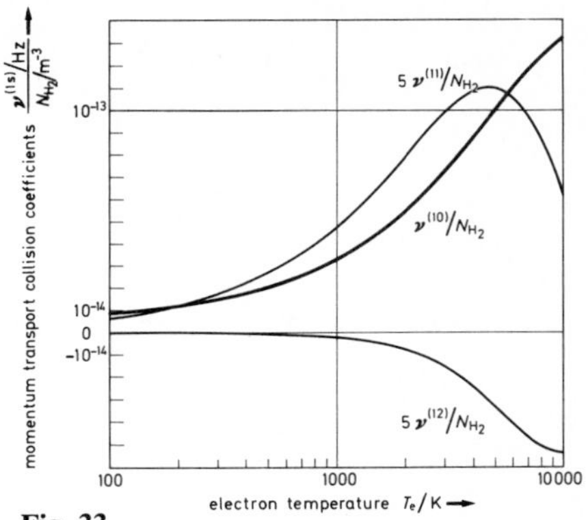

Fig. 33

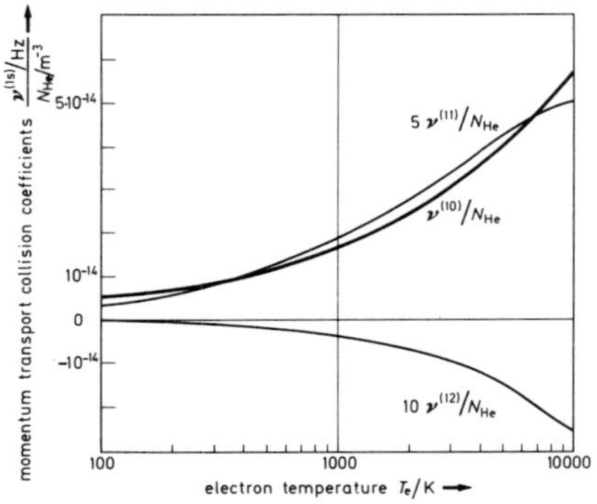

Fig. 34

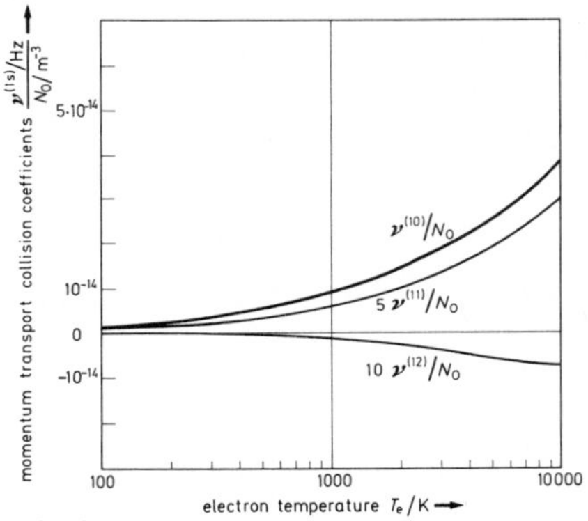

Fig. 35

Fig. 32. Momentum transport collision coefficients $\nu_{eH}^{(1s)}(T_e)/N_H$ for collisions of electrons with hydrogen atoms, using the momentum transfer cross section $Q^{(1)}(\mathscr{E})$ of Fig. 11 (Sect. 11 α)

Fig. 33. Momentum transport collision coefficients $\nu_{eH_2}^{(1s)}(T_e)/N_{H_2}$ for collisions of electrons with hydrogen molecules, using the momentum transfer cross section $Q^{(1)}(\mathscr{E})$ of Fig. 12 (Sect. 11 β)

Fig. 34. Momentum transport collision coefficients $\nu_{eHe}^{(1s)}(T_e)/N_{He}$ for collisions of electrons with helium atoms, using the momentum transfer cross section $Q^{(1)}(\mathscr{E})$ of Fig. 13 (Sect. 11 γ)

Fig. 35. Momentum transport collision coefficients $\nu_{eO}^{(1s)}(T_e)/N_O$ for collisions of electrons with oxygen atoms, using the momentum transfer cross section $Q^{(1)}(\mathscr{E})$ of Fig. 14 (Sect. 11 ε)

With the values of Eqs. (11.1) to (11.3) for the coefficients $Q_p^{(1)}$ listed at the end of Sect. 18 ζ, we obtain for collisions

$$\mathrm{e} + \mathrm{N}_2: \quad \frac{\nu_m}{\mathrm{Hz}} = 1.022 \cdot 10^{-17} \frac{T_e}{\mathrm{K}} \frac{N_{N_2}}{\mathrm{m}^{-3}} \quad \text{for } 100 < \frac{T_e}{\mathrm{K}} < 1000, \tag{21.2a}$$

$$\mathrm{e} + \mathrm{O}_2: \quad \frac{\nu_m}{\mathrm{Hz}} = 2.619 \cdot 10^{-18} \left(\frac{T_e}{\mathrm{K}}\right)^{\frac{15}{14}} \frac{N_{O_2}}{\mathrm{m}^{-3}} \quad \text{for } 100 < \frac{T_e}{\mathrm{K}} < 500, \tag{21.2b}$$

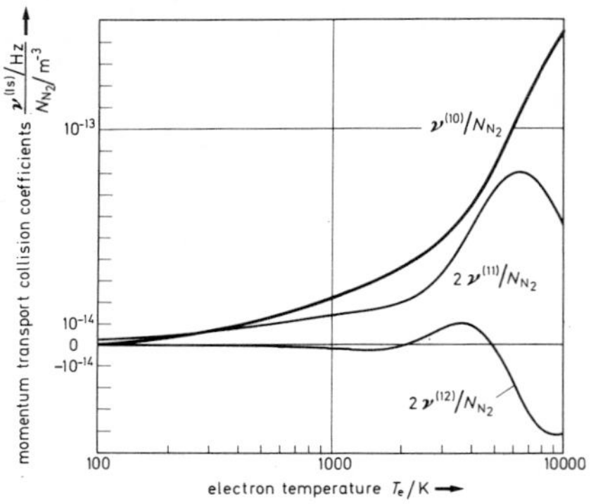

Fig. 36

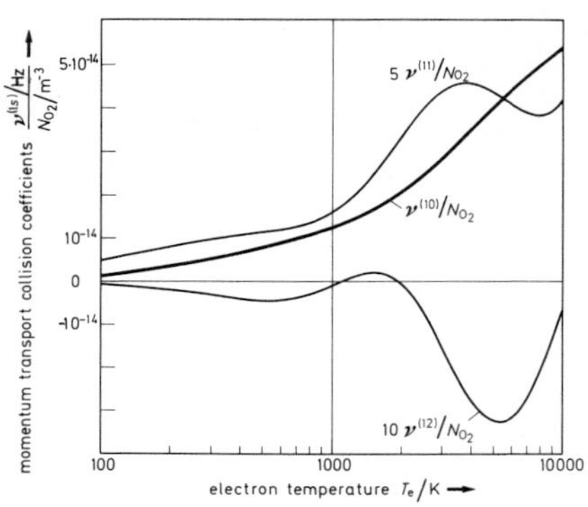

Fig. 37

Fig. 36. Momentum transport collision coefficients $\nu_{eN_2}^{(1s)}(T_e)/N_{N_2}$ for collisions of electrons with nitrogen molecules, using the momentum transfer cross section $Q^{(1)}(\mathscr{E})$ of Fig. 15 (Sect. 11 ζ)

Fig. 37. Momentum transport collision coefficients $\nu_{eO_2}^{(1s)}(T_e)/N_{O_2}$ for collicions of electrons with oxygen molecules, using the momentum transfer cross section $Q^{(1)}(\mathscr{E})$ of Fig. 16 (Sect. 11 η)

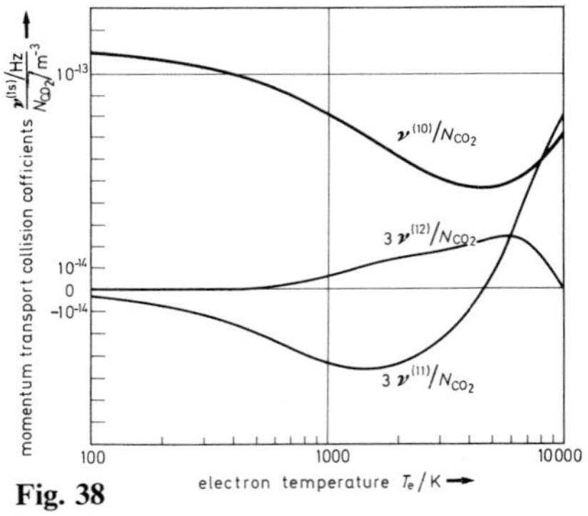

Fig. 38

Fig. 38. Momentum transport collision coefficients $\nu_{eCO_2}^{(1s)}(T_e)/N_{CO_2}$ for electron collisions with carbon dioxide molecules, using the momentum transfer cross section $Q^{(1)}(\mathscr{E})$ of Fig. 17 (Sect. 11 ϑ)

$$e + CO_2: \quad \frac{\nu_m}{\text{Hz}} = 1.01 \cdot 10^{-13} \frac{N_{CO_2}}{\text{m}^{-3}} \quad \text{for } 100 < \frac{T_e}{\text{K}} < 1000. \tag{21.2c}$$

The corresponding low-temperature momentum transport collision frequencies are

$$\nu^{(1s)}(\hat{T}) = \nu_m \frac{(p+2)!\,(p+\frac{1}{2})!}{(\frac{3}{2}+s)!\,(p+\frac{1}{2}-s)!} \qquad [19.8\text{d}]$$

with $p+\frac{1}{2}=1, \frac{15}{14}, 0$ for N_2, O_2, CO_2, respectively.

β) Analytical expressions for transport collision frequencies. If the transfer cross sections $Q^{(l)}(\mathscr{E})$ and hence the transfer collision frequencies $\nu^{(l)}(\mathscr{E}) \sim \sqrt{\mathscr{E}}\, Q^{(l)}(\mathscr{E})$, Eq. (4.14), are given by analytic expressions of the form

$$Q^{(l)}(\mathscr{E})=\sum_p Q_p^{(l)}\,\mathscr{E}^{*p}+Q_{p_l,\ln}^{(l)}\,\mathscr{E}^{*p_l}\ln\mathscr{E}^*+Q_{\ln^2}^{(l)}(\ln\mathscr{E}^*)^2, \qquad [18.3]$$

$$\nu^{(l)}(\mathscr{E})=\sum_p \nu^{(l)}\,\mathscr{E}^{*p+\frac{1}{2}}+\nu_{p_l,\ln}^{(l)}\,\mathscr{E}^{*p_l+\frac{1}{2}}\ln\mathscr{E}^*+\nu_{\ln^2}^{(l)}\,\mathscr{E}^{*\frac{1}{2}}(\ln\mathscr{E}^*)^2 \qquad [18.4a]$$

where the coefficients are related by

$$\nu_p^{(l)}=\sqrt{2\mathscr{E}_R/\mu}\,Q_p^{(l)}\,N_k \quad \text{etc.} \qquad [18.5]$$

and $\mathscr{E}_R=\mathscr{E}/\mathscr{E}^*$ is a reference energy, then the transport collision frequencies

$$\nu^{(ls)}(\hat{T}):=\frac{(-1)^s\,s!}{(l+\frac{1}{2}+s)!}\int_0^\infty \mathrm{d}w\,\mathrm{e}^{-w}\,\mathrm{L}_s^{(l+\frac{1}{2})}(w)\,w^{l+\frac{1}{2}}\,\nu^{(l)}(k\hat{T}w) \qquad [19.3]$$

with $w:=\mathscr{E}/k\hat{T}$, Eq. (19.1b), can be calculated in closed form.

With

$$\mathscr{E}^*=\frac{\mathscr{E}}{\mathscr{E}_R}=\frac{\mathscr{E}}{k\hat{T}}\,\frac{k\hat{T}}{\mathscr{E}_R}=w\,\hat{T}^*$$

we write Eq. (18.4a) in the form

$$\begin{aligned}\nu^{(l)}(k\hat{T}w)=&\sum_p \nu_p^{(l)}\,\hat{T}^{*p+\frac{1}{2}}\,w^{p+\frac{1}{2}}\\&+\nu_{p_l,\ln}^{(l)}\,\hat{T}^{*p_l+\frac{1}{2}}\,w^{p_l+\frac{1}{2}}(\ln\hat{T}^*+\ln w)\\&+\nu_{\ln^2}^{(l)}\,\hat{T}^{*\frac{1}{2}}\,w^{\frac{1}{2}}[(\ln\hat{T}^*)^2+2\ln\hat{T}^*\ln w+(\ln w)^2].\end{aligned} \qquad (21.3)$$

For inverse power potentials the reference energy was already chosen in Eq. (18.8a) as the thermal energy $k\hat{T}$. Therefore in this particular case we have $\hat{T}^*=1$.

If the transfer cross sections and the transfer collision frequencies are expressed in normalized forms $Q^{(l)*}$ and $\nu^{(l)*}$, respectively, then the normalized coefficients of the expansions Eqs. (18.6) and (18.7) are equal.

If we put the representation Eq. (21.3) for $\nu^{(l)}(k\hat{T}w)$ into the integral of Eq. (19.3) for $\nu^{(ls)}(\hat{T})$, we have to calculate the following three types of integrals:

$$\begin{pmatrix} l+\frac{1}{2} & p+\frac{1}{2} \\ & s \end{pmatrix}:=\frac{(-1)^s\,s!}{(l+\frac{1}{2}+s)!}\int_0^\infty \mathrm{d}w\,\mathrm{e}^{-w}\,\mathrm{L}_s^{(l+\frac{1}{2})}(w)\,w^{l+\frac{1}{2}+p+\frac{1}{2}}, \qquad (21.4a)$$

$$\begin{bmatrix} l+\frac{1}{2} & p+\frac{1}{2} \\ & s \end{bmatrix}:=\frac{(-1)^s\,s!}{(l+\frac{1}{2}+s)!}\int_0^\infty \mathrm{d}w\,\mathrm{e}^{-w}\,\mathrm{L}_s^{(l+\frac{1}{2})}(w)\,w^{l+\frac{1}{2}+p+\frac{1}{2}}\ln w, \qquad (21.4b)$$

$$\begin{Bmatrix} l+\frac{1}{2} & \frac{1}{2} \\ & s \end{Bmatrix}:=\frac{(-1)^s\,s!}{(l+\frac{1}{2}+s)!}\int_0^\infty \mathrm{d}w\,\mathrm{e}^{-w}\,\mathrm{L}_s^{(l+\frac{1}{2})}(w)\,w^{l+1}(\ln w)^2. \qquad (21.4c)$$

In Appendix Bδ closed expressions are derived for these integrals and they are tabulated for those values of l, p, s which are needed for the results of Sect. 18.

With the coefficients Eqs. (21.4) the transport collision frequencies $\nu^{(ls)}(\hat{T})$ for transfer collision frequencies $\nu^{(l)}(\mathscr{E})$, Eq. (21.3), can be written as

$$\nu^{(ls)}(\hat{T})=\sum_p \nu_p^{(l)}\,\hat{T}^{*p+\frac{1}{2}}\begin{pmatrix} l+\frac{1}{2} & p+\frac{1}{2} \\ & s \end{pmatrix}$$
$$+\nu_{p_l,\ln}^{(l)}\,\hat{T}^{*p_l+\frac{1}{2}}\left\{\begin{pmatrix} l+\frac{1}{2} & p_l+\frac{1}{2} \\ & s \end{pmatrix}\ln\hat{T}^*+\begin{bmatrix} l+\frac{1}{2} & p_l+\frac{1}{2} \\ & s \end{bmatrix}\right\}$$
$$+\nu_{\ln^2}^{(l)}\,\hat{T}^{*\frac{1}{2}}\left\{\begin{pmatrix} l+\frac{1}{2} & \frac{1}{2} \\ & s \end{pmatrix}(\ln\hat{T}^*)^2+2\begin{bmatrix} l+\frac{1}{2} & \frac{1}{2} \\ & s \end{bmatrix}\ln\hat{T}^*+\begin{Bmatrix} l+\frac{1}{2} & \frac{1}{2} \\ & s \end{Bmatrix}\right\}. \quad (21.5)$$

Since the transport collision frequencies $\nu^{(ls)}(\hat{T})$ are in general analytic functions[4] of the (combined) temperature $\hat{T}$, the power terms with integer p and all the logarithmic terms are valid only in a certain temperature range.

22. Survey of the methods for the determination of transport collision frequencies. In Sects. 19 through 21 the calculation of transport collision frequencies $\nu^{(ls)}(\hat{T})$, Eq. (19.3), has been described according to the shape of the corresponding transfer cross section $Q^{(l)}(\mathscr{E})$, Eq. (4.8). We distinguished results for model potentials of the collisional interaction (Sect. 20), measured or theoretical values for electron-neutral collisions (Sect. 21α), and analytical expressions (Sect. 21β). Now we collect the results and order them according to the species of the colliding particles:

α) Electron-neutral collisions,
β) Collisions between charged particles,
γ) Collisions between light atomic ions and light atoms,
δ) Other collisions between ions and neutrals,
ε) Collisions of ions in their parent gas,
ζ) Collisions between neutral particles.

α) Electron-neutral collisions. The transport collision frequencies $\nu^{(1s)}(T_e)$ are shown in Figs. 32 to 38 of Sect. 21α. The low-temperature limits for collisions of electrons with N_2, O_2, CO_2 can roughly be described by power laws listed in Eqs. (18.21) and (21.2).

β) Collisions between charged particles. If the interaction is described with sufficient accuracy by a screened Coulomb potential, Eq. (12.1), with the Debye-Hückel length, Eq. (12.2), as screening length, the normalized transport collision frequencies $\nu_{jk}^{(ls)*}(\hat{T})$ are shown in Figs. 31 of Sect. 20ϑ.

The normalization length

$$r_N := \left[4\pi\sum_c \frac{q_c^2}{|q_j q_k|}\frac{\hat{T}_{jk}}{T_c}N_c\right]^{-\frac{1}{3}} \quad [12.6a]$$

is a mean distance between charged particles; the (very small) normalization energy

$$\mathscr{E}_N := \frac{C_1}{r_N} = \frac{\mathrm{u}\,q_j q_k}{4\pi\varepsilon_0}\left[4\pi\sum_c \frac{q_c^2}{|q_j q_k|}\frac{\hat{T}_{jk}}{T_c}N_c\right]^{\frac{1}{3}} \quad [12.7]$$

[4] Weinert, U. (1978): Z. Naturforsch. Teil A *33*, 1428–1431, eq. 1.15

is the Coulomb energy between the colliding charged particles of species j and k, taken at the normalization length r_N, Eq. (12.6a).

The analytic expression for the normalized transfer cross sections

$$Q^{(l)*}(\mathscr{E}^*)=\frac{l(l+1)}{2}\mathscr{E}^{*-2}\left[\ln\left(\frac{4}{e^{\gamma}}\sqrt{\hat{T}^*}\,\mathscr{E}^*\right)-\frac{l}{2}\right]$$

$$=Q^{(l)*}_{-2}\,\mathscr{E}^{*-2}+Q^{(l)*}_{-2,\ln}\,\mathscr{E}^{*-2}\ln\mathscr{E}^* \qquad [12.9]$$

with

$$Q^{(l)*}_{-2}=\frac{l(l+1)}{2}\left(\ln 4\sqrt{\hat{T}^*}-\gamma-\frac{l}{2}\right),\qquad Q^{(l)*}_{-2,\ln}=\frac{l(l+1)}{2} \qquad [18.15]$$

yields as representation for the normalized transfer collision frequencies

$$\nu^{(l)*}(\mathscr{E}^*)=Q^{(l)*}_{-2}\,\mathscr{E}^{*-2+\frac{1}{2}}+Q^{(l)*}_{-2,\ln}\,\mathscr{E}^{*-2}\ln\mathscr{E}^*. \qquad [18.7a]$$

From the general result, Eq. (21.5), for the transport collision frequencies we obtain

$$\frac{\nu^{(ls)*}(\hat{T}^*)}{\hat{T}^{*-\frac{3}{2}}}=Q^{(l)*}_{-2,\ln}\begin{pmatrix}l+\frac{1}{2} & -\frac{3}{2}\\ & s\end{pmatrix}\ln\hat{T}^*$$
$$+Q^{(l)*}_{-2}\begin{pmatrix}l+\frac{1}{2} & -\frac{3}{2}\\ & s\end{pmatrix}+Q^{(l)*}_{-2,\ln}\begin{bmatrix}l+\frac{1}{2} & -\frac{3}{2}\\ & s\end{bmatrix}. \qquad (22.1\text{a})$$

According to Eq. (B.8c) in Appendix B δ the last coefficient can be decomposed. This leads together with the expressions (18.15) for the coefficients $Q^{(l)*}_{-2}$ and $Q^{(l)*}_{-2,\ln}$ and Eqs. (B.7c) (B.8d) for the bracket coefficients in Eq. (22.1a) to

$$\frac{\nu^{(ls)*}(\hat{T}^*)}{\hat{T}^{*-\frac{3}{2}}}=\frac{l(l+1)}{2}\begin{pmatrix}l+\frac{1}{2} & -\frac{3}{2}\\ & s\end{pmatrix}\left\{\ln\left(\frac{4}{e^{2\gamma}}\hat{T}^{*\frac{3}{2}}\right)-\frac{l}{2}\right\}$$
$$+\frac{l(l+1)}{2}\begin{bmatrix}l+\frac{1}{2} & -\frac{3}{2}\\ & s\end{bmatrix}_0. \qquad (22.1\text{b})$$

To eliminate the normalization we have to multiply $\nu^{(ls)*}(\hat{T}^*)$ with $\sqrt{2|\mathscr{E}_N|/\mu}\,\pi r_N^2 N_k$, Eq. (19.13). We obtain, together with $\mathscr{E}_N r_N=C_1:=\mathrm{u}\,q_j q_k/4\pi\varepsilon_0$, Eq. (12.7), the result

$$\frac{\nu^{(ls)}_{jk}(\hat{T})}{\pi\sqrt{2/\mu}\,C_1^2 N_k(k\hat{T})^{-\frac{3}{2}}}=\frac{l(l+1)}{2}\begin{pmatrix}l+\frac{1}{2} & -\frac{3}{2}\\ & s\end{pmatrix}\left\{\ln\left[\frac{4}{e^{2\gamma}}\left(\frac{k\hat{T}}{|\mathscr{E}_N|}\right)^{\frac{3}{2}}\right]-\frac{l}{2}\right\}$$
$$+\frac{l(l+1)}{2}\begin{bmatrix}l+\frac{1}{2} & -\frac{3}{2}\\ & s\end{bmatrix}_0. \qquad (22.2)$$

The bracket coefficients are tabulated in Tables 12 and 13, Appendix B δ.

γ) For *collisions between light atomic ions and light atoms*, i.e., $H^+ + H$, $H^+ + He$, $He^+ + H$, $He^+ + He$, the interaction potentials are listed in Table 8 (Sect. 14). They are Morse potentials, Eq. (7.1), repulsive exponential potentials, Eq. (8.2), and a Lennard-Jones (12, 4) potential, Eq. (7.14). The corresponding normalized transport collision frequencies $\nu^{(ls)*}(\hat{T}^*)$ are shown in Figs. 25, 30 and 27a, resp., of Sect. 20. For the Lennard-Jones (12, 4) potential the limit for low temperature is given in Eq. (20.2a) with $n=12$, for high temperature in Eq. (20.2c) with $n=12$, $\gamma=0$.

The normalizations are eliminated with

$$\nu^{(ls)}(\hat{T})=\sqrt{2\mathscr{E}_R/\mu}\,\pi r_R^2 N_k \nu^{(ls)*}(\hat{T}^*) \qquad [19.13]$$

and

$$\hat{T}=\mathscr{E}_R \hat{T}^*/k, \qquad [18.18]$$

where the reference energies $\mathscr{E}_R$ and the reference lengths r_R are taken from Table 8, Sect. 14.

For a *repulsive exponential (Born-Mayer) potential* the normalized transfer cross section can be written as

$$\frac{Q^{(l)}(\mathscr{E})}{\pi r_{(l)}^2}=\left(\ln\frac{\mathscr{E}}{\mathscr{E}_{(l)}}\right)^2 \quad \text{for } 2\cdot 10^{-4}\lesssim\frac{\mathscr{E}}{\mathscr{E}_{(l)}}\lesssim 2\cdot 10^{-1} \qquad [8.3\text{c}]$$

with the following relations between normalization and reference quantities:

$$\frac{r_{(1)}}{r_R}=0.977,\quad \frac{r_{(2)}}{r_R}=1.064,\quad \frac{\mathscr{E}_{(1)}}{\varphi_R}=0.464,\quad \frac{\mathscr{E}_{(2)}}{\varphi_R}=0.754. \qquad [8.3\text{e}]$$

This is a special case of the general expression Eq. (18.6) with the only nonvanishing coefficient

$$Q_{\ln^2}^{(l)*}=1. \qquad [18.13]$$

The normalized transport collision frequencies $\nu^{(ls)*}(\hat{T}^*)$ can be taken from Eq. (21.5) with $\nu_{\ln^2}^{(l)*}=Q_{\ln^2}^{(l)*}=1$ [see Eqs. (18.7a) (18.13)] as

$$\frac{\nu^{(ls)*}(\hat{T}^*)}{\sqrt{\hat{T}^*}}=\begin{pmatrix} l+\frac{1}{2} & \frac{1}{2} \\ s & \end{pmatrix}(\ln\hat{T}^*)^2+2\begin{bmatrix} l+\frac{1}{2} & \frac{1}{2} \\ s & \end{bmatrix}\ln\hat{T}^*+\begin{Bmatrix} l+\frac{1}{2} & \frac{1}{2} \\ s & \end{Bmatrix} \qquad (22.3\text{a})$$

$$=\begin{pmatrix} l+\frac{1}{2} & \frac{1}{2} \\ s & \end{pmatrix}\left[\left(\ln\frac{\hat{T}^*}{e^\gamma}\right)^2+\frac{\pi^2}{6}\right]+2\begin{bmatrix} l+\frac{1}{2} & \frac{1}{2} \\ s & \end{bmatrix}_0\ln\frac{\hat{T}^*}{e^\gamma}+\begin{Bmatrix} l+\frac{1}{2} & \frac{1}{2} \\ s & \end{Bmatrix}_0 \qquad (22.3\text{b})$$

with $\gamma:=0.5772$ and $\exp\gamma=1.7811$. The bracket coefficients are given in Eqs. (B.7) (B.8) (B.9) in Appendix Bδ and their values are listed in Tables 12, 13, 14 in this appendix. The normalization of $\nu^{(ls)*}(\hat{T}^*)$ is removed by multiplication with $\sqrt{2\mathscr{E}_{(l)}/\mu_{jk}}\,\pi r_{(l)}^2 N_k$, Eq. (19.13), and the substitution $\hat{T}^*=\hat{T}k/\mathscr{E}_{(l)}$, Eq. (18.18).

δ) Other collisions between ions and neutrals. For collisions of ions with atomic oxygen an attractive Maxwell potential $\varphi(r)=C_4/r^4$ with

$$C_4=-\frac{1}{2}Z_c^2\frac{\mathrm{u}\,q_e^2}{4\pi\varepsilon_0}\frac{\mathrm{u}\,\alpha_o}{4\pi\varepsilon_0} \qquad [9.2]$$

must be used, where the polarizability α_o is taken from Table 4 (Sect. 9) as

$$\frac{\mathrm{u}\,\alpha_o}{4\pi\varepsilon_0}=5.19\,a_0^3=0.77\cdot 10^{-30}\ \mathrm{m}^3.$$

The transfer collision frequencies $\nu^{(l)}$, Eq. (18.10a), do not depend on the energy for Maxwell interaction. In this special case they are equal to the monoenergetic transfer collision frequencies $\nu^{(l)}$ and to the transport collision frequencies $\nu^{(l\,0)}$,

$$\nu_{jk}^{(l)} = \pi\sqrt{2/\mu_{jk}}\,|C_4|^{\frac{1}{2}}\,Q^{(l)*}(4,-)\,N_k = \nu_{jk}^{(l\,0)}, \qquad [18.10\text{b}]\ [19.8\text{f}]$$

which in turn do not depend on the temperature. The factors

$$Q^{(1)*}(4,-) = 2.2092, \qquad Q^{(2)*}(4,-) = 2.3076$$

are taken from Table 1 (Sect. 6β).

If at least one of the collision partners for ion-neutral collisions is not a light particle like H, H^+, He, He^+ and the neutral particle is not an oxygen atom, a Sutherland (∞, 4) potential, Eq. (7.16), or a Mason-Schamp (12, 6, 4) potential Eq. (7.10), can be used. For the Sutherland (∞, 4) potential the normalized transport collision frequencies $\nu^{(ls)*}(\hat{T}^*)$ are shown in Fig. 29 (Sect. 20ζ), their values for low and high temperatures are given in Eqs. (20.3). The reference length r_m for the collision partners A^+ and B or A^+ and A must be calculated with

$$r_m(A^+ + B) = \tfrac{1}{2}[r_0(A) + r_0(B)] \qquad [13.1\text{a}]$$

$$r_m(A^+ + A) = \sqrt{2}\, r_0(A) \qquad [13.1\text{b}]$$

with the values for $r_0(A)$ and $r_0(B)$ taken from Table 7 (Sect. 13β). The reference energy is $-\varphi_m = -C_4/r_m^4$ with

$$C_4 = -\frac{1}{2} Z_c^2 \frac{\mathrm{u} q_e^2}{4\pi\varepsilon_0} \frac{\mathrm{u}\alpha_n}{4\pi\varepsilon_0}. \qquad [9.2]$$

The values for the polarizabilities α_n are listed in Table 4 (Sect. 9). The normalization of $\nu^{(ls)*}(\hat{T}^*)$ is eliminated by multiplication with $\sqrt{2|\varphi_m|/\mu_{jk}}\,\pi r_m^2 N_k$ and the substitution $\hat{T}^* = \hat{T}k/|\varphi_m|$. The same holds for the normalized transport collision frequencies $\nu^{(ls)*}(\hat{T}^*)$ for the Mason-Schamp (12, 6, 4) potential, Eq. (7.10), shown in Fig. 27 (Sect. 20γ). The determination of the reference quantities r_m and φ_m and of the parameter γ is quite involved and described in detail in Sect. 13γ.

ε) For *collisions of ions in their parent gas* twice the charge exchange cross section $Q^{ex}(\mathscr{E})$ has to be combined with the classical momentum transfer cross section $Q^{(1)}_{class}(\mathscr{E})$ [see Eqs. (16.3) (16.4) (16.5)]. Its normalized form is

$$\frac{2Q^{ex}(\mathscr{E})}{\pi(r_\Delta/2)^2} = \left(\ln\frac{\mathscr{E}}{\mathscr{E}_\Delta}\right)^2 \qquad [15.17\text{a}]$$

with the normalization quantities r_Δ and $\mathscr{E}_\Delta$ listed in Table 9 (Sect. 15δ). It is included in the general expression Eq. (18.6) with the only nonvanishing coefficient $Q^{(l)*}_{\ln^2} = 1$, Eq. (18.13). The normalized transport collision frequencies $\nu^{(ls)*}(\hat{T}^*)$ for the charge exchange contribution are given in Eq. (22.3). Their normalization is removed by multiplication with $\sqrt{2\mathscr{E}_\Delta/\mu_{jk}}\,\pi(r_\Delta/2)^2 N_k$ and the substitution $\hat{T}^* = \hat{T}k/\mathscr{E}_\Delta$, Eq. (18.18). For the combination of $\nu^{(1s)}_{\text{class}}(\hat{T})$, calculated from $Q^{(1)}_{\text{class}}(\mathscr{E})$, with $2\nu^{(\text{ex}\,s)}$, calculated from $Q^{\text{ex}}(\mathscr{E})$, BANKS[1] proposed

$$\nu^{(1s)}(\hat{T}) = \nu^{(1s)}_{\text{class}}(\hat{T})\,\theta(\nu^{(1s)}_{\text{class}} - 2\nu^{(\text{ex}\,s)}) + 2\nu^{(\text{ex}\,s)}(\hat{T})\,\theta(2\nu^{(\text{ex}\,s)} - \nu^{(1s)}_{\text{class}}) \tag{22.4}$$

with a smoothing of the noncontinuity over about half a decade in $\hat{T}$.

ζ) Collision between neutral particles. In Table 10 (Sect. 17) the interaction potential for collisions between aeronomic neutrals are listed. For collisions between hydrogen, helium, and oxygen atoms the potential curves are splitted into several branches. The corresponding transport collision frequencies $\nu^{(ls)}$ have to be added with statistical weights described in Sect. 17.

The interaction potentials in Table 10 (Sect. 17) are Morse potentials, Eq. (7.1), Buckingham (exp β, 6) potentials, Eq. (7.5), Lennard-Jones (12, 6) potentials, Eq. (7.14), and repulsive exponential potentials, Eq. (8.2). The normalized transport collision frequencies $\nu^{(ls)*}(\hat{T}^*)$ are shown in Figs. 25, 26, 27d, 30, resp., (Sect. 20). For the Buckingham (exp β, 6) potential the low temperature limit is given by Eq. (20.1), for the Lennard-Jones (12, 6) potential by Eq. (20.2a) with $n=12$. The high-temperature limit of this potential can be taken from Eq. (20.2c) with $n=12$, $\gamma=0$. For the repulsive exponential potential the normalized transport collision frequencies $\nu^{(ls)*}(\hat{T}^*)$ are given by Eq. (22.3).

The normalizations are generally eliminated with

$$\nu^{(ls)}(\hat{T}) = \sqrt{2\mathscr{E}_{\text{R}}/\mu}\,\pi r_{\text{R}}^2 N_k\,\nu^{(ls)*}(\hat{T}^*) \qquad [19.13]$$

and

$$\hat{T} = \mathscr{E}_{\text{R}}\hat{T}^*/k \qquad [18.18]$$

with the reference quantities $\mathscr{E}_{\text{R}}$ and r_{R} taken from Table 10 (Sect. 17). For the repulsive exponential potential the normalization quantities $\mathscr{E}_{(l)}$ and $r_{(l)}$ must be used in Eqs. (19.13) and (18.20) instead of $\mathscr{E}_{\text{R}}$ and r_{R}. Their relations to the reference quantities φ_{R} and r_{R} are listed in Eq. (8.3e).

For collisions not listed in Table 10 (Sect. 17) the reference quantities of the interaction potential can be calculated from those for collisions between neutrals of the same species. Empirical combining laws for the reference quantities must be used. They are given in Sect. 17.

[1] Banks, P. (1966): Planet. Space Sci. *14*, 1105–1122

C. The moment method

I. System of balance equations for momentum and heat flux

23. Momentum and heat flux balances. With the transport collision frequencies $\nu^{(ls)}$ of Eq. (19.3) the results of the moment method of the kinetic theory (see Sect. 3α) can be written down. This method yields an infinite system of coupled balance equations. One of these is the momentum balance, which is also called "equation of motion". It replaces Eq. (1.13) of the mean-collision-frequency method. Another one is the heat flux balance, replacing Eq. (1.15).

α) In the approximation used in this contribution the following quantities are needed in the two kinds of *balance equations.*

The *mean velocities* of particles of species j

$$\boldsymbol{v}_j := \overline{\boldsymbol{c}_j} \tag{23.1}$$

yield, when averaged with the mass densities $N_j m_j$ as weights, the *mass velocity*

$$\boldsymbol{v} := \frac{\sum_j N_j m_j \boldsymbol{v}_j}{\sum_j N_j m_j} \tag{23.2}$$

of the whole gas; the *mass fluxes* (or "diffusion flows") are then

$$\boldsymbol{J}_j := N_j m_j (\boldsymbol{v}_j - \boldsymbol{v}) \tag{23.3}$$

and the *partial heat fluxes*

$$\boldsymbol{Q}_j := N_j \frac{m_j}{2} \overline{|\boldsymbol{c}_j - \boldsymbol{v}|^2 (\boldsymbol{c}_j - \boldsymbol{v})} - \frac{5}{2} \frac{k T_j}{m_j} \boldsymbol{J}_j, \tag{23.4}$$

the upper bar meaning velocity averaging with the velocity distribution functions as weights.

The *partial pressure tensors* are obtained as averages of tensorial products:

$$\mathbf{p}_j := N_j m_j \overline{(\boldsymbol{c}_j - \boldsymbol{v})(\boldsymbol{c}_j - \boldsymbol{v})}, \tag{23.5a}$$

and the *partial hydrostatic pressures* read therefore

$$p_j := N_j m_j \tfrac{1}{3} \overline{|\boldsymbol{c}_j - \boldsymbol{v}|^2} = N_j k T_j. \tag{23.5b}$$

The circular *gyro* (or *cyclotron*) *frequencies* are

$$\omega_{B_j} := -\frac{q_j \boldsymbol{B}}{m_j c_0 \sqrt{\varepsilon_0 \mu_0}} \tag{23.6}$$

for particles of species j gyrating under the influence of a homogeneous magnetic field $\boldsymbol{B}$. Furthermore $\boldsymbol{g}_0$, Eq. (1.2), is the gravitational field and $\boldsymbol{\omega}$, Eq. (1.3), the angular velocity causing the Coriolis force (in a rotating frame of reference).

From the definitions Eq. (23.2) for the mass velocity $\boldsymbol{v}$ and Eq. (23.3) for the mass fluxes $\boldsymbol{J}_j$ there follows the linear relation

$$\sum_j \boldsymbol{J}_j = 0. \tag{23.7}$$

Hence the mass fluxes $\boldsymbol{J}_j$ are not linearly independent. For a binary mixture there holds $\boldsymbol{J}_1 + \boldsymbol{J}_2 = 0$.

Another procedure to establish balance equations uses mass fluxes, partial heat fluxes, partial pressures, and partial temperatures with definitions different from those in Eqs. (23.3), (23.4), (23.5). Instead of the mass velocity $\boldsymbol{v}$, Eq. (23.2), of the whole gas, the partial mean velocities $\boldsymbol{v}_j$ of the particle species are used as reference velocities in the definitions Eqs. (23.3), (23.4), (23.5). A review of different approaches used in the moment method is given by SCHUNK[1].

β) The *momentum balance* for particles of species j is now given[2,3] under the conditions that the mass velocity $\boldsymbol{v}$ of the whole gas is homogeneous and stationary, i.e., for $(\partial/\partial \boldsymbol{r})\boldsymbol{v} = 0 = \partial \boldsymbol{v}/\partial t$, and that the spatial variation of the partial *stress tensors* $\mathbf{p}_j - p_j \mathbf{U}$ can be neglected against the variations of the hydrostatic pressures p_j

$$\begin{aligned}&\left(\frac{\partial}{\partial t} + \boldsymbol{v}\cdot\frac{\partial}{\partial \boldsymbol{r}}\right)\boldsymbol{J}_j - (\boldsymbol{\omega}_{B_j} - 2\boldsymbol{\omega}) \times \boldsymbol{J}_j \\ &\quad + N_j \sum_k \mu_{jk}\left[\nu_{jk}^{(10)}\left(\frac{\boldsymbol{J}_j}{N_j m_j} - \frac{\boldsymbol{J}_k}{N_k m_k}\right) + \frac{\nu_{jk}^{(11)}}{k\hat{T}_{jk}/\mu_{jk}}\left(\frac{\boldsymbol{Q}_j}{N_j m_j} - \frac{\boldsymbol{Q}_k}{N_k m_k}\right) + O\left(\frac{\nu^{(l2)}}{\nu^{(l0)}}\right)\right] \\ &\quad = N_j m_j \left[\frac{q_j \boldsymbol{E} + m_j \boldsymbol{g}_0}{m_j} + (\boldsymbol{\omega}_{B_j} - 2\boldsymbol{\omega}) \times \boldsymbol{v}\right] - \frac{\partial p_j}{\partial \boldsymbol{r}}.\end{aligned} \tag{23.8}$$

The neglected terms $O(\nu^{(l2)}/\nu^{(l0)})$ contain fluxes of more complicated character than $\boldsymbol{J}_j$ and $\boldsymbol{Q}_j$. The order of the transport collision frequencies $\nu^{(ls)}$, Eq. (19.3), has been used to indicate the order of magnitude of the terms neglected. Since $|\nu^{(ls)}|$ decreases with increasing s due to the use of orthogonal polynomials $\mathrm{L}_s^{(l+\frac{1}{2})}$ in the definition Eq. (19.3) of $\nu^{(ls)}$, this indication is justified.

Likewise it is observed from the momentum balance Eq. (23.8) that the *coupling* of the heat fluxes $\boldsymbol{Q}_j$ to the mass fluxes $\boldsymbol{J}_j$ is of the order $\nu^{(11)}/\nu^{(10)}$. Here also the order s in $\nu^{(ls)}$ serves to indicate a certain order of magnitude.

If $|\nu^{(1s)}| \ll \nu^{(10)}$, then mass fluxes and heat fluxes are uncoupled and the rigorous momentum balance Eq. (23.8) coincides with the momentum balance

---

[1] Schunk. R.W. (1977): Rev. Geophys. Space Phys. *15*. 429–445

[2] Suchy, K.: See [*8*], eqs. 85.1 and 85.2

[3] McCormack. F.J. (1973): Phys. Fluids *16*. 2095–2105. eqs. 16. 20. 21

Eq. (1.13) from the mean-collision-frequency theory, provided $m\nu$ is replaced by $\sum_k \mu_{jk}\nu_{jk}^{(10)}$ and all $\boldsymbol{v}_k$ are small against $\boldsymbol{v}_j$.

It must be stressed that the transport collision frequencies $\nu_{jk}^{(ls)}$ in the equations of the rigorous kinetic theory are always multiplied with the reduced masses μ_{ik} and not with the mass m_j. Because of their definition, Eq. (19.3), they obey the relation

$$N_j\nu_{jk}^{(ls)}=N_k\nu_{kj}^{(ls)}. \qquad [4.15]$$

If one had written $m_j\nu_{jk}$ instead of $\mu_{jk}\nu_{jk}^{(ls)}$, the corresponding relation would be

$$N_j\nu_{jk}m_k=N_k\nu_{kj}m_j. \qquad (23.9)$$

This different mass factor must be kept in mind if one compares the momentum balance Eq. (23.8) with a momentum balance into which other mass factors of "collision frequencies" (or better: "friction coefficients") were introduced.

γ) Putting the mass velocity $\boldsymbol{v}=$const., the *heat flux balance* is obtained with omissions similar to the omissions of the stress tensor in the momentum balance, Eq. (23.8),

$$\left(\frac{\partial}{\partial t}+\boldsymbol{v}\cdot\frac{\partial}{\partial \boldsymbol{r}}\right)\boldsymbol{Q}_j-(\boldsymbol{\omega}_{B_j}-2\boldsymbol{\omega})\times\boldsymbol{Q}_j+\frac{5}{2}N_j\sum_k\frac{\mu_{jk}^2}{m_j^2}k\hat{T}_{jk}\nu_{jk}^{QJ}\left(\frac{\boldsymbol{J}_j}{N_jm_j}-\frac{\boldsymbol{J}_k}{N_km_k}\right)$$

$$+\boldsymbol{Q}_j\sum_k\frac{\mu_{jk}^3}{m_j^2m_k}\nu_{jk}^{[m_j/m_k]}+\sum_k\frac{\mu_{jk}^3}{m_j^2m_k}\nu_{jk}^{[-1]}\boldsymbol{Q}_k+O\left(\frac{\nu^{(l1)}}{\nu^{(l0)}}\right)=-\frac{5}{2}\frac{p_jk}{m_j}\frac{\partial T_j}{\partial \boldsymbol{r}} \qquad (23.10\text{a})$$

with[4]

$$\nu_{jk}^{QJ}:=\nu_{jk}^{(11)}-\tilde{\tau}_{jk}[4\nu_{jk}^{(10)}-\tfrac{4}{3}\nu_{jk}^{(20)}]+\tilde{\tau}_{jk}^2 3\nu_{jk}^{(11)}, \qquad (23.10\text{b})$$

$$\nu_{jk}^{[m_j/m_k]}:=\tfrac{4}{3}\nu_{jk}^{[20]}+\frac{m_k}{m_j}\nu_{jk}^{[10]}+3\frac{m_j}{m_k}\nu_{jk}^{(10)}, \qquad (23.10\text{c})$$

$$\nu_{jk}^{[-1]}:=\tfrac{4}{3}\nu_{jk}^{[20]}-\nu_{jk}^{[10]}-3\nu_{jk}^{(10)}, \qquad (23.10\text{d})$$

$$\nu_{jk}^{[20]}:=\nu_{jk}^{(20)}+\tilde{\tau}_{jk}\tfrac{33}{4}\nu_{jk}^{(11)}, \qquad (23.10\text{e})$$

$$\nu_{jk}^{[10]}:=\nu_{jk}^{(10)}+2\nu_{jk}^{(11)}+\tfrac{7}{2}\nu_{jk}^{(12)}-\tilde{\tau}_{jk}[5\nu_{jk}^{(11)}-\tfrac{14}{3}\nu_{jk}^{(21)}]+\tilde{\tau}_{jk}^2\tfrac{21}{2}\nu_{jk}^{(12)}, \qquad (23.10\text{f})$$

$$\tilde{\tau}_{jk}:=\frac{\mu_{jk}}{m_k}\frac{T_j-T_k}{\hat{T}_{jk}}. \qquad (23.10\text{g})$$

For uniform temperature of all species ($T_j-T_k\sim\tilde{\tau}_{jk}=0$) the collision terms are given by McCormack[5] in terms of the omega integrals $\Omega^{(l,s)}$, Eq. (19.1). Their connection with the transport collision frequencies $\nu^{(ls)}$, Eq. (19.3), is given by Eqs. (19.5) and (19.6).

---

[4] Suchy, K. (1973): Proc. Eleventh Intern. Conf. on Phenomena in Ionized Gases, Prague, 272–272, eq. 2. Prague: Czechoslovak Academy of Sciences, Institute of Physics

[5] McCormack, F.J. (1973): Phys. Fluids *16*, 2095–2105, eqs. 19, 21, 24, 25

SCHUNK and WALKER[6] calculated the balance for the *partial energy flux*

$$\boldsymbol{q}_j := N_j \frac{m_j}{2} \overline{|\boldsymbol{c}_j - \boldsymbol{v}|^2 (\boldsymbol{c}_j - \boldsymbol{v})} = \boldsymbol{Q}_j + \frac{5}{2} \frac{kT_j}{m_j} \boldsymbol{J}_j \tag{23.11}$$

and therefore obtained collision terms different from Eqs. (23.10b–g).

For different temperatures of the species ($T_j \neq T_k$) the powers of the *normalized temperature difference* $\tilde{\tau}_{jk}$ correspond in a certain way to the order s in the $\nu^{(ls)}$. That means that the "coupling" collision frequency ν_{jk}^{QJ}, Eq. (23.10b), [which couples the mass fluxes $\boldsymbol{J}_j$ to the heat fluxes $\boldsymbol{Q}_j$ in the heat flux balance Eq. (23.10a)] is of the order $\nu^{(11)}$ for $\tilde{\tau}_{jk}=0$ (i.e., $T_j = T_k$) or of the order $\tilde{\tau}^1 \nu^{(10)}$ for $\tilde{\tau} \neq 0$ (i.e., $T_j \neq T_k$). With this equivalence one may say that the coupling is of "first order". Note that different temperatures of the species is the normal condition for most natural and artificial plasmas.

The neglected quantities meant by $O(\nu^{(l1)}/\nu^{(l0)})$ in the heat flux balance Eq. (23.10a) are again "fluxes" of complicated nature. The error committed by neglecting these terms has yet to be investigated.

For $|\nu^{(l1)}| \ll \nu^{(l0)}$ and $|\tilde{\tau}| \ll 1$ heat fluxes $\boldsymbol{Q}_j$, Eq. (23.4), and mass fluxes $\boldsymbol{J}_j$, $\boldsymbol{J}_k$ are uncoupled. However, energy fluxes $\boldsymbol{q}_j$, Eq. (23.11), and mass fluxes $\boldsymbol{J}_j$, $\boldsymbol{J}_k$ are uncoupled only under the additional condition[7] $m_j \approx m_k$. If all $\boldsymbol{Q}_k (k \neq j)$ are small compared with $\boldsymbol{Q}_j$, then the heat flux balance, Eq. (23.10a), coincides with the energy flux balance, Eq. (1.15), derived with the mean-collision-frequency theory (if $\boldsymbol{F} = 0 = \partial N / \partial \boldsymbol{r}$ in the latter). The mean-collision frequency ν in Eq. (1.15) has to be replaced by

$$\sum_k \frac{\mu_{jk}^3}{m_j^2 m_k} \nu_{jk}^{[m_j/m_k]}$$

and the factor 3/2 of $\partial T / \partial \boldsymbol{r}$ by 5/2.

24. Balance equations in matrix notation

α) Column matrices and matrix vectors. The momentum balances, Eq. (23.8), for a multispecies plasma ($j = 1, 2, \ldots; k = 1, 2 \ldots$) and the heat flux balances, Eq. (23.10), form a system of linear equations for the unknown partial fluxes $\boldsymbol{J}_j$, $\boldsymbol{Q}_j$. The "thermodynamic forces" on the right-hand sides of the balance equations are the "sources" for the fluxes (in a physical sense), or the "inhomogeneities" of the equation (in a mathematical sense). Since a linear system is most conveniently expressed by matrices, we shall follow this compact notation. Instead introducing new letters for the matrices, we express them by their elements, for short. In this sense the set of partial pressures p_j at the right-hand side of the momentum balances, Eq. (23.8), are the elements of a column matrix, which we denote also by p_j. Then the set of partial pressure

[6] Schunk, R.W., Walker, J.C.G. (1971): J. Geophys. Res. *76*, 6159–6171, eqs. 11 and B1 through B9

[7] Schunk, R.W., Walker, J.C.G. (1971): J. Geophys. Res. *76*, 6159–6171, eqs. 11, B4, B5

gradients $\partial p_j/\partial \boldsymbol{r}$ are the elements of a *matrix vector*, or vector (column) matrix [1,2], as well as the sets of fluxes $\boldsymbol{J}_j$ and $\boldsymbol{Q}_j$.

β) For the *momentum balances*, Eq. (23.8), the transport collision frequencies $\nu_{jk}^{(10)}$ as factors of the $\boldsymbol{J}_j$ and the differential operator $\frac{\partial}{\partial t}+\boldsymbol{v}\cdot\frac{\partial}{\partial \boldsymbol{r}}$ are combined in the elements R_{jk}^0 of a square matrix as follows [3]:

$$R_{jj}^0 := \frac{\partial}{\partial t}+\boldsymbol{v}\cdot\frac{\partial}{\partial \boldsymbol{r}}+\sum_{k\neq j}\frac{\mu_{jk}}{m_j}\nu_{jk}^{(10)} \tag{24.1a}$$

$$R_{jk}^0 := -\frac{\mu_{jk}}{m_k}\nu_{kj}^{(10)} \quad \text{for } j\neq k. \tag{24.1b}$$

Note the inverted sequence of the subscripts of $\nu_{kj}^{(10)}$ in the definition Eq. (24.1b) of R_{jk}^0. Use has been made of the relation

$$N_j\nu_{jk}^{(ls)} = N_k\nu_{kj}^{(ls)}. \tag{4.15}$$

The projections $\boldsymbol{\omega}_{B_j}\cdot\hat{\boldsymbol{B}}$ of the vectorial gyrofrequencies $\boldsymbol{\omega}_{B_j} := -q_j\boldsymbol{B}/m_jc_0\sqrt{\varepsilon_0\mu_0}$, Eq. (23.6), are positive (negative) for negative (positive) charges q_j. They represent the elements of a diagonal matrix [3]

$$R_{jk}^- := \delta_{jk}\, i\boldsymbol{\omega}_{B_j}\cdot\hat{\boldsymbol{B}} = \begin{cases} i\boldsymbol{\omega}_{B_j}\cdot\hat{\boldsymbol{B}} & \text{for } j=k \\ 0 & \phantom{\text{for }} j\neq k. \end{cases} \tag{24.2}$$

The (first order) transport collision frequencies $\nu_{jk}^{(11)}$ appear as factors of the $\boldsymbol{Q}_j$ in the momentum balances, Eq. (23.8), and are collected in a square matrix S_{jk} with the elements

$$S_{jj} := \sum_{k\neq j}\frac{\mu_{jk}}{m_j}\frac{\mu_{jk}}{k\hat{T}_{jk}}\nu_{jk}^{(11)} \tag{24.3a}$$

$$S_{jk} := -\frac{\mu_{jk}}{m_k}\frac{\mu_{jk}}{k\hat{T}_{jk}}\nu_{kj}^{(11)} \quad \text{for } j\neq k. \tag{24.3b}$$

Without the Coriolis force $\sim\boldsymbol{\omega}\times\boldsymbol{v}$ and with the electric field

$$\boldsymbol{E}_v := \boldsymbol{E}+\boldsymbol{v}\times\boldsymbol{B}/c_0\sqrt{\varepsilon_0\mu_0} \tag{24.4}$$

in the center-of-mass frame moving with the mass velocity $\boldsymbol{v}\ll c_0$, the system of momentum balances, Eqs. (23.8), can now be written in matrix notation as [2]

$$\sum_k (R_{jk}^0\mathbf{U}+R_{jk}^- i\hat{\boldsymbol{B}}\times\mathbf{U})\cdot\boldsymbol{J}_k+\sum_k S_{jk}\boldsymbol{Q}_k = N_jq_j\boldsymbol{E}_v+N_jm_j\boldsymbol{g}_0-\partial p_j/\partial\boldsymbol{r}. \tag{24.5}$$

---

[1] Köhler, W.E., Raum, H.H. (1972): Z. Naturforsch. Teil A *27*, 1383–1393, chapt. III

[2] Suchy, K. (1973): Proc. Eleventh Intern. Conf. on Phenomena in Ionized Gases, Prague, 272–272. Prague: Czechoslovak Academy of Sciences, Institute of Physics

[3] Al'pert, Ya.L., Budden, K.G., Moiseyer, B.S., Stott, G.F. (1983): Phil. Trans. Roy. Soc. London A *309*, 503–557, eqs. A.20 and A.21

The matrix $S_{jk} \sim \nu^{(11)}$ describes the linear coupling between the mass fluxes $\boldsymbol{J}_k$ and the heat fluxes $\boldsymbol{Q}_k$.

Analogous to the column matrices p_j, $N_j q_j$, $N_j m_j$ and the matrix vectors $\partial p_j/\partial \boldsymbol{r}$, $\boldsymbol{J}_k$, $\boldsymbol{Q}_k$, introduced in Subsection α, we denote the quantities $R^0_{jk}\mathbf{U}$ and $R^-_{jk} i\hat{\boldsymbol{B}} \times \mathbf{U}$ as *matrix tensors* with square matrices R^0_{jk} and R^-_{jk} as components[2,4]. They are special cases of the class of *axial matrix tensors*, whose properties are discussed in Appendix A. For the inversion of matrix tensors their diagonal representation is important with the *matrix eigenvalues* as factors of the projectors (see Appendix A α, δ).

The sum

$$\mathbf{R}_{jk} := R^0_{jk}\mathbf{U} + R^-_{jk} i\hat{\boldsymbol{B}} \times \mathbf{U} \tag{24.6}$$

of the two axial matrix tensors of Eq. (24.5) is a particular axial matrix tensor, too. The particularity is the equality of one of the three matrix eigenvalues, viz. R^0_{jk}, with the arithmetic mean

$$R^+_{jk} := \tfrac{1}{2}(R^{+1}_{jk} + R^{-1}_{jk}) \tag{24.7a}$$

of the two others, $R^{\pm 1}_{jk}$ (see Appendix A ε). The matrix R^-_{jk} is equal to $\frac{1}{2}(R^{+1}_{jk} - R^{-1}_{jk})$ and thus the matrix eigenvalues $R^{\pm 1}_{jk}$ are given by [see Eq. (A.16) in Appendix A ε]

$$R^{\pm 1}_{jk} = R^0_{jk} \pm R^-_{jk}. \tag{24.7b}$$

The linear system of momentum balances, Eq. (24.5), can now be written as

$$\sum_k \mathbf{R}_{jk} \cdot \boldsymbol{J}_k + \sum_k S_{jk} \boldsymbol{Q}_k = N_j q_j \boldsymbol{E}_v + N_j m_j \boldsymbol{g}_0 - \partial p_j/\partial \boldsymbol{r}. \tag{24.8}$$

γ) For the heat flux balances, Eqs. (23.10), the linear combinations $\nu^{[m_j/m_k]}_{jk}$ and ν^{QJ}_{jk} of transport collision frequencies, appearing as factors of the $\boldsymbol{Q}_j$ and $\boldsymbol{J}_j$, are combined in the elements of square matrices $\breve{R}^0_{jk}$ and $\breve{S}_{jk}$ together with the differential operator $\dfrac{\partial}{\partial t} + \boldsymbol{v} \cdot \dfrac{\partial}{\partial \boldsymbol{r}}$. The gyrofrequencies $\boldsymbol{\omega}_{B_j}$ appear again in the elements of a diagonal matrix $\breve{R}^-_{jk} = R^-_{jk}$, Eq. (24.2). The matrix elements read as follows:

$$\breve{R}^0_{jj} := \frac{\partial}{\partial t} + \boldsymbol{v} \cdot \frac{\partial}{\partial \boldsymbol{r}} + \frac{1}{3} \nu^{(20)}_{jj} + \sum_{k \neq j} \frac{\mu^3_{jk}}{m^2_j m_k} \nu^{[m_j/m_k]}_{jk}, \tag{24.9a}$$

$$\breve{R}^0_{jk} := -\frac{\mu^3_{jk}}{m^2_j m_k} \nu^{[-1]}_{jk} \quad \text{for } j \neq k, \tag{24.9b}$$

$$\breve{R}^-_{jk} := \delta_{jk} i \boldsymbol{\omega}_{B_j} \cdot \hat{\boldsymbol{B}} = R^-_{jk}, \tag{24.9c}$$

$$\breve{S}_{jj} := \sum_{k \neq j} \frac{5}{2} \frac{\mu^3_{jk}}{m^3_j} \frac{k\hat{T}_{jk}}{\mu_{jk}} \nu^{QJ}_{jk}, \tag{24.10a}$$

---

[4] Köhler, W.E., Raum, H.H. (1972): Z. Naturforsch. Teil A *27*, 1383–1393, eq. 3.14

$$\breve{S}_{jk} := -\frac{5}{2}\frac{\mu_{jk}^3}{m_j^2 m_k}\frac{k\hat{T}_{jk}}{\mu_{jk}} v_{kj}^{JQ}. \tag{24.10b}$$

with [cf. Eqs. (23.10b, g)]

$$v_{kj}^{JQ} := v_{kj}^{(11)} + \tilde{\tau}_{kj}[4 v_{kj}^{(10)} - \tfrac{4}{3} v_{kj}^{(20)}] + \tilde{\tau}_{kj}^2 \, 3 v_{kj}^{(11)}. \tag{24.10c}$$

Without the Coriolis contribution $\sim\boldsymbol{\omega}$ the system of heat flux balances, Eqs. (23.10), can now be written in matrix notation[2] analogous to Eq. (24.8) as

$$\sum_k \breve{\mathbf{R}}_{jk} \cdot \boldsymbol{Q}_k + \sum_k \breve{S}_{jk} \boldsymbol{J}_k = -\frac{5}{2}\frac{p_j k}{m_j}\frac{\partial T_j}{\partial \boldsymbol{r}}. \tag{24.11a}$$

The axial matrix tensor

$$\breve{\mathbf{R}}_{jk} := \breve{R}_{jk}^0 \mathbf{U} + \breve{R}_{jk}^- i \hat{\boldsymbol{B}} \times \mathbf{U} \tag{24.11b}$$

has the matrix eigenvalues [see Eq. (24.7)]

$$\breve{R}_{jk}^0 \quad \text{and} \quad \breve{R}_{jk}^{\pm 1} = \breve{R}_{jk}^0 \pm \breve{R}_{jk}^-. \tag{24.11c}$$

The matrix $\breve{S}_{jk} \sim v^{JQ}$, Eqs. (24.10), (23.10b), represents the coupling between the heat fluxes $\boldsymbol{Q}_j$ and the mass fluxes $\boldsymbol{J}_j$. For equal temperatures of the species, $T_j = T_k$, its elements are of the order $v^{(11)}$; for different temperatures, $T_j \neq T_k$, they are of the order $\tilde{\tau}^1 v^{(10)}$, Eq. (23.10b), with

$$\tilde{\tau}_{jk} := \frac{\mu_{jk}}{m_k}\frac{T_j - T_k}{\hat{T}_{jk}}. \qquad [23.10g]$$

δ) For a *sinusoidal wave variation* of the fluxes, i.e., for

$$\boldsymbol{J} \sim \exp i(\boldsymbol{k} \cdot \boldsymbol{r} - \omega t) \sim \boldsymbol{Q}, \tag{24.12}$$

the differential operator $\frac{\partial}{\partial t} + \boldsymbol{v} \cdot \frac{\partial}{\partial \boldsymbol{r}}$ in the diagonal elements R_{jj}^0, Eq. (24.1a) and $\breve{R}_{jj}^0$, Eq. (24.9a), is replaced by

$$-i(\omega - \boldsymbol{k} \cdot \boldsymbol{v}) =: -i\omega_{\mathrm{D}} \tag{24.13}$$

with ω_{D} as the Doppler shifted circular frequency ω. With this replacement the matrices $R_{jk}^0(\omega_{\mathrm{D}})$ and $\breve{R}_{jk}^0(\omega_{\mathrm{D}})$ are algebraic expressions.

Note that the matrix $R_{jk}^0(\omega_{\mathrm{D}})$ has a special feature. From the expressions Eqs. (24.1) for $R_{jk}^0(\omega_{\mathrm{D}})$ one obtains the two relations

$$\sum_j R_{jk}^0(\omega_{\mathrm{D}}) = -i\omega_{\mathrm{D}} \qquad \sum_k \frac{N_k m_k}{N_j m_j} R_{jk}^0(\omega_{\mathrm{D}}) = -i\omega_{\mathrm{D}}. \tag{24.14}$$

Thus for the direct current (DC) case ($\omega_{\mathrm{D}} = 0$) rows and columns of the matrix $R_{jk}^0(0)$ are linearly dependent, hence $R_{jk}^0(0)$ is a singular matrix (without an inverse). But for the calculation of transport coefficients the matrix R^0_{jk} must be inverted. In the DC case this can be achieved with the following trick[5].

We observe that R^0_{jk} appears in the momentum balances, Eqs. (24.5), only in the combination $\sum_k R^0_{jk}\boldsymbol{J}_k$. Because of the linear relation $\sum_k \boldsymbol{J}_k=0$, Eq. (23.7), among the mass fluxes $\boldsymbol{J}_k$ one may add (or subtract) to each element R^0_{jk} in a row (with fixed j) the same arbitrary quantity without altering the result of $\sum_k R^0_{jk}\boldsymbol{J}_k$. If we subtract

$$R^0_{jj}(0)=\sum_{k\neq j}\frac{\mu_{jk}}{m_j}\nu^{(10)}_{jk} \tag{24.15}$$

from each element $R^0_{jk}(\omega_D)$, Eq. (24.1), in the j-th row, no particle species is distinguished. The new matrix $\bar{R}^0_{jk}$ with matrix elements

$$\bar{R}^0_{jj}:=R^0_{jj}(\omega_D)-R^0_{jj}(0)=-i\omega_D \tag{24.16a}$$

$$\bar{R}^0_{jk}:=R^0_{jk}-R^0_{jj}(0) \quad \text{for } j\neq k \tag{24.16b}$$

is regular, i.e., has an inverse. It can be used instead of R^0_{jk} in the DC case ($\omega_D=0$) and, if so desired, also for $\omega_D\neq 0$.

II. Transport equations

25. Transport equations in zeroth approximation. Ohm's and Fourier's laws

α) Transport equations in zeroth approximation. In the system of balance equations

$$\sum_k \mathbf{R}_{jk}\cdot\boldsymbol{J}_k+\sum_k S_{jk}\boldsymbol{Q}_k=N_jq_j\boldsymbol{E}_v+N_jm_j\boldsymbol{g}_0-\frac{\partial p_j}{\partial \boldsymbol{r}} \tag*{[24.8]}$$

$$\sum_k \breve{S}_{jk}\boldsymbol{J}_k+\sum_k \breve{\mathbf{R}}_{jk}\cdot\boldsymbol{Q}_k=-\frac{5}{2}\frac{p_jk}{m_j}\frac{\partial T_j}{\partial \boldsymbol{r}} \tag*{[24.11]}$$

the "forces" $\boldsymbol{E}_v$, $\boldsymbol{g}_0$, $-\partial p_j/\partial\boldsymbol{r}$, $-\partial T_j/\partial\boldsymbol{r}$ are expressed by the fluxes $\boldsymbol{J}_k$ and $\boldsymbol{Q}_k$. To calculate transport coefficients, the fluxes must be expressed by the forces. To achieve this the system of balance equations must be inverted. This is analogous to the solution Eq. (2.8) of the momentum balance Eq. (1.13) in the mean-collision-frequency method.

Due to the tensorial character of the coefficients $\mathbf{R}_{jk}$, $\breve{\mathbf{R}}_{jk}$, this inversion is somewhat complicated. To perform it as lucidly as possible, we neglect for a zeroth approximation the coupling coefficients S_{jk}, Eq. (24.3), and $\breve{S}_{jk}$, Eq. (24.9, 10), between the two types of fluxes $\boldsymbol{J}_k$ and $\boldsymbol{Q}_k$. This approximation is justified because $\breve{S}_{jk}$ is of order $\nu^{(11)}$ and $\breve{S}_{jk}$ of order $\nu^{(11)}$ or $\tilde{\tau}^1\nu^{(10)}$ in comparison with $\mathbf{R}_{jk}$ and $\breve{\mathbf{R}}_{jk}$, which are both of order $\nu^{(10)}$.

Neglecting the coupling between the $\boldsymbol{J}_k$ and the $\boldsymbol{Q}_k$, we have the two *transport equations*

$$\boldsymbol{J}_k=\sum_j \mathbf{b}^{[0]}_{kj}\cdot\left(N_jq_j\boldsymbol{E}_v+N_jm_j\boldsymbol{g}_0-\frac{\partial p_j}{\partial\boldsymbol{r}}\right) \tag{25.1a}$$

[5] Schaber, A., Obermeier, E. (1975): Wärme- u. Stoffübertrag. *8*, 23–31, sect. 2

$$\boldsymbol{Q}_k = -\sum_j \check{\mathbf{b}}_{kj}^{[0]} \cdot \frac{5}{2} \frac{p_j k}{m_j} \frac{\partial T_j}{\partial \boldsymbol{r}} =: +\sum_j \boldsymbol{\kappa}_{kj}^{[0]} \cdot \frac{\partial T_j}{\partial \boldsymbol{r}} \tag{25.1b}$$

with the *mobility matrix tensors* $\mathbf{b}_{kj}$ and $\check{\mathbf{b}}_{kj}$ in zeroth approximation as the inverses of $\mathbf{R}_{kj}$ and $\check{\mathbf{R}}_{kj}$, respectively:

$$\sum_j \mathbf{b}_{kj}^{[0]} \cdot \mathbf{R}_{jl} = \delta_{kl} \mathbf{U}, \qquad \sum_j \check{\mathbf{b}}_{kj}^{[0]} \cdot \check{\mathbf{R}}_{jl} = \delta_{kl} \mathbf{U}. \tag{25.2}$$

The inversion of the matrix tensors $\mathbf{R}_{kj}$ and $\check{\mathbf{R}}_{kj}$ is described in Appendix Aδ. Their matrix eigenvalues are $R_{jk}^{0,\pm 1}$, Eq. (24.7), and $\check{R}_{jk}^{0,\pm 1}$, Eq. (24.11c); the matrix eigenvalues of the mobility tensors $\mathbf{b}_{kj}$ and $\check{\mathbf{b}}_{kj}$ are in zeroth approximation the inverses of $R_{jk}^{0,\pm 1}$ and $\check{R}_{jk}^{0,\pm 1}$, respectively:

$$\sum_j b_{kj}^{0,\pm 1[0]} R_{jl}^{0,\pm 1} = \delta_{kl}, \qquad \sum_j \check{b}_{kj}^{0,\pm 1[0]} \check{R}_{jl}^{0,\pm 1} = \delta_{kl}. \tag{25.3}$$

Since both $\mathbf{b}_{kj}$ and $\check{\mathbf{b}}_{kj}$ are nondiagonal in kj, a gradient $\partial p_j/\partial \boldsymbol{r}$ or $\partial T_j/\partial \boldsymbol{r}$ of *any* partial quantity p_j or T_j causes[1] a diffusion flux $\boldsymbol{J}_k$ or a heat flux $\boldsymbol{Q}_k$, respectively, of *all* species k.

β) Diffusion tensor. The proportionality between the mass fluxes $\boldsymbol{J}_k$ and the gradients of the partial pressures $\partial p_j/\partial \boldsymbol{r}$ is usually not expressed by a mobility tensor $\mathbf{b}_{kj}$, but by a *diffusion matrix tensor*

$$\mathbf{D}_{kj} := \frac{1}{N_k m_k} \mathbf{b}_{kj} p_j. \tag{25.4}$$

The definition Eq. (25.4) corresponds to Eq. (2.10) in the mean-collision-frequency method.

For $\boldsymbol{E}_v = 0 = \boldsymbol{g}_0$ the transport equation (25.1a) can now be written with Eq. (23.2) as

$$\frac{1}{N_k m_k} \boldsymbol{J}_k = \boldsymbol{v}_k - \boldsymbol{v} = -\sum_j \mathbf{D}_{kj} \cdot \frac{1}{p_j} \frac{\partial p_j}{\partial \boldsymbol{r}}. \tag{25.5}$$

This is a generalization of *Fick's diffusion law.*

There exist a variety of possibilities for the definition of diffusion tensors in multicomponent media. Their connections are discussed by DE GROOT-MAZUR[2] and CONDIFF[3]. We use here that definition of $\mathbf{D}_{kj}$ which relates the diffusion tensor in the simplest way to the mobility tensor $\mathbf{b}_{kj}$ of the kinetic theory.

γ) Ohm's law. To obtain an expression for the *electric conduction current density*

$$\boldsymbol{j} - \rho \boldsymbol{v} := \sum_k N_k q_k (\boldsymbol{v}_k - \boldsymbol{v}) = \sum_k \frac{q_k}{m_k} \boldsymbol{J}_k \tag{25.6}$$

---

[1] St. Maurice. J.-P.. Schunk. R.W. (1976): Planet. Space Sci. *25*. 907–920

[2] deGroot, S.R., Mazur, P.: See [*4*], chapt. XI, §2

[3] Condiff. D.W. (1969): J. Chem. Phys. *51*. 4209–4212

the transport equations (25.1a) for the diffusion fluxes $\boldsymbol{J}_k$ are premultiplied with q_k/m_k and added, yielding *Ohm's law*

$$\boldsymbol{j}-\rho\boldsymbol{v}=\boldsymbol{\sigma}^{[0]}\cdot\boldsymbol{E}_v+\sum_k\frac{q_k}{m_k}\sum_j\mathbf{b}_{kj}^{[0]}\cdot\left(N_j m_j \boldsymbol{g}_0-\frac{\partial p_j}{\partial\boldsymbol{r}}\right) \tag{25.7}$$

with the *electrical conductivity tensor*

$$\boldsymbol{\sigma}:=\sum_k\frac{q_k}{m_k}\sum_j\mathbf{b}_{kj}N_j q_j. \tag{25.8a}$$

This is a "bilinear form" for the "variables" q_k/m_k and $N_j q_j$. The corresponding relations for the three eigenvalues are

$$\sigma_{0,\pm1}=\sum_k\frac{q_k}{m_k}\sum_j b_{kj}^{0;\pm1}N_j q_j. \tag{25.8b}$$

For a three-component plasma (consisting of electrons and one species of ions and neutrals) the quantities $\sigma_0^{[0]}$ and $\sigma_\pm^{[0]}=\frac{1}{2}(\sigma_{+1}^{[0]}+\sigma_{-1}^{[0]})$ have been discussed for various frequency regions by SEEGER[4], for the direct current (DC) case ($\omega_D=0$) by GUREVICH and TSEDILINA[5] and by MÖHLMANN[6]. For a five-component plasma (two ion species and two neutral species) these quantities are given by SCHLÜTER and SCHÜRGER[7].

The electrical conductivity tensor

$$\boldsymbol{\sigma}=\sigma_0\hat{\boldsymbol{B}}\hat{\boldsymbol{B}}+\sigma_+(\mathbf{U}-\hat{\boldsymbol{B}}\hat{\boldsymbol{B}})+\sigma_-i\hat{\boldsymbol{B}}\times\mathbf{U} \tag{25.9a}$$

with

$$\sigma_\pm:=\tfrac{1}{2}(\sigma_{+1}\pm\sigma_{-1}) \tag{25.9b}$$

is an axial tensor of the general form discussed in Appendix A with the direction $\hat{\boldsymbol{B}}$ of the ambient magnetic field as the generating axial vector $\hat{\boldsymbol{t}}$. The two parts $\sigma_+(\mathbf{U}-\hat{\boldsymbol{B}}\hat{\boldsymbol{B}})$ and $\sigma_-i\hat{\boldsymbol{B}}\times\mathbf{U}$ of this axial tensor give rise to particular names associated with current densities related to them. They are called *Pedersen* and *Hall currents*, respectively. The Hall current is perpendicular to both the directions of $\hat{\boldsymbol{B}}$ and of the generating field $\boldsymbol{E}_v$.

δ) Fourier's law. The partial heat fluxes $\boldsymbol{Q}_k$ are related to the partial temperature gradients $\partial T_j/\partial\boldsymbol{r}$ by the *heat conductivity matrix tensor* $\boldsymbol{\kappa}_{kj}$ as

$$\boldsymbol{Q}_k=-\sum_j\boldsymbol{\kappa}_{kj}^{[0]}\cdot\frac{\partial T_j}{\partial\boldsymbol{r}}. \qquad [25.1\text{b}]$$

The total heat flux $\boldsymbol{Q}:=\sum_k\boldsymbol{Q}_k$ is obtained by an addition of all transport equations (25.1b) for the partial heat fluxes $\boldsymbol{Q}_k$, yielding a *generalized Fourier's*

[4] Seeger, G. (1961): Z. Angew. Phys. *13*, 551–560
[5] Gurevich, A.V., Tsedilina, E.E. (1967): Space Sci. Rev. *7*, 407–450, eq. 0.26
[6] Möhlmann, D. (1974): Gerlands Beitr. Geophys. *83*, 270–274
[7] Schlüter, H., Schürger, G. (1975): Z. Naturforsch. Teil A *30*, 1600–1605

law

$$Q = -\sum_k \sum_j \boldsymbol{\kappa}_{kj}^{[0]} \cdot \frac{\partial T_j}{\partial \boldsymbol{r}}. \qquad (25.10)$$

In contrast to the electrical conductivity tensor $\boldsymbol{\sigma}$ in Ohm's law (25.8) a comparable heat conductivity tensor $\boldsymbol{\kappa}$ can not be introduced except for the case of uniform temperatures of the species $T_j = T_k$. In this case we obtain *Fourier's law*

$$Q = -\boldsymbol{\kappa} \cdot \frac{\partial T}{\partial \boldsymbol{r}} \qquad (25.11\text{a})$$

with the *heat conductivity tensor* [see Eq. (25.1b)]

$$\boldsymbol{\kappa} := \sum_k \sum_j \boldsymbol{\kappa}_{kj} = \sum_k \sum_j \check{\mathbf{b}}_{kj} \frac{5}{2} \frac{p_j k}{m_j}. \qquad (25.11\text{b})$$

The three eigenvalues are

$$\kappa_{0,\pm 1} = \sum_k \sum_j \check{b}_{kj}^{0,\pm 1} \frac{5}{2} \frac{p_j k}{m_j}. \qquad (25.11\text{c})$$

The axial heat conductivity tensor is

$$\boldsymbol{\kappa} = \kappa_0 \hat{\boldsymbol{B}}\hat{\boldsymbol{B}} + \kappa_+ (\mathbf{U} - \hat{\boldsymbol{B}}\hat{\boldsymbol{B}}) + \kappa_- i\hat{\boldsymbol{B}} \times \mathbf{U} \qquad (25.12\text{a})$$

with

$$\kappa_\pm := \tfrac{1}{2}(\kappa_{+1} \pm \kappa_{-1}). \qquad (25.12\text{b})$$

The heat flux associated with the part $\kappa_- i\hat{\boldsymbol{B}} \times \mathbf{U}$ is called the *Righi-Leduc flux*, with the direction perpendicular to both $\hat{\boldsymbol{B}}$ and $\partial T/\partial \boldsymbol{r}$.

26. Transport equations in first approximation. Thermal diffusion and diffusion thermo effect. In Sect. 25 the balance equations (24.8) (24.11) for momentum and heat flux were solved in zeroth approximation neglecting the coupling terms S_{kj} and $\check{S}_{kj}$ between the mass fluxes $\boldsymbol{J}_k$ and the heat fluxes $\boldsymbol{Q}_k$. Both balance equations were uncoupled in this approximation and could be solved independently of each other. In the first approximation the coupling is retained and the two sets of balance equations must be solved simultaneously. This is done by first eliminating the $\boldsymbol{Q}_k$ and then the $\boldsymbol{J}_k$ and solving the two resulting equations for $\boldsymbol{J}_k$ and $\boldsymbol{Q}_k$, respectively.

α) Thermal diffusion. Soret effect. To eliminate the matrix vector $\boldsymbol{Q}_k$, the set of heat flux balances, Eq. (24.11), is premultiplied with the inverse of $\check{\mathbf{R}}_{ij}$, i.e., with $\check{\mathbf{b}}_{ij}^{[0]}$, Eq. (25.2). The result is premultiplied with S_{li} and subtracted from the set of momentum balances, Eq. (24.8) with j replaced by l. The solution for the matrix vector $\boldsymbol{J}_k$ is the transport equation

$$\boldsymbol{J}_k = \sum_j \mathbf{b}_{kj}^{[1]} \cdot \left(N_j q_j \boldsymbol{E}_v + N_j m_j \boldsymbol{g}_0 - \frac{\partial p_j}{\partial \boldsymbol{r}} \right)$$

$$+ \sum_l \mathbf{b}_{kl}^{[1]} \cdot \sum_i S_{li} \sum_j \boldsymbol{\kappa}_{ij}^{[0]} \cdot \frac{\partial T_j}{\partial \boldsymbol{r}} \qquad (26.1\text{a})$$

with [see Eq. (25.1b)]

$$\boldsymbol{\kappa}_{ij}^{[0]} = \breve{\mathbf{b}}_{ij}^{[0]} \frac{5}{2} \frac{p_j k}{m_j} \tag{26.1b}$$

and the mobility matrix tensor $\mathbf{b}_{kj}$ in first approximation as the inverse of

$$\mathbf{R}_{jk} - \sum_i S_{ji} \sum_l \mathbf{b}_{il}^{[0]} \breve{S}_{lk},$$

i.e.,

$$\sum_j \mathbf{b}_{kj}^{[1]} \cdot \left(\mathbf{R}_{jm} - \sum_i S_{ji} \sum_l \mathbf{b}_{il}^{[0]} \breve{S}_{lm}\right) = \delta_{km} \mathbf{U}. \tag{26.2}$$

Comparison with $\mathbf{b}_{kj}^{[0]}$, Eq. (25.2), shows the improvement $\sim S\breve{S}$ in the expression for the mobility. Comparison of the transport equation (26.1) with the corresponding Eq. (25.1a) in zeroth approximation shows the addition of the term proportional to the temperature gradients $\partial T_j/\partial \boldsymbol{r}$. The proportionality between the mass fluxes $\boldsymbol{J}_k$ and these temperature gradients is usually expressed by the *thermal diffusion matrix tensor*

$$\mathbf{D}_{kj}^T := -\frac{1}{N_k m_k} \sum_l \mathbf{b}_{kl} \cdot \sum_i S_{li} \boldsymbol{\kappa}_{ij}^{[0]} T_j \tag{26.3}$$

with the same physical dimension as the diffusion matrix tensor $\mathbf{D}_{kj}$, Eq. (25.4). Since it is proportional to the coupling quantities $S_{li} \sim \nu^{(11)}$, Eq. (24.3), it is a correction term compared with the diffusion matrix tensor $\mathbf{D}_{kj}$. The appearance of mass fluxes $\boldsymbol{J}_k$ caused merely by temperature gradients $\partial T_j/\partial \boldsymbol{r}$ is named *thermal diffusion* or *Soret effect*. It affects the height distribution of minor ions in the topside ionosphere[1-6].

Since $\mathbf{D}_{kj}^T$ is nondiagonal in kj, a temperature gradient $\partial T_j/\partial \boldsymbol{r}$ of *any* species j causes[7] thermal diffusion fluxes $\boldsymbol{J}_k$ of *all* species k.

We intend now to determine the *thermal diffusion current* or *Soret current*, i.e., the electrical current density caused by the temperature gradients $\partial T_j/\partial \boldsymbol{r}$. In order to do so, the set of transport equations (26.1) has to be premultiplied with q_k/m_k and added:

$$(\boldsymbol{j} - \rho \boldsymbol{v})_{\text{Soret}} = -\sum_k N_k q_k \sum_j \mathbf{D}_{kj}^T \frac{1}{T_j} \frac{\partial T_j}{\partial \boldsymbol{r}}. \tag{26.4}$$

This expression for the thermal diffusion flux is analogous to Eq. (25.10) for the total heat flux $\boldsymbol{Q}$ in the case of a multitemperature plasma. Merely for a

---

[1] Walker, J.C.G. (1967): Planet. Space Sci. *15*, 1151–1156
[2] Schunk, R.W., Walker, J.C.G. (1969): Planet. Space Sci. *17*, 853–868
[3] Schunk, R.W., Walker, J.C.G. (1970): Planet. Space Sci. *18*, 535–557, 1319–1334
[4] Bauer, S.J.: See [*21*], end of sect. V.1
[5] Bauer, S.J., Donahue, T.M., Hartle, R.E., Taylor, H.A. (1979): Science *205*, 109–112, table 3
[6] Nakada, M.P., Sullivan, E.C. (1980): J. Geophys. Res. *85*, 171–176
[7] St.Maurice, J.-P., Schunk, R.W. (1976): Planet. Space Sci. *25*, 907–920

plasma with a uniform temperature $T := T_j = T_k$ a *thermal diffusion tensor*

$$\mathbf{d} := \sum_k N_k q_k \sum_j \mathbf{D}^T_{kj} \tag{26.5}$$

can be introduced. Then the thermal diffusion current, Eq. (26.4), can be written as

$$(\boldsymbol{j} - \rho \boldsymbol{v})_{\text{Soret}} = -\mathbf{d} \cdot \frac{1}{T} \frac{\partial T}{\partial \boldsymbol{r}}. \tag{26.6}$$

The thermal diffusion tensor

$$\mathbf{d} = d_0 \hat{\boldsymbol{B}}\hat{\boldsymbol{B}} + d_+ (\mathbf{U} - \hat{\boldsymbol{B}}\hat{\boldsymbol{B}}) + d_- i \hat{\boldsymbol{B}} + \mathbf{U} \tag{26.7}$$

is, of course, an axial tensor with the eigenvalues

$$d_{0,\pm 1} = \sum_k N_k q_k \sum_j D^{T0,\pm 1}_{kj}. \tag{26.8}$$

The three components of the thermal diffusion current, Eq. (26.6), associated with the three parts $d_0 \hat{\boldsymbol{B}}\hat{\boldsymbol{B}}$, $d_+(\mathbf{U} - \hat{\boldsymbol{B}}\hat{\boldsymbol{B}})$, $d_- i\hat{\boldsymbol{B}} \times \mathbf{U}$ of the thermal diffusion tensor $\mathbf{d}$, Eq. (26.7), are named after *Seebeck*, *Ettingshausen-Nernst*, and *Nernst*, respectively[8].

In moderately and strongly ionized plasmas, where Coulomb interactions dominate, the coupling matrices S_{jk}, Eq. (24.3), and $\breve{S}_{jk}$, Eq. (24.10), are of the same order of magnitude as the matrices R^0_{jk}, Eq. (24.1), and $\breve{R}^0_{jk}$, Eq. (24.9) (see Fig. 23, Sect. 19β). Therefore the "coupling" tensors $\mathbf{d}$, Eq. (26.5), and $\breve{\mathbf{d}}$, Eq. (26.13), are comparable with the conductivity tensors $\boldsymbol{\sigma}$, $\boldsymbol{\kappa}$ in these cases, and thus the electric and heat fluxes associated with the coupling terms may rise to the same order of magnitude[9,10] as the fluxes related to $\boldsymbol{\sigma}$ and $\boldsymbol{\kappa}$. In laboratory plasmas the thermal diffusion flux $-\mathbf{d} \cdot \partial T / \partial \boldsymbol{r}$, Eq. (26.6), has already been employed to explain remarkable effects[11], in particular the Nernst current[12–14] $-d_- i\hat{\boldsymbol{B}} \times \partial T / T \partial \boldsymbol{r}$.

β) Diffusion thermo effect. Peltier heat flux. To eliminate the matrix vector $\boldsymbol{J}_k$, the set of momentum balances, Eq. (24.8), is premultiplied with the inverse of $\mathbf{R}_{ij}$, i.e., with $\mathbf{b}^{[0]}_{ij}$, Eq. (25.2). The result is premultiplied with $\breve{S}_{li}$ and subtracted from the set of heat flux balances, Eq. (24.11). The solution for the matrix vector $\boldsymbol{Q}_k$ is the transport equation

$$\begin{aligned} \boldsymbol{Q}_k = &-\sum_j \breve{\mathbf{b}}^{[1]}_{kj} \cdot \frac{5}{2} \frac{p_j k}{m_j} \frac{\partial T_j}{\partial \boldsymbol{r}} \\ &-\sum_l \breve{\mathbf{b}}^{[1]}_{kl} \cdot \sum_i \breve{S}_{li} \sum_j \mathbf{b}^{[0]}_{ij} \cdot \left(N_j q_j \boldsymbol{E}_v + N_j m_j \boldsymbol{g}_0 - \frac{\partial p_j}{\partial \boldsymbol{r}} \right) \end{aligned} \tag{26.9}$$

---

[8] deGroot, S.R., Mazur, P.: See [*4*], chapt. XIII, §§6 and 7
[9] Chapman, S. (1958): Proc. Phys. Soc. *72*, 353–362
[10] Schunk, R.W., Walker, J.C.G. (1970): Planet. Space Sci. *18*, 1535–1550, figs. 4 and 5
[11] Reader, J., Wirtz, S. (1968): Z. Naturforsch. Teil A *23*, 1695–1706, fig. 4
[12] Klüber, O. (1967): Z. Naturforsch. Teil A *22*, 1599–1612, eqs. 13b, 16, 18
[13] Lehnert, B. (1974): Plasma Phys. *16*, 341–350, eq. 15
[14] Uhlenbusch, J. (1976): Physica B+C *82*, 61–85, p. 84

with a mobility matrix tensor $\check{\mathbf{b}}_{kj}$ in first approximation defined by

$$\sum_j \check{\mathbf{b}}_{kj}^{[1]} \cdot \Big(\check{\mathbf{R}}_{jm} - \sum_i \check{S}_{ji} \sum_l \mathbf{b}_{il}^{[0]} S_{lm}\Big) := \delta_{km} \mathbf{U}. \tag{26.10}$$

The term $\sim \check{S}S$ gives the improvement of $\check{\mathbf{b}}_{kj}^{[1]}$ compared with $\check{\mathbf{b}}_{kj}^{[0]}$, Eq. (25.2). The second term in the transport equation (26.9), of the order $\check{S}$, yields the heat flux due to the *diffusion thermo effect*, or *Dufour effect*, which was absent in the zeroth approximation, Eq. (25.1b). It can be of the same order of magnitude as the conducted heat flux in the topside ionosphere at high altitudes[15] and affects the temperature of minor ions in the topside ionosphere[16]. Its coefficient

$$\check{\mathbf{D}}_{kj}^{T} := -\sum_l \check{\mathbf{b}}_{kl}^{[1]} \sum_i \check{S}_{li} \mathbf{b}_{ij}^{[0]} \tag{26.11}$$

is the *diffusion thermo matrix tensor*, having the same physical dimension as the diffusion matrix tensor $\mathbf{D}_{kj}$, Eq. (25.4), and the thermal diffusion matrix tensor $\mathbf{D}_{kj}^{T}$, Eq. (26.3).

To obtain the *Peltier heat flux*, i.e., the heat flux caused by an electric field $\boldsymbol{E}_v$ even in the absence of temperature gradients $\partial T_j/\partial \boldsymbol{r}$, the transport equations (26.9) must be added

$$\boldsymbol{Q}_{\text{Peltier}} = -\check{\mathbf{d}} \cdot \boldsymbol{E}_v \tag{26.12}$$

with the *Peltier tensor*

$$\check{\mathbf{d}} := \sum_k \sum_j \check{\mathbf{D}}_{kj}^{T} N_j q_j. \tag{26.13}$$

This, again, is an axial tensor

$$\check{\mathbf{d}} = \check{d}_0 \hat{\boldsymbol{B}}\hat{\boldsymbol{B}} + \check{d}_+ (\mathbf{U} - \hat{\boldsymbol{B}}\hat{\boldsymbol{B}}) + \check{d}_- i\hat{\boldsymbol{B}} \times \mathbf{U} \tag{26.14}$$

with the eigenvalues

$$\check{d}_{0,\pm 1} = \sum_k \sum_j \check{D}_{kj}^{T\,0,\pm 1} N_j q_j. \tag{26.15}$$

The part $-\check{d}_- i\hat{\boldsymbol{B}} \times \boldsymbol{E}_v$ of the Peltier heat flux is named after *Ettingshausen*[17].

The combinations $\check{d}_\pm = \frac{1}{2}(\check{d}_{+1} \pm \check{d}_{-1})$ of the eigenvalues, Eq. (26.15), have been calculated by RAEDER and WIRTZ[18] for a three-component plasma (consisting of electrons and one species of ions and neutrals).

γ) Mobilities in higher-order approximation. The transport equations (25.1) in zeroth approximation have been obtained while completely neglecting the coupling between mass fluxes and heat fluxes. For the transport equations (26.1) (26.9), in first approximation the coupling between these two kinds of fluxes was retained. But since higher-order fluxes were neglected in the balance equations for momentum, Eq. (23.8), and heat flux, Eq. (23.10), the coupling

---

[15] Rees, M.H., Jones, R.A., Walker, J.C.G. (1971): Planet. Space Sci. *19*, 313–325, figs. 2, 3, 5

[16] Schunk, R.W., Raitt, W.J., Nagy, A.F. (1978): Planet. Space Sci. *26*, 189–191

[17] deGroot, S.R., Mazur, P.: See [4], chapt. XII, §7.I.a.3

[18] Reader, J., Wirtz, S. (1968): Z. Naturforsch. Teil A *23*, 1695–1706, sect. 3.1

between mass and heat fluxes on the one hand and higher-order fluxes on the other hand was neglected in the first approximation for the transport equations. By taking into account these higher-order fluxes and the corresponding coupling terms, the mobilities $\mathbf{b}_{kj}$ and $\breve{\mathbf{b}}_{kj}$ can be obtained with higher accuracy than given by $\mathbf{b}_{kj}^{[0]}$, $\breve{\mathbf{b}}_{kj}^{[0]}$, Eq. (25.2), and $\mathbf{b}_{kj}^{[1]}$, Eq. (26.2), $\breve{\mathbf{b}}_{kj}^{[1]}$, Eq. (26.10).

27. Onsager-Casimir relations. Generalized Bridgman relation

α) Onsager relations. For mixtures with uniform temperatures there exist certain symmetry relations for the matrices b_{kj}^M, $\breve{b}_{kj}^M$ $(M=0, \pm 1)$ and between the matrices D_{kj}^{TM} and $\breve{D}_{kj}^{TM}$, the so-called *Onsager relations* (see Sect. 2δ). To derive them we write the balance equations (24.8) (24.11) for $\boldsymbol{J}_k$ and $\boldsymbol{Q}_k$ in the form

$$\sum_k \mathbf{A}_{jk} \boldsymbol{J}_k + \sum_k B_{jk} \boldsymbol{Q}_k = \boldsymbol{X}_j \tag{27.1a}$$

$$\sum_k \breve{B}_{jk} \boldsymbol{J}_k + \sum_k \breve{\mathbf{A}}_{jk} \cdot \boldsymbol{Q}_k = \boldsymbol{Y}_j \tag{27.1b}$$

with

$$\mathbf{A}_{jk} := \frac{1}{N_j m_j} \mathbf{R}_{jk}, \qquad B_{jk} := \frac{1}{N_j m_j} S_{jk}, \tag{27.2a}$$

$$\breve{\mathbf{A}}_{jk} := \frac{2m_j}{5p_j k T_j} \breve{\mathbf{R}}_{jk}, \qquad \breve{B}_{jk} := \frac{2m_j}{5p_j k T_j} \breve{S}_{jk}, \tag{27.2b}$$

$$\boldsymbol{X}_j := \frac{1}{N_j m_j} \left(N_j q_j \boldsymbol{E}_v + N_j m_j \boldsymbol{g}_0 - \frac{\partial p_j}{\partial \boldsymbol{r}} \right), \tag{27.2c}$$

$$\boldsymbol{Y}_j := -\frac{1}{T_j} \frac{\partial T_j}{\partial \boldsymbol{r}}. \tag{27.2d}$$

From the definitions Eqs. (24.1) (24.2) (24.7) for the elements of the matrix eigenvalues R_{jk}^0 and $R_{jk}^{\pm 1} = R_{jk}^0 \pm R_{jk}^-$ one obtains the symmetry

$$A_{jk}^M = A_{kj}^M \qquad (M=0, \pm 1) \tag{27.3}$$

of the matrix eigenvalues A_{jk}^M. The definitions for the elements of S_{jk}, Eqs. (24.3), yield the symmetry

$$B_{jk} = B_{kj}. \tag{27.4}$$

Analogous symmetries for $\breve{A}_{jk}^M$ and $\breve{B}_{jk}$ hold only for uniform temperatures:

$$\left.\begin{aligned} \breve{A}_{jk}^M &= \breve{A}_{kj}^M \\ \breve{B}_{jk} &= \breve{B}_{kj} \end{aligned}\right\} \quad \text{for } T_j = T_k. \tag{27.5}$$

Under the same restriction the equality

$$\breve{B}_{jk} = B_{jk} \qquad \text{for } T_j = T_k \tag{27.6}$$

holds. Combination of Eqs. (27.3) to (27.6) yields

$$\begin{bmatrix} A_{jk}^M & B_{jk} \\ \breve{B}_{jk} & \breve{A}_{jk}^M \end{bmatrix} \quad \text{is symmetric for } T_j = T_k. \tag{27.7}$$

The solution of the linear system Eq. (27.1) is given by the transport equations (26.1) (26.9) in first approximation, which are written in the form

$$\boldsymbol{J}_k = \sum_j \mathbf{b}_{kj} N_j m_j \cdot \boldsymbol{X}_j + N_k m_k \sum_j \mathbf{D}_{kj}^T \cdot \boldsymbol{Y}_j \tag{27.8a}$$

$$\boldsymbol{Q}_k = \sum_j \breve{\mathbf{D}}_{kj}^T N_j m_j \cdot \boldsymbol{X}_j + \sum_j \breve{\mathbf{b}}_{kj} \frac{5}{2} \frac{p_j k T_j}{m_j} \cdot \boldsymbol{Y}_j. \tag{27.8b}$$

The symmetry of the three matrices Eq. (27.7) leads to

$$\begin{bmatrix} b_{kj}^M N_j m_j & N_k m_k D_{kj}^{TM} \\ \breve{D}_{kj}^{TM} N_j m_j & \breve{b}_{kj}^M \frac{5}{2} \frac{p_j k T}{m_j} \end{bmatrix} \quad \text{is symmetric for } T_j = T_k. \tag{27.9a}$$

This symmetry is a particular case of the Onsager relations[1].

For the elements of the four matrices Eq. (27.9a) the symmetry means

$$b_{kj}^M N_j m_j = N_k m_k b_{jk}^M, \qquad D_{kj}^{TM} = \breve{D}_{jk}^{TM}, \tag{27.9b, c}$$

$$\breve{b}_{kj}^M \frac{p_j}{m_j} = \frac{p_k}{m_k} \breve{b}_{jk}^M, \qquad N_k D_{kj}^M = N_j D_{jk}^M. \tag{27.9d, e}$$

Equation (27.9e) is a consequence of Eq. (27.9b) for $T_j = T_k$ as a result of the definition

$$D_{kj}^M := \frac{1}{N_k m_k} b_{kj}^M p_j \qquad [25.3]$$

of the diffusion matrix tensor. The relation between the eigenvalues of $\mathbf{D}_{kj}^T$ and $\breve{\mathbf{D}}_{jk}^T$, Eq. (27.9c), connects the thermal diffusion (Soret) effect with the diffusion thermo (Dufour) effect.

The symmetry relations Eqs. (27.9) can serve as a useful check for numerical computations involving the inversion of the matrices occurring in $\mathbf{b}_{kj}^{[1]}$, $\breve{\mathbf{b}}_{kj}^{[1]}$, Eqs. (26.2) (26.10), and in $\mathbf{D}_{kj}^T$, $\breve{\mathbf{D}}_{kj}^T$, Eqs. (26.3) (26.11). But this holds only in the case of uniform temperatures $T_j = T_k$.

β) Bridgman relation. From the definitions for the eigenvalues d_M, Eq. (26.8), and $\breve{d}_M$, Eq. (26.15), there follows with the symmetry relation Eq. (27.9c) the *generalized Bridgman relation*

$$d_M = \breve{d}_M, \quad \text{i.e.,} \quad \mathbf{d} = \breve{\mathbf{d}} \quad \text{for } T_j = T_k. \tag{27.10}$$

---

[1] Suchy, K.: See [*8*], sect. 95

The particular relation $d_- = \check{d}_-$ between the Nernst effect and the Ettingshausen effect is known as *Bridgman relation*[2]. Because of $d_- := \frac{1}{2}(d_{+1} - d_{-1})$, this is a special case of the generalization Eq. (27.10). The additional relations $d_0 = \check{d}_0$, $d_+ = \check{d}_+$ between the Seebeck and Ettingshausen-Nernst effects on the one side and the Peltier effects on the other side are also contained in the generalized Bridgman relation, Eq. (27.10).

γ) Casimir's generalization of Onsager's symmetry relations in the presence of an ambient magnetic field $\boldsymbol{B}$ is satisfied by the symmetric character of the three matrices Eq. (27.9a) since they are matrix eigenvalues of an axial matrix tensor. An axial tensor $\boldsymbol{\tau}(\hat{\boldsymbol{t}})$ is invariant under the simultaneous operations of transposition and reversal of the axis, $\boldsymbol{\tau}(\hat{\boldsymbol{t}}) = \boldsymbol{\tau}^T(-\hat{\boldsymbol{t}})$ [see Appendix A, Eq. (A.1)]. Therefore if one combines the transposition of the three matrix eigenvalues Eq. (27.9a) with the transposition of the axial tensor, then the direction $\hat{\boldsymbol{B}}$ of its axis must be reversed to keep the whole matrix tensor invariant. The separation of the tensor character from the collection of the contributions of different particle species in matrix eigenvalues permits the separation of Onsager's symmetry relations from Casimir's generalization.[3]

III. Particular transport coefficients

28. Transport coefficients for charged particles and for electrons alone. The improvement of the transport equations and the mobilities from their zeroth-order approximation in Sect. 25 to the first-order approximation in Sect. 26 was achieved by taking into account the coupling between the mass fluxes $\boldsymbol{J}_k$ and the heat fluxes $\boldsymbol{Q}_k$ in the balance equations. Now two other types of coupling should be investigated: 1) coupling between the fluxes $\boldsymbol{J}_c$, $\boldsymbol{Q}_c$ of charged particles with the fluxes $\boldsymbol{J}_n$, $\boldsymbol{Q}_n$ of the neutral particles, and 2) coupling between the electron fluxes $\boldsymbol{J}_e$, $\boldsymbol{Q}_e$ with the fluxes $\boldsymbol{J}_h$, $\boldsymbol{Q}_h$ of the heavy particles (ions and neutrals).

α) Coupling between the fluxes of charged particles and neutrals. For a first estimate the heat fluxes $\boldsymbol{Q}_k$ are neglected in the momentum balances Eq. (24.5) for the mass fluxes $\boldsymbol{J}_k$. A periodic electric field $\boldsymbol{E}_v$ acts primarily on the charged particles. Immediately after the field is switched on, the dominant periodic motion of the charged particles is their gyration around the magnetic field lines. The relevant relaxation time is of the order of the reciprocal transport collision frequencies $\nu_{ck}^{(10)}$. After this very short time has elapsed, the charged particles perform a harmonic oscillation with the angular frequency ω_D of the periodic electric field.

The neutrals are not directly influenced by the electric field but only indirectly via their collisions with the charged particles. Thus the mass fluxes $\boldsymbol{J}_c$ of the charged particles act as driving forces for the neutrals. The rise time for the neutrals until they reach a harmonic motion like the charged particles is of the order of the reciprocal transport collision frequencies $\nu_{nc}^{(10)} = (N_c/N_n)\,\nu_{cn}^{(10)}$, Eq. (4.15). Therefore this rise time is N_n/N_c times higher than the relaxation time mentioned above. In the dilute plasma of the Earth's ionosphere

---

[2] deGroot, S.R., Mazur, P.: See [4], chapt. XII, eq. 143

[3] Garrod, C., Hurley, J. (1983): Phys. Rev. *A 27*, 1487–1490, eq. 16

this factor is, in general, more than three orders of magnitude. Thus for angular frequencies ω_D appreciably higher than $\nu_{nc}^{(10)}$ the influence of the mass fluxes of the neutrals on those of the charged particles can be neglected. Phenomena with $\omega_D \approx \nu_{nc}^{(10)}$ are discussed by KOHL[1,2].

To improve these estimates the contribution of the heat fluxes $\boldsymbol{Q}_k$ should be taken into account. In the balance equations (24.11) for the heat fluxes, the mass fluxes $\boldsymbol{J}_c$ of the charged particles are the driving forces. Because of the similarity of the heat flux balances with the momentum balances, the same arguments hold for the different excitation of the harmonic oscillations of the $\boldsymbol{Q}_c$ compared with the $\boldsymbol{Q}_n$. Thus for

$$\omega_D \gg \nu_{nc}^{(10)} \equiv \frac{N_c}{N_n} \nu_{cn}^{(10)}, \tag{28.1}$$

the fluxes $\boldsymbol{J}_n$ and $\boldsymbol{Q}_n$ can be completely neglected in the systems of balance equations (24.5) and (24.11) for the charged particles. The only contribution of the neutrals to the transport coefficients is via the transport collision frequencies $\nu_{cn}^{(10)}$ in the expression for R_{cc}^0, Eq. (24.1a), and via the $\nu_{cn}^{(ls)}$ with $l=1,2$ and $s=0,1,2$ in the expression for $\breve{R}_{cc}^0$, Eq. (24.9a).

β) Coupling between the fluxes of electrons and heavy particles. To decouple the electrons from the heavy particles, one neglects all terms of order $m_e/m_n \ll 1$ and higher, where m_h are the masses of heavy particles, e.g., neutrals or ions. This means

$$R_{ee}^0 \gg R_{eh}^0, \quad S_{ee} \gg S_{eh}, \quad \breve{R}_{ee}^0 \gg \breve{R}_{eh}^0, \quad \breve{S}_{ee} \gg \breve{S}_{eh}. \tag{28.2}$$

Thus the set of momentum balances Eqs. (24.5) reduces to the momentum balance for the electrons

$$R_{ee}^0 \boldsymbol{J}_e - \boldsymbol{\omega}_{B_e} \times \boldsymbol{J}_e + S_{ee} \boldsymbol{Q}_e = N_e q_e \boldsymbol{E}_v + N_e m_e \boldsymbol{g}_0 - \frac{\partial p_e}{\partial \boldsymbol{r}} \tag{28.3}$$

with

$$R_{ee}^0 = \frac{\partial}{\partial t} + \boldsymbol{v} \cdot \frac{\partial}{\partial \boldsymbol{r}} + \nu^{(10)} = -i\omega_D + \nu^{(10)}, \tag{28.4a}$$

$$S_{ee} = \frac{m_e}{kT_e} \nu^{(11)}, \qquad \nu^{(ls)} := \sum_h \nu_{eh}^{(ls)}. \tag{28.4b, c}$$

Up to the term $S_{ee} \boldsymbol{Q}_e$, this is equivalent to the momentum balance Eq. (1.13) of the mean-collision-frequency method if the mean collision frequency ν is replaced by $\nu^{(10)}$.

The set of heat flux balances Eqs. (24.11) reduces to the heat flux balance of the electrons

$$\breve{R}_{ee}^0 \boldsymbol{Q}_e - \boldsymbol{\omega}_{B_e} \times \boldsymbol{Q}_e + \breve{S}_{ee} \boldsymbol{J}_e = -\frac{5}{2} \frac{p_e k}{m_e} \frac{\partial T_e}{\partial \boldsymbol{r}} \tag{28.5}$$

---

[1] Kohl, H. (1960): Arch. Elektr. Übertr. *14*, 169–176

[2] Kohl, H. (1963): Proc. Intern. Conf. on the Ionosphere, London, pp. 198–202. London: The Institute of Physics and the Physical Society

with

$$\check{R}^0_{\mathrm{ee}}=\frac{\partial}{\partial t}+\boldsymbol{v}\cdot\frac{\partial}{\partial \boldsymbol{r}}+\frac{1}{3}\,\nu^{(20)}_{\mathrm{ee}}+\nu^{[10]}=-i\,\omega_{\mathrm{D}}+\tfrac{1}{3}\,\nu^{(20)}_{\mathrm{ee}}+\nu^{[10]}, \tag{28.6a}$$

$$\check{S}_{\mathrm{ee}}=\frac{5}{2}\frac{kT_{\mathrm{e}}}{m_{\mathrm{e}}}\,\nu^{(11)}=\frac{5}{2}\left(\frac{kT_{\mathrm{e}}}{m_{\mathrm{e}}}\right)^2 S_{\mathrm{ee}}, \tag{28.6b}$$

$$\frac{5}{2}\,p_{\mathrm{e}}\frac{kT_{\mathrm{e}}}{m_{\mathrm{e}}}=N_{\mathrm{e}}\,m_{\mathrm{e}}\frac{\check{S}_{\mathrm{ee}}}{S_{\mathrm{ee}}}, \tag{28.6c}$$

$$\nu^{[10]}:=\sum_{\mathrm{h}}\nu^{[10]}_{\mathrm{eh}}=\sum_{\mathrm{h}}\left(\nu^{(10)}_{\mathrm{eh}}+2\,\nu^{(11)}_{\mathrm{eh}}+\tfrac{7}{2}\,\nu^{(12)}_{\mathrm{eh}}\right). \tag{28.6d}$$

Even with the connection

$$\boldsymbol{Q}_j=\boldsymbol{q}_j-\frac{5}{2}\frac{kT_j}{m_j}\boldsymbol{J}_j \qquad [23.11]$$

between heat flux $\boldsymbol{Q}_j$, Eq. (23.4), and energy flux $\boldsymbol{q}_j$, Eq. (1.14), the above heat flux balance Eq. (28.5) differs in several respects from an analogous balance for $\boldsymbol{Q}=\boldsymbol{q}-\frac{5}{2}\frac{kT}{m}N\,m\,\boldsymbol{v}$ derived from Eqs. (1.13) and (1.15). This is the reason why the mean-collision-frequency method does not yield the correct connection between the diffusion coefficient D, the thermal conductivity κ, and the thermal diffusion coefficient D_T, as mentioned in Sect. 2γ.

29. Electrical and heat conductivity for an electron plasma. To solve the combined balance equations (28.3) and (28.5), the methods of Sects. 25 and 26 can be applied. The matrix eigenvalues R^M_{jk} and $\check{R}^M_{jk}$ ($M=-1, 0, +1$), Eq. (24.7), reduce to scalar eigenvalues

$$R_M=R^0_{\mathrm{ee}}+M R^-_{\mathrm{ee}}=R^0_{\mathrm{ee}}+i\,M\,\omega_{B_{\mathrm{e}}}=-i(\omega_{\mathrm{D}}-M\,\omega_{B_{\mathrm{e}}})+\nu^{(10)}, \tag{29.1a}$$

$$\check{R}_M=\check{R}^0_{\mathrm{ee}}+M\check{R}^-_{\mathrm{ee}}=\check{R}^0_{\mathrm{ee}}+i\,M\,\omega_{B_{\mathrm{e}}}=-i(\omega_{\mathrm{D}}-M\,\omega_{B_{\mathrm{e}}})+\tfrac{1}{3}\,\nu^{(20)}_{\mathrm{ee}}+\nu^{[10]}, \tag{29.1b}$$

and with the corresponding resistivity tensors $\mathbf{R}$, $\check{\mathbf{R}}$ the momentum balance Eq. (28.3) and the heat flux balance Eq. (28.5) become

$$\mathbf{R}\cdot\boldsymbol{J}_{\mathrm{e}}+S\,\boldsymbol{Q}_{\mathrm{e}}=N_{\mathrm{e}}\,q_{\mathrm{e}}\,\boldsymbol{E}_v+N_{\mathrm{e}}\,m_{\mathrm{e}}\,\boldsymbol{g}_0-\frac{\partial p_{\mathrm{e}}}{\partial \boldsymbol{r}} \tag{29.2a}$$

$$\check{\mathbf{R}}\cdot\boldsymbol{Q}_{\mathrm{e}}+\check{S}\boldsymbol{J}_{\mathrm{e}}=-\frac{5}{2}\frac{p_{\mathrm{e}}k}{m_{\mathrm{e}}}\frac{\partial T_{\mathrm{e}}}{\partial \boldsymbol{r}}. \tag{29.2b}$$

α) The *eigenvalues for electrical and heat conductivity,*

$$\sigma_M=\frac{N_{\mathrm{e}}\,q_{\mathrm{e}}^2}{m_{\mathrm{e}}}\,b_M \qquad [25.8\,\mathrm{b}]$$

and

$$\kappa_M=\frac{5}{2}\frac{p_{\mathrm{e}}k}{m_{\mathrm{e}}}\,\check{b}_M \qquad [25.11\,\mathrm{c}]$$

are given in zeroth approximation for negligible coupling between current $\boldsymbol{j}_e - \rho_e \boldsymbol{v}$ and heat flux $\boldsymbol{Q}_e$ with $\mathbf{b}^{[0]} = \mathbf{R}^{-1}$, $\breve{\mathbf{b}}^{[0]} = \breve{\mathbf{R}}^{-1}$, Eq. (25.2), as

$$\frac{N_e q_e^2}{m_e \sigma_M^{[0]}} = \frac{1}{b_M^{[0]}} =: R_M^{[0]} = R_M = -i(\omega_D - M\,\omega_{B_e}) + \nu^{(10)}, \tag{29.3a}$$

$$\frac{5}{2}\frac{p_e k}{m_e \kappa_M^{[0]}} = \frac{1}{\breve{b}_M^{[0]}} =: \breve{R}_M^{[0]} = \breve{R}_M = -i(\omega_D - M\,\omega_{B_e}) + \tfrac{1}{3}\nu_{ee}^{(20)} + \nu^{[10]}. \tag{29.3b}$$

The simple structure of the zeroth-order resistivity eigenvalues (29.3) yields immediately $R_- := \frac{1}{2}(R_{+1} - R_{-1}) = i\,\omega_{B_e} = \breve{R}_-$. Therefore the zeroth-order approximations $\boldsymbol{\sigma}^{[0]}$ and $\boldsymbol{\kappa}^{[0]}$ for the electrical and heat conductivity are obtained with Eqs. (25.2) (24.6) (24.11b) (29.1) as[1]

$$\left(\frac{m_e}{N_e q_e^2}\boldsymbol{\sigma}^{[0]}\right)^{-1} = (\mathbf{b}^{[0]})^{-1} = \mathbf{R} = R_0\,\mathbf{U} + R_-\,i\hat{\boldsymbol{B}} \times \mathbf{U}$$
$$= (-i\,\omega_D + \nu^{(10)})\,\mathbf{U} - \boldsymbol{\omega}_{B_e} \times \mathbf{U}, \tag{29.4a}$$

$$\left(\frac{2}{5}\frac{m_e}{p_e k}\boldsymbol{\kappa}^{[0]}\right)^{-1} = (\breve{\mathbf{b}}^{[0]})^{-1} = \breve{\mathbf{R}} = \breve{R}_0\,\mathbf{U} + \breve{R}_-\,i\hat{\boldsymbol{B}} \times \mathbf{U}$$
$$= (-i\,\omega_D + \tfrac{1}{3}\nu_{ee}^{(20)} + \nu^{[10]})\,\mathbf{U} - \boldsymbol{\omega}_{B_e} \times \mathbf{U}. \tag{29.4b}$$

The combined momentum transport collision frequency $\nu^{(10)}$, Eq. (28.4c), is sometimes named "effective electron collision frequency for $\boldsymbol{\sigma}^{[0]}$."

According to Eqs. (26.1) (26.2) (26.9) (26.10), the first approximation is obtained by eliminating $\boldsymbol{Q}_e$, $\boldsymbol{J}_e$ from the two balance equations (29.2):

$$\mathbf{R}^{[1]} \cdot \boldsymbol{J}_e = N_e q_e \boldsymbol{E}_v + N_e m_e \boldsymbol{g}_0 - \frac{\partial p_e}{\partial \boldsymbol{r}} + S\breve{\mathbf{R}}^{-1} \cdot \frac{5}{2}\frac{p_e k}{m_e}\frac{\partial T_e}{\partial \boldsymbol{r}}, \tag{29.5a}$$

$$\breve{\mathbf{R}}^{[1]} \cdot \boldsymbol{Q}_e = -\breve{S}\mathbf{R}^{-1} \cdot \left(N_e q_e \boldsymbol{E}_v + N_e m_e \boldsymbol{g}_0 - \frac{\partial p_e}{\partial \boldsymbol{r}}\right) - \frac{5}{2}\frac{p_e k}{m_e}\frac{\partial T_e}{\partial \boldsymbol{r}} \tag{29.5b}$$

with

$$\mathbf{R}^{[1]} = \mathbf{R} - S\breve{\mathbf{R}}^{-1}\breve{S} = \mathbf{R} \cdot (\mathbf{U} - S\breve{S}\mathbf{R}^{-1} \cdot \breve{\mathbf{R}}^{-1}), \tag{29.5c}$$

$$\breve{\mathbf{R}}^{[1]} = \breve{\mathbf{R}} - \breve{S}\mathbf{R}^{-1}S = \breve{\mathbf{R}} \cdot (\mathbf{U} - \breve{S}S\breve{\mathbf{R}}^{-1} \cdot \mathbf{R}^{-1}). \tag{29.5d}$$

Therefore the first-order resistivity eigenvalues $R_M^{[1]}$, $\breve{R}_M^{[1]}$ are obtained by multiplication of the zeroth-order eigenvalues with $(1 - S\breve{S}/R_M \breve{R}_M)$ [see Eqs. (26.2) (26.10)]. For the electrical conductivity we obtain[2-4]

$$\frac{N_e q_e^2}{m_e \sigma_M^{[1]}} = \frac{1}{b_M^{[1]}} =: R_M^{[1]} = R_M - \frac{S\breve{S}}{\breve{R}_M}$$
$$= -i(\omega_D - M\,\omega_{B_e}) + \nu^{(10)} - \frac{\frac{5}{2}(\nu^{(11)})^2}{-i(\omega_D - M\,\omega_{B_e}) + \frac{1}{3}\nu_{ee}^{(20)} + \nu^{[10]}}. \tag{29.6a}$$

---

[1] Suchy, K.: See [25], p. 23–33, eqs. 5.3 and 5.4

[2] Landshoff, R. (1949): Phys. Rev. *76*, 904–909

[3] Landshoff, R. (1951): Phys. Rev. *82*, 442

[4] Wyller, A.A. (1961) in: Proc. Fifth Intern. Conf. on Ionization Phenomena in Gases. München, vol. 1. Maecker, H. (ed.) pp. 940–954. Amsterdam: North-Holland

The heat conductivity becomes

$$\frac{5}{2}\frac{p_e k}{m_e \kappa_M^{[1]}} = \frac{1}{\breve{b}_M^{[1]}} =: \breve{R}_M^{[1]} = \breve{R}_M - \frac{\breve{S}S}{R_M}$$

$$= -i(\omega_D - M\,\omega_{B_e}) + \tfrac{1}{3}\nu_{ee}^{(20)} + \nu^{[10]} - \frac{\tfrac{5}{2}(\nu^{(11)})^2}{-i(\omega_D - M\,\omega_{B_e}) + \nu^{(10)}} \tag{29.6b}$$

with $\nu^{[10]}$ given by Eq. (28.6d). Comparison between the first approximations, Eqs. (29.6), and the zeroth approximations, Eqs. (29.3), yields as a condition for the validity of the latter

$$(\nu^{(11)})^2 \ll |-i(\omega_D - M\,\omega_{B_e}) + \nu^{(10)}|^2. \tag{29.7}$$

β) Representation of the conductivity and resistivity eigenvalues. Because of the simple expressions for the resistivity and conductivity eigenvalues $R_M^{[0]} \sim 1/\sigma_M^{[0]}$, $\breve{R}_M^{[0]} \sim 1/\kappa_M^{[0]}$ in zeroth approximation, Eqs. (29.3), it seemed desirable to express higher-order approximations – like Eqs. (29.6) – in a similar analytic form. For the resistivity eigenvalues $R_M^{[n]}$ this was done by SHKAROFSKY[5]. He wrote

$$\frac{N_e q_e^2}{m_e \sigma_M^{[n]}} = \frac{1}{b_M^{[n]}} =: R_M^{[n]} = -i(\omega_D - M\,\omega_{B_e})\,h^{[n]} + g^{[n]}\,\nu^{(10)}. \tag{29.8}$$

The *Shkarofsky functions*

$$g^{[n]}\left(\frac{\omega_D - M\,\omega_{B_e}}{\nu^{(10)}}\right) := \frac{1}{\nu^{(10)}}\,\mathrm{Re}\,R_M^{[n]} = \frac{1}{\nu^{(10)}}\,\mathrm{Re}\,\frac{1}{b_M^{[n]}}$$

$$= g^{[n]}\left(\frac{M\,\omega_{B_e} - \omega_D}{\nu^{(10)}}\right), \tag{29.9a}$$

$$h^{[n]}\left(\frac{\omega_D - M\,\omega_{B_e}}{\nu^{(10)}}\right) := \frac{-1}{\omega_D - M\,\omega_{B_e}}\,\mathrm{Im}\,R_M^{[n]} = h^{[n]}\left(\frac{M\,\omega_{B_e} - \omega_D}{\nu^{(10)}}\right) \tag{29.9b}$$

are even functions with respect to the variable $(\omega_D - M\,\omega_{B_e})/\nu^{(10)}$ [see Eqs. (29.6) for example]. Moreover, they depend on the fractions $\nu_{eh}^{(10)}/\nu^{(10)}$ and on ratios $\nu_{eh}^{(ls)}/\nu_{eh}^{(l0)}$. Of course, in zeroth approximation, Eqs. (29.3), we have

$$g^{[0]} = 1 = h^{[0]} \tag{29.9c}$$

for all values of $(\omega_D - M\,\omega_{B_e})/\nu^{(10)}$. For high values of this variable, comparison of Eqs. (29.9) with (29.6) for $R_M^{[1]}$, $\breve{R}_M^{[1]}$ and Eqs. (29.6) for $R_M^{[0]}$, $\breve{R}_M^{[0]}$ yields $g^{[1]}(\infty) = 1 = h^{[1]}(\infty)$. This *weak absorption limit*

$$g^{[n]}(\infty) = 1 = h^{[n]}(\infty) \tag{29.9d}$$

holds for all orders n of the approximation.

---

[5] Shkarofsky, I.P. (1961): Can. J. Phys. *39*, 1619–1703, eq. 19

For $n=3$ and for power law transfer collision frequencies

$$\sum_n \nu_{\text{en}}^{(l)}(\mathscr{E}) = \nu^{(l)}(kT_{\text{e}})\, w^{p+\frac{1}{2}} \quad \text{with} \quad 2p+1=0, \pm 1, \pm 2, \pm 3 \qquad [18.20\text{a}]$$

for collisions between electrons and neutrals, curves and tables of $g^{[3]}$ and $h^{[3]}$ are given by SHKAROFSKY[5,6] and SHKAROFSKY et al.[7,8].

Since $g^{[n]}$ and $h^{[n]}$ are even with respect to $\omega_{\text{D}} - M\,\omega_{B_{\text{e}}}$, the resistivity eigenvalues $R_M^{[n]}$ and their linear combinations $R_{\pm}^{[n]} = \frac{1}{2}(R_{+1}^{[n]} \pm R_{-1}^{[n]})$ become rather simple expressions in the direct current (DC) case:

$$R_0^{[n]} = \nu^{(10)}\, g^{[n]}(0), \qquad R_+^{[n]} = \nu^{(10)}\, g^{[n]}\left(\frac{\omega_{B_{\text{e}}}}{\nu^{(10)}}\right), \qquad R_-^{[n]} = i\,\omega_{B_{\text{e}}}\, h^{[n]}\left(\frac{\omega_{B_{\text{e}}}}{\nu^{(10)}}\right) \quad \text{for } \omega_{\text{D}}=0. \qquad (29.10\text{a})$$

The resistivity tensor $\mathbf{R}^{[n]}$ takes the form [see Eq. (A1) in Appendix Aα]

$$\begin{aligned}\mathbf{R}^{[n]} &= (\mathbf{b}^{[n]})^{-1} \\ &= \nu^{(10)}\, g^{[n]}(0)\, \hat{\boldsymbol{B}}\hat{\boldsymbol{B}} + \nu^{(10)}\, g^{[n]}\left(\frac{\omega_{B_{\text{e}}}}{\nu^{(10)}}\right)(\mathbf{U} - \hat{\boldsymbol{B}}\hat{\boldsymbol{B}}) - h^{[n]}\left(\frac{\omega_{B_{\text{e}}}}{\nu^{(10)}}\right)\boldsymbol{\omega}_{B_{\text{e}}} \times \mathbf{U} \quad \text{for } \omega_{\text{D}}=0. \qquad (29.10\text{b})\end{aligned}$$

The tensor

$$\boldsymbol{\nu}_{\text{e}}^{[n]} := \mathbf{R}^{[n]} + \boldsymbol{\omega}_{B_{\text{e}}} \times \mathbf{U} \quad \text{for } \omega_{\text{D}}=0 \qquad (29.11\text{a})$$

was called "electron collision frequency tensor"[9,10]. In zeroth approximation with $g^{[0]}=1=h^{[0]}$, Eq. (29.9c), it becomes

$$\boldsymbol{\nu}_{\text{e}}^{[0]} = \nu^{(10)}\,\mathbf{U}, \qquad (29.11\text{b})$$

and this might be the reason for the nomenclature.

GINZBURG and GUREVIČ[11] introduced two functions $K_\sigma^{[n]}$ and $K_\varepsilon^{[n]}$ and wrote

$$\begin{aligned}\frac{m_{\text{e}}}{N_{\text{e}}\, q_{\text{e}}^2}\,\sigma_M^{[n]} &= b_M^{[n]} = K_\sigma^{[n]}\,\text{Re}\, b_M^{[0]} + i\,K_\varepsilon^{[n]}\,\text{Im}\, b_M^{[0]} \\ &= K_\sigma^{[n]}\frac{\nu^{(10)}}{(\nu^{(10)})^2 + (\omega_{\text{D}} - M\,\omega_{B_{\text{e}}})^2} + i\,K_\varepsilon^{[n]}\frac{\omega_{\text{D}} - M\,\omega_{B_{\text{e}}}}{(\nu^{(10)})^2 + (\omega_{\text{D}} - M\,\omega_{B_{\text{e}}})^2}. \qquad (29.12\text{a})\end{aligned}$$

The *Ginzburg-Gurevič functions*

$$\begin{aligned}K_\sigma^{[n]}\left(\frac{\omega_{\text{D}} - M\,\omega_{B_{\text{e}}}}{\nu^{(10)}}\right) &:= \frac{\text{Re}\, b_M^{[n]}}{\text{Re}\, b_M^{[0]}} \\ &= \left[1 + \left(\frac{\omega_{\text{D}} - M\,\omega_{B_{\text{e}}}}{\nu^{(10)}}\right)^2\right]\frac{g^{[n]}}{(g^{[n]})^2 + \left(\dfrac{\omega_{\text{D}} - M\,\omega_{B_{\text{e}}}}{\nu^{(10)}}\,h^{[n]}\right)^2}, \qquad (29.12\text{b})\end{aligned}$$

---

6 Shkarofsky, I.P. (1962): Can. J. Phys. *40*, 49–60

7 Shkarofsky, I.P., Bernstein, I.B., Robinson, B.B. (1963): Phys. Fluids *6*, 40–47

8 Shkarofsky, I.P., Johnston, T.W., Bachynski, M.P.: See [9], figs. 8-2 through 8-8

9 Gurevich, A.V., Tsedilina, E.E. (1967): Space Sci. Rev. *7*, 407–450, eqs. 0.20 and 0.21

10 Hill, R.J., Bowhill, S.A. (1977): J. Atmos. Terr. Phys. *39*, 803–811, eq. 24

11 Ginzburg, V.L., Gurevič, A.V. (1960): Usp. Fiz. Nauk *70*, 201–246, 393–428; English transl.: Soviet Phys. Usp. *3*, 115–146, 175–194; German transl.: Fortschr. Phys. *8*, 97–189, eq. 2.33

$$K_\varepsilon^{[n]}\left(\frac{\omega_D - M\,\omega_{B_e}}{\nu^{(10)}}\right) := \frac{\operatorname{Im} b_M^{[n]}}{\operatorname{Im} b_M^{[0]}} = \frac{h^{[n]}}{g^{[n]}}\, K_\sigma^{[n]} \tag{29.12c}$$

are related to the Shkarofsky functions $g^{[n]}$, Eq. (29.9a), and $h^{[n]}$, Eq. (29.9b), in a rather simple manner[12]. They can therefore be calculated if the latter are known. The weak absorption limit

$$K_\sigma^{[n]}(\infty) = 1, \qquad K_\varepsilon^{[n]}(\infty) = 1, \tag{29.12d}$$

obtained from Eqs. (29.12b) with (29.9d), holds for all degrees n of the approximation.

γ) Accuracy of the different approximations. A comparison between the conductivity formulas in zeroth approximation, Eqs. (29.3), and in first approximation, Eqs. (29.6), reveals that the influence of the coupling terms $\sim(\nu^{(11)})^2$ is most important in the *resonance cases* near $\omega_D = 0$ (DC case) and $\omega_D = \omega_{B_e}$ (*gyroresonance*). The influence becomes negligible for

$$(\nu^{(11)})^2 \ll |-i(\omega_D - M\,\omega_{Be}) + \nu^{(10)}|^2. \qquad [29.7]$$

In these particular cases the conductivity eigenvalues can be given exactly if electron-electron collisions can be completely neglected ($\nu_{ee}^{(20)} \ll \nu^{[10]}$) and if the sum $\sum_h \nu_{eh}^{(1)}(\mathscr{E})$ of momentum transfer collision frequencies can be approximated by a single power law

$$\sum_h \nu_{eh}^{(1)}(\mathscr{E}) = \nu^{(1)}(k T_e)\, w^{p+\frac{1}{2}} \qquad [18.20a]$$

with integer or half-integer p.

The matching of a sum of momentum transfer collision frequencies $\sum_h \nu_{eh}^{(1)}(\mathscr{E})$ by a single power law $\mathscr{E}^{p+\frac{1}{2}}$ is extensively discussed by SUCHY and RAWER[13] and by SCHUNK and WALKER[14].

Under the two assumptions, the first approximation $\sigma_M^{[1]}$, Eq. (29.6a), of the electrical conductivity is given using Eq. (19.8e) by

$$\frac{m_e}{N_e q_e^2}\,\sigma_M^{[1]} = b_M^{[1]} = \frac{1}{\nu^{(10)}}\left[1 + \frac{(p+\frac{1}{2})^2}{3+p}\right]. \tag{29.13}$$

In the brackets we find the first partial sum of a hypergeometric series representing the exact expression

$$\begin{aligned} \frac{m_e}{N_e q_e^2}\,\sigma_M = b_M &= \frac{1}{\nu^{(10)}}\, \mathrm{F}(p+\tfrac{1}{2}, p+\tfrac{1}{2}; p+3; 1) \\ &= \frac{1}{\nu^{(10)}} \frac{(p+2)!\,(1-p)!}{\frac{3}{2}!\;\frac{3}{2}!} \equiv \frac{1}{\nu^{(10)}} \frac{\Gamma(p+3)\,\Gamma(2-p)}{\Gamma(\frac{5}{2})\,\Gamma(\frac{5}{2})} \end{aligned} \tag{29.14}$$

[12] Gurevich, A.V., Tsedilina, E.E. (1967): Space Sci. Rev. 7, 407–450, footnote on p. 411
[13] Suchy, K., Rawer, K. (1971): J. Atmos. Terr. Phys. 33, 1853–1868, sect. 4.4
[14] Schunk, R.W., Walker, J.C.G. (1970): Planet. Space Sci. 18, 1535–1550, sect. 3

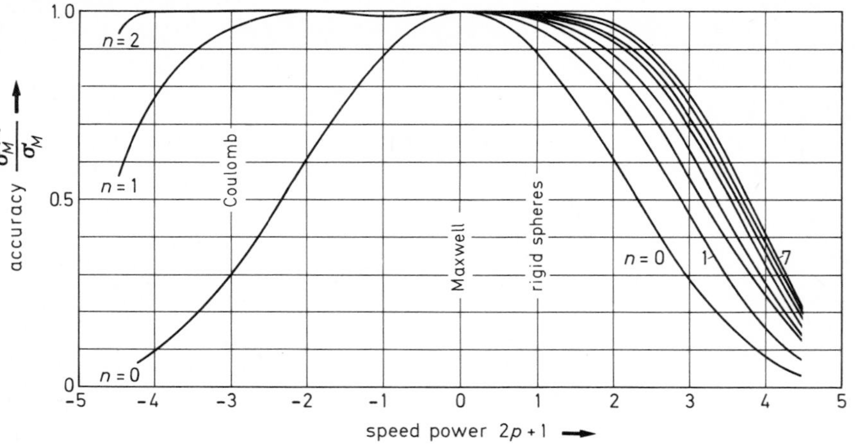

Fig. 39. Ratios of approximated conductivity eigenvalues $\sigma_M^{[n]}$, Eq. (29.13), to their exact expressions σ_M. Eq. (29.14). for transfer collision cross sections $Q^{(l)}(\mathscr{E})$. Eq. (4.8). proportional to E^p (after SCHIRMER and FRIEDRICH[17])

for b_M in the resonance cases[15,16]. Its reciprocal is sometimes named "effective electron collision frequency in the resonance approximation".

The higher-order approximations $\sigma_M^{[n]}$ for $n>1$ are the higher partial sums. A plot of the ratios $\sigma_M^{[n]}/\sigma_M$ in Fig. 39 shows the accuracy of the several approximations in the resonance cases (see also MITCHNER, KRUGER[18]). SHKAROFSKY[5] used $\sigma_M^{[3]}$ in his calculations of the alternating current (AC) conductivity.

The satisfying accuracy of $\sigma_M^{[1]}$ for Coulomb interaction ($p=-2$) is not yet a sufficient justification to use $\sigma_M^{[1]}$ of Eq. (29.6a) for a high degree of ionization. Since the aeronomic plasmas are mostly quasi neutral ($N_e=N_i$), the occurrence of electron-electron collisions is of the same order of magnitude as the electron-ion collisions. But only the latter are taken into account in the hypergeometric series for the exact expression of σ_M used in Fig. 39. However, it has been shown[17,19] that for a fully ionized plasma $\sigma_M^{[1]}$, Eq. (29.6a) in the resonance case differs only by about 1% from an expression derived by SPITZER and HÄRM[20] with a completely different method.

In the higher ionosphere where Coulomb collisions with $p=-2$ are dominant, the first approximation $\boldsymbol{\sigma}^{[1]}$, $\boldsymbol{\kappa}^{[1]}$ is very accurate, while in the lower ionosphere with dominant electron-air collisions [$p\approx\frac{1}{2}$, see Eqs. (11.1) (11.2)] even the Shkarofsky approximation $\boldsymbol{\sigma}^{[3]}$, $\boldsymbol{\kappa}^{[3]}$ is not very satisfying. The Lorentz method of kinetic gas theory yields better results in this case, where the electron-neutral collisions are dominant (Sects. 31 to 35).

---

[15] Chapman, S., Cowling, T.G.: See [*12*], 2nd. ed., eq. 10.53,10
[16] Friedrich, F.J. (1965): Dissertation. Berlin: Freie Universität
[17] Schirmer, H., Friedrich, J. (1958): Z. Phys. *151*, 375–384, fig. 1 and eq. 29
[18] Mitchner, M., Kruger, C.H.: See [*24*], chapt. VIII, fig. 4
[19] Shkarofsky, I.P., Johnston, T.W., Bachynski, M.P.: See [*9*], p. 324
[20] Spitzer, L., Härm, R. (1953): Phys. Rev. *89*, 977–981

The condition for neglecting electron-electron collisions is

$$\tfrac{1}{2}\nu_{ee}^{(20)} \ll \sum_{h} \nu_{eh}^{(10)} \tag{29.15}$$

from Eq. (29.1b). This is the condition for the applicability of the Lorentz method (see Sect. 3β). In the lower ionosphere it is certainly satisfied[21].

30. Thermal diffusion tensor and Peltier tensor for an electron plasma. The eigenvalues D_M^T, $\check{D}_M^T$ of the thermal diffusion tensor $\mathbf{D}_T$ and the diffusion thermo tensor $\check{\mathbf{D}}_T$ are given by Eqs. (26.3) and (26.11):

$$D_M^T = -\frac{b_M S \frac{5}{2} p_e k T}{m_e^2 N_e \check{R}_M} \qquad \check{D}_M^T = -\frac{\check{b}_M \check{S}}{R_M} \qquad (M=0, \pm 1). \tag{30.1}$$

From Eq. (28.6c) one obtains $S\frac{5}{2}p_e k T_e/m_e^2 N_e = \check{S}$; furthermore, Eq. (25.3) yields

$$\frac{b_M^{[0]}}{\check{R}_M} = \frac{1}{R_M \check{R}_M} = \frac{\check{b}_M^{[0]}}{R_M}.$$

Thus $D_M^{T[0]} = \check{D}_M^{T[0]}$, as required by the Onsager symmetry relation Eq. (27.9c). The first approximation is obtained by multiplication with $(1 - S\check{S}/R\check{R})$ [see Eqs. (29.5c, d)]. Thus also $D_M^{T[1]} = \check{D}_M^{T[1]}$ as required by Eq. (27.9c), which should hold for all orders n of approximations. With

$$d_M^T = N_e q_e D_M^T, \qquad \check{d}_M^T = \check{D}_M^T N_e q_e, \qquad [26.5]\ [26.13]$$

S from Eq. (28.4b), $\check{S}$ from Eq. (28.6b), R_M from Eq. (29.1a), $\check{R}_M$ from Eq. (29.1b), one has

$$\begin{aligned} \frac{d_M^{T[0]}}{p_e q_e/m_e} &= \frac{D_M^{T[0]}}{kT_e/m_e} = \frac{\check{D}_M^{T[0]}}{kT_e/m_e} = \frac{\check{d}_M^{T[0]}}{p_e q_e/m_e} = -\frac{m_e}{kT_e} \frac{\check{S}}{R_M \check{R}_M} \\ &= -\frac{\frac{5}{2}\nu^{(11)}}{\{-i(\omega_D - M\omega_{B_e}) + \nu^{(10)}\}\{-i(\omega_D - M\omega_{B_e}) + \frac{1}{3}\nu_{ee}^{(20)} + \nu^{[10]}\}} \end{aligned} \tag{30.2}$$

$$\frac{p_e q_e/m_e}{d_M^{T[1]}} = -\frac{kT_e}{m_e} \frac{R_M \check{R}_M}{\check{S}} + \frac{kT_e}{m_e} S = \frac{p_e q_e/m_e}{d_M^{T[0]}} + \nu^{(11)}. \tag{30.3}$$

Comparison between both approximations again yields the validity condition Eq. (29.7) for the zeroth approximation $d_M^{T[0]}$.

In the resonance cases $\omega_D = M\omega_{B_e}$ $(M=0, \pm 1)$ one obtains for

$$\sum_h \nu_{eh}^{(1)}(\mathscr{E}) \sim \mathscr{E}^{p+\frac{1}{2}}, \qquad [18.20a]$$

[21] Schunk, R.W., Walker, J.C.G. (1970): Planet. Space Sci. *18*, 1535–1550, fig. 1

using Eqs. (19.8e) and (29.3)

$$\frac{d_M^{T[1]}}{p_e q_e/m_e} = -\frac{1}{\nu^{(10)}} \frac{5}{4} \frac{2p+1}{p+3} = -b_M^{[0]} \frac{5}{4} \frac{2p+1}{p+3}. \tag{30.4}$$

The next higher approximation yields

$$\frac{d_M^{T[2]}/b_M^{[1]}}{p_e q_e/m_e} = -\frac{5}{4} \frac{2p+1}{p+3} \left[1 + \frac{(p+\frac{1}{2})(p+\frac{3}{2})}{p+4}\right]. \tag{30.5}$$

The bracketed term is the first partial sum of a hypergeometric series which represents the exact expression[1]:

$$\begin{aligned} \frac{d_M^T/b_M}{p_e q_e/m_e} &= -\frac{5}{4} \frac{2p+1}{p+3} \mathrm{F}(p+\tfrac{1}{2}, p+\tfrac{3}{2}; p+4; 1), \\ &= -\frac{5}{4} \frac{2p+1}{p+3} \frac{(p+3)!\,(1-p)!}{\frac{5}{2}!\ \frac{3}{2}!}, \\ &= -(p+\tfrac{1}{2}) \frac{(p+2)!}{\frac{3}{2}!} \frac{(1-p)!}{\frac{3}{2}!} = -(p+\tfrac{1}{2})\, \nu^{(10)}\, b_M. \end{aligned} \tag{30.6}$$

For the last equality Eq. (29.14) was used.

D. The Lorentz method

31. Representation of transport coefficients

α) Transport coefficients as energy averages of transfer collision tensors. For low degrees of ionization where electron-neutral and ion-neutral collisions dominate the Coulomb collisions, the moment method (Chapt. C) gives satisfying results only for rather high orders of approximations if the momentum transfer cross sections $Q_{\mathrm{eh}}^{(1)}(\mathscr{E})$ increase with increasing kinetic energy $\mathscr{E}$, i.e., if for a power law approximation

$$\sum_{\mathrm{h}} Q_{\mathrm{eh}}^{(1)}(\mathscr{E}) = Q^{(1)}(kT_e) \left(\frac{\mathscr{E}}{kT_e}\right)^p \tag{18.19a}$$

the power p is positive (see Fig. 39, Sect. 29γ). Only if the motions of the electrons contribute appreciably to the electric and heat fluxes is the so-called Lorentz method better suited for describing transport phenomena. The smallness of the ratio of the electron mass and the masses of the neutrals is taken account from the very beginning (not at the end of the calculations as in the

[1] Chapman, S., Cowling, T.G.: See [*12*], 2nd ed., following eq. 10.53,10

moment method, Sect. 28β). Electron-electron collisions, however, must have negligible effect compared with collisions between electrons and heavy particles. This was not required in the moment method, e.g., the occurrence of $\nu_{ee}^{(20)}$ in the expression Eq. (29.3b) for the heat conductivity $\boldsymbol{\kappa}$. From Eqs. (28.6a, d) one obtains the condition

$$\tfrac{1}{3}\nu_{ee}^{(20)} \ll \sum_{h} \left(\nu_{eh}^{(10)} + 2\nu_{eh}^{(11)} + \tfrac{7}{2}\nu_{eh}^{(12)}\right) \tag{31.1}$$

for the applicability of the Lorentz method. This condition is satisfied where the degree of ionization remains small.

The subscript h indicates species of heavy particles, i.e., ions and neutrals, the subscript e identifies electrons.

Contrary to the moment method used in Chapt. C, the Lorentz method does not yield coupled balance equations for $\boldsymbol{J}_e$ and $\boldsymbol{Q}_e$ – like Eqs. (28.3) (28.5) – rather than equivalents to transport equations, like Eqs. (26.1) (26.9). Mass flux $\boldsymbol{J}_e = N_e m_e(\boldsymbol{v}_e - \boldsymbol{v})$ and heat flux

$$\boldsymbol{Q}_e = \boldsymbol{q}_e - \frac{5}{2}\frac{kT_e}{m_e}\boldsymbol{J}_e \qquad [23.4]$$

are approximately given (for vanishing gravitation $\boldsymbol{g}_0$) by[1]

$$\boldsymbol{J} = -\frac{4\pi}{3}\int_0^\infty \mathrm{d}C\, C^3 \boldsymbol{\beta}^{(1)}(\mathscr{E}) \cdot \left(\frac{\partial f}{\partial C} q\boldsymbol{E}_v + mC\frac{\partial f}{\partial \boldsymbol{r}}\right), \tag{31.2}$$

$$\boldsymbol{Q} = -\frac{4\pi}{6}\int_0^\infty \mathrm{d}C\, C^3 \left(C^2 - 5\frac{kT}{m}\right) \boldsymbol{\beta}^{(1)}(\mathscr{E}) \cdot \left(\frac{\partial f}{\partial C} q\boldsymbol{E}_v + mC\frac{\partial f}{\partial \boldsymbol{r}}\right). \tag{31.3}$$

The subscript e for electrons will be omitted in the following.

Here, cf. Eq. (1.11),

$$\boldsymbol{C} := \boldsymbol{c} - \boldsymbol{v} \tag{31.4}$$

is the *intrinsic* or *peculiar velocity* of an electron, i.e., the velocity relative to the mean velocity $\boldsymbol{v}$ of the whole plasma. The function $f(\boldsymbol{r}, t, C)$ is the isotropic part of the velocity distribution function of the electrons. The tensor $\boldsymbol{\beta}^{(1)}(\mathscr{E})$ is a special case of energy dependent *transfer collision tensors*[2]

$$\boldsymbol{\beta}^{(l)}(\mathscr{E}) := \{(-i\omega_D + \nu^{(l)}(\mathscr{E}))\,\mathbf{U} - \boldsymbol{\omega}_B \times \mathbf{U}\}^{-1} \tag{31.5a}$$

with the three eigenvalues [cf. Eqs. (29.3) (29.8) (28.4c)]

$$\beta_M^{(l)}(\mathscr{E}) = \{-i(\omega_D - M\omega_B) + \nu^{(l)}(\mathscr{E})\}^{-1}, \qquad M = 0, \pm 1, \tag{31.5b}$$

where

$$\nu^{(l)}(\mathscr{E}) := \sum_h \nu_{eh}^{(l)}(\mathscr{E}) \tag{31.6}$$

---

[1] Suchy, K.: See [*8*], eqs. 105.1 and 105.2

[2] Suchy, K.: See [*25*], pp. 23–33, eqs. 6.1 and 6.2

is the sum of the transfer collision frequencies $\nu_{\mathrm{eh}}^{(l)}(\mathscr{E})$, Eq. (4.14), of electrons with heavy particles.

In general the isotropic part $f(\boldsymbol{r}, t, C)=f(\boldsymbol{r}, t, \sqrt{2\mathscr{E}/m})$ of the electron distribution function depends on several parameters, e.g., the number density N, temperature T, field strength $\boldsymbol{E}_v$, etc., which all may vary with $\boldsymbol{r}, t$. We assume in the following that the spatial variation of N and T dominates the spatial variations of the remaining parameters and write

$$\frac{\partial f}{\partial \boldsymbol{r}} \approx \frac{f}{N}\frac{\partial N}{\partial \boldsymbol{r}}+\frac{\partial f}{\partial T}\frac{\partial T}{\partial \boldsymbol{r}}=\frac{f}{p}\frac{\partial p}{\partial \boldsymbol{r}}+T\frac{\partial}{\partial T}\left(\frac{f}{T}\right)\frac{\partial T}{\partial \boldsymbol{r}}. \tag{31.7}$$

The gas law $p=NkT$ was used for the second equality.

Writing Eqs. (31.2) (31.3) in a form similar to the transport equations (26.1) (26.9) which may be written as

$$\boldsymbol{J}=\mathbf{b}\cdot Nq\boldsymbol{E}_v-\frac{Nm}{p}\mathbf{D}\cdot\frac{\partial p}{\partial \boldsymbol{r}}-\frac{Nm}{T}\mathbf{D}^T\cdot\frac{\partial T}{\partial \boldsymbol{r}}, \tag{31.8}$$

$$\boldsymbol{Q}=\frac{p}{Nm}\check{\mathbf{b}}^T\cdot Nq\boldsymbol{E}_v-\check{\mathbf{D}}^T\cdot\frac{\partial p}{\partial \boldsymbol{r}}-\boldsymbol{\kappa}\cdot\frac{\partial T}{\partial \boldsymbol{r}}, \tag{31.9}$$

the transport coefficients are expressed by

$$\left.\begin{matrix}\mathbf{b}\\ \check{\mathbf{b}}^T\end{matrix}\right\}=\pm\frac{4\pi}{3}\int_0^\infty \mathrm{d}C\,C^3\frac{-\partial}{\partial C}\left(\frac{f}{N}\right)\boldsymbol{\beta}^{(1)}(\mathscr{E})\left\{\begin{matrix}\mathrm{L}_0\\ \mathrm{L}_1^{(3/2)}\left(\dfrac{\mathscr{E}}{kT}\right),\end{matrix}\right. \tag{31.10a, b}$$

$$\left.\begin{matrix}\mathbf{D}\\ \check{\mathbf{D}}^T\end{matrix}\right\}=\pm\frac{4\pi}{3}\int_0^\infty \mathrm{d}C\,C^4\frac{f}{N}\boldsymbol{\beta}^{(1)}(\mathscr{E})\left\{\begin{matrix}\mathrm{L}_0\\ \mathrm{L}_1^{(3/2)}\left(\dfrac{\mathscr{E}}{kT}\right),\end{matrix}\right. \tag{31.11a b}$$

$$\left.\begin{matrix}\mathbf{D}^T\\ \dfrac{\boldsymbol{\kappa}}{Nk}\end{matrix}\right\}=\pm\frac{4\pi}{3}\int_0^\infty \mathrm{d}C\,C^4T^2\frac{\partial}{\partial T}\left(\frac{f}{NT}\right)\boldsymbol{\beta}^{(1)}(\mathscr{E})\left\{\begin{matrix}\mathrm{L}_0\\ \mathrm{L}_1^{(3/2)}\left(\dfrac{\mathscr{E}}{kT}\right)\end{matrix}\right. \tag{31.12a, b}$$

with the generalized Laguerre (or Sonine) polynomials

$$\mathrm{L}_0=1,\qquad \mathrm{L}_1^{(3/2)}(w)=\tfrac{5}{2}-w. \tag{31.13}$$

β) Transport coefficients represented by transport tensors. The mathematical form of the speed dependence of $f(\boldsymbol{r}, C)$ can only be given if an equation for $f(\boldsymbol{r}, C)$ has first been solved which is also provided by the Lorentz method[3,4]. For many cases a *local Maxwell distribution*

$$f^M(N(\boldsymbol{r}), T(\boldsymbol{r}), C)=\frac{N\,\mathrm{e}^{-w}}{\sqrt{2\pi kT/m^3}}\quad\text{with } w=\frac{mC^2}{2kT}=\frac{\mathscr{E}}{kT} \tag{31.14}$$

[3] Allis, W.P.: this Encyclopedia, vol. *21*, pp. 383–444, sects. 32 to 36

[4] Wilhelm, J., Wallis, G.: See [*17*], sects. 2.2.3.2, 3.1.2, 3.1.3

represents the isotropic part of the velocity distribution function with sufficient accuracy. The analytic form of f^M and the appearance of generalized Laguerre polynomials $L_s^{(3/2)}(w)$ in Eqs. (31.10) (31.11) (31.12) suggests a representation of the transport coefficients for a Maxwell distribution in terms of *transport tensors*

$$\boldsymbol{\beta}^{(ls)}(T) := \frac{(-1)^s s!}{(l+\frac{1}{2}+s)!} \int_0^\infty \mathrm{d}w\, \mathrm{e}^{-w} w^{l+\frac{1}{2}} L_s^{(l+\frac{1}{2})}(w)\, \boldsymbol{\beta}^{(l)}(kTw) \tag{31.15}$$

which are the expansion coefficients of the transfer collision tensors

$$\boldsymbol{\beta}^{(l)}(\mathscr{E}) = \sum_{s=0}^{\infty} (-1)^s \boldsymbol{\beta}^{(ls)}(T)\, L_s^{(l+\frac{1}{2})}(w) \tag{31.16a}$$

$$= \boldsymbol{\beta}^{(l0)} \text{ for energy independent } \boldsymbol{\beta}^{(l)}. \tag{31.16b}$$

This expansion is completely equivalent to the expansion of the transfer collision frequencies $\nu^{(l)}(\mathscr{E})$, Eq. (19.4), in terms of transport collision frequencies $\nu^{(ls)}(\hat{T})$, Eq. (19.3). The form of the expansion coefficients $\boldsymbol{\beta}^{(ls)}$, Eq. (31.15), is derived from Eq. (31.16) by means of identical relations for the *transport eigenvalues* $\beta_M^{(ls)}(T)$, $\beta_M^{(l)}(\mathscr{E})$ ($M=0, \pm 1$):

$$\beta_M^{(ls)}(T) = \frac{(-1)^s s!}{(l+\frac{1}{2}+s)!} \int_0^\infty \frac{\mathrm{d}w\, \mathrm{e}^{-w} w^{l+\frac{1}{2}} L_s^{(l+\frac{1}{2})}(w)}{-i(\omega_\mathrm{D} - M\omega_B) + \nu^{(l)}(kTw)} \tag{31.17}$$

$$\beta_M^{(l)}(\mathscr{E}) = \sum_{s=0}^{\infty} (-1)^s \beta_M^{(ls)}(T)\, L_s^{(l+\frac{1}{2})}(w). \tag{31.18}$$

For the transport coefficients, Eqs. (31.10, 11, 12), one obtains with a local Maxwell distribution $f^M \sim \exp - w$, Eq. (31.14),

$$\mathbf{b} = \boldsymbol{\beta}^{(10)}, \quad \mathbf{D} = \frac{kT}{m}\boldsymbol{\beta}^{(10)}, \quad \mathbf{D}^T = \frac{kT}{m}\frac{5}{2}\boldsymbol{\beta}^{(11)}, \tag{31.19a}$$

$$\check{\mathbf{b}}^T = \frac{5}{2}\boldsymbol{\beta}^{(11)}, \quad \check{\mathbf{D}}^T = \frac{kT}{m}\frac{5}{2}\boldsymbol{\beta}^{(11)}, \tag{31.19b}$$

$$\boldsymbol{\kappa} = \frac{kp}{m}\frac{5}{2}\left(\boldsymbol{\beta}^{(10)} + 2\boldsymbol{\beta}^{(11)} + \frac{7}{2}\boldsymbol{\beta}^{(12)}\right). \tag{31.19c}$$

When calculating $\boldsymbol{\kappa}$, the decomposition

$$(-L_1^{(3/2)})^2 = 2L_2^{(3/2)} - 2L_1^{(3/2)} + \tfrac{5}{2}L_0^{(3/2)} \tag{31.20}$$

of the square of the generalized Laguerre polynomial $L_1^{(3/2)}$ was used.

If the isotropic part f of the electron distribution can be approximated by a sum of two or more Gaussian distributions of different widths, i.e., Maxwellians [Eq. (31.14)], with different temperatures, then every transport tensor $\boldsymbol{\beta}^{(ls)}(T)$, Eq. (31.15), in the expressions for the transport coefficients, Eqs. (31.19), must be replaced by a sum of transport tensors with the same superscripts ls, but with different temperatures.

According to the Lorentz method the *Einstein relations*

$$\mathbf{D}=\frac{kT}{m}\mathbf{b}, \qquad \check{\mathbf{D}}^T=\frac{kT}{m}\check{\mathbf{b}}^T \tag{31.21}$$

hold exactly only for Maxwell isotropic distributions. The same holds for the *Onsager symmetry relation*[5]

$$\check{\mathbf{D}}^T=\mathbf{D}^T, \tag{31.22}$$

cf. Eq. (27.9c). This demonstrates that the validity of these relations is restricted to those cases where the isotropic distribution function is the local Maxwellian representing local thermal equilibrium (LTE). Thus these relations hold only for those weak deviations from LTE which are manifested by the anisotropic part of the distribution function alone.

32. Eigenvalues of the transport tensors expressed by Dingle integrals. If the sum

$$\nu^{(l)}(\mathscr{E}):=\sum_{\mathrm{h}}\nu^{(l)}_{\mathrm{eh}}(\mathscr{E}) \qquad [31.6]$$

of transfer collision frequencies $\nu^{(l)}_{\mathrm{eh}}(\mathscr{E})$ appearing in the transfer collision tensors $\boldsymbol{\beta}^{(l)}(\mathscr{E})$, Eq. (31.5), can be approximated by a single power law

$$\nu^{(l)}(\mathscr{E})=\nu^{(l)}(kT)\,w^{p+\frac{1}{2}}, \qquad [18.20\mathrm{a}]$$

then the eigenvalues $\beta^{(ls)}_M(T)$ $(M=0, \pm 1)$, Eq. (31.17), of the transport tensors $\boldsymbol{\beta}^{(ls)}(T)$, Eq. (31.15), can be represented by a certain class of definite integrals, the so-called Dingle integrals. The case $l=1$ was already used in Sects. 29γ and 30. The factors $\nu^{(l)}(kT)$ are called *monoenergetic transfer collision frequencies.* With Eq. (18.20a) the three eigenvalues $\beta^{(ls)}_M$, Eq. (31.17), of the transport tensors are

$$\beta^{(ls)}_M(T)=\frac{1}{\nu^{(l)}(kT)}\,\frac{(-1)^s s!}{(l+\frac{1}{2}+s)!}\int_0^\infty \frac{\mathrm{d}w\,\mathrm{e}^{-w}\,w^{l+\frac{1}{2}}\,\mathrm{L}^{(l+\frac{1}{2})}_s(w)}{w^{p+\frac{1}{2}}-i\,\dfrac{\omega_{\mathrm{D}}-M\,\omega_B}{\nu^{(l)}(kT)}} \qquad (M=0, \pm 1). \tag{32.1}$$

α) Normalized transport integrals and transport functions. Introducing the *normalized transport integrals*

$$\beta^{(ls)*}_M(p+\tfrac{1}{2};\,y):=\frac{(-1)^s s!}{(l+\frac{1}{2}+s)!}\int_0^\infty \frac{\mathrm{d}w\,\mathrm{e}^{-w}\,w^{l+\frac{1}{2}}\,\mathrm{L}^{(l+\frac{1}{2})}_s(w)}{w^{p+\frac{1}{2}}-i\,y}, \tag{32.2}$$

we write the transport eigenvalues as

$$\beta^{(ls)}_M(T)=\frac{1}{\nu^{(l)}(kT)}\,\beta^{(ls)*}_M(p+\tfrac{1}{2};\,y^{(l)}_M) \tag{32.3a}$$

[5] Mitchner, M., Kruger, C.H.: See [24], eqs. 2.38

with the normalized frequencies

$$y_M^{(l)} := \frac{\omega_D - M\,\omega_B}{\nu^{(l)}(kT)}. \tag{32.3b}$$

For energy-independent transfer collision frequencies, i.e., for $p=-\frac{1}{2}$, the denominator in the normalized transport integrals, Eq. (32.2), does not depend on the integration variable $w=\mathscr{E}/kT$. Then, according to the orthogonality of the generalized Laguerre (or Sonine) polynomials $\mathrm{L}_s^{(l+\frac{1}{2})}(w)$, all normalized transport integrals $\beta_M^{(ls)*}(0, y)$ vanish for $s>0$:

$$\beta_M^{(ls)*}(0, y) = \frac{\delta_{s0}}{1-iy} \qquad \text{for } p=-\tfrac{1}{2}, \tag{32.4}$$

hence

$$\beta_M^{(ls)}(T) = \frac{\delta_{s0}}{-i(\omega_D - M\,\omega_B)+\nu^{(l)}} \qquad \text{for } p=-\tfrac{1}{2}. \tag{32.5}$$

δ_{st} is the Kronecker symbol which is unity for $s=t$ and zero for $s\neq t$.

For $p\neq -\frac{1}{2}$ the polynomials are expanded[1]:

$$\mathrm{L}_s^{(l+\frac{1}{2})}(w) = \sum_{n=0}^{s} \binom{l+\frac{1}{2}+s}{s-n} \frac{(-w)^n}{n!}, \tag*{[19.2]}$$

since the Dingle integrals, Eqs. (32.8, 9, 10), contain powers of w in the numerator instead of generalized Laguerre polynomials.

Introducing the *transport functions*

$$B_{q,p}(y) := \frac{1}{(q+\frac{1}{2})!} \int_0^\infty \frac{\mathrm{d}w\, w^{q+\frac{1}{2}}\,\mathrm{e}^{-w}}{w^{p+\frac{1}{2}} - iy}, \tag{32.6}$$

we can write the normalized transport integrals Eq. (32.2) with the expansion for the generalized Laguerre polynomials, Eq. (19.2), as

$$\beta_M^{(ls)*}(p+\tfrac{1}{2}; y) = \sum_{n=0}^{s} (-1)^{n+s} \binom{s}{n} B_{l+n,p}(y). \tag{32.7}$$

The following *Dingle integrals* have been defined, thoroughly discussed, and tabulated by DINGLE et al.[2], DINGLE[3,4]:

$$\mathfrak{A}_q(z) := \frac{1}{q!} \int_0^\infty \frac{\mathrm{d}w\, w^q\, \mathrm{e}^{-w}}{w+z} =: \frac{1}{z}\Lambda(z) = \mathrm{U}(1, 1-q, z), \tag{32.8}$$

---

[1] Abramowitz, M., Stegun, I.A. (eds.) (1965): Handbook of mathematical functions, eq. 22.3.9. New York: Dover

[2] Dingle, R.B., Arndt, D., Roy, S.K. (1957): Appl. Sci. Res. sect. B *6*, 144–154, 155–164, 245–252

[3] Dingle, R.B. (1958): Proc. R. Soc. London Ser. A *244*, 456–475, 476–483, 484–490

[4] Dingle, R.B. (1958): Proc. R. Soc. London Ser. A *249*, 270–283, 284–292, 293–295

$$\mathfrak{C}_q(z) := \frac{1}{q!}\int_0^\infty \frac{dw\, w^q\, e^{-w}}{w^2+z^2} =: \frac{1}{z^2}\,\Pi_q(z), \tag{32.9}$$

$$\mathfrak{E}_q(z) := \frac{1}{q!}\int_0^\infty \frac{dw\, w^q\, e^{-w}}{1+z\,w^3}. \tag{32.10}$$

The function U(a, b, z) in Eq. (32.8) is a confluent hypergeometric function sometimes denoted[5] by $\Psi(a, b, z)$. Some relations derived by GILARDINI[6] could be of interest for the discussion of the Dingle integrals. Extensions of the original tabulations of $\mathfrak{A}_q$ and $\mathfrak{C}_q$ were made by BROWN and HINDLEY[7].

The transport functions $B_{q,p}(y)$ for integer $2p+1=\pm 1, \pm 2, \pm 3$ can be expressed by the Dingle integrals in the following way[8]:

$$B_{q,0}(y) = \frac{(q+1)!}{(q+\frac{1}{2})!}\,\mathfrak{A}_{q+1}(y^2) + i\,y\,\mathfrak{A}_{q+\frac{1}{2}}(y^2), \tag{32.11a}$$

$$B_{q,-1}(y) = \frac{(q+1)!}{(q+\frac{1}{2})!}\,\frac{1}{y^2}\,\mathfrak{A}_{q+1}\left(\frac{1}{y^2}\right) + \frac{i}{y}\,(q+\tfrac{3}{2})\,\mathfrak{A}_{q+\frac{3}{2}}\left(\frac{1}{y^2}\right), \tag{32.11b}$$

$$B_{q,+\frac{1}{2}}(y) = (q+\tfrac{3}{2})\,\mathfrak{C}_{q+\frac{3}{2}}(y) + i\,y\,\mathfrak{C}_{q+\frac{1}{2}}(y), \tag{32.11c}$$

$$B_{q,-\frac{3}{2}}(y) = (q+\tfrac{3}{2})\,\frac{1}{y^2}\,\mathfrak{C}_{q+\frac{3}{2}}\left(\frac{1}{y}\right) + \frac{i}{y}\,\frac{(q+\frac{5}{2})!}{(q+\frac{1}{2})!}\,\mathfrak{C}_{q+\frac{5}{2}}\left(\frac{1}{y}\right), \tag{32.11d}$$

$$B_{q,+1}(y) = \frac{(q+2)!}{(q+\frac{1}{2})!}\,\frac{1}{y^2}\,\mathfrak{E}_{q+2}\left(\frac{1}{y^2}\right) + \frac{i}{y}\,\mathfrak{E}_{q+\frac{1}{2}}\left(\frac{1}{y^2}\right), \tag{32.11e}$$

$$B_{q,-2}(y) = \frac{(q+2)!}{(q+\frac{1}{2})!}\,\mathfrak{E}_{q+2}(y^2) + i\,y\,\frac{(q+\frac{7}{2})!}{(q+\frac{1}{2})!}\,\mathfrak{E}_{q+\frac{7}{2}}(y^2). \tag{32.11f}$$

With these transport functions $B_{q,p}(y)$ all normalized transport integrals $\beta_M^{(ls)*}(y)$, Eq. (32.2), can be expressed by Eq. (32.7) for $2p+1=\pm 1, \pm 2, \pm 3$.

Fig. 40 shows plots of

$$\nu^{(10)}\,\beta_M^{(10)} = \frac{\nu^{(10)}}{\nu^{(1)}(kT)}\,\beta_M^{(10)*} = \frac{\nu^{(10)}}{\nu^{(1)}(kT)}\,B_{1,p} = \frac{(p+2)!}{\frac{3}{2}!}\,B_{1,p} \tag{32.12}$$

however as functions of

$$y_M^{(10)} := \frac{\omega_D - M\,\omega_B}{\nu^{(10)}} \tag{32.13}$$

instead of

$$y_M^{(1)} := \frac{\omega_D - M\,\omega_B}{\nu^{(1)}(kT)}. \qquad [32.3b]$$

---

[5] Abramowitz, M., Stegun, I.A. (eds.) (1965): Handbook of mathematical functions, sect. 13.1. New York: Dover

[6] Gilardini, A.L.: See [*19*], appendix B

[7] Brown, R.M., Hindley, N.K. (1964): Engineering experiment station circular, No. 84. Urbana: University of Illinois

[8] Suchy, K.: See [*8*], eqs. 108.5 through 108.10

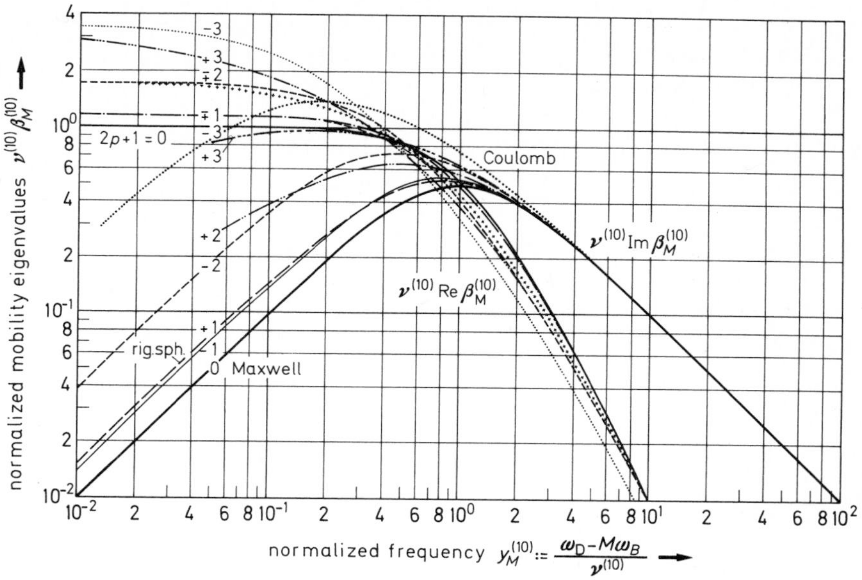

Fig. 40. Normalized mobility eigenvalues $\nu^{(10)}\beta_M^{(10)}$, Eq. (32.12), as functions of $y_M^{(10)} := (\omega_D - M\omega_B)/\nu^{(10)}$ for transfer collision cross sections $Q^{(1)}(\mathscr{E})$, Eq. (4.8), proportional to $\mathscr{E}^p$ (after Suchy[9]. Parameter of the curves is $2p+1$

Plots[10] of $\beta_M^{(10)*}(p+\frac{1}{2};\, y_M^{(1)})$ versus $y_M^{(1)}$ do not show a common asymptote of $\mathrm{Re}\,\beta_M^{(10)*}$ for high values of $y_M^{(1)}$. The appearance of a common asymptote of $\nu^{(10)}\,\mathrm{Re}\,\beta_M^{(10)}$ in Fig. 40 is due to the "high-frequency approximation", Eq. (33.2a). The limiting values for vanishing $y_M^{(10)}$, viz.,

$$\lim_{y\to 0} \nu^{(10)}\,\mathrm{Re}\,\beta_M^{(10)} = K_\sigma^{[10]}(0), \tag{32.14a}$$

$$\lim_{y\to 0} \nu^{(10)}\,\mathrm{Im}\,\beta_M^{(10)} = y_M^{(10)}\,K_\varepsilon^{[10]}(0), \tag{32.14b}$$

are given by the resonance values $K_\sigma^{[10]}(0)$, $K_\varepsilon^{[10]}(0)$, Eq. (34.7) [see Table 11, (Sect. 34)] of the Ginzburg-Gurevič functions K_σ, K_ε, which are discussed in Sect. 34.

The normalization with $\nu^{(10)}$, Eq. (32.13), instead of $\nu^{(1)}(kT)$ allows the following conclusion to be drawn from Fig. 40. Increasing $|p+\frac{1}{2}|$ means increasing energy dependence of the transfer collision frequency $\nu^{(1)}(\mathscr{E}) \sim \mathscr{E}^{p+\frac{1}{2}}$. Starting with the energy independent case $p+\frac{1}{2}=0$, the deviation of $\nu^{(10)}\beta_M^{(10)}$ increases with increasing values of $y_M^{(10)}$. This corresponds to the condition

$$\left(\frac{\nu^{(11)}}{\nu^{(10)}}\right)^2 \ll |-i\,y_M^{(10)}+1|^2 \qquad [29.7]$$

[9] Suchy, K.: See [8], fig. 6b
[10] Suchy, K.: See [8], fig. 6a

which we established in Sect. 29 for the validity of the zeroth approximation $\sigma_M^{[0]}$ of the conductivity eigenvalues calculated with the moment method. Here the small ratio

$$\left(\frac{\nu^{(11)}}{\nu^{(10)}}\right)^2 = \left(\frac{p+\frac{1}{2}}{5/2}\right)^2 \qquad [19.8\,e]$$

increases with increasing $|p+\frac{1}{2}|$ and therefore decreases the accuracy of $\sigma_M^{[0]}$.

β) Approximation of Dingle integrals by rational functions. For computing purposes it is desirable to have approximations for the Dingle integrals which can easily be handled by a computer. HARA[11] gave the approximations

$$\mathfrak{C}_{\frac{3}{2}}(x) \approx \frac{p_4(x)}{p_6(x)}, \qquad \mathfrak{C}_{\frac{5}{2}}(x) \approx \frac{p_3(x)}{p_5(x)} \qquad (32.15)$$

with polynomials

$$p_k(x) = \sum_{\kappa=0}^{k} a_\kappa^{(k)} x^\kappa \qquad (32.16)$$

and has given the values of the coefficients $a_\kappa^{(k)}$ with eight decimals. The maximum error is less than 0.7% over the whole range $0 \leqq x < \infty$.

For low-energy collisions of electrons with nitrogen molecules the value $p=1/2$ must be taken [see Eq. (11.1)]. The value of the monoenergetic momentum transfer collision frequency $\nu^{(1)}(kT) \equiv \nu_m$ is listed in Eqs. (18.21a) and (21.2a).

With Eqs. (31.19) (32.12) (32.11c) the conductivity eigenvalues become

$$\frac{m}{N q^2}\sigma_M = b_M = \beta_M^{(10)} = \frac{1}{\nu^{(1)}(kT)}\beta_M^{(10)*}(1;\, y_M^{(1)}) = \frac{1}{\nu^{(1)}(kT)} B_{1,\frac{1}{2}}(y_M^{(1)})$$

$$= \frac{1}{\nu^{(1)}(kT)}\{\tfrac{5}{2}\mathfrak{C}_{\frac{5}{2}}(y_M^{(1)}) + i\, y_M^{(1)}\,\mathfrak{C}_{\frac{3}{2}}(y_M^{(1)})\} \qquad \text{for } p=\tfrac{1}{2}. \qquad (32.17)$$

With the approximations Eqs. (32.15) for the two Dingle integrals $\mathfrak{C}_{\frac{5}{2}}$ and $\mathfrak{C}_{\frac{3}{2}}$ the conductivity eigenvalues σ_M can easily be calculated with a computer. It would be very desirable to have approximations of the Dingle integrals $\mathfrak{A}(x)$ and $\mathfrak{E}(x)$ by rational functions, too.

33. Binomial approximations of the transport eigenvalues

α) High-frequency approximation. The values of the transfer collision frequencies $\nu^{(l)}(\mathscr{E})$, Eq. (4.14), are bounded, i.e., they remain finite for all values of the electron energy $\mathscr{E} = kTw$. It is therefore justified to calculate the transport eigenvalues $\beta_M^{(ls)}(T)$, $M=0, \pm 1$, Eq. (31.17), and also linear combinations of them under the *high-frequency condition*

$$\nu^{(l)}(\mathscr{E}) \ll |\omega_D - M\,\omega_B| \qquad (33.1)$$

by means of a binomial expansion[1] of the denominator of the integrand of $\beta_M^{(ls)}(T)$. One obtains for linear combinations of $\beta_M^{(ls)}$ with different s, needed to represent real and imaginary parts of the transport coefficients Eqs. (31.19),

[11] Hara, E.H. (1963): J. Geophys. Res. *68*, 4388–4389

[1] Budden, K.G. (1965): J. Res. Natl. Bur. Stand. Sect. D *69*, 191–211, eq. 58

$$\sum_s a_s \operatorname{Re} \beta_M^{(ls)}(T) = \frac{\sum_s a_s \nu^{(ls)}}{(\omega_D - M\,\omega_B)^2} + O\left(\frac{\nu^{3(ls)}}{(\omega_D - M\,\omega_B)^4}\right), \tag{33.2a}$$

$$\sum_s a_s \operatorname{Im} \beta_M^{(ls)}(T) = \frac{a_0}{\omega_D - M\,\omega_B} + O\left(\frac{\nu^{2(ls)}}{(\omega_D - M\,\omega_B)^3}\right). \tag{33.2b}$$

Combining both equations, we obtain the complex expression

$$\sum_s a_s \beta_M^{(ls)}(T) \approx \frac{a_0}{-i(\omega_D - M\,\omega_B) + \frac{1}{a_0}\sum_s a_s \nu^{(ls)}}. \tag{33.2c}$$

In the remainders of the series in Eq. (33.2) the terms $(-1)^s \nu^{n(ls)}$ are the expansion coefficients of powers $(\nu^{(l)}(\mathscr{E}))^n$ into generalized Laguerre (Sonine) polynomials analogous to Eq. (19.4) for $n=1$.

For *energy-independent* transfer collision frequencies $\nu^{(ls)} = \delta_{s0}\,\nu^{(l)}$, Eq. (19.8d), holds. Hence

$$\sum_s a_s \beta_M^{(ls)}(T) \approx \frac{a_0}{-i(\omega_D - M\,\omega_B) + \nu^{(l)}} \quad \text{for } \nu^{(ls)} = \delta_{s0}\,\nu^{(l)}. \tag{33.3}$$

In this particular case the high-frequency approximation is exact, as shown by a comparison with the exact expression Eq. (31.5).

For *energy-dependent* transfer collision frequencies $\nu^{(l)}(\mathscr{E})$, Eq. (33.2c) yields, in the particular case $a_s = \delta_{s0}$,

$$\beta_M^{(l0)}(T) \approx \frac{1}{-i(\omega_D - M\,\omega_B) + \nu^{(l0)}} \tag{33.4a}$$

and for $a_0 = 1$, $a_1 = 2$, $a_2 = 7/2$, needed for the heat conductivity $\boldsymbol{\kappa}$, Eq. (31.19c),

$$\beta_M^{(10)} + 2\beta_M^{(11)} + \tfrac{7}{2}\beta_M^{(12)} \approx \frac{1}{-i(\omega_D - M\,\omega_B) + \nu^{[10]}} \tag{33.4b}$$

with

$$\nu^{[10]} := \nu^{(10)} + 2\nu^{(11)} + \tfrac{7}{2}\nu^{(12)}. \tag{28.6d}$$

Since the structure of the transport eigenvalues under the high-frequency condition Eq. (33.1) is so simple, one can write down the transport tensors $\mathbf{b}$, $\boldsymbol{\kappa}$, Eqs. (31.19a, c), for a Maxwell distribution in the same way as the tensors $\mathbf{b}^{[0]}$, $\breve{\mathbf{b}}^{[0]}$ in Eq. (29.4) were obtained from the eigenvalues Eqs. (29.3):

$$\frac{m}{N q^2}\boldsymbol{\sigma}_{HF} = \frac{m}{kT}\mathbf{D}_{HF} = \mathbf{b}_{HF} = \{(-i\,\omega_D + \nu^{(10)})\,\mathbf{U} - \boldsymbol{\omega}_B \times \mathbf{U}\}^{-1} \tag{33.5}$$

$$\frac{2}{5}\frac{m}{kp}\boldsymbol{\kappa}_{HF} = \{(-i\,\omega_D + \nu^{[10]})\,\mathbf{U} - \boldsymbol{\omega}_B \times \mathbf{U}\}^{-1}. \tag{33.6}$$

Since the transport collision frequency $\nu^{(10)}$ appears in the high-frequency approximation, Eq. (33.4a), of the electric conductivity, it is sometimes named "effective collision frequency" and denoted[2] by ν_{eff}.

The expressions in Eq. (33.5) for $\boldsymbol{\sigma}_{\text{HF}}$ and Eq. (33.6) for $\boldsymbol{\kappa}_{\text{HF}}$ are identical with the zeroth approximations – under the condition Eq. (29.7) – $\boldsymbol{\sigma}^{[0]}$ and $\boldsymbol{\kappa}^{[0]}$ in Eqs. (29.4). This was used by SUCHY and RAWER[3] to obtain correct matching conditions for the power law approximation of a sum of momentum transfer collision frequencies $\nu^{(1)}(\mathscr{E}) := \sum_{\text{h}} \nu_{\text{eh}}^{(1)}(\mathscr{E})$ as

$$\nu^{(l)}(\mathscr{E}) = \nu^{(l)}(kT)\, w^{p+\frac{1}{2}}. \qquad [18.20\text{a}]$$

For $\nu^{(l)}(\mathscr{E})$ obeying Eq. (18.20a) with $2p+1 = \pm 1, \pm 2, \pm 3$, the transport eigenvalues $\beta_M^{(ls)}$, Eq. (31.17), can be expressed by Dingle integrals (see Sect. 32α). If the high-frequency limits of the Dingle integrals are compared with the limits Eq. (33.2) which were calculated with a binomial expansion, one finds that the limit values after Eqs. (33.2) are the same as those obtained from the Dingle integrals; but the order of the first additional term is not always expressed by the orders given in Eqs. (33.2a, b). It turns out that these orders are correct only in the ranges

$$-2 < 2p+1 \leqq +3 \quad \text{for } \operatorname{Re} \beta_M^{(ls)}(T), \qquad (33.7\text{a})$$

$$-3 \leqq 2p+1 \leqq +3 \quad \text{for } \operatorname{Im} \beta_M^{(ls)}(T). \qquad (33.7\text{b})$$

For $|2p+1| > 3$ there exist up to now no closed expressions of the transport eigenvalues $\beta_M^{(ls)}(T)$ by Dingle integrals. Thus the validity range may exceed $+3$ for $\operatorname{Re} \beta_M^{(ls)}$ and ± 3 for $\operatorname{Im} \beta_M^{(ls)}$; but for $2p+1 = -2$ the first neglected order in Eq. (33.2a) is certainly lower than $\nu^{3(ls)}/(\omega_{\text{D}} - M\omega_B)^4$.

β) Resonance approximations. In the vicinity of the resonance values $\omega_{\text{D}} = M\omega_B$ ($M = 0, \pm 1$), i.e., for the *resonance approximation*

$$|\omega_{\text{D}} - M\omega_B| \ll \nu^{(l)}(\mathscr{E}) \equiv \sum_{\text{h}} \nu_{\text{eh}}^{(l)}(\mathscr{E}), \qquad (33.8)$$

another binomial expansion is employed. To express the results, it is convenient to introduce *reciprocal transfer collision frequencies*, i.e., time intervals which are called *transfer collision intervals*

$$\tau^{(l)}(\mathscr{E}) := \frac{1}{\nu^{(l)}(\mathscr{E})}. \qquad (33.9)$$

They express time intervals between successive collisions. Expanded after generalized Laguerre polynomials (Sonine polynomials) $\mathrm{L}_s^{(l+\frac{1}{2})}(w)$, the expansion coefficients are – up to the sign $(-1)^s$ – the *transport collision intervals*

$$\tau^{(ls)}(T) := \frac{(-1)^s\, s!}{(l+\frac{1}{2}+s)!} \int_0^\infty \mathrm{d}w\, \mathrm{e}^{-w}\, w^{l+\frac{1}{2}}\, \mathrm{L}_s^{(l+\frac{1}{2})}(w)\, \tau^{(l)}(kTw). \qquad (33.10)$$

The expansion coefficients of powers $\{\tau^{(l)}(\mathscr{E})\}^n$ are denoted by $(-1)^s\, \tau^{n(ls)}$.

---

[2] Budden, K.G. (1965): J. Res. Natl. Bur. Stand. Sect. D *69*, 191–211, eq. 60

[3] Suchy, K., Rawer, K. (1971): J. Atmos. Terr. Phys. *33*, 1853–1868

For linear combinations of the transport eigenvalues $\beta_M^{(ls)}$, Eq. (31.17), one obtains the expansions for real and imaginary parts

$$\sum_s a_s \operatorname{Re}\beta_M^{(ls)}(T) = \sum_s a_s \tau^{(ls)} + O(\tau^{3(ls)}(\omega_\mathrm{D} - M\,\omega_B)^2), \tag{33.11a}$$

$$\sum_s a_s \operatorname{Im}\beta_M^{(ls)}(T) = (\omega_\mathrm{D} - M\,\omega_B) \sum_s a_s \tau^{2(ls)} + O(\tau^{4(ls)}(\omega_\mathrm{D} - M\,\omega_B)^3). \tag{33.11b}$$

The combination of both results gives

$$\sum_s a_s \beta_M^{(ls)}(T) \approx \frac{(\sum_s a_s \tau^{(ls)})^2}{-i(\omega_\mathrm{D} - M\,\omega_B)\sum_s a_s \tau^{2(ls)} + \sum_s a_s \tau^{(ls)}}. \tag{33.11c}$$

The two particular cases corresponding to Eqs. (33.4) are

$$\beta_M^{(ls)}(T) \approx \frac{(\tau^{(ls)})^2}{-i(\omega_\mathrm{D} - M\,\omega_B)\,\tau^{2(ls)} + \tau^{(ls)}}, \tag{33.12a}$$

$$\beta_M^{(l0)} + 2\beta_M^{(l1)} + \tfrac{7}{2}\beta_M^{(l2)} \approx \frac{(\tau^{[l0]})^2}{-i(\omega_\mathrm{D} - M\,\omega_B)\,\tau^{2[l0]} + \tau^{[l0]}} \tag{33.12b}$$

with

$$\tau^{n[l0]} := \tau^{n(l0)} + 2\tau^{n(l1)} + \tfrac{7}{2}\tau^{n(l2)}. \tag{33.12c}$$

Formally, one could write down expressions for the transport tensors which would be as simple as those in Eqs. (33.4) (33.5) (33.6), for the high frequency approximation, Eq. (33.1). But this would give no meaningful expressions because of the difference of the three resonance conditions Eqs. (33.8) for the three eigenvalues $M = -1, 0, +1$. For instance, near the gyroresonance $\omega_\mathrm{D} \approx +1\omega_B$, i.e. $M = +1$, the resonance condition Eq. (33.8) is not satisfied for the other two eigenvalues $\beta_M^{(ls)}$ with $M = 0, -1$.

For a power law approximation of $v^{(l)}(\mathscr{E})$, Eq. (18.20a), one can compare the limit values, Eq. (33.11), obtained with a binomial expansion, with the limit values of closed expressions found by Dingle integrals (Sect. 32α). It turns out that the order of the first neglected term as indicated in (33.11a, b) is correct only in the ranges

$$-3 \leqq 2p+1 \leqq +2 \quad \text{for } \operatorname{Re}\beta_M^{(ls)}(T), \tag{33.13a}$$

$$-3 \leqq 2p+1 \leqq +1 \quad \text{for } \operatorname{Im}\beta_M^{(ls)}(T). \tag{33.13b}$$

For a power law approximation the transport collision intervals $\tau^{(ls)}(T)$, Eq. (33.10), can be obtained from the expressions of the transport collision frequencies, Eq. (19.8d), by replacing $\frac{1}{2}+p$ by $-(\frac{1}{2}+p)$ and $v^{(l)}(kT)$ by $1/v^{(l)}(kT)$:

$$\tau^{(ls)}(p;T) = \frac{1}{v^{(l)}(kT)} \frac{(-1)^s (l-p)!}{(l+\frac{1}{2}+s)!} \frac{(p-\frac{1}{2}+s)!}{(p-\frac{1}{2})!}, \tag{33.14}$$

$$\tau^{2(ls)}(p;T) = \frac{1}{v^{(l)}(kT)} \tau^{(ls)}(2p+\tfrac{1}{2};T). \tag{33.15}$$

γ) Two-sided Padé approximation. A rather simple algebraic approximation for the whole frequency range can be obtained by a combination of the high-frequency approximation Eq. (33.2) and the resonance approximation Eq. (33.11) in an analogous algebraic manner as Planck combined Wien's and Rayleigh-Jeans' radiation laws:

$$\sum_s a_s \operatorname{Re} \beta_M^{(ls)} \approx \frac{1}{\dfrac{1}{\sum_s a_s \tau^{(ls)}} + \dfrac{(\omega_D - M\omega_B)^2}{\sum_s a_s \nu^{(ls)}}}, \tag{33.16a}$$

$$\sum_s a_s \operatorname{Im} \beta_M^{(ls)} \approx \frac{\omega_D - M\omega_B}{\dfrac{1}{\sum_s a_s \tau^{2(ls)}} + \dfrac{(\omega_D - M\omega_B)^2}{a_0}}. \tag{33.16b}$$

For power-law transfer collision frequencies $\nu^{(l)}(\mathscr{E}) \sim w^{p+\frac{1}{2}}$, Eq. (18.20a) with $2p+1 = \pm 1, \pm 2, \pm 3$, the validity ranges Eqs. (33.7) and (33.13) for the high frequency approximation and the resonance approximation must be combined. One obtains the combined validity ranges

$$-2 < 2p+1 \leqq +2 \quad \text{for } \operatorname{Re} \beta_M^{(ls)}, \tag{33.17a}$$

$$-3 \leqq 2p+1 \leqq +1 \quad \text{for } \operatorname{Im} \beta_M^{(ls)}. \tag{33.17b}$$

Unfortunately, the value $2p+1=2$ for low-energy collisions between electrons and nitrogen molecules [see Eq. (11.1)] is outside the validity range for $\operatorname{Im} \beta_M^{(ls)}$, but it is still inside the range $|2p+1| \leqq 3$ where the transport eigenvalues $\beta_M^{(ls)}$ can be represented by Dingle integrals (Sect. 32α).

34. The Ginzburg-Gurevič representation of transport eigenvalues. Representations of (linear combinations of) transport eigenvalues can be introduced analogous to the representations for σ_M with higher approximations of the moment method, Eqs. (29.8) (29.12). We start with the representation[1]

$$\sum_s a_s \beta_M^{(ls)} = K_\sigma \operatorname{Re} \frac{1}{-i(\omega_D - M\omega_B) + \sum_s a_s \nu^{(ls)}} + i K_\varepsilon \operatorname{Im} \frac{1}{-i(\omega_D - M\omega_B) + \sum_s a_s \nu^{(ls)}} \tag{34.1}$$

which is similar to the high-frequency approximation Eq. (33.2) and analogous to Eq. (29.12). The *Ginzburg-Gurevič functions*

$$K_\sigma\left(\frac{\omega_D - M\omega_B}{\sum_s a_s \nu^{(ls)}}\right) := \left[1 + \left(\frac{\omega_D - M\omega_B}{\sum_s a_s \nu^{(ls)}}\right)^2\right] \left(\sum_s a_s \nu^{(ls)}\right) \operatorname{Re}\left(\sum_s a_s \beta_M^{(ls)}\right), \tag{34.2a}$$

$$K_\varepsilon\left(\frac{\omega_D - M\omega_B}{\sum_s a_s \nu^{(ls)}}\right) := \left[\frac{\sum_s a_s \nu^{(ls)}}{\omega_D - M\omega_B} + \frac{\omega_D - M\omega_B}{\sum_s a_s \nu^{(ls)}}\right] \left(\sum_s a_s \nu^{(ls)}\right) \operatorname{Im}\left(\sum_s a_s \beta_M^{(ls)}\right) \tag{34.2b}$$

---

[1] Ginzburg. V.L., Gurevič. A.V. (1960): Usp. Fiz. Nauk *70*, 201–246, 393–428; English transl.: Soviet Phys. Usp. *3*, 115–146, 175–194; German transl.: Fortschr. Phys. *8*, 97–189, eq. 2.34

depend not only on $(\omega_{\mathrm{D}}-M\,\omega_B)/\sum_s a_s\,\nu^{(ls)}$ but also on the numbers a_s and on the way in which the monoenergetic transfer collision frequencies $\nu^{(l)}(kTw)$ in the expression for $\beta_M^{(ls)}$, Eq. (31.17), depend on the energy $\mathscr{E}=kTw$. Their *high-frequency limit*

$$K_\sigma(\infty)=1, \qquad K_\varepsilon(\infty)=\sum_s a_s \tag{34.3}$$

is obtained by inserting the high-frequency expressions for $\beta_M^{(ls)}$, Eqs. (33.2a, b), into the right-hand sides of Eqs. (34.2a, b). Their *resonance limit*

$$K_\sigma(0)=(\sum_s a_s\,\nu^{(ls)})\,(\sum_s a_s\,\tau^{(ls)}), \tag{34.4a}$$

$$K_\varepsilon(0)=(\sum_s a_s\,\nu^{(ls)})^2\,(\sum_s a_s\,\tau^{2(ls)}) \tag{34.4b}$$

is obtained by inserting Eqs. (33.11a, b) into Eqs. (34.2a, b).

For power law transfer collision frequencies $\nu^{(l)}(\mathscr{E})\sim w^{p+\frac{1}{2}}$, Eq. (18.20a), the transport eigenvalues $\beta_M^{(ls)}$, Eq. (31.17), in the definitions for the Ginzburg-Gurevič functions K_σ, K_ε, Eqs. (34.2), can be expressed with the transport functions $B_{q,p}$, Eq. (32.6), by means of Eqs. (32.3) (32.7) as

$$\beta_M^{(ls)}=\frac{1}{\nu^{(l)}(kT)}\sum_{n=0}^{s}(-1)^{n+s}\binom{s}{n}\,B_{l+n,p}\left(\frac{\omega_{\mathrm{D}}-M\,\omega_B}{\nu^{(l)}(kT)}\right). \tag{34.5}$$

The monoenergetic transfer collision frequencies $\nu^{(l)}(kT)$ must then be replaced by the transport collision frequencies $\nu^{(ls)}$ by means of

$$\frac{\nu^{(ls)}}{\nu^{(l)}(kT)}=\frac{(l+1+p)!\,(p+\frac{1}{2})!}{(l+\frac{1}{2}+s)!\,(p+\frac{1}{2}-s)!}\equiv\frac{\Gamma(l+2+p)\,\Gamma(p+\frac{3}{2})}{\Gamma(l+\frac{3}{2}+s)\,\Gamma(p+\frac{3}{2}-s)}. \qquad [19.8\,\mathrm{d}]$$

For $2p+1=\pm1$, ±2, ±3 the transport functions $B_{q,p}$ can be expressed by Dingle integrals, Eqs. (32.11).

If the sums over s in the definitions Eqs. (34.2) of the Ginzburg-Gurevič functions K_σ, K_ε consist of one term only, say $a_s=1$ (i.e., for $a_s=\delta_{ss'}$), one obtains with the expressions for $\nu^{(ls)}$, Eq. (19.8d), and Eqs. (33.14) (33.15) for $\tau^{(ls)}$, $\tau^{2(ls)}$ the *resonance limits*

$$\begin{aligned} K_\sigma^{[ls]}(0)&=\nu^{(ls)}\,\tau^{(ls)}\\ &=(-1)^s\,\frac{2p+1}{2}\,\frac{(p-\frac{1}{2}+s)!}{(p+\frac{1}{2}-s)!}\,\frac{(l+1+p)!\,(l-p)!}{\{(l+\frac{1}{2}+s)!\}^2}, \end{aligned} \tag{34.6a}$$

$$\begin{aligned} K_\varepsilon^{[ls]}(0)&=(\nu^{(ls)})^2\,\tau^{2(ls)}\\ &=(-1)^s\,\frac{\{(l+1+p)!\}^2\,(l-2p-\frac{1}{2})!}{\{(l+\frac{1}{2}+s)!\}^3}\left\{\frac{(p+\frac{1}{2})!}{(p+\frac{1}{2}-s)!}\right\}^2\frac{(2p+s)!}{(2p)!}. \end{aligned} \tag{34.6b}$$

We have written s for s'. The ls dependence of K_σ, K_ε is indicated by the superscript $[ls]$.

The particular values

$$K_\sigma^{[10]}(0)=\nu^{(10)}\,\tau^{(10)}=\frac{(p+2)!\,(1-p)!}{\{\frac{3}{2}!\}^2}, \tag{34.7a}$$

$$K_\varepsilon^{[10]}(0)=(\nu^{(10)})^2\,\tau^{2(10)}=\frac{\{(p+2)!\}^2\,(\frac{1}{2}-2p)!}{\{\frac{3}{2}!\}^3} \tag{34.7b}$$

were used by BUDDEN[2].

For energy-independent transfer collision frequencies (with $2p+1=0$) the Ginzburg-Gurevič functions K_σ, K_ε in Eq. (34.1) must be unity for $a_s=\delta_{ss'}$ because of Eq. (32.5). With increasing deviation from this case, i.e., for increasing $|2p+1|$, the values of both functions become more and more different from unity. For the resonance values $K_\sigma^{[10]}(0)$, $K_\varepsilon^{[10]}(0)$ this can be seen from Table 11.

Table 11. Resonance values of the Ginzburg-Gurevič functions $K_\sigma^{[10]}$, $K_\varepsilon^{[10]}$, Eqs. (34.7), for power-law momentum transfer collision frequencies $\nu^{(1)}(\mathscr{E})\sim\mathscr{E}^{p+\frac{1}{2}}\sim g^{2p+1}$

$2p+1$	$K_\sigma^{[10]}(0)$	$K_\varepsilon^{[10]}(0)$	$2p+1$	$K_\sigma^{[10]}(0)$	$K_\varepsilon^{[10]}(0)$
-3	$\frac{32}{3\pi}$	$\frac{140}{\pi}$	0	1	1
-2	$\frac{5}{3}$	$\frac{35}{9}$	$+1$	$\frac{32}{9\pi}$	$\frac{128}{27\pi}$
-1	$\frac{32}{9\pi}$	$\frac{40}{9\pi}$	$+2$	$\frac{5}{3}$	$\frac{25}{3}$
0	1	1	$+3$	$\frac{32}{3\pi}$	$-\frac{512}{3\pi}$

The range of the independent variable $(\omega_D-M\,\omega_B)/\sum_s a_s\,\nu^{(ls)}$, where the functions K_σ, K_ε differ from unity, increases with increasing $|2p+1|$. This can be seen from Fig. 40 (Sect. 32α), where the deviations of the curves of $\nu^{(10)}\beta_M^{(10)}$ from the curve for $2p+1=0$ indicate the deviations of K_σ, K_ε from unity. The limit values of the curves in Fig. 40 for $\omega_D-M\,\omega_B\to 0$ indicate the resonance values $K_\sigma^{[10]}(0)$, $K_\varepsilon^{[10]}(0)$ of Table 11.

If only the term $a_0=1$ exists in the sums over s for the definitions of the Ginzburg-Gurevič functions, Eqs. (34.2), i.e., for $a_s=\delta_{s0}$, Eq. (34.5) becomes

$$\beta_M^{(l0)}=\frac{1}{\nu^{(l0)}}\frac{(l+1+p)!}{(l+\frac{1}{2})!}\,B_{l,p}\left(\frac{\omega_D-M\,\omega_B}{\nu^{(l)}(kT)}\right). \tag{34.8}$$

The Ginzburg-Gurevič functions K_σ, K_ε, Eqs. (34.2), are now

$$K_\sigma^{[l0]}\left(\frac{\omega_D-M\,\omega_B}{\nu^{(l0)}}\right)=\left[1+\left(\frac{\omega_D-M\,\omega_B}{\nu^{(l0)}}\right)^2\right]\frac{(l+1+p)!}{(l+\frac{1}{2})!}$$
$$\cdot\,\mathrm{Re}\,B_{l,p}\left(\frac{(l+1+p)!}{(l+\frac{1}{2})!}\,\frac{\omega_D-M\,\omega_B}{\nu^{(l0)}}\right), \tag{34.9a}$$

[2] Budden, K.G. (1965): J. Res. Natl. Bur. Stand. Sect. D 69, 191–211, eq. 64

$$K_\varepsilon^{[l0]}\left(\frac{\omega_D - M\omega_B}{\nu^{(l0)}}\right) = \left[\frac{\nu^{(l0)}}{\omega_D - M\omega_B} + \frac{\omega_D - M\omega_B}{\nu^{(l0)}}\right]\frac{(l+1+p)!}{(l+\frac{1}{2})!}$$
$$\cdot \operatorname{Im} B_{l,p}\left(\frac{(l+1+p)!}{(l+\frac{1}{2})!}\,\frac{\omega_D - M\omega_B}{\nu^{(l0)}}\right). \tag{34.9b}$$

For $l=1$, tables and curves of $K_\sigma^{[10]}$, $K_\varepsilon^{[10]}$ are given by GINZBURG and GUREVIČ[3] for the cases $p=0$ (rigid sphere interaction) and $p=-2$ (Coulomb interaction). For the case $p=\frac{1}{2}$ (low-energy collisions with nitrogen molecules) curves of $K_\sigma^{[10]} \equiv K_\mathrm{i} \equiv K_\mathrm{I}$ and $K_\varepsilon^{[10]} \equiv K_\mathrm{r} \equiv K_\mathrm{R}$ are plotted by BUDDEN[4] and SMITH[5].

35. The Shkarofsky representation of transport eigenvalues. A representation similar to that of GINZBURG and GUREVIČ (Sect. 34) was given by SHKAROFSKY[1], who introduced two functions, g and h, writing

$$\sum_s a_s \beta_M^{(ls)} = \frac{1}{-i(\omega_D - M\omega_B)\,h + g\sum_s a_s \nu^{(ls)}} \tag{35.1}$$

instead of Eq. (34.1). The *Shkarofsky functions*

$$g\left(\frac{\omega_D - M\omega_B}{\sum_s a_s \nu^{(ls)}}\right) := \frac{1}{\sum_s a_s \nu^{(ls)}} \operatorname{Re} \frac{1}{\sum_s a_s \beta_M^{(ls)}}$$
$$= \left[1 + \left(\frac{\omega_D - M\omega_B}{\sum_s a_s \nu^{(ls)}}\right)^2\right] \frac{K_\sigma}{K_\sigma^2 + \left(\frac{\omega_D - M\omega_B}{\sum_s a_s \nu^{(ls)}}\right)^2 K_\varepsilon^2}, \tag{35.2a}$$

$$h\left(\frac{\omega_D - M\omega_B}{\sum_s a_s \nu^{(ls)}}\right) := \frac{1}{\omega_D - M\omega_B} \operatorname{Im} \frac{-1}{\sum_s a_s \beta_M^{(ls)}} = \frac{K_\varepsilon}{K_\sigma} g \tag{35.2b}$$

have simple algebraic relations with the Ginzburg-Gurevič functions K_σ, K_ε, Eqs. (34.2). Their *high-frequency limit* is obtained from that of $K_\sigma(\infty)$, $K_\varepsilon(\infty)$, Eq. (34.3), as

$$g(\infty) = \frac{1}{K_\varepsilon^2(\infty)} = \frac{1}{(\sum_s a_s)^2}, \qquad h(\infty) = \frac{1}{K_\varepsilon(\infty)} = \frac{1}{\sum_s a_s}. \tag{35.3}$$

---

[3] Ginzburg, V.L., Gurevič, A.V. (1960): Usp. Fiz. Nauk *70*, 201–246, 393–428; English transl.: Soviet Phys. Usp. *3*, 115–146, 175–194; German transl.: Fortschr. Phys. *8*, 97–189, table 2, figs. 4 and 5

[4] Budden, K.G. (1965): J. Res. Natl. Bur. Stand. Sect. D *69*, 191–211, fig. 3

[5] Smith, M.S. (1975): Proc. R. Soc. London Ser. A *343*, 133–153, fig. 7

[1] Shkarofsky, I.P. (1961): Can. J. Phys. *39*, 1619–1703, eq. 19

In the same way we obtain their *resonance limit* from Eq. (34.4) as

$$g(0)=\frac{1}{K_\sigma(0)}=\frac{1}{(\sum_s a_s\nu^{(ls)})(\sum_s a_s\tau^{(ls)})},\tag{35.4a}$$

$$h(0)=\frac{K_\varepsilon(0)}{K_\sigma^2(0)}=\frac{\sum_s a_s\tau^{2(ls)}}{(\sum_s a_s\tau^{(ls)})^2}.\tag{35.4b}$$

For power law transfer collision frequencies $\nu^{(l)}(\mathscr{E})\sim\mathscr{E}^{p+\frac{1}{2}}$ the Ginzburg-Gurevič functions K_σ, K_ε, Eqs. (34.2), can be expressed via the transport eigenvalues $\beta_M^{(ls)}$, Eqs. (34.5), and the transport functions $B_{q,p}$, Eq. (32.6), by Dingle integrals, Eq. (32.11), for $2p+1=\pm1$, ±2, ±3. Since the Shkarofsky functions g, h are related to the Ginzburg-Gurevič functions K_σ, K_ε, they can be calculated in these cases too using Dingle integrals. For $a_s=\delta_{s0}$, $l=1$, this has been done by Shkarofsky et al.[1-4].

36. Some applications and extensions. It was our intention to give a state-of-the-art review of present knowledge about basic features of the definition, calculation, computation, and measurement of collision cross sections and transport coefficients. A few applications of these results shall be summarized in this section together with some extensions and clarifications of the results presented on the foregoing pages.

α) Ambipolar conditions. In several situations of transport processes, restrictions regarding the electric current density are necessary. There are situations where no electric current can exist. This is achieved by an "ambipolar field" $\boldsymbol{E}_v^{\text{amb}}$ which compensates the contribution of $N_j m_j \boldsymbol{g}_0-\partial p_j/\partial\boldsymbol{r}$ in Ohm's law, Eq. (25.7), plus the thermal diffusion (Soret) current of Eq. (26.4). Insertion of this field $\boldsymbol{E}_v^{\text{amb}}$ into the transport equations (26.1) (26.9) for the mass fluxes $\boldsymbol{J}_k$ and the heat fluxes $\boldsymbol{Q}_k$ changes the contributions of $\boldsymbol{g}_0$, $\partial p_j/\partial\boldsymbol{r}$, and $\partial T_j/\partial\boldsymbol{r}$ to $\boldsymbol{J}_k$ and $\boldsymbol{Q}_k$ and therefore modifies the mobility matrix tensors $\mathbf{b}_{kj}$ and $\breve{\mathbf{b}}_{kj}$. The so-modified diffusion tensor $\mathbf{D}_{kj}$, Eq. (25.4), is usually named "ambipolar diffusion tensor", the modified heat conductivity tensor $\boldsymbol{\kappa}$, Eq. (25.11b), is denoted as "effective" heat conductivity tensor[1,2] or – less satisfyingly – "usual" heat conductivity tensor[3]. Conversion formulas between the different heat conductivity tensors for an electron plasma can be found in these references and in Schunk and Walker[4].

A more complicated situation exists if one requires that only one vector component of the electric current density vanishes. This happens in the ionosphere where the vanishing of the vertical component at the lower boundary is

[2] Shkarofsky, I.P. (1962): Can. J. Phys. *40*, 49–60
[3] Shkarofsky, I.P., Bernstein, I.B., Robinson, B.B. (1963): Phys. Fluids *6*, 40–47
[4] Shkarofsky, I.P., Johnston, T.W., Bachynski, M.P.: See [*9*], figs. 8-2 through 8-8
[1] Wyller, A.A. (1963): Astrophys. Norv. *8*, 53–77, 79–98
[2] Wyller, A.A. (1973): Astrophys. J. *184*, 517–538
[3] Mitchner, M., Kruger, C.H.: See [*24*], eq. VIII 4.14
[4] Schunk, R.W., Walker, J.C.G. (1970): Planet. Space Sci. *18*, 1535–1550, eqs. 2, 3, 5, 6d, 6f, fig. 7

required. An analogous situation applies for the radial current density in a cylindrical gas discharge. If the ambient magnetic field is parallel or perpendicular to the vanishing component, then the constraint can be dealt with analogously to the above procedure[5,6]. But for an arbitrary angle between the ambient magnetic field and the direction of the vanishing current component, a corresponding procedure has still to be worked out.

β) The *viscosity* of an anisotropic medium is a tensor of fourth rank (as mentioned in Sect. 1). Its evaluation is much more complicated than that of the transport tensors of second rank, which was described in Chapts. C and D. But the methods of the kinetic theory can, of course, also be applied for deriving the viscosity tensor. Some results are given by BURGERS[7] and by SCHUNK[8]. A description of the tensorial behavior in contrast to that of second rank tensors is given by STURHANN and SUCHY[9].

If the plasma is collision dominated or in a high-frequency limit, it is almost isotropic (see end of Subsect. 2α). The viscosity becomes a scalar in these cases. Some applications for aeronomic isotropic plasmas have been discussed by SCHUNK[10] and CONRAD and SCHUNK[11] and SCHUNK and ST.-MAURICE[12].

γ) For *strong deviations from local partial thermodynamic equilibrium* the methods of this contribution are not applicable. For these cases the anisotropy of the velocity distribution functions cannot be represented by small corrections to a local Maxwellian. For plasmas in strong magnetic fields the (unperturbed isotropic) Maxwellian is often replaced by an uniaxial Gaussian with a temperature parallel to the magnetic field much different from the temperature perpendicular to the magnetic field ("temperature anisotropy"). Transport equations for these plasmas were derived by CHODURA and POHL[13], DEMARS and SCHUNK[14], and BARAKAT and SCHUNK[15], SCHUNK and WATKINS[16].

δ) *Nonlinear phenomena* were not dealt with in this review. They play an important rôle for the energy balances of the particle species[17] and for diffusion processes under the influence of strong electric fields[18-20].

---

[5] Bauer, S.J.: See [*21*], sect. V.1

[6] Krall, N.A., Trivelpiece, A.W.: See [*22*], sect. 6.11

[7] Burgers, J.M.: See [*11*], sect. 39

[8] Schunk, R.W. (1975): Planet. Space Sci. *23*, 437–485

[9] Sturhann, U., Suchy, K. (1976): Z. Naturforsch. Teil A *31*, 1514–1516

[10] Schunk, R.W. (1978): Planet. Space Sci. *26*, 605–610

[11] Conrad, J.R., Schunk, R.W. (1979): J. Geophys. Res. *84*, 5355–5360

[12] Schunk, R.W., St.-Maurice, J.-P. (1981): J. Geophys. Res. *86*, 4823–4827

[13] Chodura, R., Pohl, F. (1971): Plasma Phys. *13*, 645–658

[14] Demars, H.G., Schunk, R.W. (1979): J. Phys. D *12*, 1051–1077

[15] Barakat, A.R., Schunk, R.W. (1981/2): J. Phys. D *14*, 421–428/D *15*, 1195–1216; (1982): Plasma Phys. *24*, 389–418

[16] Schunk, R.W., Watkins, D.S. (1981/2): J. Geophys. Res. *86*, 91–102/*87*, 171–180

[17] Weinert, U. (1975): Proc. Twelfth Intern. Conf. on Phenomena in Ionized Gases, Eindhoven, 266–266. Amsterdam: North-Holland

[18] Viehland, L.A., Mason, E.A. (1975): Ann. Phys. N.Y. *91*, 499–533

[19] Viehland, L.A., Mason, E.A. (1975): J. Chem. Phys. *63*, 2913–2915

[20] Stubbe, P. (1981): Radio Sci. *16*, 417–425

ε) The *determination of interaction potentials* has attracted considerable attention in recent years. Originally the rigid-sphere potential was widely used in kinetic gas theory as well as the Maxwell potential $\sim r^{-4}$; later improvements lead to the Sutherland (∞, n) potential, Eq. (7.16), and the Lennard-Jones (12, 6) potential, Eq. (7.14c). The "softening" of the repulsive core from the rigid sphere type of the Sutherland (∞, n) potential to the r^{-12} type of the Lennard-Jones (12, 6) potential is regarded as not yet sufficient (ZUNGER and HULER[21]; EVANS[22]). As a result of combined theoretical and experimental efforts, the present trend is towards a generalization of the Buckingham $(\exp \beta, n)$ potential, Eq. (7.5)[23-25]. It describes the core by a repulsive exponential potential which seems to be the best fit of experimental results for high kinetic energies[26]. The tail is chiefly described by a linear combination of inverse power potentials like those for the Mason-Schamp potential, Eq. (7.10). Cross sections calculated for such types of (generalized Buckingham) potentials are highly recommended so as to allow the most recent results obtained for interaction potentials to be applied at the computation of transport coefficients. A discussion of a variety of analytical interaction potentials was given by PAULY[27].

ζ) The *measurements of differential cross sections* became possible by beam and swarm experiments in recent years. Therefore the theoretical efforts in the field of cross section investigations concentrated upon the calculation of these differential cross sections as needed for the interpretation of these experiments. This is particularly true for electron-neutral collisions, but some interest begins to show for other collisions, too. Up to now such measurements and calculations were mostly done for moderate and high energies, i.e., above 10 eV. However, for transport processes in aeronomic and most laboratory plasmas, data below 10 eV are urgently needed.

η) *Higher-order approximations* for the calculation of transport coefficients are scarcely available. Only those of zeroth and first order are presented in this contribution. However, for the calculation of the coefficients of the thermo diffusion and diffusion thermo effects (Sect. 26) higher orders are desirable. The formulas necessary for the calculating of higher-order moments for multitemperature plasmas were worked by WEINERT and SUCHY[28] and by WEINERT[29,30]. Their application to the higher-order approximations of transport coefficients is an important task for the near future.

Acknowledgements. The first draft of this contribution was written during the author's first short stay at the Physics Department of the Israel Institute of Technology (Technion) in Haifa.

---

[21] Zunger, A., Huler, E. (1975): J. Chem. Phys. *62*, 3010–3020

[22] Evans, D.J. (1977): Mol. Phys. *33*, 979–986

[23] Buck, U. (1975): Adv. Chem. Phys. *30*, 313–388, chapt. IV

[24] Ellis, R.L. (1976): J. Chem. Phys. *64*, 342–348

[25] van Hemert, M.C., Berns, R.M. (1982): J. Chem. Phys. *76*, 354–361, fig. 2

[26] Foreman, P.B., Lees, A.B., Rol, P.K. (1976): Chem. Phys. *12*, 213–224, tables 1 and 3

[27] Pauly, H. (1979) in: Atom-molecule collision theory. Burke, P.G., Kleinpoppen, H. (eds.), pp. 111–199. New York: Plenum

[28] Weinert, U., Suchy, K. (1977): Z. Naturforsch. Teil A *32*, 390–400

[29] Weinert, U. (1978): Z. Naturforsch. Teil A *33*, 480–492

[30] Weinert, U. (1979): J. Math. Phys. N.Y. *20*, 2339–2346

This was made possible by financial support from that department and from the Deutsche Forschungsgemeinschaft. Discussions with Prof. Altman and Dr. Fijalkow are greatly appreciated. For the plotting of most of the diagrams the author is indebted to his collaborators at the Institute for Theoretical Physics of the University of Düsseldorf, especially to Dr. Weinert. The final version was completed during the author's stay at the Cavendish Laboratory of the University of Cambridge, while holding an Overseas Visiting Fellowship of St. John's College, Cambridge.

Appendix A. Axial tensors and axial matrix tensors

If the anisotropy of a medium is caused merely by an axial vector, say $\hat{\boldsymbol{t}}$, then a tensor of second rank representing this particular kind of anisotropy has the general form

$$\boldsymbol{\tau}(\hat{\boldsymbol{t}}) := \tau_0 \hat{\boldsymbol{t}}\hat{\boldsymbol{t}} + \tau_+ (\mathbf{U} - \hat{\boldsymbol{t}}\hat{\boldsymbol{t}}) + \boldsymbol{\tau}_-\, i\hat{\boldsymbol{t}} \times \mathbf{U} = \boldsymbol{\tau}^T(-\hat{\boldsymbol{t}}). \qquad \text{(A.1)}$$

Gibbs[1] gave it the name "cyclotonic". In analogy to an axial vector we denote it as an *axial tensor*. The axial vector is named the *gyration vector*, the corresponding anisotropic medium a special case of *gyrotropic medium*[2,3].

α) Eigenvalues, eigenvectors, projectors. Let the gyration vector $\hat{\boldsymbol{t}}$ together with unit vectors $\hat{\boldsymbol{s}} \perp \hat{\boldsymbol{t}}$ and $\hat{\boldsymbol{r}} := \hat{\boldsymbol{s}} \times \hat{\boldsymbol{t}}$ form a right-handed triple of unit vectors. Then the (right) *eigenvectors* of the axial tensor $\boldsymbol{\tau}(\hat{\boldsymbol{t}})$, Eq. (A.1), are

$$\boldsymbol{t}_{-1} = \frac{\hat{\boldsymbol{r}} - i\hat{\boldsymbol{s}}}{\sqrt{2}}, \qquad \boldsymbol{t}_0 = \hat{\boldsymbol{t}}, \qquad \boldsymbol{t}_{+1} := \frac{\hat{\boldsymbol{r}} + i\hat{\boldsymbol{s}}}{\sqrt{2}} \qquad \text{(A.2a)}$$

with

$$\boldsymbol{t}_M \cdot \boldsymbol{t}_{M'} = \delta_{M,-M'} \boldsymbol{t}_M \qquad (M = 0, \pm 1), \qquad \text{(A.2b)}$$

and the corresponding *eigenvalues* are

$$\tau_{-1} = \tau_+ - \tau_-, \qquad \tau_0, \qquad \tau_{+1} = \tau_+ + \tau_-. \qquad \text{(A.3)}$$

The corresponding orthogonal *projectors*

$$\mathbf{P}_0 = \boldsymbol{t}_0 \boldsymbol{t}_0 = \hat{\boldsymbol{t}}\hat{\boldsymbol{t}}, \qquad \text{(A.4a)}$$

$$\mathbf{P}_{\pm 1} = \boldsymbol{t}_{\pm 1} \boldsymbol{t}_{\mp 1} = \tfrac{1}{2}(\mathbf{U} - \hat{\boldsymbol{t}}\hat{\boldsymbol{t}} \pm i\hat{\boldsymbol{t}} \times \mathbf{U}) \qquad \text{(A.4b)}$$

with

$$\mathbf{P}_M \cdot \mathbf{P}_{M'} = \delta_{MM'} \mathbf{P}_M \qquad \text{(A.4c)}$$

[1] Gibbs, J.W. (1906): The scientific papers of J. Willard Gibbs, vol. II, p. 70. New York: Dover

[2] Sommerfeld, A. (1959): Optik, 2nd edn, § 29. Leipzig: Akademische Verlagsgesellschaft

[3] Landau, L.D., Lifšic, E.M.: Elektrodinamika splosnyh sred. Moskva 1957. English transl.: Electrodynamics of dense media. London: Pergamon 1958. German transl.: Elektrodynamik der Kontinua. Berlin: Akademie-Verlag 1967, § 82

allow the *diagonal representation* of the axial tensor $\boldsymbol{\tau}(\hat{\boldsymbol{t}})$ as

$$\begin{aligned}\boldsymbol{\tau}(\hat{\boldsymbol{t}}) &= \tau_{-1}\mathbf{P}_{-1} + \tau_0\mathbf{P}_0 + \tau_{+1}\mathbf{P}_{+1} && \text{(A.5a)}\\ &= \tau_{-1}\boldsymbol{t}_{-1}\boldsymbol{t}_{+1} + \tau_0\boldsymbol{t}_0\boldsymbol{t}_0 + \tau_{+1}\boldsymbol{t}_{+1}\boldsymbol{t}_{-1}. && \text{(A.5b)}\end{aligned}$$

With the linear combinations

$$\tau_{\pm} = \tfrac{1}{2}(\tau_{+1} \pm \tau_{-1}) \qquad \text{(A.6)}$$

of the eigenvalues and the corresponding ones

$$\begin{aligned}\mathbf{U} - \hat{\boldsymbol{t}}\hat{\boldsymbol{t}} &= 2\,\mathrm{Re}\ \mathbf{P}_{\pm 1} = \mathbf{P}_{+1} + \mathbf{P}_{-1},\\ i\hat{\boldsymbol{t}} \times \mathbf{U} &= \pm 2i\,\mathrm{Im}\ \mathbf{P}_{\pm 1} = \mathbf{P}_{+1} - \mathbf{P}_{-1} && \text{(A.7)}\end{aligned}$$

of the projectors, the original representation of the axial tensor $\boldsymbol{\tau}(\hat{\boldsymbol{t}})$, Eq. (A.1), is regained.

β) Group property. All axial tensors $\boldsymbol{\tau}(\hat{\boldsymbol{t}})$ with the same gyration vector $\hat{\boldsymbol{t}}$ have the same projectors $\mathbf{P}_M$, Eq. (A.4). Therefore a sum of two such axial tensors $\boldsymbol{\tau}'(\hat{\boldsymbol{t}})$ and $\boldsymbol{\tau}''(\hat{\boldsymbol{t}})$ resulting in

$$\boldsymbol{\tau}'''(\hat{\boldsymbol{t}}) := \boldsymbol{\tau}'(\hat{\boldsymbol{t}}) + \boldsymbol{\tau}''(\hat{\boldsymbol{t}}) \quad \text{is equivalent to } \tau'''_M = \tau'_M + \tau''_M, \qquad \text{(A.8a)}$$

the sum of the corresponding eigenvalues ($M=0, \pm 1$). Because of the orthogonality of the projectors $\mathbf{P}_M(\hat{\boldsymbol{t}})$, Eq. (A.4c), the same holds for the scalar multiplication:

$$\boldsymbol{\tau}'''(\hat{\boldsymbol{t}}) := \boldsymbol{\tau}'(\hat{\boldsymbol{t}}) \cdot \boldsymbol{\tau}''(\hat{\boldsymbol{t}}) \quad \text{is equivalent to } \tau'''_M = \tau'_M \tau''_M. \qquad \text{(A.8b)}$$

Because of Eqs. (A.8a, b) any relation among axial tensors with the same gyration vector $\hat{\boldsymbol{t}}$ can be replaced by the same relation between corresponding eigenvalues. Therefore these axial tensors form a group.

γ) Cartesian base and eigenbase. If the right-handed triple $\hat{\boldsymbol{r}}, \hat{\boldsymbol{s}}, \hat{\boldsymbol{t}}$ of unit vectors is used as a *cartesian base*, then the representation of $\boldsymbol{\tau}(\hat{\boldsymbol{t}})$ in this base can be written with Eq. (A.1)

$$\boldsymbol{\tau}(\hat{\boldsymbol{t}}) = [\hat{\boldsymbol{r}} \quad \hat{\boldsymbol{s}} \quad \hat{\boldsymbol{t}}]\begin{bmatrix}\tau_+ & -i\tau_- & 0\\ +i\tau_- & \tau_+ & 0\\ 0 & 0 & \tau_0\end{bmatrix}\begin{bmatrix}\hat{\boldsymbol{r}}\\ \hat{\boldsymbol{s}}\\ \hat{\boldsymbol{t}}\end{bmatrix}. \qquad \text{(A.9)}$$

The eigenvalue τ_0, Eq. (A.3), and the linear combinations

$$\tau_{\pm} = \tfrac{1}{2}(\tau_{+1} \pm \tau_{-1}) \qquad \text{[A.6]}$$

of the other two eigenvalues $\tau_{\pm 1}$ are the components of $\boldsymbol{\tau}(\hat{\boldsymbol{t}})$ in this cartesian base.

The triple $\boldsymbol{t}_{-1}$, $\boldsymbol{t}_0$, $\boldsymbol{t}_{+1}$ of right eigenvectors, Eq. (A.2), together with its reciprocal triple $\boldsymbol{t}^{-1}=\boldsymbol{t}_{+1}$, $\boldsymbol{t}^0=\boldsymbol{t}_0$, $\boldsymbol{t}^{+1}=\boldsymbol{t}_{-1}$ of left eigenvectors is the *eigenbase* of the axial tensor $\boldsymbol{\tau}(\hat{\boldsymbol{t}})$. The representation of $\boldsymbol{\tau}(\hat{\boldsymbol{t}})$ in this base can be taken from the diagonal representation Eq. (A.5):

$$\boldsymbol{\tau}(\hat{\boldsymbol{t}})=[\boldsymbol{t}_{-1} \quad \boldsymbol{t}_0 \quad \boldsymbol{t}_{+1}]\begin{bmatrix}\tau_{-1} & 0 & 0\\ 0 & \tau_0 & 0\\ 0 & 0 & \tau_{+1}\end{bmatrix}\begin{bmatrix}\boldsymbol{t}_{+1}\\ \boldsymbol{t}_0\\ \boldsymbol{t}_{-1}\end{bmatrix}. \tag{A.10}$$

The sequence $\boldsymbol{t}_{+1}=\boldsymbol{t}^{-1}$, $\boldsymbol{t}_0=\boldsymbol{t}^0$, $\boldsymbol{t}_{-1}=\boldsymbol{t}^{+1}$ of the reciprocal base vectors in the column matrix corresponds to the sequence $\boldsymbol{t}_{-1}$, $\boldsymbol{t}_0$, $\boldsymbol{t}_{+1}$ of the original base in the row matrix[4].

δ) The *reciprocal*. To obtain the reciprocal $\boldsymbol{\tau}^{-1}$ of $\boldsymbol{\tau}(\hat{\boldsymbol{t}})$, two methods can be used. The first method employs the general law that the eigenvalues of the reciprocal tensor $\boldsymbol{\tau}^{-1}$ are the reciprocals of the eigenvalues τ_0, $\tau_{\pm 1}$, Eq. (A.3), of $\boldsymbol{\tau}$. This general law holds even if the eigenvalues are matrices[5]. Starting with the diagonal representation of $\boldsymbol{\tau}$, Eq. (A.5), one obtains the following representation of its reciprocal:

$$\boldsymbol{\tau}^{-1}(\hat{\boldsymbol{t}})=\tau_{-1}^{-1}\mathbf{P}_{-1}+\tau_0^{-1}\mathbf{P}_0+\tau_{+1}^{-1}\mathbf{P}_{+1}, \tag{A.11a}$$

$$=[\boldsymbol{t}_{-1} \quad \boldsymbol{t}_0 \quad \boldsymbol{t}_{+1}]\begin{bmatrix}\tau_{-1}^{-1} & 0 & 0\\ 0 & \tau_0^{-1} & 0\\ 0 & 0 & \tau_{+1}^{-1}\end{bmatrix}\begin{bmatrix}\boldsymbol{t}_{+1}\\ \boldsymbol{t}_0\\ \boldsymbol{t}_{-1}\end{bmatrix}, \tag{A.11b}$$

$$=\tau_0^{-1}\hat{\boldsymbol{t}}\hat{\boldsymbol{t}}+\frac{\tau_{+1}^{-1}+\tau_{-1}^{-1}}{2}(\mathbf{U}-\hat{\boldsymbol{t}}\hat{\boldsymbol{t}})+\frac{\tau_{+1}^{-1}-\tau_{-1}^{-1}}{2}i\hat{\boldsymbol{t}}\times\mathbf{U}, \tag{A.11c}$$

$$=[\hat{\boldsymbol{r}} \quad \hat{\boldsymbol{s}} \quad \hat{\boldsymbol{t}}]\begin{bmatrix}\dfrac{\tau_{+1}^{-1}+\tau_{-1}^{-1}}{2} & -i\dfrac{\tau_{+1}^{-1}-\tau_{-1}^{-1}}{2} & 0\\ +i\dfrac{\tau_{+1}^{-1}-\tau_{-1}^{-1}}{2} & \dfrac{\tau_{+1}^{-1}+\tau_{-1}^{-1}}{2} & 0\\ 0 & 0 & \tau_0^{-1}\end{bmatrix}\begin{bmatrix}\hat{\boldsymbol{r}}\\ \hat{\boldsymbol{s}}\\ \hat{\boldsymbol{t}}\end{bmatrix}. \tag{A.11d}$$

Another method for the calculation of $\boldsymbol{\tau}^{-1}$ uses $\boldsymbol{\tau}$ as a tensor operator in a linear system

$$\boldsymbol{y}=\boldsymbol{\tau}\cdot\boldsymbol{x}=\tau_0\hat{\boldsymbol{t}}\hat{\boldsymbol{t}}\cdot\boldsymbol{x}+\tau_+(\mathbf{U}-\hat{\boldsymbol{t}}\hat{\boldsymbol{t}})\cdot\boldsymbol{x}+\tau_-i\hat{\boldsymbol{t}}\times\boldsymbol{x} \tag{A.12}$$

and solves the system for $\boldsymbol{x}$. To do this, three relations are calculated by scalar multiplication of Eq. (A.12) from the left with $\hat{\boldsymbol{t}}\hat{\boldsymbol{t}}$, $(\mathbf{U}-\hat{\boldsymbol{t}}\hat{\boldsymbol{t}})$, and $i\hat{\boldsymbol{t}}\times\mathbf{U}$, respectively. Then the first of these three relations is multiplied with τ_0^{-1}, the second one is multiplied from the left with $(\tau_++\tau_-)^{-1}\,\tau_+$ and from the right with $(\tau_+-\tau_-)^{-1}$, the third one is multiplied from the left with $(\tau_++\tau_-)^{-1}$ and from the right with $-\tau_-(\tau_+-\tau_-)^{-1}$. The sum of these three equations yields

$$\begin{aligned}\boldsymbol{\tau}^{-1}(\hat{\boldsymbol{t}})=\tau_0^{-1}\hat{\boldsymbol{t}}\hat{\boldsymbol{t}}&+(\tau_++\tau_-)^{-1}\tau_+(\tau_+-\tau_-)^{-1}(\mathbf{U}-\hat{\boldsymbol{t}}\hat{\boldsymbol{t}})\\ &-(\tau_++\tau_-)^{-1}\tau_-(\tau_+-\tau_-)^{-1}i\hat{\boldsymbol{t}}\times\mathbf{U}.\end{aligned} \tag{A.13}$$

---

[4] Rawer, K., Suchy, K.: this Encyclopedia, vol. 49/2, pp. 1–546, eq. 6.10

[5] Suchy, K. (1973): Proc. Eleventh Intern. Conf. on Phenomena in Ionized Gases, Prague, p. 272–272. Prague: Czechoslovak Academy of Sciences, Institute of Physics

This is equivalent with Eq. (A.11c) even if the eigenvalues of $\boldsymbol{\tau}$ do not commute. This happens for matrix eigenvalues (see Subsect. ζ below). In such cases the sequence of multiplication is important in Eq. (A.13). This was the reason for the multiplication sequence during the derivation of Eq. (A.13). For commuting eigenvalues of $\boldsymbol{\tau}$ one can write

$$(\tau_+ + \tau_-)^{-1}\tau_\pm(\tau_+ - \tau_-)^{-1} = (\tau_+^2 - \tau_-^2)^{-1}\tau_\pm \quad \text{for } [\tau_+, \tau_-] = 0. \tag{A.14}$$

ε) Special cases. For an axial tensor of the form

$$\boldsymbol{\tau}(\hat{\boldsymbol{t}}) = \tau_0 \mathbf{U} + \tau_- i\hat{\boldsymbol{t}} \times \mathbf{U} \tag{A.15}$$

the eigenvalue τ_0 must be equal to the linear combination $\tau_+ = \frac{1}{2}(\tau_{+1} + \tau_{-1})$ [cf. Eqs. (A.1), (A.6)]. Because of $\tau_- = \frac{1}{2}(\tau_{+1} - \tau_{-1})$, Eq. (A.6), one has

$$\tau_{\pm 1} = \tau_0 \pm \tau_- \quad \text{for } \tau_+ = \tau_0. \tag{A.16}$$

If $\tau_- \ll \tau_0 = \tau_+$, then the representation Eq. (A.13) for the reciprocal yields

$$\boldsymbol{\tau}^{-1} = \tau_0^{-1}\mathbf{U} - \tau_0^{-1}\tau_-\tau_0^{-1} i\hat{\boldsymbol{t}} \times \mathbf{U} - \tau_0^{-1}\tau_-\tau_0^{-1}\tau_-\tau_0^{-1}(\mathbf{U} - \hat{\boldsymbol{t}}\hat{\boldsymbol{t}}) + O(\tau_-^3\tau_0^{-4}). \tag{A.17}$$

For the other limit one obtains

$$\boldsymbol{\tau}^{-1} = \tau_0^{-1}\hat{\boldsymbol{t}}\hat{\boldsymbol{t}} + \tau_-^{-1} i\hat{\boldsymbol{t}} \times \mathbf{U} - \tau_-^{-1}\tau_0\tau_-^{-1}(\mathbf{U} - \hat{\boldsymbol{t}}\hat{\boldsymbol{t}}) + O(\tau_0^2\tau_-^{-3}). \tag{A.18}$$

In the first limiting case the anisotropy due to $\hat{\boldsymbol{t}}$ has almost no influence on the reciprocal $\boldsymbol{\tau}^{-1}(\hat{\boldsymbol{t}})$, in the second limit, however, it has a very strong influence.

ζ) Axial matrix tensors

$$\boldsymbol{\tau}_{jk}(\hat{\boldsymbol{t}}) := \tau_{jk}^0 \hat{\boldsymbol{t}}\hat{\boldsymbol{t}} + \tau_{jk}^+(\mathbf{U} - \hat{\boldsymbol{t}}\hat{\boldsymbol{t}}) + \tau_{jk}^- i\hat{\boldsymbol{t}} \times \mathbf{U} \tag{A.19}$$

are axial tensors with square matrices

$$\tau_{jk}^{-1} = \tau_{jk}^+ - \tau_{jk}^-, \quad \tau_{jk}^0, \quad \tau_{jk}^{+1} = \tau_{jk}^+ + \tau_{jk}^- \tag{A.20}$$

as eigenvalues. The eigenvectors $\boldsymbol{t}_0$, $\boldsymbol{t}_{\pm 1}$, Eq. (A.2), the projectors $\mathbf{P}_0$, $\mathbf{P}_{\pm 1}$, Eq. (A.4), and their linear combinations $\mathbf{U} - \hat{\boldsymbol{t}}\hat{\boldsymbol{t}}$, $i\hat{\boldsymbol{t}} \times \mathbf{U}$, Eq. (A.7), are the same as for axial tensors. Therefore the diagonal representation Eq. (A.5) reads

$$\boldsymbol{\tau}_{jk}(\hat{\boldsymbol{t}}) = \tau_{jk}^{-1}\mathbf{P}_{-1} + \tau_{jk}^0\mathbf{P}_0 + \tau_{jk}^{+1}\mathbf{P}_{+1}. \tag{A.21}$$

For the calculation of the reciprocal the representations corresponding to Eqs. (A.11a), (A.11c), and (A.13) must be employed, where no commutation between the matrix eigenvalues has been used. The simplifying relation Eq. (A.14) can in general not be used, since the matrix eigenvalues do not in general commute.

All axial matrix tensors with the same gyration vector $\hat{\boldsymbol{t}}$ form a group. Relations between them can be replaced by the same relations between corresponding matrix eigenvalues.

Appendix B. Some definite integrals and special functions

α) Integrals needed for the calculation of transport collision frequencies. We calculate transport collision frequencies

$$\nu^{(ls)}(\hat{T}) := \frac{(-1)^s s!}{(l+\frac{1}{2}+s)!} \int_0^\infty dw\, e^{-w} L_s^{(l+\frac{1}{2})}(w)\, w^{l+\frac{1}{2}} \nu^{(l)}(k\hat{T}w) \qquad [19.3]$$

from transfer collision frequencies

$$\nu_{jk}^{(l)}(\mathscr{E}) = N_k \sqrt{2\mathscr{E}/\mu}\, Q^{(l)}(\mathscr{E}). \qquad [4.14]$$

The latter are proportional to transfer cross sections $Q^{(l)}(\mathscr{E})$, Eq. (4.8), which are calculated or approximated analytically. The following integrals are then needed:

$$\int_0^\infty dw\, e^{-w} w^\alpha = \alpha! \equiv \Gamma(\alpha+1) \quad \text{for } \operatorname{Re}\alpha > -1. \qquad (B.1a)$$

This result is a special case for $L_0^{(0)} = 1$ of the more general integral[1]

$$\int_0^\infty dw\, e^{-w} w^{\alpha+\lambda} L_s^{(\lambda)}(w) = \frac{(\alpha+\lambda)!}{s!} \frac{(s-\alpha-1)!}{(-\alpha-1)!} = \frac{(\alpha+\lambda)!}{s!}(-1)^s \frac{\alpha!}{(\alpha-s)!}$$

$$\text{for } \lambda > -1 \text{ and } \alpha > -1-\lambda \qquad (B.1b)$$

with $L_s^{(\lambda)}(w)$ as generalized Laguerre (or Sonine) polynomials. If the power α is an integer or a half-integer, the result can also be written as

$$\int_0^\infty dw\, e^{-w} w^{\alpha+\lambda} L_s^{(\lambda)}(w) = \frac{(\alpha+\lambda)!}{s!}(-1)^s \frac{(2\alpha)!!}{2^s(2\alpha-2s)!!}. \qquad (B.1c)$$

The results of the integrals[2,3]

$$\int_0^\infty dw\, e^{-w} w^\alpha \ln w = \alpha!\, \psi(\alpha+1) \quad \text{for } \alpha > -1 \qquad (B.2)$$

and

$$\int_0^\infty dw\, e^{-w} w^\alpha (\ln w)^2 = \alpha!\, [\psi^2(\alpha+1) + \zeta(2, \alpha+1)] \qquad (B.3)$$

are expressed by the digamma function $\psi(z+1)$ and the generalized zeta function $\zeta(z, \alpha)$. The parameter α in the last two integrals Eqs. (B.2) (B.3) is an integer in the applications.

[1] Erdélyi, A. (ed.) (1954): Tables of integral transforms, vol. 1, eq. 4.11.29. New York: McGraw-Hill

[2] Erdélyi, A. (ed.) (1954): Tables of integral transforms, vol. 1, eq. 4.6.7. New York: McGraw-Hill

[3] Gradshteyn, I.S., Ryzhik, I.M. (1980): Table of integrals, series, and products, eq. 4.358.2. New York: Academic Press

β) The *generalized zeta function* is defined by[4]

$$\zeta(z,\alpha) := \sum_{k=0}^{\infty} \frac{1}{(k+\alpha)^z} \quad \text{for } \operatorname{Re} z>0 \text{ and } \alpha \neq 0, -1, -2 \ldots \tag{B.4a}$$

The special value $\alpha=0$, viz.,

$$\zeta(z,1) = \sum_{k=1}^{\infty} \frac{1}{k^z} =: \zeta(z), \tag{B.4b}$$

yields Riemann's zeta function $\zeta(z)$. For the integral Eq. (B.3) the zeta function $\zeta(2,n)=\sum_{k=n}^{\infty} k^{-2}$ can be decomposed as the difference between

$$\sum_{k=1}^{\infty} \frac{1}{k^2} = \zeta(2) = \frac{\pi^2}{6} \tag{B.4c}$$

and a finite sum over k^{-2}, viz.,

$$\zeta(2,n) = \frac{\pi^2}{6} - \sum_{k=1}^{n-1} \frac{1}{k^2}. \tag{B.4d}$$

The following particular cases are used:

$$\begin{aligned} &\zeta(2,1)=\frac{\pi^2}{6}, &&\zeta(2,2)=\frac{\pi^2}{6}-1, &&\zeta(2,3)=\frac{\pi^2}{6}-\frac{5}{4}, &&\zeta(2,4)=\frac{\pi^2}{6}-\frac{49}{36} \\ &\zeta(2,5)=\frac{\pi^2}{6}-\frac{205}{144}, &&\zeta(2,6)=\frac{\pi^2}{6}-\frac{5269}{3600}, &&\zeta(2,7)=\frac{\pi^2}{6}-\frac{5369}{3600}, &&\zeta(2,8)=\frac{\pi^2}{6}-\frac{266781}{176400}. \end{aligned} \tag{B.4e}$$

γ) The *digamma function* is defined by[5]

$$\psi(z+1) := \frac{\mathrm{d}}{\mathrm{d}z} \ln(z!) \equiv \frac{\mathrm{d}}{\mathrm{d}z} \ln \Gamma(z+1). \tag{B.5a}$$

For integer values of z one obtains[6]

$$\psi(n+1) = -\gamma + \sum_{k=1}^{n} \frac{1}{k} \quad \text{with } \gamma = 0.5772 \tag{B.5b}$$

with the special values

$$\begin{aligned} &\psi(1)=-\gamma, &&\psi(2)=-\gamma+1, &&\psi(3)=-\gamma+\frac{3}{2}, &&\psi(4)=-\gamma+\frac{11}{6}, \\ &\psi(5)=-\gamma+\frac{25}{12}, &&\psi(6)=-\gamma+\frac{137}{60}, &&\psi(7)=-\gamma+\frac{49}{20}, &&\psi(8)=-\gamma+\frac{363}{140}. \end{aligned} \tag{B.5c}$$

---

[4] Erdélyi. A. (ed.) (1953): Higher transcendental functions. vol. 1. eq. 1.10.1. New York: McGraw-Hill

[5] Abramowitz, M., Stegun, I.A. (eds.) (1964): Handbook of mathematical functions, eq. 6.3.1. New York: Dover

[6] Abramowitz, M., Stegun, I.A. (eds.) (1964): Handbook of mathematical functions, eq. 6.3.2. New York: Dover

The real part of the digamma function for imaginary argument is also needed with the following properties[7]:

$$\mathrm{Re}\,\psi(iy) = \mathrm{Re}\,\psi(1+iy) = \ln y + \frac{1}{12y^2} + \frac{1}{120\,y^4} + O\left(\frac{1}{y^6}\right), \tag{B.6a}$$

$$= -\gamma + y^2 \sum_{k=1}^{\infty} \frac{1}{k(k^2+y^2)}. \tag{B.6b}$$

The last relation can be expanded for small y and yields with Eq. (B.4b)

$$\mathrm{Re}\,\psi(iy) = -\gamma + y^2\zeta(3) + O(y^4) \qquad \text{with } \zeta(3) = 1.2021. \tag{B.6c}$$

δ) For the *calculation of transport collision frequencies* $\nu^{(ls)}(\hat{T})$ from transfer collision frequencies $\nu^{(l)}(\mathscr{E})$ of the form Eq. (18.4a), three types of integrals are needed. First, we calculate

$$\begin{pmatrix} \lambda & \kappa \\ & s \end{pmatrix} := \frac{(-1)^s s!}{(\lambda+s)!} \int_0^\infty \mathrm{d}w\, e^{-w}\, \mathrm{L}_s^{(\lambda)}(w)\, w^{\lambda+\kappa}. \tag{B.7a}$$

We insert the representation

$$\mathrm{L}_s^{(\lambda)}(w) = \sum_{r=0}^{s} \begin{pmatrix} \lambda+s \\ s-r \end{pmatrix} \frac{(-1)^r}{r!} w^r \tag*{[19.2a]}$$

for the generalized Laguerre polynomials $\mathrm{L}_s^{(\lambda)}(w)$ and obtain, using Eq. (B.1a) with $\alpha = \kappa + r$ and Eq. (B.1b) with $\alpha = \kappa$,

$$\begin{pmatrix} \lambda & \kappa \\ & s \end{pmatrix} = \sum_{r=0}^{s} (-1)^{s+r} \begin{pmatrix} s \\ r \end{pmatrix} \frac{(\lambda+\kappa+r)!}{(\lambda+r)!} = \frac{(\lambda+\kappa)!}{(\lambda+s)!} \frac{\kappa!}{(\kappa-s)!}. \tag{B.7b}$$

For the application in Eq. (21.4a) we need

$$\begin{pmatrix} l+\frac{1}{2} & p+\frac{1}{2} \\ & s \end{pmatrix} = \frac{(l+1+p)!}{(l+\frac{1}{2}+s)!} \frac{(p+\frac{1}{2})!}{(p+\frac{1}{2}-s)!} \tag{B.7c}$$

with integer l and s, while p take integer and half-integer values. Then Eq. (B.7c) can be written as[8]

$$\begin{pmatrix} l+\frac{1}{2} & p+\frac{1}{2} \\ & s \end{pmatrix} = \frac{(l+1+p)!}{(l+\frac{1}{2})!} \frac{(2l+1)!!}{(2l+1+2s)!!} \frac{(2p+1)!!}{(2p+1-2s)!!}. \tag{B.7d}$$

This expression vanishes for half-integer p, i.e., $2p+1$ even, if $2s > 2p+1$, i.e., $s > p + \frac{1}{2}$, in particular

$$\begin{pmatrix} l+\frac{1}{2} & 0 \\ & s \end{pmatrix} = \delta_{s0} \equiv \begin{cases} 1 & \text{for } s=0 \\ 0 & \phantom{\text{for }} s \neq 0. \end{cases} \tag{B.7e}$$

---

[7] Abramowitz, M., Stegun, I.A. (eds.) (1964): Handbook of mathematical functions, eqs. 6.3.10, 6.3.19, 6.3.17. New York: Dover

[8] Suchy, K., Rawer, K. (1971): J. Atmos. Terr. Phys. *33*, 1853–1868, eqs. 4.2 and 4.3

Table 12 presents values of the coefficients Eq. (B.7d) for $l=1,2$; $s=0, 1, 2$; p integer and half-integer.

Table 12. Values of the coefficients $\binom{l+\frac{1}{2}\ p+\frac{1}{2}}{s}$, Eq. (B7d)

$2p+1$	$\binom{\frac{3}{2}\ p+\frac{1}{2}}{0}$	$\binom{\frac{3}{2}\ p+\frac{1}{2}}{1}$	$\binom{\frac{3}{2}\ p+\frac{1}{2}}{2}$	$\binom{\frac{5}{2}\ p+\frac{1}{2}}{0}$	$\binom{\frac{5}{2}\ p+\frac{1}{2}}{1}$
5	$\frac{32}{\sqrt{\pi}}$	$\frac{32}{\sqrt{\pi}}$	$\frac{160}{7\sqrt{\pi}}$	$\frac{64}{\sqrt{\pi}}$	$\frac{320}{7\sqrt{\pi}}$
4	$\frac{35}{4}$	7	2	$\frac{63}{4}$	9
3	$\frac{8}{\sqrt{\pi}}$	$\frac{24}{5\sqrt{\pi}}$	$\frac{24}{35\sqrt{\pi}}$	$\frac{64}{5\sqrt{\pi}}$	$\frac{192}{35\sqrt{\pi}}$
2	$\frac{5}{2}$	1	0	$\frac{7}{2}$	1
1	$\frac{8}{3\sqrt{\pi}}$	$\frac{8}{15\sqrt{\pi}}$	$-\frac{8}{105\sqrt{\pi}}$	$\frac{16}{5\sqrt{\pi}}$	$\frac{16}{35\sqrt{\pi}}$
0	1	0	0	1	0
-1	$\frac{4}{3\sqrt{\pi}}$	$-\frac{4}{15\sqrt{\pi}}$	$\frac{4}{35\sqrt{\pi}}$	$\frac{16}{15\sqrt{\pi}}$	$-\frac{16}{105\sqrt{\pi}}$
-2	$\frac{2}{3}$	$-\frac{4}{15}$	$\frac{16}{105}$	$\frac{2}{5}$	$-\frac{4}{35}$
-3	$\frac{4}{3\sqrt{\pi}}$	$-\frac{4}{5\sqrt{\pi}}$	$\frac{4}{7\sqrt{\pi}}$	$\frac{8}{15\sqrt{\pi}}$	$-\frac{8}{35\sqrt{\pi}}$

The second integral is

$$\begin{bmatrix}\lambda & \kappa\\ s\end{bmatrix} := \frac{(-1)^s s!}{(\lambda+s)!}\int_0^\infty \mathrm{d}w\, e^{-w} \mathrm{L}_s^{(\lambda)}(w)\, w^{\lambda+\kappa} \ln w. \tag{B.8a}$$

Using again the series expansion Eq. (19.2a) for the generalized Laguerre polynomials $\mathrm{L}_s^{(\lambda)}(w)$, we obtain with Eq. (B.2), putting $\alpha=\lambda+\kappa+r$,

$$\begin{bmatrix}\lambda & \kappa\\ s\end{bmatrix} = \sum_{r=0}^{s} (-1)^{s+r} \binom{s}{r} \frac{(\lambda+\kappa+r)!}{(\lambda+r)!} \psi(\lambda+\kappa+r+1). \tag{B.8b}$$

Now the digramma function $\psi(n+1)$ is decomposed in two parts according to Eq. (B.5b), the first part $-\gamma=-0.5772$ being independent of the argument of the digamma function. Therefore it is not influenced by the summation in Eq. (B.8b) which can be performed using Eq. (B.7b). The result is an expression linear in γ, viz.,

$$\begin{bmatrix}\lambda & \kappa\\ s\end{bmatrix} = -\gamma \binom{\lambda\ \ \kappa}{s} + \begin{bmatrix}\lambda & \kappa\\ s\end{bmatrix}_0 \tag{B.8c}$$

with

$$\begin{bmatrix} \lambda & \kappa \\ & s \end{bmatrix}_0 := \sum_{r=0}^{r} (-1)^{s+r} \binom{s}{r} \frac{(\lambda+\kappa+r)!}{(\lambda+r)!} \sum_{k=1}^{\lambda+\kappa+r} \frac{1}{k}. \tag{B.8d}$$

The applications in Eq. (21.4b) require $\lambda = l+\frac{1}{2}$, $\kappa = p+\frac{1}{2}$, and $s=0,1,2$ with $l=1,2$ and $p=1,0,-2$ (see Table 13).

Table 13. Values of the coefficients $\begin{bmatrix} l+\frac{1}{2} & p+\frac{1}{2} \\ & s \end{bmatrix}_0$, Eq. (B.8d)

$2p+1$	$\begin{bmatrix} \frac{3}{2} & p+\frac{1}{2} \\ & 0 \end{bmatrix}_0$	$\begin{bmatrix} \frac{3}{2} & p+\frac{1}{2} \\ & 1 \end{bmatrix}_0$	$\begin{bmatrix} \frac{3}{2} & p+\frac{1}{2} \\ & 2 \end{bmatrix}_0$	$\begin{bmatrix} \frac{5}{2} & p+\frac{1}{2} \\ & 0 \end{bmatrix}_0$	$\begin{bmatrix} \frac{5}{2} & p+\frac{1}{2} \\ & 1 \end{bmatrix}_0$
5	$\frac{200}{3\sqrt{\pi}}$	$\frac{1192}{15\sqrt{\pi}}$	$\frac{216}{5\sqrt{\pi}}$	$\frac{2192}{15\sqrt{\pi}}$	$\frac{368}{3\sqrt{\pi}}$
3	$\frac{44}{3\sqrt{\pi}}$	$\frac{12}{\sqrt{\pi}}$	$\frac{108}{35\sqrt{\pi}}$	$\frac{80}{3\sqrt{\pi}}$	$\frac{528}{35\sqrt{\pi}}$
1	$\frac{4}{\sqrt{\pi}}$	$\frac{28}{15\sqrt{\pi}}$	$-\frac{4}{35\sqrt{\pi}}$	$\frac{88}{15\sqrt{\pi}}$	$\frac{184}{105\sqrt{\pi}}$
-1	$\frac{4}{3\sqrt{\pi}}$	$\frac{4}{15\sqrt{\pi}}$	$-\frac{4}{21\sqrt{\pi}}$	$\frac{24}{15\sqrt{\pi}}$	$\frac{8}{105\sqrt{\pi}}$
-3	0	$\frac{8}{15\sqrt{\pi}}$	$-\frac{64}{105\sqrt{\pi}}$	$\frac{8}{15\sqrt{\pi}}$	$-\frac{8}{105\sqrt{\pi}}$

The last integral

$$\begin{Bmatrix} \lambda & \kappa \\ & s \end{Bmatrix} := \frac{(-1)^s s!}{(\lambda+s)!} \int_0^\infty \mathrm{d}w \, \mathrm{e}^{-w} \, \mathrm{L}_s^{(\lambda)}(w) \, w^{\lambda+\kappa} (\ln w)^2 \tag{B.9a}$$

is dealt with analogously to Eq. (B.8a) and becomes with Eq. (B.3)

$$\begin{Bmatrix} \lambda & \kappa \\ & s \end{Bmatrix} = \sum_{r=0}^{s} (-1)^{s+r} \binom{s}{r} \frac{(\lambda+\kappa+r)!}{(\lambda+r)!} [\psi^2(\lambda+\kappa+r+1) + \zeta(2, \lambda+\kappa+r+1)]. \tag{B.9b}$$

The digamma function $\psi(n+1)$ is again decomposed with Eq. (B.5b) and the generalized zeta function $\zeta(2,n)$ with Eq. (B.4d). The result is

$$\begin{Bmatrix} \lambda & \kappa \\ & s \end{Bmatrix} = \left(\gamma^2 + \frac{\pi^2}{6}\right) \begin{pmatrix} \lambda & \kappa \\ & s \end{pmatrix} - 2\gamma \begin{bmatrix} \lambda & \kappa \\ & s \end{bmatrix}_0 + \begin{Bmatrix} \lambda & \kappa \\ & s \end{Bmatrix}_0 \tag{B.9c}$$

with

$$\begin{Bmatrix} \lambda & \kappa \\ & s \end{Bmatrix}_0 := \sum_{r=0}^{s} (-1)^{s+r} \binom{s}{r} \frac{(\lambda+\kappa+r)!}{(\lambda+r)!} \left[\left(\sum_{k=1}^{\lambda+\kappa+r} \frac{1}{k} \right)^2 - \sum_{k=1}^{\lambda+\kappa+r} \frac{1}{k^2} \right]. \tag{B.9d}$$

For the application in Eq. (21.4c) we need $\lambda = l+\frac{1}{2}$ with $l=1,2$; $\kappa=\frac{1}{2}$, and $s=0,1,2$ (see Table 14).

Table 14. Values of the coefficients $\begin{Bmatrix} l+\frac{1}{2} & \frac{1}{2} \\ s & \end{Bmatrix}_0$, Eq. (B.9d)

$\begin{Bmatrix} \frac{3}{2} & \frac{1}{2} \\ 0 & \end{Bmatrix}_0$	$\begin{Bmatrix} \frac{3}{2} & \frac{1}{2} \\ 1 & \end{Bmatrix}_0$	$\begin{Bmatrix} \frac{3}{2} & \frac{1}{2} \\ 2 & \end{Bmatrix}_0$	$\begin{Bmatrix} \frac{5}{2} & \frac{1}{2} \\ 0 & \end{Bmatrix}_0$	$\begin{Bmatrix} \frac{5}{2} & \frac{1}{2} \\ 1 & \end{Bmatrix}_0$
$\frac{8}{3\sqrt{\pi}}$	$\frac{56}{15\sqrt{\pi}}$	$\frac{8}{15\sqrt{\pi}}$	$\frac{32}{5\sqrt{\pi}}$	$\frac{64}{15\sqrt{\pi}}$

Appendix C. Conversion of units[1]

u = 1 in rationalized but = 4π in nonrationalized systems of units.
Units of length:

Bohr radius

$$\frac{4\pi\,\varepsilon_0}{\mathrm{u}\,q_e^2}\,\frac{\hbar^2}{m_e} =: a_0 = 5.292\cdot 10^{-11}\,\mathrm{m}$$

$$a_0^2 = 2.800\cdot 10^{-21}\,\mathrm{m}^2$$

$$\pi\,a_0^2 = 8.797\cdot 10^{-21}\,\mathrm{m}^2$$

$$a_0^3 = 1.482\cdot 10^{-31}\,\mathrm{m}^3.$$

Units of energy, see Eq. (5.5b):

$$1\,\text{joule} = 6.241\cdot 10^{18}\,\mathrm{eV}$$

$$1\,\mathrm{eV} = 1.6022\cdot 10^{-19}\,\mathrm{J}$$

$$\frac{\hbar^2}{2m_e a_0^2} =: 1\,\text{rydberg} = 13.606\cdot 10^0\,\mathrm{eV} = 21.799\cdot 10^{-19}\,\mathrm{J} = 1.5790\cdot 10^5\,k\,\text{kelvin}$$

$$1\,k\,\text{kelvin} = 8.617\cdot 10^{-5}\,\mathrm{eV} = 1.3807\cdot 10^{-23}\,\mathrm{J} = 6.333\cdot 10^{-6}\,\text{rydberg}$$

$$1\,\frac{\text{kcal}}{N_A\,\text{mol}} = 4.339\cdot 10^{-2}\,\mathrm{eV} = 6.953\cdot 10^{-21}\,\mathrm{J}.$$

General references

Books and Review Articles

[*1*] Lorentz, H.A. (1915): The theory of electrons. Leipzig: Teubner
[*2*] Herzberg, G. (1950): Spectra of diatomic molecules. Princeton: van Nostrand
[*3*] Present, R.D. (1958): Kinetic theory of gases. New York: McGraw-Hill
[*4*] deGroot, S.R., Mazur, P. (1962): Non-equilibrium thermodynamics. Amsterdam: North-Holland
[*5*] Hirschfelder, J.O., Curtiss, C.F., Bird, R.B. (1964): Molecular theory of gases and liquids, 2nd ed. New York: Wiley & Sons
[*6*] McDaniel, E.W. (1964): Collision phenomena in ionized gases. New York: Wiley & Sons
[*7*] Montgomery, D.C., Tidman, D.A. (1964): Plasma kinetic theory. New York: McGraw-Hill
[*8*] Suchy, K. (1964): Ergeb. Exakten Naturwiss. *35*, 103–294
[*9*] Shkarofsky, I.P., Johnston, T.W., Bachynski, M.P. (1966): The particle kinetics of plasmas. Reading, Mass.: Addison-Wesley
[*10*] Bernstein, R.B., Muckermann, J.T. (1967): Adv. Chem. Phys. *12*, 389–486
[*11*] Burgers, J.M. (1969): Flow equations for composite gases. New York: Academic
[*12*] Chapman, S., Cowling, T.G. (1970): The mathematical theory of non-uniform gases, 3rd edn. Cambridge: Cambridge University Press
[*13*] Takayanagi, K., Itikawa, Y. (1970): Adv. At. Mol. Phys. *6*, 105–151
[*14*] Mason, E.A., Marrero, T.R. (1970): Adv. At. Mol. Phys. *6*, 155–232
[*15*] Bederson, B., Kieffer, L.J. (1971): Rev. Mod. Phys. *43*, 601–640
[*16*] Massey, H., Burhop, E.H., Gilbody H.B. (1971): Electronic impact phenomena. vol. III: Slow collisions of heavy particles. London: Oxford University Press
[*17*] Wilhelm, J., Wallis, G. (1971): Ergeb. Plasmaphys. Gaselektronik *2*, 155–252
[*18*] Ferziger, J.H., Kaper, H.G. (1972): Mathematical theory of transport processes. Amsterdam: North-Holland
[*19*] Gilardini, A.L. (1972): Low energy electron collisions in gases. New York: Wiley & Sons
[*20*] Banks, P.M., Kockarts, G. (1973): Aeronomy. New York: Academic Press
[*21*] Bauer, S.J. (1973): Physics of planetary ionospheres. Berlin, Heidelberg, New York: Springer
[*22*] Krall, N.A., Trivelpiece, A.W. (1973): Principles of plasma physics. New York: McGraw-Hill
[*23*] McDaniel, E.W., Mason, E.A. (1973): The mobility and diffusion of ions in gases. New York: Wiley & Sons
[*24*] Mitchner, M., Kruger, C.H. (1973): Partially ionized gases. New York: Wiley & Sons
[*25*] Rawer, K. (ed.) (1974): Methods of measurements and results of lower ionosphere structure. Berlin: Akademie-Verlag
[*26*] Kleinpoppen, H., McDowell, M.R.C. (eds.) (1976): Electron and photon interactions with atoms. New York: Plenum
[*27*] Thomas, L., Rishbeth, H. (eds.) (1977): Photochemical and transport processes in the upper atmosphere. London: Pergamon
[*28*] Joachain, C.J. (1978): Quantum collision theory. Amsterdam: North-Holland
[*29*] Bernstein, R.B. (ed.) (1979): Atom-molecule collision theory. New York: Plenum
[*30*] Brown, S.C. (ed.) (1979): Electron-molecule scattering. New York: Wiley & Sons
[*31*] Duderstadt, J.J., Martin, W.R. (1979): Transport theory. New York: Wiley & Sons
[*32*] Hasted, J.B. (1979): Adv. At. Mol. Phys. *15*, 205–232
[*33*] Rescigno, T., McKoy, V., Schneider, B. (eds.) (1979): Electron-molecule and photon-molecule collisions. New York: Plenum
[*34*] Nesbet, R.K. (1980): Variational methods in electron-atom scattering theory. New York: Plenum
[*35*] Maitland, G.C., Rigby, M., Smith, E.B., Wakeham, W.A. (1981): Intermolecular forces. Oxford: Clarendon
[*36*] Hinze, J. (ed.) (1983): Electron-atom and electron-molecule collisions. New York: Plenum
[*37*] Bransden, B.H. (1983): Atomic Collision Theory, 2nd edn. Reading, Mass.: Benjamin/Cummings
[*38*] Morrison, M.A. (1983): Aust. J. Phys. *36*, 239–286

[1] Villena, L. (ed.) (1980): Symbols, units and nomenclature in physics. Physica A *93*, 1–60

Modelling of Neutral and Ionized Atmospheres

By

K. RAWER

With 224 Figures

Dedicated to Yves Rocard, Paris, at the occasion of his 80th anniversary

1. Fundamental relations

α) The atmosphere around a planet is held together by the gravitational force. In the *stationary state* the weight of a column of air above the height level z is supported by the pressure force at its base; from this reasoning we get the *hydrostatic equation* describing the pressure profile:

$$p(z) = \int_z^\infty \mathrm{d}z\, g(z) \sum_j m_j(z) \cdot n_j(z) \tag{1.0}$$

where g is the acceleration of gravity and m_j and n_j are the mass and numerical density of constituent j. It follows that the pressure p monotonically increases with decreasing height.

Under non-stationary conditions, however, NEWTON's dynamic equation must be used. It may be written separately for each constituent:

$$m_j n_j \frac{\mathrm{d}\boldsymbol{v}_j}{\mathrm{d}t} = m_j n_j \cdot (\boldsymbol{g} + \boldsymbol{f}) + \frac{\partial}{\partial \boldsymbol{r}} p \tag{1.1}$$

where $\boldsymbol{v}_j$ is the bulk velocity, $\boldsymbol{f}$ acceleration due to volume forces other than gravity $\boldsymbol{g}$ (including collisional friction with other constituents) and p is total pressure. In Eq. (1.0) only the first and third members on the right-hand side were taken into account. Since Eq. (1.1) is a partial differential equation in space and time, from the stationary solution it may also admit wave-like solutions, e.g. sound and gravity waves[1]. Under such conditions Eq. (1.0) is not rigorously applicable to instantaneous values. Wherever Eq. (1.0) applies it may also be used in differential form (the sign is due to the downward direction of $\boldsymbol{g}$):

$$\frac{\mathrm{d}p}{\mathrm{d}z} = -g(z) \cdot \sum_j m_j(z) \cdot n_j(z). \tag{1.2}$$

[1] Jones, W.L. (1976): This Encyclopedia, vol. 49/5, pp. 177–216

Pressure and density are related by the ideal gas equation [see SUCHY's contribution in this volume, his Eq. (1.12)]:

$$p=\sum_j n_j k T_j, \tag{1.3}$$

where T is absolute temperature and k is BOLTZMANN's constant ($1.38\cdot 10^{-23}$ J/K).

In most planetary atmospheres the different neutral constituents are in thermal equilibrium so that $T_j \equiv T$. Ionized constituents are often at a higher temperature than the neutrals. However, their part in the total density is so small that they can be neglected in relations (1.0–1.3). Thus the pressure may be written as

$$p=kT\sum_j n_j = kTn \tag{1.3a}$$

with $n\equiv\sum_j n_j$ the total numerical density.

We introduce the average molecular mass (at height z) by

$$\bar{m}=\sum_j m_j n_j \Big/ \sum_j n_j \equiv \sum_j m_j n_j/n \equiv \frac{\rho}{n}, \tag{1.4}$$

$\rho\equiv\sum_j m_j n_j$ being the mass density.

By inserting Eqs. (1.3a) and (1.4) into Eq. (1.2) we obtain a linear differential equation for the pressure p:

$$\frac{\mathrm{d}\log p}{\mathrm{d}z} \equiv \frac{1}{p}\frac{\mathrm{d}p}{\mathrm{d}z} = -\bar{m}g(z)/kT(z), \tag{1.5}$$

(log being natural logarithm). Eq. (1.5) is called the *barometrical differential equation*[2]. It is immediately resolved by

$$\log(p/p_0) = -\int_{z_0}^{z} \mathrm{d}z \frac{\bar{m}g(z)}{kT(z)}.$$

Now writing $1/H(z)$ for the integrand, i.e. defining:

$$H(z) \equiv -1 \Big/ \frac{\mathrm{d}}{\mathrm{d}z}\log p = \frac{kT(z)}{\bar{m}(z)g(z)} = \textit{scale height} \tag{1.6}$$

(more precisely *pressure* scale height), we obtain

$$p=p_0 \exp\left(-\int_{z_0}^{z} \frac{\mathrm{d}z}{H(z)}\right), \tag{1.7}$$

the *barometrical formula.*

[2] In this Encyclopedia log means natural (NEPER's) logarithm. In order to simplify we write $\mathrm{d}\log x/\mathrm{d}z$ for $(1/x)\,\mathrm{d}x/\mathrm{d}z$

The denominator depends primarily on z for the height-dependent temperature, but also for the average molecular mass $\bar{m}$, which, in a diffusive atmosphere, decreases with increasing height. Thus diffusive separation has a similar effect to a temperature increase. Finally, the acceleration of gravity, g, decreases slowly with increasing height, which gives a small effect in the same direction. It goes with the inverse cube of the radius so that at an altitude of 1,000 km above Earth g has decreased by 36% against its value at ground level.

β) The *average molecular mass*[3], $\bar{m}$, is 28.96 au near the ground and is almost constant up to 90 km [*12*]. Above that altitude it decreases as a result of photodissociation of molecular oxygen by solar ultraviolet radiation [*35*].

The time constants of the counter processes are long enough to have conditions which, up to 125 km, show no serious diurnal variation. However, due to contraction of the cooler night atmosphere, day and night values of $\bar{m}$ are largely different at heights above 200 km, with lower night values. In the day/night average profile the steepest decrease of $\bar{m}$ occurs just above 100 km and is about 1 au/6 km. The same average profile has values of 24, 20, 18 and 16 au at about 130, 200, 250 and 400 km. However, the extreme values which may occur at 400 km are 3.8 and 18.3 au [*12*]. Further decrease, above 400 km, is due to the increasing part of helium and hydrogen by diffusive separation (see YONEZAWA's contribution in vol. 49/6).

Below 90 km vertical exchange by up and down going turbulence elements ("eddy diffusion") is so strong that the relative densities of all stable constituents are constant. Ozone is an important exception since it is produced in chemical processes following the photodissociation of molecular oxygen and then propagates downwards by eddy transport [*35*]. Ozone remains, however, a minor constituent.

Turbulence[4] remains important up to a limiting level, called *turbopause*, which, above Earth, is found around 110 km but may vary by some kilometres (see YONEZAWA's contribution in vol. 49/6).

Below it mixing is decisive and $\bar{m}$ remains constant. This height range is called the homosphere. Neglecting the small variation of the acceleration of gravity, g, the temperature profile alone determines pressure and density as functions of height.

While the pressure is given by Eq. (1.7), the *total numerical density* follows from Eq. (1.3a). Under full mixing conditions the relative densities have invariable, i.e. height-independent ratios. Only some minor constituents produced in chemical or photochemical reactions do not follow this rule, which is true up to the so-called *homopause* (in the terrestrical atmosphere situated around 90 km). The nomenclature of atmosphere levels and regions is summarized in Fig. 1.

Above the turbopause, as also shown in YONEZAWA's contribution in vol. 49/6, diffusion takes over. Thus the density of heavier constituents decreases more rapidly with increasing height than that of lighter ones. In the full diffusion regime each individual constituent is distributed according to its own scale height [Eq. 1.6)]:

$$H_j(z) = \frac{k \cdot T_j(z)}{m_j \cdot g(z)} = \frac{k \cdot T(z)}{m_j \cdot g(z)}. \tag{1.8}$$

[3] The international symbol au for atomic mass numbers should not be confused with the "astronomical unit" a.u. (which is equal to the average distance between Earth and Sun)

[4] Schlichting, H. (1959): Entstehung der Turbulenz. This Encyclopedia, vol. 8/1, pp. 351–450; Corrsin, S. (1963): Turbulence. Experimental methods. This Encyclopedia, vol. 8/2, pp. 525–590; Lin, Ch.Ch. (1963): Turbulent flow, theoretical aspects. This Encyclopedia, vol. 8/2, pp. 438–523

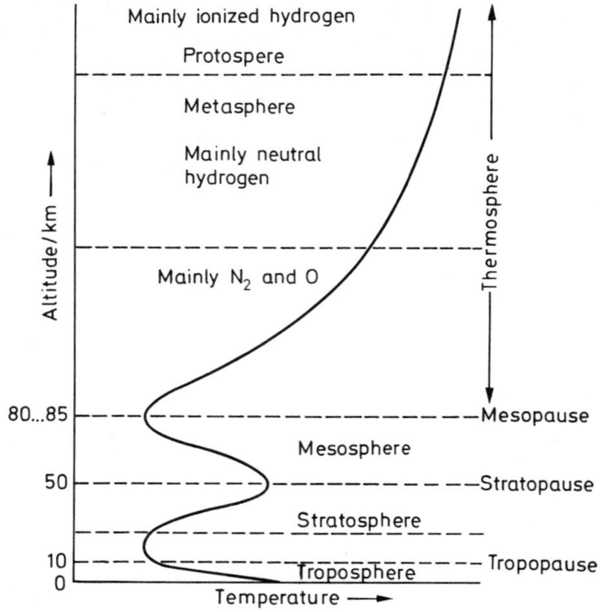

Fig. 1. Upper atmosphere nomenclature after CHAPMAN in [*38*]. (The transition range to the exosphere, at top, may be called 'thermopause')

(The right-hand term is valid if the neutrals are in thermal equilibrium; this can always be taken for granted.) Atomic oxygen has twice the scale height of the molecular species such that, in the terrestrial as in the Venusian atmosphere, it is prevailing constituent in the upper thermosphere. In this height range, as in the homosphere, the temperature profile is decisive for the large scale structure of the atmosphere.

γ) At very large heights the density becomes so small that the mean free path of a molecule is of the order of the scale height. In this condition, molecules with a larger upward component of velocity may "jump" upwards (like a trout jumps above the water surface), and we have a dynamical equilibrium [*34*]. This "fringe region" of the atmosphere is called the *exosphere*. For neutral constituents, though still belonging to the atmosphere, it is the transition zone towards interplanetary space. Molecules with velocities above a critical value may escape into it (see Sect. 31 of YONEZAWA's contribution). This is, however, quite different for the ionized components which, at least in the presence of an internal planetary magnetic field, remain linked with the planetary atmosphere in a much more extended range called the *magnetosphere*[5].

δ) As we have learned above, the temperature profile, $T(z)$, mainly characterizes the conditions in an atmosphere. In the ideal case, without energy advection or interchange in the atmosphere, the temperature profile is ruled by the condition of *adiabatic equilibrium*. It means that the temperature decreases with increasing height in the same measure as an upward transported air

[5] Poeverlein, H. (1972): This Encyclopedia, vol. 49/4, pp. 7–113

element would cool down by pressure decrease. Thus, in the adiabatic condition, an atmosphere is stable against vertical displacements. The condition is equated by expressing the temperature gradient as a function of the pressure (or density) gradient and also applying the adiabatic equation of state.

From the pressure equation, Eq. (1.3a), we obtain by logarithmic differentiation[2]

$$\frac{\mathrm{d}}{\mathrm{d}z}\log p=\frac{\mathrm{d}}{\mathrm{d}z}\log T+\frac{\mathrm{d}}{\mathrm{d}z}\log n. \tag{1.9}$$

Adiabatic gas behaviour is characterized by a particular relation between pressure p and density ρ:

$$p=\text{const}\cdot\rho^{\kappa} \tag{1.10}$$

where κ is the ratio of the specific heats at constant pressure and volume. For ideal gases $\kappa=1.66$, 1.4 and 1.3 for mono-, di- and tri-atomic molecules, and these values apply to the major atmospheric constituents. With the aid of Eq. (1.4) we replace mass density ρ by numerical density n and find (with height-independent mean molecular mass $\bar{m}$)

$$p=\text{konst}\cdot n^{\kappa} \tag{1.10a}$$

and by logarithmic differentiation[2]

$$\frac{\mathrm{d}}{\mathrm{d}z}\log p=\kappa\cdot\frac{\mathrm{d}}{\mathrm{d}z}\log n.$$

After inserting, Eq. (1.9) reads

$$\frac{\mathrm{d}}{\mathrm{d}z}\log T=(\kappa-1)\frac{\mathrm{d}}{\mathrm{d}z}\log n=\frac{\kappa-1}{\kappa}\frac{\mathrm{d}}{\mathrm{d}z}\log p. \tag{1.11}$$

Finally, with the definition of the scale height, Eq. (1.6), we get

$$\frac{\mathrm{d}}{\mathrm{d}z}\log T\equiv\frac{1}{T}\frac{\mathrm{d}T}{\mathrm{d}z}=-\frac{\kappa-1}{\kappa}\Big/H,$$

or

$$-\left(\frac{\mathrm{d}T}{\mathrm{d}z}\right)_{\mathrm{ad}}=\frac{\kappa-1}{\kappa}\frac{\bar{m}g}{k}. \tag{1.12}$$

This is the *adiabatic temperature gradient;* numerical values are given in Table 1.

Table 1. Adiabatic temperature gradient in different gases

Species	H	He	O	N_2	Air	O_2	A	CO_2
Gradient (negative)	0.47	1.88	7.50	9.44	9.76	10.78	18.78	11.98 K/km

Such large negative gradients occur nowhere in the real atmosphere since there is always some energy interchange. Isothermal behaviour, which is often dealt with as another typical case, is also not a realistic condition but for a few quite narrow ranges of altitude near peaks of temperature.

ε) A more general case which is quite helpful in many applications is that of the *constant height gradient* of temperature or, better, of *scale height* $H(z)$. Let us introduce with NICOLET[6]

$$H = H_0 \cdot (1 + \beta z) = H_0 + B \cdot z. \tag{1.13}$$

Neglecting the very small variations of molecular mass and gravity with height, Eq. (1.13) after Eq. (1.6) means a constant gradient of temperature

$$\frac{dT}{dz} = \frac{\bar{m} g}{k} H_0 \cdot \beta, \tag{1.14}$$

or

$$\frac{d}{dz} \log T = H_0 \cdot \beta / H = B/H. \tag{1.14a}$$

Comparison with Eq. (1.12) shows that in the special case of adiabatic behaviour

$$B_{ad} \equiv B_0 \cdot \beta_{ad} = -(\kappa - 1)/\kappa,$$

i.e. −0.3976, −0.2857 and −0.2308 respectively for mono-, di- and tri-atomic molecules. In real atmospheres β takes positive and negative values as well. In the terrestrial atmosphere (see Fig. 1) it is negative in the troposphere and mesosphere but positive in the strato- and thermosphere. Typical values are shown in Table 2 (near-ground estimates for the main planets[7] and upper atmospheric values for Earth).

Introducing Eq. (1.14a) into the general relation Eq. (1.9) we find

$$\frac{d}{dz} \log p - \frac{d}{dz} \log n = \frac{\beta}{H} = \frac{\beta}{1 + \beta \cdot z}.$$

Table 2. *Adiabatic atmospheres:* Temperature gradient Γ, specific heat ratio κ, polytropic index q, scale height gradient B (after D.M. HUNTEN[7])

Planet[a]	Γ/K km^{-1}	κ	q	B_{ad}
Venus, 100 bar	8.0	(1.205)	4.87	−0.170
0.1 bar	11.0	1.32	3.13	−0.242
Earth, 1 bar, wet	6.5	(1.23)	4.35	−0.187
dry	10.0	1.4	2.5	−0.286
Mars	5.0	1.37	2.70	−0.270
Jupiter	1.9	1.4	2.5	−0.286

[a] 1 bar = 10^5 Pa ≈ 750 Torr

[6] Nicolet, M. in: Kuiper, G.P. (ed.) (1954): The Earth as a planet. Chicago: Univ. Chicago Press, Chap. 13; Nicolet, M., Bossy, L. (1949): Ann. Géophys. *4*, 275

[7] Hunten, D.M. (1971): Space Sci. Rev. *12*, 539

Thus, similarly to the adiabatic case, we have a relation between pressure and numerical density of shape [from Eqs. (1.6, 1.9, 1.14a)]

$$p = \text{const}_1 \cdot n^{1/(1+B)}, \tag{1.15a}$$

and, correspondingly, between density and temperature

$$n = \text{const}_2 / T^{(1+B)/B}, \tag{1.15b}$$

and finally, between temperature and pressure

$$T = \text{const}_3 / p^B. \tag{1.15c}$$

ζ) The above relations can be derived in the most general way from *statistical kinetic theory*, for example, from the BOLTZMANN equation[8]. In full equilibrium the gas molecules follow a statistical distribution which is called Maxwellian. It is either written in full, i.e. as a distribution function, f^{M}, in three-dimensional velocity space with coordinates c_x, c_y, c_z (Fig. 2a), or just for the absolute value $|\boldsymbol{c}|$ of the velocity taking together the different directions of $\boldsymbol{c}$ since isotropy obtains. We designate this latter distribution function by $f^{\mathrm{M}*}$ (Fig. 2b). As can be seen by comparison of both figures, $f^{\mathrm{M}*}$, which is zero at the origin, has a clear maximum at the most frequent molecular velocity, C_{w}, while f^{M} takes its maximum value at the origin and, towards higher velocities, is decreasing everywhere. If m is molecular mass, T temperature, n total numerical density and $\mathscr{E}$ particle energy, it can be shown that the *three-dimensional distribution function* f^{M} (Fig. 2a) is[9]

$$\frac{\mathrm{d}n}{n} \equiv \frac{f^{\mathrm{M}} \cdot \mathrm{d}^3 c}{n} = \left(\frac{m}{2\pi k T}\right)^{3/2} \cdot \exp\left(-\frac{\mathscr{E}}{kT}\right) \cdot \mathrm{d}c_x \cdot \mathrm{d}c_y \cdot \mathrm{d}c_z \tag{1.16}$$

while the *one-dimensional distribution function* is[9]

$$\frac{\mathrm{d}n}{n} \equiv \frac{f^{\mathrm{M}*} \cdot \mathrm{d}c}{n} = \frac{4}{\sqrt{\pi}\, C_{\mathrm{w}}} \cdot \left(\frac{c}{C_{\mathrm{w}}}\right)^2 \cdot \exp\left(-\left(\frac{c}{C_{\mathrm{w}}}\right)^2\right) \cdot \mathrm{d}c. \tag{1.17}$$

The most frequent molecular velocity is[10]

$$C_w = \left(\frac{2kT}{m}\right)^{1/2}. \tag{1.18}$$

This velocity is indicated for $m = 16$ au in Fig. 2 by a bold arrow tip; it appears right at the peak of the one-dimensional distribution functions $f^{\mathrm{M}*}$. The different abscissa scales in Fig. 2 allow

[8] Rawer, K., Suchy, K. (1967): This Encyclopedia, vol. 49/2, p. 8, Eq. (2.3) from which the Maxwellian distribution f^{M}, their Eq. (2.4), is derived as a "trivial" solution. It is the same as our Eq. (1.6)

[9] Waldmann, L. (1962) in: Physikalisches Taschenbuch. Ebert, H. (ed.), p. 205. Braunschweig: Friedr. Vieweg und Sohn

[10] Though the order of magnitude is the same, C_{w} must not be confounded [9] with the mean velocity, $\bar{c} = 2 \cdot (2kT/\pi \bar{m})^{1/2}$, or with the root of the mean squared velocity $\overline{c^2} = 3kT/\bar{m}$

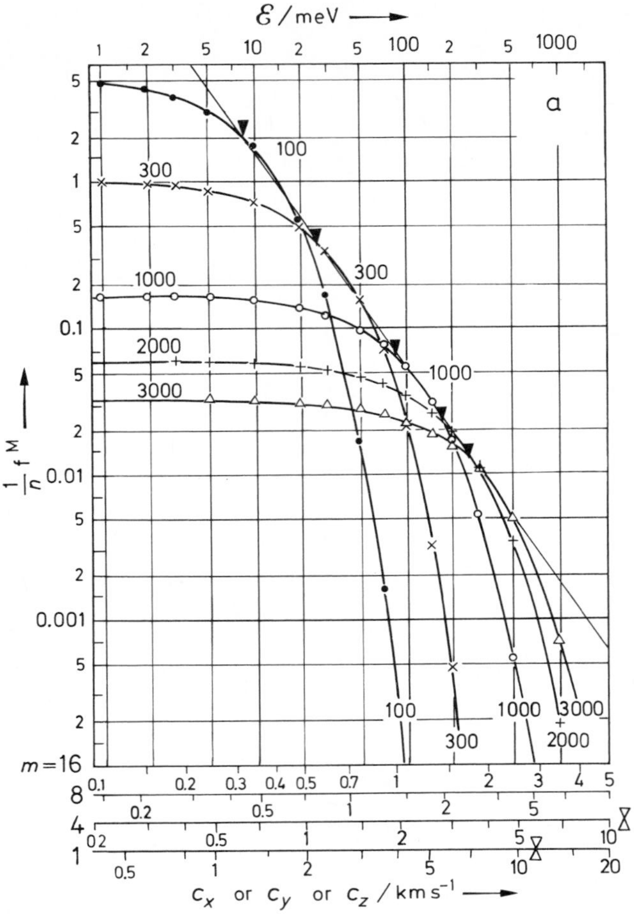

Fig. 2 a, b. Maxwellian velocity distribution functions f^{M} and $f^{\mathrm{M}*}$ for five temperatures (parameter: ● 100 K, × 300 K, ○ 1,000 K, +2,000 K, △ 3,000 K) and four mass numbers: 16, 8, 4 and 1 au; see abscissa scales (bottom). The ordinate scale is correct for $m=16$ au (atomic oxygen). For the other mass numbers an ordinate shift is needed to get correct absolute values, but the shape of the curves is exactly the same anyway. Logarithmic scales on both axes. Abscissa: velocity (additional energy scale on top); ordinates: statistical density (i.e. ratio of unit cell occupation number to total numerical density n). (**a**) $1/n \cdot f^{\mathrm{M}}$ refers to unit cell $(1\,\mathrm{km/s})^3 = \mathrm{d}^3\boldsymbol{c}$ in three-dimensional, Cartesian velocity space [Eq. (1.16); ordinate is $\mathrm{d}n/n \cdot \mathrm{d}c_x \cdot \mathrm{d}c_y \cdot \mathrm{d}c_z$]. (**b**) $1/n \cdot f^{\mathrm{M}*}$ refers to unit cell $1\,\mathrm{km/s} = |\mathrm{d}\boldsymbol{c}|$, one-dimensional after integration over the two angular coordinates in a spherical velocity space [Eq. (1.17); ordinate is $\mathrm{d}n/n \cdot |\mathrm{d}\boldsymbol{c}|$]

these to be applied not only to atomic oxygen ($m=16$ au) but also the helium ($m=4$ au) and atomic hydrogen ($m=1$ au). The case $m=8$ au is of some interest for a simple approximation describing a plasma of electron and O^+ ions as found near the peak of the terrestrial ionosphere[11].

[11] The charged particles of both signs are held together by strong electrostatic forces so that the plasma as whole behaves approximately like a gas with the average molecular mass and average temperature of the charged particles of both signs

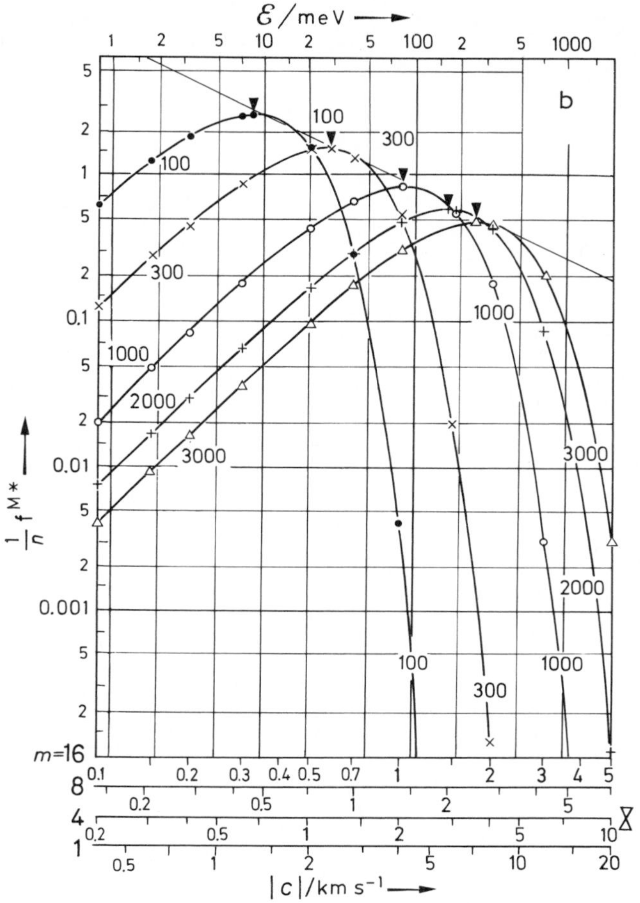

Fig. 2b

Equation (1.16) is, of course, the more general one. In isotropic space one identifies the energy $\mathscr{E}$ with the kinetic energy of the particle [as done in Eq. (1.17)]. One may, however, introduce a more general expression for $\mathscr{E}$, for example, one which also admits some space-dependent potential energy, $\mathscr{E}_p$[9],

$$\mathscr{E} = \tfrac{1}{2} m |\boldsymbol{c}|^2 + \mathscr{E}_p. \tag{1.19}$$

Thus, Eq. (1.16) can immediately be used to find the effect of a gravitational field and derive the barometrical formula, Eq. (1.7), above.

The statistical distribution of velocities is, of course, very important for all diffusion problems since the faster molecules promote the particle flow more efficiently than the slower ones (see YONEZAWA's contribution in volume 49/6, his Sect. 3). As shown in the same contribution (his Sect. 31), it is of prime importance for the *escape problem* in the *exosphere*. The velocity needed for escaping from the Earth's gravitational field is about 11 km/s (in a vertical direction). As seen from Eq. (1.18) the most frequent velocities are greater for lighter molecules. We see from Fig. 2a that for He ($m=4$ au) and *a fortiori* for H ($m=1$ au) this velocity has non-negligible probability at temperatures of 1,000 K and more as are found in the exosphere[12]. (These values are marked by a spindle-shaped symbol in Fig. 2a.)

η) *A general system of equations* must be used when the physical nature of the atmospheric structure is to be considered. While for empirical modelling it is sufficient to adopt an empirically obtained temperature profile as basic input, by theoretical modelling one should hope to find out why such a profile occurred. Thus, energy interchange must be considered. Above we had used the "equation of state" of the gas, Eq. (1.3), and the assumed hydrostatic equilibrium as in Eq. (1.2). Instead of the latter we now have to admit motions, i.e. use the general dynamic equation, Eq. (1.1), take account of production and loss processes for each species, and consider all energy interchange processes. A fairly general set of equations including the effect of charged particles is the following one:

(i) *Equation of motion* (dimension $\mathrm{N\,m^{-3}}$):

$$\rho\left(\frac{\partial \boldsymbol{v}}{\partial t}+\boldsymbol{v}\frac{\partial \boldsymbol{v}}{\partial \boldsymbol{r}}\right)+\frac{\partial}{\partial \boldsymbol{r}}p+\frac{\partial}{\partial \boldsymbol{r}}\cdot\Pi=-\rho g\boldsymbol{u}_z-2\rho\boldsymbol{\omega}_\mathrm{E}\times\boldsymbol{v}+\tfrac{1}{2}\rho_\mathrm{i}\nu_\mathrm{in}(\boldsymbol{v}_\mathrm{i}-\boldsymbol{v}). \tag{1.20}$$

Π is the tensor of viscous stress (dimension $\mathrm{N\,m^{-2}}$), $\boldsymbol{\omega}_\mathrm{E}$ is the rotation frequency (pulsation) of Earth which provokes the CORIOLIS force, ρ_i mass density of the ionized component, $\boldsymbol{v}_\mathrm{i}$ its bulk motion and ν_in the transport collision frequency for momentum transfer from the ionized component to the neutrals (ion drag) (see SUCHY's contribution in this volume). $\boldsymbol{u}_z$ is the unit vector in the vertical direction.

Equation (1.20) is often simplified to EULER's equation:

$$\frac{\partial \boldsymbol{v}}{\partial t}+\left(\boldsymbol{v}\cdot\frac{\partial}{\partial \boldsymbol{r}}\right)\boldsymbol{v}=-\frac{1}{\rho}\frac{\partial}{\partial \boldsymbol{r}}p-\boldsymbol{g}. \tag{1.20a}$$

(ii) *Overall continuity equation* (dimension $\mathrm{kg\,m^{-3}\,s^{-1}}$):

$$\frac{\partial \rho}{\partial t}+\frac{\partial}{\partial \boldsymbol{r}}\cdot(\rho\boldsymbol{v})=0, \tag{1.21}$$

requests conservation of mass.

(iii) *Set of individual continuity equations for the components* (dimension $\mathrm{m^{-3}\,s^{-1}}$):

$$P_j-L_j=\frac{\partial n_j}{\partial t}+\frac{\partial}{\partial \boldsymbol{r}}\cdot(n_j(\boldsymbol{v}+\boldsymbol{v}_j)-n_j'\boldsymbol{v}_j') \tag{1.22}$$

[12] One might argue that electrons should escape most easily due to their small mass. See, however [11]

P_j, L_j production and loss ($m^{-3}s^{-1}$). $\boldsymbol{v}_j$ is the relative velocity of species j due to diffusion. The apostrophe identifies a drifting population of species j. Note that the divergence term on the right-hand side corrects the production/loss balance for the effect of motions of all kinds.

(iv) *Energy transfer equation* (dimension $J\,m^{-3}s^{-1}$)

$$\rho\, c_v \left(\frac{\partial T}{\partial t} + \boldsymbol{v} \cdot \frac{\partial T}{\partial \boldsymbol{r}}\right) + \frac{\partial}{\partial \boldsymbol{r}} \cdot \boldsymbol{q} + p \frac{\partial}{\partial \boldsymbol{r}} \cdot \boldsymbol{v} + \Pi \cdot\cdot \frac{\partial \boldsymbol{v}}{\partial \boldsymbol{r}}$$
$$= Q_R - \sum_j m_j K_{jk} (U_j - \tfrac{1}{2} v_j^2) + A(T_e - T) + B(T_i - T) \qquad (1.23)$$
$$+ Q_A + \tfrac{1}{2} \rho_i \nu_{in} (\boldsymbol{v}_i - \boldsymbol{v})^2$$

where c_v is specific heat at constant volume and $\boldsymbol{q}$ is the heat flux vector (conduction and eddy transport). The right-hand side describes the sources and sinks of heat transfer due to heating by absorbed radiation (Q_R), by dissipation of potential (U_j) and kinetic energy, by heat flux from the electronic (suffix e) and from the ionic (suffix i) constituents, by radiation and by collisional effects of the ions. See Sect. 16 for a detailed discussion.

For a detailed discussion of the energy equation see, for example, P.R. BLUM in [*68*], where solutions are also indicated, for steady state as well as a time-dependent model.

A. Measurements

2. Air density determinations from satellite drag. With the advent of low-orbiting artificial Earth satellites, at altitudes, above 200 km, the total atmospheric density became measurable in quite a simple way.

α) *Aerodynamic braking* is in fact quite easily determined by its systematic influence on the orbital parameters of a satellite. Braking makes the excentricity of the original orbit decrease, mainly by a decrease of the apogee while the perigee remains almost unchanged (Fig. 3). Other effects are a small reduction of the orbital period[1].

The *braking force* is, of course, proportional to the mass density of the air in the bracking zone, i.e. to $\sum n_j m_j = \rho$ in a multicomponent atmosphere. Minor constituents are, therefore, of no importance in the process. Since satellites move supersonically in a rarified gas the braking force is, to first-order approximation, given by the momentum gain of the air mass captured by the satellite

$$F_b \approx \rho \cdot A \cdot v_S^2 = M v_S \qquad (2.1\,a)$$

[1] Rather easily observable by optical as well as by radio wave observations. The latter are described in Sect. 53 of Rawer, K., Suchy, K. (1967): This Encyclopedia, vol. 49/2. The COSPAR World list of tracking stations [COSPAR Transactions] contains both types of stations

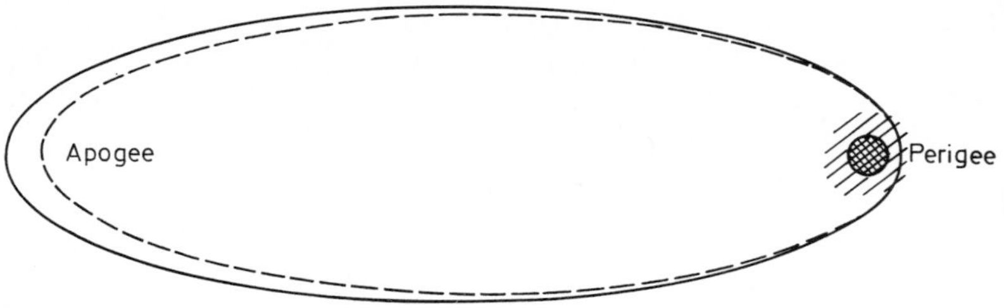

Fig. 3. Satellite braking in the dense atmosphere near perigee: the orbit (original: full line) in the course of time reaches smaller excentricity (broken line)

where A is the cross section perpendicular to the velocity $\boldsymbol{v}_S$ and $M=\rho A v_S$ is the total air mass captured at the satellite front surface per time unit; it is the speeding up of this mass to satellite velocity v_S which causes the drag.

To obtain a second-order approximation one may introduce a corrective factor. $C_D/2$, the numerical value of which is, of course, not very far from one. For the drag coefficient C_D, 2.2 is an often applied estimate [52]. Instead of Eq. (2.1a) we have then

$$\boldsymbol{F}_b = -\frac{\rho}{2}\, C_D A\, v_S \boldsymbol{v}_S, \tag{2.1}$$

the minus sign indicating that, in an inertia system, the drag force is opposed to the velocity of the satellite. (In the satellite system, however, drag force and air flow have the same direction.)

Bodies without spherical symmetry present at different times different *cross sections* to the incoming air stream. For random orientation the average cross section is one-fourth the total surface area provided it is convex everywhere[2]. However, random tumbling is not very likely with elongated satellites. For spinning satellites the position of the instantaneous rotation axis is often unknown. A satellite of elongated shape may be approximated by a cylinder of length L and diameter d. The extreme modes of rotation around an axis of maximum moment of inertia are propellerwise and tumbling end-over-end. In the first case the average cross section is $\bar{A}=L\cdot d$, but it is $Ld\cdot(2/\pi+d/2L)$ in the second case [52]. As the axis of rotation is unknown, the mean value of the two extremes is often taken as an estimate; the maximum error made with this guess[3] is about 20%.

Stabilized satellites make a much better estimate possible. For example, for a spin-stabilized satellite with artificially enforced solar orientation the variation of the effective cross section A can be geometrically determined and enables us to assess the variation of A along the satellite orbit[4].

β) The *drag coefficient* C_D depends not only on the geometrical characteristics of the satellite but also on its configuration and on the specifics of the interaction between the vehicle and the gas density the determination of which is intended. The mean free path of the air molecules, λ_0, as compared with the dimensions of the vehicle, L, determines the KNUDSEN number K_K. The gas is rarified enough to admit free molecular flow when K_K is much greater than one. This is the case at higher levels in the upper atmosphere. For small K_K we obtain the other extreme case where (supersonic) continuum mechanics[5] must be applied. While this case is not important with free-flying satellites, the intermediate regime can be of some importance for large vehicles at satellite

[2] Radzijewskij, V.V., Razbitnaja, E.I. (1953): Astron. Žurnal *30*, 616
[3] King-Hele, D.G., Walker, D.M.C. (1961): Space Res. *II*, 918
[4] Roemer, M., Richter, E., Slowey, J.W. (1979): J. Geomag. Geoelectr. *31* (Suppl.), 63
[5] (1960): This Encyclopedia, vol. 9 [Schiffer, M., pp. 1–161, Cabannes, H., pp. 162–224]

altitudes lower than 200 km. The prevalance of the free molecular flow condition is also due to the fact that reflected or re-emitted molecules create a zone of increased density in front of the vehicle[6]. All these mechanisms are taken into account by the drag coeffcient C_D which should still be dependent on the energy accommodation coefficient α which is defined by $\alpha=(\mathscr{E}_i-\mathscr{E}_r)/(\mathscr{E}_i-\mathscr{E}_w)$ where $\mathscr{E}_i$ and $\mathscr{E}_r$ are the average kinetic energies of incident and reflected molecules and $\mathscr{E}_w$ is the average thermal energy corresponding to the surface temperature T_w. Unfortunately, due to the incoming air molecules, this latter is often increased at lower altitudes where the air is rather dense. As a function of α the drag coefficient C_D varies[6] between about (2.55 ± 0.07) for $\alpha=0.5$ and 2.08 for $\alpha=1.0$.

γ) At higher altitudes where the air density is small *other braking forces* might be deduced before using a relation like Eq. (2.1) in order to determine the density ρ. Three mechanisms of *non-neutral drag* must be considered [52]. First, Coulomb drag provoked by an interaction of atmospheric ions with the charged vehicle[7]. Second, an electromagnetic braking force produced by currents in the vehicle skin which are induced when the satellite moves through the variable magnetic field of Earth. Third, though the neutral density is too small for acoustic waves to be excited, oscillations and waves of electromechanical structure [8] may be created in the ionospheric plasma and so provoke energy loss[7]. Anyway, non-neutral drag is negligible up to 700 km of altitude and probably not a source of major error in the determination of atmospheric densities [*52*].

Solar radiation pressure, however, is not negligible at low densities [*52*]. The total wave energy radiated from the Sun at 1 a.u. is $S=1.4\,\mathrm{kW\,m^{-2}}$ so that the force onto a surface A is directed away from the Sun and has for absolute value

$$F_0 = A \cdot S/c_0 = 4.7 \cdot 10^{-6}\,\mathrm{N} \cdot \frac{A}{m^2}. \tag{2.2}$$

δ) *Drag affects the orbital parameters* in different ways. For an elliptical orbit the perigee height is decreased by drag and is more sensitive than that at apogee. But since the orbital period can be measured very accurately it is the indicator normally used. Drag decreases the orbital period since it decreases the orbital energy, $\mathscr{E}$, of the satellite. Along an orbital element ds the work $\boldsymbol{F}_b \cdot \mathrm{d}\boldsymbol{s}$ is performed by the braking force, and the vector $\boldsymbol{F}_b$ is parallel to d$\boldsymbol{s}$. The total loss of energy during one orbit, $\Delta\mathscr{E}$, is then obtained by integration. With Eq. (2.1) one obtains

$$\Delta\mathscr{E} = -\frac{A}{2}\oint \mathrm{d}\boldsymbol{s}\, C_D \rho v_S \cdot \boldsymbol{v}_S. \tag{2.3}$$

For a spherical satellite in a circular orbit, radius R is constant and the acceleration of gravity $g_0(R_{♁}/R)^3$ equals the centripetal acceleration v_S^2/R so that $v_S^2 = g_0 R_{♁}^3/R^2$ and

$$\Delta\mathscr{E} = -\pi A\, C_D g_0 \rho R_{♁}^3/R = -7.966\,\mathrm{J} \cdot \frac{\rho}{10^{-15}\,\mathrm{kg\,m^{-3}}}\,\frac{A/\mathrm{m}^2}{R/\mathrm{Mm}}\,C_D. \tag{2.4}$$

---

[6] Cook, G.E. (1965): Planet. Space Sci. *13*, 929–946; (1966): Ann. Géophys. *22*, 53
[7] Al'pert, J.A. (1976): This Encyclopedia, vol. 49/5, Chap. B
[8] Ginzburg, V.L., Ruhadze, A.A. (1972): This Encyclopedia, vol. 49/4, Chaps. C, E

In the case of an excentric orbit, due to the exponential decrease of ρ with increasing radius R, the contributions from higher parts of the orbit are negligible against those from the neighbourhood of perigee[9]. In the general case, taking account of a relation between the anomalistic period P and the orbital energy $\mathscr{E}$ given by the Keplerian theory one obtains

$$\dot{p} \equiv \frac{\Delta P}{P} = -\frac{3}{2}\frac{Aa}{M_S \mu}\rho_{\text{per}} \oint \mathrm{d}s\, C_D \frac{\rho}{\rho_{\text{per}}} v_S \cdot \boldsymbol{v}_S \tag{2.5}$$

where M_S is the mass of the satellite, index per designates perigee, a is the semimajor axis of the orbit and μ is the constant of the terrestrial gravitational field. Unfortunately, the determination of $\dot{p}$, the rate of change of the period, is a rather difficult operation. Applying a particularly accurate procedure[10] to optical satellite observations, an overall accuracy of only 5% could be reached, in spite of a much lower relative error in $\dot{p}$ within the analysis.

ε) Assuming an exponential decrease of ρ with altitude corresponding to the density scale height H and a Keplerian orbit, Eq. (2.5) can be evaluated and the *density at perigee* be determined [*52*]:

$$\rho_{\text{per}} = -\frac{\dot{p}}{-3C_D}\cdot\frac{M}{A}\left(\frac{2e}{\pi a H}\right)^{1/2}\cdot\left[1-2e-\frac{H}{8ae}+0\left(e^2,\frac{H^2}{a^2e^2}\right)\right], \tag{2.6}$$

e being the numerical excentricity of the orbit. This equation is obtained by an asymptotic expansion and is valid for not too small excentricity e. Corrections for taking account of a linear variation of H with height[11] and for other influences including atmospheric rotation have been derived [*52*].

Unfortunately, atmospheric density depends on local time so that the assumption of a fixed density profile around the whole orbit is not correct. This influence can be taken into account by using the more general Eq. (2.5) and resolving it by iteration[12].

Due to the integrating procedure involved the determination of densities from orbital decay has rather low resolution in position and time. A time resolution of 2.5 h (for relative values only) is only rarely reached[13] so that short-term effects cannot be seen. Also the accuracy in position is of the order of 30° only. A comparison with in-situ determination was, however, not so bad and gave an average systematic difference of the order of 20% only[14].

3. In-situ composition measurements. In rockets and satellites mass spectrometers have quite often been used for determining relative abundances of the different neutral constituents, or of the ions in the ionospheric plasma. Rather generally, the temperature profile is inferred from the partial density profiles or from their sum. There are, however, devices now available for directly measuring the neutral temperature and which can at least be used aboard satellites.

[9] Note that, even near perigee, ρ is always less then 10^{-15} kg m^{-3}
[10] Jacchia, L.G. (1963): Rev. Mod. Phys. *35*, 973–991
[11] Jacchia, L.G. (1960): J. Geophys. Res. *65*, 2775
[12] Roemer, M. (1966): Smithson. Astrophys. Obs. Spec. Rep. No. 199
[13] Roemer, M. (1967): Phil. Trans. Roy. Soc. *262*, 185
[14] Von Zahn, U. (1970): J. Geophys. Res. *75*, 5517

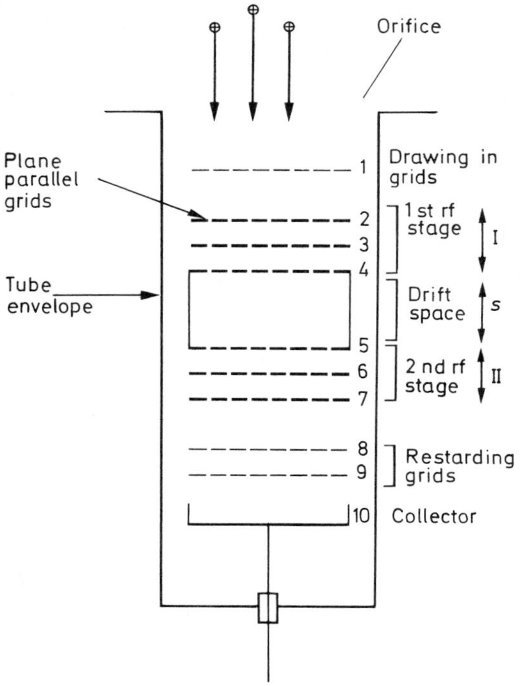

Fig. 4

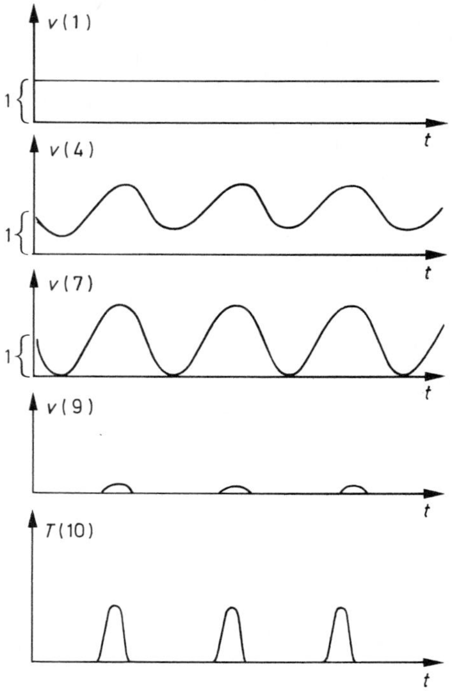

Fig. 5

Fig. 4. Time-of-flight mass spectrometer, schematic. Velocity modulation is obtained in the two "filters" (I and II)

Fig. 5. Time-of-flight mass spectrometer: velocities and current (well adjusted). Numbers refer to Fig. 4

As for total densities in the upper atmosphere variations have been determined with accelerometers aboard low-altitude satellites. Ionization vacuum gauges have also been used, particularly aboard rockets.

a) Mass spectrometers. These instruments are used either for neutral or for ionic composition measurements. Both types necessarily contain some kind of mass analyzer; if applied to neutrals a suitable ion source must be put in front of the aperture of the analyzer. We shall first discuss a few types of mass analyzers (cf. THOMAS' contribution in vol. 49/6, Sect. 19).

α) Time-of-flight radio frequency spectrometer. Figure 4 is a schematic drawing of such an instrument, which is often called a 'Bennett spectrometer'[1–4].

[1] Johnson, C.Y., Holmes, J.C. (1960): Space Res. *I*, 417; Taylor, H.A., Brinton, H.C. (1960): J. Geophys. Res. *66*, 2587

[2] Istomin, V.G. (1960): Iskusstv. Sputniki zemli *4*, 171; Istomin, V.G., Pohemkov, A.A. (1963): Space Res.*III*, 117

[3] Taylor, H.A., Brace, L.H., Brinton, H.C., Smith, C.R. (1963): J. Geophys. Res. *68*, 5339; Taylor, H.A., Brinton, H.C., Smith, C.R. (1965): J. Geophys. Res. *70*, 5769. – A modern instrument of this type is described in: Thomas, L. (1981): This Encyclopedia, vol. 49/6, his Fig. 31 and p. 74

[4] Istomin, V.G. (1965) in: All Union Conference on Space Research, Moskva. Skuridin, G.A., Al'pert, Ja.L., Krassovskij, U.I., Sarev, V.V. (eds.), p. 259

Such instruments contain two radio frequency driven accelerator/decelerator devices. Each of these consists of two grids at fixed (zero) potential with a third grid in between to which a radio frequency (f) voltage is applied. If an ion stream of constant velocity v goes through such a device, the velocity at the output is modulated and is thus a function of time. Putting two such devices in series at distance s one obtains a kind of velocity filter. Maximum velocity modulation is obtained when the beam finds the same radio frequency phase at the second device as it had encountered in the first device. If, however, opposed phase are encountered, the second device just compensates for the effect of the first, and no modulation appears at its output. Selection of the fastest particles is obtained by a suitable retarding potential (Fig. 5) applied to the retarding grids which are mounted between the second filter and the collector. If this (positive) potential is high enough, only those ions which have obtained maximum acceleration arrive at the collector (Fig. 5). The condition for this optimum is that the time of flight s/v equals just one radio frequency period, $1/f$, or an integer multiple of it:

$$v = \frac{1}{n} s f. \tag{3.1}$$

If v is expressed by the accelerating voltage U applied to the entrance aperture of the whole instrument, we have

$$v^2 = 2 \frac{q}{m} U \tag{3.2}$$

q and m being charge and mass of an ion. From Eqs. (3.1) and (3.2) we obtain the mass selection condition

$$\frac{m}{q} = \frac{2n^2}{s^2 f^2} U \qquad (n = 1, 2, 3 \ldots). \tag{3.3}$$

For a detailed analysis see [5].

β) *Magnetic deflection mass spectrometers.* This classical type[6] uses magnetic velocity selection. In a magnetic field B, ions follow a curved path of radius $r = mv/qB$. Ions of different ratios will have different curvature radii r after

$$\frac{m}{q} = \frac{r^2 B^2}{2U}. \tag{3.4}$$

By variation of either the magnetic field B or the accelerating voltage U any mass number can be brought to a given radius r to which the collector is adjusted. The classical type achieves focusing by combination with an electric

---

[5] Johnson, C.Y. (1960) in: The Encyclopedia of Spectroscopy. Clark, S.L. (ed.), p. 587. New York: Reinhold

[6] Aston, F.W. (1919): Phil. Mag. (London) *38*, 709; Aston, F.W., Fowler, R.H. (1922): Phil. Mag. (London) *43*, 514; Aston, F.W. (1924): Isotopes. London: E. Arnold & Co.

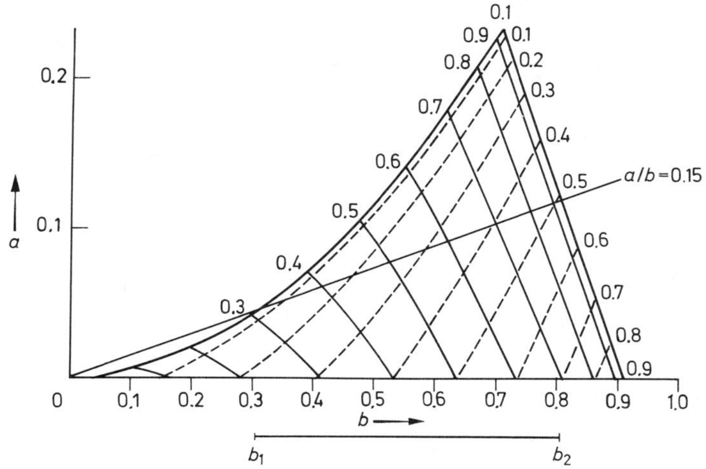

Fig. 6. Quadrupole mass spectrometer: stability diagram[9] (see text); the straight line marked $a/b = 0.15$ is shown as an example of a working line

(transverse) field[6]. Instruments were flown in quite a few rockets but also on some satellites[7]. Since a strong magnetic field is needed, stray fields are difficult to avoid and may disturb other experiments or require volume- and weight-consuming protecting devices. The different more recent instruments using magnetic deflection are discussed by THOMAS (vol. 49/6, Sect. 19(a), α, β).

γ) *Quadrupole mass spectrometer*[8]. Fundamentally, a "*Massenfilter*" uses an oscillating electric field to provoke transverse oscillations of the ions in a beam with increasing amplitude. Only under particular conditions does their amplitude remain limited, and this gives an opportunity for selecting specific ions. In Fig. 32 of THOMAS' contribution (in vol. 49/6) the transverse potential applied to the four rods[9] creates a hyperbolic potential distribution $\varphi = \varphi_0(x^2 - y^2)/r_0^2$. The potential is a combination of d.c. ('direct current') and radio frequency (ω) with amplitudes U_{dc} and U_{rf}.

The motion of the ions is described by MATTHIEU's differential equation[10], from which follows a rather limited stability range as shown in Fig. 6. The coordinates used are proportional to the two applied voltages:

$$a = 8q\,U_{dc}/mr_0^2\omega^2; \qquad b = 4q\,U_{rf}/mr_0^2\omega^2.$$

By choosing the ratio $a/b \equiv 2U_{dc}/U_{rf}$ one determines the working line in the stability diagram. Stability can only be obtained with $a/b < 0.3357$. Best mass resolution is obtained slightly below this value (>0.325 is typical). Even when there is stability the amplitude of the oscillation may become larger than the rods allow. In order to avoid this one has to watch out for a lateral beam limitation[9].

---

[7] Pelz, D.T., Newton, G.P., Kasprzak, W.T., Clem, T.D. (1973): NASA Doc. GSFC X-623-73-142; Zbinden, P.A., Hidalgo, M.A., Eberhardt, P., Geiss, J. (1975): Planet. Space Sci. *23*, 1621

[8] Paul, W., Steinwedel, H. (1953): Z. Naturforsch. *8a*, 448; Paul, W., Raether, M. (1955): Z. Physik *140*, 262

[9] Paul, W.H., Reinhard, P., von Zahn, U. (1958): Z. Physik *152*, 143

[10] Its solutions are discussed by Thomas, L. (1981): This Encyclopedia, vol. 49/6, p. 75/76 with schematic Fig. 32. A detailed drawing of a modern instrument is also shown there as Fig. 33

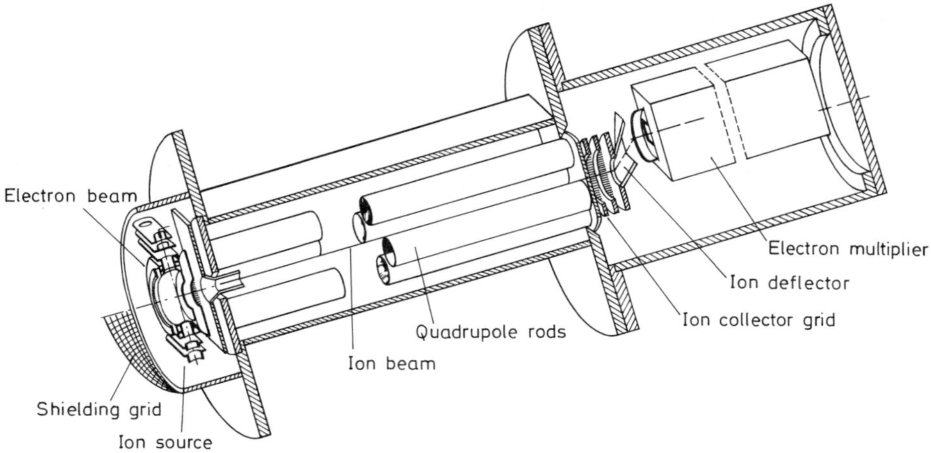

Fig. 7. Quadrupole mass spectrometer as flown in the AEROS satellites[12]

Under appropriate working conditions the filter selects a charge mass ratio corresponding to the midpoint of the working range (a_w, b_w) and

$$\frac{q}{m} = \frac{r_0^2 \omega^2}{8 U_{dc}} a_w = \frac{r_0^2 \omega^2}{4 U_{rf}} b_w. \tag{3.5}$$

Adjustment to a certain mass can be obtained by varying either U_{dc} and U_{rf}, with fixed ratio, or by variation of ω[11]. Figure 7 shows an instrument of this type used in the AEROS satellites[12]. Other applications are in satellite ESRO-4[13] and in quite a few USA satellites[14]. More details are given by THOMAS in vol. 49/6 (Sect. 19(a), γ).

δ) *Ion mass spectrometers.* A mass analyzer can be applied directly for determining the composition of the ions in the ionospheric plasma provided its axis is aligned with the velocity vector of the incoming particle flux. The angular aperture of most mass analyzers being quite small this condition often constitutes a serious difficulty, e.g. aboard spin-stabilized satellites. At lower altitudes a shock wave may appear in front of the vehicle, which, due to increased temperature, might alter the ion composition. Special devices are applied to have the aperture of the instrument outside the shock[15]. More recently, kryo-pumped mass spectrometers have been applied an descending rockets (or on balloons), down into the stratosphere[16]. Since below about 80 km loosely

---

[11] Bitterberg, W., Bruchausen, K., Offermann, D., von Zahn, U. (1970): J. Geophys. Res. *75*, 5528

[12] Krankowsky, D., Bonner, F., Wieder, H. (1974): J. Geophys. *40*, 601

[13] Trinks, H., von Zahn, U. (1975): Rev. Sci. Instr. *46*, 213

[14] Narcisi, R.S. (1966): Ann. Géophys. *22*, 159

[15] Arnold, F., Krankowsky, D. (1979): J. Atmos. Terr. Phys. *39*, 625

[16] Offermann, D., von Zahn, U. (1971): J. Geophys. Res. *76*, 2520, and Scholz, T.G., Offermann, D. (1974): J. Geophys. Res. *79*, 307 used a kryogenic ion source in the 85...115 km height range; Arnold, F. (1980): Nature *284*, 610 [also: ESA-SP *152*, 479] using balloons, was able to measure down to 24 km; Henschen, G., Arnold, F. (1981): Geophys. Res. Letts. *8*, 999; Arnold, F., Henschen, G., Ferguson, E.E. (1981): Planet. Space Sci. *29*, 185

bound cluster ions are abundant, it is important to hold the attractive aperture potential at quite a low voltage; otherwise, higher order clusters may be dissociated during the acceleration process[17].

ε) *Neutral mass spectrometers.* In order to obtain the composition of the neutrals these must first be ionized before entering the analyzer. Different types of *ion sources* are used. The most common system is ionization at the surface of heated wire; collisional ionization is also applied, usually with an electron beam which is perpendicular to the incoming flux[11]. The efficiency of ionization depends on the ionization energy and is therefore different for different species. Noble gases and hydrogen need a particularly high energy input. Therefore, the efficiency of ionization must be determined by an individual calibration of each instrument.

Open and closed sources (with many intermediate cases) are distinguished. In a closed source collisions inside the ion source compartment are not rare enough to prevent chemical reactions from occurring in this room, which make its composition different from that in free space. This danger is particularly serious for atomic oxygen, O, which after colliding with a wall may reappear as O_2 or NO by combining with molecules attached to the surface[18]. For this reason most atomic oxygen measurements obtained with a closed ion source must be considered as not directly 'measured' but 'inferred' determinations[19,20]. Open sources are better in this context but have a smaller collecting efficiency. Due to the rapid motion of satellites it is extremely difficult to obtain a completely open source, i.e. to avoid any wall effect, including that due to the vehicle skin itself.

ζ) *Calibration of mass spectrometers.* Neutral and ion mass spectrometers must be painstakingly calibrated. With laboratory calibration one simulates in-flight conditions by applying a molecular or ion beam. It is not easy to obtain perfect simulation. In-flight overall calibration may be obtained for neutral instruments by summing up the major constituents and comparing them with the total density determinations [*26*] (see Sect. 2).

b) *Retarding potential analyzers.* Ion mass spectrometers may be calibrated and the efficiency as function of the mass number may be determined by comparison with simultaneous *retarding potential analyzer* (RPA) measurements. This device is now often applied for in-situ determination of the main parameters of the ionospheric plasma[21], e.g. electron temperature, electron density, ion temperature, ion partial densities; certain instruments even enable the bulk plasma (ion) motion or the individual ion temperatures to be mea-

[17] Kopp, E., Eberhardt, P., Herrmann, U. (1978): Space Res. *XVIII*, 245

[18] Kasprzak, W.T., Krankowsky, D., Nier, A.O. (1968): J. Geophys. Res. *73*, 6765; von Zahn, U. (1967): J. Geophys. Res. *72*, 5933; Ackermann, M., Simon, P., von Zahn, U., Laux, U. (1974): J. Geophys. Res. *79*, 4757

[19] A solution using ion and neutral composition data, and theory is specified in: Mayr, H.G., Bauer, P., Brinton, H.C., Brace, L.H., Potter, W.E. (1976): Geophys. Res. Lett. *3*, 77

[20] Sometimes the analyzer peak corresponding to O_2^+ or NO^+ is used to deduce the original density of atomic oxygen O. See for example: Lake, L.R., Krankowsky, D. (1975): Geophys. Res. Lett. *2*, 245

[21] Serbu, G.P. (1965): Space Res. *V*, 564; Knudsen, W.C. (1966): J. Geophys. Res. *71*, 4669; Hanson, W.B., Sanatani, S. Zuccaro, D., Flowerday, T.W. (1970): J. Geophys. Res. *75*, 5483

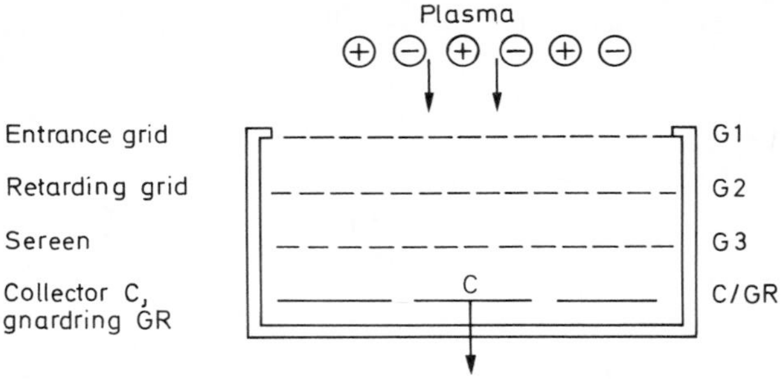

Fig. 8. Scheme of the planar retarding potential analyzer (RPA) flown aboard the AEROS satellites[22] [read 'guardring' at left hand bottom]

sured. Spherical RPAs are sometimes used for electron measurements. Planar ones have a wider application since they may be applied to electrons or ions according to the sign of applied voltages. An RPA with a large aperture[22] is shown in Fig. 8.

We shall consider the instrument in the ion mode, i.e. with a positive retarding voltage, since this is the type used for composition measurements. No acceleration of the ions is applied since these enter the instrument with vehicle velocity. One needs only to select the incoming ions according to their kinetic energy. Seen from the vehicle all ions have the same linear velocity, but different kinetic energy, according to their mass. The thermal energy part is small for ions (not at all, however, for electrons[23]), compared with that due to the supersonic vehicle velocity.

Since kinetic energy is mass-proportional, particle energies corresponding to different species obey the same rule. Energy selection is therefore obtained in the most simple manner by a retarding potential which can only be overcome by those ions which (at the instrument aperture) had an eV-energy greater than the retarding potential just applied. The device thus counts all ions above a certain minimum energy, i.e. it is an integrating measuring system. Differentiation with the sweeping retarding potential makes the determination of the abundance of the individual ion species possible.

In fact, the resolution is not very good compared with mass spectrometers. It is, for example, not possible to distinguish between the molecular ions N_2^+, NO^+, O_2^+, and it is often even difficult to distinguish between H^+ and He^+. However, the absolute accuracy of the density measurement is much better than for uncalibrated mass spectrometers. The only problem is the transparency of the entrance grid, which must be determined by simulation in the laboratory, or by comparison with redundant data from, for example, a radio frequency impedance probe [*26*][24].

The reduction of electron measurements with an RPA is quite straightforward: the temperature T_e is determined from the slope at the steepest part of the $\log I$ vs U characteristic (I being the current), see Fig. 9.

[22] Spenner, K., Dumbs, A. (1974): J. Geophys. *40*, 585; Spenner, K. (1981): Bundesmin. Forschung und Technologie, Forschungsber. W 81-007. Karlsruhe: Fachinformationszentrum Energie, Physik, Mathematik

[23] Rawer, K., Suchy, K. (1967): This Encyclopedia, vol. 49/2, p. 481

[24] Neske, E., Kist, R. (1974): J. Geophys. *40*, 593; see also: Rawer, K., Suchy, K. (1967): This Encyclopedia, vol. 49/2, p. 484

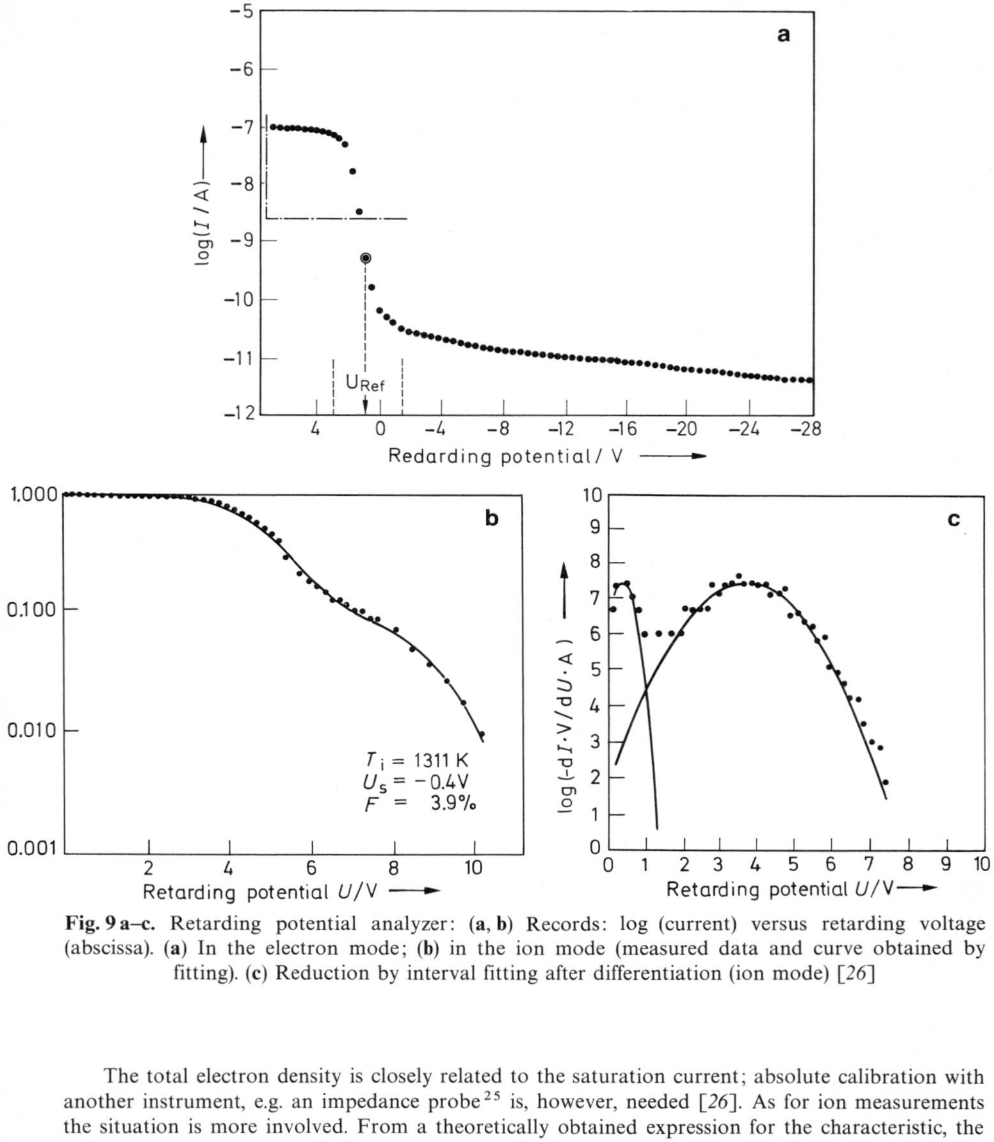

Fig. 9 a–c. Retarding potential analyzer: (**a, b**) Records: log (current) versus retarding voltage (abscissa). (**a**) In the electron mode; (**b**) in the ion mode (measured data and curve obtained by fitting). (**c**) Reduction by interval fitting after differentiation (ion mode) [*26*]

The total electron density is closely related to the saturation current; absolute calibration with another instrument, e.g. an impedance probe[25] is, however, needed [*26*]. As for ion measurements the situation is more involved. From a theoretically obtained expression for the characteristic, the unknown parameters, namely ion temperature, effective zero potential and individual ion densities, are determined by fitting with the observed characteristic (fortunately, the individual densities enter linearly into this procedure). Figure 9b shows a typical example. Another fitting method uses first differentiation of the characteristic and then individual (parabolic) fitting in U-intervals (Fig. 9c) adapted to the different species. This latter procedure is not applicable to complex characteristics.

Recent RPA designs used in planetary research have reached a high state of perfection, and automatic onboard data reduction is achieved in such instru-

[25] Neske, E., Kist, R. (1981): Bundesmin. Forschung u. Technologie, Forschungsber. W 81-008. Karlsruhe: Fachinformationszentrum Energie, Physik, Mathematik

Fig. 10. Advanced retarding potential analyzer flown in the VENUS-ORBITER mission of NASA[26]. View of the instrument (See [27] for results)

ments (Fig. 10; a few results are shown in Sect. 15, Fig. 166)[26]. The state of the art is described in [28] for plasma measuring devices and in [29] for neutral mass spectrometers.

4. Other measurements of neutral atmosphere parameters[0]

α) The very first rocket measurements in the upper atmosphere were made with *pressure gauges*[1]. Mechanical membrane gauges were successfully used to measure stagnation (ram) and static pressure down to about 200 Pa. If membrane deflections were measured using an ionization transducer, pressures down to about 0.1 Pa could even be determined[1]. However, the range of small pressures is better covered by the different types of vacuum meters which must be chosen according to the desired height range [*67*/I] [*4*].

In a PIRANI gauge the heat conductivity of the air (which is proportional to the *pressure*) is measured by its cooling effect upon a heated wire. The

[26] Knudsen, W.C., Spenner, K., Bakke, J., Novak, V. (1979): Space Sci. Instr. *4*, 351

[27] Rawer, K., Spenner, K., Knudsen, W.C. (1982): Z. Flugwiss. Weltraumforsch. (ZFW) *6*, 147

[28] Goldstein, R., Neugebauer, M.M. (1983): Adv. Space Res. *2* (10), 271

[29] Niemann, H.B., Kasprzak, W.T. (1983): Adv. Space Res. *2* (10), 261

[0] This subsection partly follows a NASA report by S.J. Bauer [*4*]

[1] LaGow, H.E., Horowitz, R., Ainsworth, J. (1958): Ann. Géophys. *14*, 117; (1960): Space Res. *I*, 164

electrical resistance of the wire varies with temperature and can easily be telemetered. In the early V2 flights this device was applied in the pressure range from 300 to 0.4 Pa, i.e. in the strato- and mesosphere. In a rapidly ascending rocket the retardation in the system must be corrected for since it corresponds to a few kilometres in altitude.

Still lower pressures can be determined with ionization gauges. In these instruments ionizing collisions between electrically charged and neutral particles are artificially provoked; the ionizing effect depends on the probability of such collisions, which is a measure of the neutral air *density*.

A straightforward application of this principle is the *alphatron:* He^{++} ions emitted from a weak alpha source ionize the air in a vessel which has an opening of suitable size towards free space and which contains two suitably arranged electrodes. A voltage is applied to these, provoking a current which is fed to an impedance-transforming amplifier; the input impedance is extremely high, between 50 MΩ and 100 GΩ [67/I]. It is an advantage of this system that it does not suffer from inertia and covers a very large density range.

The classical ionization gauges commonly used as laboratory vacuum meters are also applied. A prototype of such devices is the classical triode in the space-charge limited regime; it is extremely sensitive to a few positive ions which are produced when electrons collide with a neutral molecule. To obtain a higher sensitivity one of the different possible methods for increasing the path length of the electrons is applied.

In the PHILIPS instrument this is done by a rather strong magnetic field parallel to the symmetry axis of the triode configuration. This system can also be used with a cold cathode[2], when it covers a range from about 100 mPa down to 10 μPa. Other versions of *ionization gauges* are the BAYARD-AL'PERT hot filament thermionic gauge[3,4] and the very sensitive magnetron type "Redhead" gauge, which was very often used in the USA[4]. While the first instrument has a linear response, the latter type achieves a much higher sensitivity, in a (non-linear) range from 20 μPa down to 20 nPa; see Fig. 11.

A certain difficulty with all these gauges is the fact that the sensitivity depends on the effective cross section of the neutral gas against collisional ionization. This is particularly small for noble gases, with their high ionization potential. So the ratio of sensitivities for He to N_2 is 0.17. The main constituent at the level where the measurement is taken should therefore be known. This means in practice the N_2 over O numerical density ratio.

It is not easy to take account of another influence, namely that of the air flow around the vehicle, which is connected to that of the flow through the aperture of the measuring volume. On a spinning vehicle true ambient pressure is found at those instants for which the rocket velocity is perpendicular to the normal on the orifice plane; only at these instants is there no ram or rarification. These points are reached one-fourth of a spin cycle before and after the pressure maximum; Fig. 12 shows that on a satellite this is not always easy since the ram pressure exceeds the ambient pressure by an order of magnitude and outgassing may provoke a non-negligible disturbance. (For satellite velocities the pressure right in the wake should only be 10^{-20} of the ambient one.) The

---

[2] Danilin, B.S., Mihnevič, V.V., Repnev, A.I., Švidkovskij, E.T. (1957): Usp. Fiz. Nauk *63*, Nr 1b; Poloskov, S.M. (1960): Space Res. *I*, 95

[3] Newton, G.D., Pelz, D.T., Miller, G.E., Horowitz, R. (1963): Trans. 10th Nat. Vacuum Symposium, p. 208. New York: McMillan

[4] Spencer, N.W., Bogess, R.R., LaGow, H.E., Horowitz, R. (1959): Amer. Rocket Soc. J., 290

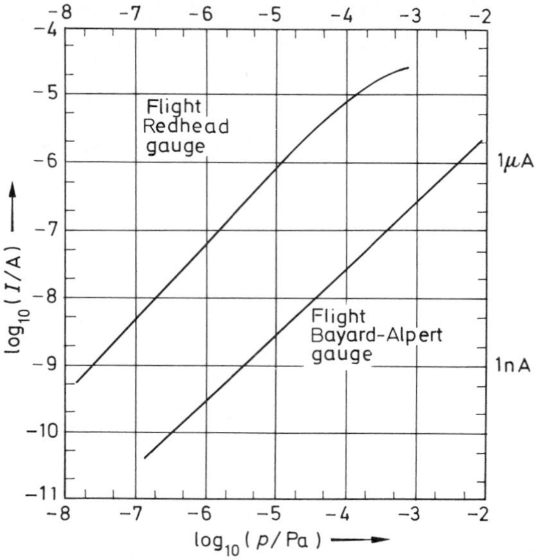

Fig. 11. Calibration curves (current I vs pressure p) of two ionization gauges showing the high sensitivity of the "Redhead" (magnetron type) device as compared with "Bayard-Alpert" instrument. Valid for molecular nitrogen[0]

following formula is usable provided the orifice is large enough to avoid hysterisis, and outgassing has become small enough; otherwise it is not valid near and inside the wake. Then the ratio of the indicated pressure p_g to the ambient pressure, p_a, is given by[5]

$$\frac{p_g}{p_a} = (T_g/T_a)^{1/2} \cdot f(S) \tag{4.1}$$

where $S = v_1/V_T = v_1/(2kT_a/m)^{1/2}$ and (from the Maxwellian distribution, see Sect. 1ε) $f(S) = \exp(-S^2) + S \cdot \pi^{1/2}(1 - \mathrm{erf}(S))$, erf being the Gaussuan error function. v_1 is the component of the vehicle velocity parallel to the normal on the orifice plane; V_T is the characteristic thermal velocity.

The first factor in Eq. (4.1) takes account of the fact that the temperature inside the gauge, T_g, is not equal to the ambient temperature, T_a, but is determined by the wall temperature which the incoming molecules reach after a few collisions.

Note that in the neighbourhood of the transition into the wake the function strongly depends on T_a; this gives a means for measuring the ambient temperature (see Sect. 4γ below). The ambient number density can be calculated after[6]

$$n_a = n_g \cdot (T_a/T_g)^{1/2} / f(S) \tag{4.2}$$

---

[5] Newton, G.P., Horowitz, R., Priester, W. (1955): Planet. Space Sci. *13*, 599

[6] Horowitz, R. (1966): Ann. Géophys. *22*, 1

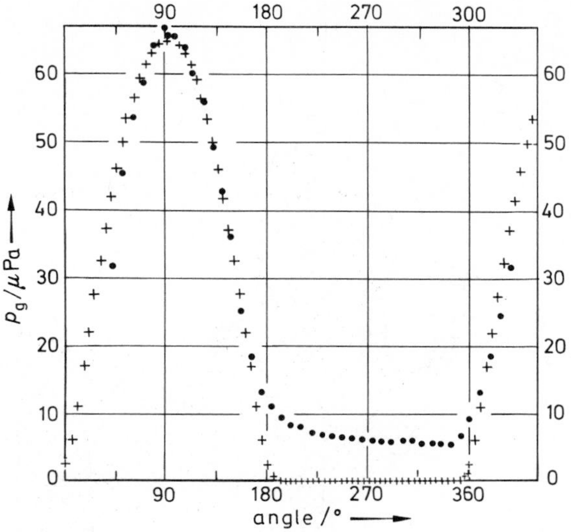

Fig. 12. "Redhead" gauge pressure p_g as function of the roll angle. Dots: values measured aboard a spinning satellite; crosses: theoretically computed values (molecular nitrogen). The difference is due to outgassing from the vehicle skin [4]

and the ambient mass density, ρ_a, from the spin modulation (Fig. 12)

$$\rho_a = (p_g(0°) - p_g(180°))/\pi^{1/2} v_1 V_T. \tag{4.3}$$

On rockets the measurement of the ambient pressure is often resolved by positioning the measuring place outside the disturbed flow, i.e. in front of the vehicle. For not too low pressure (i.e. with the orifice limitations given by the device) this can be achieved with a Pitot tube [7]. With recent improvements of this method measurements could even be achieved in the height range 50–120 km [8].

β) Direct measurement of the *mass density* were first obtained with a microphone gauge [9].

In this instrument the interaction (i.e. the momentum transfer) between a microphone ribbon and the ambient atmosphere is observed by the ram pressure at the ribbon, p_r, which is proportional to the mass density ρ:

$$p_r = K \cdot \rho \cdot v^2 \tag{4.4}$$

where v is the vehicle velocity and K an accommodation constant of the order of 1. Measurements up to an altitude of 550 km have been performed with this device [9].

The most important modern type of measurement is that with true accelerometers. In such an instrument a mass M is elastically suspended so that its deviation from zero position can be used as a measure of the acceleration force. In satellites instruments of this type measure directly the very small

[7] Horvath, J.J. (1972): Neutral atmosphere structure measurements by Pitot probe technique. Ann Arbor: Univ. of Michigan Report 05776-I-F

[8] Bäte, J., Becker, M., Niederlöhmer, U., Papanikas, D.G. (1977): J. Geophys. *44*, 147

[9] Sharp, G.W., Hanson, W.B., McKibbin, D.D. (1962): J. Geophys. Res. *67*, 1375

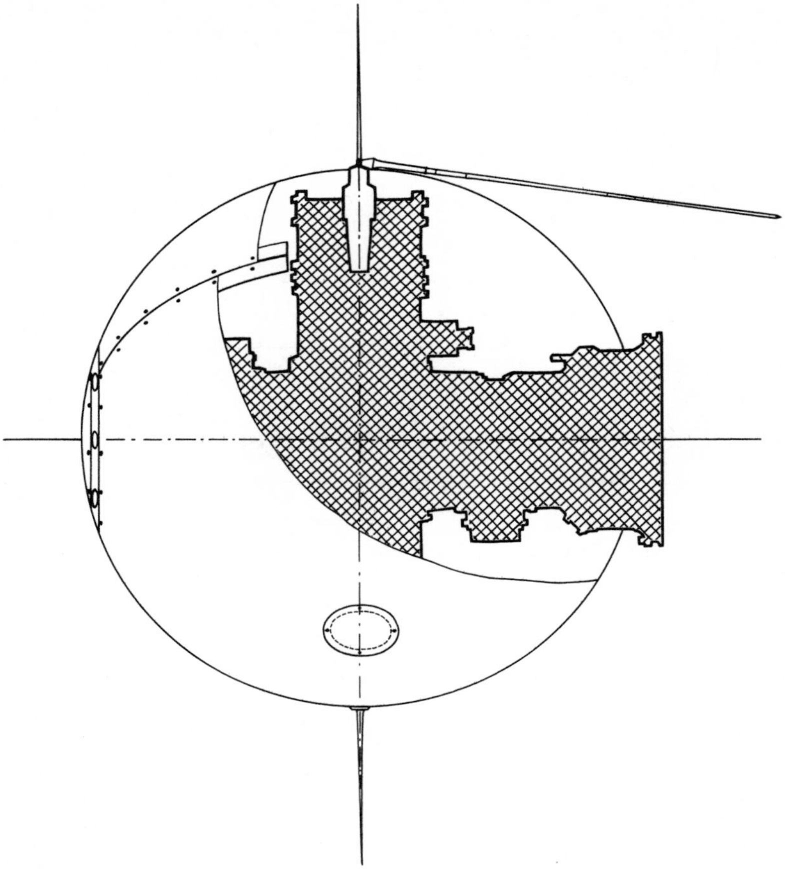

Fig. 13. The SAN MARCO drag balance system: spherical shell (partially cut) and internal mass (drum)[10]

deceleration which is due to air drag. While the principle was well known it took a long time before fully usable instruments appeared.

In the Italo-USA satellite series SAN MARCO such measurements were successfully carried out for the first time. The whole satellite is constructed as a measuring device[10].

It consists of a thin, metallic, spherical shell with all the heavy parts (more than 100 kg) arranged inside two concentric cylinders (the so-called "drum") so that their centre of gravity coincides with the centre of the shell. The latter and the drum are connected by elastic connections which are arranged as a "balance" system with the drum's rest position in the centre of the outer shell (Fig. 13). The displacement of the drum against the shell is the measured quantity. Thus, mass forces give no effect, but surface forces onto the shell do. Therefore, the system is sensitive only to drag forces. The "balance" gives a three-dimensional indication so that the direction of the drag is also determined. An absolute orientation is obtained with an astrometric device. It is important

[10] Broglio, L. (1967): Space Res. *VII*, 1135; (1968): Space Res. *VIII*, 90; (1969): Space Res. *IX*, 547

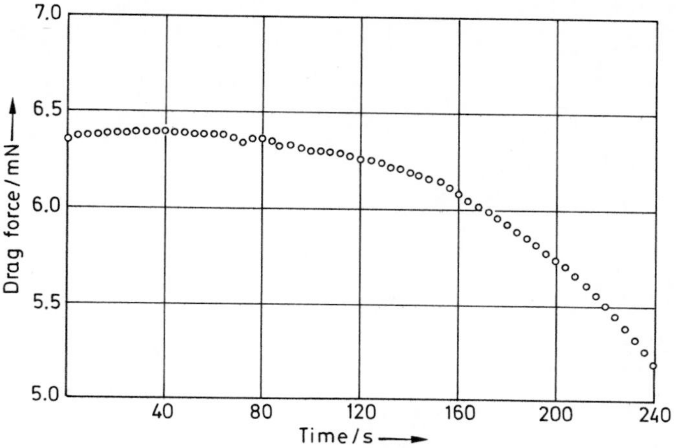

Fig. 14. Variation of the density along a satellite path of about 2,000 km measured 15 May 1967 by SAN MARCO near Nairobi/Kenya at an altitude of 207 km[10]

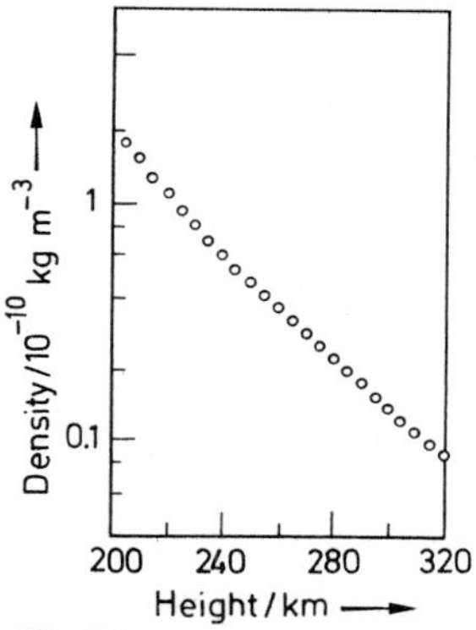

Fig. 15

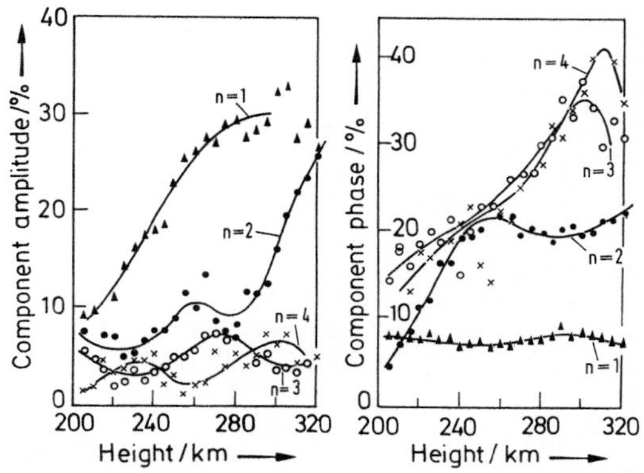

Fig. 16

Fig. 15. Diurnal average density profile after the SAN MARCO-3 accelerometer measurements[11]

Fig. 16. Average quiet day relative amplitudes (*left*) and phases (*right*) of the four first diurnal harmonics ($n=1 \ldots 4$ with periods of 24, 12, 8 and 6 h) determined after SAN MARCO-3 measurements[11]. Amplitude percentage values refer to the diurnal average (shown in Fig. 15), phase percentages to the relevant value of n

that the system is stiff enough, i.e. its fundamental frequency high enough against that of the rotation around the centre of mass. The maximum displacement is therefore quite small (1 μm). The measurable forces range between 50 mN and 250 mN, with a sensitivity of 10 μN.

The big advantage of the system is that it enables a continuous and direct record of atmospheric density to be made. Figure 14 shows typical (short-term) results about the directly measured drag force[10]. In deducing densities from the drag force the reasoning made in Sect. 2 can be applied. Figure 15 displays average density results obtained during a longer measuring period of satellite SAN MARCO-3[11]. Observed quiet day diurnal variations were fitted with

[11] Broglio, L., Arduini, C., Buongiorno, C., Ponzi, U., Ravelli, G. (1976): J. Geophys. Res. *81*, 1335 and Space Res. *XVI*, 203

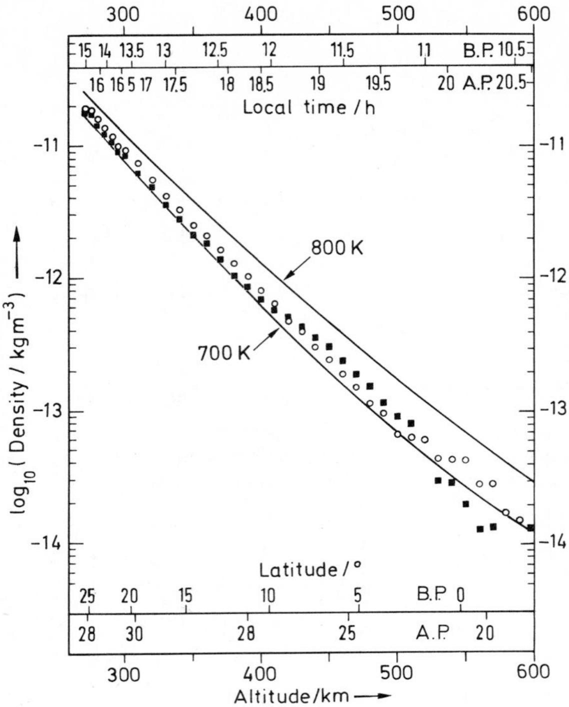

Fig. 17. Density profile deduced from CACTUS accelerometer measurements for one pass (29 June 1975, around 06 UT). The CIRA-72 model densities [*12*] for 700 and 800 K exospheric temperature are shown for comparison. Open symbols designate measurements on the inbound part, bold symbols on the outbound part of the orbit[13]

a fourth-order harmonic development, with amplitudes and phases depending on the altitude (Fig. 16)[11].

Electrostatic accelerometer systems are more sensitive and have a shorter response time[12]. (A sample record is found as Fig. 82 in Sect. 10.) A very refined version is the French accelerometer called CACTUS[13].

This has a much smaller mass (around 500 g) than the mechanical systems described above. The mass is an inner sphere of 80 mm diameter inside an exterior hollow sphere of the same diameter, with only 170 nm spacing. Both surfaces are finished to 1 μm and are gold-covered; there is no contact, the inner sphere being held in its symmetry position by electric forces with the help of six electrodes, two for each of the three spatial axes. The force is proportional to the square of the applied voltage and the displacement is measured by the relevant change of the capacity between the electrodes.

These electrostatic systems are extremely sensitive. Figure 17 shows a density profile in the 250–600 km height range from one pass of a CACTUS

[12] Such instruments were first flown on Explorer-satellites No. 54 and 55 of NASA (USA)

[13] Barlier, F., Boudon, Y., Falin, J.L., Futanlly, R., Villain, J.-P., Walch, J.J., Mainguy, A.M., Bordet, J.P. (1977): Space Res. *XVII*, 341; Bernard, A., Gay, M., Mainguy, A.M., Juillerat, R., Walch, J.J., Boudon, Y., Barlier, F., Lala, P. (1978): Space Res. *XVIII*, 163; Barlier, F., Berger, C., Bordet, J.P., Falin, J.L., Futanlly, R., Villain, J.-P. (1978): Space Res. *XVIII*, 169

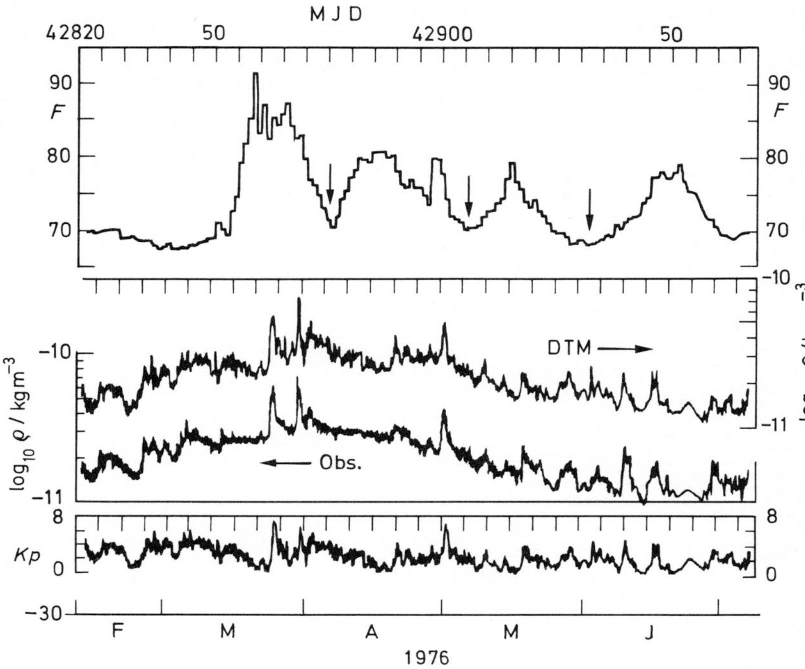

Fig. 18. Middle time-scale results (day-by-day values) of CACTUS accelerometer[13] measurements relative to 270 km during 5 months in 1976. The middle field shows the observed data and those computed according to activity figures from model DTM (=M1, [*62*]). For comparison: COVINGTON index F (top) and planetary magnetic index Kp (bottom). The satellite had a low-latitude (30° inclination) orbit[16]. (The long-term variation corresponds to that of the smoothed solar activity). The abscissa on top gives a continuous day count, the Mean Julian Date (MJD)

instrument. It achieves a very high resolution in a time (of only 2.7 s) so that rather small density structures of a few 10 km (e.g. gravity waves[14]) can be resolved[15]. Figure 18 shows an example with a larger time scale where characteristic peaks appear (in data at constant height) during periods of increased solar activity and/or magnetic activity[16].

Accelerometers have even also been used in rocket experiments[17]. The sensitive instruments were placed inside inflatible spheres and were only released after ejection (above the mesopause) of the spheres from the rocket, and inflation. The spheres then descended with a velocity depending on the neutral air drag (falling sphere method[18]); the corresponding deceleration was mea-

---

[14] Jones, W.L. (1976): This Encyclopedia, vol. 49/5, pp. 177–216

[15] Villain, J.-P. (1979): Space Res. *XIX*, 231

[16] Falin, J.L., Kockarts, G., Barlier, F. (1981): Adv. Space Res. *1* (12), 221

[17] Philbrick, C.R., Murphy, E.A., Zimmerman, S.P., Fletcher, E.T. Jr., Olasen, R.O. (1980): Space Res. *XX*, 79

[18] Smith, W.S. (1969): Meteorological Data from the falling sphere technique ... Washington D.C.: NASA Spec. Publ. No. 219

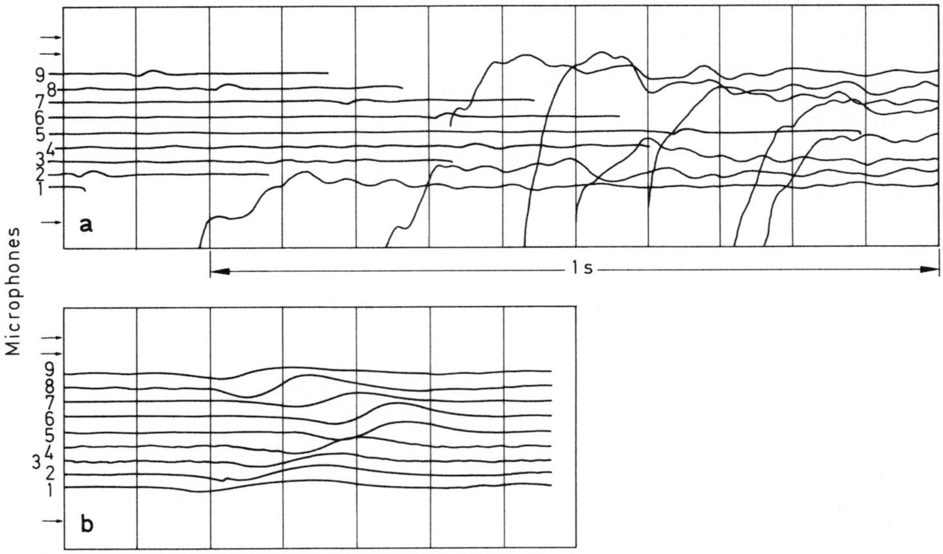

Fig. 19a, b. Sound records at the different ground stations: (**a**) of a 0.9 kg TNT explosion at 31.1 km altitude; (**b**) of 1.8 kg at 94.3 km [22]

sured. The data were used for checking existing models, and detecting turbulent regions [19].

γ) Direct *temperature measurements* are not so easy to carry out, so that in earlier days temperatures were inferred, mostly from pressure/density profiles. In the strato- and mesosphere *sound wave propagation* has long been used for determining temperatures via the velocity of sound [*10*] [*21*]

$$V_s^2 = \kappa k T / \bar{m} \tag{4.5}$$

κ is the ratio of specific heats at constant pressure and volume, k BOLTZMANN's constant, T temperature and $\bar{m}$ mean molecular weight. Since at these altitudes the composition is essentially that at ground there is no problem in determining κ and $\bar{m}$, so that T can unambiguously be derived from V_s. In the so-called grenade method [20] explosive charge are ejected from the rocket at different heights, and the arrival times are recorded at several suitably placed sound receiving stations on the ground. Account is taken of the increased speed in the neighbourhood of the explosion. Perfect synchronization between board and ground was achieved by the radio navigation system DOVAP. It is important to identify carefully the arrival times of the acoustic signal at the different stations [21]; see Fig. 19 [22].

[19] Zimmerman, S.P., Keneshea, T.J. (1976): J. Geophys. Res. *81*, 3187

[20] Stroud, W.G. Terhune, A.E., Venner, J.H., Walsh, J.R., Weiland, S. (1955): Rev. Sci. Instr. *26*, 427; Stroud, W.G., Nordberg, W., Walsh, J.R. (1956): J. Geophys. Res. *61*, 45; Ference, M., Stroud, W.G., Walsh, J.R., Weisner, A.G. (1956): J. Meteorol. *13*, 5

[21] Weisner, A.G. (1956): J. Meteorol. *13*, 30

[22] Stroud, W.G., Nordberg, W., Bandeen, W.R., Bartman, F.L., Titus, P. (1960): Space Res. *I*, 117

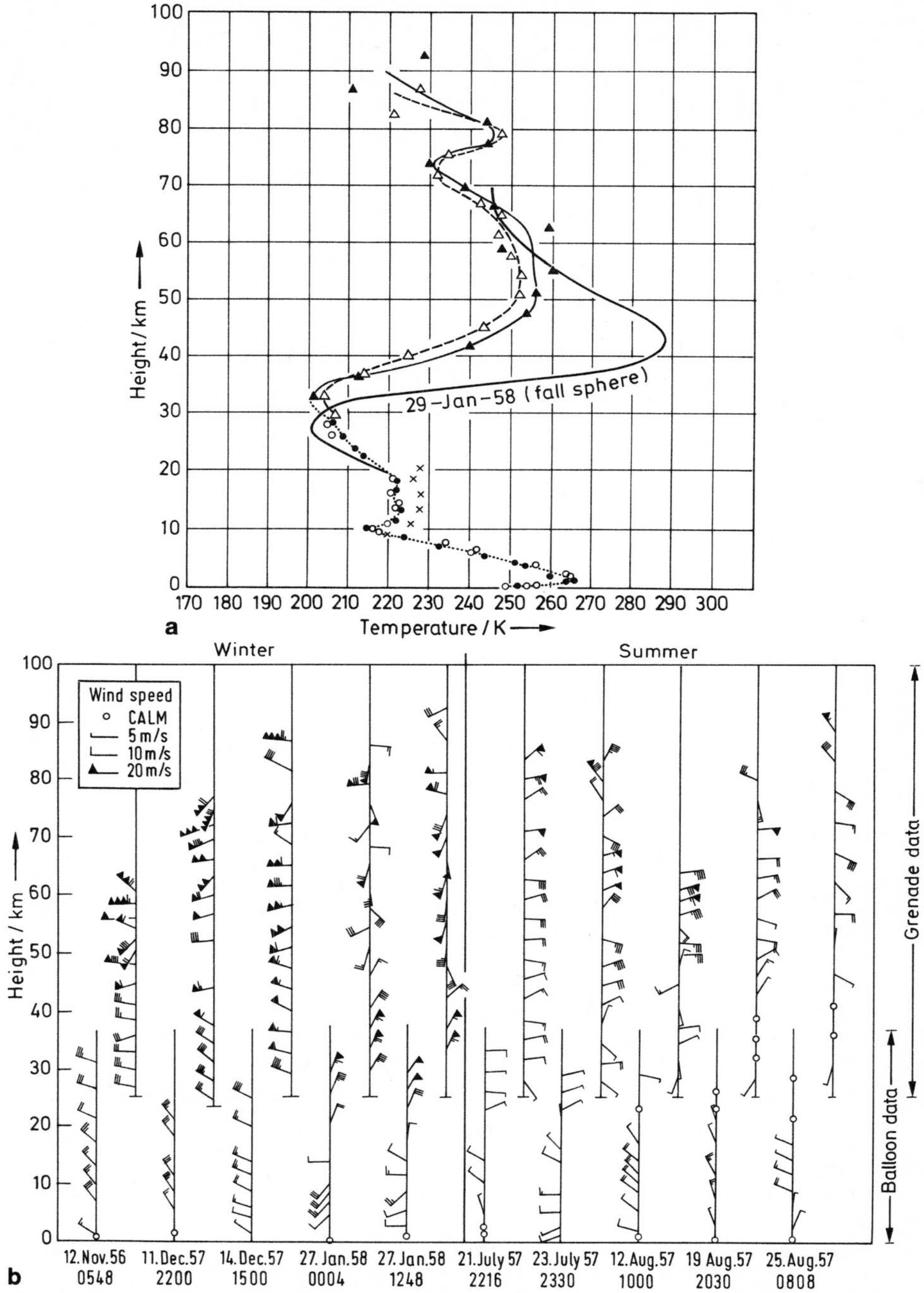

Fig. 20. (**a**) Temperatures profiles obtained by the grenade method (triangles), by meteorological balloon radiosonde (circles) and by the falling sphere method (bold curve, 29 Jan 1958). Crosses show radiosonde data 3 Feb 1958; all other data are from 27 Jan 1958. Open symbols are for midnight, bold symbols for noon[22]. (**b**) Winds up to 90 km observed with balloons (low altitudes) and with the grenade method between Nov 1956 and Jan 1958[22]

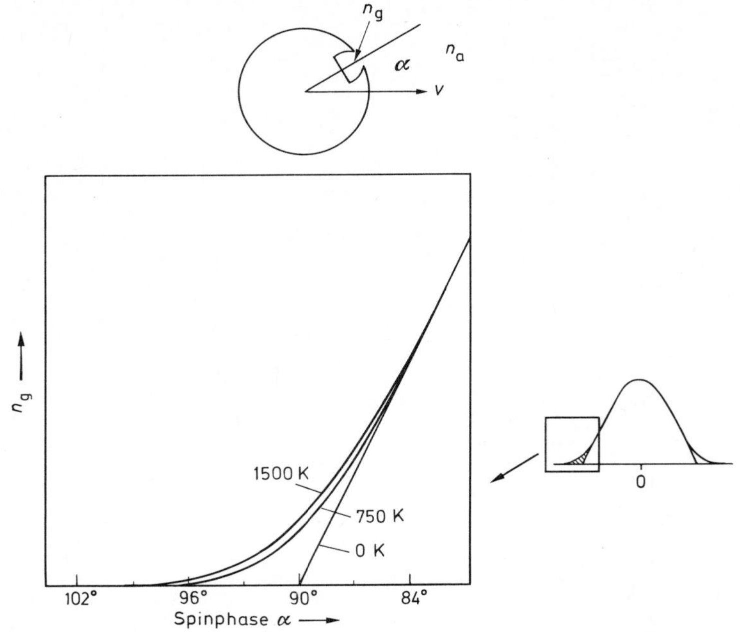

Fig. 21. Variation of the pressure in a measuring chamber with an opening towards outer space on a rotating satellite. Parameter: temperature of the environment. n_a and n_g, ambient and gauge density of the gas (the latter being indicated by the sensor); $\boldsymbol{v}$, satellite velocity vector; α, phase angle of rotation (zero = forward direction). The line for zero temperature corresponds to the chamber filling in the (hypothetical) absence of thermal motion [*26*]

On reduction of these data an initially assumed temperature profile is corrected by an optimum fit method. However, it is not possible to consider just the geometry since the sound propagation is influenced at each level by the respective wind. So the problem must finally be resolved with three a priori unknown profiles, namely those of temperature direction and velocity of wind (which is supposed to be horizontal).

Figure 20a shows temperature profiles determined with the grenade method, together with comparable profiles obtained up to 30 km by classical meteorological radiosondes (bimetal instruments), and one profile inferred from density measurements with the 'falling sphere' method (see [18] above). Figure 20b shows ten wind profiles determined by the grenade method[22].

After overcoming quite a few experimental problems direct determinations of the thermal velocity distribution[23] can now be performed and allow the temperature to be directly derived from it. Mass spectrometers (see our Sect. 3 and Sect. 19 in THOMAS' contribution in volume 49/6) were mostly used as indicating devices in a fixed mass condition, i.e. as mass filter[24].

The so-called "Neutral Atmosphere Temperature Experiment" (NATE)[25] is applicable in spinning satellites. It uses the fact that at the transition from

[23] Maxwellian distribution function. Seee Sect. 1ε, Eq. (1.16...18) and Fig. 2

[24] Hedin, A.E., Avery, C.P., Tschetter, C.D. (1964): J. Geophys. Res. *69*, 4637

[25] Spencer, N.W., Niemann, H.B., Carignan, G.R. (1973): Radio Sci. *8*, 284

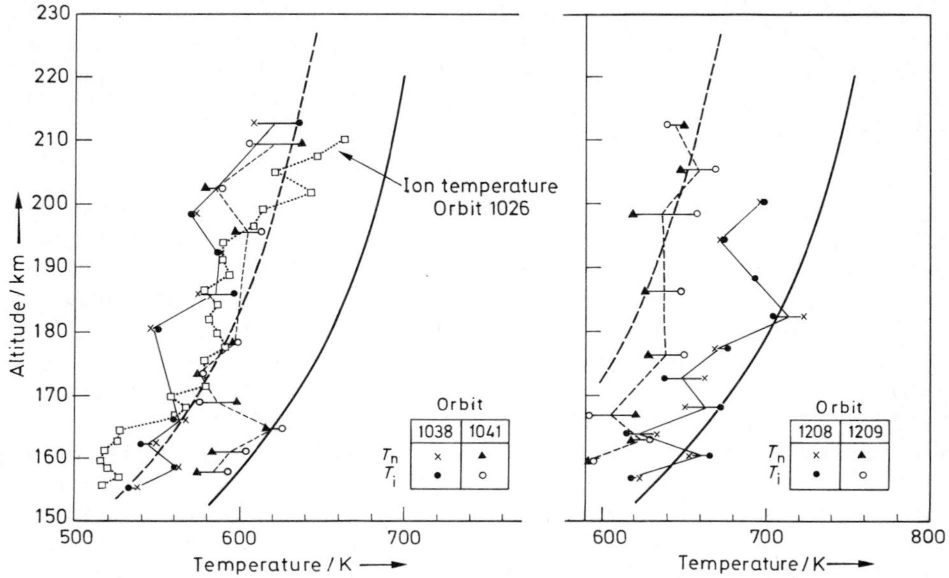

Fig. 22. Temperature profiles after measurements with the "fly-through" mode mass spectrometer aboard satellite AE-C compared with CIRA 72 (broken line, [*12*]) the MSIS-model (full line, [*20*]). Open squares are ion temperatures determined with an RPA (see Sect. 3b)[31]

the front to the rear side of the vehicle a sharp decrease of pressure occurs (see Fig. 12). If there is no outgassing effect the steepness of the pressure decrease in the wake region follows Eq. (4.1) and the ambient temperature T_a can be derived from the width of the observed transition. Aboard a spinning satellite this can be directly seen with a quick pressure indicator, the input opening of which lies an the equator of the satellite, so that the angular width can be measured as a time (since the spin period is known); see Fig. 21 [*26*].

An instrument of this type was first tried out aboard the SAN MARCO-3 satellite[26], then on AEROS-A. An improved system on AEROS-B was the first to deliver usable data with ± 50 K accuracy, at least for altitudes below 300 km and in the absence of magnetic activity[27]. An instrument flown on the AE-C satellite gave satisfactory agreement with incoherent scatter observations of ion temperature[28].

Another temperature-measuring device aboard the last-mentioned satellite uses a particular "open source" mass spectrometer (see Sect. 3), which has a special mode in which the source voltages were adjusted to discriminate against all ions that do not have the satellite velocity, i.e. against those due to outgassing or to interaction with the vehicle surfaces[29]. Focusing was arranged so as to suppress large off axis particles so that the response of this instrument

---

26 Spencer, N.W., Pelz, D.T., Niemann, H.B., Carignan, G.R., Caldwell, J.R. (1974): J. Geophys. *40*, 613

27 Chandra, S., Spencer, N.W., Krankowsky, D., Lämmerzahl, P. (1976): Geophys. Res. Lett. *3*, 718

28 Hedin, A.E., Spencer, N.W., Hanson, W.B., Bauer, P. (1976): Geophys. Res. Lett. *3*, 469

29 Nier, A.O., Potter, W.E., Kayser, D.C., Finstad, R.G. (1974): Geophys. Res. Lett. *1*, 197

to a monoenergetic molecular beam peaks sharply in the ram direction[30]. If, for example, the N_2 mass is selected, both N_2 density and temperature can be deduced from the response curve obtained on a spinning satellite, in the so-called "fly-through" mode[31]. A comparison of some data with the CIRA 72 [*12*] and the MSIS [*20*] models is shown in Fig. 22.

5. Incoherent scatter sounding[0]. Though this is a radio technique[1] originally invented for ionospheric measurements[2], it is now widely used for determining parameters of the neutral atmosphere[1]. (Compare also THOMAS, his Sect. 20 in volume 49/6.) It should be noted that the world wide network of incoherent scatter stations is rather restricted (see Fig. 23a).

a) Plasma measurements. Since the early days[3] instrumental capability and reduction techniques have been considerably improved. The main parameter obtained was originally the total power return from a certain level as provoked by a radio wave transmitted at ground with extremely high energy. The local plasma density can be derived from the power return ratio[3]. The returned spectrum has since been analyzed in much detail and a greater number of parameters can therefore be deduced. However, it should be borne in mind that only plasma parameters can be obtained directly and as "measured" quantities.

α) The *return spectrum* from a given height level (chosen by the travel time) may be understood as to stem from a mixing process betweeen the incoming (monochromatic) wave and irregular fluctuations in the plasma which might be represented by a stochastic conglomerate of sound-like waves travelling through the plasma in all directions. Since the fluctuations are provoked by the thermal motions in the plasma, this low-frequency spectrum is rather broad. Now, since the incoming wave has a high field strength, non-linear interaction occurs with the sound-like spectrum and this creates a return wave[3] (Fig. 23b, c) the spectral width of which is given by the receiver bandwidth. To each frequency inside that band corresponds a certain Doppler-shift between transmitted and returned wave. Also by the geometry of the system a specific direction is given thus selecting a certain sound wave vector (Fig. 23c).

Unlike the usual conditions in neutral gases thermal equilibrium is not necessarily valid in the plasma, but electrons often have a much higher temperature than the ions. Therefore, the width of the acoustic wave spectra should be different for ions and electrons. Since the scattering effect on the

[30] French, J.B., Reid, N.M., Nier, A.O., Hayden, J.L. (1975): Amer. Inst. Aeron. Astronautics J. *13*, 1641

[31] Kayser, D.C., Potter, W.E. (1976): Geophys. Res. Lett. *3*, 455; Kayser, D.C., Breig, E.L., Power, R.A., Hanson, W.B., Nier, A.O. (1979): J. Geophys. Res. *84*, 4321

[0] An important part of the basic information used in this Sect. was presented by J.V. Evans at the 1978 COSPAR meeting held at Innsbruck

[1] Fejèr, J. (1963) in: Advances in upper atmosphere research. Landmark, B. (ed.), p. 265. Oxford: Pergamon Press (survey paper)

[2] Bowles, K.L. (1961): J. Res. Nat. Bur. Stand. *65D*, 1; Bowles, K.L., Ochs, G.R., Green, J.L. (1962): ibidem *66D*, 395 (original papers)

[3] Rawer, K., Suchy, K. (1967): This Encyclopedia, vol. 49/2, Sect. 42δ, p. 356

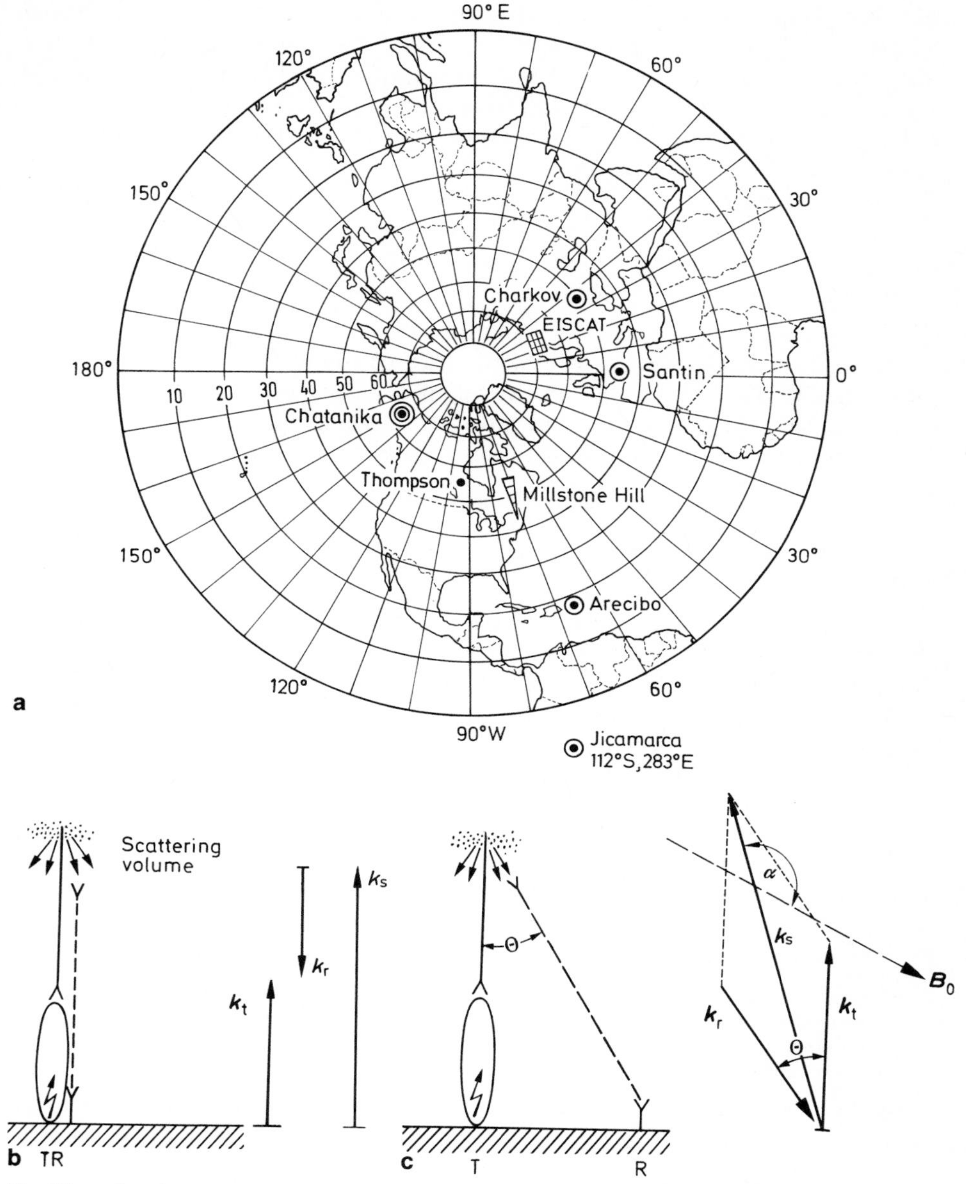

Fig. 23a–c. Incoherent scatter stations. (**a**) World map of incoherent scatter stations, 1981/1982. (**b**) Monostatic facility: the returned wave vector $\boldsymbol{k}_r$ is opposed and almost equal in absolute size to the transmitted one, $\boldsymbol{k}_t$; the sound wave vector $\boldsymbol{k}_s$ is twice that size. (Note that though it corresponds to the small frequency difference of the Doppler, it depends inversely on the very small velocity of sound such that the wave number is comparable with that of the radio waves.) (**c**) Bistatic facility: the relation between the three wave vectors is $\boldsymbol{k}_s = \boldsymbol{k}_t - \boldsymbol{k}_r$. ($\theta$ = angle between $\boldsymbol{k}_t$ and $-\boldsymbol{k}_r$)

incoming radio wave is due to the inhomogeneous and variable spatial distribution of the electrons, it was first concluded that the width of the return spectrum corresponded to the electron temperature, T_e. It was, however, found that this is not true but that the ion temperature, T_i, is decisive for the spectral width. This astonishing fact was explained by FEJER[4], who took into account the electrostatic forces which are provoked by space charges appearing when electrons and ions deviate locally from electrostatic equilibrium: The heavier ions pull the electrons with them so that finally the spatial spectrum corresponds to their temperature, T_i. Therefore, the width of the observed return spectrum corresponds to the average energy of the ions rather than that of the electrons.

β) So, first of all, the *ion temperature* can be found from the return spectrum (Fig. 24). The *plasma density*, N, might also be evaluated from the total energy return[3]. This measurement, however, needs some calibration, which is usually obtained by comparison with ionosonde measurements[5].

There exists, however, another method from which the plasma density can be found by a frequency measurement which does not need calibration. On both sides of the centre of the radiated frequency (f) appears a "plasma line" at frequency ($f \pm f_N$) where f_N is the 'plasma frequency'. Its relation with N is[6] $(f_N/\text{MHz})^2 = 80.62 \cdot 10^{-12}\ N/\text{m}^{-3}$. Apparently, suprathermal electrons must be present in a large enough quantity for the plasma lines to be excited; their relative strength increases with the deviation from thermal equilibrium. Unfortunately, the plasma lines are so weak that they are observed only rarely and under particular conditions, e.g. when the spatial distribution of the plasma is very inhomogeneous.

Measurements of the plasma line were first made in Puerto Rico[7], then in the USA[8], France[9] and Alaska[10]. The excitation of the line emission depends on local electron flux and is stronger for higher and for inhomogeneous electron density.

A detailed theory[11] after establishing the dispersion relation of plasma oscillations[12] shows that the precise condition for resonance is:

$$f^2 \approx f_N^2 + (f_B \sin\alpha)^2 + 12 \cdot \sin^2(\theta/2) \frac{k T_e}{m_e \lambda^2} \tag{5.1}$$

where f_B is the electron gyro-frequency[6], λ the radar wavelength, m_e electron mass and the angles as in Fig. 23c. If there are simultaneous observations under different angles, one may resolve not only after f_N (f_B being known by the magnetic field) but also after the electron temperature T_e. Initial measurements have been made with the tristatic station network in France[13]. Figure 25 shows typical, asymmetrical plasma line shapes. A comparison of peak electron densities derived from measured f_N values at two stations is seen in Fig. 26. Though this technique gives very precise results, its application is limited. Most routine observations continue to be made by the total energy return method.

γ) The shape of the main part of the spectrum depends on the *ratio* of *electron* temperature T_e *to ion temperature* T_i (see Fig. 35 of THOMAS' contri-

[4] Fejèr, J.A. (1960): J. Geophys. Res. *65*, 2635; Canad. J. Phys. *38*, 1114
[5] Rawer, K., Suchy, K. (1967): This Encyclopedia, vol. 49/2, Sects. 16…21, p. 199
[6] Rawer, K., Suchy, K. (1967): This Encyclopedia, vol. 49/2, Sect. 1, p. 2
[7] Yngverson, K.O., Perkins, F.W. (1968): J. Geophys. Res. *73*, 97
[8] Evans, J.V., Gastmann, I.J. (1970): J. Geophys. Res. *75*, 807
[9] Vidal Madjar, D., Kofman, W., Lejeune, G. (1975): Ann. Géophys. *31*, 227
[10] Wickwar, V.B. (1978): J. Geophys. Res. *83*, 5186
[11] Hagfors, T., Lehtinen, M. (1981): J. Geophys. Res. *86*, 119
[12] Ginzburg, V.L., Ruhadze, A.A. (1972): This Encyclopedia, vol. 49/4, Sect. 34, p. 524
[13] Kofman, W., Lejeune, G., Hagfors, T., Bauer, P. (1981): J. Geophys. Res. *86*, 6795

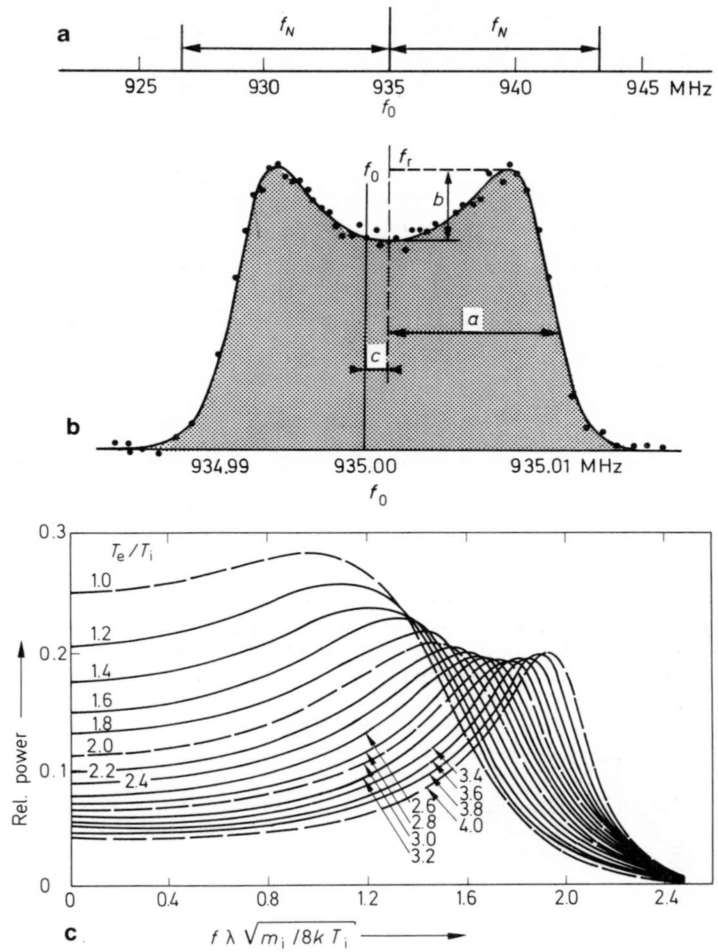

Fig. 24a–c. Return-wave spectrum. (**a**) Plasma frequencies distant from centre by $f_p(\equiv f_N)$ on either side (compare Fig. 26 below). (**b**) Main (central) return spectrum (compare THOMAS' Fig. 34, p. 81 in vol. 49/6): measured values (dots) and fitted curve. Total power (shadowed surface) is proportional to N. Width a depends on ion mass and ion temperature (see Fig. 35 on p. 82 of THOMAS' contribution in vol. 49/6 of the Encyclopedia). The Doppler shift c corresponds to the average bulk velocity of the plasma (in the spatial sounding range) [*17*]. (**c**) Spectra computed for one ion species of mass m_i and different ratios of T_e/T_i (electron to ion temperature = parameter of the curves) after[4] [see: EVANS, J.V., LOEWENTHAL, M. (1964): Planet. Space Sci. *12*, 915]

bution in vol. 49/6, p. 82). Figure 24c gives a set of theoretically computed shape curves depending on this ratio. If one has one unique ion species, or if one ion species is largely predominant, one may obtain this ratio, and T_e, by fitting with these curves. However, in many cases more than one positive ion species is present. Then the shape is also influenced by their mass ratio. In practice, one normally knows the two most prominent species and by fitting with two unknowns determines both ratios. T_i being known (see above), the

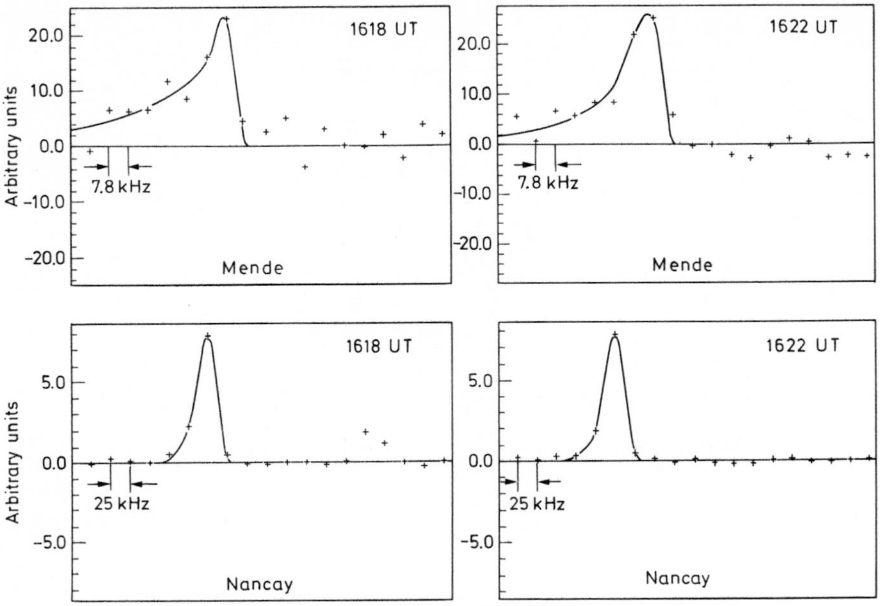

Fig. 25. Examples of plasma line shapes measured at the F2-peak (best fitting spectra). French receiving stations Mende and Nancay distant by 100 and 300 km respectively from the transmitter at St. Santin. Abscissa: frequency scale (arbitrary zero); ordinate: return signal (arbitrary units)[13]

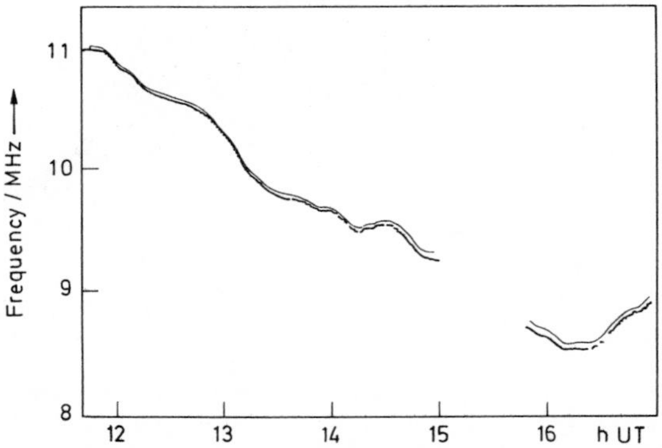

Fig. 26. Simultaneous plasma line frequency measurements at Nancay (bold) and Mende (thin line). 10 June 1980, altitude 325 km[13]

spectrum shape allows T_e and the ratio of the most prominent ion species to be determined.

δ) Finally, the centre of the return spectrum is quite often slightly shifted away from the radiated frequency (f_r instead of f_0, see Fig. 24b). This is a Doppler shift[14] due to the inward (positive) or outward (negative) motion of

[14] Gordon, W.E. (1958): Proc. Inst. Engrs. N.Y. *46*, 1824; see also Bremmer, H. (1958): This Encyclopedia, vol. 16, Sect. 92, p. 584

the plasma as whole. It gives us this *bulk velocity*, i.e. the projection of the plasma motion onto the vertical

$$w_D = -2c_0(f_r - f_0)/f_0. \tag{5.2}$$

This is so with vertical sounding (so-called monostatic arrangement, see Fig. 23b). For a bistatic facility (Fig. 23c) the velocity at angle $\theta/2$ is obtained from the Doppler shift. With three receiving stations one obtains three projections of the bulk plasma motion. The 'multistatic' incoherent scatter facility in southern France allows this measurement to be carried out[15].

No Doppler shift can be seen at the dip-equator with a monostatic facility since the plasma motions are horizontal there.

ε) *Summarizing* we may state that the most important parameters of the ionospheric plasma can be determined by the incoherent scatter technique[16]. These are: plasma density N, ion temperature T_i, ratio of the two main types of ions (provided these were correctly specified), electron temperature T_e and bulk plasma motion. At low altitudes, the collisional influence can also be determined[1]. Thus incoherent scatter stations are particularly efficient installations for ionospheric research purposes.

b) Conclusions concerning the neutral gas. Neutral parameters are not directly measurable with incoherent scatter techniques. One may, however, deduce from theory relations between neutral and plasma parameters and then determine from these the neutral ones as "inferred" parameters [*17*][17]. It is very important to bear in mind that this is not a direct measurement but a deduction depending on the applicability of the theoretical model which was used to derive the relation. This is not always possible, as shown by some examples discussed below.

α) The neutral gas is in *heat contact* with ions and electrons[17]. Due to the mass relations direct heat transfer to the neutrals occurs mainly via ion-neutral collisions. On the other hand the electrons are in fairly good heat contact with the ions since electron-ion collisions are of the Coulomb type and so have effective cross sections much larger than ion-neutral collisions. In daytime the electrons are heated up by excess energy from photo-ionization processes. First of all these create a "hot electron" population which is quite far from a Maxwellian (Fig. 2) energy distribution so that it can only approximately be characterized by an effective temperature (which would be of the order of 10^5–10^6 K). This population comprises only a small part of all electrons so that by energy transfer (in Coulomb collisions) the main electronic population is heated to a much lower temperature, T_e. Except for very special conditions where other heating influences are active, for heating by photo-ionization we have

$$T_e \geqq T_i \geqq T_n \tag{5.3}$$

[15] Bauer, P., Waldteufel, P., Vialle, C. (1974): Radio Sci. *9*, 77

[16] Evans, J.V. (1969): Proc. Inst. Electronic. Elec. Engrs. *57*, 496; Farley, D.T. (1970): J. Atmos. Terr. Phys. *32*, 597

[17] Evans, J.V. (1974): J. Atmos. Terr. Phys. *36*, 2183; Petit, M. in: Atmospheres of Earth and Planets. McCormack, B.M. (ed.) (1975), p. 159. Dordrecht: De Reidel

since the heat flux goes from hot electrons to electrons, from these to ions, and from these to neutrals. By day, there is a constant *local* heat transport in this direction and the stationary ion temperature T_i depends on this flux, but also on heat conductivities and capacities. Equating this[18] one obtains [17]:

$$T_i - T_n = \text{const} \cdot N_e^2 (T_e - T_i)/T_e^{3/2}. \tag{5.4}$$

The constant depends on the neutral partial densities and is said to be proportional[1] to:

$$[6.6 \cdot n(N_2)/\text{cm}^{-3} + 5.8 \cdot n(O_2)/\text{cm}^{-3} + 0.21 \cdot (n(O)/\text{cm}^{-3}) \cdot ((T_i - T_n)/\text{K})^{1/2}]^{-1}.$$

β) Since incoherent scatter observations yield N_e, T_e, and T_i one can "infer" the *neutral temperature*, T_n, from Eq. (5.4), provided the neutral number densities are taken from a suitable neutral atmosphere model, and spatial heat fluxes are negligible. This is not so at very high altitudes where heat and plasma transport along fieldlines becomes important and one has to use a considerably more involved relation[17,19]. On the other hand, at very low altitudes (around 100 km), heat contact is so good and the heat capacity of the neutrals so large that $T_e = T_i = T_n$ is valid. This holds also on many nights when the heat input is negligible. Anyway, in a middle range of heights up to 400 km Eq. (5.4) is applicable.

γ) The *exospheric temperature*, T_∞, is the most important parameter to be determined for comparison with neutral models (see Sects. 10–12). Two methods can be used to this end.

(i) T_n is inferred at a certain number of levels and a Bates profile is used as a fitting function between these determinations. This profile is given by[20]

$$T_n(z) = T_\infty - (T_\infty - T_n(z_0)) \exp(-s \cdot (z - z_0)) \tag{5.5}$$

and the thermospheric shape parameter s is assumed to be height independent ($0.2\,\text{km}^{-1}$ is an often used estimate). The lower boundary z_0 is often 120 km and $T_0 \equiv T(z_0)$ is then[21] about 355 K.

(ii) A normally well-justified simplification of Eq. (5.4) neglects the molecular species [$n(N_2) = n(NO) = n(O_2) = 0$] since above 300 km atomic oxygen is largely predominant (except for magnetically disturbed periods). With the remaining dependence on $n(O)$ the height dependence of the constant in Eq. (5.4) simplifies. The Bates profile is further used after Eq. (5.5) and an $n(O)$ profile deduced from this by assuming diffusive equilibrium for oxygen atoms; see Eq. (1.8). Taking a reference altitude z_0, not far from the turbopause, above which this assumption is valid to a first approximation, $n(O)$ at z_0 can be taken with good accuracy from neutral models. Solving by regression analysis, one obtains not only T_∞ but also the shape parameter s and $n(O)$ at the reference level; in Figs. 27a–c some results concerning s are presented[22]. Finally, if the data are consistent enough one may even also determine T_0[23].

---

[18] Carru, H., Petit, M., Vasseur, G., Waldteufel, P. (1967): Ann. Géophys. *23*, 455; Nisbet, J.S. (1967): J. Atmos. Sci. *24*, 586; Lejeune, G., Waldteufel, P. (1970): Ann. Géophys. *26*, 223; Waldteufel, P. (1971): Ann. Géophys. *27*, 167; Lejeune, G. (1972): Ann. Géophys. *28*, 15

[19] Bilitza, D. (1975): J. Atmos. Terr. Phys. *37*, 1219; Bilitza, D., Thiemann, H. (1981): Kleinheubacher Berichte *25*, 237. See also Annex A, p. 525, in this volume

[20] Bates, D.R. (1959): Proc. Roy. Soc. (London) *A 253*, 451

[21] Salah, J.E., Evans, J.V. (1973): Space Res. *XIII*, 267

[22] Oliver, W.L. (1977): Trans. Am. Geophys. Un. EOS *58*, 701

[23] Bauer, P., Waldteufel, P., Alcaydé, D. (1970): J. Geophys. Res. *75*, 4825; Alcaydé, D. (1979): Space Res. *XIX*, 211; Oliver, W.L. (1983): Adv. Space Res. *3* (1), 113

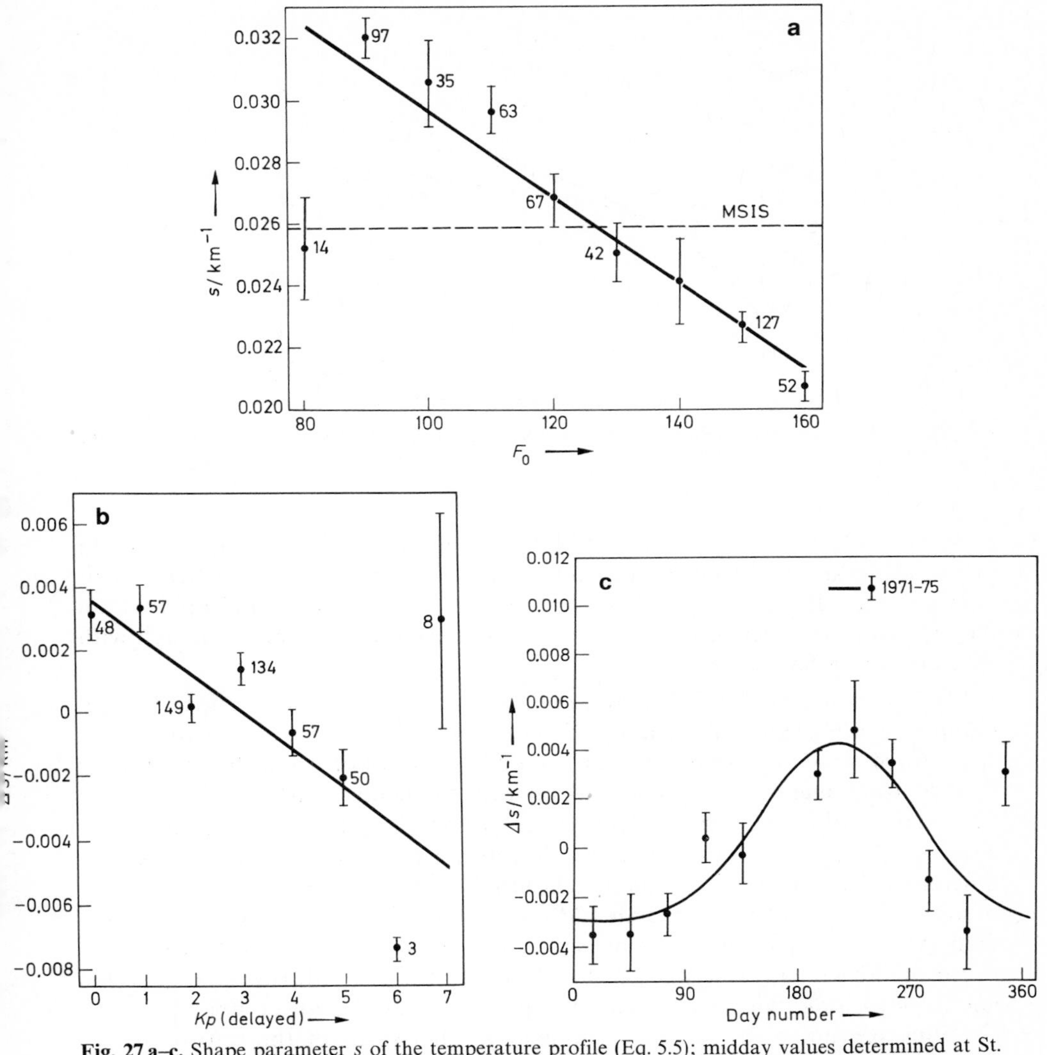

Fig. 27 a–c. Shape parameter s of the temperature profile (Eq. 5.5); midday values determined at St. Santin, France (Parameter: number).[22] (**a**) Variation with monthly averaged COVINGTON index F_0 (obtained from solar radio emission on 10.7 cm wavelength). (**b**) Dependence upon magnetic activity (planetary magnetic index Kp) after delaying by 6 h. (**c**) Seasonal variation from observations between 1971 and 1975, suppressing other influences as shown above. Bold curve: average over 5 years

In all these determinations the accuracy of the ion temperature, T_i, is decisive. Errors may be due to incorrect assumptions about the ions (e.g. if H^+ is present when O^+ alone is considered). Any kind of heat transfer other than that considered in Eq. (5.4) may, if appreciable, lead to an erroneous value of T_n. In the presence of very strong neutral winds (of more than 200 ms^{-1}) or

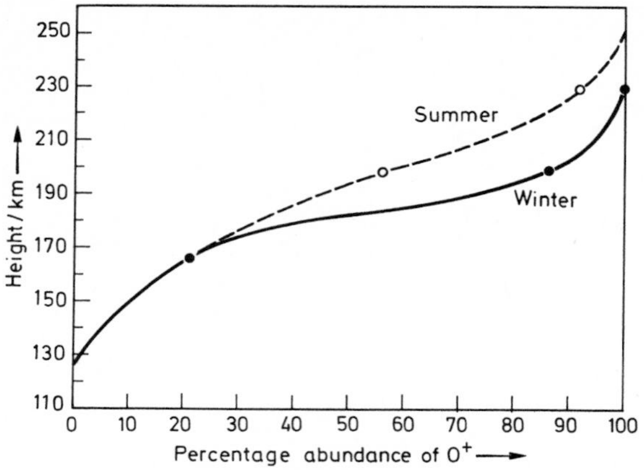

Fig. 28. Average summer and winter ionic composition profiles based on average values of the ratio $N(O^+)/N_e$ measured at Millstone Hill (USA)[30]

strong electric fields[24], frictional heating of the ions becomes important. At night, heat transport from the top side is not always negligible, particularly at high latitudes. If this is not so the electrons may cool down more rapidly than the ions and the order of the temperatures can even be inverted[25], though with differences of a few tens of K only[1].

δ) The *neutral density determinations* are apparently less certain[1]. The reliability of $n(O)$ values determined by regression is still under discussion[23,26], since these show a considerably larger diurnal variation than do the models derived from mass-spectrometric in-situ measurements[1]. This may be due to frictional heating, but this influence should only explain part of the difference[27]. Also the heat exchange coefficients may be in error and somewhat different ones were used in different groups[22,27]. Nevertheless, the abundance of atomic oxygen at 475 km, 14 h LMT was shown[28] to be substantially lower than given by the (meanwhile replaced) CIRA 1965 model [*11*] (see Sect. 10).

In the height range between 130 and 300 km there is a drastic change in the *ionic composition*, with molecular ions prevailing at the lower boundary and O^+ ions prevailing at the upper boundary. The width of the return spectra depends on the ionic composition and it is difficult to distinguish this influence from that of the temperatures. By modelling the variation of one of the latter a regression solution for the ratio of atomic to molecular ions can be obtained and, finally, by ion chemical relations, the $n(O)/n(N_2)$ ratio[29]. In these determinations different assumptions were used concerning the temperature balance and the applicability of Eq. (5.5). A large seasonal variation of the ratio

[24] McClure, J.P. (1971): J. Geophys. Res. *76*, 3106
[25] Mazaudier, C., Bauer, P. (1976): J. Geophys. Res. *81*, 3447
[26] Alcaydé, D., Bauer, P., Jaeck, C., Falin, J.L. (1972): J. Geophys. Res. *76*, 7814
[27] Alcaydé, D., Bauer, P. (1977): Ann. Géophys. *33*, 305
[28] Mahajan, K.K. (1971): J. Geophys. Res. *76*, 4621
[29] Evans, J.V., Oliver, W.L., Salah, J.E. (1979): J. Atmos. Terr. Phys. *41*, 259

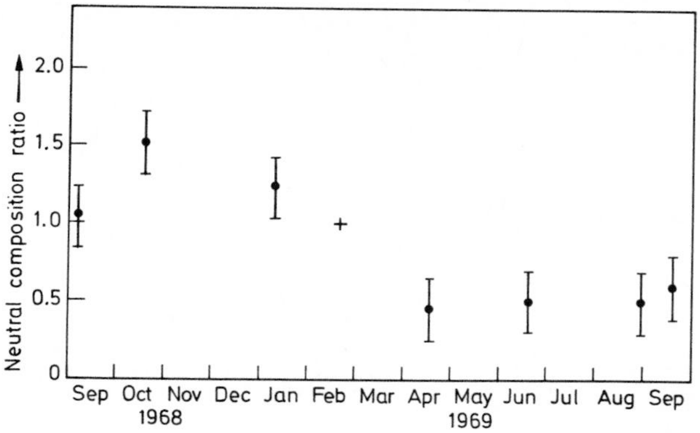

Fig. 29. Annual relative variation of the ratio $n(\mathrm{O})/n(\mathrm{N_2})$ at height 198 km, inferred from $N(\mathrm{O^+})/N(\mathrm{N_2^+})$ spectral measurements. Ordinate = ratio to value of 19 Feb 1969 (as arbitrary reference). Solar flux dependence was eliminated. Millstone Hill (USA) measurements[30]

was obtained[30], supposing Eq. (5.4) to apply and $T_n = T_i$. Figure 28 shows the basic (averaged) measurements of the ionic composition from which the neutral ratio $n(\mathrm{O})/n(\mathrm{N_2})$ was inferred using photochemical theory[31]. It shows a considerable seasonal variation visible in Fig. 29. Compare also Sect. 9 below.

ε) At very low altitudes *collisions* affect the return spectrum. With a suitable theory their effect could be determined. It depends on the parameter

$$\nu_{in}/\sqrt{2}\,k_s\,V_i; \qquad V_i = (k\,T_i/m_i)^{1/2}, \tag{5.6}$$

where ν_{in} is the ion-neutral momentum transfer collision frequency (see SUCHY, in this volume, Sect. 4γ, p. 11). Since FEJÈR (1960)[4] the relevant theory has been developed and improved[32–34].

Measurements were first made in Puerto Rico with the large 430 MHz radar[35]. A particular, simplified autocorrelation technique[36] was applied and the most recent fluid equation approach was used at the interpretation. Some results obtained at altitudes between 75 and 100 km are shown in Fig. 30a.

Doppler spectra found with this method, as well as electron density and collision frequency profiles are displaid in Figs. 30, 31[37]. Applying kinetic theory these latter were even used to deduce a neutral number density profile (top scale in Fig. 30b) down to 70 km. More recent determinations[38] with the

[30] Evans, J.V., Cox, L.P. (1970): J. Geophys. Res. *75*, 159
[31] Cox, L.P., Evans, J.V. (1970): J. Geophys. Res. *75*, 6271
[32] Dougherty, J.P., Farley, D.T. Jr. (1963): J. Geophys. Res. *68*, 5473
[33] Tannenbaum, B.S. (1968): Phys. Rev. *171*, 215
[34] Hagfors, T., Brockelmann, R.A. (1971): Phys. Fluids *14*, 1143
[35] Tepley, C.A., Mathews, J.D. (1978): J. Geophys. Res. *83*, 3299
[36] Hagen, J.B., Farley, D.T. Jr. (1973): Radio Sci. *8*, 775
[37] Tepley, C.A., Mathews, J.D., Ganguly, S. (1981): J. Geophys. Res. *86*, 1130
[38] Schlegel, K., Kohl, H., Rinnert, K. (1980): J. Geophys. Res. *85*, 710

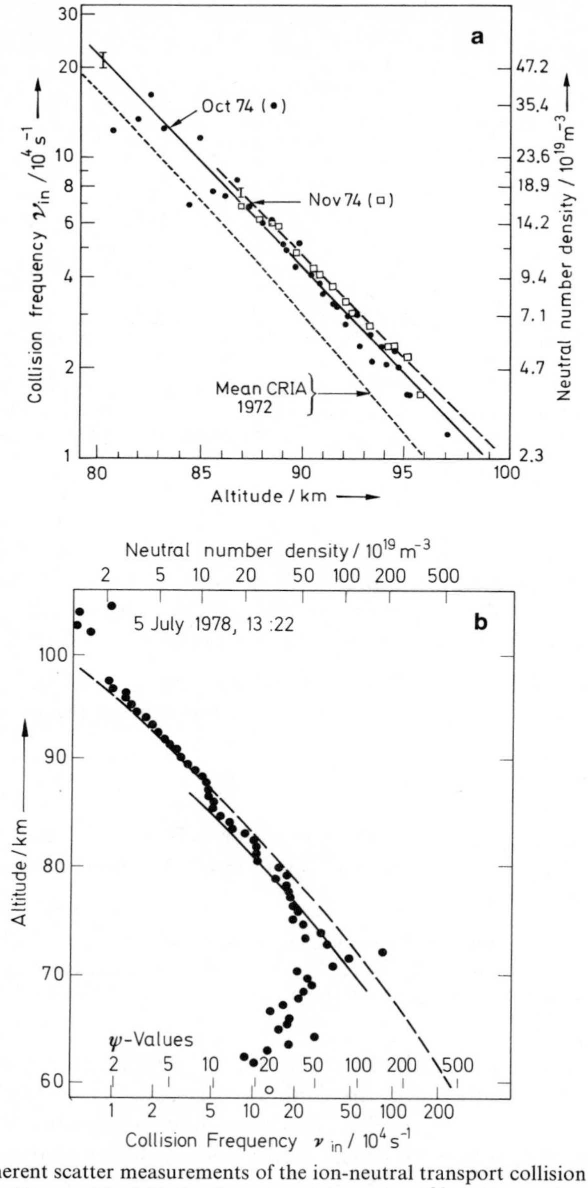

Fig. 30a, b. Incoherent scatter measurements of the ion-neutral transport collision frequency ν_{in} made with the Arecibo (Puerto Rico) Radar (wavelength $\lambda_0 = 0.7$ m)[35]. (**a**) Monthly averages of ν_{in} with values deduced from CIRA-72 [*12*] for comparison. [142 measurements in October, average ion temperature 207(+19/−16) K; 259 measurements in November, 215(+16/−15) K]. (**b**) Ion-neutral transfer collision frequency ν_{in} (lower scale) and inferred neutral number density (upper scale) [same time]. ψ is the 'normalized transfer collision frequency' $= \nu_{in} k_s/(2kT_i/m_i)^{1/2}$, k_s being the scattering wave number $4\pi/\lambda_0$. The broken curve was deduced from CIRA-72 [*12*]

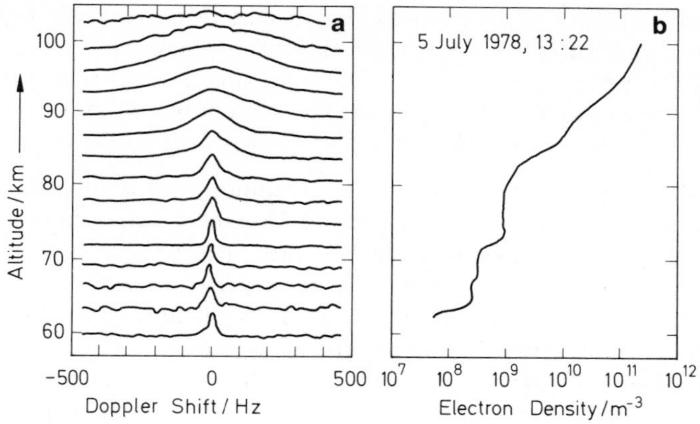

Fig. 31. (**a**) Example of Doppler spectra from which temperatures were deduced [noon measurements 5 Jul 1978]. (**b**) Electron density profile (obtained independently from the spectral measurements from observations before and after these) [same time]

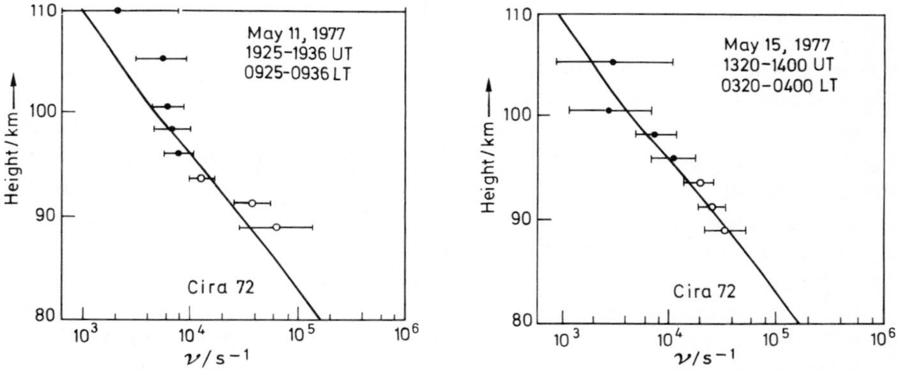

Fig. 32. Incoherent scatter measurements of ion-neutral transfer collision frequency determined by fitting of (i) three parameters (ν, T_e, T_i, dots) or (ii) two parameters (ν, $T_e = T_i$, open circles)[0]. CIRA-72 [*12*] for comparison (full curve)

Alaskan facility at Chatanika gave results (Fig. 32) which compared well with CIRA-72 [*12*].

The seasonal variations of density and temperature at heights up to 115 km were inferred from measurements of that type[39].

An interesting generalization of physical theories of incoherent scatter[40] takes account of the effects of thermally induced chemical fluctuations which are followed-up by such ionized species. These are no more negligible at heights below 70 km, where negative ions become important. The incoherent scatter cross section is then considerably increased in a ± 20 Hz band around the measuring frequency. This is one reason why current theory becomes obsolete below 70 km (compare Fig. 30b).

---

[39] Waldteufel, P. (1970): Planet. Space Sci. *18*, 741; Alcaydé, D., Bauer, P., Fontanari, J. (1974): J. Geophys. Res. *79*, 629; Salah, J.E., Evans, J.V., Wand, R.H. (1974): Radio Sci. *9*, 231

[40] Kockarts, G., Wisemberg, J. (1981): J. Geophys. Res. *86*, 5793

6. Optical methods of observation. In this section we shall only discuss methods that give results from which neutral atmosphere parameters are either directly obtained or easily inferred. The large field of optical phenomena in the upper atmosphere and detailed conclusions obtainable from these are dealt with in other volumes of this Encyclopedia[1] [*36*] [*67*/III], also [*21*].

α) *Optical density determinations.* The technique is absorption measurements, usually with the Sun as light source. The short wavelength part of the solar spectrum is strongly absorbed in the upper atmosphere. The method was developed to a high degree of perfection for observations of the ozone layer, including ground-based observations of rather weak absorption bands in the visible spectrum and the first balloon-borne experiments[2] [*36*].

With the advent of rockets and satellites measurements became possible in those wavelength ranges which are fully absorbed in the high atmosphere. The principle is shown in Fig. 33. Measured intensities I are interpreted by the absorbing effect along the ray[3]

$$I_{(z)} = I_0 \exp\left(-\int_0^\infty \sigma_\mathrm{T}\, n_j\, \mathrm{d}s\right), \tag{6.1}$$

where σ_T is the total absorption cross section and n numerical density. The integral is customarily called the *optical depth*, τ. For not too large values of the solar zenith angle χ one has $\mathrm{d}s = \mathrm{d}z \cdot \sec\chi$ and obtains directly the height integral over the absorbing constituent j with density profile $n_j(z)$. Measurements at different altitudes as directly obtained in rocket experiments can be used to determine the instantaneous profile, the lower limit of the integral being varied by the actual satellite motion[4]. The situation is more involved when the Sun is near the horizon (χ around or above $\pi/2$). In this condition $\sec\chi$ must be replaced[5] by the so-called Chapman function $\mathrm{Ch}(\chi, z)$; also horizontal density gradients may then be of some importance[4,6] and should be considered. An example of an early rocket observation is shown in Fig. 34.

In the case of satellite measurements the more involved geometry must be taken into account, and measurements obtained at different instants are usually combined in order to get an average profile, assuming not too large a change between successive measurements. Around $\chi = \pi/2$, however, one obtains a quicker sequence of data with the largely changing angle χ; see Fig. 33b. It appears that the minimum height of the ray towards the Sun increases rapidly so that the lower limit of the integral is "differentiated" by the satellite motion.

---

[1] See (1957): vol. 48: Middleton, W.E.K.: Vision through the atmosphere, pp. 254–287, and (1966): vol. 49/1: Akasofu, S.-I., Chapman, S., Meinel, A.B.: The aurora, pp. 1–158, further (1976): vol. 49/5: Vassy, A.T., Vassy, E.: La luminescence nocturne, pp. 5–116

[2] The famous record of E. and V.H. Regener's balloon flight in 1934 (up to 34 km) is found as Fig. 12, p. 383 in [*36*]

[3] See Thomas, L. (1981): This Encyclopedia vol. 49/6, Sect. 4, Eq. (4.2)

[4] Howlett, L.C., Baker, K.D., Megill, L.R., Shaw, A.W., Pendleton, W.R. (1980): J. Geophys. Res. *85*, 1291

[5] Chapman, S. (1931): Proc. Phys. Soc. (London) *43*, 26, 483; Proc. Roy. Soc. (London) *A 132*, 353; see also [14]

[6] Jones, W.L. (1976): This Encyclopedia, vol. 49/5, p. 177

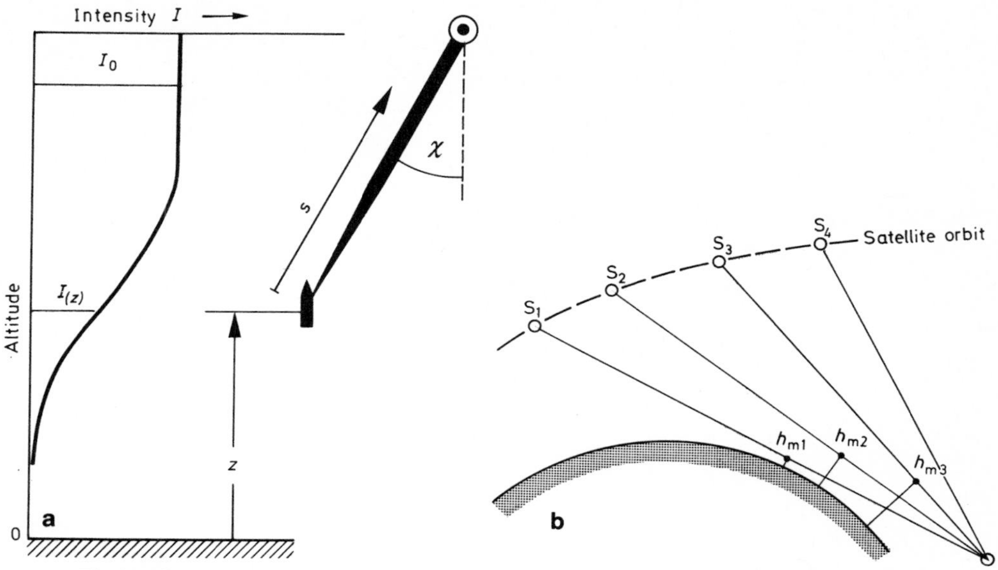

Fig. 33 a, b. Measuring atmospheric absorption: (**a**) aboard a rocket, (**b**) aboard a satellite

Such measurements are particularly useful for determining partial density profiles, under sunrise or sunset conditions of course; see Sect. 6α(ii) below.

(i) *Ionization chambers* were quite early exposed to the solar radiation *aboard rockets*. A rather large wavelength range was selected by the combined action of the filling gas ionization potential which determines the long wavelength limit, and a window determining that at the short wavelength side. The situation is sometimes involved when thin metal foils are used; one of the 'edges' in the metallic absorption spectrum is then used, but there may be ambivalence due to another edge, allowing for a second range of transmission.

The method was introduced by FRIEDMAN and co-workers at NRL[7]; they used an NO-filled photocell with an LiF window, thus isolating the range 112–134 nm in which the 121.6 nm Lα-line of hydrogen is found, the strongest emission line in the solar EUV-spectrum. Two more spectral ranges were observed, one of which (141–154 nm) is strongly absorbed by molecular oxygen (see Fig. 34). More wavelength ranges could be assessed in the soft X-ray range[8] using suitable photocathodes behind different windows; Fig. 35 shows typical calibration curves[9]. It is evident that this was a rather crude start but it was the first time that soft X-ray penetration into the atmosphere could be studied at all, and at that time existing models were checked and corrected with it. The technique has since been stabilized, but it is now used mainly in satellites for monitoring short- and long-term variations of the solar X-ray emissions in different wavelength ranges[10], the so-called SOLRAD program.

---

[7] Friedman, H., Lichtman, S.W., Byram, E.T. (1951): Phys. Rev. *83*, 1025

[8] Byram, E.T., Chubb, T., Friedman, H., Gailar, N. (1953): Phys. Rev. *91*, 1278. See also: Byram, E.T., Chubb, T., Friedman, H. (1954), p. 274 in: Rocket Exploration. Oxford: Pergamon

[9] Friedman, H. (1960) in: [*38*], 133

[10] Kreplin, R.W. (1961): Ann. Géophys. *17*, 151; Horan, D.M., Kreplin, R.W. (1980): J. Geophys. Res. *85*, 4257

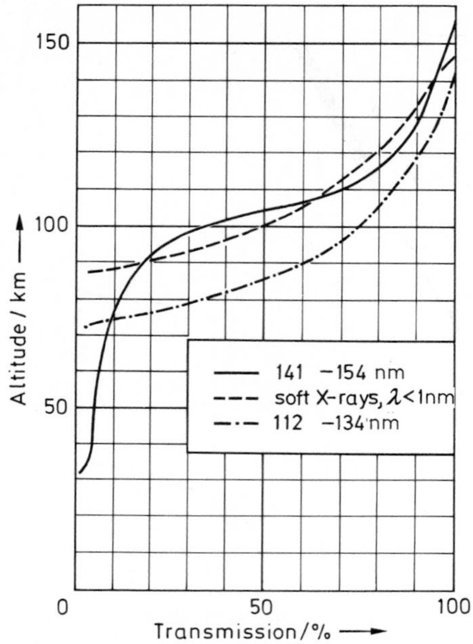

Fig. 34. Intensity variations with height in three wavelength ranges observed aboard an early sounding rocket[7]

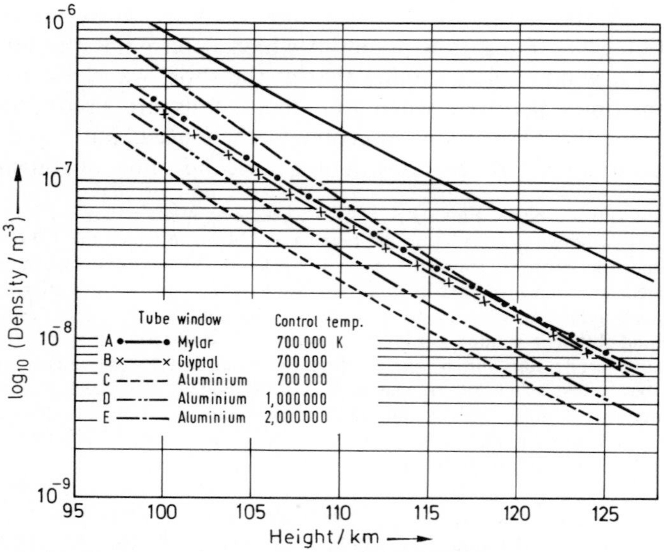

Fig. 35. Air densities deduced from X-ray intensities observed aboard a rocket behind different windows. The saw-tooth shape of the absorption spectrum makes the results depend on the spectral distribution of the solar radiation; it has been characterized here by an assumed coronal temperature. (Full line is 1952 estimate of USA rocket panel)[9]

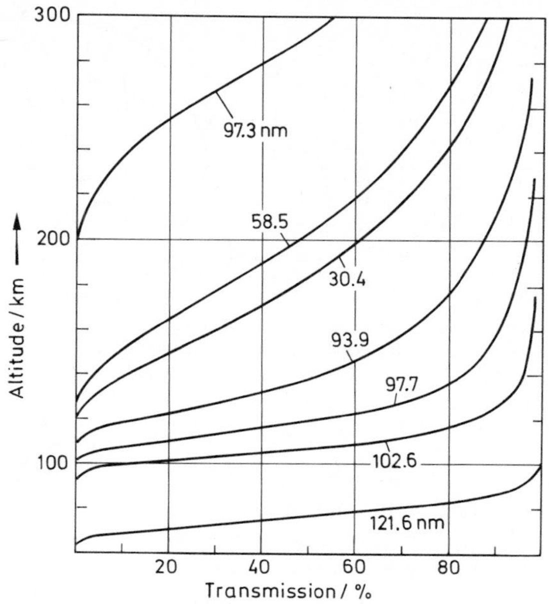

Fig. 36. Atmospheric transmission curves for different wavelengths in the EUV range, published in 1962 by AFCRL (USA)[11]

Other observations have been made with rocket-borne spectroscopes with photo-electric detectors (see NIKOL'SKIJ's contribution in vol. 49/6, his Sects. 1–6). Figure 36 shows penetration curves obtained quite early by the AFCRL group[11].

Ionization cells with LiF or MgF_2 windows and NO as filling gas have spectral response in the ranges 105–135 and 112–135 nm, respectively, and are still largely used for measuring Lα-intensities and the relevant absorption[12–15]. Such measurements are very helpful for determining density profiles of molecular oxygen O_2, which is of great interest in view of its dissociation in the height range mesopause/lower thermosphere (see YONEZAWA in vol. 49/6, his Chap. III). In fact, O_2 has a window at Lα and, though a minor constituent, NO also contributes to the absorption; this gas is of particular interest in ion chemistry (see THOMAS in vol. 49/6, his Sects. 4, 12 and 27–30).

It is important to note that at altitudes up to about 92 km the solar radiation reaching the detector is almost only Lα, thus monochromatic and with well-defined absorption cross sections for the different gases of interest. Above this altitude other contributions are no longer negligible[16].

[11] Watanabe, K., Hinteregger, H.E. (1962): J. Geophys. Res. *67*, 999

[12] Kupperian, J.E., Byram, E.T., Friedman, H. (1959): J. Atmos. Terr. Phys. *16*, 174

[13] Carver, J.H., Mitchell, P., Murray, E.L., Hunt, B.G. (1964): J. Geophys. Res. *69*, 3755; Carrer, J.H., Horton, B.H., Ilyas, M., Lewis, B.R. (1977): J. Geophys. Res. *82*, 2613

[14] Weeks, L.H., Smith, L.G. (1968): J. Geophys. Res. *73*, 4853; Weeks, L.H. (1975): J. Geophys. Res. *80*, 3655

[15] Hall, J.E. (1972): J. Atmos. Terr. Phys. *34*, 1337

[16] Iliyas, M. (1978): J. Atmos. Terr. Phys. *40*, 1065; (1980): J. Geophys. Res. *85*, 5113

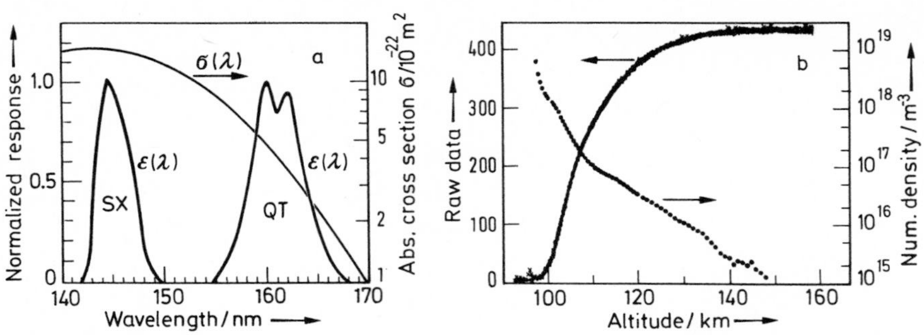

Fig. 37. Left: Spectral response curves of typical quartz-triethylamine (QT) and sapphire-xylene-(SX) ion chambers. Right: Extinction curve observed according to Fig. 33a with a QT ion chamber aboard a rocket, and derived density profile of molecular oxygen O_2[19]

Other ion chambers are mainly used in the height range above 100–110 km. One is the sapphire-xylene combination for the range 142–149 nm[14,17,18], which makes it possible to monitor the 145 nm wavelength and measure O_2 absorption; the temperature dependence of the cross section must be taken into account[18]. Another combination is quartz-triethylamine with a 155–168 nm bandwidth[17,19]; see Fig. 37.

(ii) *Absorption measurements aboard satellites* (see Fig. 33b) were preferentially executed with spectrometers of prefixed wavelength setting. Emissions on 121.6, 145 and 171 nm were simultaneously recorded aboard the ESRO-4 satellite[20] and gave information about O_2 in the 90–190 km height range; crosschecking with mass spectrometer measurements was successful[21].

Absorption measurements were made with the EUV spectrometer aboard the AEROS satellites on much shorter wavelengths, namely the helium resonance lines 58.4 nm (neutral) and 30.4 nm (ionized)[22]. Figure 38 shows typical height profiles of atomic oxygen obtained with this method (compare Fig. 33b); a correction due to N_2 absorption was needed at the evaluation[23].

With improving instrumental quality and calibration accuracy optical absorption measurements from satellites may become more important in the future, in particular when orbiting observatories are more and more used.

β) Temperature determinations by optical means. Atmospheric emissions, if in thermal equilibrium with the environment, can be used to this end.

(i) *Linewidth* measurement *from satellites* is a very suitable means. A particularly successful experiment has been flying since 1969 in the OGO-6 satellite.

---

[17] Carver, J.H., Edwards, P.J., Gough, P.L., Gregory, A.G., Rofe, B., Johnson, S.G. (1969): J. Atmos. Terr. Phys. *31*, 563

[18] Weeks, L.H. (1975): J. Geophys. Res. *80*, 3661

[19] Carver, J.H., Davis, L.A., Horton, B.H., Ilyas, M. (1978): J. Geophys. Res. *83*, 4377

[20] Ackerman, M., Simon, P. (1973): Solar Phys. *30*, 345

[21] Ackerman, M., Simon, P., von Zahn, U., Laux, U. (1974): J. Geophys. Res. *79*, 4757

[22] Schmidtke, G., Rawer, K., Fischer, Th., Lotze, W. (1974): Space Res. *XIV*, 169

[23] Schmidtke, G., Münther, Ch., Rawer, K. (1975): Space Res. *XV*, 221; see also [*26*]

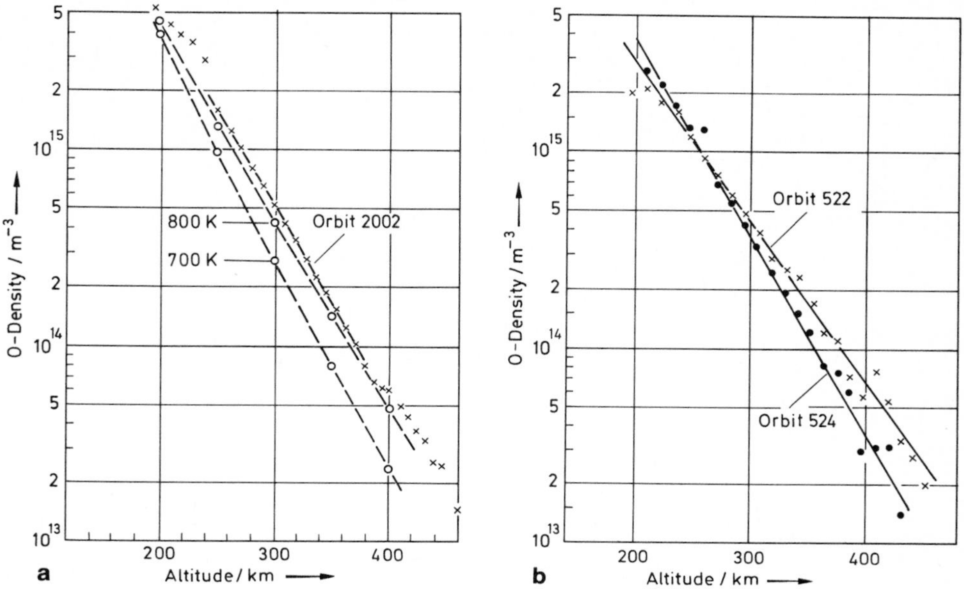

Fig. 38a, b. Height profiles of atomic oxygen O derived, according to Fig. 33b, from absorption measurements on 30.4 and 58.4 nm aboard the AEROS-A satellite [*26*]. (**a**) Observed height profile (crosses and short broken line) compared with CIRA-72 [*12*] model profiles (circles and broken lines); (**b**) Two height profiles observed with 3 h UT time difference (but same local time) on 19 Jan 1973 (dots and crosses with full lines)[23]

The profile of the forbidden line of atomic oxygen at 630 nm was almost continuously recorded and the linewidth determined. Since the excited state $^1D_2$ is metastable with a lifetime of 110 s, the atoms raised to this level reach a Maxwellian distribution by elastic collisions so that the neutral temperature T_n can be directly deduced from the linewidth[24].

The instrument used was a spherical Fabry-Perot interferometer after Connes[25]. Figure 39a shows the very heart piece of the interferometer, namely two quartz planoconcave lenses the spherical surfaces of which were coated with a dielectric reflecting film (the reflectivity was 0.80). The beam passing between both lenses was restricted to the central region by diaphragms of 2 mm diameter. The wavelength selected by the filter is given by the distance of both lenses. This distance was modulated by the action of a sawtooth tension upon the piezo-electric tube onto which one of the lenses was mounted. A wavelength modulation of 7.5 pm made it possible to cover the linewidth (see the typical record in Fig. 39b). The full instrumental arrangement is shown in Fig. 39c. A flat, moving mirror in front of the equipment is used to keep the optical axis in the wanted direction, which was a few degrees above the horizon, centered upon the emitting layer around 250 km. The light reflected at the mirror passed through a broad enough interference filter for 630 nm (of 0.7 nm width). This filter is placed between lenses L1 and L2 (field limiting) and the interferometer itself (D1, D2) between lenses L2 and L3. In order to assess the contribution of stray light, the filter was periodically replaced by another of larger bandwidth (1 nm). A trialkali photomultiplier of quantum efficiency 0.10 was used as detector. The instrument was wavelength recalibrated in flight (on ground command) with a cadmium dicharge source; no readjustment was

[24] Blamont, J.E., Luton, J.M. (1972): J. Geophys. Res. *77*, 3534

[25] Connes, P. (1956): Rev. Opt. Theor. Instr. *35*, 37; (1958): J. Phys. Radium *19*, 262

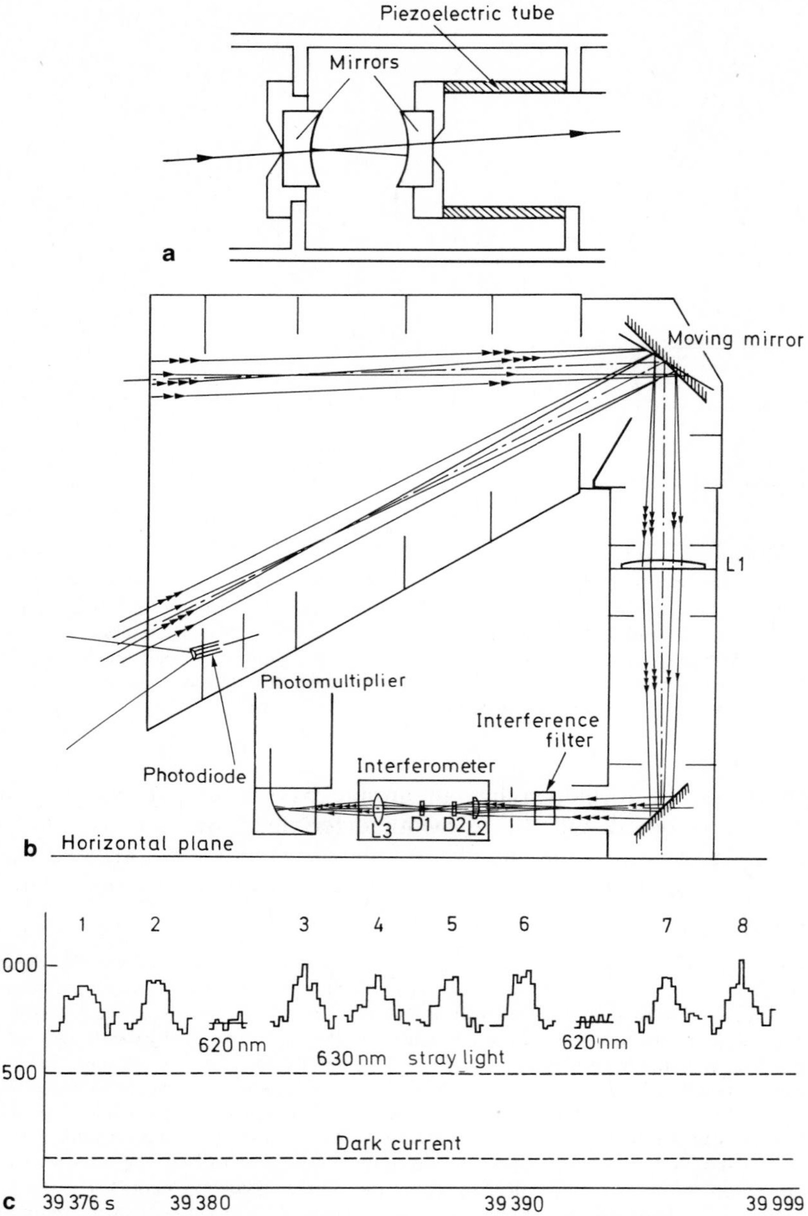

Fig. 39 a–c. Interferometric measurement of the width of the O-emission line 630 nm[24] (see text): (**a**) interferometer; (**b**) whole instrument; (**c**) data record

needed. The sensitivity of the photomultiplier, however, had to be recalibrated in flight; this was done by the light of a tungsten filament, making it possible to determine the transmission of the interference filters as well. During a full measurement cycle of 92 s the orientation was systematically changed in 2° steps so as to reach atmospheric layers of different altitudes in turn.

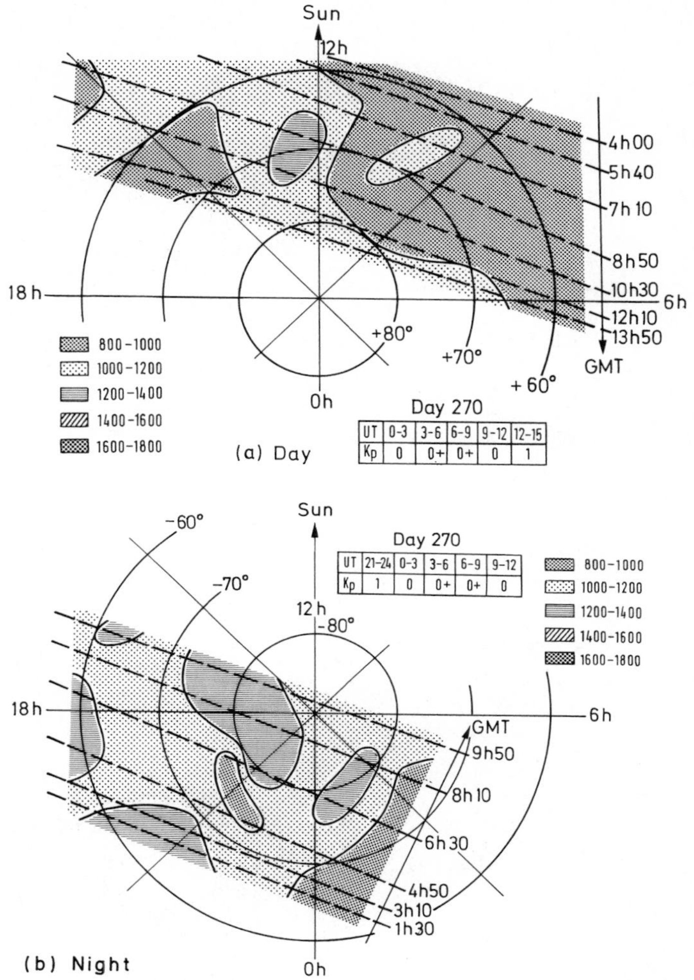

Fig. 40 a, b. Temperature maps of north polar region, day 270, 1969. (**a**) day, (**b**) night[24]

At data reduction a Doppler-shaped line with intensity distribution after

$$I(\lambda)=\frac{I_0}{w}\exp(-((\lambda-\lambda_0)/w)^2) \tag{6.2}$$

was assumed, w being the linewidth.

With this instrument the OGO-6 satellite was able to draw temperature maps at altitudes between 200 and 320 km. Figure 40 shows some typical maps for different hours during a normal day, Fig. 41 the same for a magnetically disturbed day. Due to the high inclination of the satellite orbit the polar regions were, fortunately, covered. This is of particular interest because the

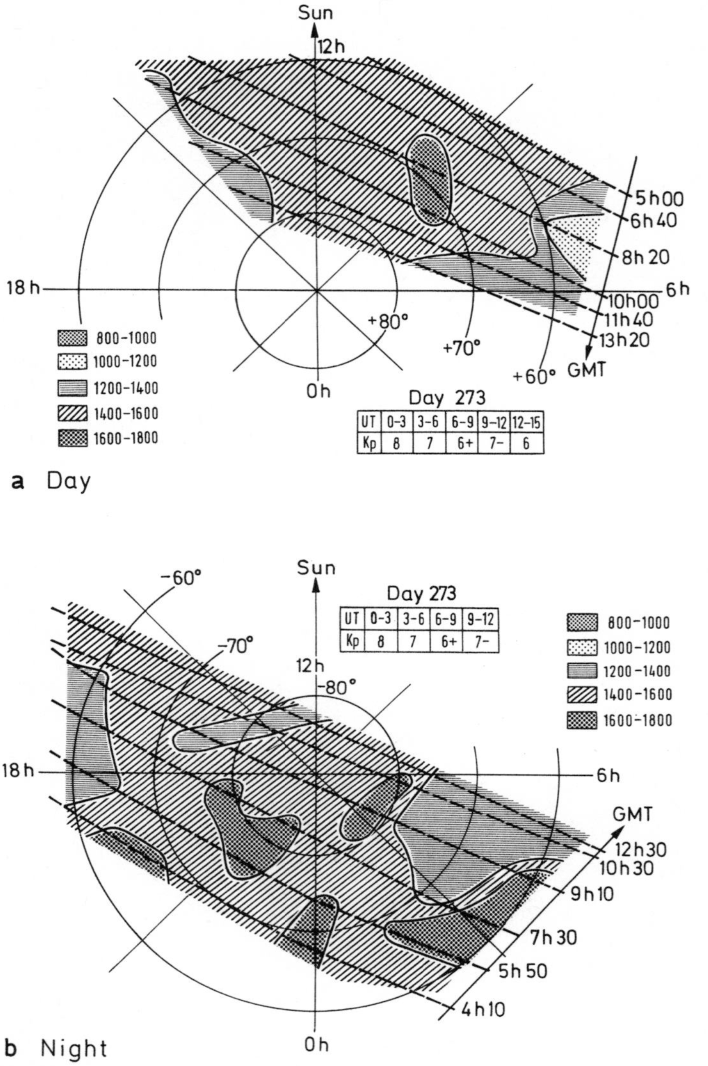

Fig. 41 a, b. Temperature maps as in Fig. 40 for magnetically disturbed day 273, 1969[24]. (**a**) day, (**b**) night

heating effects in the auroral zones could so be quantitively measured; comparing Figs. 40 and 41 the large heating effect occurring during corpuscular disturbances becomes visible. Another feature is the detection of localized warm and cold areas. The results were compared locally with incoherent scatter measurements (see Sect. 5) of ion temperature from which neutral temperatures were inferred[26]. Good agreement was reached except for the noon

[26] Carru, H., Waldteufel, P. (1969): Ann. Géophys. *25*, 485

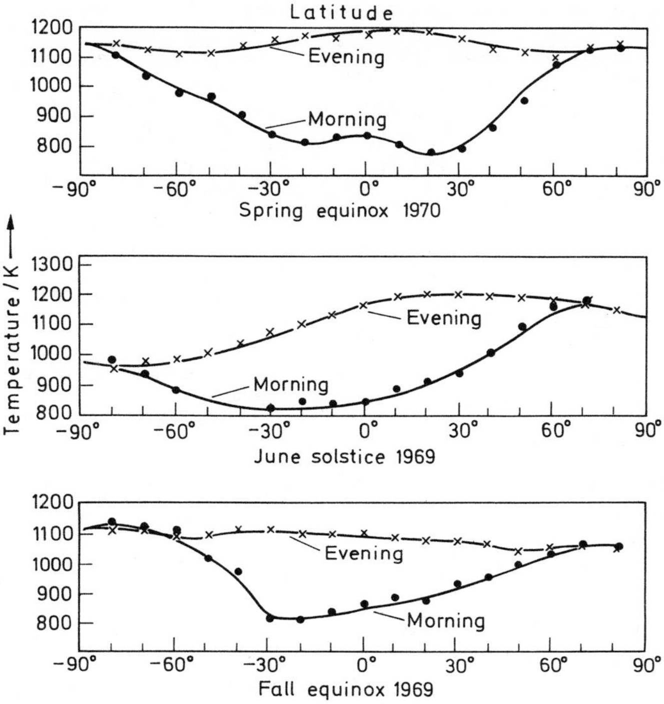

Fig. 42. Temperature at 270 km near sunrise and sunset hours deduced from optical satellite measurements[27]

values for which the values inferred from incoherent scatter were somewhat higher, by 100 K only[27]. Since, rather generally, errors in temperature measurements tend to increase the resulting temperature value, it is highly probable that the optical determination was more reliable.

While Figs. 40 and 41 were for individual days, the following figures[27] stem from averages over seasons which were deduced for the period June 1959 to August 1970. Figure 42 gives latitudinal sections through the morning and evening temperatures, i.e. measurements close to sunrise and sunset; due to the time constant of heating by photo-electrons the morning values are much lower (except for the polar regions). The maximum average morning temperature occurs at very high latitudes while in the evening the rather flat curve culminates at low latitudes. Polar temperatures are similar in South and North. As for the seasonal effect there is symmetry to the equator at the spring but not at the fall equinox. Figure 43 is a local time vs latitude map of the temperature at 270 km for the spring of 1969. The overall temperature maximum occurs near 17 h LT over the equator, but is shifted to later hours (up to 20 h LT) at higher latitudes. The morning increase is quite steep by noon.

---

[27] Blamont, J.E., Luton, J.M., Nisbet, J.S. (1974): Radio Sci. *9*, 247. [The authors put $B=\pi/12$ as if D had h for unit which, apparently, is incorrect.]

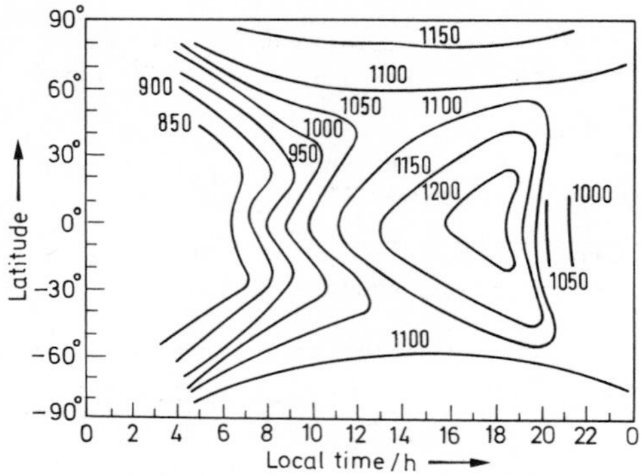

Fig. 43. Worldwide temperature distribution at 270 km deduced from optical satellite measurements, spring 1969 (equinox)[27]

From the set of observations the following empirical formula for the diurnal effect was derived:

$$T_{max}/T_{min} = 1 + A \cdot \cos\varphi \cdot [1 - \cos(B \cdot D)] \tag{6.3}$$

where $D = \arccos[(1 - 0.1738 - \sin\varphi \cdot \sin\delta)/\cos\varphi \cdot \cos\delta]$ is the 'day length angle' (at 200 km), φ being latitude and δ solar declination, $A = 0.21$[27].

THULLIER et al.[28] have represented the OGO-6 Fabry-Perot measurements in a descriptive model (see Sect. 11). It appeared that certain features of the upper atmosphere are more accurately assessed than with satellite braking (see Sect. 2) and with incoherent scatter (see Sect. 5) techniques.

(ii) *Ground-based linewidth observations* have also been made with the 630 nm forbidden emission of atomic oxygen. Techniques are similar to that described above, namely Fabry-Perot interferometers. Since distant nightglow is used as source, larger apertures are, however, needed on the ground. Also the potential perturbation by two neighbouring emissions in a band of OH must be carefully eliminated since this is an important emission in the lower parts of the upper atmosphere[29]. Measurements were executed from 1973 to 1976[30] at Fritz Peak, Colorado, USA. Apart from temperature horizontal winds were also determined with the Doppler method; see Sect. 6γ. The main results for the temperature are typically shown in Figs. 44 and 45. During magnetic disturbances a large temperature increase is observed; an example is shown in Fig. 47[31] below.

At Arecibo (Puerto Rico) similar measurements were made[32]; main emphasis was on winds; see Sect. 6γ.

---

[28] Thullier, G., Falin, J.L., Wachtel, C. (1977): J. Atmos. Terr. Phys. *39*, 399

[29] Hernandez, G., Roble, R.G. (1976): J. Geophys. Res. *81*, 2065

[30] Hernandez, G., Roble, R.G. (1977): J. Geophys. Res. *82*, 5505

[31] Hernandez, G., Roble, R.G. (1976): J. Geophys. Res. *81*, 5173

[32] Burnside, R.G., Herrero, F.A., Meriwether, J.W. Jr., Walker, J.C.G. (1981): J. Geophys. Res. *86*, 5532

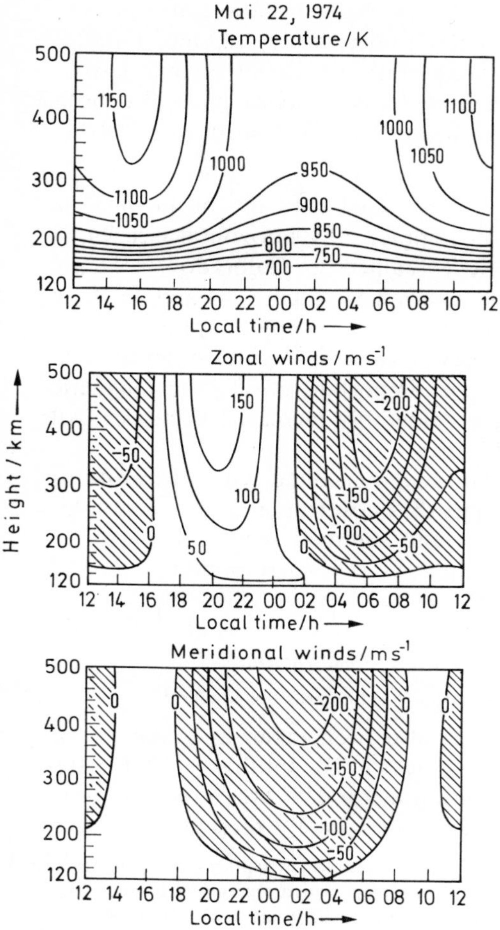

Fig. 44

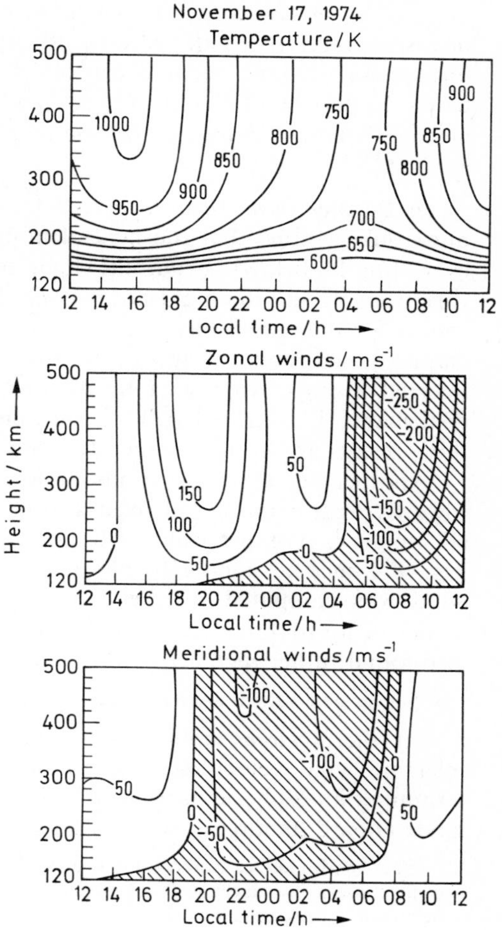

Fig. 45

Fig. 44. Computed diurnal variations of neutral temperature (top) and winds, adapting to optical observations made in USA on 22 May 1974[29]. (Zonal winds positive towards E, meridional ones towards N). Compare Fig. 69, p. 308

Fig. 45. As Fig. 44, for 17 Nov 1974[29]

(iii) *Distribution of populations* in rotational or vibrational bands can be a suitable means for temperature determination[33] [67/I].

We may mention here one active rocket-borne experiment[34] in which the first negative band system of N_2^+ ions was artificially excited by electron bombardment near the rocket and vibrational temperatures inferred from the distribution of populations[33]. A typical night profile gave a temperature increase from 800 K below 115 km to 1,500 K near 175 km[34].

[33] Vassy, A.T., Vassy, E. (1976): This Encyclopedia, vol. 49/5, Sect. 41

[34] O'Neil, R.R., Pendleton, W.R. Jr., Hart, A.M., Stair, A.T. Jr. (1974): J. Geophys. Res. *79*, 1942

γ) *Wind determination by optical means.* Narrow line airglow emissions can now be received with good optical interferometers[24,29]. Doppler effect measurements have become feasible. Ground stations are, of course, particularly helpful in this context[29,32], but satellite-based systems have also been used[35,36]. Other, more indirect methods of wind determination have also been applied[37,38].

(i) *Doppler measurements* give a fairly direct answer to the problem. One must only assume that the neutral motion is practically horizontal (see, however, Fig. 48 below). From the long series obtained in Colorado[29,30] daily variations during the night hours could be determined. Figures 44 and 45 show results for individual days; these were obtained, however, by adapting a semi-empirical aeronomical theory to the observed data[29]. Figure 46 gives such inferred average wind data by month[30]. The important effect of a magnetic disturbance is seen in Fig. 47[31], showing considerably increased velocities. All these results are from a temperate latitude site.

Another set of observations at low latitude was made with observations towards eight azimuths, at fixed zenith angle, thus covering a larger geographical range[32]. Average wind fields for spring and summer nights are shown in Fig. 48. Since a non-negligible vorticity appears the corresponding vertical wind was computed by the condition of closed flow; see Fig. 49. Average values of up to 4 ms^{-1} were found; the motion goes downwards at night[32] (as also found rather regularly with incoherent scatter observations).

Similar observations are reported also from Kwajalein (Pacific), near the equator[39].

(ii) *The apparent motion of structures* has also been taken as a vehicle for inferring wind velocities. One method is quite similar to that used for determining ionospheric drifts with a three-station method[40]. Measurements were made aboard an aircraft with an all-sky monochromatic TV presentation[41].

Other more indirect methods for inferring winds use semi-empirical relations between height and direction of the magnetically conjugate main emission centres which are shifted by the large-scale winds[36]. Peak velocities at low latitude were 110 ms^{-1} meridionally (near 20 h) and 260 ms^{-1} zonally near 22 h LT[42].

Quite recently Lidar (Laser-Radar) echoes from the atmospheric sodium layer in the 82–99 km height range were analyzed by the three-station method[40] and the displacement of its irregular structure determined[43].

Satellite measurements have been used from the OGO-4 satellite determining the intensity ratio of the O-emissions 135.6 and 630 nm[42] and of 130.4 and 135.6 nm[44] and deducing winds from semi-empirical relations. A review is found in [38].

---

[35] Chandra, S., Reed, E.I., Meier, R.R., Opal, C.B., Hicks, G.T. (1975): J. Geophys. Res. *80*, 2327

[36] Bittencourt, J.A., Tinsley, B.A. (1976): J. Geophys. Res. *81*, 3781

[37] Hicks, G.T., Chubb, T.A. (1970): J. Geophys. Res. *75*, 6233; Barth, C.A., Schaffner, S. (1970): J. Geophys. Res. *75*, 4299

[38] Meier, R.R. (1979): Rev. Geophys. Space Phys. *17*, 485

[39] Sipler, D.P., Biondi, M.A. (1978): Geophys. Res. Lett. *5*, 373

[40] Rawer, K., Suchy, K. (1967): This Encyclopedia, vol. 49/2, Sects. 50, 51

[41] Weber, E.J., Buchau, J., Eather, R.H., Mende, S.B. (1978): J. Geophys. Res. *83*, 712

[42] Tinsley, B.A., Bittencourt, J.A. (1975): J. Geophys. Res. *80*, 2333

[43] Clemesha, B.R., Kirchhoff, W.J.H., Simonich, D.M., Batista, P.P. (1981): J. Geophys. Res. *86*, 868

[44] Gerard, J.-C., Anderson, D.N., Matsushita, S. (1977): J. Geophys. Res. *82*, 1126

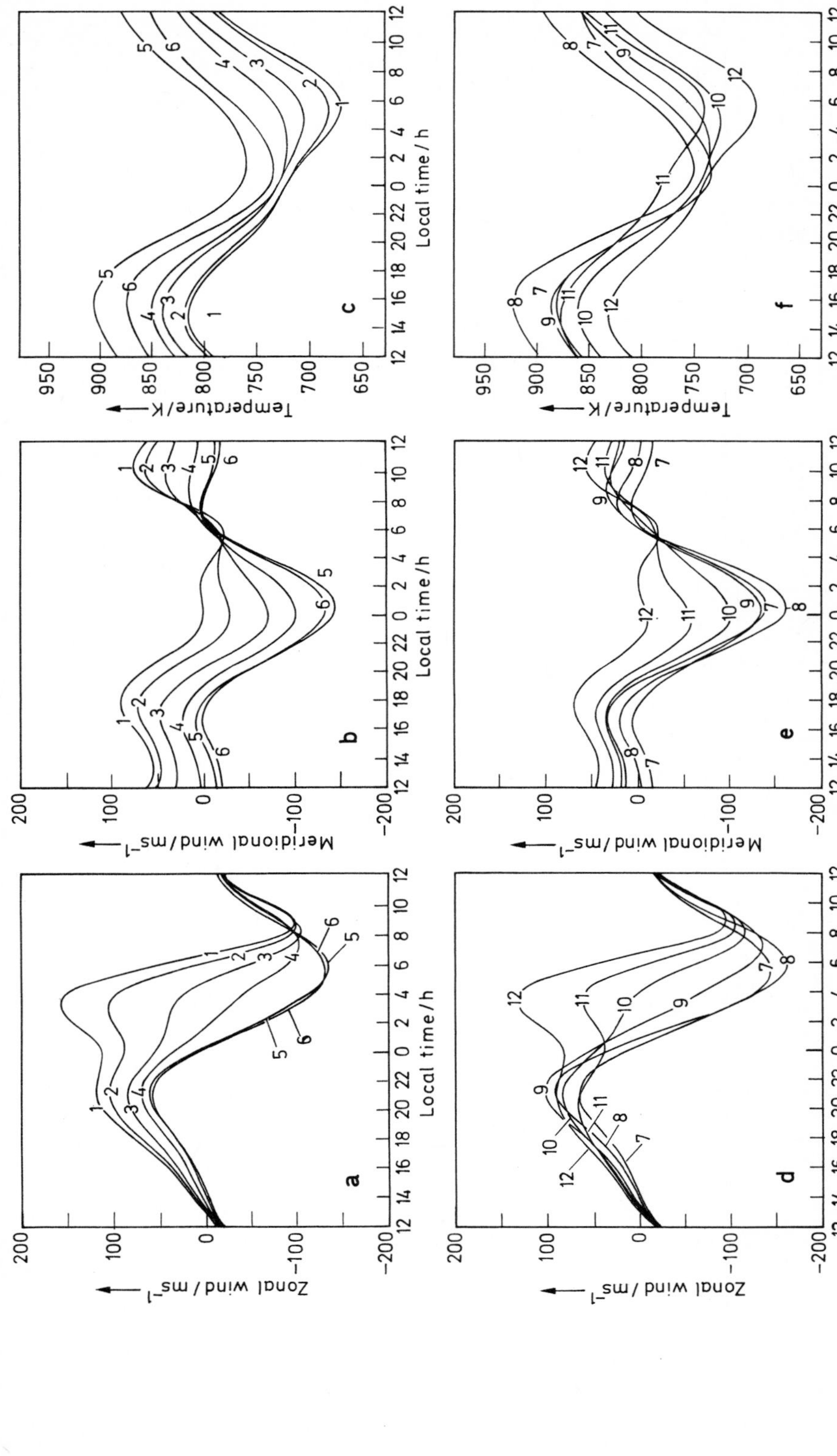

Fig. 46a–f. Neutral temperature and winds at 250 km altitude, computed with adaption to optical linewidth and Doppler-shift measurements at Fritz Peak (USA). Parameter of the curves is month[30]

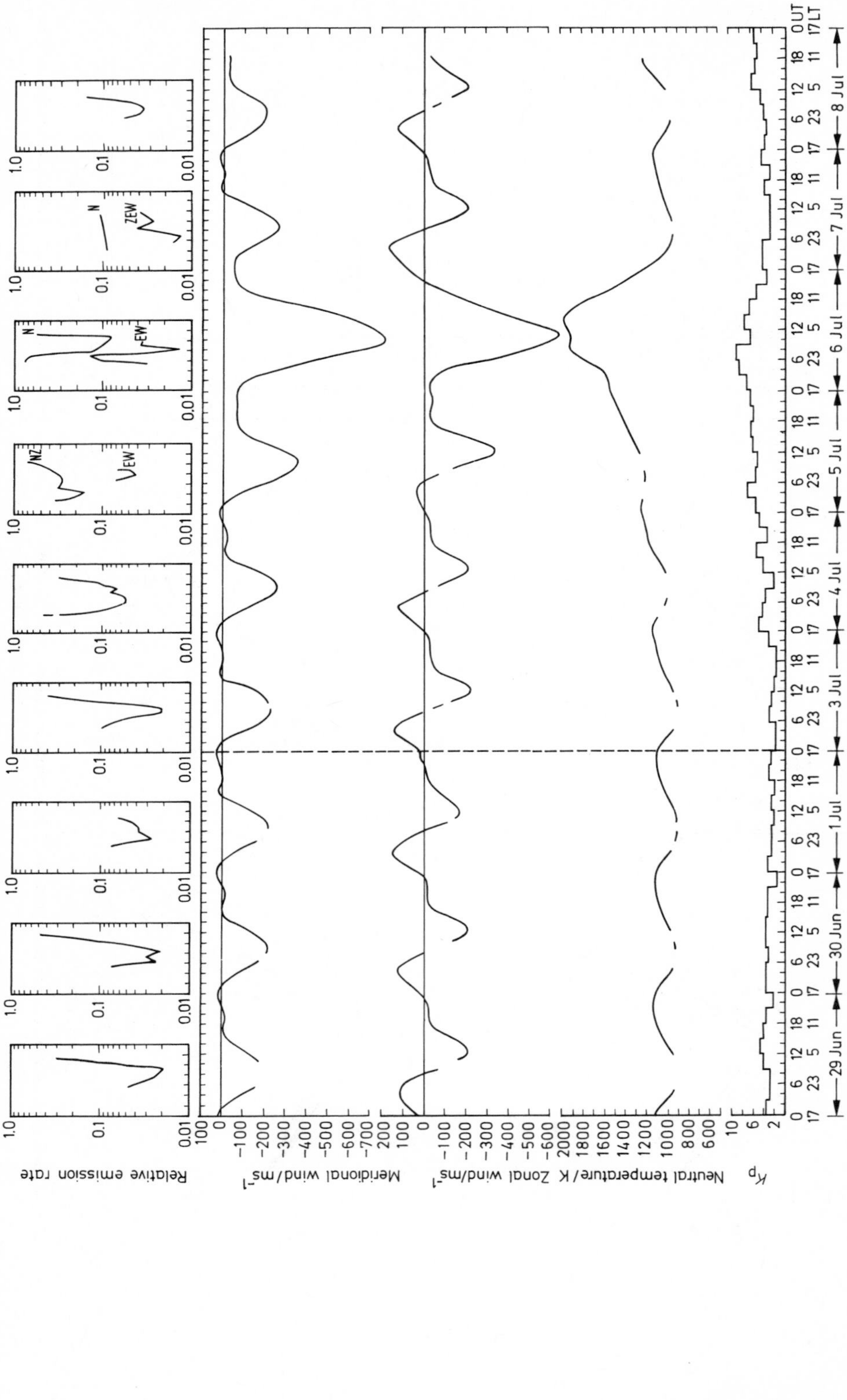

Fig. 47. Variations of emission rate on 630 nm, deduced neutral temperature, zonal and meridional winds (top to bottom) from 29 June to 8 July 1974 after observations at Fritz Peak (USA). (Lowest step curve is magnetic character figure Kp)[31]

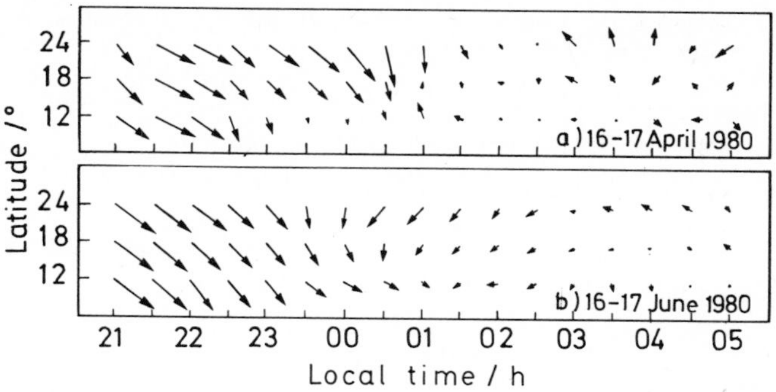

Fig. 48. Wind field inferred for typical spring (top) and summer (bottom) night conditions. Basic observations made by optical means at Arecibo (Puerto Rico)[32]

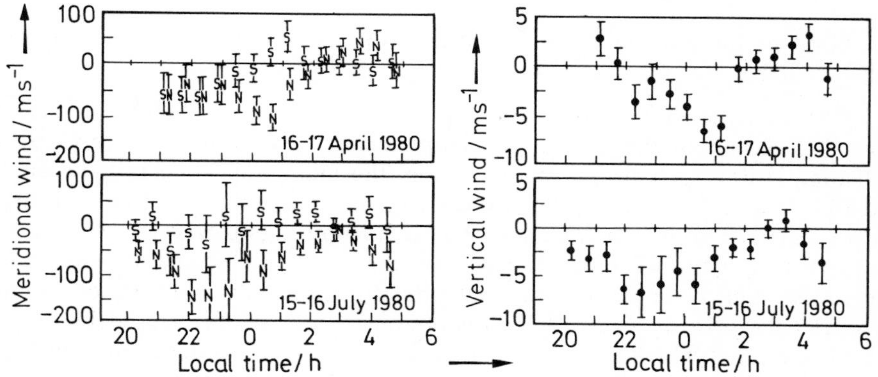

Fig. 49. Inferred meridional winds (left) and vertical wind (right) from optical observations at Arecibo (Puerto Rico). (Top: 16/17 April; bottom: 15/16 July 1980)[32]

B. Results

7. Temperature in the upper atmosphere[0]. In this section we intend to show a few remarkable results obtained with the observational techniques discussed in the preceding sections.

As we have learned in Sect. 1 the most important information about the actual situation in an atmosphere is the variation of temperature as a function of height. Inferred neutral *temperature profiles* obtained by incoherent scatter sounding (see Sect. 5) were extensively studied at the temperate latitude 'key-stations' in France (St. Santin) and in the USA (Millstone Hill). From a large number of ion temperature measurements at St. Santin, after seasonal group-

[0] In writing this section, a preprint of J.V. Evans was largely used which was distributed at the COSPAR meeting held at Innsbruck in 1978

Table 3. Coefficients of Eq. (7.1) in the 95–120 km height range[1]

Season	T_m/K	$B_2/10^{-4}\,km^{-2}$	$B_4/10^{-8}\,km^{-4}$	z_0/km
Spring	188.0	24.1	− 10.1	98.8
Summer	184.1	22.9	− 61.5	94.4
Autumn	185.8	19.7	+ 59.3	98.0
Winter	198.3	7.5	+263	98.4

These profiles all show a sharp temperature increase above 110 km.

ing, a fourth-order power series after the altitude z

$$T(z) = T_m[1 + B_2(z - z_0)^2 + B_4(z - z_0)^4] \tag{7.1}$$

was fitted to the data in the lower middle thermosphere. The coefficients are found in Table 3[1].

α) Most important is the *diurnal variation.* Figure 50 shows the typical diurnal variation of the exospheric temperature T_∞, [see Sect. 5, Eq. (5.5)] at Millstone Hill[2,3]. According to the season, the maximum appears between 15 and 18 h, near 16 h at the equinoxes; earlier atmospheric models based on total density determinations[4] assumed an earlier maximum (see Sect. 8). Many ideas have been brought forward in order to explain this discrepancy (see [3] for references), but it appears now that the diffusive redistribution of atomic oxygen is the primary cause. A two component thermospheric model with N_2 and O only, and which includes horizontal transport, reproduces the observed behaviour[5]. A very crude worldwide description of the exospheric temperature, based upon incoherent scatter observations in France and in Peru is shown in Fig. 51[6].

The total amplitude of the diurnal variation of T_∞ is about 200–300 K, thus quite appreciable. Fourier analysis[7] of data obtained in the period 1969–1972 at the key stations clearly showed a diurnal (24 h) wave which has at least three times greater amplitude than the semidiurnal (24 h) wave[8], and which culminates around 14 h LT. The diurnal 24-h amplitude is about 10% of the diurnal maximum value. Since the latter increases roughly linearly with solar

[1] Alcaydé, D., Fontanari, J., Kockarts, G., Bauer, P., Bernard, R. (1979): Ann. Géophys. *35*, 41

[2] Measurements at incoherent scatter stations are routinely made only a few days each month; thus, unfortunately, uninterrupted series of daily measurements still do not exist

[3] Evans, J.V. (1978): Rev. Geophys. Space Phys. *16*, 195 (survey paper)

[4] Jacchia, L.G. (1965): Smithsonian Contrib. Astrophys. *8*, 215

[5] Mayr, H.G., Volland, H. (1973): J. Geophys. Res. *78*, 7480; the model was experimentally checked, see: Hedin, A.E., Mayr, H.G., Reber, C.A., Carignan, G.R., Spencer, N.W. (1973): Space Res. *XIII*, 1315

[6] Waldteufel, P., McClure, J.P. (1969): Ann. Géophys. *25*, 785

[7] Salah, J.E., Evans, J.V., Alcaydé, D., Bauer, P. (1976): Ann. Géophys. *32*, 257

[8] There might exist a lunar semidiurnal influence too, which would disturb phase and amplitude determinations of the 12 h solar harmonic; due to the interruptions in the data series [2] such effects could not really be studied

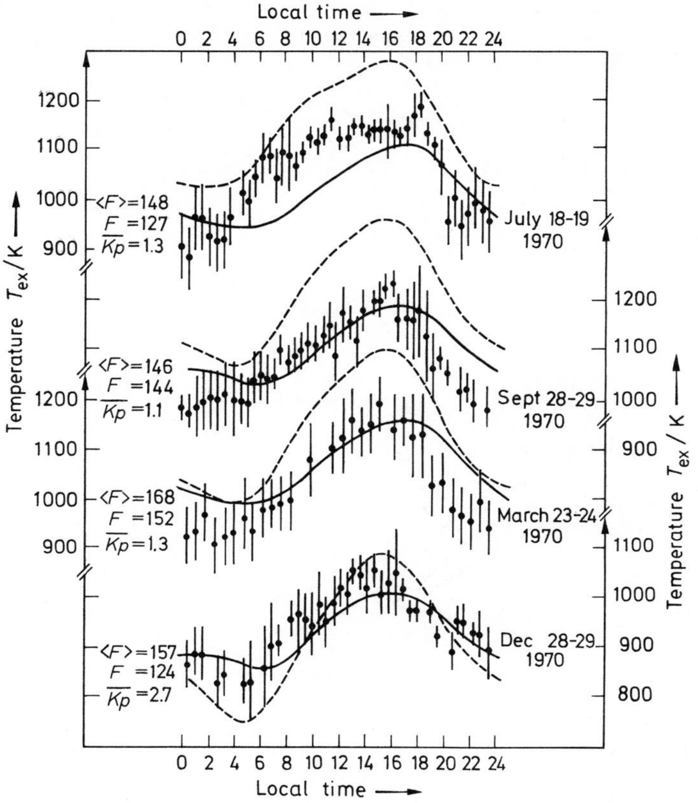

Fig. 50. Variation of exospheric temperature T_∞ with local hour for typical observational periods in summer, equinox and winter 1970[3]. Broken curves were computed from HEDIN's mass-spectrometer based model [*19*], full curves from that of THULLIER which is based on optical measurements [*62*], both made aboard the OGO-6 satellite [see Sects. 11α and β]

activity measured by the Covington index F (mostly called $F_{10.7}$ in the literature)[9], both harmonics increase in absolute value with increasing solar activity. At mid-latitude these two tidal influences were determined as

$$T_\infty = \bar{\bar{T}}_\infty + (18 + 0.64F) \cdot K \cdot \cos\left(\frac{\pi}{12}(t - t_{24})\right) + (7 + 0.14F) \cdot K \cdot \cos\left(\frac{\pi}{6}(t - t_{12})\right) \quad (7.2)$$

with phases $t_{24} \approx 14$ h and t_{12} rather variable.

[9] This is a pure number deduced from daily measurements of the solar radio noise made at Ottawa on the wavelength of 10.7 cm as specified by COVINGTON. The number is obtained by dividing the measured flux value by 10^{-22} W m^{-2} Hz^{-1}

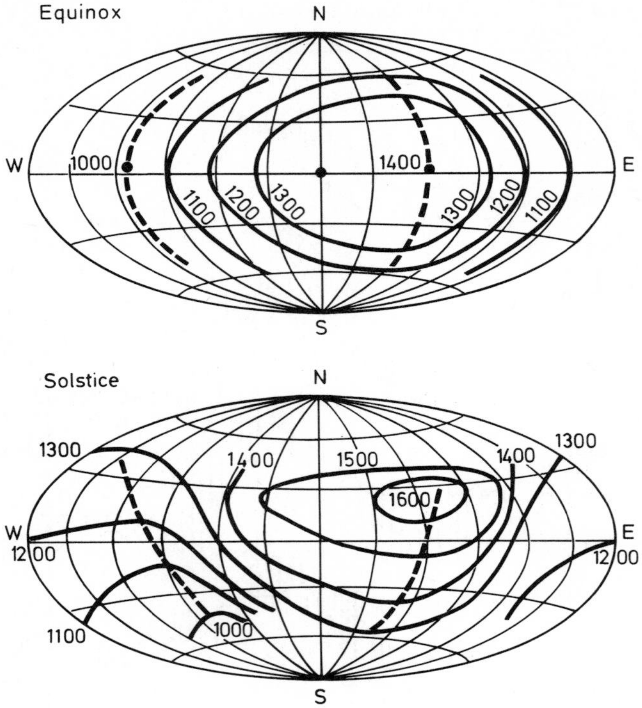

Fig. 51. Crude map of exospheric temperature, T_{∞}/K, constructed[6] from observations at St. Santin (France, 45° N) and at Jicamarca (Peru, 12° S), for equinox (top) and northern summer solstice (bottom)

β) Variations of this kind must be described as *tidal* effects provoked by solar and lunar gravity and, in particular, by heating due to solar radiation[10] [*8*, *28*, *69*]. The tides of greatest interest are, on the one hand, those generated in situ by absorption of solar XUV radiation in the thermosphere itself and, on the other hand, upward propagating [mainly (1, 1)] tidal waves which are generated at lower levels by absorption of solar radiation in the troposphere (water vapor) and in the ozonosphere[11]. In the mesosphere and lower thermosphere the wave amplitude increases with height; this is the consequence of energy flux conservation through an atmosphere with decreasing density. However, when wave energy is dissipated by eddy and molecular diffusion (see [16] and Sect. 8γ), this is no longer true. So, at some level in the thermosphere, due to viscosity (increasing with decreasing density) and to ion drag (see Sect. 16),

[10] Gravitational atmospheric tides are described in much detail in Kertz, W. (1957): This Encyclopedia, vol. 48, pp. 928–981; since that time, however, it has become evident that in the upper atmosphere, thermal excitation of tides is more important than gravitational excitation. See: Bricard, J., Kastler, A. (1944): Ann. Géophys. *1*, 1; Haurwitz, B. (1955): Bull. Amer. Meteorol. Soc. *36*, 311; Sawada, R. (1955): Geophys. Mag. *26*, 267; Siebert, M. (1956): Ber. Dtsch. Wetterdienstes *4*, 65; see also: Hvostikov [*21*], 2nd book, chap. 4, Lindzen [*28*], Forbes [*16*]

[11] Murata, H. (1974): Space Sci. Rev. *16*, 461; Groves, G.V. (1976): Proc. Roy. Soc. (London) *A351*, 437; (1983): Planet. Space Sci. *31*, 67

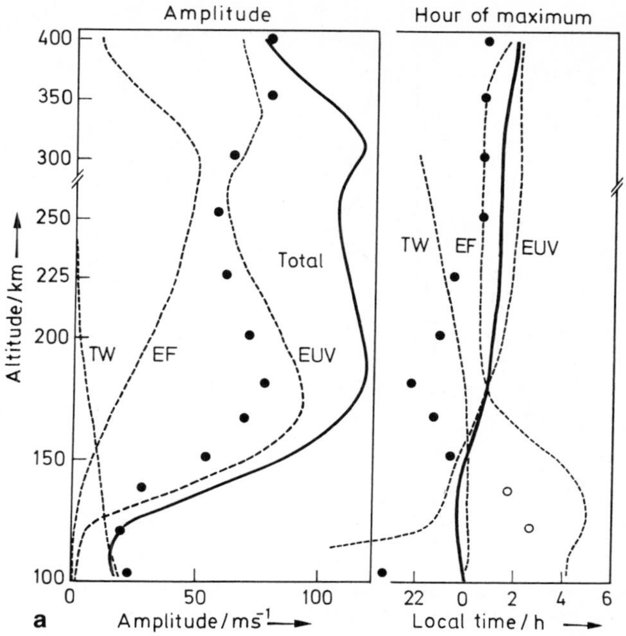

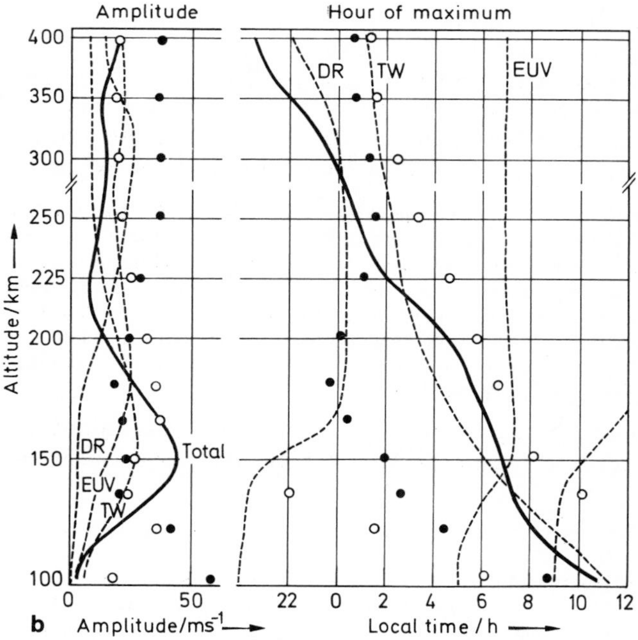

Fig. 52a, b. Diurnal and semidiurnal tides in the meridional neutral wind inferred from observed plasma motions[17]. (**a**) Diurnal trapped mode (1, −2) as theoretically predicted (bold curves: total; individual contributions are dotted: EUV-heating, electric fields and travelling waves[15]), compared with plasma wind speeds (left) and phases (right) observed in France (dots). (**b**) Semidiurnal tides: the fundamental mode (2, 2) as predicted[15] compared with tidal observations as in (**a**)

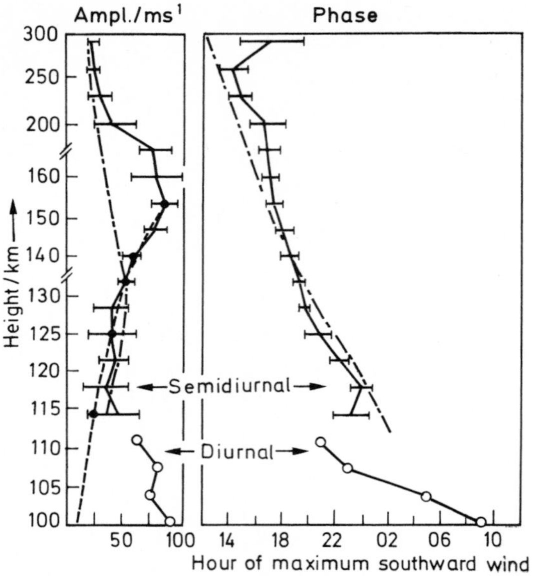

Fig. 53. Diurnal and semidiurnal tidal components identified in the incoherent scatter southward plasma wind measurements, obtained at Arecibo (Puerto Rico) in the middle of August 1974[16]; broken curves were computed theoretically after[12]

the amplification of wave amplitude is stopped so that it decreases above that level. Dissipation introduces much difficulty into the tidal computations[12] [16].

Solar heating excites a certain number of tidal oscillation modes, the latitudinal structure of which is described by Hough functions or extensions of these[13].

A mode may be excited by direct (thermal) and/or indirect influences (mode coupling), both with different phases. While the (2, 4) mode receives in phase contributions, the effect of ozone heating upon the (2, 3) mode is diminished by mode coupling. Different modes have different vertical wavelengths. The fundamental *diurnal* (1, 1) mode has a rather short wavelength of about 30 km and reaches its maximum amplitude in the 105–109 km height range[14], at low latitudes, with quite a small amplitude. Modes with greater vertical wavelength are more important in the thermosphere. The (1, −2) mode covers a large height range so that it is simultaneously excited in this whole range by solar XUV heating.

It is not a propagating but a standing ('trapped') oscillation[13, 15, 16] and was detected with incoherent scatter drift measurements at St. Santin (France) (Fig. 52)[17].

---

[12] Lindzen, R.S., Hong, S.S. (1974): J. Atmos. Sci. *31*, 1421; Hong, S.S., Lindzen, R.S. (1976): J. Atmos. Sci. *33*, 135

[13] Forbes, J.M., Garret, H.B. (1976): J. Atmos. Sci. *33*, 2226

[14] Hines, C.O. (1966): J. Geophys. Res. *71*, 1453; Kochanski, A. (1966): Mon. Weather Rev. *94*, 199; (1973): Ann. Géophys. *29*, 77; Woodrum, A., Justus, C.G., Roper, R.G. (1969): J. Geophys. Res. *74*, 4099

[15] Volland, H., Mayr, H.G. (1972): J. Atmos. Terr. Phys. *34*, 1745, 1769; (1973): Ann. Géophys. *29*, 61; (1974): Radio Sci. *9*, 263; (1977): Rev. Geophys. Space Phys. *15*, 203

[16] Mayr, H.G., Harris, I. (1977): J. Geophys. Res. *82*, 2628

[17] Amayenc, P. (1974): Radio Sci. *9*, 281

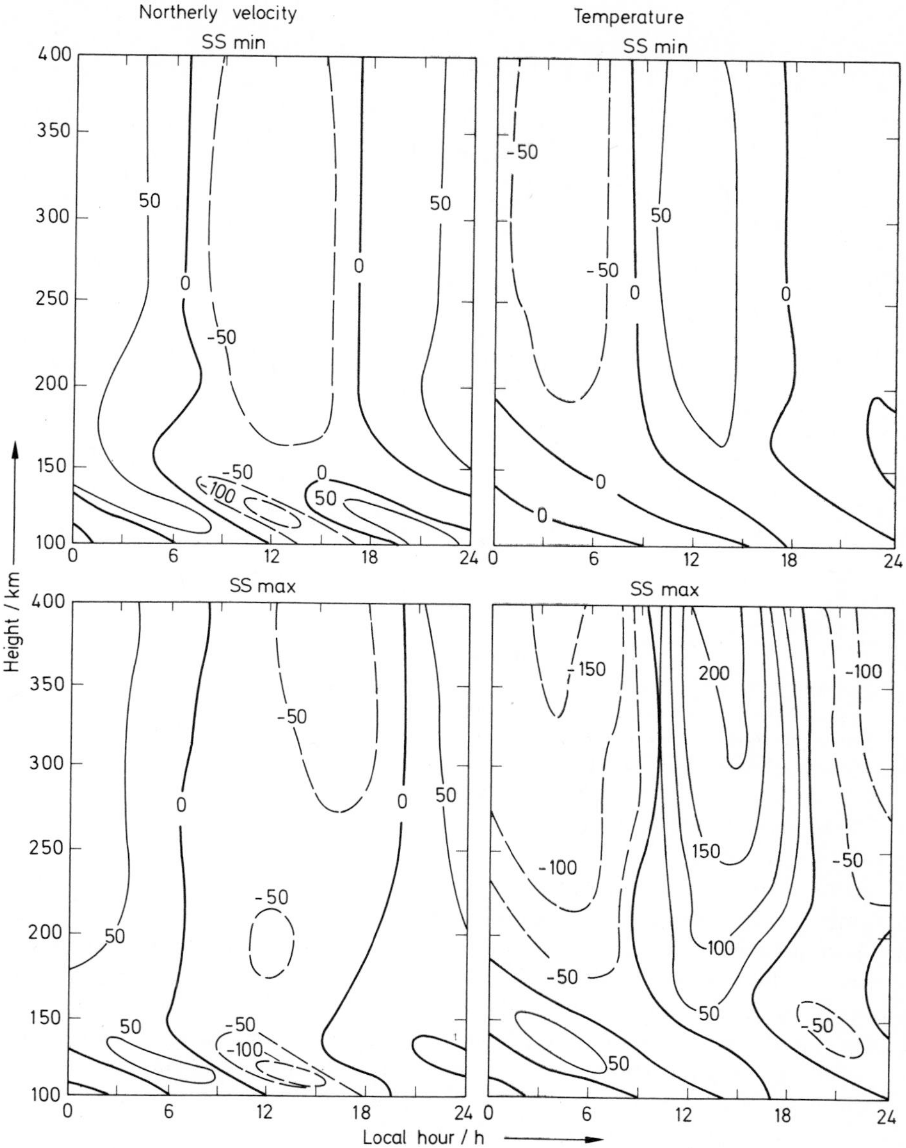

Fig. 54. Altitude/local time contours of northerly wind velocity/ms^{-1} (left) and temperature variation/K (right) for minimum (top) and maximum (bottom) solar cycle conditions computed by the GARRETT and FORBES theoretical model[21] at 45° latitude.[19]

Semidiurnal oscillations are particularly important, even dominating, in the 100–200 km height range (see Fig. 53)[18]. Classical theory of the relevant modes

[18] Harper, R.M. (1977): J. Geophys. Res. *82*, 3243

[19] Champion, K.S.W., Forbes, J.M. (1978): Survey paper presented at the XXIst COSPAR meeting Innsbruck

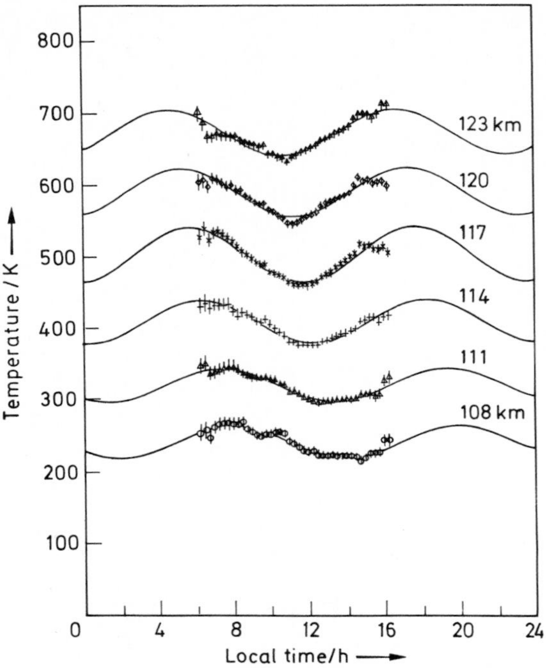

Fig. 55. Semidiurnal tide: Temperatures in the lower thermosphere, averaged from a total of 45 d of observations in 1971/72 at Millstone Hill (USA) with best fitting 12 h sinusoids at successive altitudes in the lower thermosphere (parameter). (Curves are shifted by 50 K between successive levels; ordinate scale is valid for 108 km.) Phase progression with increasing height should be due to an upward propagating component [22]

[*8*] seems to be insufficient but dissipation must be taken into account [12]. The fundamental (2, 2) mode is, apparently, trapped below the mesopause so that only a small part of the energy penetrates from below into the thermosphere directly [20]; there may, however, be some transfer by coupling with higher modes in the presence of winds [12]. This mode was detected in the plasma drift at low [18] and middle latitudes [17]. In Fig. 53 the higher modes are felt to be dissipated at heights above 130 km so that only the (2, 2) mode remains important in the upper thermosphere. In Fig. 52b the observed semidiurnal tides at St. Santin [17] are compared with theory [15] without obtaining perfect agreement. With a theoretical model [21] [*16*] the results displayed in Fig. 54 were obtained; solar cycle effects were found to be quite important.

In the lower thermosphere, between 100 and 130 km, incoherent scatter temperature measurements, or plasma drifts, might be used for determining tides; a review can be found in [3]. There is agreement that the winds rotate through about 360° in 12 h as first shown by HARNISCHMACHER with 'fading

[20] Lindzen, R.S. (1971): Geophys. Fluid Dynam. *1*, 303, *2*, 89

[21] Garrett, H.B., Forbes, J.M. (1978): J. Atmos. Terr. Phys. *40*, 657

[22] Salah, J.E., Wand, R.H. (1974): J. Geophys. Res. *79*, 4259

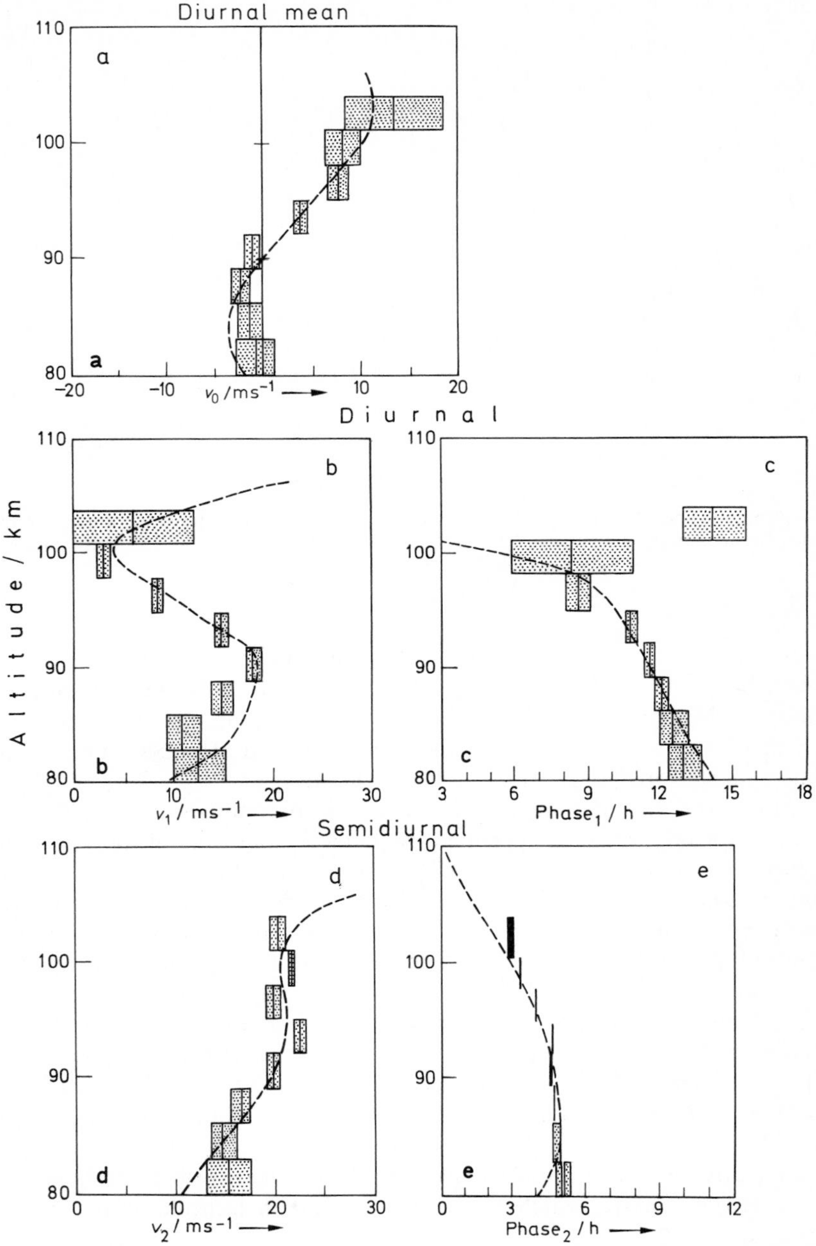

Fig. 56a–e. Height profiles of winds obtained by the meteor method (hatched rectangles). Top (**a**): diurnal average (steady component) of the northerly wind (southward wind vector) v_0. Centre (**b, c**): tidal diurnal rotating wind v_1. Bottom (**d, e**): tidal semidiurnal rotating wind v_2. Amplitude left, phase right. (Phase is the time of northward maximum.) Broken lines are third order parabolic fits[23]. Basic data: about 24,500 meteor echoes observed from 20 July to 7 Aug 1978 at Kyoto[26]

[23] Groves, G.V. (1959): J. Atmos. Terr. Phys. *16*, 344

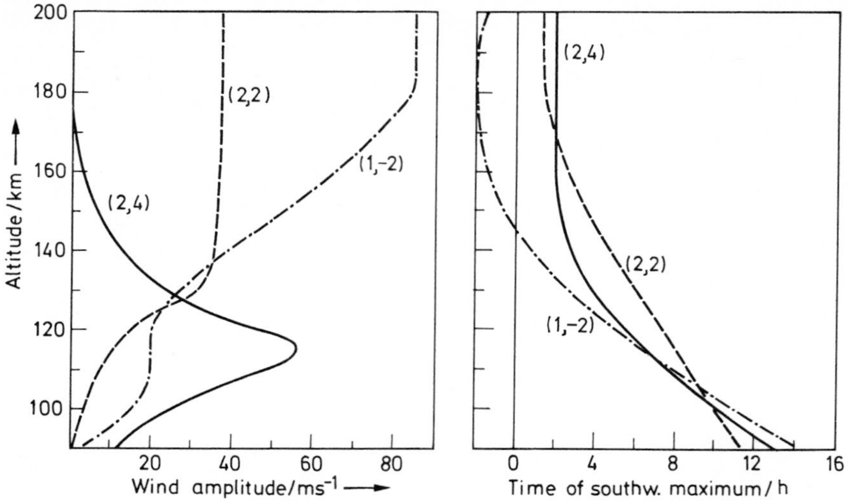

Fig. 57. Amplitude and phase of tidal winds (three principal modes) in the lower ionosphere based upon incoherent scatter observations of temperature and plasma drift at the mid-latitude key-stations in France and USA [28]

drift' observations [24]. The important day-by-day and seasonal variabilities were also first shown by this author. Figure 55 shows the daily mean temperature variations at Millstone Hill for successive levels in the lower thermosphere, fitted with 12 h sinusoids. Similar results were obtained at St. Santin [25]. At both (mid-latitude) stations the (2, 4) mode is dominant.

At still lower altitudes near and above the mesopause (in the 80–110 km height range) winds can be determined by meteor radar, one of the ionospheric drift measuring methods [*32*, *41*, *46*]. The very localized ionization produced at the evaporation of the meteorite material acts as a tracer (meteor trail), which goes with the neutral wind. So, as for artificial releases (see YONEZAWA's contribution in vol. 49/6, his Sect. 13), this is a direct determination of neutral winds. Figure 56 shows recent results obtained in Japan with a very sensitive device [26]. It appears that above 90 km the diurnal drift decreases rapidly while the semidiurnal drift increases. The phase stability obtained may be taken as proof of the high statistical significance. These results agree with earlier determinations showing greater statistical spread since they used less sensitive devices [27].

A descriptive model of tidally driven winds [28] is shown in Fig. 57. The (2, 5) HOUGH extension mode, which is antisymmetrical in latitude, is felt by some

---

[24] Harnischmacher, E., Rawer, K. (1958): Geofis. pura e appl. *39*, 216; Rawer, K., Suchy, K. (1967): This Encyclopedia, vol. 49/2, Sect. 50, p. 405; Harnischmacher, E., Rawer, K. (1968): [*41*], 242

[25] Bernard, R. (1974): J. Atmos. Terr. Phys. *36*, 1105

[26] Aso, T., Tsuda, T., Takashima, Y., Ito, R., Kato, S. (1980): J. Geophys. Res. *85*, 177

[27] Greenhow, J.S., Neufeld, E.L. (1961): Quart. J. Roy. Meteorol. Soc. *87*, 472

[28] Salah, J.E., Evans, J.V. (1977): J. Geophys. Res. *82*, 2413

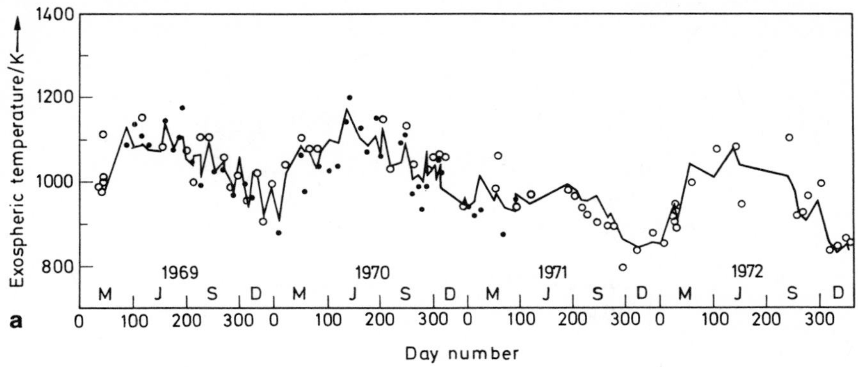

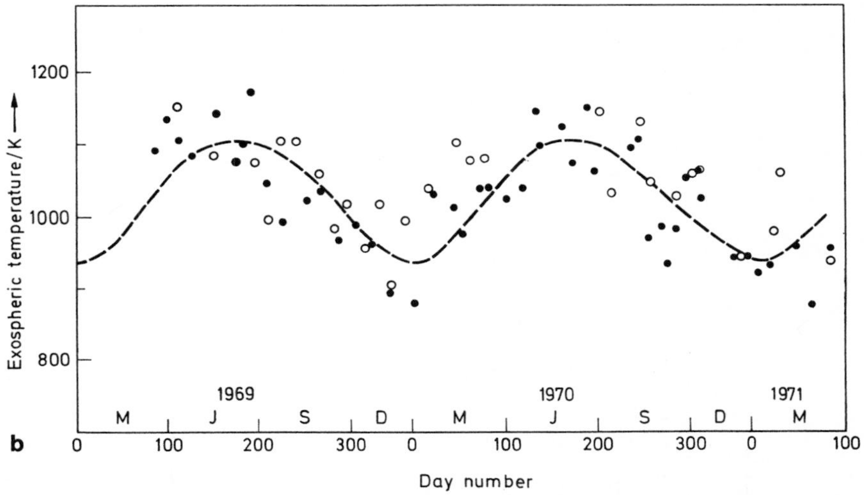

Fig. 58 a, b. Seasonal, variation of the "mean exospheric temperature" 1969–1972 obtained by averaging over all hours, month by month; stations St. Santin (France, circles) and Millstone Hill (USA, dots). (**a**) (top) Model function after Eq. (7.3a) (i.e. admitting activity changes); (**b**) (bottom) Seasonal fit with two harmonic seasonal components (year and half-year period) after removing activity influences; see Eq. (7.3b)[30]

authors to be present and must be taken account of in order to explain the observations[29].

γ) *Seasonal variations*, though less important than the diurnal ones, are not negligible. Results are shown for the mean exospheric temperature in Fig. 58, based on observations at the two mid-latitude key-stations[30]. The authors describe the variations for 45° latitudes including the solar and magnetic

[29] Lindzen, R.S. (1976): J. Geophys. Res. *81*, 2923; Wand, R.H. (1977): Trans. Amer. Geophys. Un. EOS *58*, 702

[30] Salah, J.E., Evans, J.V., Alcaydé, D., Bauer, P. (1976): Ann. Géophys. *32*, 257

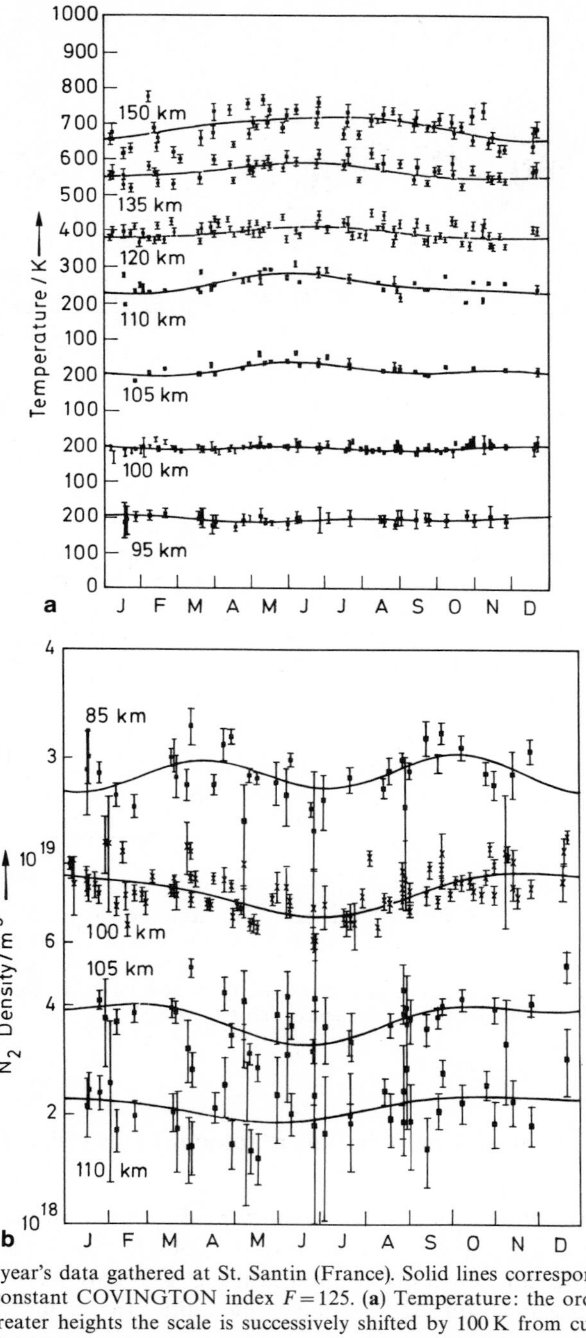

Fig. 59a, b. One year's data gathered at St. Santin (France). Solid lines correspond to a fit with two harmonics and constant COVINGTON index $F = 125$. (**a**) Temperature: the ordinate scale is valid for 110 km; at greater heights the scale is successively shifted by 100 K from curve to curve, while below 110 km the interrupted ordinate scales apply directly. (**b**) Inferred density of molecular nitrogen[1]

Table 4. Fourier analysis of inferred N_2 densities after Eq. (7.4)[1]

Height km	Average A_0	Flux coeff. $a/10^{-4}$	Amplitude $A_1/10^{-2}$	Phase ϕ_1/d	Amplitude $A_2/10^{-2}$	Phase ϕ_2/d
95	19.18	6.20	1.1	242	4.7	97
100	18.88	3.10	5.9	353	1.4	102
105	18.48	6.76	4.6	356	2.7	78
110	18.32	−0.18	3.4	326	0.94	65

activity effects (see below) by:

$$\bar{T}_\infty/\mathrm{K} = (566 + 3 \cdot F_0) \cdot \left[1 + 0.075 \cdot \cos\left(2\pi\left(\frac{d}{\mathrm{d}} - 179\right)\Big/365\right) + 0.01 \cdot \cos\left(4\pi\left(\frac{d}{\mathrm{d}} - 97\right)\Big/365\right)\right] + 1.4(F - F_0) + 15\,Kp \tag{7.3a}$$

where F is the Covington index[9], F_0 the gliding average of F over three solar rotation periods of 27 d, Kp the planetary magnetic index[31] and d day number through the year. After removing the activity influences, for $F = F_0 = 150$ and $Kp = 2$, the following relation was derived:

$$\bar{T}_\infty = 1018 + 85 \cdot \cos\left(2\pi\left(\frac{d}{\mathrm{d}} - 181\right)\Big/365\right) + 9 \cdot \cos\left(4\pi\left(\frac{d}{\mathrm{d}} - 106\right)\Big/365\right). \tag{7.3b}$$

Comparable relations with two harmonics were deduced for the lower middle thermosphere temperature from the French incoherent scatter facility[1]. Figure 59a shows all data gathered in 1 year at altitudes between 95 and 150 km. It appears that, in the lower middle thermosphere, the phase of the seasonal variations as well as the amplitudes depends strongly on the altitude. Similar descriptions applied to inferred neutral partial densities, e.g. of N_2 in Fig. 59b, show comparable behaviour but a stronger semi-annual variation. This diagram shows that the often accepted idea of 'stable conditions' near the lower limit of the thermosphere is not fulfilled. The numerical results reached by Fourier analysis of the N_2 data[1] after

$$\log_{10} \frac{n(\mathrm{N}_2)}{\mathrm{m}^{-3}} = A_0 + a\,F_0 + \sum_{j=1}^{2} A_j \cdot \cos\left[\frac{2\pi j}{365\,\mathrm{d}}(d - \phi_j)\right] \tag{7.4}$$

are summarized in Table 4.

δ) The *solar activity dependence* should be considered separately for short- and long-term variations. Priester[32] found the short-term variations of total

[31] Siebert, M. (1971): This Encyclopedia, vol. 49/3, p. 222
[32] Priester, W. (1959): Naturwiss. *46*, 197

density to be closely related to the solar activity as given by the Covington index F[9]. This effect was soon confirmed[33,34] and has since been well established. (See Sect. 11, Eq. (11.8) for a more recent descriptive formula based upon only one F-value, shifted, however, by one day.) Good correlation may also exist with other solar indices[33], but the best is obtained with the Covington index. As for the long-term variability all indices when given a sliding average yield comparable and quite good correlations.

The dependence on F has since been assumed for quite a few of the upper atmospheric parameters. These are, of course, not brought about by the solar 10.7 cm noise radiation which is used to establish the Covington index. This radiation penetrates the upper atmosphere without absorption. What really varies and is absorbed at these heights is the solar XUV radiation. Solar activity indices based on this radiation have been tentatively determined in recent years[35], but no detailed comparison with density variations has yet been established.

Dependencies on the gliding average of F (F_0, usually established over three solar rotation periods, i.e. 81 d) were specified[1] in Eqs. (7.3) and (7.4). Also the profile parameters s (see Sect. 5) was found to depend on F_0. It is now quite certain that variations of the solar XUV radiation influence not only the ionized populations of the atmosphere[36] but also very much the neutral populations. Since, at least at low and temperate latitudes, solar XUV energy is the main heating source, and since the coefficients of many molecular reactions important for aeronomy are temperature dependent, it can be understood that the very structure and composition of the upper atmosphere is considerably altered during the 11 year solar activity cycle[37].

Due to the complex nature of these reactions, even the sign of such changes is often dependent on the altitude. For example, Fig. 60 shows the variation with altitude of the solar flux dependency of the temperature in the lower and middle thermosphere[1]. The influence is weak below 110 km, but clearly negative at 120 km. Above 130 km it turns positive again and increases sharply with height. The influence of solar activity appears in all models of the upper atmosphere. Quite often it is given by two terms, depending on F and F_0, respectively, in order to separate the short-term and long-term dependencies. Mixed descriptions based upon $(F-F_0)$ and F_0 are also used (see Sects. 9 and 10 for more details).

A clear solar cycle dependence has been found for the amplitudes of the diurnal and semidiurnal tides in the exospheric temperature T_{ex} as determined by incoherent scatter techniques (see Sect. 5 above) at midlatitudes[29]. The relation of T_{ex} with F is straight forward, see Fig. 31 in SCHMIDTKE's contribution in this volume.

ε) *Magnetic activity* also has an important influence upon the upper atmosphere. During magnetic disturbances[38] intense corpuscular radiation arrives in

---

[33] Jacchia, L.G. (1959): Nature *183*, 327

[34] Paetzold, H.-K., Zschörner, H. (1959): Kleinheubacher Berichte 5, 90; (1960): Space Res. *I*, 24

[35] Schmidtke, G., Münther, Chr., Rawer, K. (1975): Space Res. *XV*, 221; see also the contribution by G. SCHMIDTKE in this volume, p. 33

[36] Rawer, K., Suchy, K. (1967): This Encyclopedia, vol. 49/2, p. 269

[37] Rawer, K. (1981): Adv. Space Res. *1* (12), 87

[38] Nagata, T. (1971): This Encyclopedia, vol. 49/3, p. 5

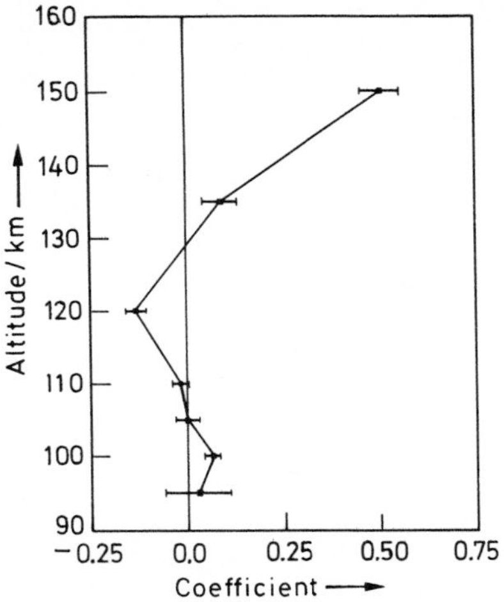

Fig. 60. Altitude profile of the coefficient describing the dependence of local temperature upon solar activity expressed by the average COVINGTON index F_0[1]

both auroral zones and finally deposits heat in the whole upper atmosphere at these high latitudes. Secondary effects are observed at middle and even low latitudes. So the magnetic activity effect depends strongly on the latitude, more precisely on magnetic latitude[39,40].

Different magnetic activity indices[41] can be used for describing such dependencies, but Kp is mostly used[30]; see for example Eq. (7.3a). The influence of magnetic activity is also included in many upper atmospheric models (see Chap. C).

8. Atmospheric structure and transport[0]. In previous contributions a few parameters were discussed which are not really 'state quantities' but have a more involved significance. One of these is the collision frequency, ν_{in}, of ions with neutrals, Eq. (5.6), which we have mentioned in Sect. 5. Another parameter, discussed in Sect. 5, is the shape parameter s of the temperature profile, Eq. (5.5), which describes the temperature increase with height occurring in the thermosphere.

α) A very important parameter is the *scale height*

$$H = kT/\bar{m}g \qquad [1.6]$$

---

[39] Hess, W.N. (1972): This Encyclopedia, vol. 49/4, Sect. 5, p. 124

[40] Due to the difference between magnetic and geographical coordinate systems, when geographical coordinates are used in the analysis the magnetic influence appears as a longitudinal effect, appearing at high latitudes in particular, see: Prölss, G.W. (1977): J. Geophys. Res. *82*, 1635; Oliver, W.L. (1977): Trans. Amer. Geophys. Un. EOS *58*, 1197

[41] Siebert, M. (1971): This Encyclopedia, vol. 49/3, p. 206

[0] Hedin, A.E. (1979): Rev. Geophys. Space Phys. *17*, 477 is a rather recent survey paper

which in the range of predominant diffusion must be applied separately to each constituent as

$$H_j = kT_j/m_j g. \qquad [1.8]$$

The scale height may be deduced from observed profiles of pressure [Eq. (1.5)] or of density and temperature [Eq. (1.3)]. Below the turbopause it may be computed after Eq. (1.6) directly, provided the mean molecular weight $\bar{m}$ is known. (Usually, it is not.) Above the turbopause, Eq. (1.8) makes it possible to determine, just from the temperature profile $T_n(z)$, the individual scale heights of those species which are in diffusive equilibrium if there is temperature equilibrium between the neutral species (which is usually granted). In the middle thermosphere, where the temperature increases rapidly with height and $\bar{m}$ decreases, the scale height also increases. In the upper thermosphere, however, where T_n is almost independent of height ($T_n \approx T_\infty$), and atomic oxygen is the dominant constituent ($m = 16$ au), the scale height is practically constant over a large height range. Therefore, scale height determinations above 250 km can be used to infer the exospheric temperature, T_∞.

The method of optical absorption sounding (see Sect. 6) is particularly suited since from the relative variations of solar light intensity with height the scale height is easily determined.

Observations of satellite braking (see Sect. 2) are mainly used to determine the total mass density ρ_{per} at the perigee of a satellite orbit. For an elliptical orbit, however, the scale height just above the perigee is of comparable importance. This follows from Eq. (2.5) where the integral contains the height-dependent density

$$\rho \approx \rho_{per} \cdot \exp(-(z - z_{per})/H), \qquad \text{see also Eq. (2.6).}$$

β) Some efforts have been made to determine an effective scale height of the *ionospheric plasma*[1]. In a simple but useful approximation the plasma is replaced by a neutral model gas with mass and temperature (kinetic energy) just obtained by averaging over positive ions and electrons. The model has $(m_i + m_e)/2 \approx m_i/2$ for particle mass and $(T_e + T_i)/2$ for temperature. Since in the F-region O^+ ions are largely predominant, the "mean molecular mass" is known and the averaged temperature might be derived from a "plasma scale height". The latter is not directly measured but mostly inferred from a thickness parameter of the F2-layer. It was noted that the peak altitude is also temperature dependent but the additional influence of neutral winds[2] is so important that this parameter cannot really be used (see STUBBE's contribution in vol. 49/6, Sect. 22, p. 285); only in the region near the dip equator this latter influence is negligible but electric fields have strong effects (ibidem Sect. 25, p. 288). Even for the layer thickness it is not easy to estimate the error due to

[1] Becker, W., Stubbe, P. (1963) in: The Ionosphere, Proc. Intern. Conf. on the Ionosphere. London: Inst. Physics and Physical Soc. Stickland, A.C. (ed.), p. 35; Stubbe, P. (1964): Kleinheubacher Berichte *10*, 147; Stubbe, P. (1964): J. Atmos. Terr. Phys. *26*, 1055

[2] Kohl, H., King, J.W. (1965): Nature *206*, 699; (1967): J. Atmos. Terr. Phys. *29*, 1045; Kohl, H., King, J.W., Eccles, J.W. (1969): J. Atmos. Terr. Phys. *31*, 1011

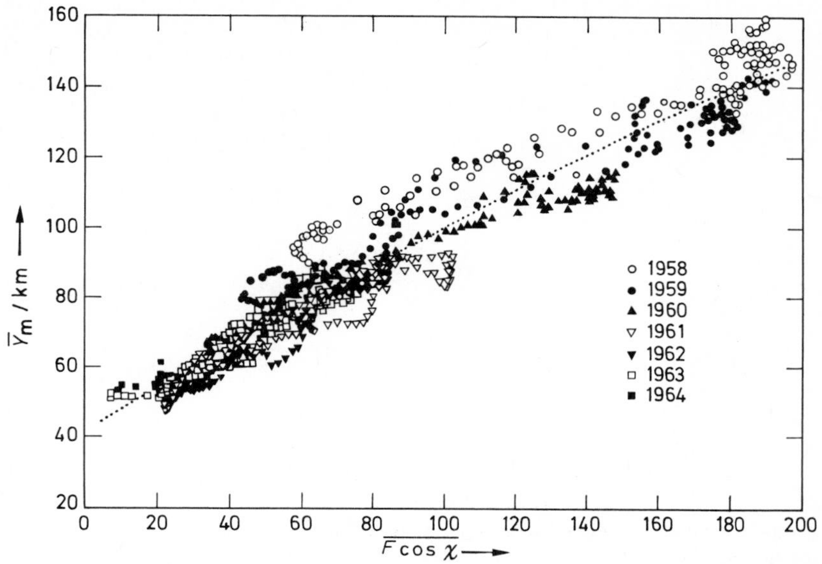

Fig. 61. Seven year correlogram between the averaged thickness parameter Y_m of the ionospheric F2-layer at Lindau (Harz), Germany, and reduced solar activity characterized by $S = F \cdot \cos\chi$ where F = COVINGTON index[5] and χ solar zenith angle. Entries are running means over the solar rotation period of 27d; the dotted line corresponds to the relation[4] $\bar{Y}_m/\mathrm{km} = 41.5 + 0.641\,S - 0.000535\,S^2$. [$Y_m$ is the half thickness of a parabolic approximation]

wind effects. By applying aeronomical theory with, however, important simplifications it could be shown that with a parabolic relation between neutral temperature and thickness parameter, by fitting the parameters of this relation with empirical (or model) temperature data, at least *relative changes of temperature* can be followed[3]. This was done in studies of day-by-day changes and, in particular, of the influence of solar activity[1]. Figure 61 shows an example[4] where, after reduction of the solar zenith angle dependence, a correlation between the thickness parameter and COVINGTON's solar activity index[5] was established.

Total electron content measurements using beacon satellites can be used to derive a representative thickness of the ionosphere, so-called "slab-thickness"; it is determined as the ratio of total (columnar) electron content and peak electron density[6–8]. After establishing "relationships" between T_e, T_i and T_n THITHERIDGE established a model from which a relation between slab thickness τ and neutral temperature T_n derives[9] (see Fig. 62). Day and night relations

[3] Rüster, R. (1969): Kleinheubacher Berichte *13*, 115

[4] Becker, W. (1973): Kleinheubacher Berichte *16*, 201

[5] See footnote [9] in Sect. 7

[6] Rawer, K., Suchy, K. (1967): This Encyclopedia, vol. 49/2, Sects. 54, 55, in particular p. 470

[7] A more recent survey is: Evans, J.V. (1977): Rev. Geophys. Space Phys. *15*, 325

[8] See the reports of meetings of the 'Beacon Satellite Group' (now in the Union Radioscientifique Internationale, URSI), most recently: [*70*]

[9] Titheridge, J.E. (1973): Planet. Space Sci. *21*, 1755

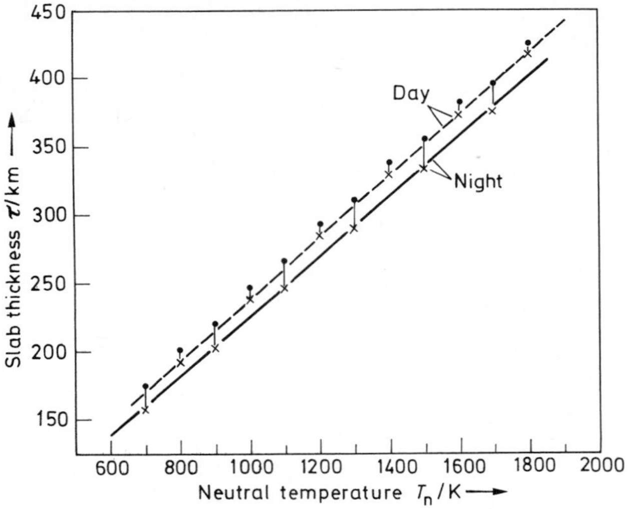

Fig. 62. Computed relation between ionospheric slab-thickness τ (mid-latitudes) and neutral temperature at the F2-peak; dots and crosses identify different assumptions about the occurrence of H^+ ions (0.1 and 0.001 %)[9]

differ because of changes in neutral and ionic composition. Large night-time increases in τ are also explained by such changes so that the day/night variation cannot simply be described by temperature change alone. In the earlier days an ionospheric electron density profile after ELIAS-CHAPMAN[10] used to be assumed; in this case, a simple, often-used relation links the slab thickness τ with the atmospheric (neutral) scale height H[11]:

$$\tau = 4.133\, H.$$

However, the assumptions under which this particular type of layer shape was derived are very far from the real conditions in the terrestrial ionosphere since there is no thermal equilibrium between ionized and neutral constituents[12] (except, approximatively, by night). Thus, slab thickness values should be taken more as indicating temperature variations than for inferring absolute temperature values. Using the relations shown in Fig. 62 seasonal variations of (neutral) exospheric temperature were determined[9] as to be as large as ± 250 K. This is too much as seen by comparison with data from incoherent scatter sounding (see Sect. 5) and satellite braking (see Sect. 2). The discrepancy is felt to be due to seasonal changes in the neutral composition of the thermosphere which markedly change the ionized layer shape[13,14].

---

[10] Elias, G.J. (1923): Tydschr. Ned. Radio Gen. *2*, 1; Chapman, S. (1931): Proc. Phys. Soc. *43*, 26, 483

[11] Bauer, S.J. (1960): J. Geophys. Res. *65*, 1685

[12] Thomas, L. (1966): J. Geophys. Res. *71*, 1357; see also his contribution in vol. 49/6 of this Encyclopedia, Sect. 3, p. 11

[13] Ivanov-Holodnyj, G.S., Kolomiytsev, O.P. (1967): Geomagn. i Aeronomija 7, 731; see also [22]

[14] Strobel, D.F., McElroy, M.B. (1970): Planet. Space Sci. *18*, 1181

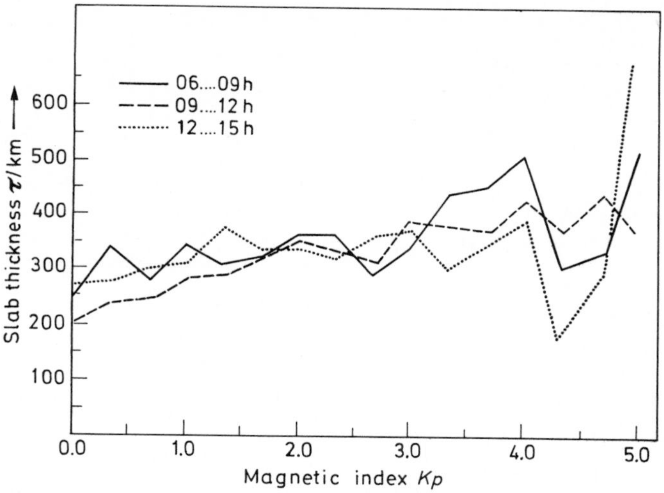

Fig. 63. Average dependence of slab-thickness τ (over three daylight hourly intervals) upon the planetary magnetic activity index Kp[16]. Beacon satellite measurements made from 1961 to 1964 at Urbana, Illinois, USA[15]

In spite of its drawbacks, slab thickness has proved to be quite helpful in studying short-term variations. Figure 63[15] shows the (averaged) influence of magnetic activity upon the slab thickness at mid-latitude. The increase is mainly due to heating of the atmosphere in the auroral zones.

γ) Transport in an atmosphere is of particular importance. While bulk motions, i.e. winds are the main vehicle for horizontal transport (see Sect. 8 δ), gravity force acts against vertical displacement of neutral air masses so that these remain small. The main vehicle for *vertical transport* is diffusion, which is either above the turbopause, due to free molecular flow, or below due to eddy transport (see Chap. II of YONEZAWA's contribution in volume 49/6). The distinction between both transport phenomena is very important. The molecular diffusion coefficient D describes the strength of diffusive separation under the influence of gravity. It can be computed from gas-kinetic theory; see contributions by YONEZAWA in vol. 49/6 (Sect. 5, p. 136) and SUCHY in this volume (Sect. 25, p. 118). On the other hand, the eddy diffusion coefficient K describes the strength of turbulent mixing against gravity; see YONEZAWA (ibidem, Sect. 8, p. 144). The turbopause is defined as the level where eddy diffusion which is dominant at lower heights just equals molecular diffusion, the importance of which increases with height; see THOMAS' contribution in vol. 49/6 (his Fig. 36, p. 86).

Our Fig. 64 shows, first of all, measured density ratios in the critical height range around 100 km where diffusive separation begins to be important. Since the molecular weights are rather

---

[15] Yeh, K.C., Flaherty, B.J. (1966): J. Geophys. Res. *71*, 4557

[16] For detailed explanation of the Kp index see: Siebert, M. (1971): This Encyclopedia, vol. 49/3, p. 222

different the pair Ar and N_2 is an appropriate choice. The measured values show rather important irregularities which are understood to be due to the occurrence of thin turbulent layers as are often observed during release experiments. These are interpreted by the different profiles of diffusion coefficients, shown at the left-hand side of Fig. 64. PHILBRICK[17], with only short-range smoothing, obtained the very variable dotted curve for K while the descriptive profile functions given by other authors[18,19] for K reproduce the data with much more smoothing. Particularly interesting is the profile with a minimum between two maxima[19] because it admits three levels where $K=D$, the turbopause condition, is satisfied. Since the levels of major turbulence are probably variable, this reasoning shows the limitations of the idea that the turbopause should always be a clearly defined level. Wherever turbulence exists in the lower thermosphere, the conditions of vertical transport are considerably more involved and variable than higher up in the middle thermosphere.

Studies of turbulence and eddy transport have not only been made with mass-spectrometric composition measurements but also when analyzing release experiments producing chemical tracers[20] and with radar type radio measurements on meteor trails and their development in time [*32*][21]. The latter give access to the height range between 80 and 100 km, i.e. the mesopause and lowest thermosphere regions[20,22]. Scale heights were also deduced from such observations. Its validity has been critically rediscussed[23]. A broad survey on electromagnetic probing of the atmosphere is found in [*6*].

Eddy diffusion is particularly important in the context of vertical heat transport. The question is whether and how the part of eddy diffusion in the atmospheric heat balance can be determined with enough accuracy. As for the lower thermosphere, it is not a priori clear whether eddy diffusion makes a positive or negative contribution. In a discussion of different planetary atmospheres IZAKOV concluded that cooling prevails in the thermospheres of Earth and Venus, whereas heating is prevalent in the thermosphere of Jupiter. For Mars both signs of the effect upon thermal interchange may occur[24].

Considering the heat balance against vertical transport a one-dimensional balance equation may be written formally as:

$$\frac{d\phi_z}{dz} = P(z) - L(z) \tag{8.1}$$

where ϕ_z is the vertical heat flux and P and L are the local heating and cooling rates. [Eq. (8.1) is often applied as a time average over 1 day then describing the gross variations of the thermal flow.] The downward heat flux ϕ_z would just be $-\lambda\, dT/dz$ if molecular heat conduction were acting alone, λ being the

[17] Philbrick, C.R., Faucher, G., Bench, P. (1978): Space Res. *XVIII*, 139

[18] Shimazaki, T. (1971): J. Atmos. Terr. Phys. *33*, 1383

[19] Blum, P.W., Beese, H.J., Schuchardt, K.G.H. (1978): Paper presented at the COSPAR meeting held at Innsbruck (Austria)

[20] Zimmerman, S., Pereira, G.P., Murphy, E.A., Theon, J. (1973): Space Res. *XIII*, 209; Zimmerman, S. (1973): J. Geophys. Res. *78*, 3927

[21] Lovell, A.B.C. (1957): This Encyclopedia, vol. 48, pp. 427–454

[22] Hess, G.C., Geller, M.A. (1976): Aeronomy Rept. No. 74. Urbana: Univ. Illinois; McIntosh, B.A., Simek, M. (1977): Bull. Astr. Insts. Čsl. *28*, 181

[23] For a discussion of the validity of the method see: Forti, G. (1958): J. Atmos. Terr. Phys. *40*, 89; Jones, J. (1970): Planet. Space Sci. *18*, 1836; Jones, J. (1975): Mon. Not. Roy. Astr. Soc. *173*, 636; Baggaley, W.J. (1979): Planet. Space Sci. *27*, 1131

[24] Izakov, M.N. (1978): Kosm. Issled. *16*, 403

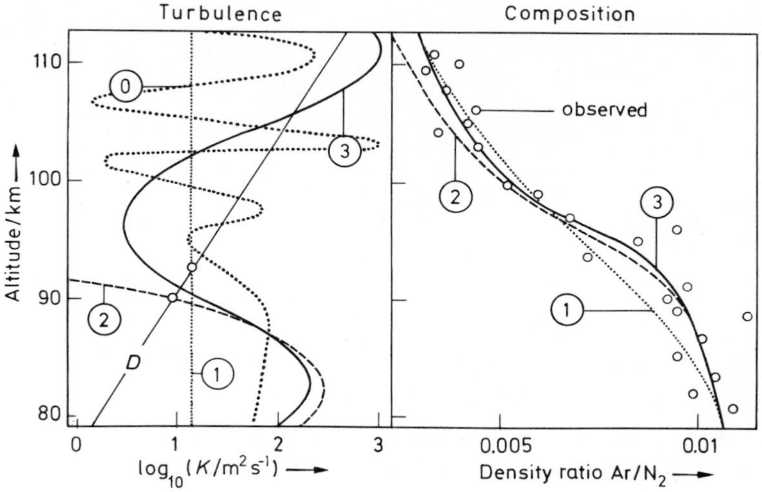

Fig. 64. Molecular (D) and eddy (K) diffusion coefficient profiles (left-hand side) compared with the relevant composition profiles (Ar/N_2, right)[19]. Circles: measured data; PHILBRICK's corresponding K-profile (dotted line) marked 0. K-profiles proposed by different authors (on left) were fitted to the data (see the relevant composition profiles on right) and marked: 1, K=const.; 2, SHIMAZAKI[18]; 3, BLUM et al.[19]

molecular heat conduction coefficient[25]. In the presence of eddy diffusion a second term appears which contains K, the eddy diffusion coefficient

$$\phi_z = -\lambda \frac{\partial T}{\partial z} - K \rho c_p \left(\frac{\partial T}{\partial z} - \frac{g}{c_p}\right) \tag{8.2}$$

where ρ is mass density, c_p the specific heat capacity and g the acceleration of gravity.

For application to atmospheric measurements it is often helpful to derive a height-integrated relation which is obtained by integrating Eq. (8.1) from z_0 to infinity. Using Eq. (8.2) for elimination of ϕ_z this gives an expression for the *eddy diffusion coefficient*[26]

$$K = \frac{\int_{z_0}^{\infty} dz(P-L) - [\partial T/\partial z]_0}{\rho_0 c_p [(\partial T/\partial z)_0 + g/c_p]}. \tag{8.3}$$

At the *mesopause* temperature minimum we have $[\partial T/\partial z]_m = 0$ so that, simply

$$K = \frac{1}{\rho_m g} \int_{z_m}^{\infty} dz(P-L), \tag{8.3a}$$

[25] See contribution by K. Suchy in this volume, Sect. 23γ, p. 69 and 25δ, p. 176

[26] Alcaydé, D., Fontanari, J., Kockarts, G., Bauer, P., Bernard, R. (1979): Ann. Géophys. *35*, 41

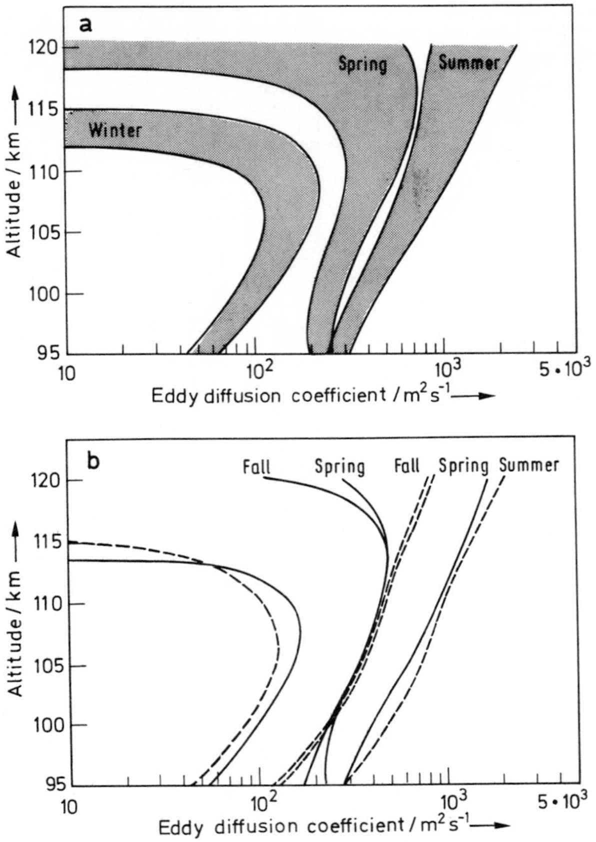

Fig. 65 a, b. Vertical profiles of the eddy diffusion coefficient deduced from aeronomical theory using incoherent scatter average temperature profiles (for St. Santin, France, latitude 45°)[26]. (**a**) Computed with aeronomical theory. The shaded ranges represent the uncertainty of the overhead EUV heat input (above 120 km) between the limits 1 and 2 mW m^{-2}. (**b**) Deduced from incoherent scatter data (full lines) and from the CIRA-72 [12] model (broken lines)

which means that K can be interpreted as the ratio of total net energy input to gravitational energy.

The gain term P contains as major contributions absorption of solar light in the SCHUMANN-RUNGE bands and continuum, and in the HERZBERG continuum, and absorption of solar EUV. Using an atmospheric model these contributions were computed and added up[26]. On the other hand, the cooling term L is, in the lower thermosphere, mainly due to infrared radiation in the 15 nm band of CO_2[27]. Since the vertical distribution of CO_2 depends on eddy transport, L depends on K and Eq. (8.3) is an implicit one which must be resolved by iteration. Figure 65 gives the results of such computations showing large seasonal effects[26].

Molecular diffusion predominates above the turbopause, i.e. in the middle and upper thermosphere. It is dealt with in vol. 49/6 in YONEZAWA's contri-

[27] Houghton, J.T. (1969): Quart. J. Roy. Met. Soc. *95*, 1

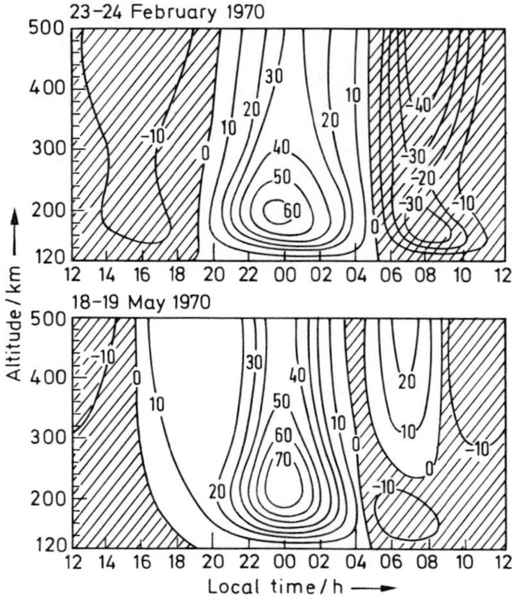

Fig. 66. Diurnal variation of the neutral wind component along the magnetic field line over Millstone Hill (USA) after semi-empirical theorey[29] (see text). Parameter of the curves: w/ms^{-1} (positive upwards). Periods characteristic for winter (top) and summer (bottom)

bution (his Sect. 5, p. 136) and, under another aspect, also by THOMAS (his Sect. 22, p. 84). Release experiments (usually of water vapour) are particularly used for determining experimentally the diffusion coefficients (and scale heights). The relevant theory is given by YONEZAWA (ibidem), where the experimental results are also discussed.

As shown by KOHL and KING[2] the ionospheric plasma moves along the magnetic field lines, driven by (mainly horizontal) neutral winds[28]. Therefore, plasma transport measurements, e.g. by the incoherent scatter technique (Sect. 5) make possible conclusions on the neutral motions, though their vertical component will often be different. In Fig. 66 the neutral velocity in the magnetic field direction was computed from a semi-empirical theory which was numerically adjusted to the data measured at Millstone Hill (USA)[29]. The pattern may be described as a mixture of diurnal and semidiurnal tides. Velocities are much higher than previously thought.

δ) *Horizontal transport* as bulk motion of neutrals is called wind. The relevant theory is found in STUBBE's contribution in vol. 49/6 (Chap. BI, in particular Sects. 2–5). The same author also gives a short survey on observed wind data (his Chap. D), mainly with results from optical (ground-based) obser-

[28] Situations are not covered where electric fields have significant effect (provoking plasma motion transverse to the magnetic field line), e.g. near the dip equator and in the auroral zones (see Sect. 18)

[29] Roble, R.G., Salah, J.E., Emery, B.A. (1977): J. Atmos. Terr. Phys. *39*, 503

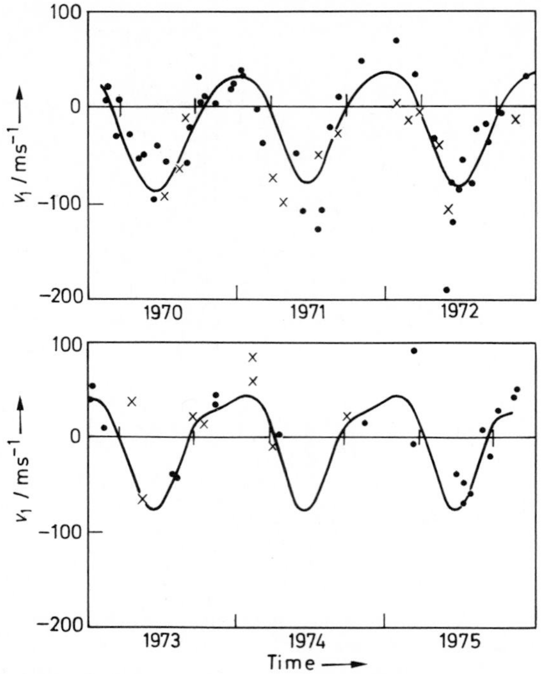

Fig. 67. Diurnally averaged meridional winds at Millstone Hill (USA) for 300 km inferred from incoherent scatter Doppler measurements by EMERY[31] (dots for quiet, crosses for magnetically disturbed days). The solid curve is a third-order harmonic fit to the quiet day results. (Note that the total number of observing days per month is rather small.)[32] Compare Figs. 180 and 181 in Sect. 17

vations of rocket-borne metallic vapour releases. In Sect. 6 are described in some detail more recent uses of the airglow method when applied aboard a satellite and its results.

In Sect. 5 a few more results of the incoherent scatter method may be found (see also [30]). This technique, of course, determines plasma motion (called "drift") which may differ from the neutral motion and probably does quite often at higher thermospheric altitudes and where electric fields are important; see Sect. 18. Data from the longest series of such measurements (Millstone Hill, Massachusetts, USA) were used as an input into a semi-empirical dynamic model for the declining portion of Sunspot cycle 20[31]. Diurnally averaged horizontal meridional wind velocities so obtained are reproduced in Fig. 67 for the period 1970–1975[32]. They show a clear seasonal variation with a mean poleward flow in winter, but a stronger, and equatorward one in summer. For comparison we show the meridional circulation as computed by ROBLE et al.

[30] Evans, J.V. (1972): J. Atmos. Terr. Phys. *34*, 175 (review paper on drift measurements with incoherent scatter)

[31] Emery, B.A. (1978): J. Geophys. Res. *83*, 5691 and 5704

[32] Babcock, R.R. Jr., Evans, J.V. (1979): J. Geophys. Res. *84*, 7348

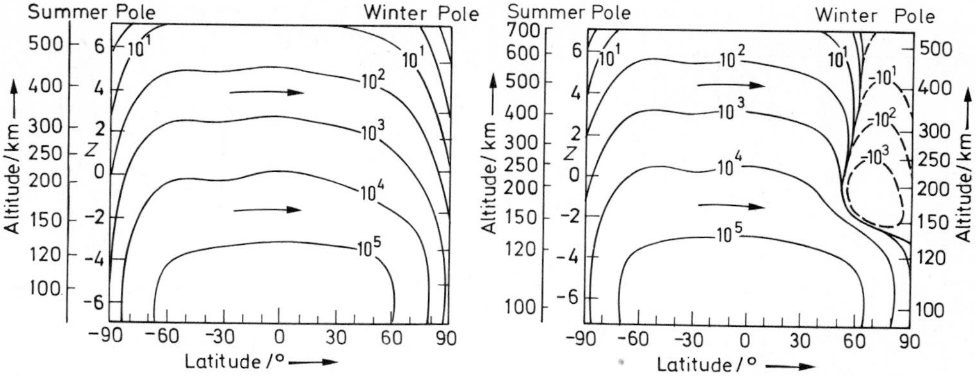

Fig. 68. *Zonal means of* thermospheric circulation computed by ROBLE et al.[33] with a semi-empirical aeronomical theory. The *meridional maps* for December solstice (left for low, right for high solar activity) give the total air transport (in kg s^{-1}) down to the relevant altitude, always from the summer towards the winter pole. The reverse cell appearing in winter (upper right corner) is driven by auroral heating (inferred from incoherent scatter data). Compare Figs. 180 and 181 in Sect. 17

in Fig. 68; these authors apply a semi-empirical aeronomical theory (see Sect. 12) with incoherent scatter results included as inputs[33].

The seasonal dependency of the diurnal variation of the neutral average winds computed[29] as for Fig. 66 is shown in Fig. 69. It may be noted that semidiurnal components (and lunar ones) appear quite regularly in ionospheric drift measurements[34–37] while they should be much smaller after the model computations. Coherent Doppler Radar (MST-radar) is more and more applied for measuring winds in the middle atmosphere, namely by day in the lower ionosphere (60–90 km) but by day and night in the tropo- and stratosphere (1–35 km).

Other radio methods for measuring plasma drifts were discussed in vol. 49/2 of this Encyclopedia, including results as known in 1967. Except for the method of observing meteor trails as tracers [*32*, *41*] the relation of plasma drifts with neutral winds remains questionable.

Fortunately, in the lower and middle thermosphere (up to 150 km) conditions are such that drifts should almost be identical with neutral winds[35–39]. Earlier results of such drift measurements are summarized in [*40*, *41*][37,40,41],

---

[33] Roble, R.G., Dickinson, R.E., Ridley, E.C. (1977): J. Geophys. Res. *82*, 5493; see also SCHMIDTKE's contribution in this volume,

[34] Czechowsky, P., Schmidt, G., Rüster, R. (1983): Radio Sci. *19*, 441

[35] Wright, R.W.H. (1968): J. Atmos. Terr. Phys. *30*, 919. This paper showed that the reduction method known as "similar fades method" [34,36,37] gives essentially correct large-scale velocities (though the individual values obtained with releases may be greater). As for the "full correlation method" [38] it gives too small values; an explanation can be found in [39]

[36] Harnischmacher, E., Rawer, K. (1958): J. Atmos. Terr. Phys. *13*, 1; (1968): J. Atmos. Terr. Phys. *38*, 871; (1968): [*41*], 242

[37] Harnischmacher, E. (1968): [*41*], 277

[38] See: Suchy, K., Rawer, K. (1967): This Encyclopedia, vol. 49/2, Sect. 51 (many references); add: Wright, J.W., Fedor, L.S. (1969): J. Atmos. Terr. Phys. *31*, 925

[39] Sprenger, K., Schminder, R. (1969): J. Atmos. Terr. Phys. *31*, 1085

[40] Sprenger, K., Greisiger, K.M., Schminder, R. (1969): Ann. Géophys. *25*, 505; Sprenger, K., Schminder, R. (1969): J. Atmos. Terr. Phys. *31*, 217; Lauter, E.A., Sprenger, K., Entzian, G. (1969): Progress in Astronautics and Aeronautics *22*, 401

[41] Kent, S.G. (1970): Rev. Geophys. Space Phys. *8*, 229

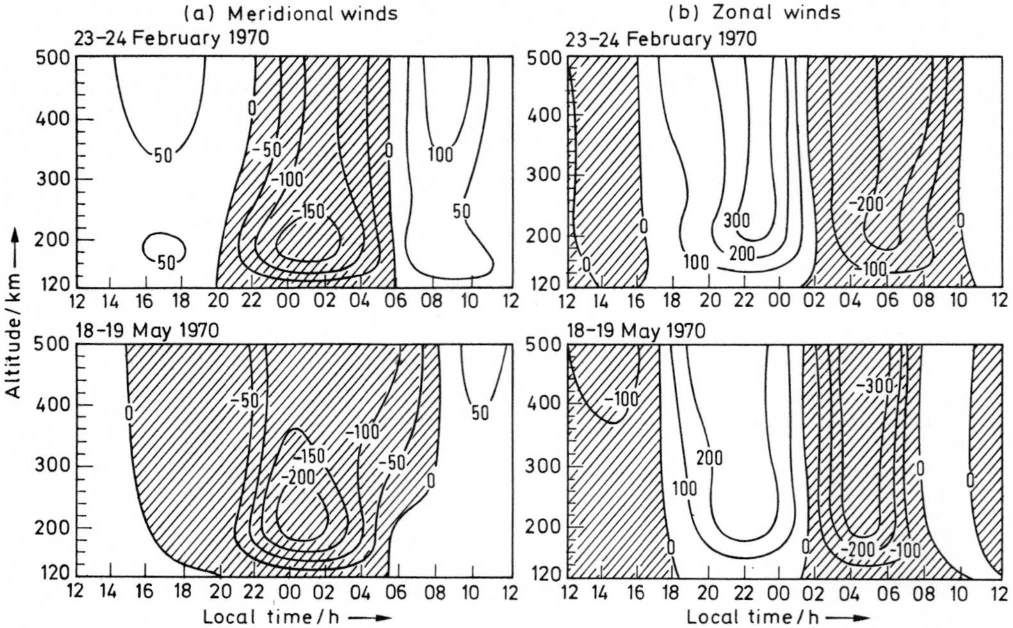

Fig. 69. Typical seasonal height profiles of thermospheric horizontal winds (left, meridional; right zonal) after semi-empirical theory adapted to the incoherent scatter data measured at Millstone Hill (USA)[29]. Compare Fig. 66 with field-aligned motion, also Figs. 44 and 45, p. 279, which were similarly established with optical measurements

more recent ones in[42,43]. The most important result of these ground-based measurements (under quiet mid-latitude conditions) is the appearance of diurnal and semidiurnal tides. Below 100 km quite regular, mainly diurnal wind patterns were found, depending on the season and to some extent on the solar cycle of 11 years[40]. In the E-region (100–120 km) semidiurnal rotating drifts are more often, seen, except in summer. Day-by-day changes are important, such as if different wind systems were in competition[36,44]; lunar semidiurnal tides, also rotating with the Sun are not negligible and produce complex phase variations[45].

9. Composition. Since the discovery of ozone resulting from photodissociation of oxygen, and of the importance of this latter reaction (see RAWER's Introduction in vol. 49/6) the role of chemistry has become more and more important in aeronomy (see the contributions by THOMAS and by YONEZAWA in vol. 49/6 and [*1*]). Composition measurements are the most adequate tool

[42] Kazimirovsky, E.S., Kokourov, V.D., Žovtij, E.I. (1979): J. Atmos. Terr. Phys. *41*, 867 (review); Kazimirovsky, E.S., Kokourov, V.D. (1979): J. Geomag. Geoelectr. *31*, 195

[43] Kazimirovsky, E.S. (1984): [*45*], 185

[44] Harnischmacher, E., Rawer, K. (1981): Arch. Elektr. Übertr. (AEÜ) *35*, 141

[45] Harnischmacher, E., Rawer, K. (1958): Geofis. Pura e Appl. *39*, 216; (1978/9): Riv. Ital. Geofiz. e Science Aff. *V*, 105

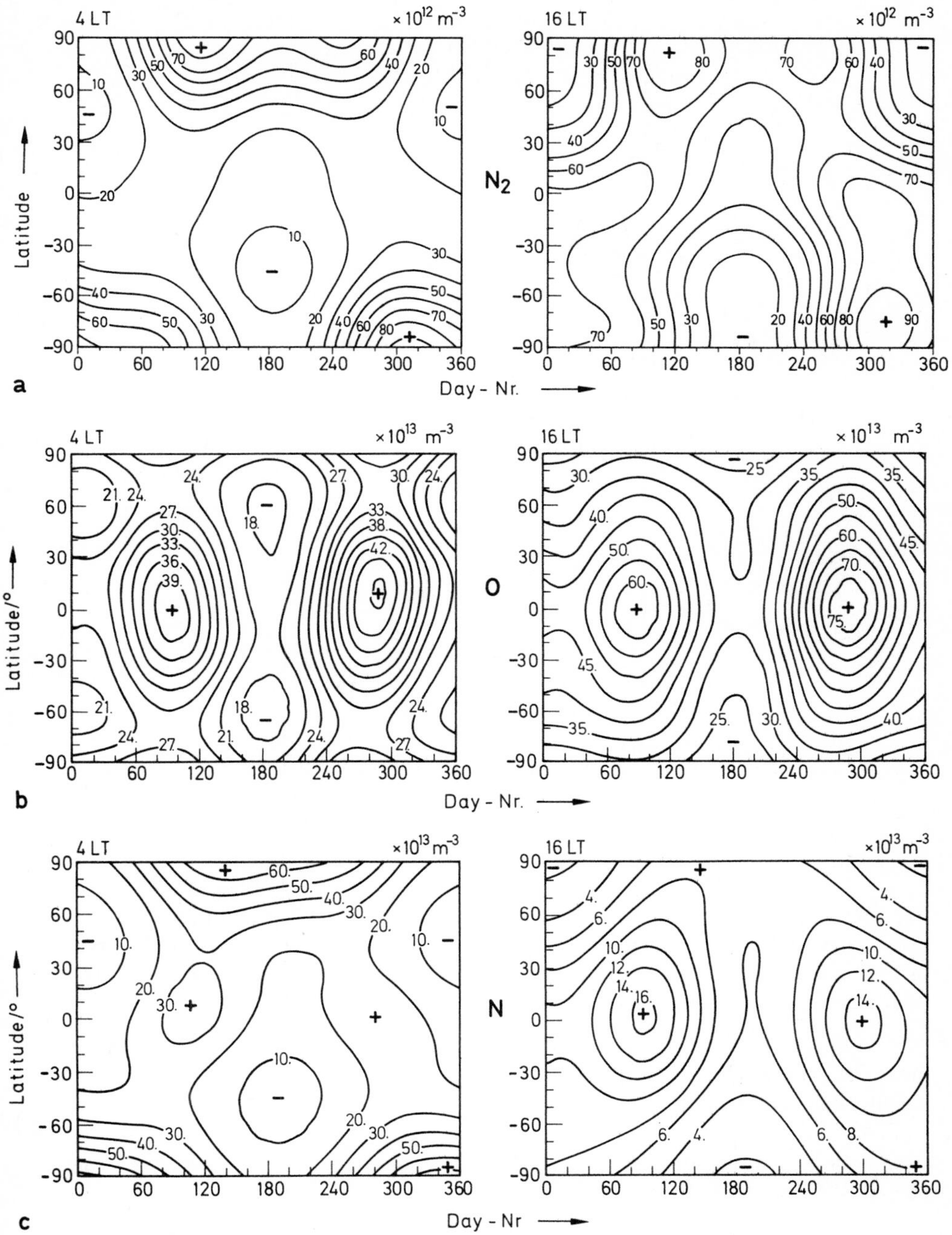

Fig. 70 a–c. Averaged latitudinal/seasonal variation of 300 km constituent densities as measured aboard the AEROS-B satellite (1974/1975). Abscissa yearly day number. The parameter given by each curve, multiplied by the upper right-hand corner exponential, gives the density in m^{-3}. The curves were drawn according to a descriptive model. Diagrams for night (04 h LT, left) and day (16 h LT, right). (**a**) Molecular nitrogen N_2; (**b**) Atomic exygen O; (**c**) Atomic nitrogen N [*26*]

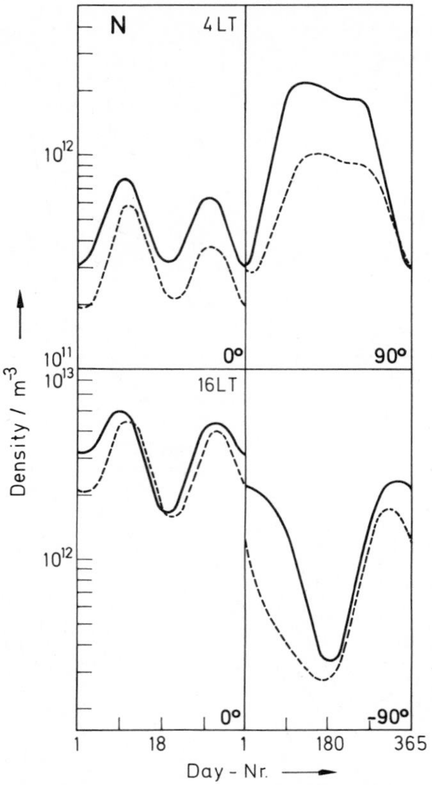

Fig. 71. Seasonal variations of atomic nitrogen density in 375 km at equatorial (0°, left) and polar (90°, right) latitudes. Top diagrams for 04, bottom ones for 16 h local time. Abscissa, yearly day number. Broken curve, model derived from AE-C; full curves from AEROS-B satellite [*26*]. [Read 180 instead of 18, left on abscissa]

for checking the results of aeronomical chemistry. Experimental methods are discussed by THOMAS in vol. 49/6 (Chap. E) and in our Sect. 3. For the major constituents the transition between molecular and atomic oxygen is of particular interest and is discussed by YONEZAWA in vol. 49/6 (Chap. III).

α) There is still some discussion about the amplitude of *seasonal variations* near 200 km, where data inferred from incoherent scatter[1] give a larger variation than mass spectrometers[2]. Near 70° N latitude the O_2 densities were

[1] Alcaydé, D., Bauer, P., Fontanari, J. (1974): J. Geophys. Res. *79*, 629

[2] Nier, A.O., Potter, W.E., Kayser, D.C. (1976): J. Geophys. Res. *81*, 17; Kayser, D.C., Potter, W.E. (1978): J. Geophys. Res. *83*, 1147

Fig. 72a–c. Semidiurnal tidal amplitudes (left) and phases (right) in the 145–400 km height range determined with the data from the neutral atmosphere composition experiment NACE aboard the satellites AE-C and SAN MARCO-3[13]. Dots designate average values, the dispersion range being given by bars. Curves, theoretical models[14] (dotted) and [15] (broken), and MSIS [*20*] model (full line). The dashed line is from an empirically adapted diffusion model. (**a**) for N_2; (**b**) for O; (**c**) for He

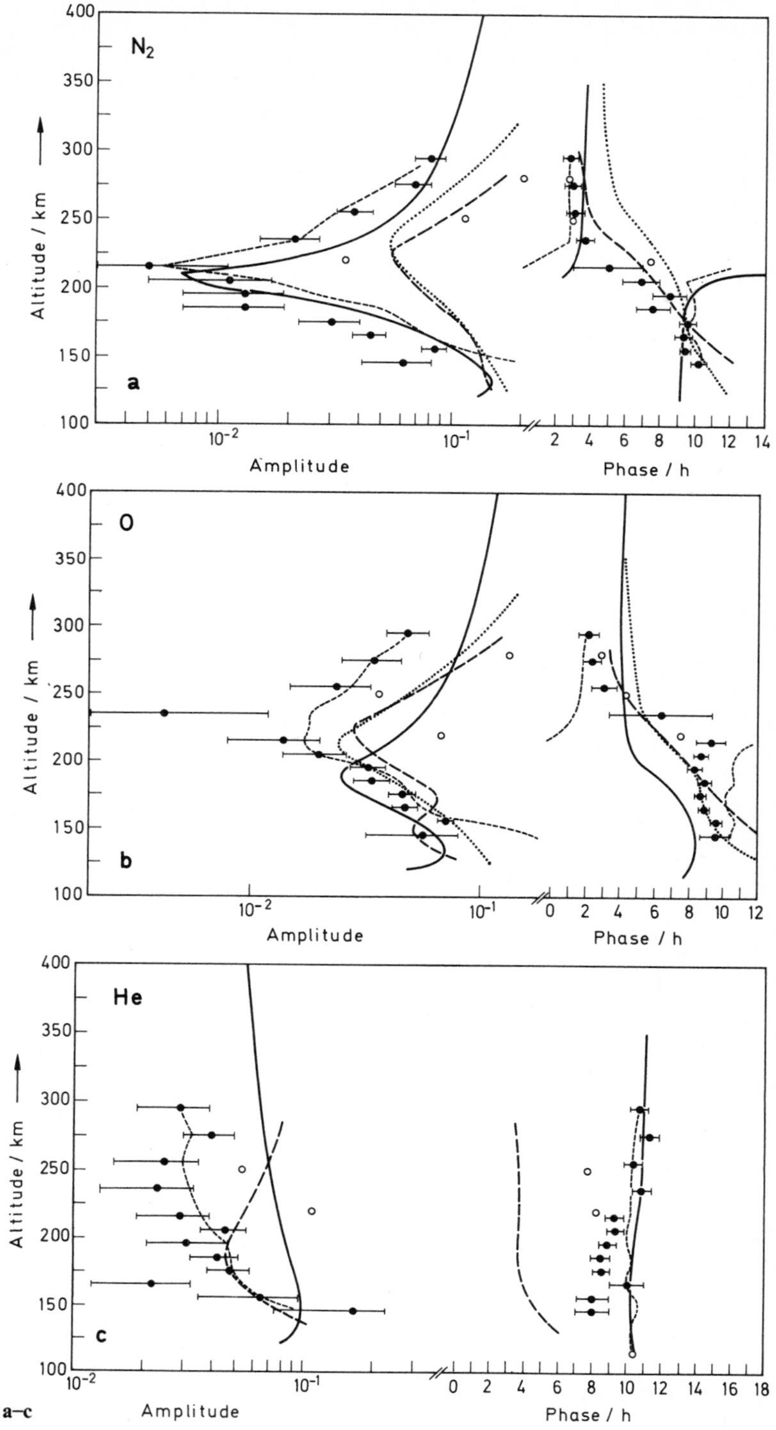

Fig. 72a–c

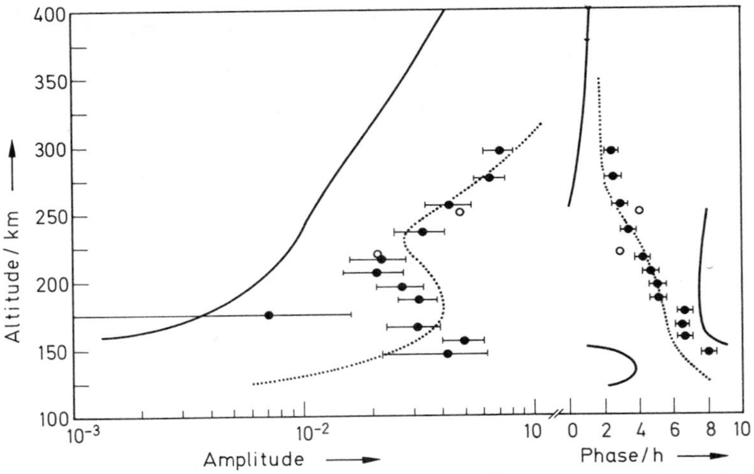

Fig. 73. Terdiurnal tidal amplitudes and phases for N_2[13]. Same presentation as in Fig. 72

greatest[2]. Measurements by the USA-satellite AE-C are also reported to show significant increases at higher geomagnetic latitudes[3]; this may be due to the increased temperature in the auroral zones (see Figs. 40–43, Sect. 6). Summary maps obtained from the AEROS-B mass-spectrometric measurements [*26*] are displayed in Fig. 70[4] for N_2, O and N.

β) As for *minor constituents* see YONEZAWA (Chap. V) and THOMAS (Sect. 23) in vol. 49/6. THOMAS also considers so-called "odd" (mostly excited) species and their important role in thermospheric aeronomy.

A few results obtained with optical methods are displayed in Sect. 6. For measuring atomic nitrogen with mass spectrometers problems with oxidation in the instrument occur. Data inferred from the observed NO peak[5,6] exhibit diurnal variation by a factor of 8 (maximum around 16 h), also latitudinal, seasonal and semi-annual variations[4,7]; see Figs. 70 and 71 [*26*]. NO also increases towards higher latitudes as was found with the USA satellite OGO-4[8], and was confirmed with AE-C measurements[9].

Noble gases, though inactive in chemical reactions, are important as a tool for studying the temperature profile and diffusion; see YONEZAWA in vol. 49/6 (Chap. V); this author also discusses the escape problem of helium and of hydrogen (his Sect. 31). As H is the lightest gas it exhibits the greatest variations in the upper thermosphere and exosphere.

γ) More recent measurements of neutral composition have been made with a particular mass spectrometer collecting data from prefixed mass numbers only[10]. This instrument was flown

---

[3] Oppenheimer, M., Dalgarno, A., Brinton, H.C. (1976): J. Geophys. Res. *81*, 4678

[4] Köhnlein, W., Krankowsky, D., Lämmerzahl, P., Joos, W., Volland, H. (1979): J. Geophys. Res. *84*, 4355

[5] Mauersberger, K., Engebretson, M.J., Kayer, D.C., Potter, W.E. (1976): J. Geophys. Res. *81*, 2413

[6] Krankowsky, D., Joos, W. (1977): Trans. Amer. Geophys. Un. EOS *58*, 457

[7] Engebretson, M.J., Mauersberger, K., Kayser, D.C., Potter, W.E., Nier, A.O. (1977): J. Geophys. Res. *82*, 461

[8] Rusch, D.W., Barth, C.A. (1975): J. Geophys. Res. *80*, 3719

[9] Gravens, T.E., Stewart, A.I. (1978): J. Geophys. Res. *83*, 2446; Stewart, A.I., Cravens, T.E. (1978): J. Geophys. Res. *83*, 2453

[10] Spencer, N.W., Niemann, H.B., Carignan, G.R. (1973): Radio Sci. *8*, 204; Pelz, D.T., Reber, C.A., Hedin, A.E., Carignan, G.E. (1973): Radio Sci. *8*, 277

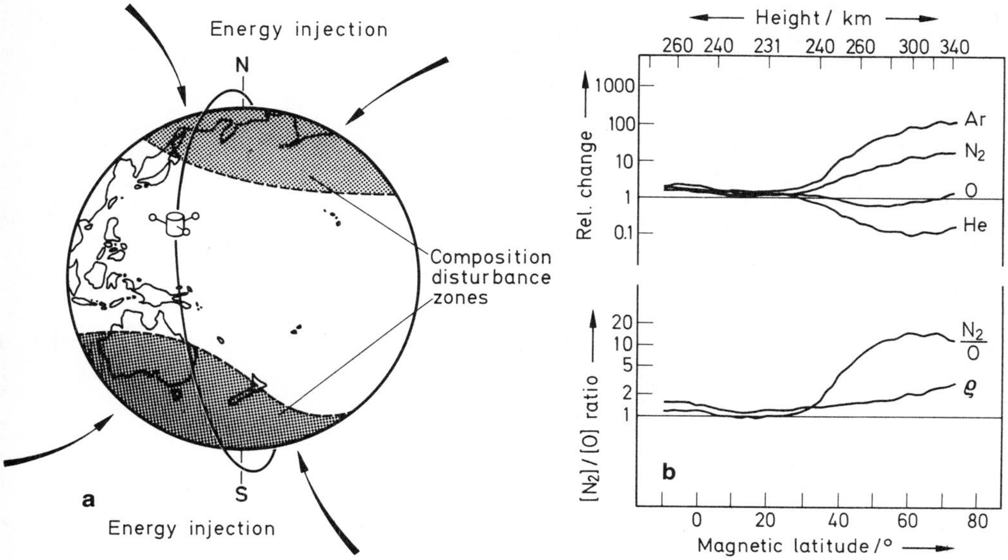

Fig. 74a, b. Effects of magnetic storms on composition: (**a**) Regions where the two types of changes occur; (**b**) Perturbance of the partial densities (upper diagram, relative values) and of total density ρ, and density ratio of N_2 to O (bottomside). Abscissa, magnetic latitude[16]

aboard satellites SAN-MARCO-3[11] and AE-C. The latter satellite had a special propulsion system which made possible the coverage of a low-height range down to 145 km. As part of a study of the *diurnal* variations[12] a large number of data were analyzed after semi- and ter-diurnal *tidal variations*[13].

Some results concerning the semidiurnal tide are shown in Fig. 72. The tidal amplitude is up to 10% with a net minimum slightly above the 200 km level.

This feature is indicated in the theoretical models of Mayr et al.[14] (dotted line) and of Forbes[15] (broken line) but not as strong as observed. The empirical MSIS model [*20*] (see Sect. 11) describes this phenomenon fairly well, except for He. A diffusion model with the *observed* temperature variations as input, however, gives full agreement. Also, except for He, there is a drastic phase change by 6 h (i.e. 180°) at the said level. Ter-diurnal variations are much smaller, a few percent only in O and Ar, but up to 8% in N_2 (see Fig. 73).

δ) Summarizing recent results allowing for a global description of *magnetic storm* activated *perturbations*[16], two zones with different characteristics are distinguished. At high and mid-latitudes the heavier constituents Ar, N_2, O_2

---

[11] Newton, G.P., Kasprzak, W.T., Curtis, S.A., Pelz, D.T. (1975): J. Geophys. Res. *80*, 2289

[12] Hedin, A.E., Spencer, N.W., Mayr, H.G., Harris, I., Porter, H.S. (1978): J. Geophys. Res. *83*, 3355

[13] Hedin, A.E., Spencer, N.W., Mayr, H.G. (1980): J. Geophys. Res. *85*, 1787

[14] Mayr, H.G., Harris, I., Spencer, N.W., Hedin, A.E., Wharton, L.E., Porter, H.S., Walker, J.C.G., Carlson, H.C. (1979): Geophys. Res. Lett. *6*, 447. See also: Mayr, H.G., Hedin, A.E., Reber, C.A., Carignan, G.R. (1974): J. Geophys. Res. *79*, 619

[15] Forbes, J.M. (1978): J. Geophys. Res. *83*, 3691

[16] Prölss, G.W. (1980): Rev. Geophys. Space Phys. *18*, 183

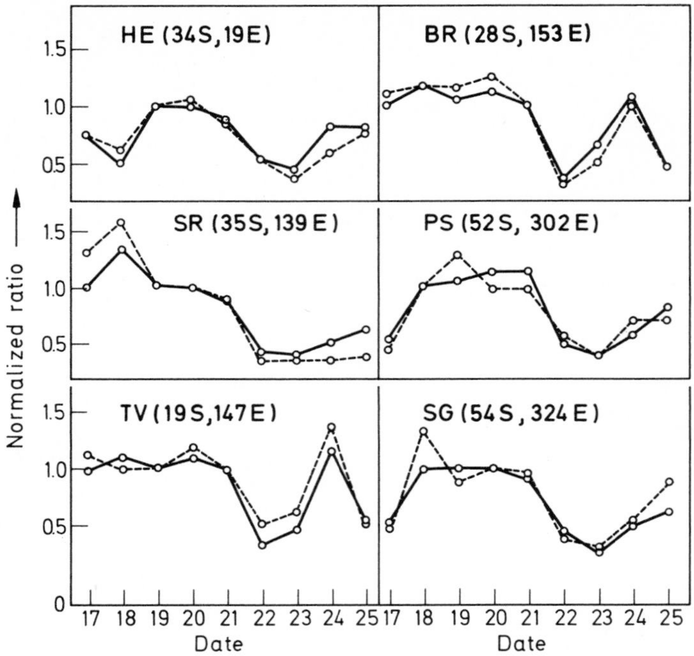

Fig. 75. Comparison of the normalized variations of (i) the composition ratio $[O]/[N_2]$ (full curve, data from ESRO-4) and (ii) the peak electron density of the ionospheric F2-layer (broken curves) during the second half of Feb. 1973. A magnetic storm occurred on Feb. 22. Southern hemisphere ionospheric sounding stations: HE, Hermanus; BR, Brisbane; SR, Salisbury; PS, Port Stanley; TV, Townsville; SG, South Georgia[16]

are increased while He decreases significantly as a consequence of heating (see Fig. 74)[16]. At low latitudes a small increase of all constituents is observed. The reaction time of the thermosphere is short, about 1 h. The observed composition changes cannot be explained with vertical diffusive transport alone[17], but also require horizontal transport on a global scale. A narrow connection was found for "negative" ionospheric stroms[17] between the peak plasma density and the composition as given by the ratio of O to N_2 (Fig. 75)[16]. This finding seems to justify the idea that the composition changes are the primary cause of those of the plasma density via ion chemistry, as demonstrated by a numerical simulation model[18]. At high latitudes, the invariant (magnetic) latitude[19] is the decisive coordinate, but some dependence on UT is also found which is explained by the rotation of the true magnetic poles[20].

---

[17] See Rawer, K., Suchy, K. (1967): This Encyclopedia, vol. 49/2, Sect. 44; see also Sect. 18 below

[18] Jung, M.J., Prölss, G.W. (1979): J. Atmos. Terr. Phys. *40*, 1347

[19] See: Hess, W.N. (1972): This Encyclopedia, vol. 49/4, p. 126

[20] Dachev, T.P., Carignan, G.R., Walker, J.C.G. (1981): preprint Space Phys. Res. Lab., Univ. of Michigan, Ann Arbor (USA)

C. Empirical (descriptive) modelling

10. International reference atmospheres. The first approaches for modelling the upper atmosphere were "standard atmospheres", i.e. tables of the height dependence for the main atmospheric parameters (see Sect. 1) which should be valid everywhere and at every time[1]. Such tables are quite helpful up to 120 km, but not at greater altitudes where diurnal variations become important. With the advent of satellites measured data became available (see Chap. A) and an international effort was undertaken.

α) The result was CIRA 1961 [*25*] established by an ad hoc committee chaired by H. KALLMANN-BIJL; it was approved only $3\frac{1}{2}$ years after the launching of SPUTNIK 1, the first artificial satellite. Basic data were the *drag measurements* (see Sect. 2) made with the rather low orbiting satellites of these first years. At the very time when CIRA-61 was accepted by COSPAR, NICOLET[2] interpreted new data from the first high-altitude satellite, ECHO-1; he found that, unlike the assumptions of the committee, the influence of diffusive separation (see YONEZAWA's contribution in vol. 49/6, Chap. II) was very important. So, just when printed, the first CIRA had become obsolete. Another drawback was the poor description of variations; with only two tables per day (14 and 04 h LT) there was no chance of reproducing the large variability of the thermosphere which was revealed more and more with the rapidly increasing number of observations.

β) Instead of an essentially empirical approach as that of CIRA 1961 [*25*] HARRIS and PRIESTER[3] in 1962 presented a model based upon *one-dimensional aeronomical theory*. These authors were the first to introduce a time-dependent energy balance including vertical heat conduction and the energy transport due to the diurnal expansion and nightly contraction of the upper atmosphere. Their balance equation reads [compare our Eq. (1.23)]:

$$\frac{\partial}{\partial z}\left(K(T)\frac{\partial T}{\partial z}\right)-\rho\, c_p\frac{\partial T}{\partial z}\cdot\int\limits_{z_0}^{z}\frac{dz'}{T^2}\frac{\partial T}{\partial z}+\sum_i Q_{\rm i}(z,t)=\rho\, c_p\frac{\partial T}{\partial t} \tag{10.1}$$

where K is thermal conductivity, ρ mass-density and c_p specific heat at constant pressure. The decisive term is that with $Q_{\rm i}$, namely the heat sources and sinks. One heat source was felt to be well known, namely solar EUV radiation (see SCHMIDTKE's contribution in this volume), which is absorbed by atomic and molecular oxygen and nitrogen. As only sink infrared reradiation of atomic oxygen by the transition $({}^3\mathrm{P}_1 \rightarrow {}^3\mathrm{P}_2)$ was considered. At the lower boundary level ($z_0 = 120$ km) temperature and densities were kept constant. For higher levels Eq. (10.1) together with that of hydrostatic equilibrium

$$n_{\rm j}(z,t)=n_{\rm j}(z_0,t)\frac{T(z_0,t)}{T(z,t)}\exp\left[-\int\limits_{z_0}^{z}dz\frac{m_{\rm j}\,g(z)}{kT(z,t)}\right] \tag{10.2}$$

[see our Eqs. (1.3a), (1.7), (1.8)] were simultaneously integrated on a large computer. It was, however, found out that the diurnal variation so obtained did not agree with the observations. In particular, the theoretical density maximum occurred at 17 h instead of 14 h LT.

---

[1] For example, U.S. Standard Atmosphere. Washington D.C.: U.S. Govnt. Printing Office

[2] Nicolet, M. (1961): Smithsonian Astrophys. Obs. Spec. Rep. No. 75

[3] Harris, I., Priester, W. (1962): J. Atmos. Sci. *19*, 286

In order to *repair* this shortcoming an additional '*second heat source*' was introduced ad hoc. By suitably fitting the source profile and intensity good agreement between observed and calculated densities could be achieved. By this method, the procedure had become semi-empirical; nevertheless, it was theoretical in the sense that no independent temperature input was used.

The real nature of the "second heat source" is not necessarily of a physical character. To a certain extent it should be better understood as a correction for the simplifications introduced in the theory. On the other hand it has been stated that gravity waves[4] by their very structure transport energy through the lower boundary which stems from motions in the dense lower layers.

γ) Anyway, the computer program established by HARRIS and PRIESTER[3] provided a sufficient number of free parameters to reproduce the observed densities. So it was finally used as a vehicle to include also the most important features of a set of empirical models established by JACCHIA[5].

These were patterned after an early, more theoretical model of NICOLET[2], in which an equilibrium temperature profile was first computed for very strong solar XUV irradiation followed by computing temperature profiles for different times after shutoff of the source.

Including these results, the Working Group now chaired by W. PRIESTER produced CIRA 1965 [*11*], more precisely the thermospheric Part III of it (120–800 km range). In Part I (prepared by CHAMPION) it presented a "mean atmosphere" from 30 to 300 km, still following the tradition of a "standard atmosphere", though it is noted that "at high altitudes (viz. above 300 km) the properties of the atmosphere varied so much with solar flux and time of day that the use of mean properties is not meaningful". In Part III of CIRA 1965 a set of ten models is given each covering 12 pages, namely one table for every second hour, each with columns for temperature, density, pressure, scale height, mean molecular weight and numerical densities of six atmospheric constituents (between H and Ar in weight). Each model is for a certain level of solar activity ranging from "extremely low" to "extremely high". With this variability CIRA 1965 was a much better solution than its predecessor.

δ) It lasted for 7 years, then CIRA 1972 [*12*] became officialized and still is[6]. With L. JACCHIA as chairman the Working Group had tackled the problem in a new way, going back to a more empirical procedure. In the meantime the first reliable, direct temperature measurements (see Chap. A) had become available and were now taken as an almost independent input. JACCHIA[7] had pointed out that the temperature profiles obtained with the two models mainly in use at that time[2,3] could be very well represented [see Eq. (5.5)] by

$$T = T_{ex} - (T_{ex} - T_{120}) \exp[-s(z - 120\,\text{km})], \tag{10.3}$$

T_{ex} being the asymptotic (exospheric) temperature. He constructed a model similar to NICOLET's[2], starting with the boundary conditions of CIRA 1965 at 120 km, with exponential temperature profiles according to (Eq. (10.3)) and approximated to the observed data by fitting the gradient parameters[7] (see Sect. 5). In 1966,

---

[4] Jones, W.L. (1976): This Encyclopedia, vol. 49/5, pp. 177–216

[5] Jacchia, L.G. (1964): Smithsonian Astrophys. Obs. Spec. Rep. No. 170; (1965): Smithsonian Contr. Astrophys. *8*, 215

[6] The next CIRA is expected to be ready for 1986

[7] Jacchia, L.G. (1970): Smithsonian Astrophys. Obs. Spec. Rep. 313

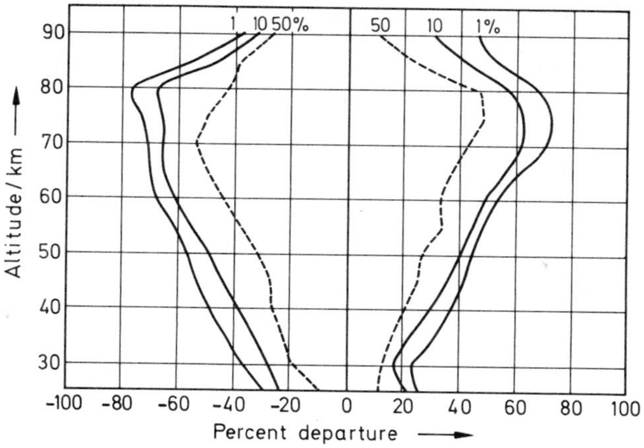

Fig. 76. Dispersion of measured densities in the 30–90 km height range[12] as compared with the CIRA-72 standard profile [*12*]

these models (sometimes called J 65) were incorporated into "COESA", the US Standard Atmosphere Supplement. Difficulties arising with the lower boundary condition at 120 km were resolved by shifting it down to 90 km; there exists a well-known isopycnic layer[7]. The exponential shape of the temperature profile [Eq. (10.3)] could not be extended down to that height. Another problem was related to oxygen dissociation (see YONEZAWA's contribution in vol. 49/6, Chap. III). Incited by new mass-spectrometric data compiled by von ZAHN[8] the model was revised again[9] in 1970 with considerable increase of the atomic oxygen density and decrease of that of molecular oxygen, so as to match the mass-spectrometric data (see Sect. 3) near 150 km[10]. A first draft of the tables for the thermosphere was published [*23*] before finally being officialized as CIRA 1972 [*12*].

ε) Again Part I of CIRA 72 ought be taken as a kind of *standard atmosphere*. Between 25 and 75 km the model of GROVES[11] is used. Above 120 km, assuming a latitude of 30° and COVINGTON index 145, diurnal, seasonal and semi-annual variations were averaged out of JACCHIA's thermosphere model [*23*], which is described below. In the region between 75 and 120 km a smooth transition was arranged starting from a composite temperature profile based on models and observed data. Figure 76[12] depicts the variability of monthly means in this range. Including higher altitudes Fig. 77 shows means of temperature and density together with the extreme values taken from Part III.

ζ) The latter is the most important one. An effort was made to take account of most of the observed *variations* in time and space, namely (see

[8] von Zahn, U. (1970): J. Geophys. Res. *75*, 5517

[9] Jacchia, L.G. (1971): J. Geophys. Res. *76*, 4602

[10] A first draft of the tables for the thermosphere was published by JACCHIA [*23*] (often referred as Jacchia 71, or simply J 71) before being finally officialized by COSPAR as CIRA 1972 [12]

[11] Groves, G.V. (1970): Seasonal and latitudinal models of atmospheric temperature, pressure and density, 25 to 110 km = Air Force Surveys in Geophys. No. 218. Cambridge (Ma., USA): AFCRL

[12] Champion, K.S.W., Schweinfurth, R.A. (1972): [*12*], 3

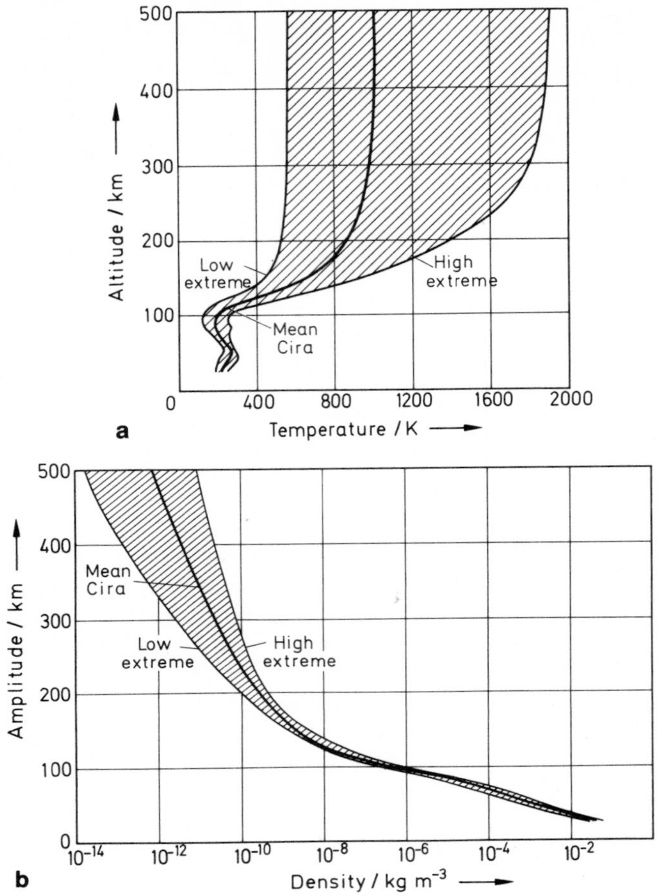

Fig. 77 a, b. CIRA-72 mean and extreme profiles of (**a**) temperature and (**b**) mass density[12] [read 'altitude' on both ordinates]

Chap. B) with (i) the solar cycle, (ii) daily change in solar activity, (iii) diurnal variation, (iv) effect of geomagnetic activity, (v) semi-annual variation and (vi) seasonal-latitudinal variations in the lower thermosphere and of helium. This is, of course, an extremely difficult task.

In order to resolve it with limited effort one particularly sensitive parameter, namely the *exospheric temperature*, T_{ex}, was chosen as intermediate variable. Apart from the diurnal variation (iii), which is accounted for by subdividing the tables (like in CIRA-65 [*11*]), the most important variable influences are only taken account of via their influence on T_{ex}, after which the main tables (two for each group) are specified for 17 groups covering the T_{ex} range from 500 to 2200 K in steps starting with 50 K and increasing with T_{ex} to, finally, 200 K. The individual influences [(i) to (vi) above] are so represented that T_{ex} depends:

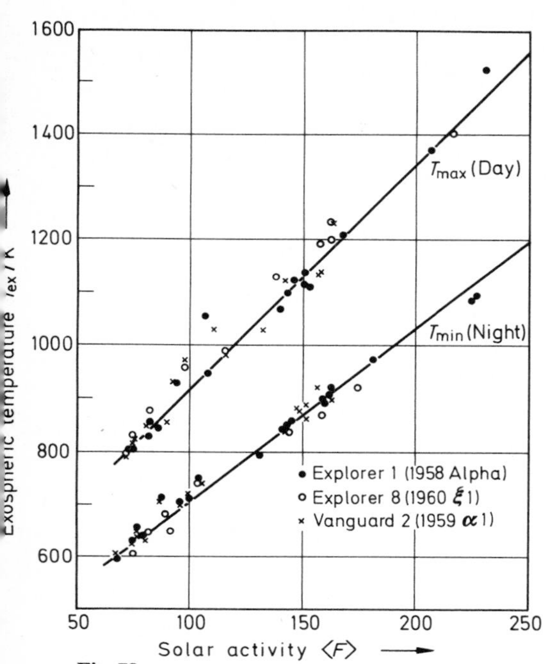

Fig. 78

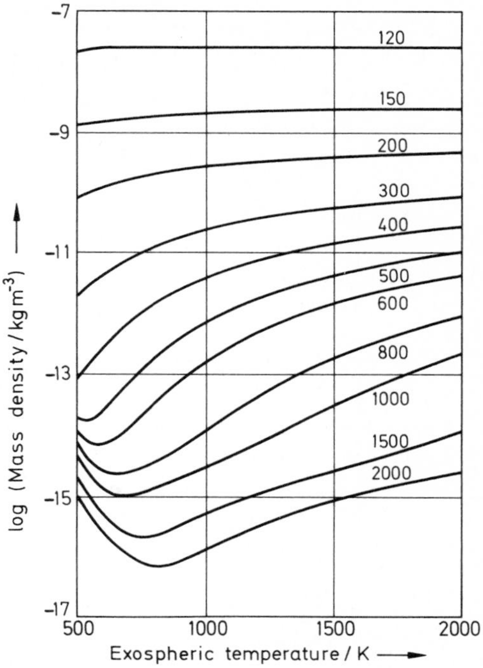

Fig. 79

Fig. 78. Night-time minimum, T_{min}, and daytime maximum exospheric temperatures measured by three satellites, plotted against the COVINGTON-index F_0. ⟨...⟩ signifies averaging over 81 d. Magnetic activity effects are excluded $[Kp=0]$[15]. The lines are given by: $T_{min}/K = 379 + 3.24 \cdot \langle F \rangle$ and $T_{max}/K = 1.30\, T_{min}$

Fig. 79. CIRA-72 model [*12*] mass-density at different altitudes/km (parameter of the curves) as a function of the exospheric temperature[15]

(i) Linearly on the long-term average $\langle F \rangle$ of the COVINGTON index[13], see Fig. 78[14].

(ii) Linearly on the difference[16] between the daily value F and $\langle F \rangle$.

(iii) In a complicated manner on the solar hour angle: Assuming a quasi-sinusoidal variation of daily maximum T_D and night-time minimum T_N of T_{ex}

---

[13] This is the numerical value of the solar radio noise flux measured on Earth at 2,800 MHz (wavelength 10.7 cm), expressed in units of $10^{-22}\,W\,m^{-2}\,Hz^{-1}$. Incited by COVINGTON, the "official" daily measurements are regularly made at Ottawa (Canada)

[14] These linear variations are applied to the night-time minimum, T_c, of the global exospheric temperature, for vanishing magnetic activity $(Kp=0)$. The increase of T_c with $\langle F \rangle$ is 3.24 K per F-unit [but 1.3 only with daily values F as considered under (ii)]

[15] Jacchia, L.G. (1972): [*12*], 227

[16] For this combination see SCHMIDTKE's Eq. (4.4), p. 31 in this volume. The longterm average is currently made over 81 d (i.e. three solar rotations), centred on day no. 40. This definition might be statistically reasonably but has the queer effect that a certain influence of *future* solar activity is assumed

after:

$$T_D = T_c(1 + A\cos^m \eta); \qquad T_N = T_c(1 + A\sin^m \theta)$$
$$\eta = \tfrac{1}{2}|\varphi - \delta|; \qquad \theta = \tfrac{1}{2}|\varphi + \delta|, \tag{10.4}$$

φ is geographical latitude, δ the solar declination. The global daily maximum is obtained at the subsolar latitude. For intermediate hours another quasi-sinusoidal variation with the hour angle is used. The exponents were empirically optimized. The amplitude A of the diurnal variation is assumed to be independent of the solar cycle[17] (see Fig. 78).

(iv) On Kp by a 'heating term' ΔT for which two relations of shape

$$\Delta T/\mathrm{K} = a \cdot Kp + b \cdot \exp(Kp) \tag{10.5}$$

are presented. With the second one, a similar relation is assumed for an increase of the logarithm of pressures[18].

The remaining influences are dealt with by additional terms to the mass density ρ. To take account of (v) a term $\Delta \log \rho$ is given as a product of rather involved functions of semi-annual phase, and of altitude. As for (vi) lower thermosphere seasonal variations derived from experiments[11,19] were matched in such a way by an additional $\Delta \log \rho$ that the resultant seasonal effect reaches a maximum between 105 and 120 km[20]. Finally, the strong increase of helium above the winter pole known as "helium bulge"[21–26] is included by an additional term in the number density of this constituent, $\Delta \log n(\mathrm{He})$, which depends on latitude and solar declination. The range of He variations at the poles is thus brought up to a factor of 4.5.

It appears that CIRA 72, though of high quality, is not easily suitable for computer use. The next CIRA[6], which is now being promoted in COSPAR, is to adopt a computerized scheme as has since been used by several 'private' models; see Sect. 11.

η) In order to describe some of *CIRA 1972 characteristic features* we reproduce a few figures from JACCHIA's paper in the official report[15]. Figure 79 shows the effect of variations in the exospheric temperature T_{ex} upon the mass density at different thermospheric heights; it is particularly strong in the exosphere. Figure 80 shows global maps for different seasons (see the subsolar

---

[17] Some doubts have since been voiced about this assumption. Also A seems to be height dependent

[18] For current use these changes are summarized in Tables 2a and 2b in [*12*]

[19] Champion, K.S.W. (1967): Space Res. *VII*, 1101

[20] The corresponding temperature variations are ignored

[21] Kasprzak, K.W.T., Krankowsky, D., Nier, A.O. (1968): J. Geophys. Res. *73*, 6765, 7291

[22] Hartmann, G., Mauersberger, K., Müller, D. (1968): Space Res. *VIII*, 940; Müller, D., Hartmann, G. (1969): J. Geophys. Res. *74*, 1287

[23] Federova, N.I. (1967): Issled. po Geomagn., Aeronomij i fizike solnca (Akad. Nauk SSSR) *13*, 53; Shefov, N.N. (1968): Planet. Space Sci. *16*, 1103

[24] Jacchia, L.G., Slowey, J. (1968): Planet. Space Sci. *16*, 509

[25] Keating, G.M., Prior, E.J. (1968): Space Res. *VIII*, 982

[26] Tinsley, B.A. (1968): Planet. Space Sci. *16*, 91

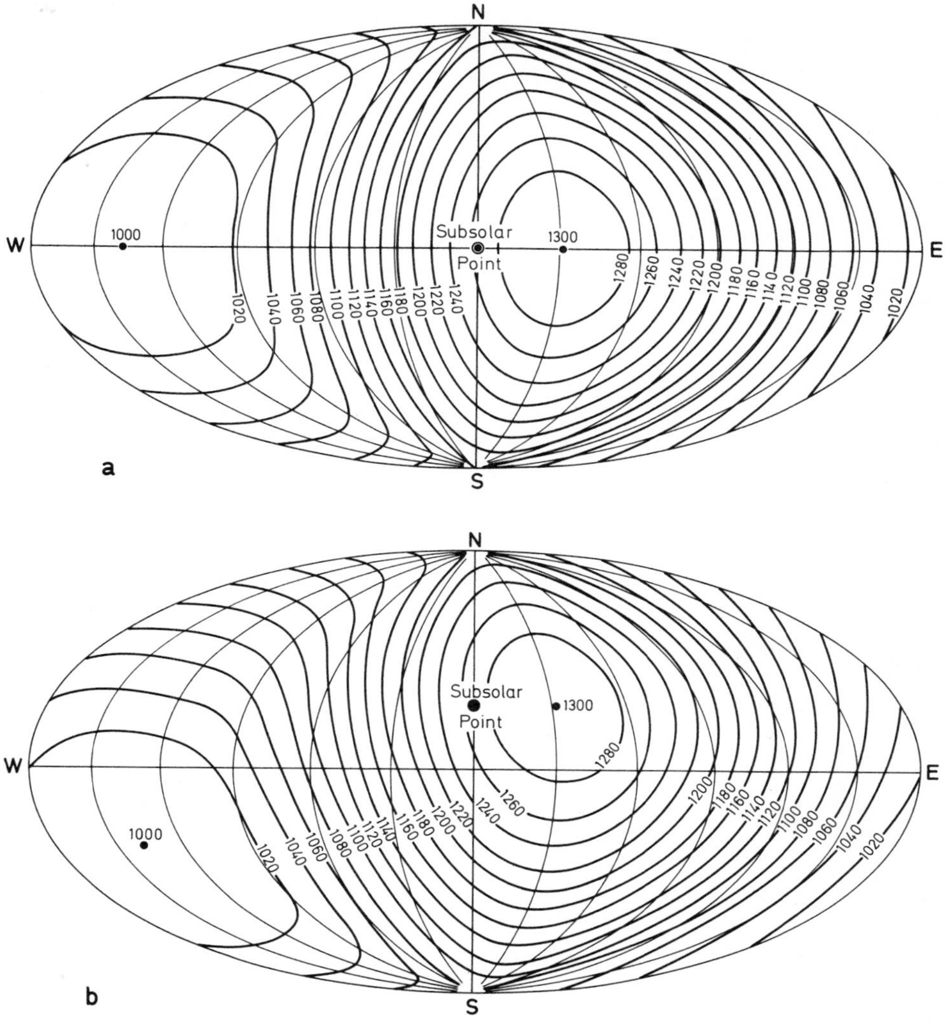

Fig. 80 a, b. Clobal maps of exospheric model temperature (isotherms labelled in K) after CIRA-72 [*12*] for a total diurnal variation of 300 K (minimum 1,000 K). (**a**) For equinox; (**b**) for northern solstice conditions[15]

point). These charts were often used for deducing thermospheric winds (see STUBBE's contribution in vol. 49/6). As compared with actual observations (see Sect. 7), they include, of course, a lot of smoothing.

ϑ) Meanwhile CIRA 1972 was *compared with* a large amount of *measured data*. The earliest comparisons are reported by ROEMER[27]. The helium bulge (Fig. 81) does not quite agree with the model[28]. Figure 82 shows wave-like

---

[27] Roemer, M. (1972): [*12*], 341 (many references)

[28] Reber, C.A., Harpold, D.N., Horowitz, R., Hedin, A.E. (1971): J. Geophys. Res. *76*, 1845

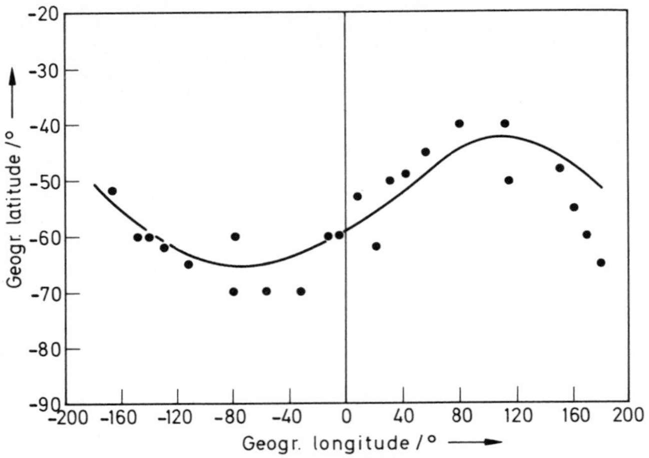

Fig. 81. Observed position of the helium peak as a function of *geographical* latitude and longitude during a 52 h period in June 1969. Comparison with the curve identifying 53° S geomagnetic latitude shows that an apparent longitude effect exists which is due to magnetic control[28]

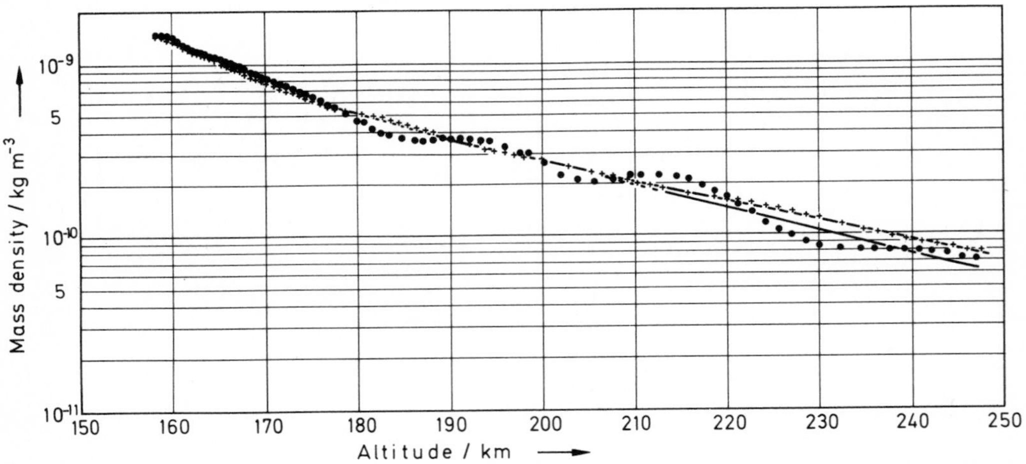

Fig. 82. Accelerometer density measurements for one orbit of satellite OV 1-15 (6 Aug 1968) showing wave-like disturbances[29]

disturbances in two instantaneous density profiles[29]; such deviations, generally due to gravity waves[4], must be admitted as a regularly occurring but impredictable feature of the real thermosphere.

There are other features which are difficult to represent in a global model, e.g. lunar tides and a midnight temperature maximum which might also be tide related and which appears in only about 50% of all observations[30].

[29] Marcos, F.A., Champion, K.S.W. (1972): Space Res. *XII*, 791

[30] Spencer, N.W., Carignan, G.R., Mayr, H.G., Niemann, H.B., Theis, R.F., Warton, L.E. (1979): Geophys. Res. Lett. *6*, 444

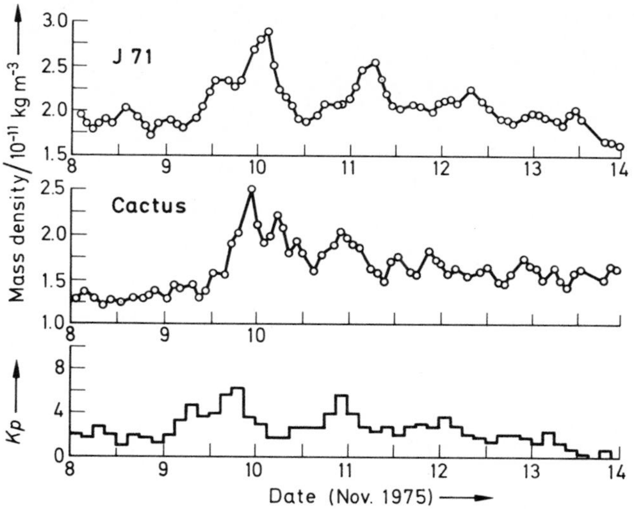

Fig. 83. Observations with the CACTUS accelerometer at 300 km during November 1975 (middle) compared with CIRA-72 [*12*] (top, labelled J 71) and with magnetic activity index *Kp* (bottom)[31]

The dependencies on actual (daily) changes on solar and magnetic activities, via indices *F* and *Kp*, are not as well established as the description for quiet periods. A comparison with satellite accelerometer measurements (see Sect. 4) in Fig. 83 shows that the effect of *Kp* is not as clear as given in the model[31]. Also CIRA-72 (identical with J 71), during perturbed conditions, considerably overestimates the density (and temperature) increase, which is caused by the perturbation, at least at low latitudes[31]. Like all neutral atmosphere models (even the more recent ones described in Sect. 12 below) CIRA uses geographical coordinates; these are not really suited to represent geomagnetic effects [see Fig. 91 in Sect. 11].

At mesospheric heights and below CIRA 1972 gives an invariable, standard model. Figure 84 shows, however, that non-negligible medium-term variations exist at low latitudes[32] which show no apparent relation with those of solar activity. On the other hand, a long-term influence upon the stratospheric temperature has been revealed[33] by radiosonde measurements (see Fig. 85).

ι) Basically, CIRA is a set of tables which needs, however, some computational work to get the right entries and to determine quite a number of corrections. Computer programming has only rarely been applied. It may be worthwhile mentioning one approach which makes possible an approximate representation of the tables by a set of simple equations. Rawer[34], adapting a transformation proposed by Booker[35] in another context, starts from the

[31] Berger, C., Barlier, F. (1981): Adv. Space Res. *1* (12), 231

[32] Lall, S., Subbaraya, B.H., Narayanan, V. (1979): Space Res. *XIX*, 147

[33] Elling, W., Schwentek, H. (1981): Jahresbericht 1980. Katlenburg-Lindau: MPI-Aeronomie, p. 31

[34] Rawer, K. (1977): J. Atmos. Terr. Phys. *39*, 753

[35] Booker, H.G. (1977): J. Atmos. Terr. Phys. *39*, 619

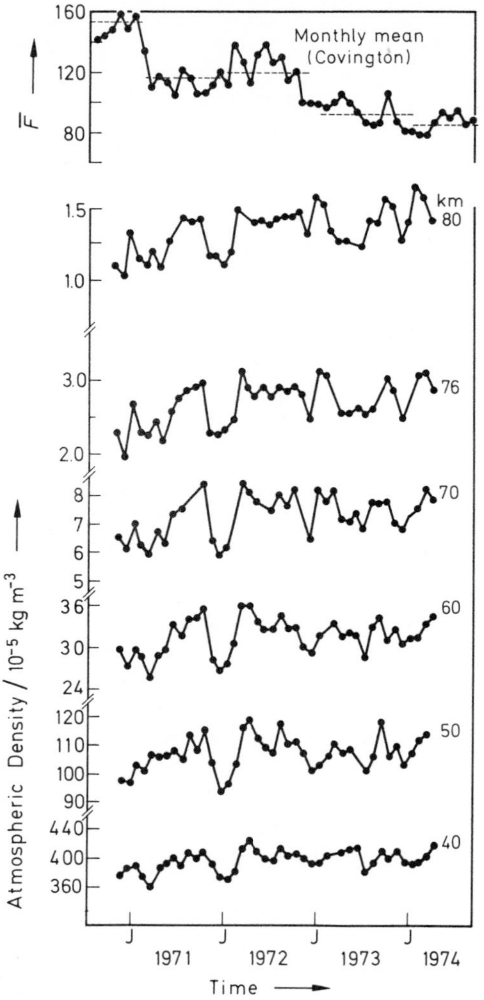

Fig. 84. Atmospheric density variations from 1971 to 1974, at a tropical site (Ahmedabad, 23° N: annual variation virtually absent). Parameter: altitude/km. Monthly mean COVINGTON index[13] on top[32]

"density scale height" (see Sect. 1)

$$H' = -(\mathrm{d}\log N/\mathrm{d}z)^{-1} \equiv -\frac{n}{\mathrm{d}n/\mathrm{d}z}, \tag{10.6}$$

which can easily be derived from the tables. For a given T_{ex}, this data set as a function of the height z is approximated by a sum of a few EPSTEIN step functions of shape[36]

[36] A graph of this function can be found in Rawer, K., Suchy, K. (1967): This Encyclopedia, vol. 49/2, Fig. 63 on p. 188

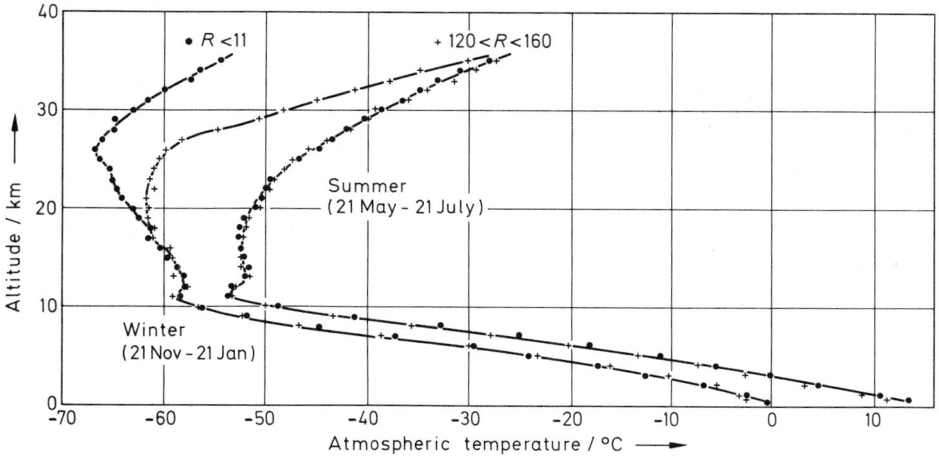

Fig. 85. Stratospheric temperature profiles for summer and winter after balloon radiosonde measurements made by the Institut für Meteorologie der Freien Universität Berlin (52.5° N). Data from 1958 to 1972 were selected after solar activity (Zürich sunspot number R) and show in winter large differences between low and high solar activity[33]

$$\mathrm{Eps}_0(x) \equiv \frac{1}{1+\exp(-x)}, \tag{10.7}$$

$$H' = H_0 + \sum_k H_k \cdot \mathrm{Eps}_0\left(\frac{z-z_k}{B_k}\right). \tag{10.8}$$

After determining the coefficients (for each T_{ex}, by optimization) the dependence on T_{ex} is approximated in the whole field by fitting each coefficient with a linear function of T_{ex}, but a parabolic one for each B_k. An integration is then needed. The number of terms needed depends on the required height range and accuracy. For representing the upper thermosphere from 200 to 400 km one term only is sufficient (see Table 5a). More than that, even the influence of the exospheric temperature can easily be taken into account by admitting a quadratic function of T_{ex} for each of the three parameters H_k, z_k, B_k. Thus, for a given component, one has ten constants (including H_0) which are determined by fitting with the H' values found from the given CIRA tables. One more constant, namely the density value at the lower boundary, is needed when deducing the partial density at altitude z from H':

$$n_j(z) = n_{j_0} \cdot \exp\left(-\int_{z_0}^{z} \mathrm{d}z/H'(z)\right). \tag{10.9}$$

Thanks to the special choice of the profile function the integral is easily obtained in closed form.

Table 5b gives a set of nine constants (to each constituent) which are needed to describe the whole CIRA H' table (in the 200–400 km height range) when allowing for quadratic functions in T_{ex} to z_1 and B_1, and for a linear function to H_1.

A better approximation, including greater heights, is obtained with two terms. In that case, it is sufficient to take z_2 constant and admit only linear dependence on T_{ex} for H_0, H_1 and H_2 so that the total number of coefficients needed to describe the H' field comes to 16. An extension to lower heights needs a third term (22 coefficients).

Such representation is particularly useful when further integration is needed, e.g. in order to calculate the extinction integral for an optical ray. Even a horizontal density gradient can easily be taken into account with this method[34].

Table 5a. *Parameters to Eq. (10.8)* obtained by fitting to the CIRA [12] density profile independently for each value of exospheric temperature T_{ex} in the 200–400 km height range. Reference height is 300 km. Set (1) was obtained for the '*density scale height*' which is defined by $H' = -dz/d(\ln(n))$, from which the density distribution can be obtained by integration. Set (2) straightforward describes the height variation of the logarithm of density by the *step scale*, H_{st}, as $\ln(n(z)) = \ln(n(z_0)) - (z - z_0)/H_{st}$ with $z_0 = 300$ km. Indications are separately given for the three major constituents O, O_2 and N_2. The slight irregularities appearing with B_1 in vertical direction are due to the fact that the CIRA tables too were independently established for each value of T_{ex}

	/	z_1/m	H_0/m	H_1/m	B_1/m	z_1/m	H_0/m	H_1/m	B_1/m	z_1/m	H_0/m	H_1/m	B_1/m
(1)	700	133,314	12,000	30,393	71,904	127,075	8,000	13,242	79,198	125,233	8,000	16,084	72,976
	800	145,084	12,000	36,452	68,364	133,774	8,000	16,185	75,113	132,288	8,000	19,589	71,913
	900	153,816	12,000	41,902	62,549	143,828	8,000	19,243	71,597	141,514	8,000	22,993	69,369
	1,000	164,830	12,000	48,367	62,406	153,430	8,000	22,140	65,929	153,964	8,000	26,250	62,384
	1,100	172,828	12,000	54,240	60,041	161,589	8,000	25,069	63,230	159,383	8,000	29,649	61,620
	1,200	179,909	12,000	59,575	57,745	167,617	8,000	28,058	61,793	168,642	8,000	32,938	58,351
	1,300	187,075	12,000	64,817	51,982	174,233	8,000	30,807	57,600	174,644	8,000	36,580	59,069
	1,400	194,355	12,000	72,080	56,085	179,515	8,000	33,861	58,251	179,342	8,000	40,161	60,654
	1,500	199,208	12,000	77,108	54,427	185,176	8,000	36,822	56,459	185,445	8,000	42,861	53,880
(2)	700	174,916	32,000	9,054	73,776	231,800	18,000	2,530	60,396	213,652	20,000	3,497	69,807
	800	146,596	32,000	14,856	79,067	167,023	18,000	5,597	86,935	156,860	20,000	6,917	86,325
	900	138,031	32,000	20,638	80,585					139,094	20,000	10,161	82,259
	1,000	136,816	32,000	26,223	78,307					128,339	20,000	13,490	86,089
	1,100	131,386	32,000	32,924	97,802	126,996	18,000	14,246	86,415	133,046	20,000	16,531	74,381
	1,200					127,144	18,000	17,014	83,359	126,851	20,000	20,075	86,326
	1,300	157,470	32,000	41,956	58,234	127,155	18,000	19,985	87,235	142,850	20,000	22,692	64,504
	1,400	152,975	32,000	48,626	76,972	132,397	18,000	22,757	83,479	135,406	20,000	26,353	80,283
	1,500	156,271	32,000	54,536	82,366	138,175	18,000	25,335	77,381	140,279	20,000	29,447	77,558
			O_1				O_2				N_2		

Table 5b. *Parameters* to describe the *density scale height* H' in the *whole* T_{ex} *range* (valid in the 200–400 km height range). The variations with T_{ex} are reproduced for z_1, B_1 and for H_1 by quadratic and linear functions, respectively. [(*i*) identifies the grade in T_{ex}]

	$z_1^{(0)}$/m	$z_1^{(1)}$/m K	$z_1^{(2)}$/m K^2	$B_1^{(0)}$/m	$B_1^{(1)}$/m K	$B_1^{(2)}$/m K^2	$H_1^{(0)}$/m	$H_1^{(1)}$/m K	H_0/m
O_1	48,593.750	143.125	−0.028	107,268.750	−64.375	0.019	−10,900.000	59.000	12,000.000
O_2	43,437.500	141.250	−0.031	128,500.000	−89.250	0.027	− 7,262.500	29.375	8,000.000
N_2	40,890.625	140.937	−0.030	104,500.000	−55.500	0.015	− 7,337.500	33.625	8,000.000

11. Computerized descriptive models of the neutral atmosphere. Soon after CIRA 1972 [*12*] had been published a large amount of data from the OGO-6 satellite mission was presented. Thus, at least in the 400–600 km height range, the variations in space and time became much better known. Two new atmospheric models were deduced which made CIRA 1972 more or less obsolete[1]: one was based on mass-spectrometric measurements, shedding new light onto the composition of the main constituents in the upper thermosphere; the other was derived from optical temperature determinations, giving direct and reliable temperature values for the first time.

α) A NASA team established a computerized *model of partial densities* applying generalized computational methods [*19*] by introducing the well-known spatial development in terms of spherical harmonics (LEGENDRE functions[2]) in order to describe the spatial variations of key parameters. In the limited height range, the height distribution was computed for three constituents (Ne, O, He) from a BATES-WALKER approximation to the temperature profile

$$T_n(z) = T_{ex} - (T_{ex} - T_n(z_0)) \cdot \exp(-s(z - z_0)), \qquad [5.5]$$

using a constant gradient parameter s throughout. The exospheric temperature T_{ex} was linearly related to the averaged COVINGTON index[3] $\langle F \rangle$ and quadratically to its daily value F. Assuming fixed boundary conditions at 120 km, temperature profiles were derived from measured N_2 densities, so minimizing differences which had shown up when comparing with temperatures determined by the incoherent scatter technique (see Sects. 5, 7).

All coefficients were determined separately for each individual constituent by fitting with the relevant measurements such that each "was free to vary in its own individual way - and these ways turned out to be quite different"[1]. A difference in behaviour of N_2 and O was now clearly appearing.

The descriptive system used in [*19*] takes the values at $z_0 = 450$ km as primary input for the partial density, the worldwide behaviour of which is described by a generalized development function G(L) after

$$n(450, L) = n(450) \cdot \exp(G(L) - 1) \qquad (11.1)$$

$n(450)$ being the worldwide average value.

L is a symbolic variable representing the geographical and geophysical parameters on which density is assumed to depend. The expansion specifying G(L) depends on associated LEGENDRE functions[4] (Table 6), which give a

[1] Jacchia, L.G. (1979): Space Res. *XIX*, 179

[2] Meixner, J. (1956): This Encyclopedia, vol. 1, p. 159. Schlögl, F. ibidem p. 320, 325. Instead of the original Legendre function P_n in our context one needs the "associated Legendre functions", P_n^m (German: *Kugelflächenfunktionen*). One defines $P_n^m(\cos\theta) = \sin^m\theta\, d^m P_n(\cos\theta)/d(\cos\theta)^m$, m being the order and n the grade. Tables can be found in: Jahnke, E., Emde, F. (1945): Tables of Functions. New York: Dover, pp. 107–125; Jahnke, E., Emde, F., Lösch, F. (1966): Tafeln höherer Funktionen [7th German, 8th Engl.]. Stuttgart: Teubner

[3] See footnote [13] in Sect. 10

[4] Stacey, F.D. (1977): Physics of the Earth [2nd]. New York: John Wiley, App. C, 319–323. Different normalization factors are, unfortunately, used in different fields. The definition as in [2] may be called 'mathematical'. [Meixner[2], however, has an additional sign factor $(-1)^m$]. Numerically different normalizations are used in analyses of the geoid on the one hand [see text to Table 6] and in geomagnetism on the other [see: Bartels, J. (1957): This Encyclopedia, vol. 48, 744]

Table 6. Legendre polynomials $P_n(\cos\theta)$ and associated polynomials $P_n^m(\cos\theta)$ after[4]

	$m=0$	$m=1$
$n=0$	$1\,(1)$	–
1	$\cos\theta\,(\sqrt{3})$	$\sin\theta\,(\sqrt{3})$
2	$\frac{1}{2}(3\cos^2\theta-1)\,(\sqrt{5})$	$3\cos\theta\sin\theta\,(\sqrt{5/3})$
3	$\frac{1}{2}(5\cos^3\theta-3\cos\theta)\,(\sqrt{7})$	$\frac{3}{2}(5\cos^2\theta-1)\sin\theta\,(\sqrt{7/6})$
4	$\frac{1}{8}(35\cos^4\theta-30\cos^2\theta+3)\,(\sqrt{9})$	$\frac{5}{2}(7\cos^3\theta-3\cos\theta)\sin\theta\,(\sqrt{9/10})$

	$m=2$	$m=3$	$m=4$
$n=0$	–	–	–
1	–	–	–
2	$3\sin^2\theta\,(\sqrt{5/12})$	–	–
3	$15\cos\theta\sin^2\theta\,(\sqrt{7/60})$	$15\sin^3\theta\,(\sqrt{7/360})$	–
4	$\frac{15}{2}(7\cos^2\theta-1)\sin^2\theta\,(\sqrt{1/20})$	$105\cos\theta\sin^3\theta\,(\sqrt{1/280})$	$105\sin^4\theta\,(\sqrt{1/2240})$

Factors in brackets convert P_n^m to p_n^m, the "fully normalized" function as now used in gravimetry. In atmospheric sciences the standard normalization is used such that these factors do not apply. A third normalization (after ADOLF SCHMIDT) is used in geomagnetism.

global distribution in latitude φ and local time t. The argument[5] is $\sin\varphi$ and $Pnm \equiv P_n^m(\sin\varphi)$. Orders of m up to 3 and grades of n up to 5 are admitted. The coefficients of these functions are assumed to depend in a rather arbitrary manner on solar and magnetic activity (given by the COVINGTON index[3] F and magnetic perturbation index[6] Ap), further on season ϑ and local solar time t[7]. The full function $G(L)$ contains ten terms:

Five symmetrical zero-order terms[8], namely: one time-independent (4th grade), one depending on solar activity (worldwide), another one (2nd grade) on magnetic activity; also an annual and a semi-annual term (both of 2nd grade) which are identical in both hemispheres

Two antisymmetrical zero-order terms[8], an annual (5th grade) and a semi-annual one (1st grade) in order to assess the main seasonal variations which are opposite in both hemispheres

Three terms to describe the *diurnal variations*, all depending on solar activity by a common factor; in detail: one (5th grade) of first order (harmonic diurnal), one (3rd grade) of second order (semidiurnal) and a last one (3rd grade) of third order (ter-diurnal).

[5] The authors use co-latitude θ; thus $\cos\theta$ ($\equiv\sin\varphi$). Choosing geographical coordinates appeared almost self-evident for an atmosphere under solar radiation influence; this is at least so during equinoctial conditions. At high latitude, however, the choice may have unwanted consequences insofar as phenomena depending on other (e.g. magnetic) coordinates might not be reproduced so well

[6] Siebert, M. (1971): This Encyclopedia, vol. 49/3, pp. 225–227. Ap is the daily average of the eight 3 hourly "activity measures" ap

[7] The author's nomenclature is t_a for day count over the year (our ϑ) and τ for the hour (our t)

[8] Zero-order terms are independent from the hour. The latitudinal variation is symmetrical for even, antisymmetrical for odd orders

As for the dependence on solar activity it is described by a second-order function of the day-by-day change of the COVINGTON index[3], i.e. of $(F - \langle F \rangle)$ where $\langle F \rangle$ is the sliding means over 81 d. The 11 year solar cycle is assessed by an additional term which is linear in $\langle F \rangle$. Apart from the worldwide and all three diurnal terms, the asymmetrical annual term is assumed to depend also on solar activity in the same manner.

The total number of coefficients is 36 for each constituent, viz. N_2, He and O; a 37th is the average density at 450 km. Numerical values can be found in Table 1 of [*19*] by which the worldwide distribution in the 450 km level is determined for these three constituents[9]. Also height profiles of the partial densities are deduced[10] starting with Eq. [5.5] (see above) for the temperature:

$$n(z, L) = n(120)\,\mathrm{D}(z, T_{ex}(L), s)$$
$$\mathrm{D}(z, T_{ex}(L), s) = \left(\frac{1-a}{1-a\cdot\exp(-\sigma\,\xi)}\right)^{\kappa} \exp(-\gamma\,\sigma\,\xi) \tag{11.2}$$

with $\kappa = 1 + \gamma + \alpha$ (α thermal diffusion coefficient); $a = 1 - T_{120}/T_{ex}(L)$; $\sigma = s + 1.5 \cdot 10^{-4}\,\mathrm{km}^{-1}$; s temperature gradient parameter of Eq. [5.5]; $\xi = (R_E + 120\,\mathrm{km}) \cdot (z - 120\,\mathrm{km})/(R_E - z)$; $\gamma = (1-a)/\sigma \cdot H_{120}$ (dimensionless, H is scale height). R_E av. Earth radius ($= 6356.77$ km).

It is important to note that the exospheric temperature T_{ex} was inferred from the molecular nitrogen data; a set of coefficients is found in Table 2 of [*19*]; these values must be used in Eq. (11.2). The numerical coefficients were determined by a best fit program based upon about 900 individual measurements for each constituent.

In Figs. 2–4 of [*19*] measured and computed data are compared. The dispersion ranges are large, in particular for N_2 (and thus for T_{ex}) going from about 0.8 to 1.4[11]. Figure 86a, b shows average T_{ex} maps for equinox and solstice conditions while Fig. 86c describes the atomic oxygen density.

This shows that either there is still some important influence $G(L)$ does not take account of, or the disturbing effects of variable solar and magnetic activities are not fully described by this function. In fact, there is some arbitrariness in the manner in which these were introduced. This point is a difficult one for all of these models; see Sect. 12γ below.

Depending on the hour and latitude there are also some systematic differences between model and measured values[12]; these are stronger[13] for solstice than for equinox conditions; see Fig. 50 (in Sect. 7).

β) Independently, a *global temperature model* based on 630 nm airglow Doppler measurements of OGO-6[14] was established by THUILLIER et al. [*62*,

---

[9] The authors give in their Table 2 similar values for 120 km as well. These are, of course, only inferred and were obtained with the diffusion method of: Jacchia, L.G. (1965): Smithsonian Contrib. Astrophys. *8*, 215

[10] Walker, J.C.G. (1965): J. Atmos. Sci. *22*, 462

[11] This is, of course, much less than the natural variability at the 450 km level

[12] Salah, J.E., Evans, J.V. (1973): Space Res. *XIII*, 268

[13] This is probably due to the choice of geographical coordinates which are best adapted to irradiation conditions at equinox

[14] See Sect. 6β(i)

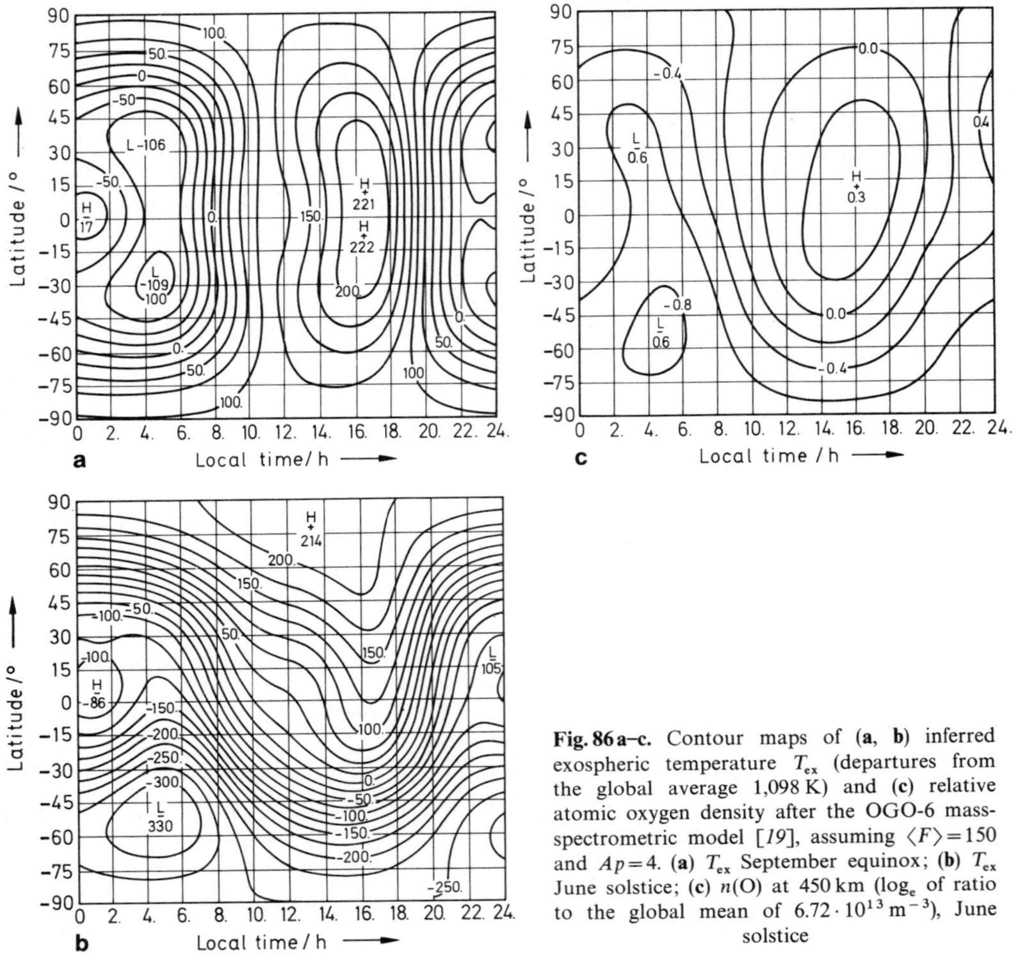

Fig. 86 a–c. Contour maps of (**a**, **b**) inferred exospheric temperature T_{ex} (departures from the global average 1,098 K) and (**c**) relative atomic oxygen density after the OGO-6 mass-spectrometric model [*19*], assuming $\langle F \rangle = 150$ and $Ap = 4$. (**a**) T_{ex} September equinox; (**b**) T_{ex} June solstice; (**c**) $n(\mathrm{O})$ at 450 km ($\log_e$ of ratio to the global mean of $6.72 \cdot 10^{13}\,\mathrm{m}^{-3}$), June solstice

63]. Almost the same development as described in Sect. 11α was used and quite similar results were obtained. However, instead of the *Ap* index Bartels' *Kp* was used[15], thus facilitating the analysis during geomagnetically disturbed days. The assessment of the influence of solar activity is almost the same as described above:

$$S \equiv 1 + F_1 = 1 + A_4(F - \langle F \rangle) + A_5(F - \langle F \rangle)^2 + A_6(\langle F \rangle - 150). \qquad (11.3)$$

However, the factor S is now admitted to *all* seasonal and diurnal terms. The full expression for $\mathrm{G}(L)$ (which gives the exospheric temperature T_{ex} in units of

[15] *Kp* is a planetary three-hourly perturbation index with quasi-logarithmic scale, see Siebert, M. (1971): This Encyclopedia, vol. 49/3, p. 212, 222

1,000.8 K) reads [*63*]:

$$\begin{aligned}G(L)=S+A_2 P_2^0+A_3 P_4^0+(A_7+A_8 P_2^0)\,Kp\\ +S\cdot\{(A_9+A_{10}P_2^0)\cdot\cos(\Omega(\vartheta-A_{11}))+(A_{12}+A_{13}P_2^0)\cdot\cos(2\Omega(\vartheta-A_{14}))\\ +(A_{15}P_1^0+A_{16}P_3^0+A_{17}P_5^0)\cdot\cos(\Omega(\vartheta-A_{18}))+A_{19}P_1^0\cdot\cos(2\Omega(\vartheta-A_{20}))\\ +A_{21}P_1^1+A_{22}P_3^1+A_{23}P_5^1+(A_{24}P_1^1+A_{25}P_2^1)\cdot\cos(\Omega(\vartheta-A_{18}))\cdot\cos\omega t\\ +A_{26}P_1^1+A_{27}P_3^1+A_{28}P_5^1+(A_{29}P_1^1+A_{30}P_2^1)\cdot\cos(\Omega(\vartheta-A_{18}))\cdot\sin\omega t\\ +(A_{31}P_2^2+A_{32}P_3^2\cos(\Omega(\vartheta-A_{18}))\cdot\cos(2\omega t)\\ +(A_{33}P_2^2+A_{34}P_3^2\cos(\Omega(\vartheta-A_{18}))\cdot\sin(2\omega t)\\ +A_{35}P_3^3\cos(3\omega t)+A_{36}P_3^3\sin(3\omega t)\ldots\},\end{aligned} \tag{11.4}$$

ω and Ω being the angular frequencies of diurnal and annual variation[16]. Table 7 gives the full set of 35 coefficients A_i ($i=2\ldots36$) for the exospheric temperature. Coefficient A_1 is a basic value, viz. exospheric average temperature: 1,000.8 K in "M1" [*62*], 999.8 K in 'M2' [*63*].

Figure 87 shows typical global temperature maps while a comparison with measured data was shown as Fig. 50 (in Sect. 7). The latitudinal variation of T_{ex} was checked with data from the AEROS-A satellite [*26*] (see Fig. 88); apparently, the observations show a larger effect than the model does.

A more detailed assessment of the magnetic activity influence at high latitudes was aimed at in a later, special analysis[17]. A three-dimensional composition model (N_2, O, He) was established by combining satellite drag data (see Sect. 2) with the [*63*] temperature model "M2" [*2*].

γ) Another model using these computational methods was established by von Zahn et al.[18] with the gas-analyzer data of satellite ESRO-4; it is valid for *low solar activity*. The particular merit of this work is the addition of argon, which is particularly suited for inferring temperatures and study disturbance phenomena. A global model of the disturbed thermosphere was also produced[19]. A later analysis[20] considered longitudinal effects; to this end, in addition to the classical analysis in latitude/hour coordinates, different longitude zones were separately analyzed. Correction terms describing this effect were derived as continuous functions of latitude and longitude. As indicated by different authors[21,22] the effect should be due to asymmetrical *corpuscular heating* in the auroral zones. In fact, these are approximately centred on the magnetic rather than the geographical poles. It can be seen from Fig. 89 that

---

[16] Functions of same grade as order are only dependent on the hour t, not on latitude φ. $\Omega=0.0172028\,\mathrm{d}^{-1}$; $\omega=6.2832\,\mathrm{d}^{-1}$

[17] Thuillier, G., Falin, J.L., Barlier, F. (1980): J. Atmos. Terr. Phys. *42*, 653

[18] von Zahn, U., Köhnlein, W., Fricke, K.H., Laux, U., Trinks, H., Volland, H. (1977): Geophys. Res. Lett. *4*, 33

[19] Jacchia, L.G., Slowey, J.W., von Zahn, U. (1977): J. Geophys. Res. *82*, 684

[20] Laux, U., von Zahn, U. (1979): J. Geophys. Res. *84*, 1942

[21] Mayr, H.G., Trinks, H. (1977): Planet. Space Sci. *25*, 607

[22] Berger, C., Barlier, F. (1981): Adv. Space Res. *1* (12), 231

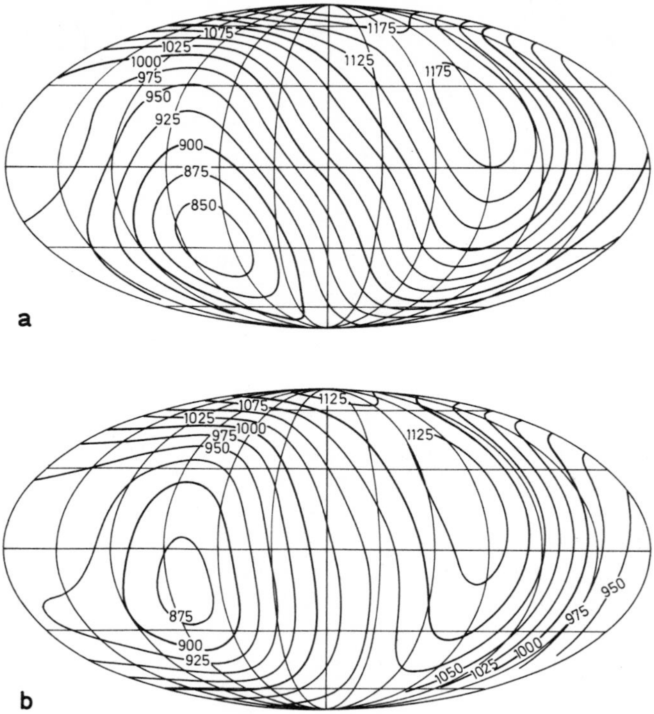

Fig. 87 a, b. Global maps of exospheric temperature T_{ex} (absolute values) after THUILLIER's optically based OGO-6 model [*62*], for low magnetic ($Kp=1$) and high solar activity ($\langle F \rangle = 150$). (**a**) for June (northern solstice); (**b**) for March (equinox)

positive corrections in the heavier constituents N_2 and Ar occur at a longitude of about 270° in the northern and 90° in the southern hemisphere, just where the auroral zone[23] is nearest to the geographical equator but are found in O and He on the opposite hemisphere.

δ) As corpuscular effects are closely related to *magnetic disturbances*, the influence of these was systematically investigated by JACCHIA et al.[19,24], resulting in a non-linear equation for the *Kp*-induced heating effect $\Delta_G T$:

$$\Delta_G T = A \cdot \sin^4 \text{Л}$$

with

$$A/\mathrm{K} = 57.3\, Kp'(1 + 0.027 \exp(0.4\, Kp')), \tag{11.5}$$

Kp' being Kp at a time somewhat earlier than the effect (by 0.1 d at high, 0.3 d at low latitudes). Since Л is the "invariant" latitude[25], the effect increases

[23] See Akasofu, S.-I., Chapman, S., Meinel, A.B. (1966): This Encyclopedia, vol. 49/1, Sect. 3, p. 2

[24] Jacchia, L.G., Slowey, J.W., von Zahn, U. (1976): J. Geophys. Res. *81*, 36

[25] See Hess, W.N. (1972): This Encyclopedia, vol. 49/4, p. 126

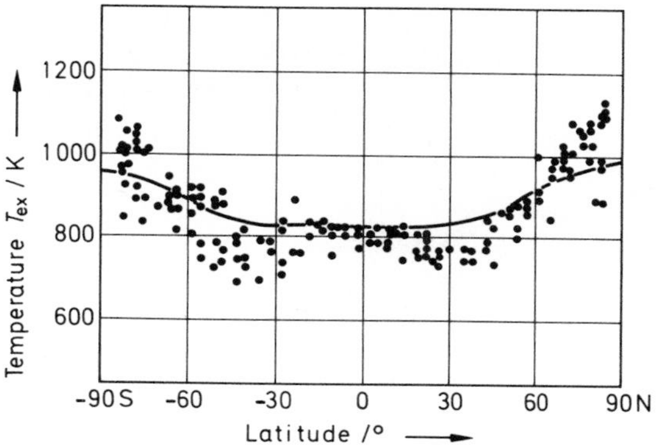

Fig. 88. Comparison of the latitudinal variation after THUILLIER's model [62] (curve) with measured data from the AEROS-A NATE experiment (see Sect. 3) [26]: dots

strongly towards the magnetic poles. As a consequence of heating the diffusive equilibrium in the thermosphere is displaced[26] towards a structure with more heavier and less lighter components; see Figs. 90 and 91.

Since magnetic disturbances are "individuals" such a global description of their effects cannot really be applied to specific events. Jacchia, in his J77 model [24], made a major effort to obtain at least a good average description based on observations from OGO-6 and ESRO-4 satellites. As in his previous model J71 [23] (and in CIRA 1972 [12]), he uses descriptive equations of the different physical processes rather than a global development. For a comparison, see Sect. 12 below.

ε) The descriptive *model* which is *most applied* at present [20] is called *MSIS* because it combines data from five satellite missions with mass spectrometers (see Sect. 3) and incoherent scatter data (see Sect. 5) from three stations. This was the largest data base available, covering a whole solar cycle of 11 years. The mathematical description follows the same lines as described in Sects. 11α and 11β, above. The descriptive function $G(L)$ is almost the same as given in Eq. (11.4) except for:

(i) An asymmetrical (hemispherical) global term with P_1^0 (which, however, is zero for T_{ex}).

(ii) A simplified dependence on solar activity by suppressing the day-by-day variability in S of Eq. (11.3), except for the very first term of Eq. (11.4). So, all other members allow only for an 11 year solar cycle effect[27].

(iii) Ap^6 not Kp^{15} is used as indicator of magnetic disturbance effects (see Sect. 12γ below).

[26] The problem of the time needed for redistribution is dealt with by Yonezawa, T. (1981): This Encyclopedia, vol. 49/6, Sect. 15, p. 169

[27] The factor is $(1+a(F-150))$ where (for T_{ex}) $a=0.0056$ to be compared with THUILLIER's $A_6=0.0027\|0.0025$. There is no solar influence in lines 2 and 3 of Eq. (11.4), except for the asymmetrical annual term (at the beginning of line 3)

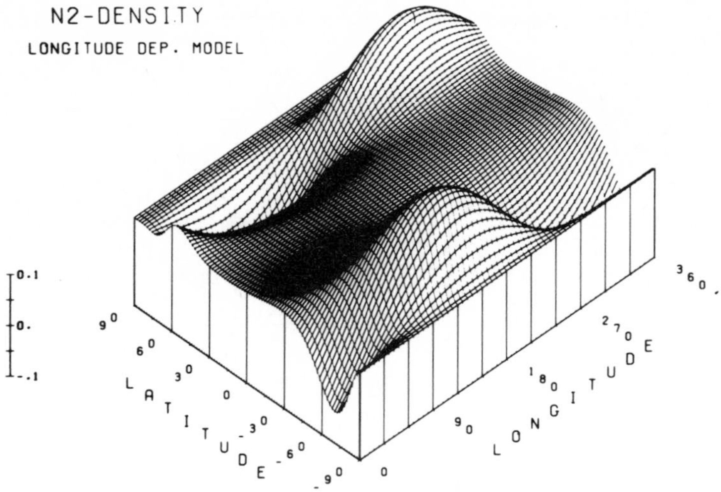
N2-DENSITY
LONGITUDE DEP. MODEL
0.1
0.
-.1
90
60
30
0
-30
-60
-90
LATITUDE
0
90
180
270
360
LONGITUDE

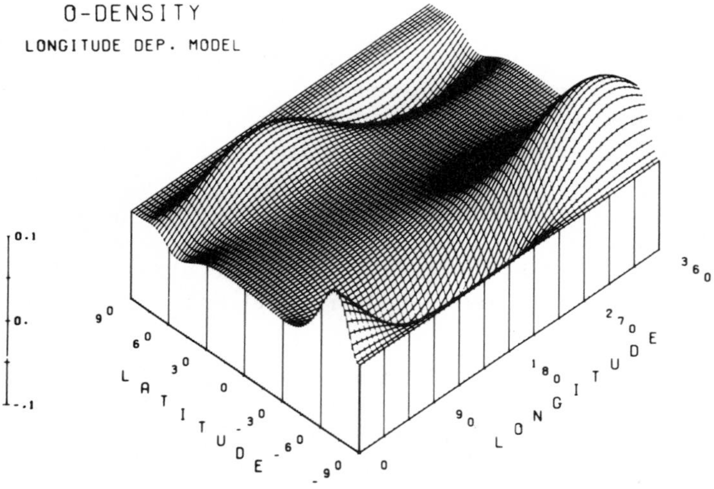
O-DENSITY
LONGITUDE DEP. MODEL
0.1
0.
-.1
90
60
30
0
-30
-60
-90
LATITUDE
0
90
180
270
360
LONGITUDE

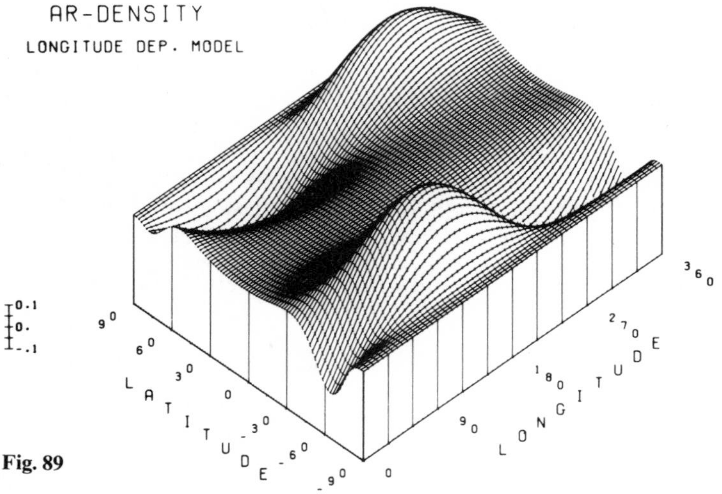
AR-DENSITY
LONGITUDE DEP. MODEL
0.1
0.
-.1
90
60
30
0
-30
-60
-90
LATITUDE
0
90
180
270
360
LONGITUDE

Fig. 89

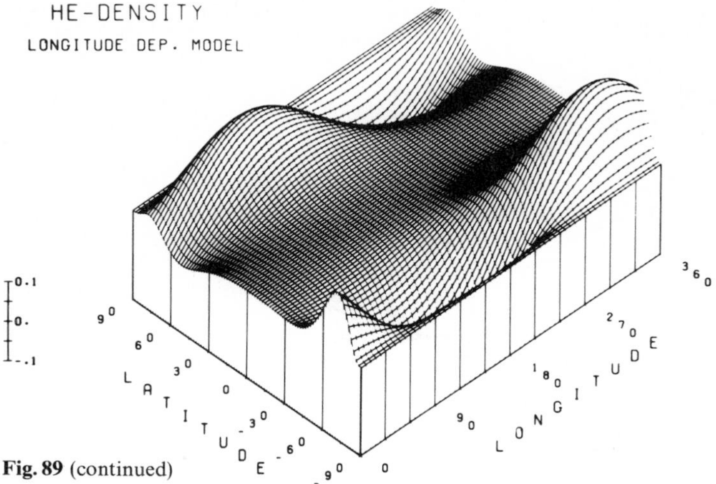

Fig. 89 (continued)

Fig. 89. Longitudinal effect on partial densities: correction against a model with only latitudinal dependence as function of latitude and longitude (see text)[20]

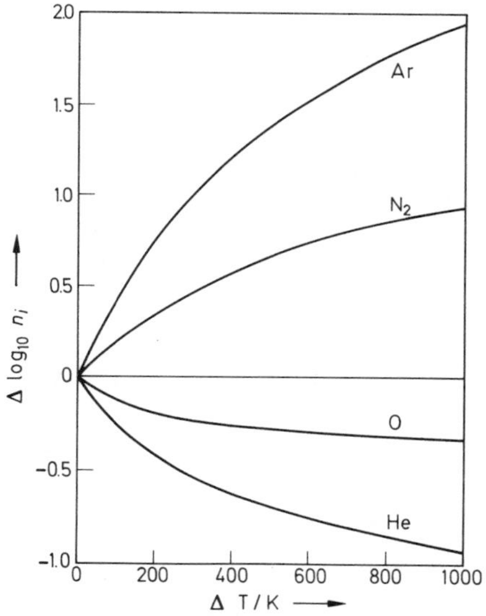

Fig. 90. Relative variations of partial densities, n_i, at 280 km as a function of geomagnetic heating, Eq. (11.5), starting with a "quiet" temperature of 900 K[22]

Table 7 shows the coefficients for the exospheric temperature T_{ex} after MSIS and makes possible comparison with the THUILLIER model (of Sect. 11 β).

The most important improvement in MSIS is a more flexible formula for the *height profiles*. Equation [5.5] is used for temperature profiles with, however, a geographically variable gradient parameter:

$$s = \bar{s} \cdot G(L). \tag{11.6}$$

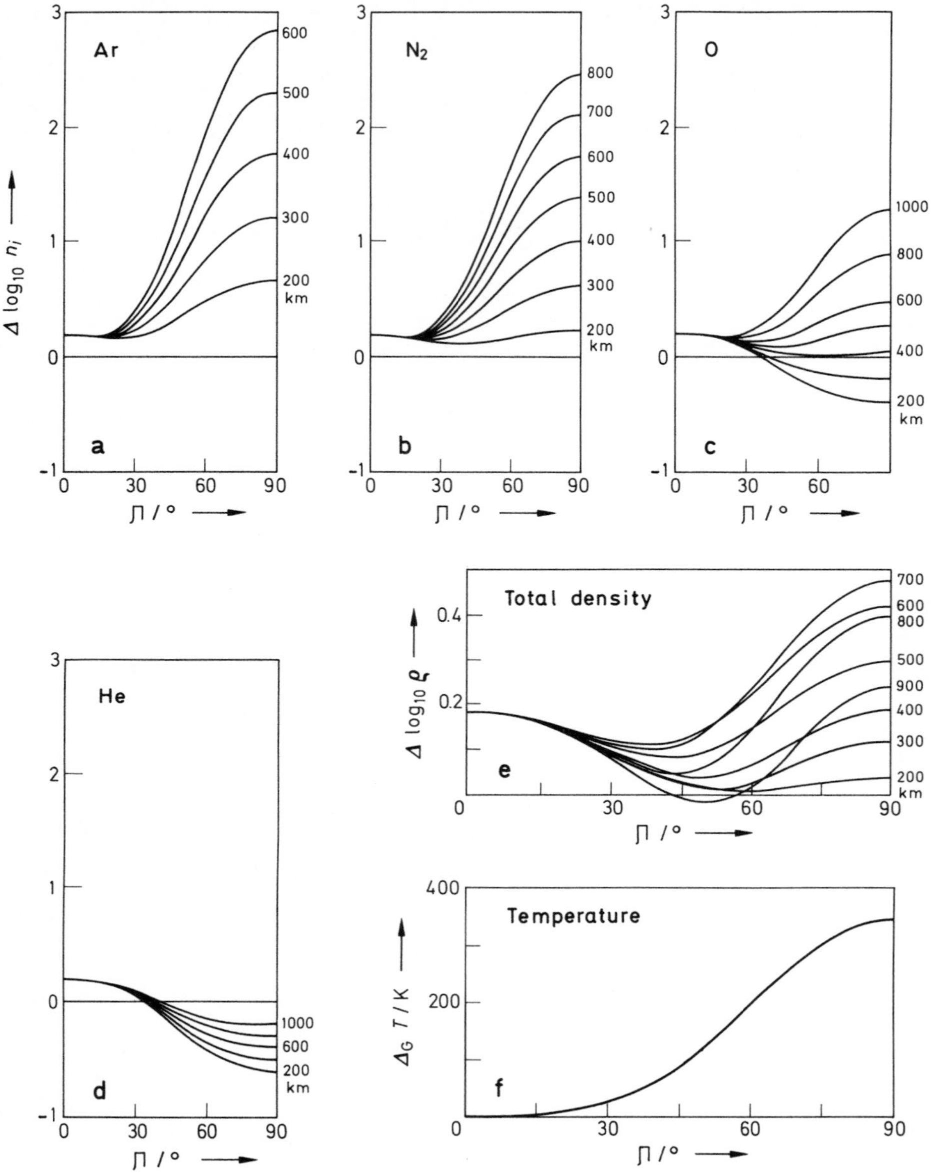

Fig. 91 a–f. Relative latitudinal changes[22] in partial (**a**–**d**) and total density (**e**) as well as temperature (**f**) provoked by a magnetic disturbance of $Kp=5$. Π is "invariant" magnetic latitude[25]. Conditions as in Fig. 90

Table 7. Set of coefficients of $G(L)$ for exospheric temperature

	THULLIER (M1) [62]	THULLIER (M2) [63]	HEDIN (1977) [20]	HEDIN (1983) *
A_{2}	−0.1328 E−2	−0.3636 E−2	+0.2599 E−2	+0.2394 E−1
A_{3}	+0.2531 E−1	+0.2459 E−1	+0.3333 E−1	−0.1616 E−2
A_{4}	+0.1486 E−2	+0.1326 E−2	+0.1198 E−2	+0.1778 E−2
A_{5}	−0.9175 E−5	−0.5623 E−5	−0.5374 E−5	−0.5460 E−5
A_{6}	+0.2705 E−2	+0.2536 E−2	+0.3197 E−2	+0.3067 E−2
A_{7}	+0.1827 E−1	+0.1766 E−1	+0.5929 E−2	+0.5975 E−2
A_{8}	+0.3373 E−1	+0.3368 E−1	+0.9056 E−2	+0.7619 E−2
A_{9}	−0.3104 E−2	−0.3764 E−2	+0.1125 E−1	+0.1098 E−1
A_{10}	+0.1830 E−1	+0.1745 E−1	0	0
A_{11}	−0.2116 E+3	−0.2115 E+3	−0.2696 E+2	+0.3762 E+2
A_{12}	−0.5880 E−2	−0.2727 E−2	+0.6276 E−2	+0.1424 E−1
A_{13}	+0.2945 E−1	+0.2747 E−1	+0.7290 E−2	0
A_{14}	−0.8596 E+2	−0.9522 E+2	+0.9902 E+2	+0.1304 E+3
A_{15}	−0.1328 E+0	−0.1337 E+0	−0.1298 E+0	−0.1561 E+0
A_{16}	−0.3123 E−1	−0.2732 E−1	−0.2522 E−1	−0.3303 E−1
A_{17}	−0.2075 E−1	−0.9673 E−2	0	0
A_{18}	−0.1602 E+2	−0.1458 E+2	−0.1143 E+2	−0.7692 E+1
A_{19}	−0.2795 E−1	−0.2747 E−1	−0.1611 E−1	−0.1853 E−1
A_{20}	−0.1770 E+3	−0.1740 E+3	+0.3213 E+2	+0.6667 E+1
A_{21}	−0.3786 E−1	−0.6657 E−1	−0.1239 E+0	−0.1120 E+0
A_{22}	−0.3846 E−2	−0.5960 E−2	+0.4381 E−2	−0.6432 E−2
A_{23}	+0.1120 E−2	+0.6745 E−2	0	0
A_{24}	−0.3405 E−1	−0.2662 E−1	+0.1123 E−1	0
A_{25}	+0.1826 E−1	+0.1469 E−1	+0.1176 E−1	+0.1146 E−1
A_{26}	−0.1069 E+0	−0.1097 E+0	−0.1094 E+0	−0.1273 E+0
A_{27}	+0.7798 E−2	+0.8870 E−2	+0.2097 E−2	+0.1191 E−3
A_{28}	+0.4333 E−2	+0.3692 E−2	0	0
A_{29}	+0.1532 E−1	+0.1222 E−1	−0.6750 E−2	0
A_{30}	−0.3789 E−2	−0.7636 E−2	−0.6396 E−2	−0.7107 E−2
A_{31}	−0.4382 E−2	−0.4489 E−2	−0.3743 E−2	−0.2736 E−2
A_{32}	+0.7232 E−3	+0.2365 E−2	+0.3977 E−2	+0.3948 E−2
A_{33}	+0.5610 E−2	+0.5057 E−2	+0.5432 E−2	+0.9537 E−2
A_{34}	+0.1214 E−2	+0.1079 E−2	+0.1339 E−2	+0.1887 E−2
A_{35}	−0.8199 E−3	−0.7161 E−3	+0.2873 E−3	+0.8287 E−3
A_{36}	+0.1098 E−2	+0.9639 E−3	+0.8585 E−3	+0.1150 E−2
A_{1} (av.)	1,000.8 K	999.8 K	1,041 K	1,035 K

Note: See Eq. (11.4) after THUILLIER et al. [63]; Model 'M1' [62] is exclusively based on optical data, model 'M2' [63] includes density data from drag analysis. The HEDIN et al. [20] formula is slightly different from Eq. (11.4), see text. See Sect. 11ε for the MSIS model entitled 'HEDIN' [20]. The more recent HEDIN model (1983) called 'revised thermospheric model' uses considerably more terms and comprises a total of almost 100 coefficients. We give only those which refer to the functional members which were earlierly used. Reference: HEDIN, A.E. (1983): J. Geophys. Res. *88*, 10170 [see Sect. 11η below].

For densities a more involved expression than that of Eq. (11.2) is introduced, namely:

$$n(z)=n_{120}\cdot(\tau/(1-(1-\tau)\exp(-\sigma\,\xi))^{1+\gamma+\alpha}\cdot\exp(-\gamma\,\sigma\,\xi) \qquad (11.7)$$

with $\tau=T_{120}/T(z)$; σ, α as in Eq. (11.2), also "virtual" height

$$\xi=(z-120\,\mathrm{km})\cdot(R_{\mathrm{E}}+120\,\mathrm{km})/(R_{\mathrm{E}}+z)$$

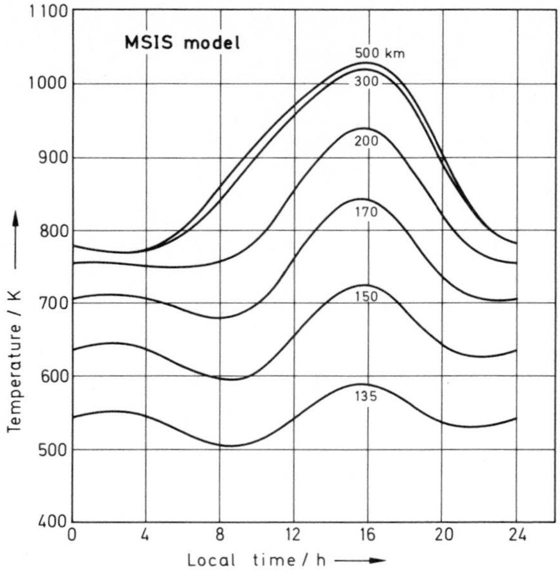

Fig. 92. Diurnal variation of temperature at different altitudes (parameter)[1] for $F=\langle F\rangle=100$, $Kp=1$, invariant latitude 0, solar declination 0 and MSIS model [20]

[where z altitude and $R_E=6{,}356{,}770$ m = Earth radius]; however,

$$\gamma=\frac{\bar{m}\,g_{120}}{kT}\equiv\frac{1}{\mathrm{H}(L)}$$

is a reciprocal standard scale height.

The diurnal variations were made variable with height, the semidiurnal tide predominating below 140 km (this is known, e.g. from the ionospheric drift observations[28], see Sect. 8); at greater heights the diurnal component takes over gradually in agreement with satellite results (see Fig. 92)[1]. Typical maps of T_{ex} and of molecular nitrogen density are shown as Fig. 93.

Tables with the sets of coefficients for the partial densities of six constituents (N_2, O, O_2, Ar, He, H) can be found in [20].

ξ) Apparently, the influences of solar and magnetic activity are not yet well assessed. Though expressions like Eq. (11.3) are currently used, the combination of an actual and an averaged measure of *solar activity* is not really satisfying.

[28] Harnischmacher, E., Rawer, K. (1968): [41], 242; see also Rawer, K., Suchy, K. (1967): This Encyclopedia, vol. 49/2, Sect. 50

Fig. 93 a–c. Global maps (local time versus latitude) after the MSIS model [20]: (**a**, **b**) exospheric temperature T_{ex} (as a difference against its global average), (**a**) at the September equinox, (**b**) at the June solstice; furthermore (**c**) the molecular nitrogen density (at 450 km (as the nat. log of its ratio to the global average), at the September solstice. $F=\langle F\rangle=150$, $Ap=4$

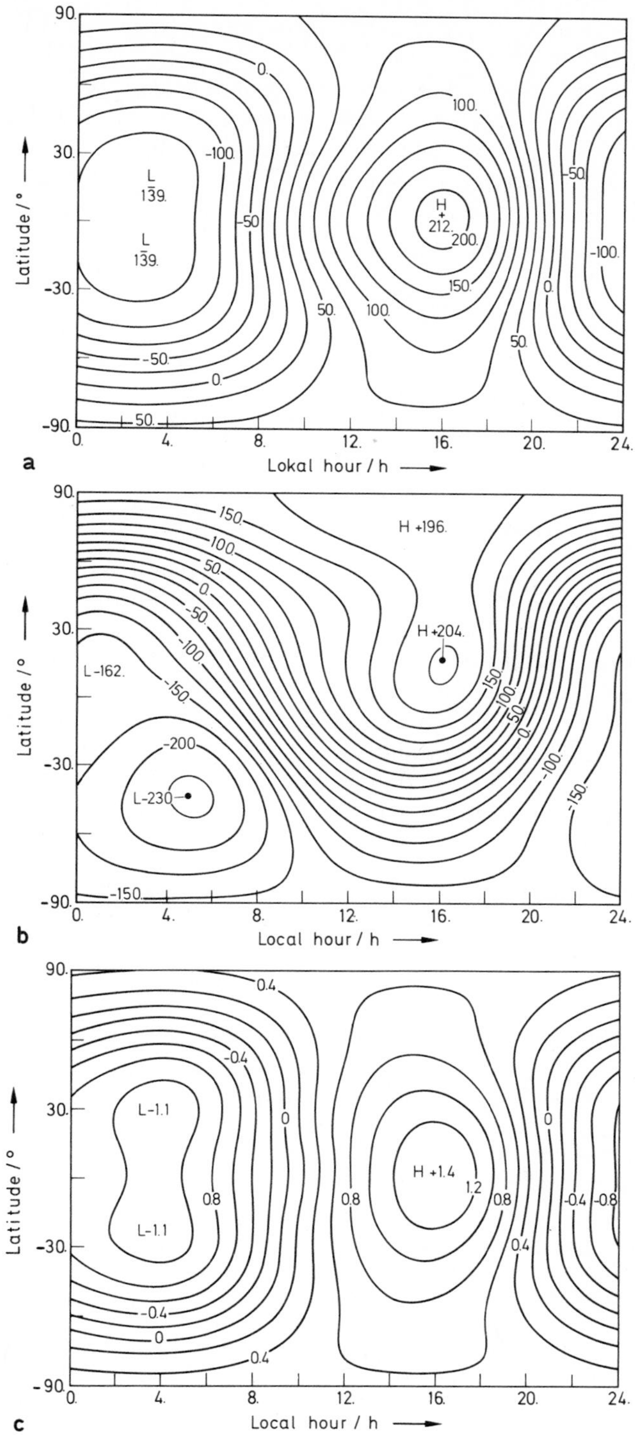
a
Latitude / °
Lokal hour / h
90.
30.
-30.
-90.
0.
4.
8.
12.
16.
20.
24.
L
139.
L
139.
H
212.
0.
100.
-100.
-50
50.
200.
150.
-50.
0.
50.
-100.
b
Latitude / °
Local hour / h
H +196.
H +204.
L -162.
L-230
150.
100.
50.
0.
-50.
-100.
-150.
-200.
c
Latitude / °
Local hour / h
L-1.1
L-1.1
H +1.4
1.2
0.8
0.4
-0.4
0
-0.8

Fig. 93 a–c

Rather, one must conclude that present measures of solar activity (like R_Z or F) are not so well suited[29].

A recent attempt is worth noting by which the actual index of the day under consideration was replaced by that of the day before. So a time shift was admitted between a solar activity change and its consequences in the terrestrial atmosphere. A better correlation than with F_0 (the index of the day itself) was reached with F_{-1} (namely that of the day before) for the exospheric temperature. A formula for T_{ex} was given which depends on solar activity only by F_{-1}[30]:

$$T_{ex}/\mathrm{K} = 190(\pm 12) \cdot F_{-1}^{0.35(\pm 0.01)}. \tag{11.8}$$

The correlation coefficient for magnetically quiet days was found to be as high as 0.85.

η) Quite recently, a new *improved version* of the *MSIS-model* [*20*] has been published[31]. It uses a considerably more involved formula with about 100 constants. See Table 7 on p. 337.

12. Intercomparison of different atmospheric models

α) The models described in Sect. 11 have been compared amongst themselves and with observed data. The most interesting parameter is the *exospheric temperature*, T_{ex}. In Fig. 94[1] the most important systematic variations are presented as given by the five models described in Sect. 11 α–ε. As for the diurnal variation (Fig. 94a) the THUILLIER model [*62*][2] (Sect. 11 β) shows the smallest diurnal effect with a peak shifted to about 18 h (the other models show one around 16 h). Appreciable differences are seen in the latitudinal variation between MSIS (Sect. 11 ε) and in all other models (Fig. 94b). Figure 95 is an example of the longterm changes in T_{ex} (daily average) with seasonal and solar cycle variations.

β) *Composition*, i.e. the partial densities of the constituents are also of great interest. A very systematic study[3] resulted in Figs. 96–98. Figure 96 describes the long-term variations: J77 (Sect. 11 δ) assumes a considerably greater slope in the dependence of T_{ex} on F than do the other models. ESRO-4 (Sect. 11 γ) has more helium at high solar activity as appears also from Fig. 97, showing the seasonal and latitudinal influences, respectively. Larger discrepancies appear in Fig. 98, where the diurnal variations of atomic oxygen partial density are compared with data inferred from incoherent scatter measurements[4] (see

---

[29] The same conclusion was reached when comparing with solar irradiation data in the extreme ultraviolet, see SCHMIDTKE's contribution in this volume. R_Z is the Zürich sunsport number

[30] Hernandez, G. (1982): J. Geophys. Res. *87*, 1623; (1983): Adv. Space Res. *3* (1), 129

[31] Hedin, A.E. (1983): J. Geophys. Res. *88*, 10170

[1] Jacchia, L.G. (1979): Space Res. *XIX*, 179

[2] It should be noted that this is the only model relying upon direct (optical) measurements of neutral temperature

[3] Barlier, F., Berger, C., Falin, J.L., Kockarts, G., Thuillier, G. (1978): J. Atmos. Terr. Phys. *41*, 527

[4] Bauer, P., Waldteufel, P., Alcaydé, D. (1970): J. Geophys. Res. *75*, 4825; Alcaydé, D., Bauer, P. (1977): Ann. Géophys. *33*, 305; Alcaydé, D., Bauer, P., Hedin, A., Salah, J.E. (1978): J. Geophys. Res. *83*, 1141

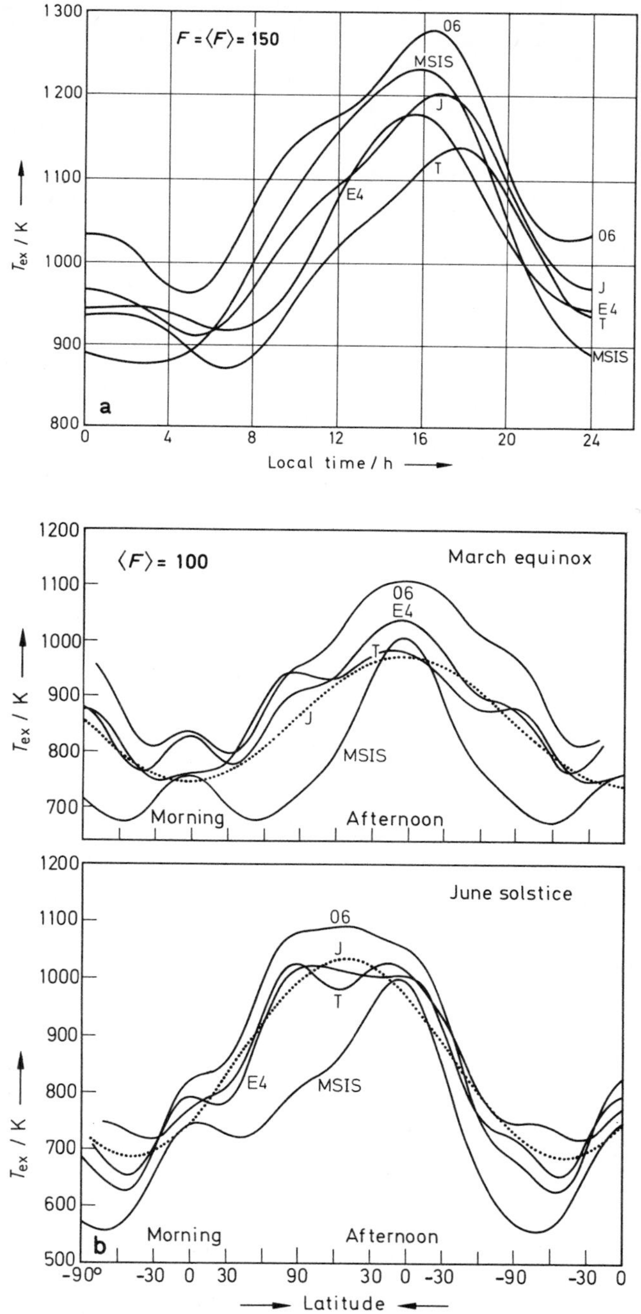

Fig. 94a, b. Exospheric temperature T_{ex} according to different models[1]: (**a**) diurnal variation ($F=\langle F\rangle=150$); (**b**) latitutinal variation along the full meridional cycle of 05/17 h. Top diagram March equinox, bottom: June solstice. ($F=\langle F\rangle=100$, $Kp=Ap=0$). Models: O6=OGO-6 (see Sect. 11α), T=THUILLIER (β); E4=ESRO-4 (γ); J=J77 (δ); MSIS (ε)

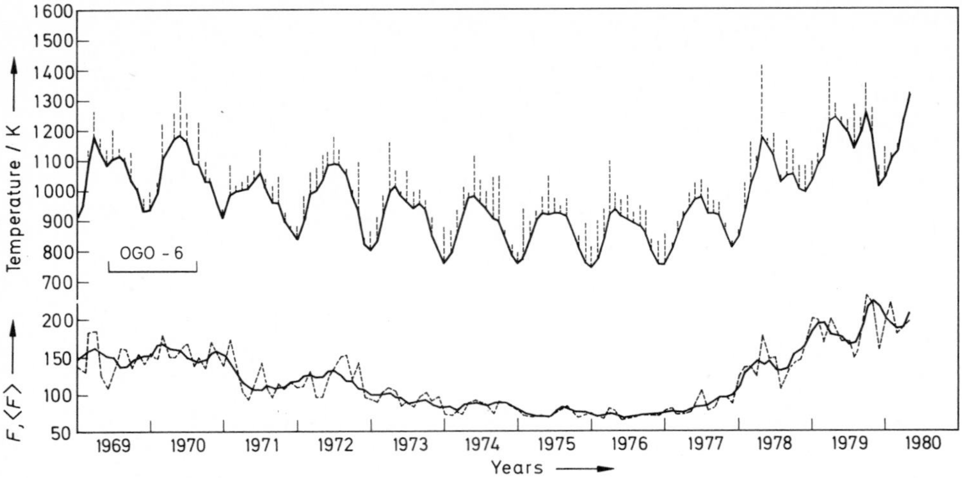

Fig. 95. Longterm variation of T_{ex} over Belgium after the OGO-6 model [*19*]. Full curve: smoothed daily average; disturbance peaks indicated by broken (vertical) lines. COVINGTON-index F (broken) and $\langle F \rangle$ below. [Courtesy: G. KOCKARTS]

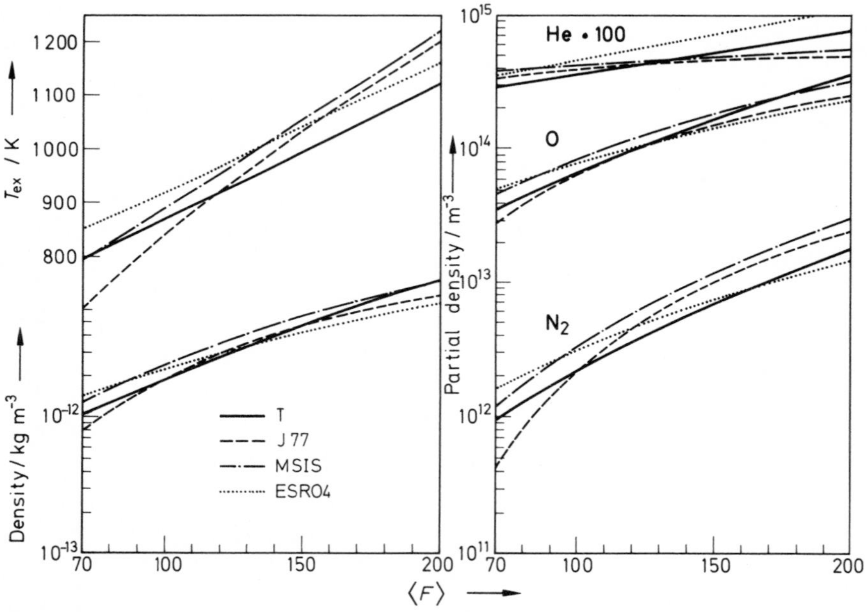

Fig. 96. Daily averaged thermospause temperature, T_{ex}, total density and partial (numerical) densities at 400 km above the equator [3] as functions of the (smoothed) COVINGTON-index $\langle F \rangle = F$. [21 Sept., $Kp=2$, $Ap=7$]

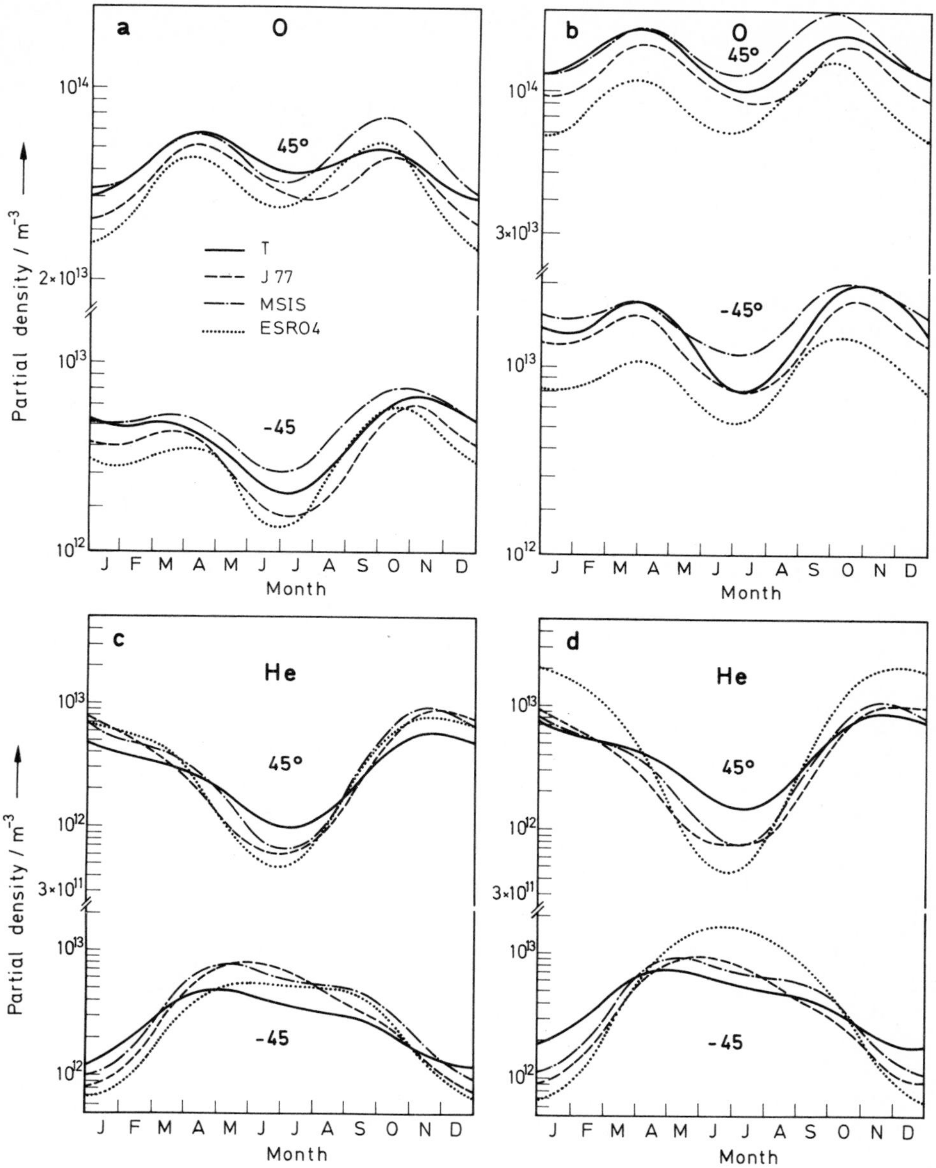

Fig. 97a–d (Fig. 97e, f see page 344)

Fig. 97a–f. Variations of daily averaged atomic oxygen and helium partial densities. (**a–d**) seasonal: (**a, c**) for low, (**b, d**) for high solar activity [$F = \langle F \rangle = 92$ and 150, respectively] at 400 km, ±45° latitude.

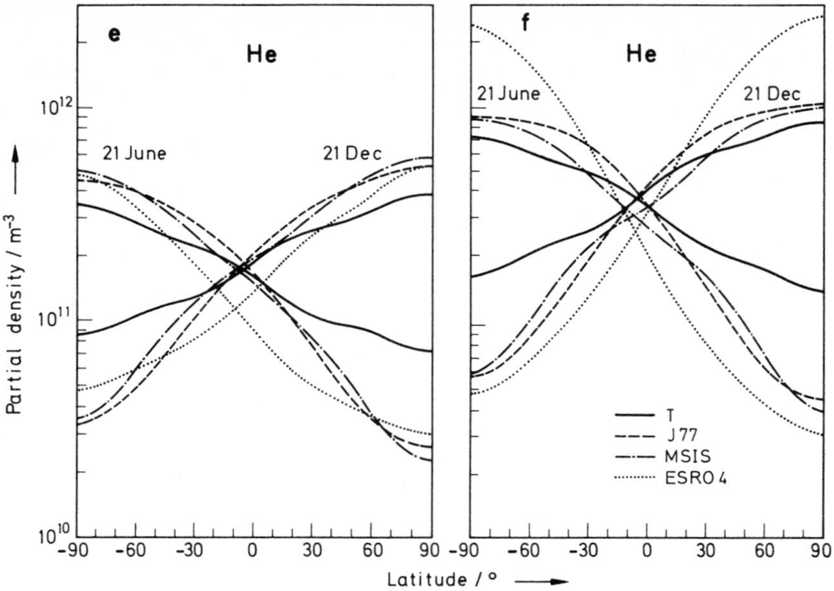

Fig. 97 e, f. Latitudinal variation of daily averaged helium partial density at 1,000 km altitude: solar activity (**e**) low [as in (**a**, **c**)], (**f**) high activity [as in (**b**, **d**)]. $Kp=2$; $Ap=7$

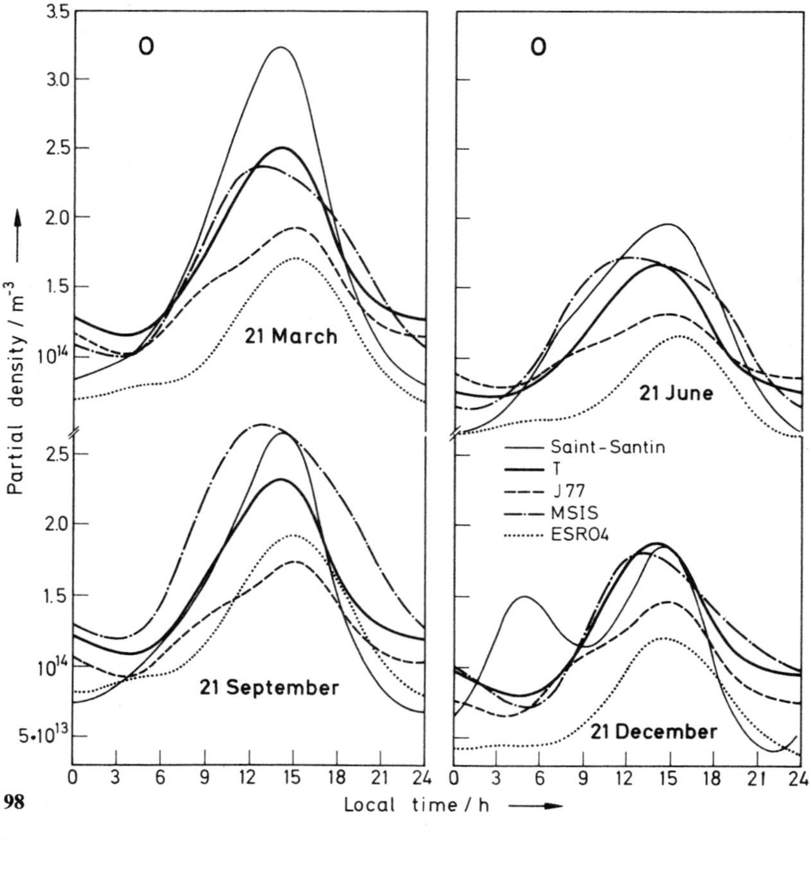

Fig. 98

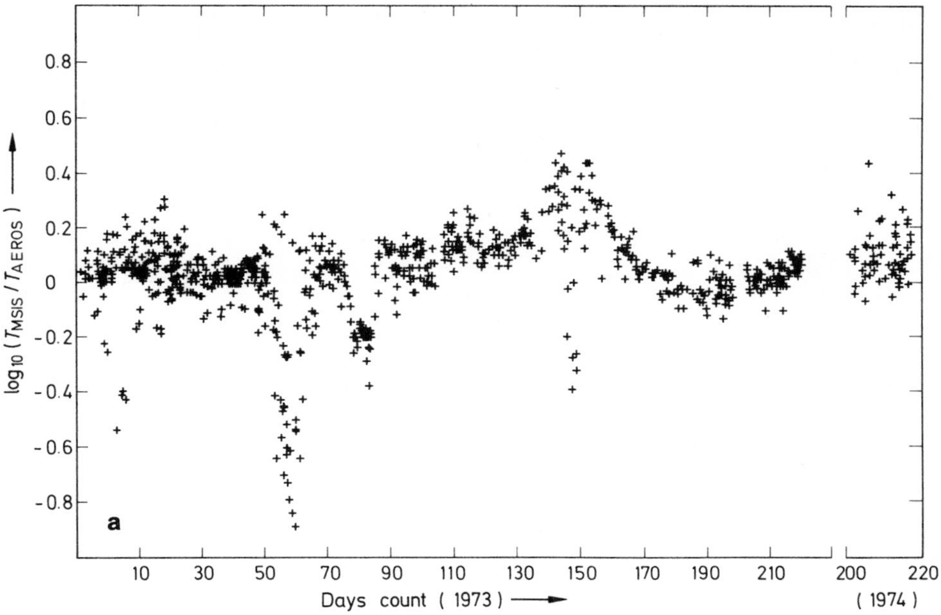

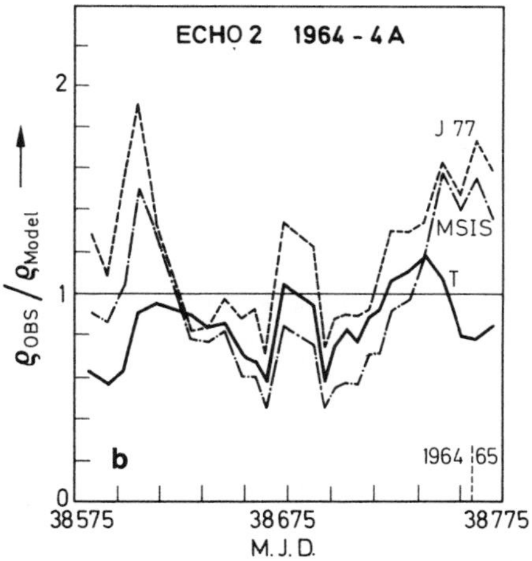

Fig. 99. (**a**) Ratio (log. ordinate) of observed (AEROS [*26*]) to model (MSIS [*20*] temperatures at great altitudes. Abscissa, day count 1973. (**b**) Total densities (lin. ordinate) derived from braking data of balloon satellite ECHO-2. Abscissa, modified Julian date. (Perigee latitude varied from 75° N to 75° S during this period). See Fig. 97, p. 343 for designations

←

Fig. 98. Diurnal variation of atomic oxygen partial density at 400 km above Saint Santin (France). Results of the local (station) model based on incoherent scatter data are given as thin line, the other curves are for models as indicated [3]. High solar activity $F = \langle F \rangle = 150$. $Kp = 2$, $Ap = 7$ [p. 344]

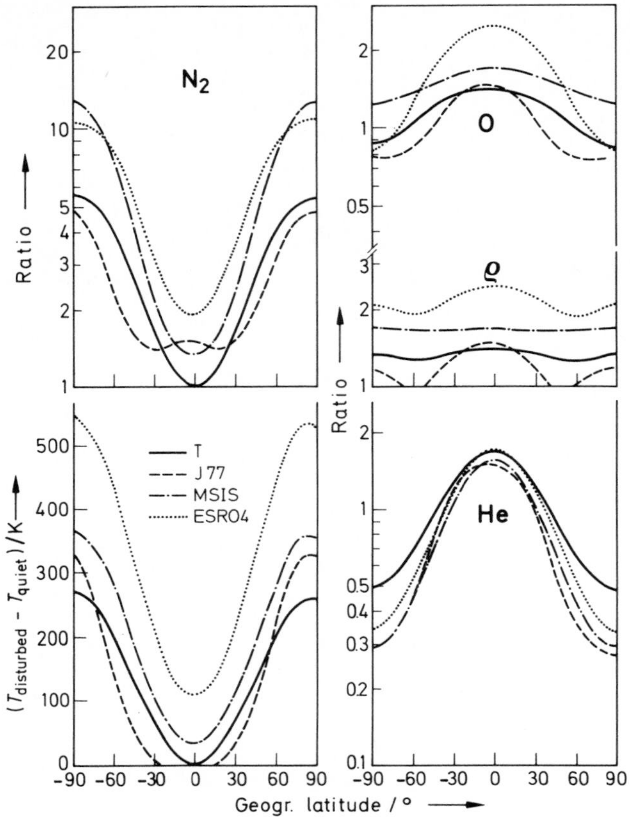

Fig. 100. Geomagnetic effect at an altitude of 400 km due to $Kp=5$ ($Ap=48$) on temperature (bottom left) and partial densities (ratio "disturbed"/"quiet") of N_2, O, He; ρ total density. 21 September, high solar activity: $F=\langle F\rangle=150$

Sect. 5). Apparently, the amplitude of the diurnal variation[5] is underestimated by the global models[6].

γ) A crucial problem is that of the effect of *magnetic disturbances*. First of all, it is certain that these are individuals, so that their average behaviour only can be described. But since all models admit a dependence on some magnetic disturbance parameter, Kp or Ap[7], they intend at least to take account of such effects. After the example shown in Fig. 99[3] THUILLIER's [*62*] Kp-based model might have the best performance in this context. The average effects themselves (considerably smoothed by averaging) are shown in Fig. 100. Typical at high

[5] The secondary peak at night (in December) is not certain since these values were extrapolated from only daytime measurements

[6] This is understandable since the fitting procedure has a smoothing tendency, in particular when measurements of different origin are combined

[7] See Siebert, M. (1971): This Encyclopedia, vol. 49/3, p. 212, 222 (for Kp), p. 225 (for Ap)

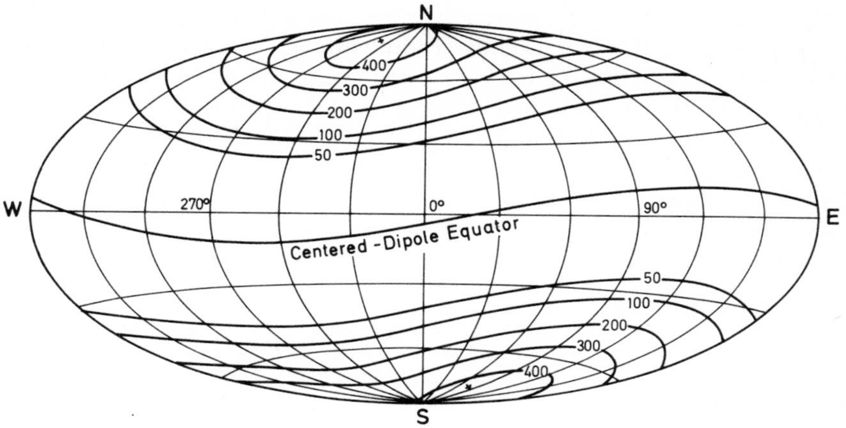

Fig. 101. Temperature increase from quiet conditions to $Kp=6$ after the J 77 [24] model[3]

latitudes are the density increases of light and decreases of heavy constituents[8]. At the same time an increase of light constituents at low latitudes appears.

JACCHIA[1] criticizes the formula used in MSIS [20] for the dependence on Ap since it becomes almost singulary at $Ap=0$. At the poles a change of Ap from 0 to 4 produces a temperature increase of 200 K, so that this model does not really cover quiet conditions. This might explain the particular behaviour of the (quiet) latitudinal variation of T_{ex} (see Fig. 94b) of this model, which even allows a secondary maximum at the equator[1]. JACCHIA proposes the following expression for the characteristic temperature increase due to magnetic activity:

$$\Delta T = c_1 \cdot Ap + c_2 \cdot [1 - \exp(c_3 \cdot Ap)]$$

and

$$Kp = 1.89\,\mathrm{Arsinh}(0.154 \cdot Ap). \qquad (12.1)$$

Figure 101 shows the global disturbance increase of T_{ex} after JACCHIA. Though it has been smoothed too much at high latitudes (the auroral ovals do not appear!), it is probably better than MSIS in this particular context. See Sect. 18 for more information about high latitude phenomena.

D. Empirical modelling of the ionosphere

13. Modelling vertical profiles

a) Electron density profile. Unlike neutral densities the electron density in the ionosphere is not a monotonically decreasing function of height but increases at lower altitudes and decreases at higher altitudes, so that a peak

[8] The ESRO-4 model, established from data at solar cycle minimum should not be considered at high levels of solar activity

usually appears somewhere between 250 and 400 km of height. The two main regions, called E and F, are usually separated by a "valley", which makes the height profile rather complicated[1]. Also the ionization in the regions D, E and F (with separation heights of about 90 and 140 km) is greatly variable in space and time. All this makes it difficult to obtain a satisfying description of the electron density distribution as function of height, geographical coordinates and time. Furthermore, since ionospheric electron density models are of particular interest for radio wave propagation purposes a reasonably accurate description of the different features in the height profile is often important.

α) In *early studies* of ionospheric radio wave propagation geometrical optics was applied and the profile was approximated either by a parabolic layer or one of the ELIAS-CHAPMAN type[2]. Wave optical computations were also made with the parabolic model[3] and with the EPSTEIN type of layer[4]. The latter model provides a unique description covering all altitudes by one analytical function. RAWER[5] has shown that full wave solutions can be obtained for a whole class of profiles which derive from an EPSTEIN "layer"[6] or "step"[4] by suitable analytical transformations (which, however, must be monotonic[5]).

In the following we discuss different types of more detailed profile functions as have been used more recently, mostly for radio wave propagation purposes. Some of these cover only the bottom side (i.e. sub-peak altitudes) since this is sufficient when the waves are only reflected from the ionosphere.

β) A combination of parabolic models enables the *main layers* E and F2 to be reproduced quite well by *segments* which are assumed parabolic in electron density, and a simple linear transition in between (Fig. 102a)[7]. An improved transition between the regions has recently been introduced (Fig. 102b)[8] by replacing the descriptive function for the F2-layer (which is parabolic in [7]) by a $\cos^2$-function so that the plasma frequency f_N as function of altitude z reads:

F_2 segment:

$$f_N(z)/foF2 = \cos\left(\frac{hmF2 - z}{ymF2}\right). \tag{13.1a}$$

f_N^2 is proportional to the electron density N_e. A transition function is then assumed in the

$E-F$ segment:

$$f_N(z)/foE = \sec\left(\frac{z - hmE}{\beta \cdot ymF2}\right) \tag{13.1b}$$

where $foF2$, foE, $hmF2$ and hmE are the internationally agreed designations of peak plasma frequencies and altitudes and ym is the (parabolic, half layer)

[1] Rawer, K., Suchy, K. (1967): This Encyclopedia, vol. 49/2, Sect. 30, p. 259

[2] Ibidem, Sect. 9, p. 97, Eqs. (9.25) and (9.26). See also: Yonezawa, T. (1982): This Encyclopedia, vol. 49/6, p. 200 [Eq. (25.5)] and Fig. 13, p. 201. The original publications may be found as [10] in Sect. 8

[3] Rawer, K., Suchy, K. (1967): This Encyclopedia, vol. 49/2, Sect. 15β, p. 183 ff.

[4] Ibidem, Sect. 15δ, p. 187 ff. and Fig. 63. The 'layer' is the derivative of the 'step', Eq. [10.7]

[5] Rawer, K. (1939): Ann. Physik *35*, 385

[6] Such a layer shape, multiplied by a 'modulation function', has been used by: Kerblai, T.S., Nosova, G.N. (1976): [*15*], 104, in order to simulate large-scale horizontal structures

[7] Bradley, P.A., Dudeney, J.R. (1973): J. Atmos. Terr. Phys. *35*, 2131

[8] Dudeney, J.R. (1978): J. Atmos. Terr. Phys. *40*, 195

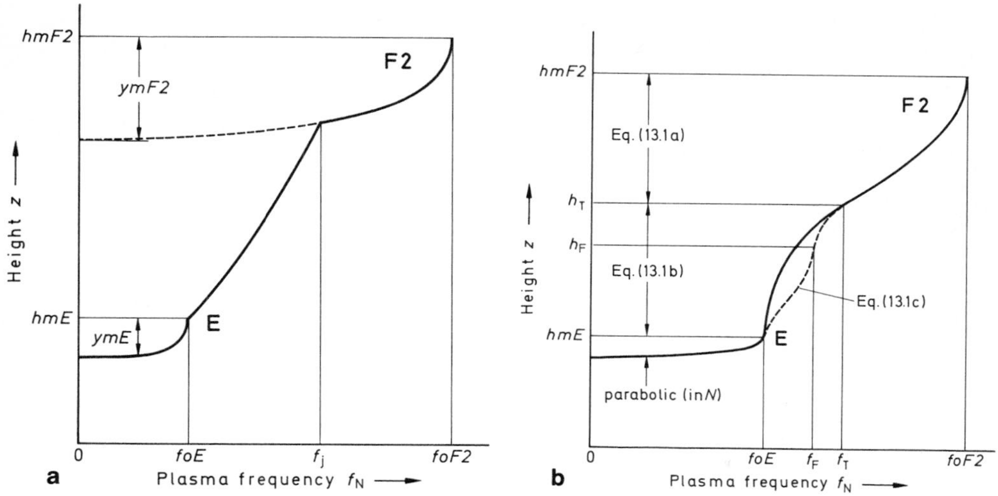

Fig. 102 a, b. Parabolic electron density profile models: (**a**) with linear connection[7], (**b**) with a connection of continuous slope[8]. Abscissa, plasma frequency f_N (proportional to the square root of the electron density N_e)

thickness[9]. β is determined by numerically solving two simultaneous equations involving f_T (see Fig. 102b) so as to obtain the desired continuity of the derivative. The model can be extended to include a ledge representing the F1-layer[10] by generalizing[8] Eq. (13.1b):

$$f_N(z)/foE = \sec\left(\frac{z-hmE}{\beta \cdot ymF2}\right) + A_F \cdot \left[1-\cos\left(2\pi\frac{z-hmE}{h_T - hmE}\right)\right] \qquad (13.1\text{c})$$

the amplitude A_F being chosen so that the inflexion (or peak) in plasma frequency occurs at the desired plasma frequency $foF1$[10]. This is, again, obtained by resolving two transcendental equations[8].

While the model is quite adequate for reproducing day-time profiles obtained around noon it is not so good for other hours where a distinct valley appears between regions E and F[11].

Another bottom-side model[12], called IONCAP, allows for a valley of constant electron density with a (linear or parabolic) transition segment towards the main parabola which represents the F2-layer. The transition admits a "kink" at the characteristic plasma frequency[13] $foF1$ (Fig. 103). As for the

[9] Piggott, W.R., Rawer, K. (1961/1976): [*37*]; see also Rawer, K., Suchy, K. (1967): This Encyclopedia, vol. 49/2, Chap. C, p. 220ff.

[10] Ibidem, Sect. 36, p. 285

[11] This feature sometimes allows particular propagation modes; see Woyk (Chvojkova), E. (1959): J. Atmos. Terr. Phys. *16*, 124

[12] Lloyd, J.L., Haydon, G.W., Lucas, D.L., Teters, L.R. (1978): Estimating the performance of telecommunication systems using the ionospheric transmission channel. Rep. Boulder, Co., USA: Nat. Telecomm. and Information Administration

[13] Rawer, K., Suchy, K. (1967): This Encyclopedia, vol. 49/2, Chap. C, p. 220ff.

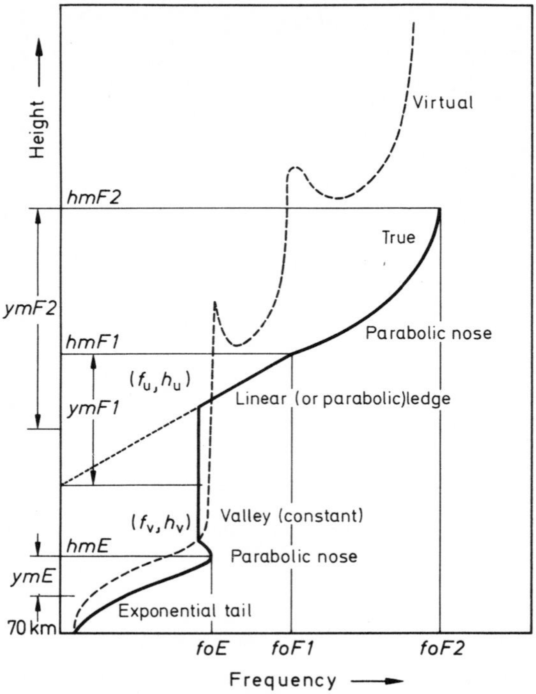

Fig. 103. Electron density model "IONCAP". The broken curve gives the virtual height trace[13] computed from the profile[12,14]

D-layer an exponential tail with smooth transition is assumed below the parabola representing the E-layer[14,15].

While the choice of a parabola as fundamental layer shape makes computations fairly easy some authors feel that the ELIAS-CHAPMAN shape[2] is more appropriate to the physics of the ionosphere: It reads in general form:

$$N_e(Z)/N_{e\,max} = \exp(a(1 - Z - e^{-Z})); \qquad Z = \frac{z - hm}{2q} \tag{13.2}$$

(peak at $z = hm$, thickness parameter q).

Just one layer of this type is assumed in KÖHNLEIN's ionospheric model[16]. Three superposed layers representing the E-, F1- and F2-layers, are applied in the model of CHING and CHIU[17]; quite

[14] The "IONCAP" model is an improved version of an earlier one with just three parabolic layers, E, F1 and F2, used in the ITS-78 ionospheric prediction system

[15] Barghausen, A.L., Finney, J.W., Proctor, L.L., Schultz, L.D. (1969): Predicting long-term operational parameters of high-frequency sky-wave telecommunication systems. ESSA Tech. Rep. ERL 110-ITS 78

[16] Köhnlein, W. (1978): Rev. Geophys. Space Phys. *16*, 341; the author proposes future parametrization of Z in order to allow for more complex layer shapes. The paper also gives a survey of models used up to 1977. The same profile formula was used before in: Köhnlein, W., Raitt, W.J. (1977): Space Res. *XVII*, 439

[17] Ching, B.K., Chiu, Y.T. (1973): J. Atmos. Terr. Phys. *35*, 1615; compare, however, the later version [63]

a similar model has been used for displaying quicklook maps of ionospheric measurements[18]. Models like these present great advantages in the context of wave optical computations since they avoid segmentation.

Another model profile with particular emphasis on the top side has been established in view of refraction effects[19] occurring with radio waves penetrating the ionosphere (BENT model)[20]. The empirical data base was a very large number of bottom- and top-side ionograms[21], the latter mainly from the ALOUETTE top-side sounders[22]. Since only refraction effects were intended to be assessed the bottom-side profile was very roughly replaced by just one (fourth order) parabola. The main merit of the model is, however, its top-side profile. It consists of a (second order) parabolic segment just above the peak which is matched with three segments with exponentially decaying electron density; these are connected continuously with each other at prefixed altitudes, with discontinuous transition of the "scale height", however. The parameters were determined by fitting a large number of observed profiles, separately for different latitude regions, seasons and values of solar activity (see Sect. 13γ and Fig. 104).

γ) The profile used for the "*International Reference Ionosphere*" (IRI) [*43*, *44*] derives in its top-side part indirectly from the large data base used in the BENT model. However, in order to obtain a unique profile BENT's discontinuously segmented description was replaced by a fully continuous one.

BOOKER[23], feeling a need to represent the electron density by one analytical function covering all altitudes, has introduced a "skeleton function" which gives a 'scale height' H defined by BOOKER as

$$-\frac{d\log_e N_e}{dz} \equiv -\frac{1}{N_e}\frac{dN_e}{dz} = \frac{1}{H}. \tag{13.3}$$

An exponential increase or decrease corresponds to a constant negative or positive scale height H. In order to obtain continuous transitions BOOKER applied EPSTEIN step functions[24]; for a transition between two "scale height" values H_1 and H_2 he writes:

$$-\frac{1}{N_e}\frac{dN_e}{dz} = \frac{1}{H_1} + \frac{\frac{1}{H_2}-\frac{1}{H_1}}{1+\exp(-(z-a)/b)} = \frac{1}{H_1} + \frac{H_1-H_2}{H_1\cdot H_2}\,\mathrm{Eps}_0(z,b,a) \tag{13.4}$$

---

[18] Flattery, T.W., Tascione, T.F., Secan, J.A., Taylor, J.W. Jr. (1979): A four-dimensional ionospheric model. Unpublished report of AFGWC (USA)

[19] Rawer, K., Suchy, K. (1967): This Encyclopedia, vol. 49/2, Chap. F I, p. 443

[20] Llewellyn, S.K., Bent, R.B. (1973): Documentation and description of the Bent ionospheric model. Rept. AFCRL-TR-73-0657. Hanscom, Ma.: AF Geophys. Labs.; a shorter version is: Bent, R.B., Llewellyn, S.K., Nesterzyk, G., Schmid, P.E. (1975) in: Proc. symposium on the effects of the ionosphere on space systems and communications. p. 559. Washington, D.C., USA: Naval Research Lab.

[21] The Bent model is the only easily accessible condensation of the enormous mass of top-side profiles gathered during solar cycle 20 by the Alouette satellites

[22] Rawer, K., Suchy, K. (1967): This Encyclopedia, vol. 49/2, Sect. 58, p. 503

[23] Booker, H.G. (1977): J. Atmos. Terr. Phys. *39*, 619

[24] Eq. [10.7]; a drawing of the 'step' function can be found in vol. 49/2 of this Encyclopedia as Fig. 63, p. 188. The Epstein 'transition', $\mathrm{Eps}_{-1}(x)$, is obtained by integrating the 'step', Eq. [10.7]

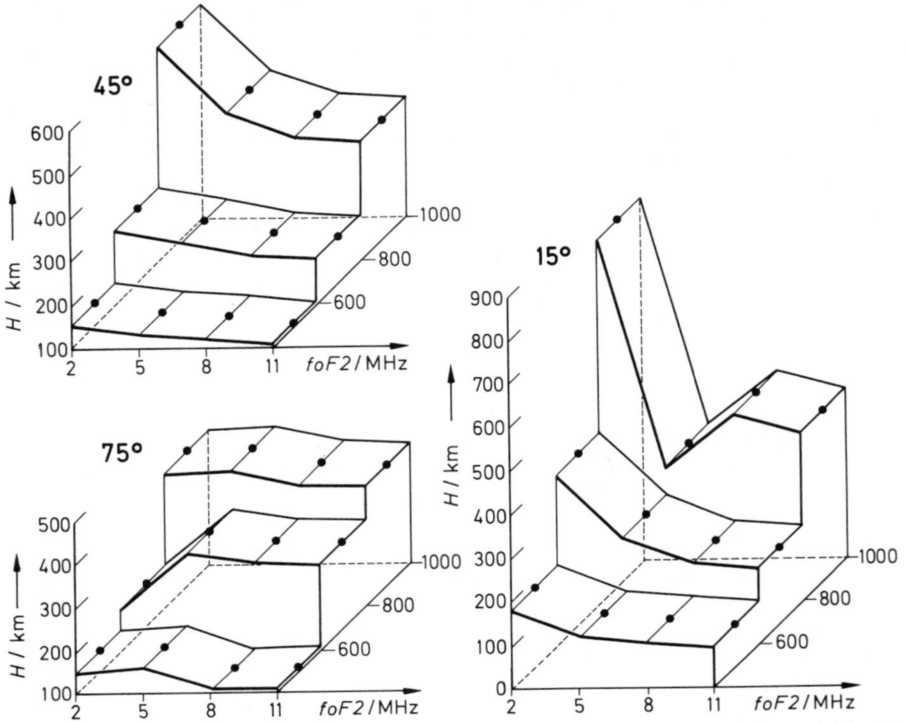

Fig. 104. Original skeleton functions of 'ionospheric scale height' H [Eq. (13.3)] as obtained by LLEWELLYN and BENT[20] from the ALOUETTE topside sounder data set by fitting *independently* in preselected ranges of height and F2 peak density ($N_{\mathrm{e\,max}}$ is proportional to $foF2^2$). Parameter: (American) latitude

where b determines the width of the transition range around $z=a$. In order to represent a number of height segments analytically a "skeleton" profile is established which admits a sequence of "scale heights" $H_0 \dots H_{\mathrm{m}}$ with continuous EPSTEIN transitions after Eq. (13.4). From this function the electron density profile is obtained by integration leading to

$$N_{\mathrm{e}}(z)/N_{\mathrm{e}0} = \exp\left\{-\left[\frac{z-z_0}{H_0} + \sum_{n=1}^{m} b_n \cdot \left(\frac{1}{H_n} - \frac{1}{H_{n-1}}\right) \cdot \log_{\mathrm{e}}\left(\frac{1+\mathrm{e}^{Z_n}}{1+\mathrm{e}^{Z_{0n}}}\right)\right]\right\}; \quad (13.5)$$

$$Z_n = (z-z_n)/b_n; \qquad Z_{0n} = (z_0 - z_n)/b_n.$$

The exponent may also be expressed with the "EPSTEIN transition function" Eps_{-1}

$$[\dots] \equiv \frac{1}{H_0}(z-z_0) + \sum_{n=1}^{m} b_n \cdot \frac{H_{n-1}-H_n}{H_n \cdot H_{n-1}} \cdot \{\mathrm{Eps}_{-1}(z; b_n, z_n) - \mathrm{Eps}_{-1}(z_0; b_n, z_n)\}.$$

Asymptotically it becomes a linear function of z for large and small values of z. Each member of the sum may be interpreted to be a smooth transition

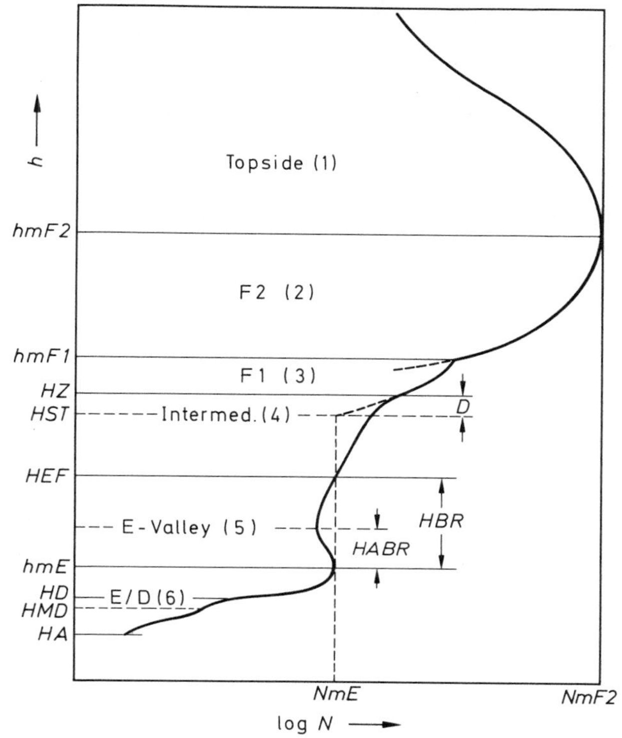

Fig. 105. IRI-1978 and 1979 segmented profile [*43*, *44*]

between two segments each with linear variation. We call Eps_{-1} an *Epstein transition.* Thus the profile $N_e(z)$ might be represented by a sequence of quasi-exponential segments with fully continuous transitions of width b_n at altitude z_n.

(i) At the present time IRI uses this kind of description only for the *top side*, i.e. above the F2-peak (with $m=2$, i.e. three segments). The parameters were determined by fitting[25] with the BENT model[20], i.e. essentially with the ALOUETTE data.

It may, however, be seen from Fig. 104 that results are rather irregular in their dependence on peak electron density [proportional to $(foF2)^2$], and on height. Therefore considerable smoothing had to be applied when replacing BENT's discontinuous description by a fully continuous one after Eq. (13.5).

Quite good agreement with electron density data observed by different techniques was reported at a 1980 critical discussion symposium on IRI [*45*].

(ii) As for the bottom side, the present-time IRI still applies segmentation in five height ranges (Fig. 105). The shape of the bottom-side F2-layer [segment

[25] Rawer, K., Bilitza, D., Ramakrishnan, S., Sheikh, N.M. (1978): [*57*], vol. 1, 6-1

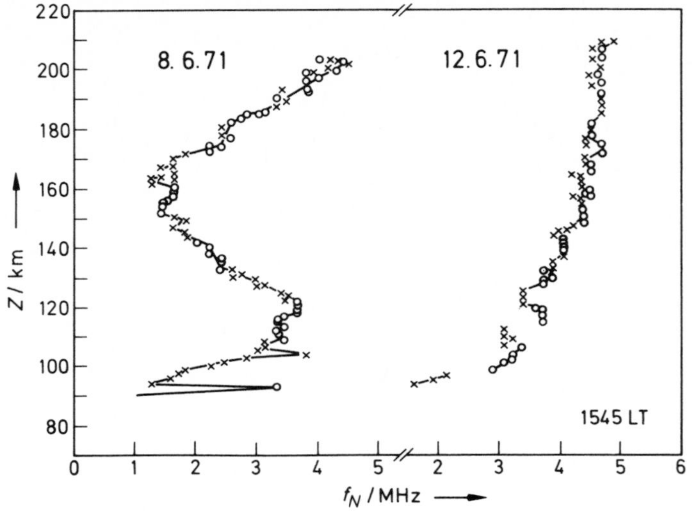

Fig. 106. Electron density profiles in the 'valley'. Both were obtained with an impedance probe aboard a rocket at exactly the same hour (15 h 45 LT) with 4 d time lag at Kourou, French Guyana (4.5° N)[30] [compare also Fig. 310, p. 480 in vol. 49/2 of this Encyclopedia]

(2)] is given as:

$$N_e/N_{e\,max} = \exp(-x^3)/\cosh(x); \qquad x = \frac{hmF2 - z}{B_0}, \tag{13.6}$$

$N_{e\,max}$, $hmF2$ being peak density and height, B_0 a thickness parameter. The latter was determined by fitting with empirical profile data[26] obtained from ionograms[27]. The F1-layer, if present, is obtained by adding below $NmF1//hmF1$ [segment (3)] a term proportional to $(hmF1 - z)^{1/2}$.

(iii) The *valley* [segment (4), smoothly connected upwards via segment (5)] is of some importance in propagational applications[11] but also presents geophysical interest. Around noon, it is often completely filled up, in particular at low latitudes. [In such condition segment (5) is omitted in the model.] It is, however, quite a regular feature when the Sun's altitude is lower, and at night.

This results from all rocket measurements[28, 29]. From the aeronomical point of view it is important to note that the valley may be quite different at

[26] For temperate latitudes after: Becker, W. (1973): Kleinheubacher Ber. *16*, 201

[27] See Rawer, K., Suchy, K. (1967): This Encyclopedia, vol. 49/2, Chap. C II, p. 252

[28] Compilations by Maeda, K.-I. (1969): J. Geomag. Geoelect. *21*, 557; (1970): ibidem, *22*, 551; (1971): ibidem, *23*, 133; (1972): Space Res. *XII*, 1229

[29] Soboleva, T.N. (1972): Model'nye profil sutočnogo raspredelenija electronnoj koncentracii spokojnoj ionosfery na srednih sirotah (average profile of the electron density distribution in the quiet ionosphere over the whole day). Rep. Moscow: IZMIRAN. At night, this model gives a much deeper valley (down to 7%) than [28], which achieves only 23% of the (night) NmE value (which is quite small). Booker has put forward arguments in favour of [28], which was therefore used in [*43*, *44*]

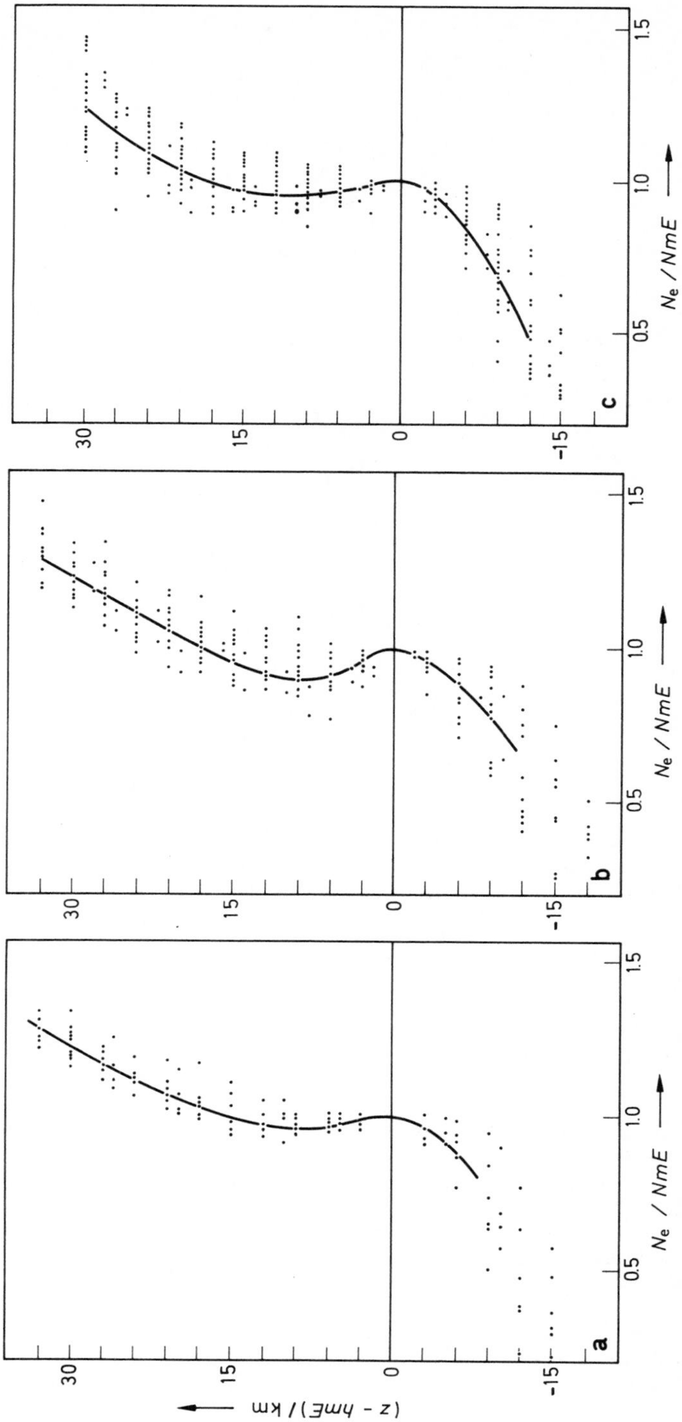

Fig. 107 a–c. Electron densities in the valley region obtained by incoherent scatter observations made in England[31]. Full line, fifth-order polynomial approximation. **(a)** Summer; **(b)** winter; **(c)** equinox, all noon[32]

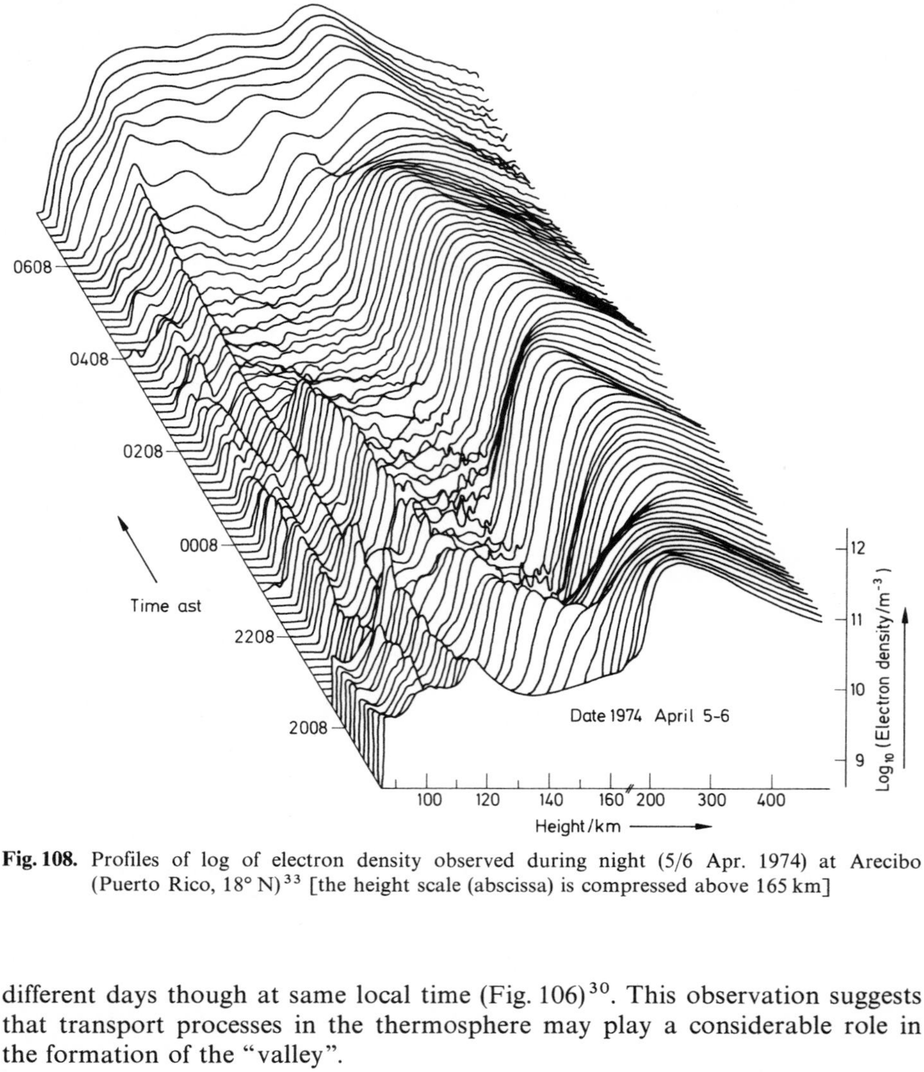

Fig. 108. Profiles of log of electron density observed during night (5/6 Apr. 1974) at Arecibo (Puerto Rico, 18° N)[33] [the height scale (abscissa) is compressed above 165 km]

different days though at same local time (Fig. 106)[30]. This observation suggests that transport processes in the thermosphere may play a considerable role in the formation of the "valley".

At mid-latitudes incoherent scatter observations are available[31], showing a rather shallow valley even at noon; see Fig. 107[32]. A polynomial in z was fitted with such data in order to model the daytime valley. Its depth is made to increase with increasing solar zenith angle so that at night a rather deep valley is assumed (descending to about a quarter of the E-layer peak density).

The broader valley occurring at night often shows layering with structures which are irregularly variable. Descending layers have been reported from incoherent scatter measurements (see Sect. 5) at low latitudes (Fig. 108)[33]. These features which cannot be represented in average models may be of some importance when instantaneous observations are considered.

[30] Neske, E., Kist, R. (1973): Space Res. *XIII*, 485

[31] Measurements were made at Malvern (UK) by G.N. Taylor

[32] Ramakrishnan, S., Rawer, K. (1974): Ann. Géophys. *30*, 347

[33] Shen, J.A., Swartz, W.E., Fairley, D.T., Harper, R.M. (1976): J. Geophys. Res. *81*, 5517

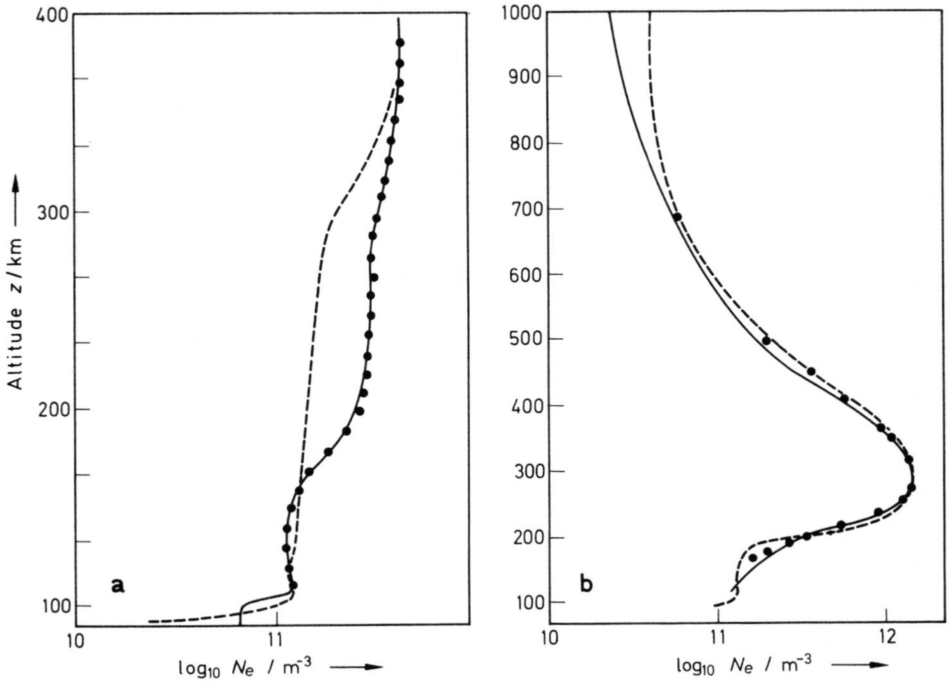

Fig. 109 a, b. Comparison of model profile shapes with individual observed profiles derived[37] from ionograms (full curve). Broken curve, peak-adapted IRI profile [*43*]; dots: BOOKER profile according to Eq. (13.5) fitted at several points. (**a**) Bottomside daytime profile (with F1-layer) [Moscow 4 Apr. 1979, 08 h 45 LT]. Disturbed conditions. (**b**) Top- and bottomside profile without F1-layer [Yamagawa, Japan, 16 Mar 1976, 11 h 00 LT][35]

(iv) When *critically discussing* the sub-peak IRI profiles [*45*] it was stated that rather often the ionization in the F1-layer and below can be smaller than given in the IRI profile[34]. This seems to be particularly so in cases where the F1-feature is particularly well seen in the ionogram. GULYAEVA has presented typical examples in [*45*]; she finds, however, also cases where the opposite is true (Fig. 109). When applying the BOOKER method to the sub-peak profile, she obtains a composite profile after fitting the parameters in Eq. (13.5) at a few selected points of the observed profile, in particular at inflection points[35,36]. While it is certain that predicted models cannot be adapted in such a way, it might be a good idea to replace the segmented scheme of Fig. 105 by one after Eq. (13.5), fitting with suitably chosen points which can a priori be determined, e.g. the F1 inflection point.

In that case, however, another difficulty may arise since when fully applying BOOKER's method a larger number, *m*, of terms in Eq. (13.5) will be needed. Now, when using standard fitting

---

[34] See, for example, Zelenova, T.I., Legen'ka, A.D., Fatkulin, M.N. (1975): [*14*], 7; (1976): [*15*], 6. Also: Gulyaeva, T.L. as well as Chasovithin, Ju.K. et al. in (1984): [*45*], p. 22 and 38

[35] Gulyaeva, T.L. (1981): Radio Sci. *16*, 135 and (1984): [*45*], 22

[36] Note that IRI gives average profiles for prediction purposes. When fitting the parameters in Fig. 105 with observed data as done in [34] better agreement would surely be achieved

[37] Methods of derivation are described in: McNamara, L. (ed.) (1978): A comparative study of methods of electron density profile analysis. Rep. UAG-68. Boulder, C., USA: NOAA, World Data Center-A (STP). See also Rawer, K., Suchy, K. (1967): This Encyclopedia, vol. 49/2, Chap. C II, p. 252ff.

procedures with many unknowns there is always the risk that the functions which are being adapted in one height range disturb the good representation in another. A way out from this was indicated by RAWER, who recently[38] proposed replacing Eq. (13.5) by:

$$N_e(z)/N_{e0} = \sum_{l=1}^{3} \mathrm{Fi}_l(z) \cdot \exp\{-[\ldots m_l \ldots]\}, \tag{13.6a}$$

$[\ldots m_l \ldots]$ being the expression in the brackets of Eq. (13.5) with $m = m_l$.

Fi(z) is a continuous filter function; for $l=1$ and 3 one may take an EPSTEIN step function, for $l=2$ the difference of two of these:

$$\mathrm{Fi}_2(z) = \mathrm{Eps}_0(z; d, hmE) - \mathrm{Eps}_0(z; d, hmF2), \tag{13.6b}$$

with a transition width d of a few kilometres only. With this scheme fitting could be performed independently in each subrange without disturbing the others. It is intended to fit at up to three points in the bottom-side profile; these must, of course, be determined a priori (e.g. the F1 inflection point, top and bottom of the valley).

(v) In the *lowest ionosphere*, below the E-peak [E+D region, segment (6) in Fig. 105] basic profile information stems from in-situ rocket measurements mainly [*42*]. These were compiled by different authors[28,29]. While the subpeak thickness of the E-region (d_E) can be thus established fairly well, the lower part (D-region) was for a long time largely unknown, different methods of observation ending up with different profiles [*42*]. Both regions are separately represented in IRI [*43*, *44*] by exponentials

$$\begin{aligned} N_e(z)/N_{E\,max} &= \exp\left(-\frac{(hmE-z)^k}{d_E}\right); \qquad z > hD \\ N_e(z)/N_{D\,max} &= \exp\left(\sum_{i=1}^{3} F_{pi} \cdot x^i\right); \qquad z \leqq hD, \end{aligned} \tag{13.7}$$

and $x = z - hmD$. The coefficients are determined so as to have an inflection point at hmD and continuity including the derivative at hD, and be in agreement with the measurements. These were summarized by MECHTLY and BILITZA[39], but other sources were also used (see, for example, Fig. 111 below)[40].

The inversion of propagation experiments from ground is too uncertain to get reliable profile data [*45*]; see Fig. 110 [and (vi) below], but the data might be used for checking a once

---

38 Rawer, K. (1982): Adv. Space Res. *2* (10), 183

39 Mechtly, E.A., Bilitza, D. (1974): Models of D-region electron concentrations. Report IPW-WB2. Freiburg i.Br.: Fraunhofer Inst. physikal. Weltraumforschung

40 Gnalalingam, S., Kane, J.A. (1978): J. Atmos. Terr. Phys. *40*, 629 provided data from low latitudes which were used in [*43*, *44*]. A more recent compilation covering a good number of *disturbed* D-region profiles in: McNamara, L. (1978): Selected disturbed D-region electron density profiles. Rep. UAG-69. Boulder, Co. USA: NOAA, World Data Center-A (STP). The data of this latter report were not available when the IRI-profile [*44*] was established

established profile[41]; doing this with a large set of ionospheric absorption measurements by the A1-method[42], SINGER et al.[43], after careful analysis, concluded that for certain conditions their data seemed to disagree with the profiles from the in-situ measurements (as seen from Fig. 110b).

In Fig. 111 some D-region profiles[39] are displayed. As a permanent feature [which is reproduced in Eq. (13.7)] an inflection point occurs near 80 km altitude (the examples with a maximum, Figs. 111c, d, are rarer cases). The corresponding density NmD, under non-disturbed conditions, seems to be quite well defined [*44*][39].

(vi) Unlike conditions with decametric radio waves, ionospheric *propagation of longer radio waves* is not only dependent on the profile of electron density but also on that of collision frequency[44]. In fact, *conductivity* is the important parameter[45] which is, essentially, the product of both parameters. Since the lowest part of the ionosphere is alone effective at this type of propagation, a rather simple profile shape is most often used. Assuming exponential laws for both, such a law is also obtained for the conductivity[46] [*33*]. The parameters may be obtained by fitting field-strength data, e.g. measured as a function of the distance aboard an aeroplane, with propagation theory results. Up to a frequency of 300 kHz (by day – at least 60 kHz by night)[47] the waveguide model is appropriate for such computations.

Figure 111e, f shows exponential electron density and collision frequency profiles determined in this manner [*33*]. Earlier determined profiles of different shape are also shown[48] (see also Fig. 110). Though the exponential approach achieves good agreement with the measurements it must be noted that the product of both parameters is only determined. The conclusion is therefore that quite a small height range is decisive in such experiments in which, of course, conductivity varies almost exponentially with height.

In a recent investigation statistical analysis[49] was applied to some 700 experimental D-region profiles obtained by any method. Dependencies were derived for five geophysical influences, namely: the solar zenith angle χ; modified dip μ [Modip, see Eq. (14.7)]; solar activity R; magnetic activity; season.

Probably the most important finding is that there are different dependencies at different height levels. Given the large dispersion of the inputs the following representation is proposed for the 75–90 km range (detailed tables

---

[41] Ramanamurthy, Y.V. (1984): [*45*], 9

[42] Rawer, K. (ed.) (1976): Absorption manual. Rep. UAG-57. Boulder, Co., USA: NOAA, World Data Center-A (STP). See also Rawer, K., Suchy, K. (1967): This Encyclopedia, vol. 49/2, Sect. 39, p. 318

[43] Singer, W., Taubenheim, J., Bremer, J. (1979): J. Atmos. Terr. Phys. *42*, 241; also (1983): [*45*], 1

[44] See SUCHY's contribution in this volume

[45] Bremer, H. (1949): Terrestrial radio waves p. 217. Amsterdam: Elsevier

[46] Bickel, J.E., Ferguson, J.A., Stanley, G.V. (1970): Radio Sci. *5*, 19

[47] Morfitt, D.G. (1977): Effective electron density distributions describing vlf/lf propagation data = Techn. Rep. 141; Pappert, R.A. (1981): Lf daytime earth-ionosphere waveguide calculations = Techn. Rep. 647. San Diego, Ca., USA: Naval Ocean Systems Center

[48] Bain, W.C., Harrison, M.D. (1972): Proc. IEEE *119*, 790; Thomas, L., Harrison, M.D. (1970): J. Atmos. Terr. Phys. *32*, 1

[49] McNamara, L.F. (1979): Radio Sci. *14*, 1156

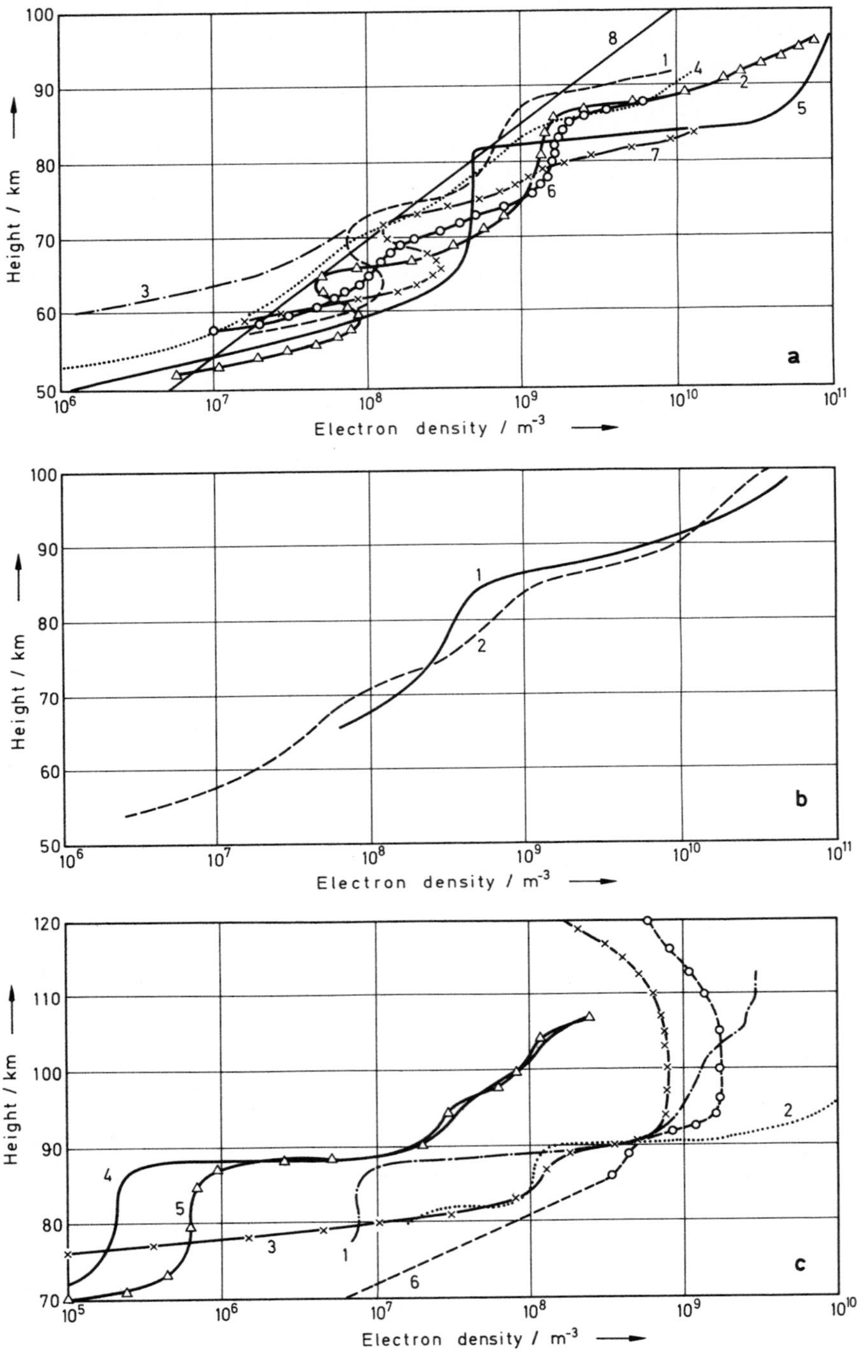

Fig. 110a–c

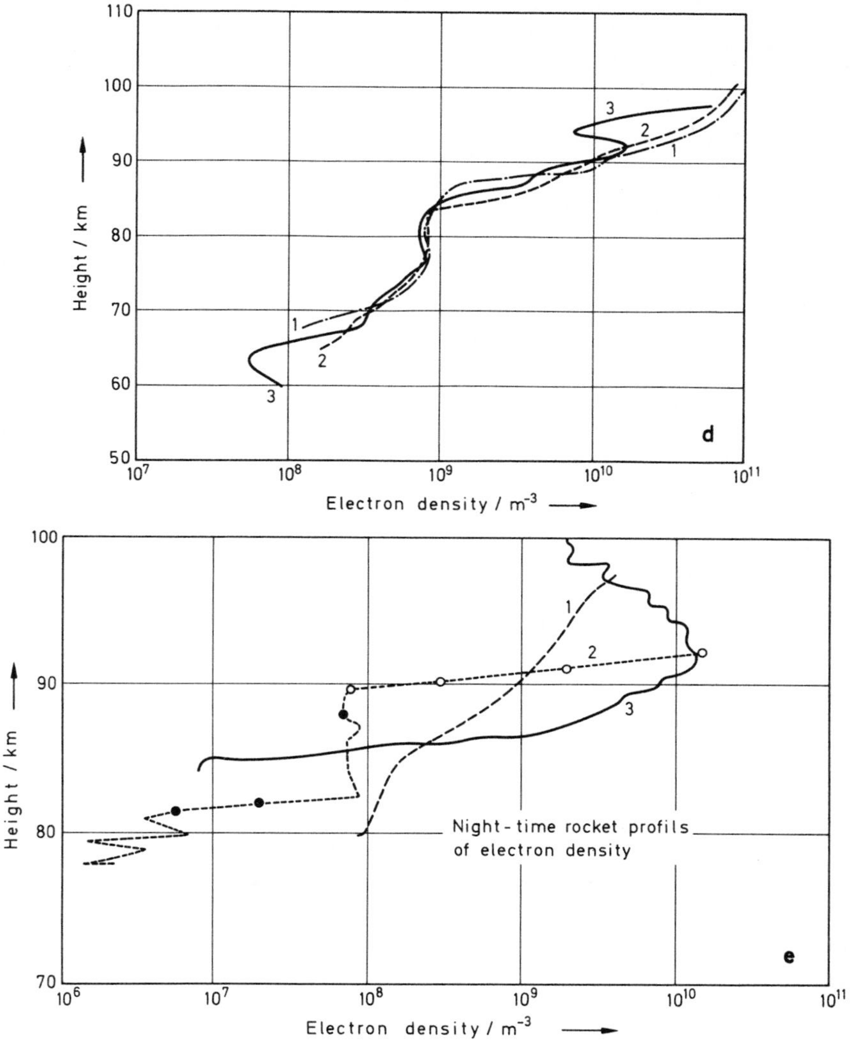

Fig. 110a–e. Compilation of D-region profiles determined by different techniques [courtesy of Y.V. Ramanamurty]. (**a**) Noon, $\chi<60°$, radio wave propagation: 1 --- Deeks (vlf, lf); 2 —▵— Bain and Harrison (vlf, lf) 3 -·- Foley et al. (vlf "omega"); 4 .... Bremer and Singer (lf, mf abs); 5 —— Behroozi - T. and Booker (elf) 6 —o— Krasnuškin and Kolen'ikov (vlf, lf); 7 —×— Krasnuškin and Kolen'ikov (HSA); 8 —— Wait and Spies (exponential). (**b**) $\chi=60°$ [$\varphi=50°$ N, $\lambda=10°$ E, June, $R=10$], 1 —— Rawer et al. (IRI-1978) [*43*]; 2 -- Bremer and Singer (1977), lf, mf); (**c**) Night, radio wave propagation: 1 -·- Deeks (vlf, lf); 2 .... Thomas and Harrison (vlf, lf); 3 —×—×— Behroozi - T. and Booker (elf); 4 —— Cairo et al. (vlf satell.): summer; 5 —▵— Cairo et al. (vlf satell.): winter; 6 —o— Rawer et al. (IRI-1978) [*43*]. (**d**) Day, medium solar activity, in-situ measurements: 1—·—N. England ($\chi=44°$); 2 --- Sardinia ($\chi=41°$); 3 —— S. India ($\chi=48°$) [from MITRA and Somayajulu. 1979]. (**e**) Night, ground-rocket propagation and in situ): 1 —— Kane 1972 ($\chi=98°$); 2 ----- Mechtly and Smith 1968 ($\chi=108°$) [Langmuir probe]; 3 —— Prakash et al. 1970 ($\chi=152°$)

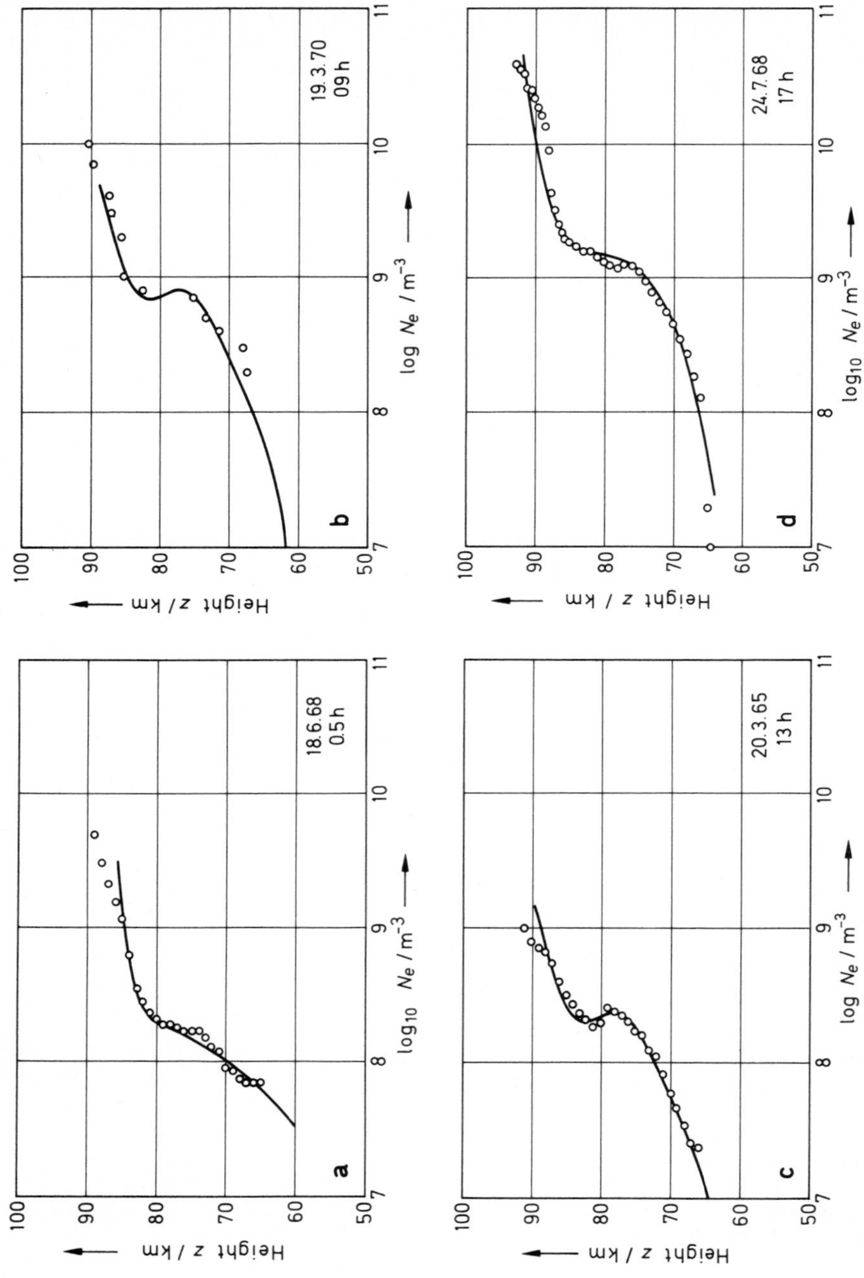
a
18.6.68
0.5 h
b
19.3.70
09 h
c
20.3.65
13 h
d
24.7.68
17 h
Height z / km
log₁₀ Ne / m⁻³
log Ne / m⁻³

Fig. 111 a–d

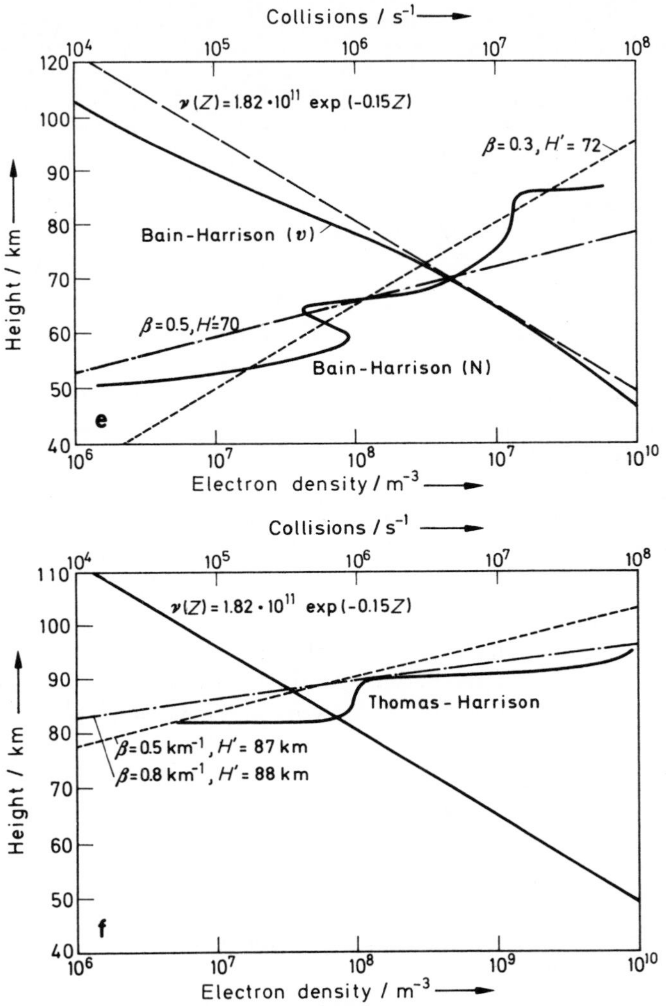

Fig. 111a–f. Typical D-region electron density profiles during daytime: (**a–d**) probe measurement checked with ground to rocket propagation experiment (circles). Fitting curve as third-order polynomial. (**a**) 05 h; (**b**) 09 h; (**c**) 13 h; (**d**) 17 h LT [39] (**e, f**) exponential and other profiles inferred from radio wave fieldstrength data. Exponential profiles (straight lines) following $N_e = N_{e0} \cdot \exp[\beta(z - H')]$. Collision frequency model profiles $\nu(z)$ are also shown. (**e**) Day; (**f**) night [48]

can be found in [50]):

$$\log N_e = C_0 + C_1 \cos\chi + C_2 \sin(2\mu) + C_3 R.$$

Tables of the constants are given. The seasonal effect is found to be small but that of magnetic disturbances is very important and increases the density [50].

[50] McNamara, L.F. (1978): Ionospheric D-region profiles data base ... = Rep. UAG-67; Selected disturbed D-region electron density profiles = Rep. UAG-69. Boulder, Co., USA: World Data Center-A for Solar Terrestrial Physics

(vii) Figure 112a–k shows a few examples of vertical profiles of electron density and temperatures as presented in IRI-1979 [*44*].

δ) The *parameter values* needed when establishing profiles of electron density depend on time and geographical location (see Sect. 14 below). We discuss only the most important ones:

(i) Peak altitudes of the *lower layers* D, E and even F1 are not very variable so that they can be described by simple diurnal functions [51].

As for the peak electron densities, IRI [*44*] has for the D-region inflection point (Fig. 111 a–d):

$$NmD/\mathrm{m}^{-3} = (7.47 + 0.2 \cdot R_{12}) \cdot 10^8 \cdot \cos^{2.3}\chi \tag{13.8}$$

(χ solar zenith angle, R_{12} gliding average of the Zürich sunspot number over 1 year). This equation was obtained as the average relation from a large number of measurements [52].

The peak density of the normal E-layer [53] depends very closely on the solar zenith angle and on solar activity *R*. IRI applies a formula [54] in which the different influences controlling the peak plasma frequency foE appear separated:

$$foE/\mathrm{MHz} = (A + B + C + D)^{1/4}, \tag{13.9}$$

A depending on solar activity (COVINGTON index *F* [see [13] Sect. 10]), *B* on the noon value of χ, *C* on the latitude and *D* specifying the variation with the local hour taking account of the decrease at sunset and of the variation during night.

Sporadic Es [55] cannot be taken account of in global models since it is too variable.

Similarly for the F1-layer [56] IRI gives [*44*]:

$$foF1/\mathrm{MHz} = K_{F1}(\cos\chi)^m. \tag{13.10}$$

K_{F1} is linearly interpolated between the values for $R=0$ and 100:

$$K_{F1\,0} = 4.35 + \frac{\mu}{100}(0.58 - 0.012 \cdot \mu); \qquad K_{F1\,100} = 5.348 + \frac{\mu}{100}(1.1 - 0.023 \cdot \mu);$$

$100 \cdot m = 9.3 + \mu \cdot (0.46 - 0.0054 \cdot \mu) + 0.03R$; μ is the "modified dip latitude" [see Eq. (14.7)].

(ii) The *peak altitude of the F-region, hmF2*, can be directly determined with in-situ or with incoherent scatter measurements (see Sect. 5 above). Unfor-

---

[51] IRI [*44*] uses EPSTEIN step functions, Eq. (13.4), of appropriate transition width

[52] Bilitza, D. (1981): [*44*], 7; compare [39]

[53] Rawer, K., Suchy, K. (1967): This Encyclopedia, vol. 49/2, Sect. 35, p. 278 and Eqs. (35.1–3)

[54] Kouris, S.S., Muggleton, L.M. (1973): J. Atmos. Terr. Phys. *35*, 133, and contribution No. 6/3/07 in CCIR rept. 252-2. Genève: Comité Consultatif International des Radio-communications (CCIR)

[55] See [*56*], also: Rawer, K., Suchy, K. (1967): This Encyclopedia, vol. 49/2, Sect. 40, p. 332

[56] Ibidem, Sect. 36, p. 285 and Eq. (36.1)

tunately, since this altitude is quite variable a worldwide picture cannot be deduced from such measurements which are made at a few locations only. The direct measurements are, however, helpful for calibrating the indirect methods which have been in use for many years.

These use the "*M*-factor" called *M(3000)F2* [*37*], which is routinely deduced from ionograms[57]. By comparing with parabolic model fits SHIMAZAKI[58] statistically established a linear relation[57] of the peak altitude *hmF2* with $1/M$. More recently, his formula was improved by taking account of a few other parameters which are of some importance in determining *M* or for the layer shape[59]. The relation finally used[60] in IRI reads [*43*, *44*]:

$$\begin{aligned} hmF2/\mathrm{km} &= 1490/(M+\Delta M)-176, \\ \Delta M &= a_1 + a_2 \cdot [1-(R/150)\cdot \exp(-\psi_\mathrm{m}^2/a_3)]/(x-a_4), \end{aligned} \tag{13.11}$$

the coefficient $a_1 \ldots a_4$ depending on solar activity R; ψ_m is magnetic "dip-latitude" derived from magnetic dip ψ [$\psi_\mathrm{m} = \arctan(\frac{1}{2}\tan\psi)$]. Empirical relations giving *hmF2* directly as a function of hour and geographical coordinates have also been given.

(iii) The *peak electron density* of the ionosphere, *NmF2*, or its equivalent the peak plasma frequency *foF2*, cannot be so easily described since it admits rather complicated variations in time and space[61]. A rather involved formula giving the worldwide pattern by a particular LEGENDRE-type development is dealt with in Sect. 14. IRI [*43*, *44*] recommends that this program[62] is used.

An extremely simple empirical formula is given in [63], admitting a second-order harmonic function of the local hour with coefficients depending on geographical and magnetic coordinates, season and, last but not least, solar activity indicated by the Zürich sunspot number *R*. It might be used with IRI as a very provisional guess but only when the CCIR method is not available.

The dependence on solar activity is an important feature when predicting the ionospheric propagation conditions[64]. Experience seems to show that this relation is not identical for different solar cycles or even in the rising and falling period of one (11 year) cycle[65]. A secular variation over four cycles might be the reason[66].

---

[57] Ibidem, Sect. 38, p. 307

[58] Shimazaki, T. (1955): J. Radio Res. Lab. (Tokyo) *2*, 85; compare Figs. 139, 140 in [57]

[59] Viz. the ratio of the peak plasma frequencies, $x = foE/foF2$, (DUDENEY), further solar activity (sunspot number *R*) and magn. dip-latitude μ [see Eq. (14.7)] (EYFRIG)

[60] Bilitza, D., Eyfrig, R. (1978): Kleinheubacher Berichte *21*, 167. Compares with (direct) determinations by incoherent scatter technique. See also Bradley, P.A. (1979): AGARD Lecture Series *99*, 9-1

[61] Rawer, K., Suchy, K. (1967): This Encyclopedia, vol. 49/2, Sect. 37, p. 288

[62] The Comité Consultatif International des Radiocommunications (CCIR) by its Rep. 340 [*9*] made this program 'official' for *international* prediction purposes

[63] Chiu, Y.T. (1975): J. Atmos. Terr. Phys. *37*, 1563. His formula has dependence on geomagnetic latitude, solar declination angle, solar activity and on the hour (*1*st order harmonic only) [one printing error was corrected in IRI]

[64] Rawer, K. (1975): Radio Sci. *10*, 669

[65] Smith, P.A., King, J.W. (1981): J. Atmos. Terr. Phys. *43*, 1057

[66] Muggleton, L.M. (1969): J. Atmos. Terr. Phys. *31*, 1413

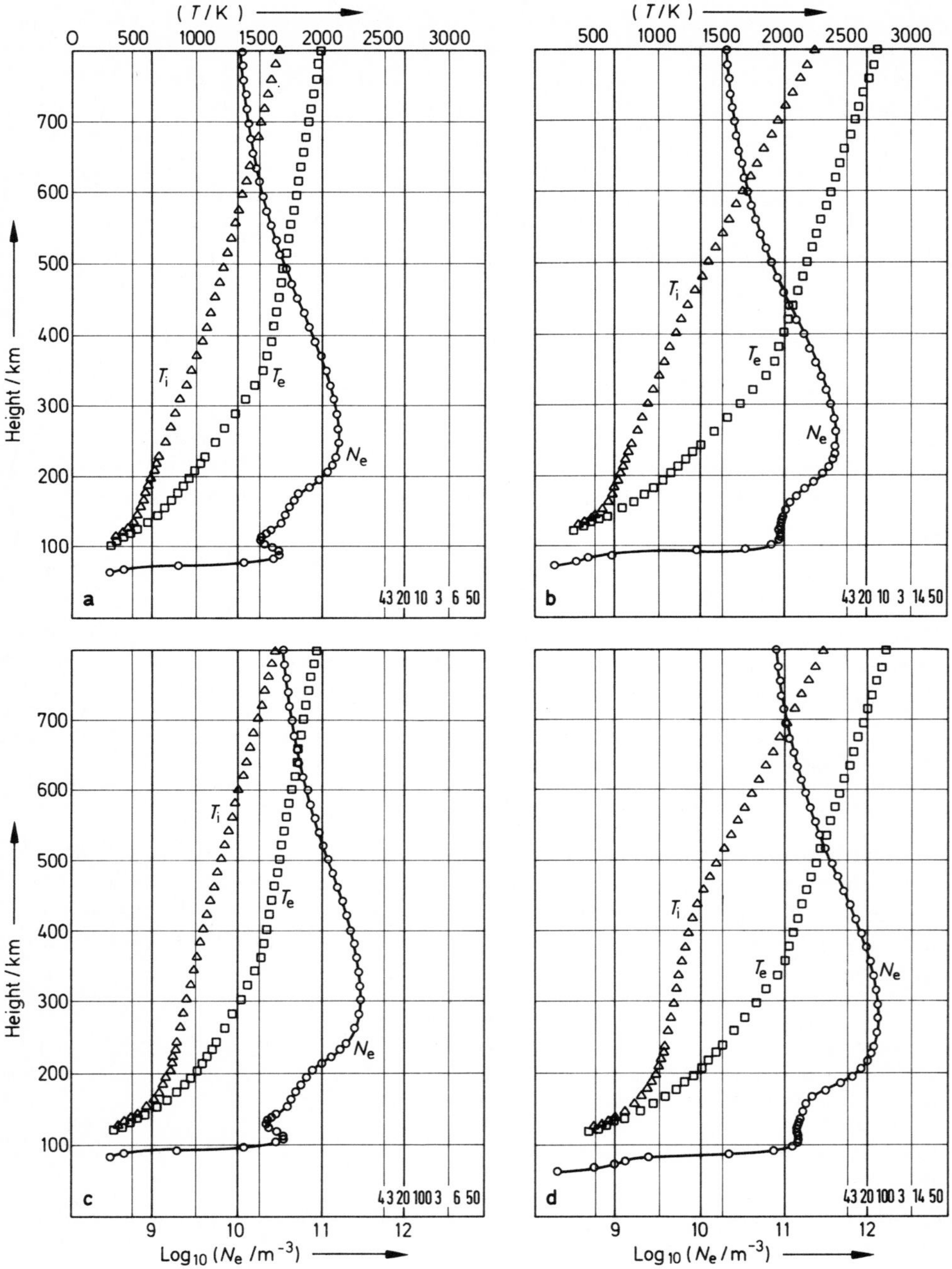

Fig. 112a–d

Fig. 112a–h. A selection of IRI-79 profiles of electron density [curves through circles] and plasma temperatures (electron temperature, squares; ion temperature, triangles). All for March; longitude 20° E; **(a–d)** modip 50° (latitude 43° N); **(e–h)** modip 0° (magnetic equator, latitude 10° N) [*44*]

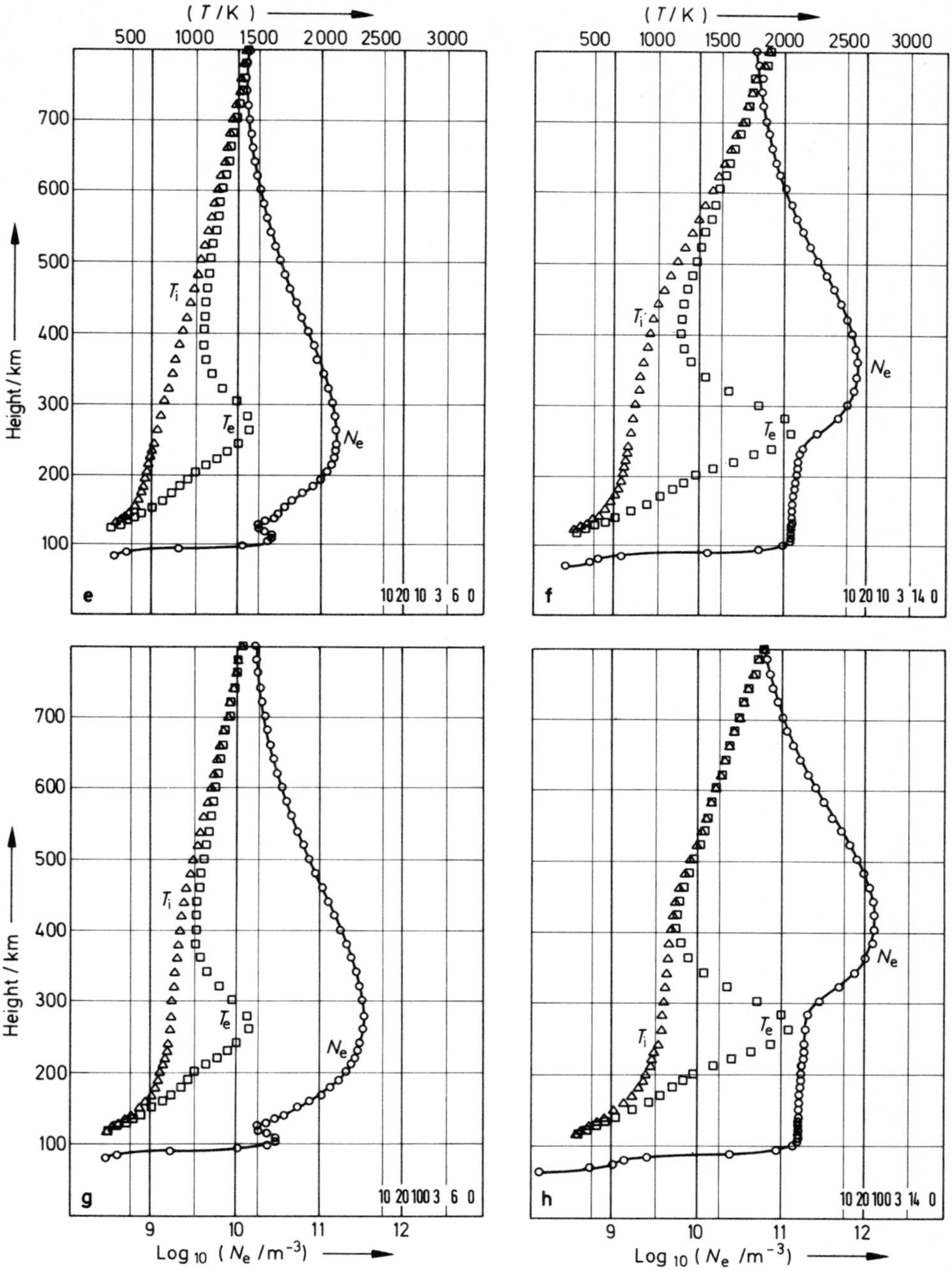

Fig. 112 e–h

	a, e	b, f	c, g	d, h
Hour LMT	06	14	06	14
Sunspot number *R*	10	10	100	100

[Lower right hand labels identify: latitude/°, longitude/°, *R*, month, hour, Modip/°]

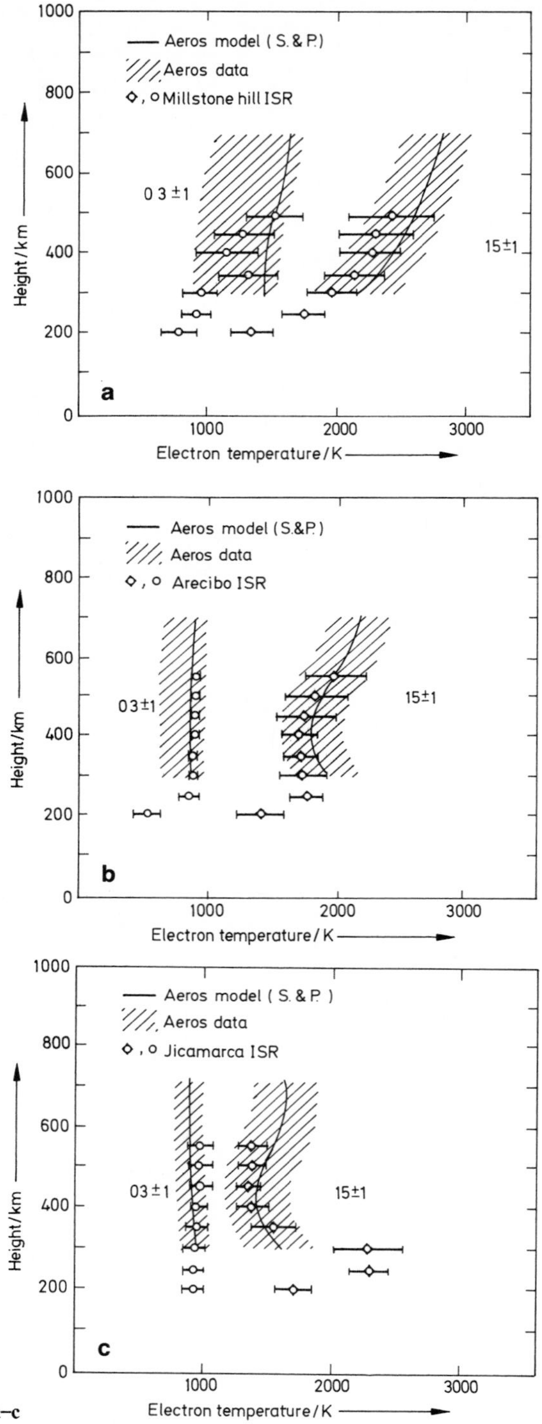
Aeros model (S.&P.)
Aeros data
◇, ○ Millstone hill ISR
03±1
15±1
a
Height/km
Electron temperature/K
Aeros model (S.&P.)
Aeros data
◇, ○ Arecibo ISR
03±1
15±1
b
Height/km
Electron temperature/K
Aeros model (S.&P.)
Aeros data
◇, ○ Jicamarca ISR
03±1
15±1
c
Height/km
Electron temperature/K

Fig. 113 a–c

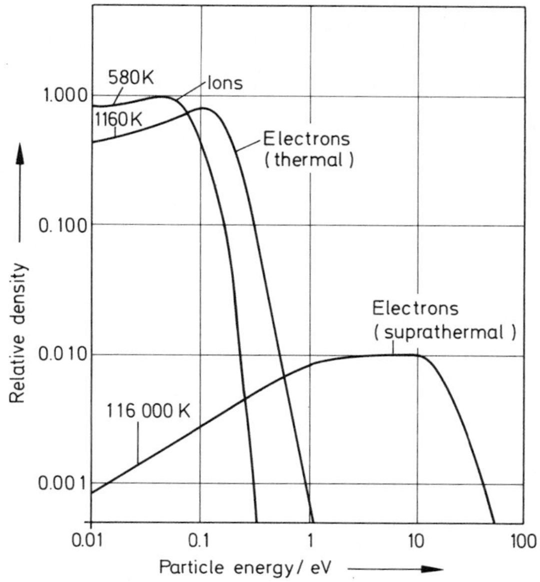

Fig. 114. Example showing energy spectra of plasma components measured by SPENNER with a retarding potential analyzer aboard satellite AEROS-B [*26*]. Temperatures were determined from the slope of the relevant characteristic curves (see Sect. 3b). The non-Maxwellian energy spectrum of the electrons was approximated by a combination of two largely different quasi-Maxwellian populations

b) Plasma temperatures

ε) Since full thermal equilibrium does not exist very often in the ionosphere, *electron and ion temperatures*, T_e and T_i, must be considered separately (see Sect. 5b α). There are mainly two *sources of information*, namely incoherent scatter (see Sect. 5) and direct (probe) measurements with the help of rockets or satellites[67]. Important discrepancies between the results of both techniques have been noted[68]. Reliable, in-situ measured T_e data have only been available for a few years and from a few satellites (Fig. 113).

[67] See Rawer, K., Suchy, K. (1967): This Encyclopedia, vol. 49/2, Sect. 56, p. 475ff.; also Maeda, K.-I. (ed.) (1967): Electron density and temperature measurements in the ionosphere (COSPAR Techn. Manual Series). Paris: Committee on Space Research

[68] See [*6*] (1970) wherein T_e values obtained from almost all in-situ measurements known at that time were found too high (by up to 1,000 K)

[69] Spenner, K., Bilitza, D., Plugge, R. (1979): J. Geophys. *46*, 57

Fig. 113a–c. Statistical intercomparison of electron temperature measurements obtained by an RPA [see Sect. 3b] aboard AEROS-B [*26*] (hatched, curves from model equation[83], 03 and 15 h LT) with local profile measurements by incoherent scatter sounding (bars) [see Sect. 5]. **(a)** Millstone Hill (43° N, 72° W); **(b)** Arecibo (18° N, 67° W); **(c)** Jicamarca (12° S, 77° W)[69] [p. 368]

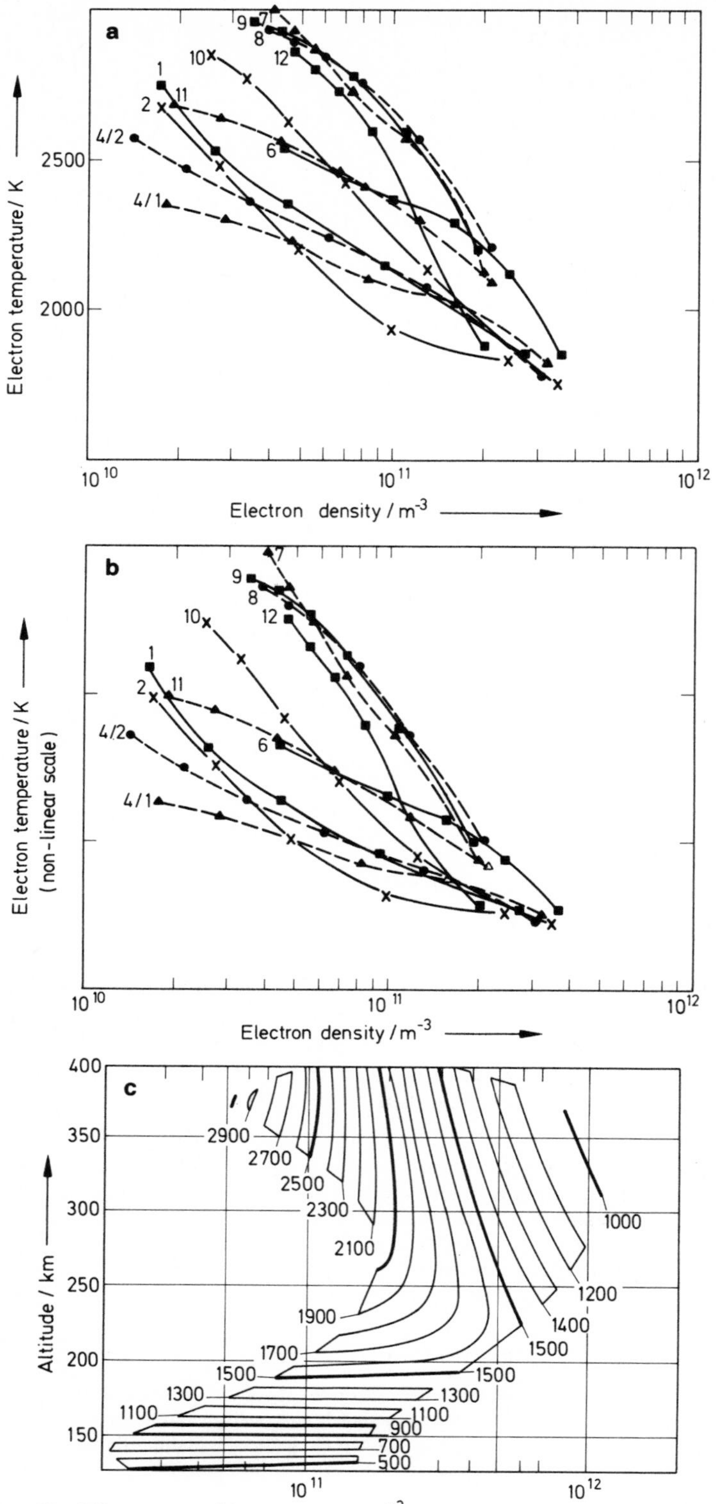
a
Electron temperature / K
2500
2000
Electron density / m^-3
1
2
4/2
4/1
11
10
9
7
8
12
6
b
Electron temperature / K (non-linear scale)
Electron density / m^-3
c
Altitude / km
400
350
300
250
200
150
2900
2700
2500
2300
2100
1900
1700
1500
1300
1100
900
700
500
1000
1200
1400
Plasma density / m^-3

Fig. 115 a–c

Vertical profiles of T_e and T_i are quite well known for the few locations of incoherent scatter sounders. In most cases (but not always[70]) there is temperature equilibrium ($T_e = T_i = T_n$) up to 120 km. Above this altitude, during daylight hours the gradient in T_e becomes considerably greater than for T_i and T_n. This is a consequence of heating by photo-electrons which form a hot electron population. These heat the main (thermal) electron population, which in turn is the heat source for the ions (Fig. 114). The energy loss by electron-ion collisions is faster the greater the plasma density. [It goes with N_e^2, see Eq. (5.4).] So T_e and N_e must behave complementarily and an inverse relation between T_e and N_e will result. This is observed regularly in daytime at altitudes between about 200 and 600 km; see Fig. 115a, b. A linear relation with negative slope was at first proposed for describing the facts (by day)[71,72].

In the upper part of this range heat transfer is, however, not restricted to local gain and loss but is also brought about by heat conduction in the electron gas[73]; since this also increases with N_e^2, the inverse relation between T_e and N_e is preserved. Only at heights above 600 km does another influence become important, namely collisional heating by trapped charges[74], at higher latitudes also by precipitating electrons. Since such effects occur fairly irregularly and with important variations the above relation will be less representative at such great altitudes[75]. The problem is discussed in more detail in Appendix A, p. 525.

The complementary behaviour of N_e and T_e had first been demonstrated with incoherent scatter data[71], only later with in-situ measurements[76]. BRACE and THEIS[77] from a large number of satellite plasma measurements established an empirical relation (for daylight hours).

$$T_e/1000\,\mathrm{K} = 1.05 + (1.7 \cdot (z/100\,\mathrm{km}) - 2.75) \cdot \exp[6.09 \cdot 10^{-12}\, N_e/\mathrm{m}^{-3} - (z/100\,\mathrm{km}) \cdot (0.051 + 3.35 \cdot 10^{-12}\, N_e/\mathrm{m}^{-3})]; \tag{13.12}$$

see Fig. 115c.

[70] Oyama, K.I., Hirao, K., Banks, P.M., Williamson, P.R. (1979): Planet. Space Sci. *28*, 207; these authors assert that $T_e = T_n$ at 120 km 'is quite rare'

[71] Lejeune, G., Waldteufel, P. (1970): Ann. Géophys. *26*, 223; Waldteufel, P. (1971): ibidem, *27*, 167; Lejeune, G. (1972): ibidem, *28*, 15. See also Lejeune, G., Taylor, G.N. (1972): Planet. Space Sci. *20*, 1061

[72] See also: Bilitza, D. (1975): J. Atmos. Terr. Phys. *37*, 1219; Mahajan, K.K., Pandey, V.K. (1978): Ind. J. Radio and Space Phys. *7*, 305; (1980): J. Geophys. Res. *85*, 213

[73] Mainly along the magnetic field lines as a consequence of magnetic control of the free motions of charges. See, for example, Petit, M. (1971): [*68*], 111

[74] See Hess, W.N. (1972): This Encyclopedia, vol. 49/4, p. 115

[75] Schunk, W.R., Walker, J.C.G. (1973) in: Progress in high temperature physics and chemistry, vol. 5. Rouse, C.A. (ed.). New York: Pergamon Press, p. 2; see also [*54*]

[76] Spenner, K., Wolf, H. (1975): Space Res. *XV*, 321

[77] Brace, L.H., Theis, R.F. (1978): Geophys. Res. Lett. *5*, 275

Fig. 115a–c. Relations between electron temperature and density. (**a, b**) Empirical monthly mean relations [each from a few days of incoherent scatter measurements at Millstone Hill (USA) in 1965]. Numbers indicate the month. (**b**) As (**a**) but with power 7/2 on the ordinate [for checking theories[71,72] claiming linear relations for this combination]. (**c**) After BRACE and THEIS from satellite in-situ measurements Eq. (13.12): curves of constant T_e/K (parameter) in plasma density versus altitude diagram[77]. [Standard deviation in the 150–350 km range is said to be less than 10%]

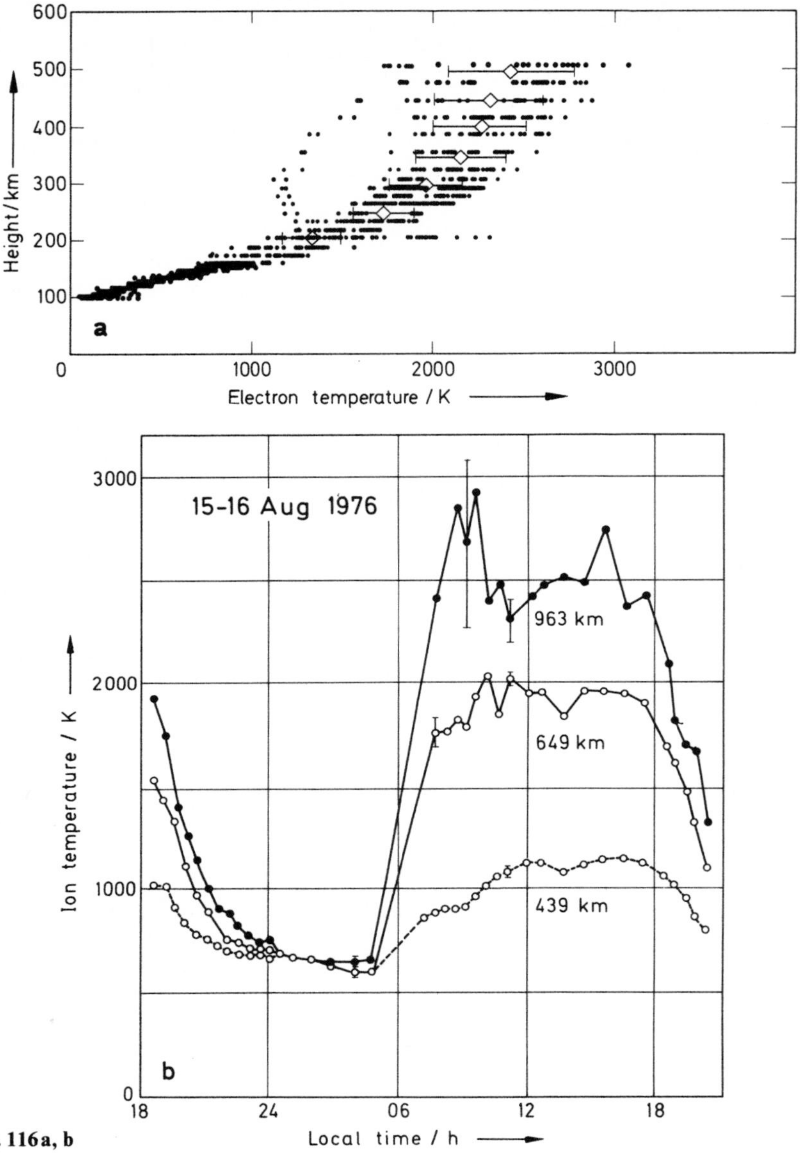

Fig. 116a, b

In a preliminary edition of IRI[78] another[79] unequivocal relation between T_e and N_e was used so that T_e could be determined directly from N_e. This proved unsatisfactory afterwards because there are a few other geophysical parameters which have some importance.

[78] Rawer, K., Ramakrishnan, S., Bilitza, D. (1975): Preliminary reference profiles of electron and ion densities and temperatures. Rep. WB-2. Freiburg i.Br.: Fraunhofer-Inst. f. physikal. Weltraumforschung

[79] Bilitza, D. (1975): J. Atmos. Terr. Phys. *37*, 1219

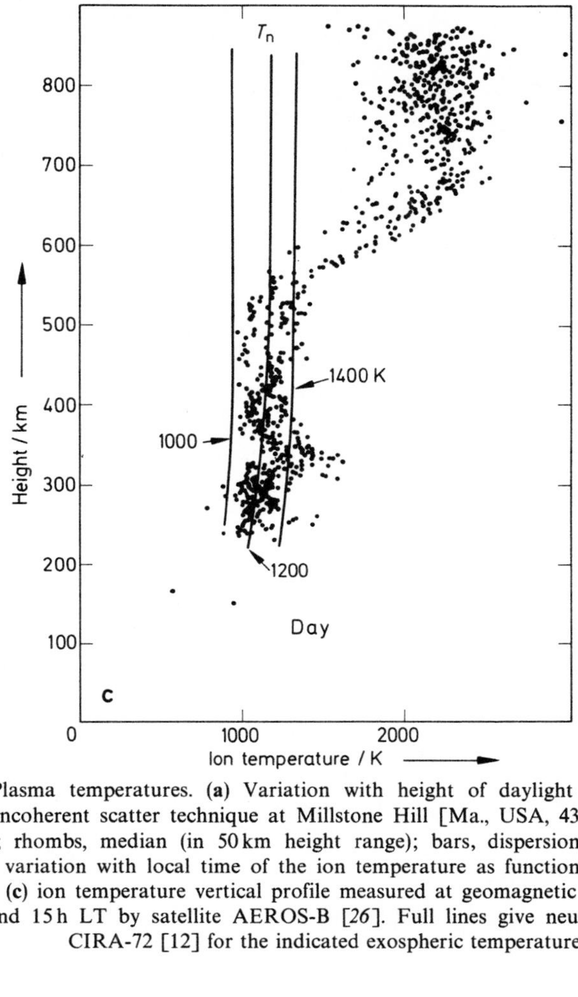

Fig. 116a–c. Plasma temperatures. (**a**) Variation with height of daylight electron temperature measured by incoherent scatter technique at Millstone Hill [Ma., USA, 43° N]. Dots, individual measurements; rhombs, median (in 50 km height range); bars, dispersion range (standard deviation)[81]; (**b**) variation with local time of the ion temperature as function of hour and altitude (parameter)[81]; (**c**) ion temperature vertical profile measured at geomagnetic latitudes between 12° and 65°, around 15 h LT by satellite AEROS-B [*26*]. Full lines give neutral temperature after CIRA-72 [12] for the indicated exospheric temperature

An investigation using a large set of satellite data was recently made by THIEMANN, from which[80] the following relation was derived. [Eqs. (10.7; 13.4) give the definition of the function Eps_0.]

$$T_e/\mathrm{K} = P_1 + (P_2 - P_1) \cdot \mathrm{Eps}_0(\log_{10}(N_e/\mathrm{m}^{-3});\ P_4, P_3) \tag{13.13}$$

with

$P_1 = \mathbf{220.3} + 8.1(z/\mathrm{km}) - 0.846(z/\mathrm{km})^2 - 258.3(\mu/°) + 106.5(\mu/°)^2;$

$P_2 = \mathbf{35.7} - 6.06(z/\mathrm{km}) + 0.556(z/\mathrm{km})^2 + 9.0(\mu/°) - 65.1(\mu/°)^2;$

$P_3 = \mathbf{9.865} - 0.872(\mu/°) + 3.304(\mu/°)^2;$

$P_4 = \mathbf{0.446} + 0.285(\mu/°) + 0.957(\mu/°)^2;$

---

[80] Bilitza, D., Thiemann, H. (1982): Kleinheubacher Berichte *25*, 237

[81] Data from: Evans, J.V. (1975): Proc. IEEE *63*, 1636 and from Vikrey, J.F., Swarts, W.E., Farley, D.T. (1979): J. Geophys. Res. *84*, 7307; compare also Fig. 5, p. 13 in vol. 49/6 of this Encyclopedia

μ is Modip, see Eq. (14.7). This formula gives no longer a one-to-one connection of T_e with N_e. It might be appropriate for a future IRI.

ζ) When *independently modelling* T_e as a function of time and coordinates one necessarily obtains a rather large dispersion (due to the above effect of the plasma density variations[71,72]). Nevertheless it is quite feasible to model T_e and T_i if a large amount of basic data is at hand. Figure 116 shows medium latitude incoherent scatter data[81], from which a monotonous profile can easily be deduced[82], rather simple profile functions being sufficient. The main data bases used in IRI have been summarized by Spenner and Plugge[83], who used a third-grade polynomial (cf. Fig. 113); day and night measurements were separately represented. However, the Epstein transition functions[24] of Eq. (13.5) are more adequate, and therefore are used in IRI [*43*, *44*], see Fig. 112, p. 366.

As for the *ion temperature* the diurnal variation is easily obtained by incoherent scatter measurements (see Sect. 5). An example is shown in Fig. 116b. Typical satellite data are shown in Fig. 116c. At altitudes below about 400 km the ion temperature, T_i, is not too far from that of the neutrals, T_n, but above that level the ion temperature increases again while T_n has reached its exospheric limiting value. Thus, the T_i profile must match the T_n profile in the lower thermosphere, but is independent from it at great altitudes (compare Appendix A).

At night there is no heating by photo-electrons; other heating sources, in particular heat transfer along the magnetic field line, may reach the ionospheric top side when the magnetically conjugate point is sunlit. Also, at higher latitudes, corpuscular heating is almost always present (see Figs. 140b, d, 142, Sect. 14). Otherwise the temperatures of the different constituents tend towards thermal equilibrium, and the differences between T_e, T_i and T_n become small (see Appendix A and Fig. 137b in Sect. 14).

c) *Ion composition*

η) The *relative importance* of the different *positive* ion species in the ionosphere depends less on the direct photo-ionization processes than on the subsequent chemical reactions which are different in different height ranges[84]. Thus the ion composition depends heavily on the altitude. In the terrestrial ionosphere[85,86] there are molecular ions prevailing below a *lower transition level* at about 160 km. Recent evidence has shown that below this level NO^+ ions are about twice as many as O_2^+ while N_2^+ ions are still less frequent. Above the transition level, i.e. in the upper thermosphere the part of O^+ increases more and more, and this species reaches almost 100% between 300

---

[82] A secondary peak is reported to occur below 300 km in the $T_e(z)$ profile near noon (at lower latitudes only). See Annex A

[83] Spenner, K., Plugge, R. (1978): Space Res. *XVIII*, 241; (1979): J. Geophys. *46*, 43

[84] These are explained in detail by Thomas, L. (1982): This Encyclopedia, vol. 49/6, p. 7

[85] Data from: Johnson, C.Y. (1966): J. Geophys. Res. *71*, 330. See also Fig. 6, p. 14 in vol. 49/6 of this Encyclopedia

[86] As for the conditions in other planetary atmospheres see [*3*, *31*, *71*] and, most recently [*59*]. The very distinct conditions in the ionosphere of Venus are summarized in: Rawer, K., Spenner, K., Knudsen, W.C. (1982): Zeitschr. f. Flugwiss. Weltraumforschung *6*, 147

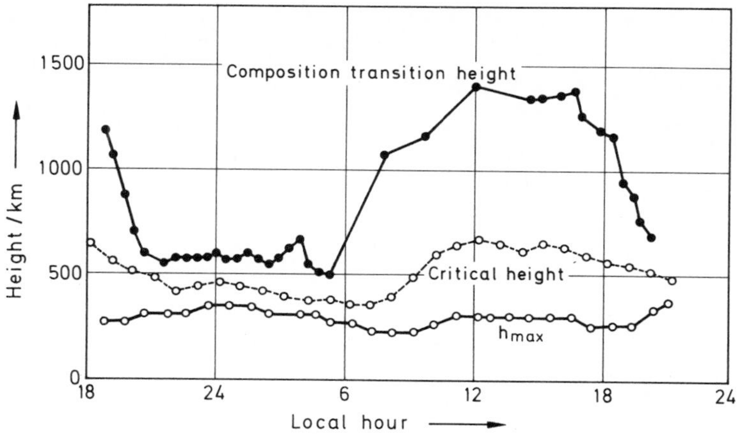

Fig. 117. Ion composition described by 'transition height': upper curve (dots); upper transition height for the O^+ ion, i.e. the height where $N(O^+)=N_i/2$ (dots). The (full) curve with open circles shows the peak altitude of the F2-layer

and 400 km. Higher up, there is another transition level from O^+ to the light ions He^+ and H^+, with H^+ normally prevailing[87]. This *upper transition level*, see Fig. 117, is usually found slightly below 1,000 km by night but much higher up by day. These are the most important ion species and their relative importance is described in IRI mainly by functions of the kind shown in Eq. (13.5). Figure 118 shows a few examples. Basic data used were a compilation of rocket data by Danilov and Semonov[88], RPA data from AEROS [*26*] and results of most incoherent scatter stations[89].

It must be acknowledged that the data base for these models is still rather poor[90]. Numerical indications could be given for the major species only, though minor species do exist[85] and are of considerable aeronomical interest. In particular, in the lower ionosphere, in the E- and D-regions, the ion chemistry is quite involved and this shows up in the minor ions; see, for example, Fig. 119.

The appearance of *metallic ions* in the lower thermosphere is a regular feature. Most authors agree that they are mainly of meteoric origin. Since their ionization energy is so much smaller than that of the atmospheric gases, if once a metallic atom has been ionized there is no way back for its energy to

---

[87] The ratio of He^+ lies between 10% and 20%; IRI-79 [*44*] applies 10% throughout

[88] Danilov, A.D., Semenov, V.K. (1978): J. Atmos. Terr. Phys. *40*, 1093

[89] For example, Giraud, A. (1970): Space Res. *XI*, 1057; Schunk, R.W. (1972): Planet. Space Sci. *20*, 581; see also Sect. 9 above

[90] At routine reduction of satellite measurements, unfortunately only *absolute* ion densities are determined. Exceptions are quite rare. What is needed for modelling purposes are, however, percentages of the total plasma density. Yet, up to now in the satellite height range, most published data stem from a period of low solar activity. The situation is worse for the rocket height range, where almost all data are from medium northern latitudes and from sunlit or early evening hours

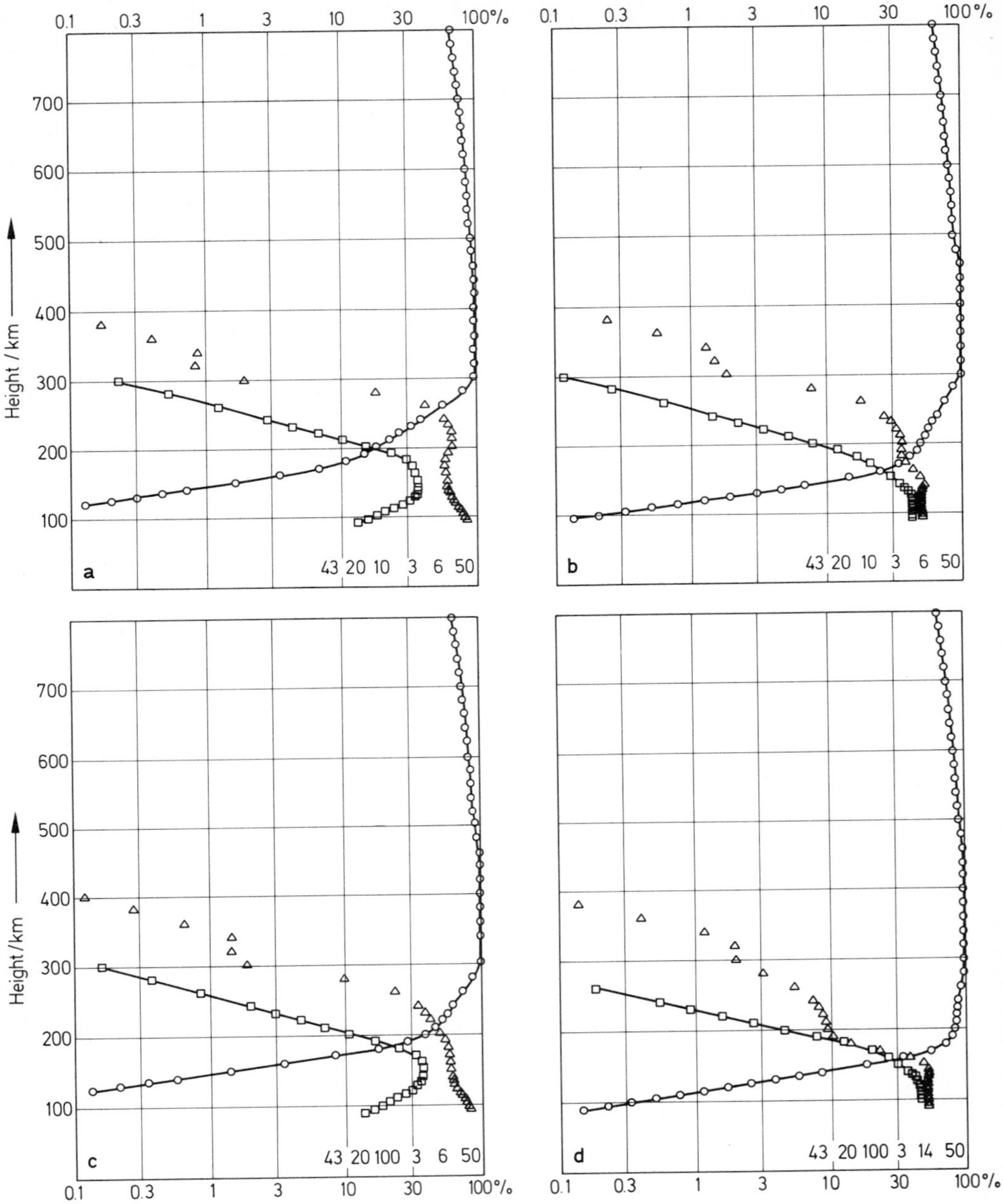
0.1
0.3
1
3
10
30
100%
Height / km
100
200
300
400
500
600
700
a
b
c
d
43 20 10 3 6 50
43 20 100 3 6 50
43 20 100 3 14 50

Fig. 118 a–d

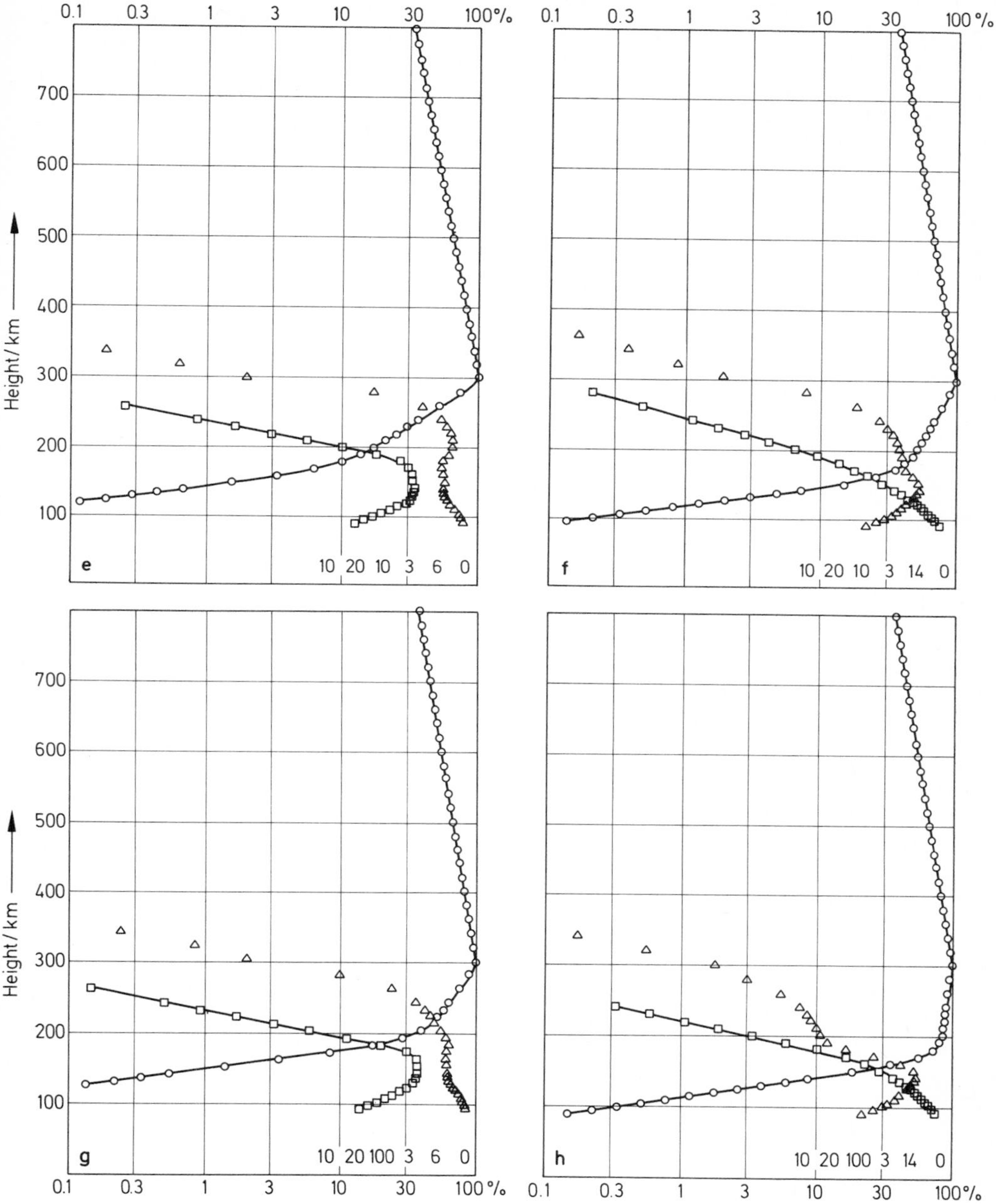

Fig. 118a–h. Ion composition diagrams (in per cent) after IRI [*44*]. Same choice of examples as in Fig. 112. Symbols: square, O_2^+; triangle NO^+; circle O^+

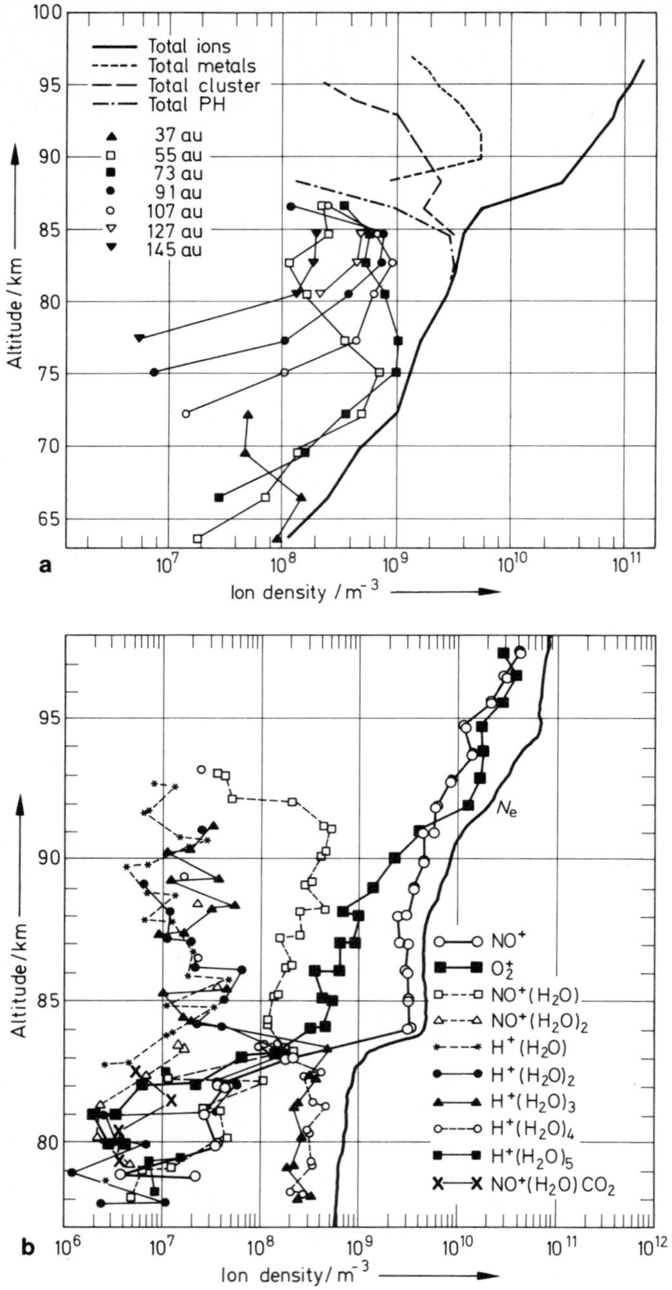
Total ions
Total metals
Total cluster
Total PH
37 au
55 au
73 au
91 au
107 au
127 au
145 au
Altitude / km
Ion density / m^{-3}
a
N_e
NO^+
O_2^+
$NO^+(H_2O)$
$NO^+(H_2O)_2$
$H^+(H_2O)$
$H^+(H_2O)_2$
$H^+(H_2O)_3$
$H^+(H_2O)_4$
$H^+(H_2O)_5$
$NO^+(H_2O)CO_2$
b

Fig. 119 a, b

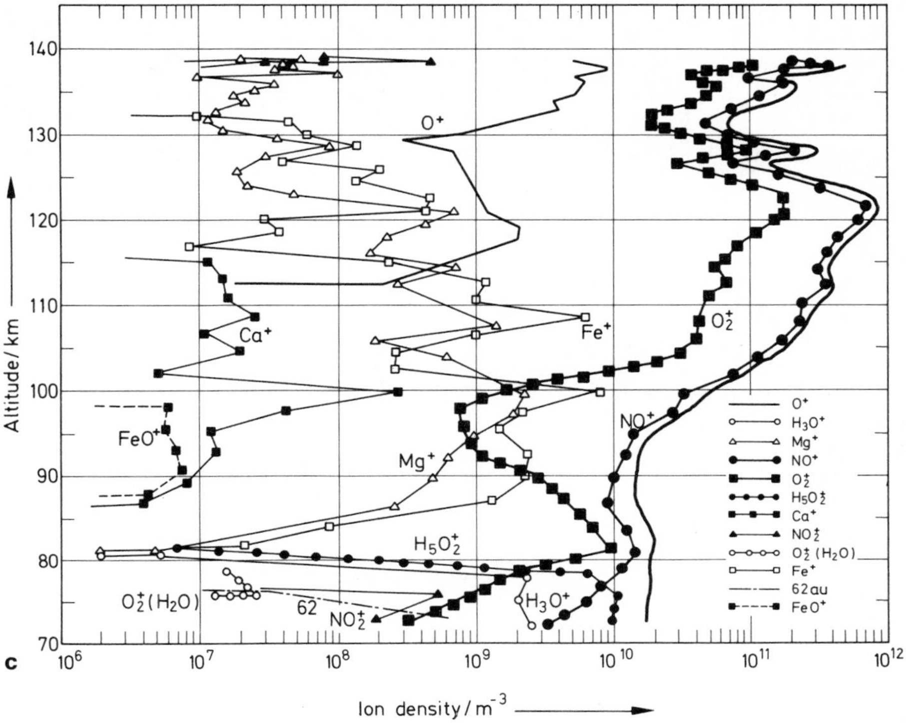

Fig. 119a–c. Positive ions in the E- and D-regions. (**a, b**) Summer days. (**a**) [10 June 1975, 13h19 UT] in northern Europe[91] (Andoya, 69° N); (**b**) [12 Aug 1976; $\chi = 28.1°$] at temperate latitude (Wallops Isld.)[95]; (**c**) night [3 Nov 1969, 07h30 UT] in the polar cap (Ft. Churchill, Canada) under the particular conditions of a "polar cap absorption" and auroral event[92]

ionize a molecule by collisional energy transfer. So dissociative recombination is impossible. On the other hand radiative recombination is very rare such that a metallic ion has a very long life expectance. The accumulation of metallic ions in a thin layer is at the origin of 'sporadic E' layers, currently called Es[55].

The situation is still more involved in the *D-region* where below about 85 km *cluster* ions become the dominant species[93], see Figs. 120 [*44*] and 121[91,94,95]. There appear even *negative ions* below about 80 km by day, 75 km

---

[91] Arnold, F., Krankowsky, D. (1977): [*18*], 93; Arnold, F., Joos, W. (1979): Geophys. Res. Lett. *6*, 763

[92] Narcisi, R.S. (1974): [*42*], 207 [another example from the same author is Fig. 7, p. 15 in vol. 49/6 of this Encyclopedia]. PCA events as such are described in: Rawer, K., Suchy, K. (1967): This Encyclopedia, vol. 49/2, Sect. 47, p. 388

[93] See Thomas, L. (1982): This Encyclopedia, vol. 49/6, Sects. 27, 28, p. 100

[94] Arnold, F. (1981): [*44*], 19

[95] Kopp, E. (1984): [*45*], 140 [many references]; Kopp, E., Herrmann, U. (1984): Ann. Geophysicae *2*, 83

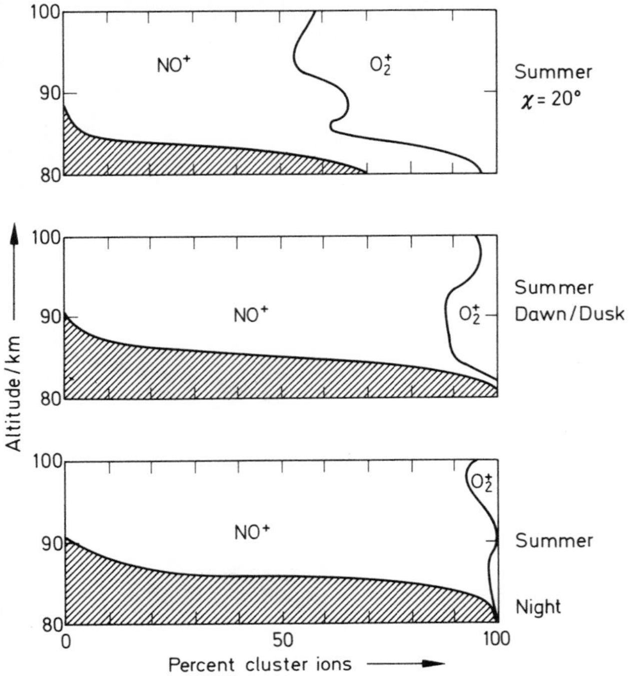

Fig. 120. Percentage of cluster and molecular positive ions between 80 and 100 km (at mid-latitude, solar activity $F \approx 70$). Summary graph from A.D. DANILOV [44]

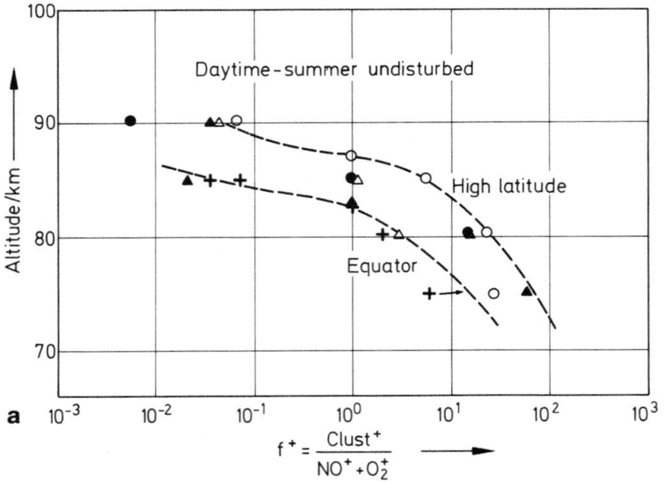

Fig. 121a–d. Positive cluster ions in the lower ionosphere. (**a–c**) Ratio f^+ of positive cluster to molecular ions[95]. (**a**) Undisturbed summer days, different northern latitudes [cross 8°; triangles 38–39°; circles 69°]; (**b**) mid-latitude days with winter anomaly; (**c**) high-latitude winter nights: —●●— undisturbed; - - - aurora (weak); other curves for stronger auroras or even PCA (triangles); (**d**) composition of positive cluster ions [winter day: Spain, 37° N; 4 Jan 1976, 14h30 UT; $\chi = 66°$][91]

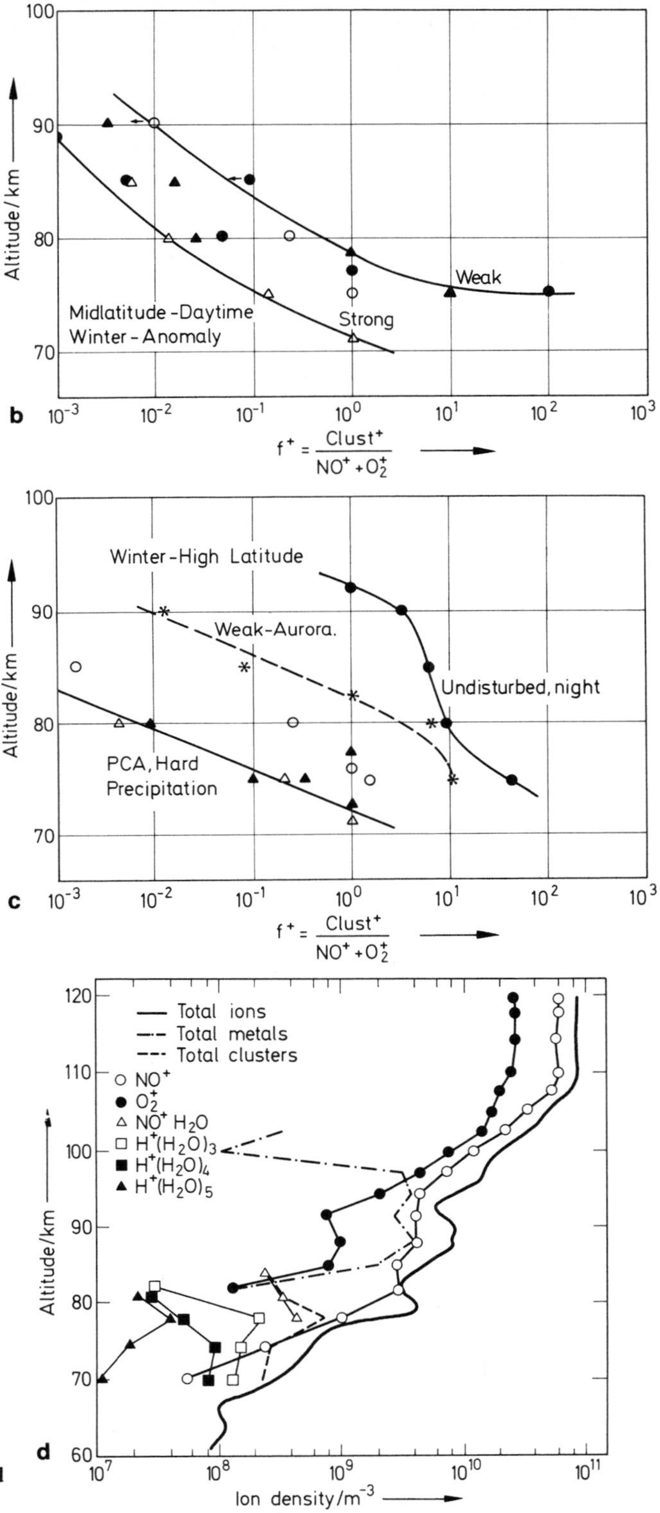

b
Altitude/km
Midlatitude-Daytime
Winter-Anomaly
Strong
Weak
$f^+ = \frac{Clust^+}{NO^+ + O_2^+}$
c
Altitude/km
Winter-High Latitude
Weak-Aurora.
Undisturbed, night
PCA, Hard
Precipitation
$f^+ = \frac{Clust^+}{NO^+ + O_2^+}$
d
Altitude/km
Total ions
Total metals
Total clusters
NO^+
O_2^+
$NO^+ H_2O$
$H^+(H_2O)_3$
$H^+(H_2O)_4$
$H^+(H_2O)_5$
Ion density/m^{-3}

Fig. 121 b–d

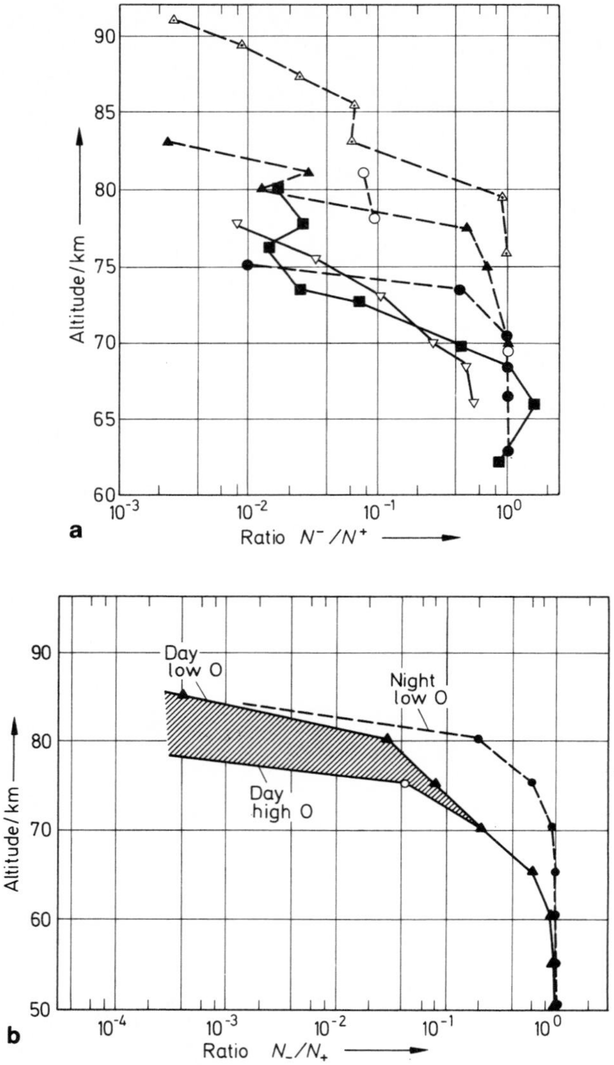

Fig. 122a, b

by night, see Fig. 122[94–96]. So far empirical modelling of these particular features has not been undertaken.

On the other hand, in the outermost ionosphere and in the plasmasphere the conditions are very variable, disturbances producing enormous changes over several days. Ions of ionospheric origin, in particular O^+, appear during the early phase of magnetic storms[97], see Sect. 18.

---

[96] See also Fig. 8, p. 15 in vol. 49/6 of this Encyclopedia

[97] Lennartsson, W., Sharp, R.D., Shelley, E.G., Johnson, R.G., Balsiger, H. (1981): J. Geophys. Res. *86*, 4628

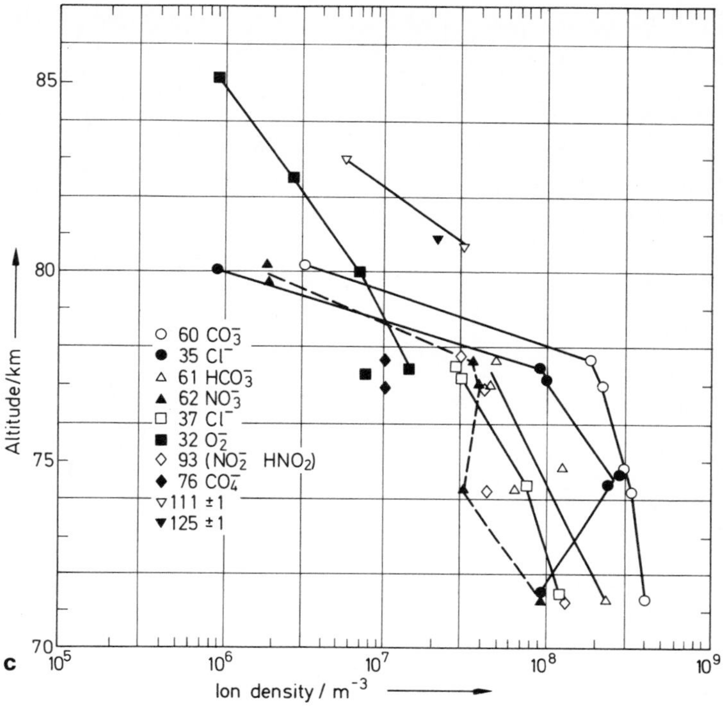

Fig. 122a–c. Negative ions in the lower ionosphere. (**a, b**) Ratio of total negative ions (excluding electrons) to total positive ions. (**a**) Rocket mass spectrometer data from northern Scandinavia by night (full square and open triangle) and day (other symbols)[94]; (**b**) computed for different conditions by a 3 ion model, classification after atomic oxygen density (inscriptions)[95]; (**c**) night time mass spectrometer observations of the negative ion composition aboard a rocket in northern Scandinavia[91,94]

14. Worldwide aspect of the ionosphere. The fundamental parameters of the vertical profiles discussed in Sect. 13 depend on the geographical location. However, geographical coordinates are not necessarily the best choice for representing ionospheric parameters worldwide. Rather, one must find out the most important physical influence and choose the coordinates accordingly. Thus, for the lower layers D, E and F1, which are under the overwhelming influence of the Sun's height (i.e. the solar zenith angle χ), the ideal coordinates would be Sun-oriented (e.g. spherical coordinates with the subsolar point as pole). On the other hand, for the F2-layer and the top-side ionosphere, since they are under magnetic field control, the coordinates should be adapted to the real magnetic field, e.g. to the magnetic inclination (dip)[1,2]. Finally, far out from Earth in the magnetosphere the field is essentially of dipole character but

[1] Rawer, K. (1951): Archiv elektr. Übertragung (A.E.Ü.) *5*, 154
[2] Rawer, K. (1963): [*27*], 221

compressed by the solar wind so that McIlwain's L-coordinate (or the equivalent "invariant magnetic latitude")[3] is adequate.

a) *Electron density*

α) *The lower layers* D, E and F1 are quite well describable by rather simple formulas depending mainly upon the solar zenith angle χ; see Sect. 13γ. Thus purely empirical mapping is not really needed.

(i) Nevertheless, the numerical mapping system which is in use for the F2-parameters (see Sect. 14γ below) has been applied to the *peak electron density* of the E- and F1-layers[4,5], the numerical coefficients being based on ground ionosonde data.

In our opinion this is a weak point since the true accuracy of these data is not as good as usually expected [*37*]; in particular, systematic calibration errors and different practices at data reduction are appearing on these maps.

Since the dispersion range of $f\circ E$ and $f\circ F1$ (as a function of true local time) is quite small; empirically proved formulas[6,7] depending on the solar zenith angle χ are probably more reliable.

As for the sporadic E-layer, Es[8], the ITS-78 model[9] had average numerical maps, unfortunately without considering the occurrence probability. The more recent IONCAP model[10] gives median and upper and lower decile.

IRI [*43, 44*] has more involved expressions for the peak (or inflection point) densities of the lower layers; see Eqs. [13.8–13.10], all of which were derived from critically analyzed ionosonde data. The inflection point density of the F1-layer, for example, is given by the corresponding plasma frequency as[11]

$$f\circ F1 = f_{ss}\cdot(\cos\chi)^m \qquad [13.10]=(14.1)$$

where the exponent is found from

$$10^3\cdot m = 93+\mu\cdot(4.6-0.054\,\mu)+0.3\,\bar{R} \qquad (14.1\,\text{a})$$

with μ modip [see Eq. (14.7) in Sect. 14γ]; $\bar{R}$ (monthly average) Zürich sunspot number. The subsolar characteristic frequency f_{ss} is obtained by linear interpolation between the values for $\bar{R}=0$ and 100:

$$\begin{aligned} f_s(0)/\text{MHz} &= 4.35+\mu\cdot(5.8-0.12\,\mu)/10^3,\\ f_s(100)/\text{MHz} &= 5.348+\mu(11-0.23\,\mu)/10^3. \end{aligned} \qquad (14.1\,\text{b})$$

As for the E-layer peak density see Eq. [13.9].

The relation between plasma frequency f_N and plasma density N is

$$\begin{aligned} f_N/\text{MHz} &= 0.8979\cdot 10^{-5}(N/\text{m}^{-3})^{1/2};\\ N/\text{m}^{-3} &= 1.24\cdot 10^{10}(f_N/\text{MHz})^2. \end{aligned} \qquad (14.2)$$

---

[3] See Hess, W.N. (1972): This Encyclopedia, vol. 49/4, Sect. 5, p. 124; an *L*-map at ionospheric altitude is found as Fig. 333 in vol. 49/2

[4] Leftin, M. (1976): Numerical representation of monthly median critical frequencies of the regular E-region ($f\circ E$). Rep. 76-88. Washington, D.C.: U.S. Govnt. Printing Off.

[5] Rosich, R.K., Jones, W.B. (1973): The numerical representation of the critical frequency of the F1 region of the ionosphere. OT-Rep. 73-22. Washington, D.C.: U.S. Govnt. Printing Off.

[6] Rawer, K., Suchy, K. (1967): This Encyclopedia, vol. 49/2, Sect. 35, p. 278

[7] Ibidem, Sect. 36, p. 285

[8] Ibidem, Sect. 40, p. 332

[9] Barghausen, A.L., Finney, J.W., Proctor, L.L., Schultz, L.D. (1969): Predicting long-term operational parameters of high-frequency sky-wave telecommunication systems. Rep. ERL 110-ITS 78. Washington, D.C.: U.S. Govnt. Printing Off.

[10] Flattery, T.W., Tascione, T.F., Secan, J.A., Taylor, J.W. Jr. (1979): A four-dimensional ionospheric model. Unpublished report of AFGWC (USA)

[11] Ducharme, E.D., Petrie, E., Eyfrig, R. (1971): Radio Sci. *6*, 369; (1973): ibidem, *8*, 837

Another empirical description is given by CHING and CHIU's[12] model (see Sect. 14β below, where the symbols are explained):

$$NmF1/10^{11}\,\mathrm{m}^{-3} = 2.44 \cdot \underset{\text{(cycle)}}{(1+1.24\rho+0.25\rho^2)^{1/2}} \cdot \underset{\text{(diurnal)}}{\mathrm{D}(1;0.5)} \cdot \underset{\text{(hour; latitude)}}{\exp(-0.25\,\Theta)} \tag{14.3}$$

with $\rho \equiv R/100$, φ geographical latitude, δ solar declination, τ local hour angle and

$$\Theta = \cos(\varphi + \delta\cos\tau) - \cos\varphi.$$

The diurnal variation D is given by an involved exponential function:

$$\mathrm{D}(a;b) = \exp\{[a + b\log_e(1+30\rho)] \cdot [\mathrm{sign}(\cos\chi) \cdot (\cos|\chi|)^{1/2} - 1]\}$$

which depends mainly on $(\cos|\chi|)^{1/2}$.

Similarly the same authors give for the E-region:

$$NmE/10^{11}\,\mathrm{m}^{-3} = 1.36 \cdot (1+1.15\rho)^{1/2} \cdot \mathrm{D}(2;0) \cdot \exp(-0.4\,\Theta). \tag{14.4}$$

(ii) Rather simple formulas can be used for *altitude and thickness* of the lower layers. Invariable D- and E-region characteristics have earlier been used for daytime [110 km for hmE and 20 km thickness; D by an experimental tail down to 70 km[9]; constant ratio of thickness $ymF1$ to altitude $hmF1$ of 1:4][9,10].

IRI [*43, 44*] admits continuous diurnal variations of the heights, but these are taken as being only dependent on sunrise and sunset hours, thus only indirectly on the geographical position, and quite independent from solar activity. The range of variation for the D-region inflection, hmD (Fig. 105), is between 81 km (night) and 88 km (day) and for its lower borderline HA is 65 km (by day) and 80 km (by night). Similarly for the height of the E-peak, hmE, IRI admits a small but continuous diurnal variation between 105 km as night value and 110 km by day. A more involved description (Fig. 105 and Sect. 13γ) is used for the F1-layer. It is important that conditions with and without the F1-layer are distinguished. This is done according to criteria depending on season ϑ, noon solar zenith angle χ_m and latitude φ.

CHING and CHIU[12] have fixed heights (of 180 and 110 km, respectively) and thickness parameters (of 34 and 10 km, respectively) for the F1- and E-layers.

β) The decisive *parameters of the F2-layer* show quite complicated variations in space and time. It is therefore not feasible to describe them just as functions of a few independent variables as done above for the lower layers. The problem was tackled in different ways which afford different computational effort. The simplest way is *a direct development* of the physical parameters themselves; this is the *phenomenological method.* After preliminary work by different authors[13] a real piece of pioneer work was presented by YONEZAWA[14]. One rather recent effort along his lines is due to CHING and CHIU[12]. As explained in Sect. 13β (and like YONEZAWA[14]), these authors superpose three layers of the ELIAS-CHAPMAN type, Eq. (13.2), which represent the E-, F1- and F2-layers. The peak density is given by an amplitude A and a "layer peak function", which itself consists of a polar function P and a non-polar function U which are combined by a "folding factor" f after:

$$N_m = A \cdot \{f \cdot \mathrm{P} + (1-f) \cdot \mathrm{U}\} \tag{14.5a}$$

[12] Ching, B.K., Chiu, Y.T. (1973): J. Atmos. Terr. Phys. *35*, 1615

[13] References can be found in: Rawer, K. (1969): Ann. Intern. Geophys. Y. *5*, 97, and in King, J.W., ibidem, 131; see also Yonezawa, T. (1966): Space Sci. Rev. *5*, 3

[14] Yonezawa, T. (1971): J. Atmos. Terr. Phys. *33*, 889

where f depends only on latitude, φ, and geomagnetic latitude, ϕ, i.e. on the location on Earth, while P and U depend on a few more variables, viz.:

ϑ = season (in angular measure) from midwinter;
τ = local time (in angular measure) from midnight [i.e. $\tau = 2\pi t/24$ where t is local time/h]; χ solar zenith angle; δ the Sun's declination;
$\rho \equiv R/100$ = reduced solar activity (R = Zürich sunspot number).

The non-polar function U is separated after the different geophysical influences[12]:

$$\mathrm{U} = \underset{\text{cycle}}{\mathrm{S}(\rho)} \cdot \underset{\text{diurnal}}{\mathrm{D}(\tau, \chi, \delta, \phi; \rho)} \cdot \underset{\text{latitude}}{\mathrm{L}(\phi, \tau)} \cdot \underset{\text{annual}}{\mathrm{T}(\vartheta, \varphi, \phi, \tau; \rho)} \cdot \underset{\text{equatorial}}{\mathrm{E}(\phi, \tau; \rho)}. \qquad (14.5\text{b})$$

In the following we reproduce most formulas from the more recent version[15]. The folding function f has a strong peak around the geomagnetic poles so that there the polar function P takes over from U[15]:

$$f = \exp(-\{[2.4 + (0.4 + 0.1\rho)\sin\phi] \cdot \cos\phi\}^6). \qquad (14.5\text{c})$$

On the other hand the equatorial function $\mathrm{E}(\phi, \tau; \rho)$ introduces the equatorial anomaly[16]. (Details can be found in the appendix of [12] or, better [15].) The worldwide behaviour[1] is mainly determined by the geomagnetic latitude ϕ:

$$\mathrm{L} = (0.7 + 0.5\sin^2\phi) \cdot \exp(1.5 \cdot \cos[\phi(\sin\varphi - 1)]. \qquad (14.5\text{d})$$

The model coefficients were determined from the monthly averaged results of 68 ionosonde stations during the period 1957–1970.

As for the height of the F2-peak it is given by[16]

$$hmF2/\text{km} = 240 + 75\rho + 83\rho\sin\delta\sin\phi\cos\phi - 30\cos(\tau - 4.5|\phi|) + 10\cos\phi \cdot \cos(2\vartheta - 3\pi/2) \qquad (14.6\text{a})$$

and its thickness parameter (twice the 'scale height' in the Elias-Chapman function) by [12,15]

$$ymF2 = 40\,\text{km} + 0.2 \cdot hmF2. \qquad (14.6\text{b})$$

Chiu[15] improving[12] extended the analytical description, Eq. (14.5b), by including a term with the geomagnetic longitude and further took the dip latitude as an additional input. His model is programmed in IRI [*43*, *44*] as a second option (the first option being the CCIR-map described in Sect. 14γ below). Clearly it cannot describe the worldwide distribution in much detail since functions L and E cannot represent small-scale variations. On the other hand the computational effort is small.

Figure 123 compares electron density contours as function of local time as obtained from the model[12] with the results of another model and with measurements from incoherent scatter stations (see Sect. 5). Apparently, the model[12] introduces considerable smoothing as a consequence of the rather simple formula used for the diurnal variation. In Fig. 124 the seasonal and long-term

---

[15] Chiu, Y.T. (1975): J. Atmos. Terr. Phys. *37*, 1563

[16] Rawer, K., Suchy, K. (1967): This Encyclopedia, vol. 49/2, Sects. 37, p. 288 and 58, p. 503

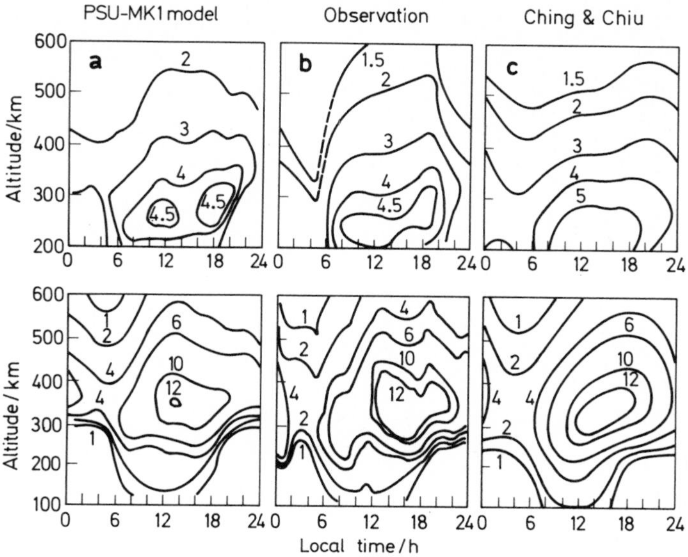

Fig. 123a–c. Contours of constant electron density in height versus local time fields. Parameter: plasma frequency/MHz [Eq. (14.2)]. (**a**) PSU-MK1 model (see Sect. 14δ). (**b**) Average of incoherent scatter observations (a few days only). (**c**) CHING and CHIU model[12]. (*Top*) Millstone Hill [Ma., USA, 43° N], July 1964; (*Bottom*) Arecibo [Puerto Rico, 18° N], June 1968

variations are checked. The agreement is much better in solar cycle 19 than for cycle 20. This might be due to the secular variation mentioned in Sect. 13.

Another phenomenological model due to KÖHNLEIN[17] allows only one ELIAS-CHAPMAN-like profile depending on a reduced height coordinate $x=(z-hmF2)/H$. One assumes, however, dependencies of the parametrized 'scale height' H and of $hmF2$ on geographical latitude φ, longitude λ, season ϑ, local time τ and on the COVINGTON index F[18]. The general expression of the plasma density is

$$N=A\cdot N_0\cdot \exp[1-x-\exp(-x)], \tag{14.6}$$

where the amplitude function $A(x, \dots)$ represents the horizontal density pattern for the altitude level x. In fact, this model directly applies a global development of A in terms of LEGENDRE functions[19], as they have been used in other disciplines of geophysics, in particular for describing the neutral atmosphere; see Sect. 11 above and Eq. (11.4). Like these, it develops in terms of geographical coordinates (see our remarks at the beginning of this Sect.). The time-independent terms are expanded to the fifth grade, the coefficients themselves still depending on τ, ϑ and F. About 50 coefficients are needed altogether. The peak altitude is developed in the same way. The model reproduces the observed data with some smoothing.

The F1-layer peak (with a fixed altitude of 180 km) is not described in the same way but by an expansion in spherical harmonics of position, local time, season, and depends on solar activity F. The average noon-to-midnight ratio is about 60.

The top-side ionosphere is cut into four layers (at 500, 1,000, 2,000 and 3,500 km), each layer being expanded into spherical harmonics as was done for the F2-layer. The intermediate space is dealt with by deriving the altitude dependence of the coefficients from *actual* profiles. This layered description, though similar to that of BENT (Sect. 13β), is continuous in the derivative.

---

[17] Köhnlein, W. (1978): Rev. Geophys. Space Phys. *16*, 341; models of other authors are summarized in this review paper. For the Elias-Chapman profile see [2] in Sect. 13

[18] This is the numerical value of the solar radio wave flux on a wavelength of 10.7 cm in units of $10^{-22}\,\mathrm{W\,m^{-2}\,Hz^{-1}}$

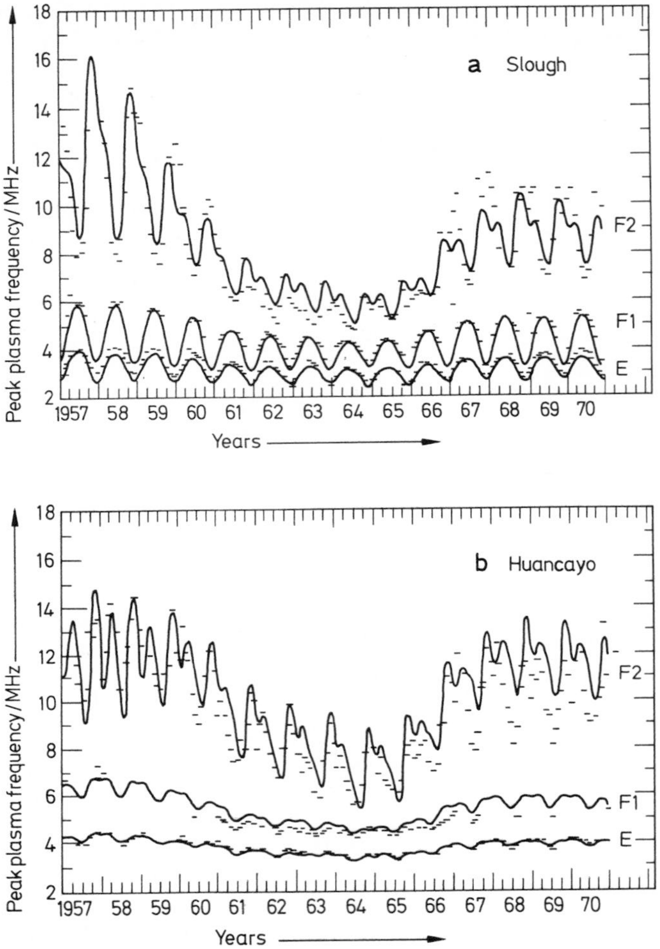

Fig. 124a, b. Long-term variation of peak plasma frequencies (noon) from 1957 to 1970. Dashes, ionosonde monthly median data; curve, CHING and CHIU model[12]. **(a)** Slough [England, 52° N]; **(b)** Huancayo [Peru, 12° S]

γ) The "*numerical mapping*" procedure was introduced after earlier efforts with direct development[19] (see Sect. 14β) by LEGENDRE functions in space and harmonic ones in time (like later [17]). These had not been satisfactory. It was found that the very characteristic and locally different diurnal variations[15] were smoothed out too much at direct mapping. This is an important point for propagational applications.

(i) JONES, applying in a first step *harmonic analysis* to the monthly average *diurnal curves* of $f_\circ F2$ (up to 11th order), plotted the coefficients obtained from

[19] The first to do this was R. GALLET (in the 1950s) at the former "Service de Prévision Ionosphérique Militaire" (SPIM)

this against geographical latitude[20]. JONES and GALLET[21] took the decisive second step by applying worldwide LEGENDRE development *separately* for each Fourier coefficient up to the 8th order[22], i.e. for 17 coefficients. The final map was then constructed by recomposing $f \circ F2$ from the harmonics with the amplitudes obtained for each geographical position by the relevant 15 sets of LEGENDRE coefficients.

Due to the irregular distribution of ionospheric sounding stations special methods had to be applied at the determination of the coefficients. SCHMIDT-GRAHAM orthogonalization[23] and least square fitting were used. However, the first trials produced negative values in the Pacific – as a consequence of the total lack of stations in that region. Therefore, in order to fill-up the main gaps, so called "*screen points*" were introduced. At these points (where there were no data) the authors[21] introduced coefficients from the overall average in longitude at the relevant latitude. These artificial curves are not correct but they are needed in order to stabilize the pattern in the gap. The final analysis is then made including the original and the screen points as well.

The "geographical functions" $G(\varphi, \lambda)$ are built up by several sets of terms (LEGENDRE functions), the first one being dependent on φ only (thus independent on longitude λ):

$$1; \sin\varphi; \sin^2\varphi; \dots \sin^9\varphi \qquad \text{(zero order in } \lambda\text{)}.$$

This gives the average latitudinal variation. The other terms are obtained (as couples) by multiplication of the zero-order functions with: $\cos\varphi \cdot \cos\lambda$ and $\cos\varphi \cdot \sin\lambda$ (first order in λ), then with $\cos^2\varphi \cdot \cos 2\lambda$ and $\cos^2\varphi \cdot \sin 2\lambda$ (2nd order), etc. (cf. Table 8). An example of a numerical map after this original method[21] is found in Fig. 125a,c.

(ii) These maps were not fully satisfactory either. It must be seen that the data basis is by far not overwhelming (as usually assumed in purely mathematical considerations). The screening procedure has a few drawbacks which are quite annoying: first of all, it produces longitudinal smoothing (see below); furthermore, it introduces a kind of *interpolation* along the circles of constant geographical latitude, which is not the right way to tackle the equatorial ionosphere, which is strongly dependent on the *magnetic dip*[16]. Therefore, the choice of the coordinates in which the development is finally executed is not at all insignificant[24] (as a mathematician might suppose).

In order to interpolate between maps of ionization ($f \circ F2$ maps) established for different continents, RAWER[2] has proposed a new coordinate, now called "modified dip" or "modip" μ, which is defined by

$$\tan\mu = \psi/(\cos\varphi)^{1/2}, \tag{14.7}$$

ψ being the true magnetic dip in the ionosphere (usually at 300 km).

---

[20] Jones, W.B. (1962): Atlas of Fourier coefficients of diurnal variation of $f \text{o} F2$. NBS Techn. Note 142. Washington, D.C.: U.S. Govnt. Printing Off.

[21] Jones, W.B., Gallet, R.M. (1962): Telecomm. J. *29*, 129; (1965): ibidem, *32*, 18

[22] Higher orders appear "noisy" as a consequence of limited sampling and irregularly appearing geophysical disturbances

[23] See: Davis, P., Rabinowitz, P. (1954): J. Assoc. for Computing Machinery *1*, 183

[24] Rawer, K. (1975): Radio Sci. *10*, 669

In the equatorial zone the lines of constant μ are practically identical to those of magnetic inclination ψ, but with increasing latitude they deviate and come nearer to those of constant geographical latitude φ. The poles ($\mu=\pi/2$) are identical to the geographical ones (see Fig. 126). Of course, the high-latitude regions (auroral zones and polar caps) need separate treatment for mapping by any procedure, and no satisfactory solution is available yet. But for low and mid-latitudes two major efforts with checking of different coordinates[25,26] came out with a clear preference for μ[27]. It can be seen from Fig. 125b that the difference is very important[25]; in particular, the well-known structure in the equatorial region with the "trench" at the true magnetic equator[16] clearly appears in the μ-based map only.

Finally, CCIR accepted the JONES and GALLET system with RAWER's coordinate as the official "numerical map" for radio propagation purposes [9]. Table 8 shows the final set of LEGENDRE functions; the FOURIER analysis in time is cut after the 7th order so that 15 coefficients have to be developed all around the globe. The total number of coefficients is almost 1,000 per month; a full set of 12 months and low and high solar activity affords about 20,000 coefficients[28,29].

Table 8. CCIR geographic coordinate functions $G_k(\varphi,\lambda)$ (X is a function of φ and λ, m is the maximum order in longitude) $q_0=k_0$; q_i $(i=1,m)=\dfrac{k_i-k_{i-1}-2}{2}$

k	Main latitude variation	k	First order longitude	k	Second order longitude	…	k	mth order longitude
0	1	k_0+1	$\cos\varphi\cos\lambda$	k_1+1	$\cos^2\varphi\cos 2\lambda$	…	$k_{m-1}+1$	$\cos^m\varphi\cos m\lambda$
1	$\sin X$	k_0+2	$\cos\varphi\sin\lambda$	k_1+2	$\cos^2\varphi\sin 2\lambda$	…	$k_{m-1}+2$	$\cos^m\varphi\sin m\lambda$
2	$\sin^2 X$	k_0+3	$\sin X\cos\varphi\cos\lambda$	k_1+3	$\sin X\cos^2\varphi\ \cos 2\lambda$	…	$k_{m-1}+3$	$\sin X\cos^m\varphi\ \cos m\lambda$
·		k_0+4	$\sin X\cos\varphi\sin\lambda$	k_1+4	$\sin X\cos^2\varphi\ \sin 2\lambda$	…	$k_{m-1}+4$	$\sin X\cos^m\varphi\ \sin m\lambda$
⋮		⋮		⋮			⋮	
k_0	$\sin^{q_0} X$	k_1-1	$\sin^{q_1} X\cos\varphi\ \cos\lambda$	k_2-1	$\sin^{q_2} X\cos^2\varphi\ \cos 2\lambda$	…	k_m-1	$\sin^{q_m} X\cos^m\varphi\ \cos m\lambda$
		k_1	$\sin^{q_1} X\cos\varphi\ \sin\lambda$	k_2	$\sin^{q_2} X\cos^2\varphi\ \sin 2\lambda$	…	k_m	$\sin^{q_m} X\cos^m\varphi\ \sin m\lambda$

Explicitly written the numerical map Ω is described by:

$$\Omega(\varphi,\lambda,T)=\sum_{k=0}^{K} U_{0,k}\cdot G_k(\varphi,\lambda)+\sum_{l=1}^{H}\left[\cos(lT)\cdot\sum_{k=0}^{K} U_{2l,k}\cdot G_k(\varphi,\lambda)+\sin(lT)\cdot\sum_{k=0}^{K} U_{2l-1,k}\cdot G_k(\varphi,\lambda)\right]$$

$K=k_m$ = highest geodetical order; H highest Fourier order. φ geodetic latitude, λ geodetic longitude, ψ magnetic inclination,

$$\mu=\varphi/\cos^{\frac{1}{2}}\varphi\ \ (\textit{Modip}),\qquad X=\arctan\mu.$$

The coefficients U were determined by analysis of a large number of monthly mean data from sounding stations and can be obtained on magnetic tape from CCIR at Geneva.

[25] Jones, W.B., Graham, P., Leftin, M. (1966): Advances in ionospheric mapping by numerical methods. ESSA Techn. Rep. ERL 107-ITS 78. Washington, D.C.: U.S. Govnt. Printing Off.

[26] Another effort was undertaken in Australia by W.G. BAKER

[27] For which the smallest residues were obtained

[28] These are distributed on magnetic tape through national telecommunication administrations

[29] An effort to reduce the number by adopting continuous seasonal variations was undertaken by: Jones, W.B., Obitts, D.L. (1970): Global representation of annual and solar cycle variation of $foF2$ monthly median 1954–1958. OT/ITS Res. Rep. 3. Washington, D.C.: U.S. Govnt. Printing Off.

A similar development[30] for *M(3000)F2* which allows *hmF2* to be derived (see Sect. 13δ) is also contained in [9][31]. A listing of conversion formulas given by different authors, and the relevant data base can be found in [32].

A three-dimensional mapping approach was undertaken by JONES and STEWART using LANCZOS' method[33] (a modified trigonometric interpolation) for representing vertical profiles. (The latter were obtained by inversion technique from ionograms, the top-side profile being obtained by an extrapolation procedure.) At each of 36 stations all over the world, and every second hour LT, the profile was reproduced by a spatial FOURIER interpolation series (32 coefficients), with the altitude range 100–900 km as basic interval. Applying a worldwide analysis in time and in one coordinate (modip) only, a four-dimensional representation for each of the 32 coefficients was obtained, with about 10^4 coefficients per month, however (!); see Fig. 127 for an example.

(iii) *Checking* of the CCIR model was first undertaken with ground-based ionosonde data, later with satellite measurements.

The latter have much better worldwide coverage. In-situ measurements, unfortunately, do not produce peak density data; these can only be derived indirectly when a certain profile shape is assumed. With top-side sounding, on the other hand, the actual height of the satellite should be irrelevant. Unfortunately, most of the earlier top-side ionograms[34] were technically unsuitable for deducing the peak plasma frequency $f \circ F2$. Also for satellites in higher orbits the effective sounding direction is often off-vertical.

The problem was resolved only with the Japanese "critical frequency sounder"[35] from the measurements of which worldwide critical frequency maps were empirically established; see Fig. 128.

The main problem with satellite data is that the different longitudes are reached at times which may differ by several days such that day-by-day changes cannot be separated from longitudinal ones. In the published maps some averaging therefore had to be applied.

In order to establish these maps data of at least three months had to be rassembled; thus quite a bit of averaging was applied in the procedure. Figure 128 shows the peak density by the average critical frequency $f \circ F2$ as function of latitude and longitude for a given hour *Universal Time* (UT). i.e. a "worldwide view" of the ionosphere[35a]. If there were no longitude effect at all the maps for different hours should be identical except for a shift in longitude. Since, however, the F-region is under strong geomagnetic control the different maps Figs. 128a–c are far from being identical. The Figs. 129a–e were constructed from the same data set but each of these maps is valid for constant *local time* (LT) so that the longitude effect is directly appearing[35a]. (If there were no such effect, these maps should exhibit a pattern of horizontal lines only.) Anyway, comparison with Fig. 128 clearly shows that interpolation is much more adequate in LT than in UT. This is, besides, the procedure which was applied wen establishing the CCIR-maps (see above).

---

[30] Needing about half that number of coefficients

[31] Certainly not all coefficients are significant for the result

[32] Bilitza, D., Sheikh, N.M., Eyfrig, R. (1979): Telecomm. J. *46*, 549

[33] Jones, W.B., Stewart, F.G. (1970): Radio Sci. *5*, 773; the mathematical method being due to: Lanczos, C. (1966): Discourse on Fourier Series. New York: Hafner

[34] Rawer, K., Suchy, K. (1967): This Encyclopedia, vol. 49/2, Sect. 58, p. 503

[35] Reports on ISS-b satellite results obtained by CCIR numerical mapping procedure [9]: (a) Matuura, N. (1979): Atlas of ionospheric critical frequencies (fo$F2$) obtained from Ionosphere Sounding Satellite-b observation; (b) – (1981): World-wide maps of electron density and temperature, mean ion mass, ion temperature and ion composition obtained from ...; (c) – (1982): Atlas of proton, helium and oxygen ion densities obtained from .... Tokyo: Radio Res. Labs.

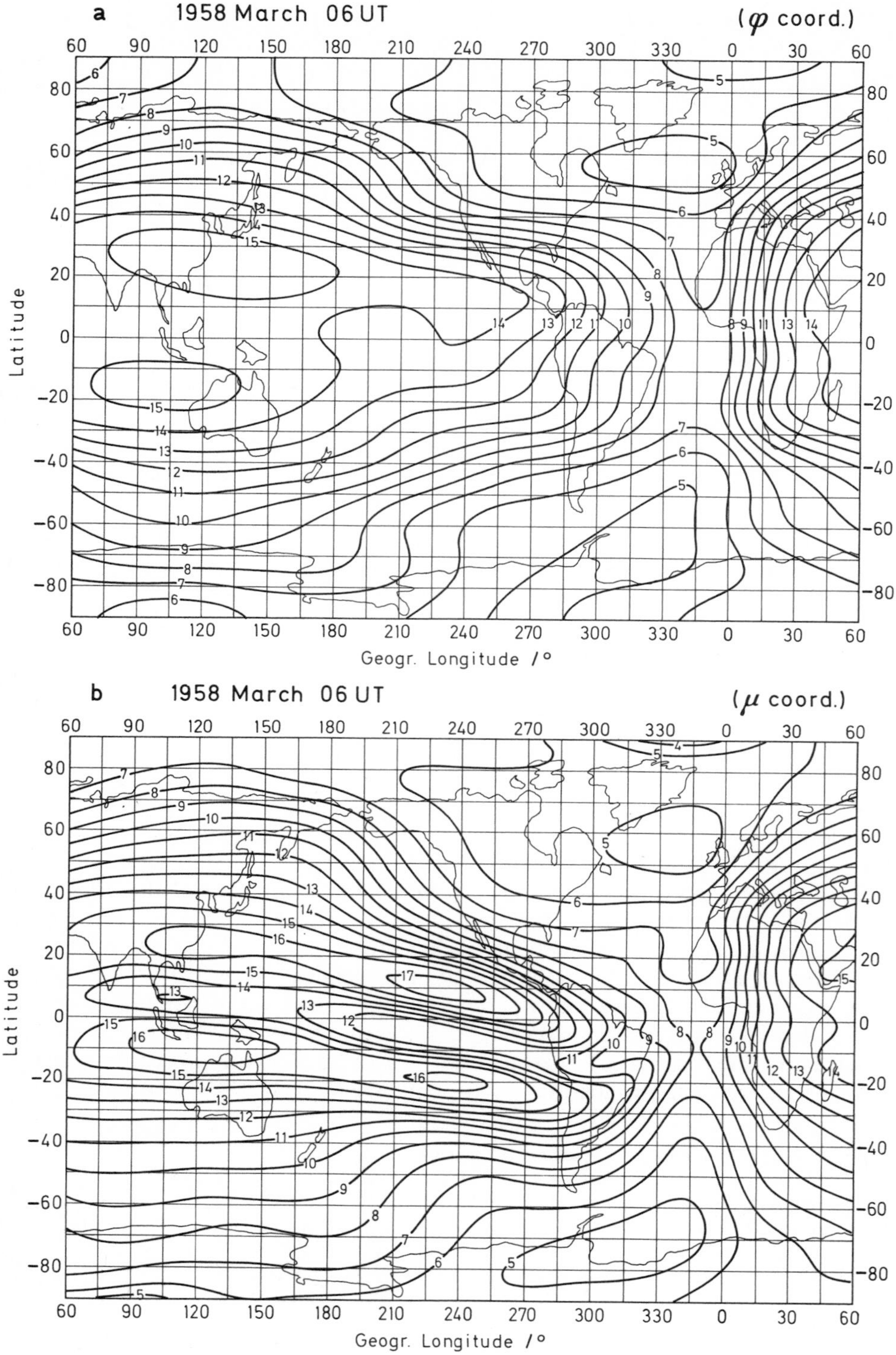

Fig. 125a–d. World maps of F2-layer peak plasma frequency, $foF2$/MHz, for March 1958, 06 (**a**, **b**) and 18 h UT (**c**, **d**). [Time analysis was made in LMT]. (**a**, **c**) Original JONES and GALLET

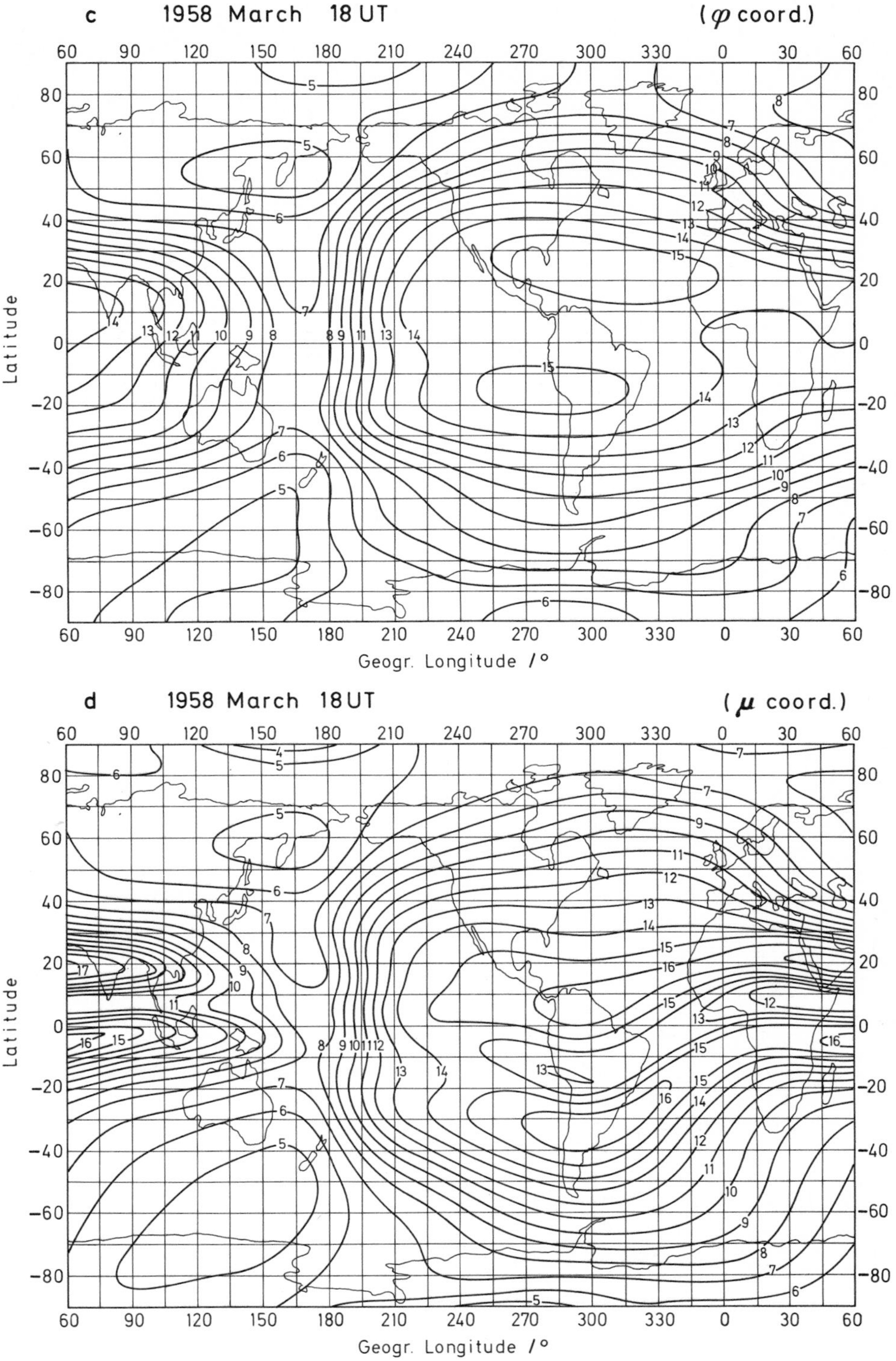

method with analysis in geographical latitude as coordinate. (**b, d**) Improved CCIR method with analysis in modip coordinate[25]

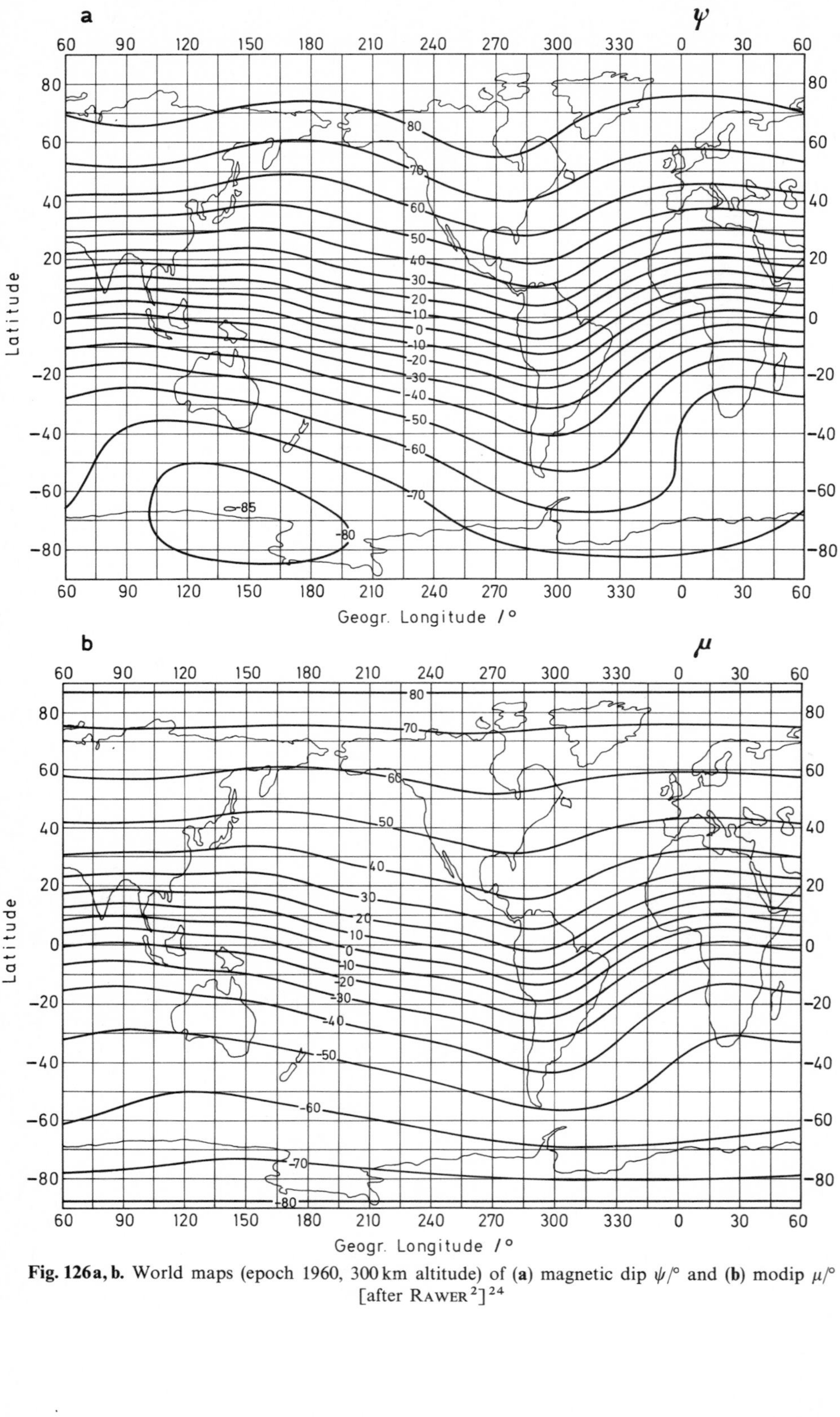

Fig. 126a, b. World maps (epoch 1960, 300 km altitude) of (**a**) magnetic dip $\psi/°$ and (**b**) modip $\mu/°$ [after RAWER[2]][24]

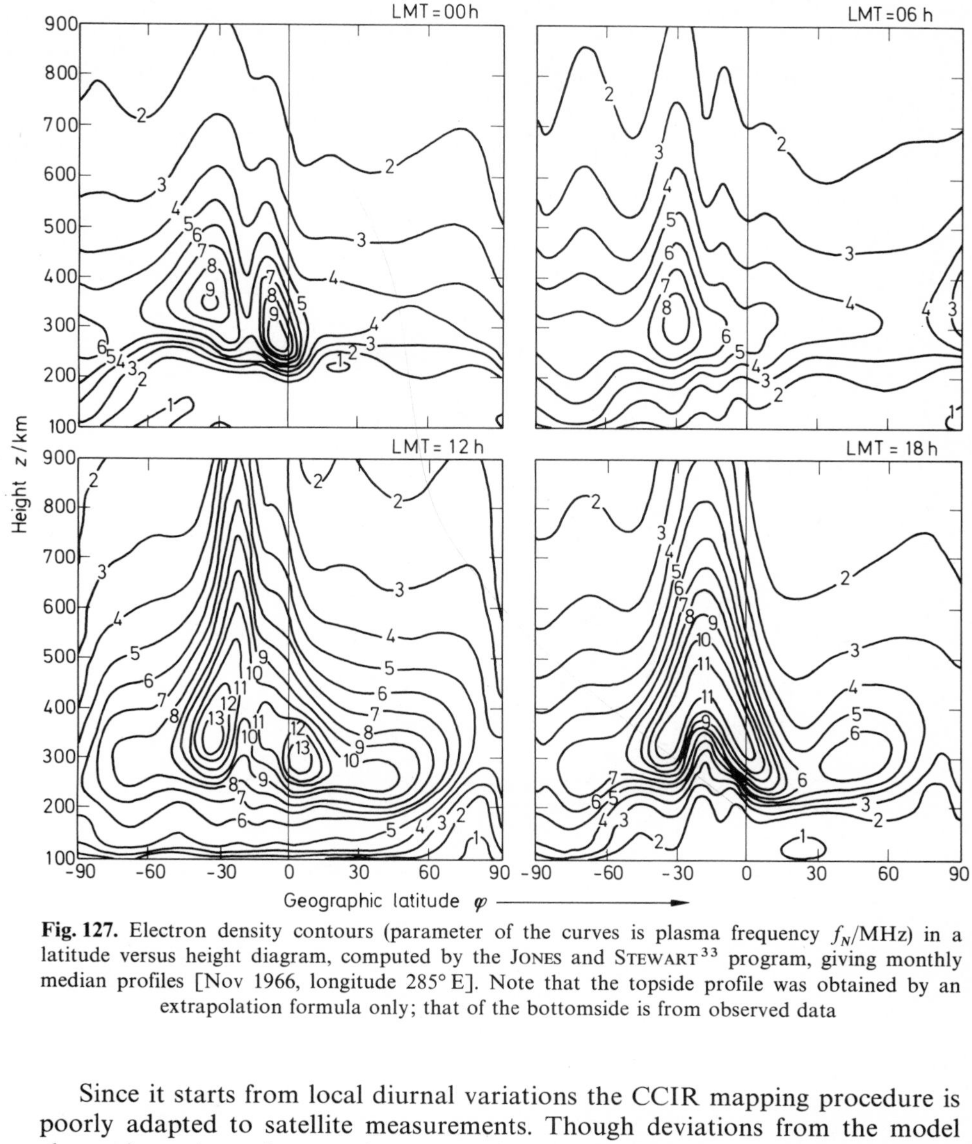

Fig. 127. Electron density contours (parameter of the curves is plasma frequency f_N/MHz) in a latitude versus height diagram, computed by the JONES and STEWART[33] program, giving monthly median profiles [Nov 1966, longitude 285° E]. Note that the topside profile was obtained by an extrapolation formula only; that of the bottomside is from observed data

Since it starts from local diurnal variations the CCIR mapping procedure is poorly adapted to satellite measurements. Though deviations from the model charts have been known for some time, it is difficult to incorporate the valuable information obtained from in-situ measuring satellites. One successful effort was carried out by N.M. SHEIKH[36], who used data from the AEROS satellites which were orbiting at almost constant local time. Transferring in-situ measurements from the top side to the peak with BENT's profile (see Sect. 13β) he obtained the actual longitudinal/latitudinal variation along the projection of the orbit. Comparing these with the CCIR model, very considerable deviations

[36] Sheikh, N.M. (1980): Introduction of satellite data into F2-layer models (PhD Thesis). Lahore (Pakistan): University of Eng. & Technol.

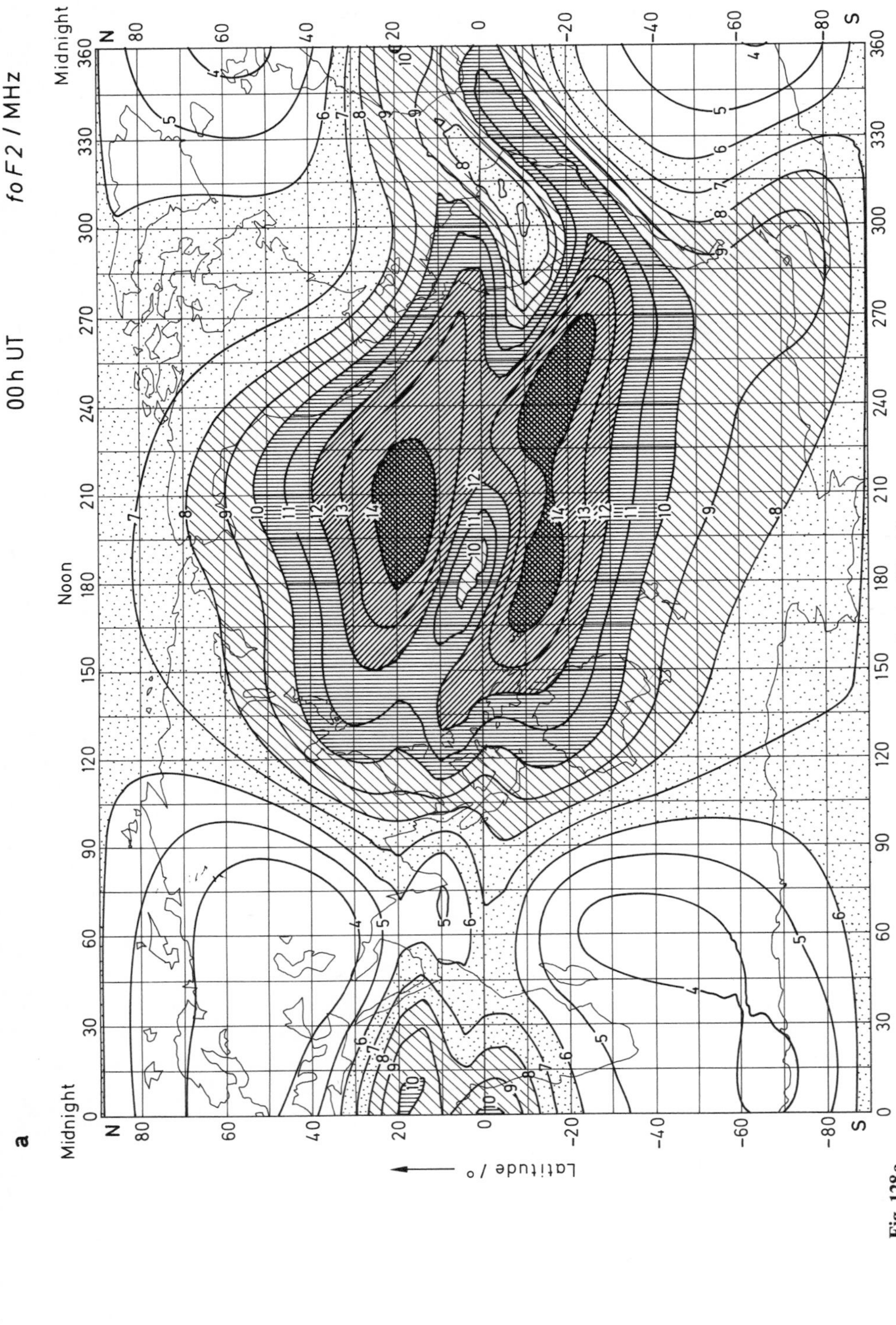
a
00h UT
foF2 / MHz
Midnight
Noon
Midnight
Longitude (E) / °
Latitude / °
N
S

Fig. 128a

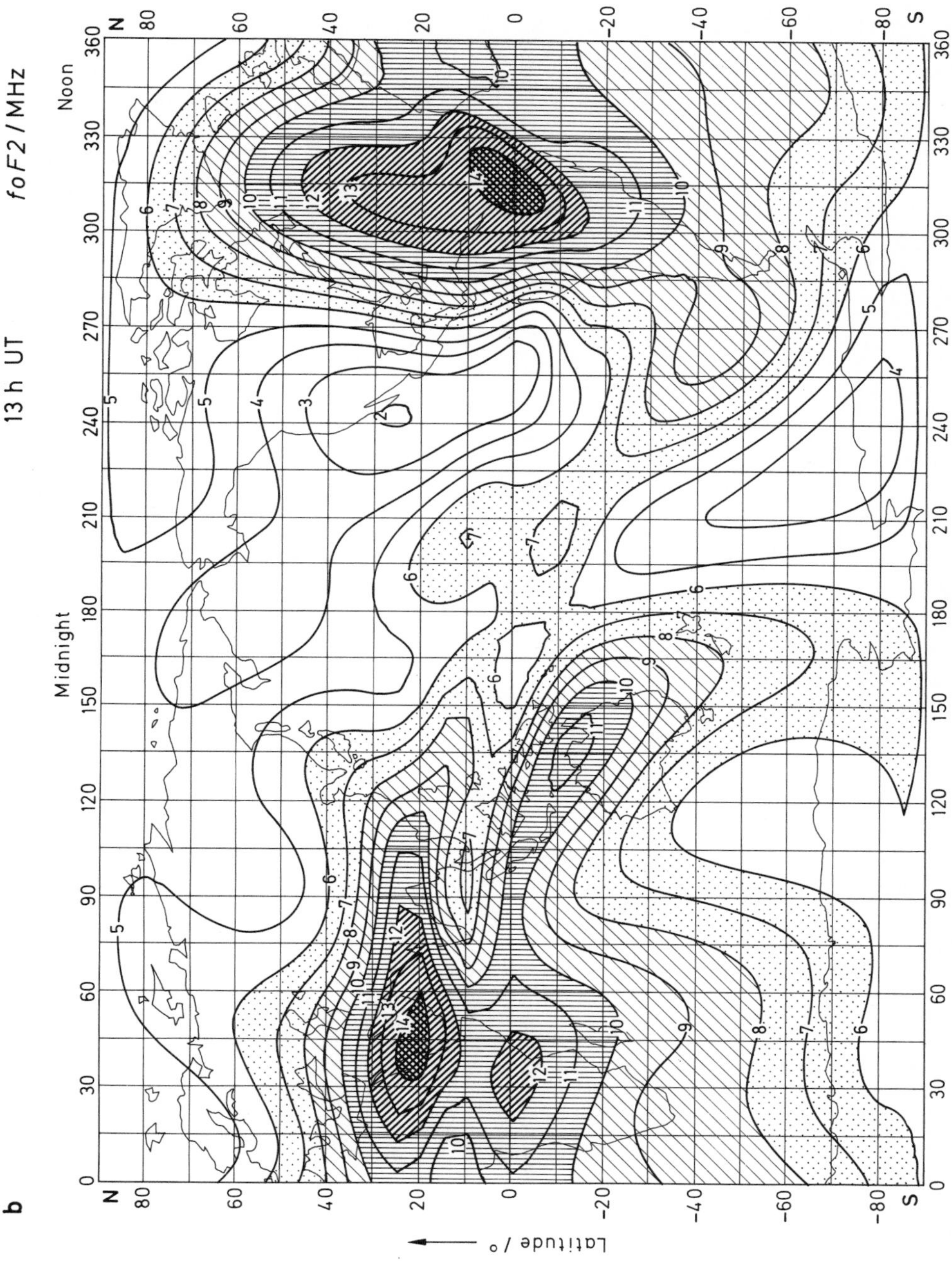

Fig. 128b (Legend see page 398)

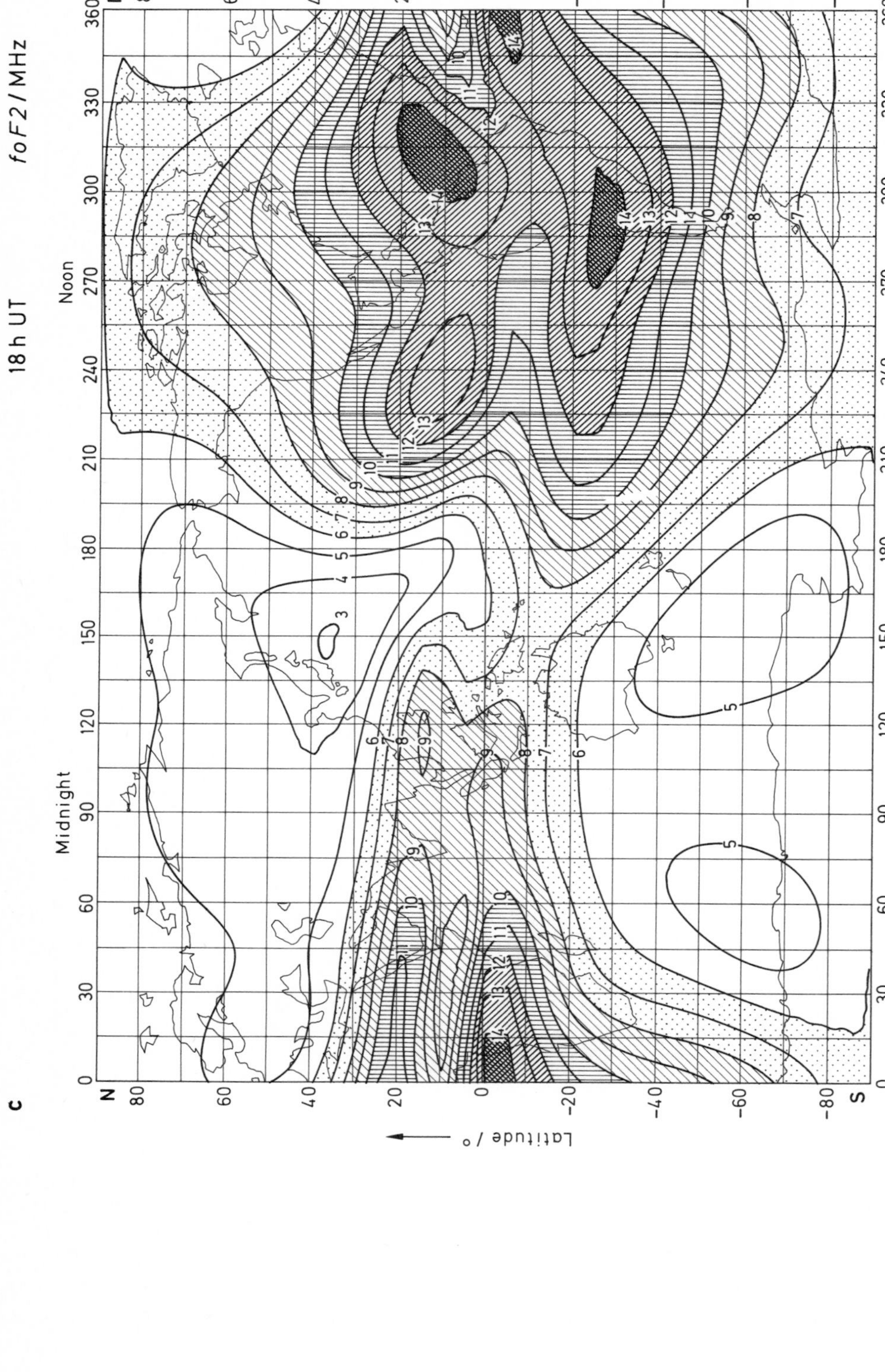

Fig. 128a–c. World maps of "critical" $foF2$ each for a given *Universal Time*, obtained from the Japanese "critical frequency sounder" satellite[35a]. Abscissa is longitude (and so Local Time), ordinate latitude. **(a)** 00 h UT (noon at centre); **(b)** 13 h UT (noon at right); **(c)** 18 h UT (sunrise at centre)

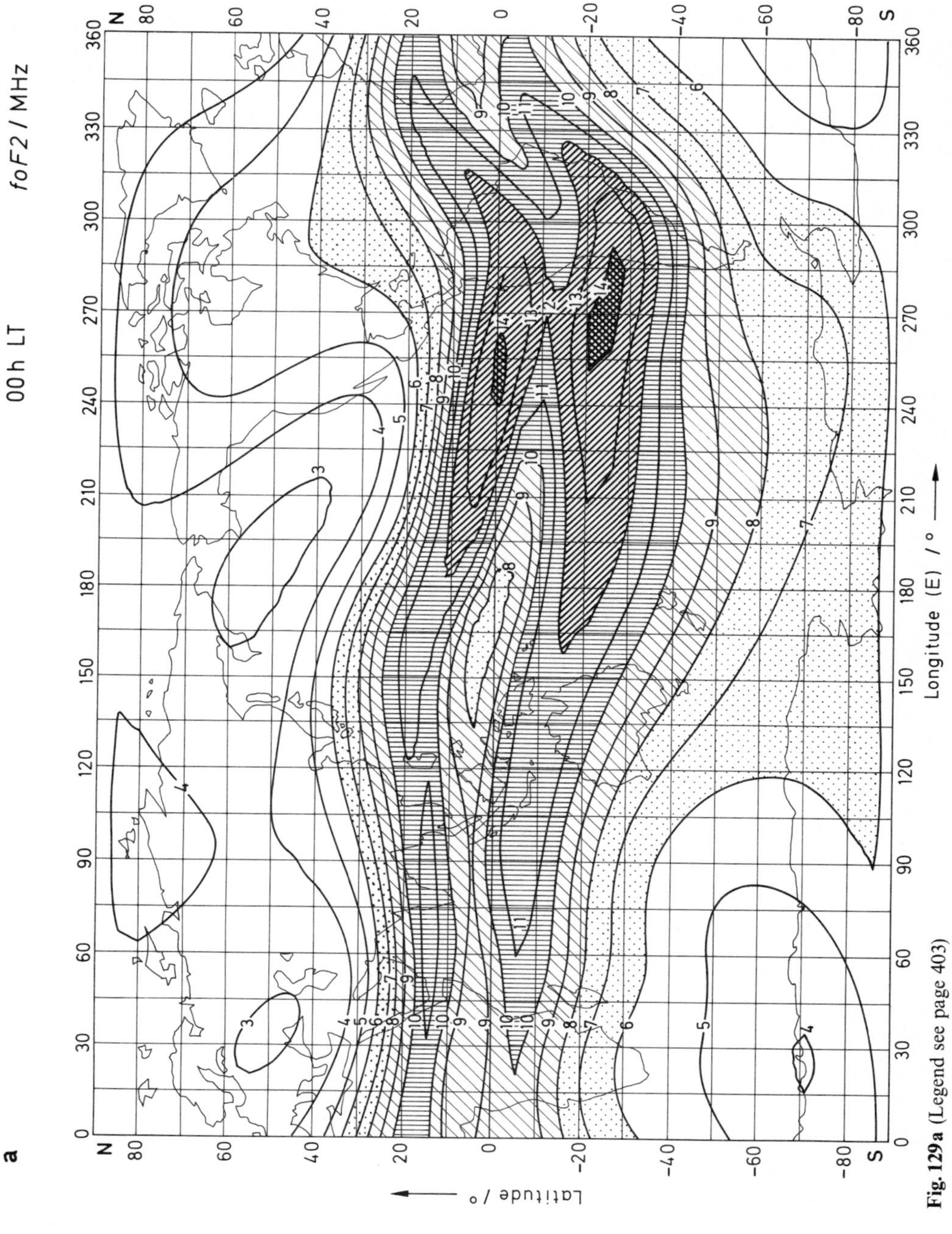

Fig. 129a (Legend see page 403)

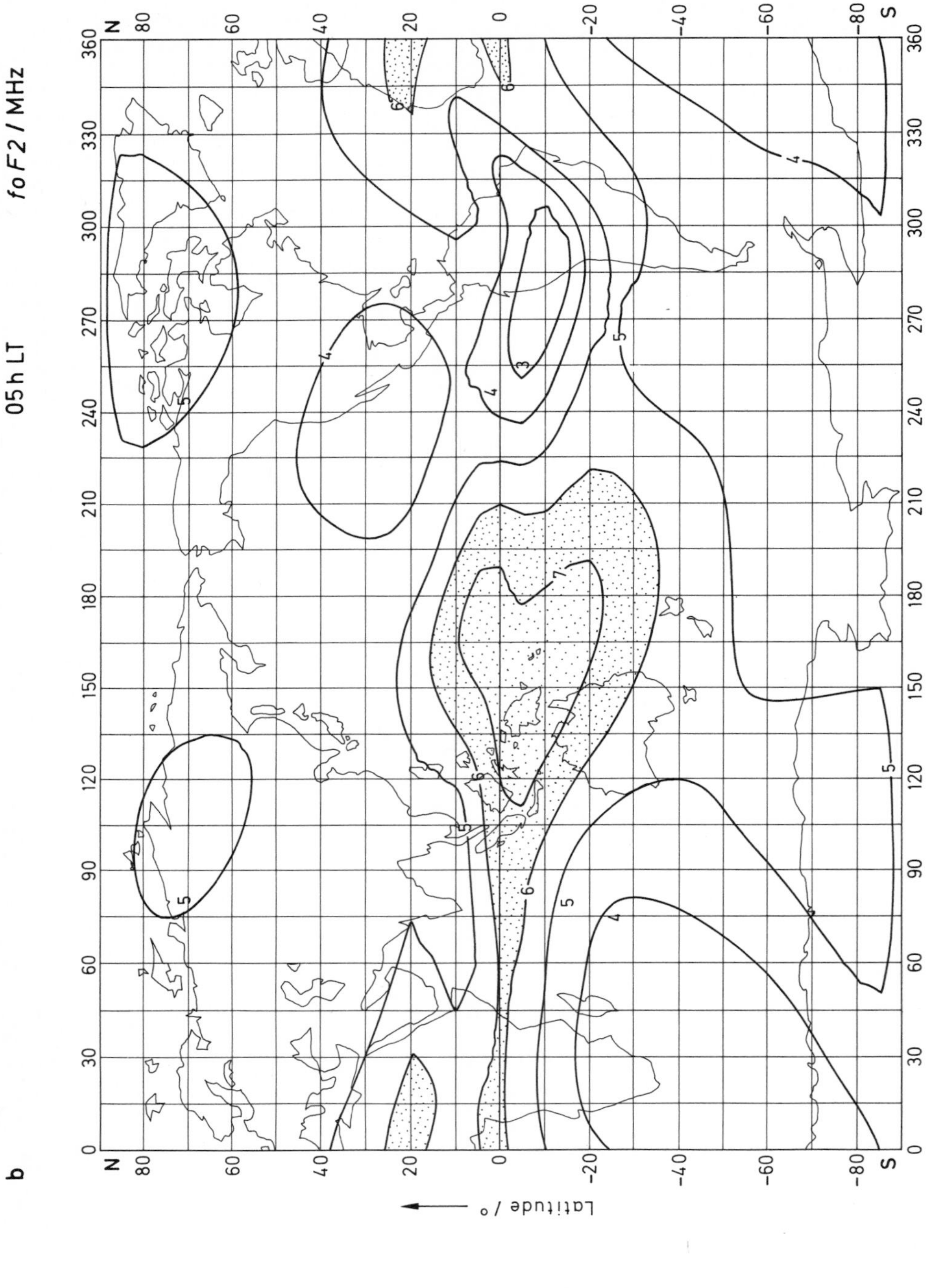

Fig. 129b

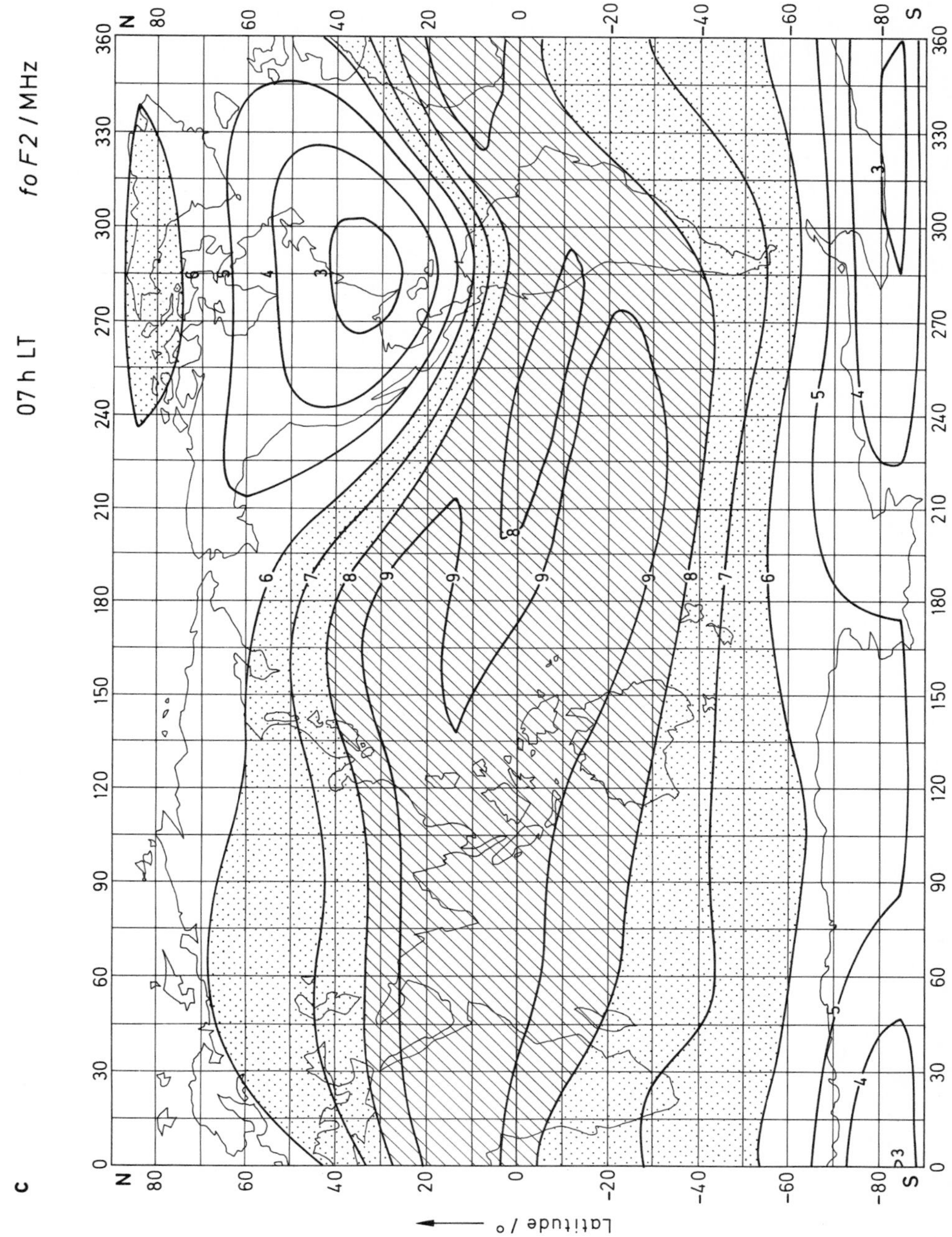

Fig. 129c (Legend see page 403)

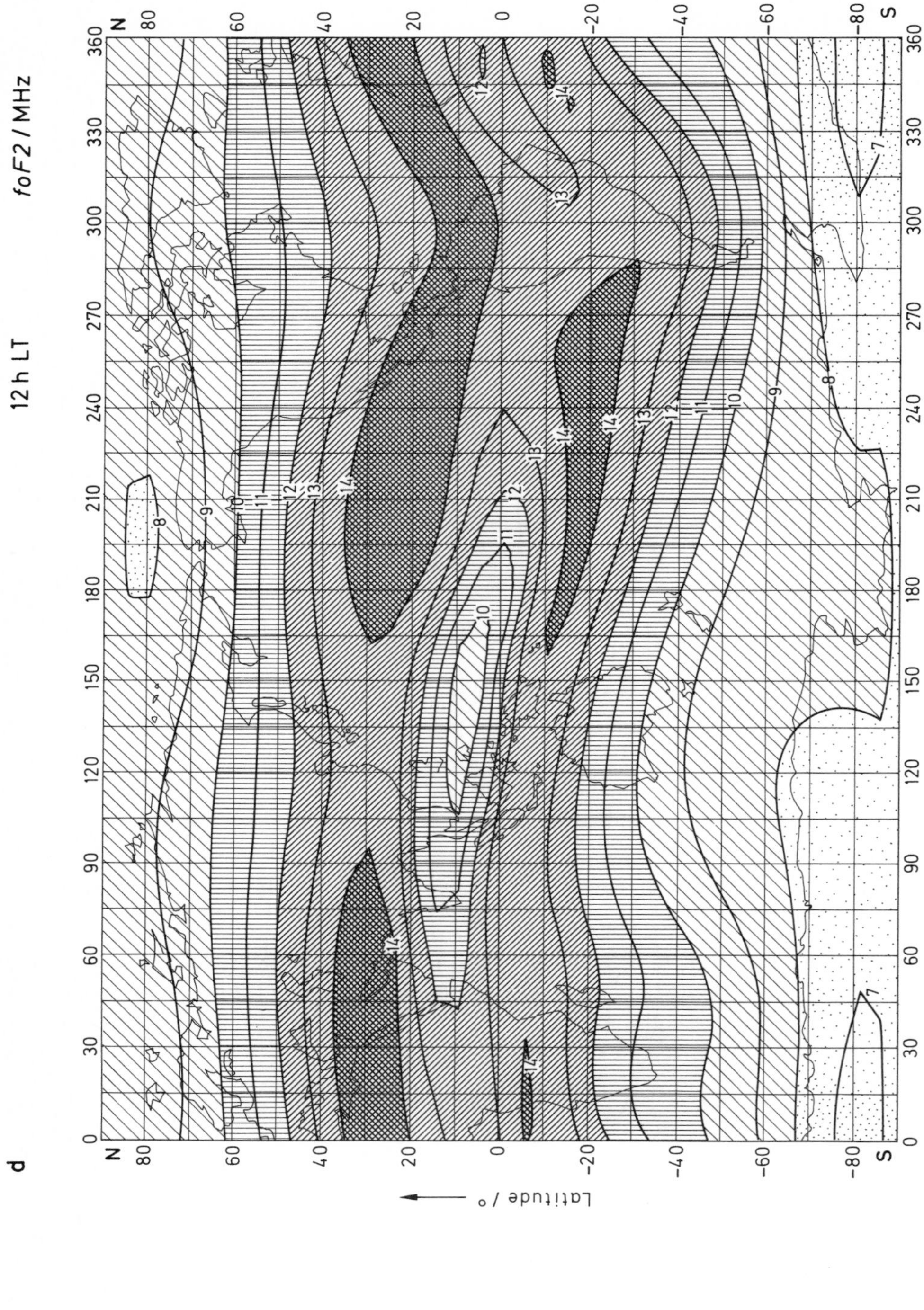

Fig. 129d

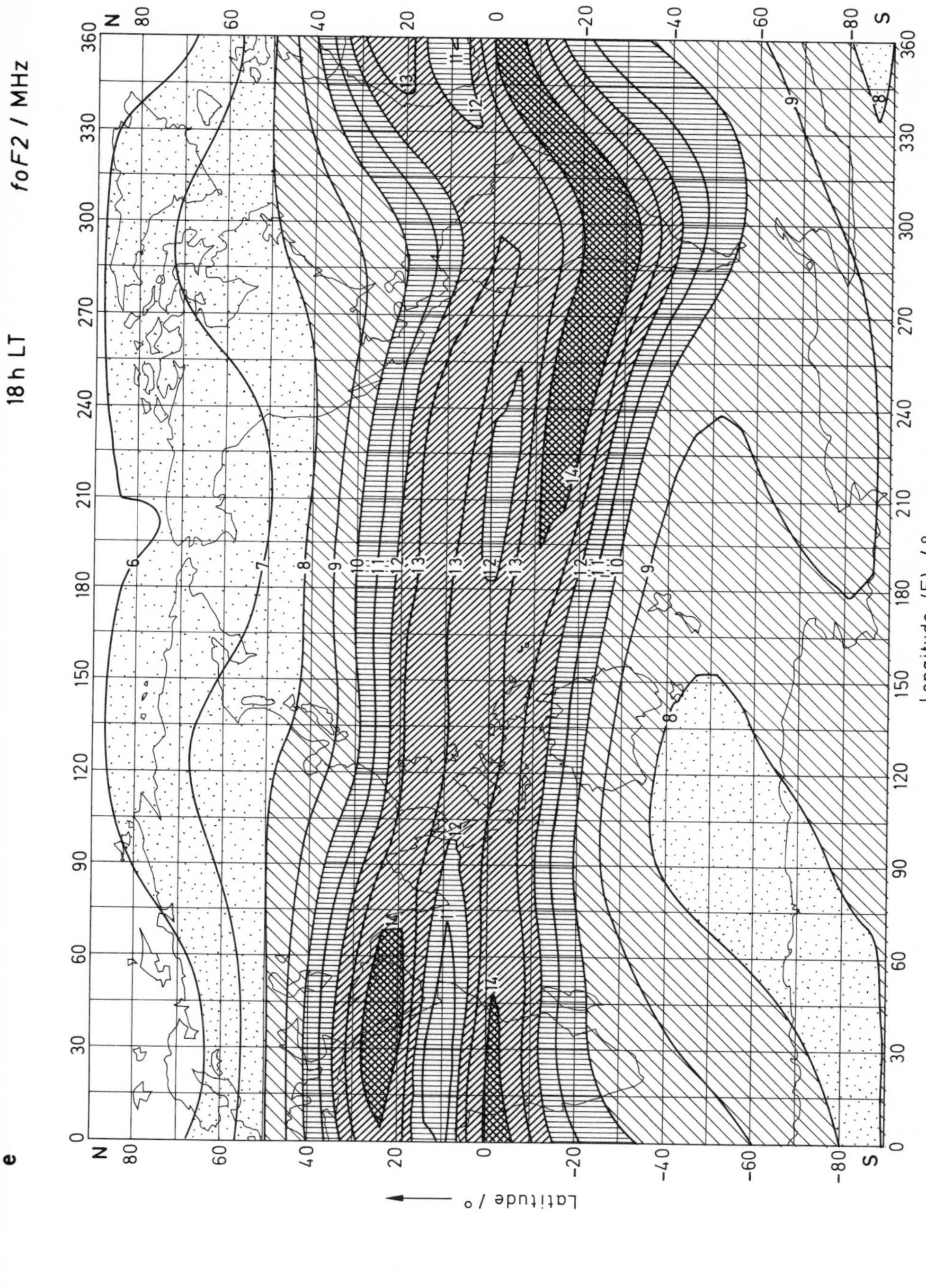

Fig. 129a–e. World maps of "critical frequency" each for a given *Local Time*, obtained from the Japanese "critical frequency sounder" satellite [35a]. Abscissa is longitude (and so Universal Time), ordinate latitude. **(a)** 00 h LT; **(b)** 05 h LT; **(c)** 07 h LT; **(d)** 12 h LT; **(e)** 18 h LT

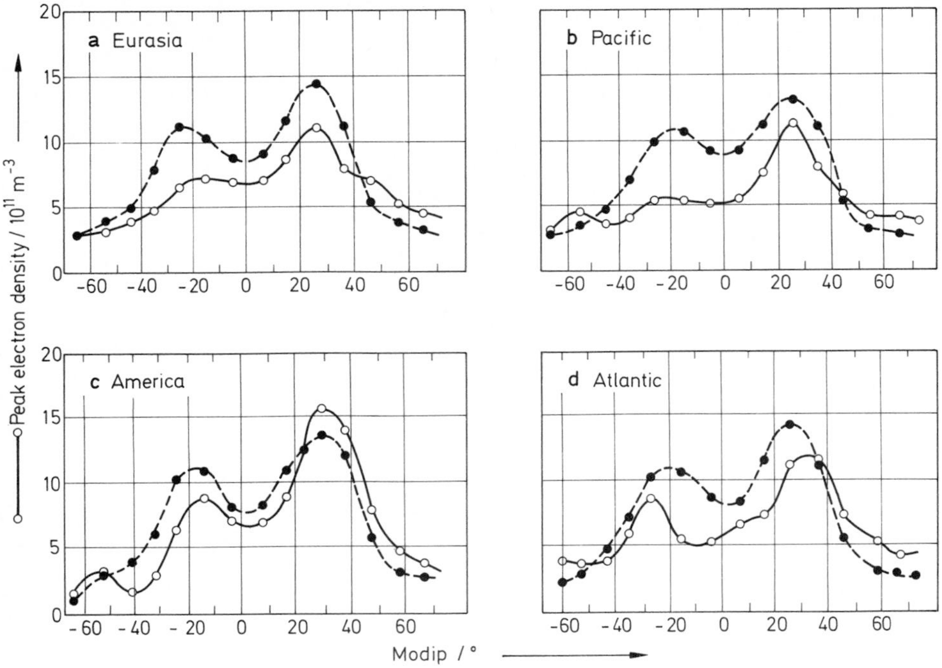

Fig. 130a–d. Latitudinal dependence of peak electron density for 16.4 h LT, August 1974 comparing in-situ probe measurements by satellite AEROS-B [*26*][36–38] (full curves) with the (activity adapted) CCIR prediction [*9*] (broken curves); (**a–d**) latitudinal dependence after the "modip" coordinate [Eq. (14.7)] for different longitudes as indicated

were found (Figs. 130, 131) in the southern hemisphere and, quite generally, in the Pacific. These are the regions where the station coverage is poor so that "screen points" might be decisive in the CCIR maps. Also the true longitudinal variations are observed much larger than given by the models. This is certainly due to oversmoothing in these (which was inevitable by the lack of stations in certain areas).

Sheikh[36] first computed the peak density value by extrapolating from the satellite altitude down to the peak with Bent's top-side profile (see Sect. 13 β). Then he applied a harmonic analysis in longitude to the satellite data, on the one hand, and to those computed from the model on the other. Oversmoothing in the latter was particularly apparent in the first two harmonics.

For a given local time he then determined a correction factor K, which is developed after coordinates as in the second step of the CCIR procedure (with a restricted number of orders, however). (The bold curve in Fig. 131 shows the improvement which could thus be obtained.) A second attempt, trying to improve the amplitude of the diurnal variation, was also undertaken. To this end two correction factors were determined, one for the fundamental and another one for all (longitudinal) harmonics.

---

[37] Rawer, K., Rebstock, C., Sheikh, N.M., Bilitza, D., Neske, E. (1978): Space Res. *XVIII*, 229; Sheikh, N.M., Neske, E., Rawer, K., Rebstock, C. (1978): Telecomm. J. *45*, 225

[38] Rawer, K., Bilitza, D., Ramakrishnan, S., Sheikh, N.M. (1978): [*57*] I, 6-1

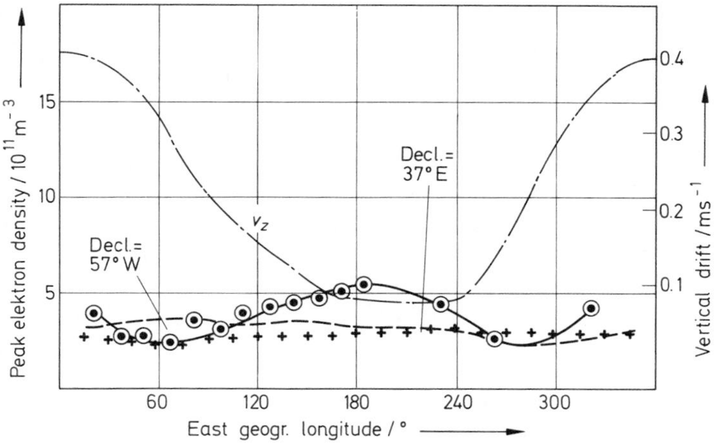

Fig. 131. Longitudinal dependence for a fixed modip (60° S), dots (in circles) are deduced from satellite measurements, the CCIR prediction is shown by the broken line. The crosses show results with the CHING and CHIU model[12]. In order to explain the longitudinal variation the thin dash-dot curve (with right hand ordinate) identifies the vertical drift velocity of the plasma computed[36] according to the KOHL and KING theory[39]

After worldwide analysis at *constant local time* SHEIKH finally came to a combination of both data sets (in a given month) by assuming the latitudinal variation of CCIR to be good enough but obtaining the longitudinal variation from the satellite data. These, derived from direct longitudinal analysis, are thus superimposed on a longitudinal average determined by the CCIR model.

(iv) A semi-theoretical solution of the 'ocean problem' has recently been proposed[40]. Aeronomically[39] computed peak values were fitted with the observed data from those regions on Earth which are covered with stations by adapting the meridional and zonal components of the neutral wind. The same wind model was then applied in the 'empty' regions and the theoretical result was taken there (instead of 'screen points'). A realistic model of the terrestrial magnetic field is used and the declination effect is taken account of. Also different neutral wind diurnal variations are admitted in the northern and southern hemispheres. It appeared that EYFRIG's effect of magnetic declination is extremely important.

(v) The behaviour at *low latitudes* is much better described in magnetic dip than in geographical coordinates (see above). Nevertheless, this region is a difficult one for global mapping since there are not only very strong variations, with the dip-latitude appearing in both peak density and altitude (Fig. 132a–d) but also the profile shape (Fig. 133); consequently, the total electron content

[39] Stubbe, P. (1981): This Encyclopedia, vol. 49/6, p. 247

[40] Rush, C.M., PoKempner, M., Anderson, D.N., Stewart, F.G., Perry, J. (1983): Radio Sci. *18*, 95

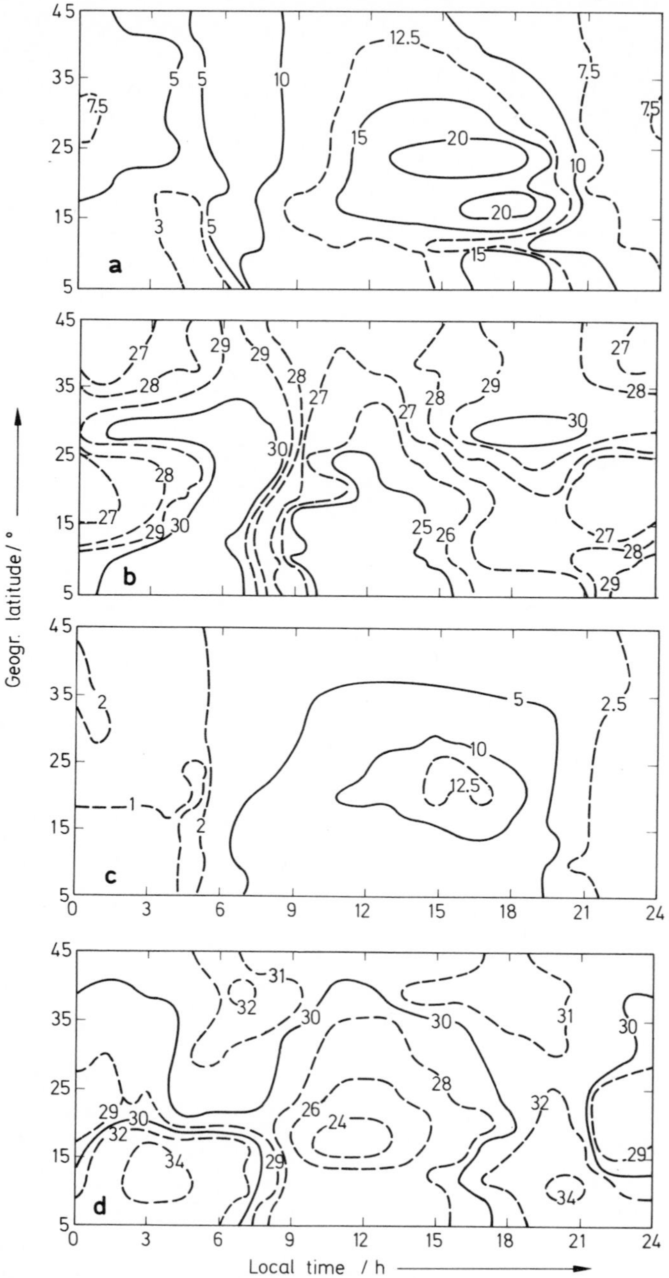

Fig. 132a–d. Low latitude pecularities (results from Indian stations, in summer)[41] influence of solar activity: (**a, b**) 1968, high solar activity; (**c, d**) 1965, low activity; (**a, c**) peak electron density/ $10^{11}\,m^{-3}$; (**b, d**) $10 \cdot M(3000)F2$ [from which the peak altitude can be inferred]

[41] Somayajulu, Y.V., Tyagi, T.R., Bhatnagar, V.P. (1965): Space Res. *V*, 641; Somayajulu, Y.V., Ghosh, A.B. (1979): Ind. J. Radio Space Phys. *8*, 47

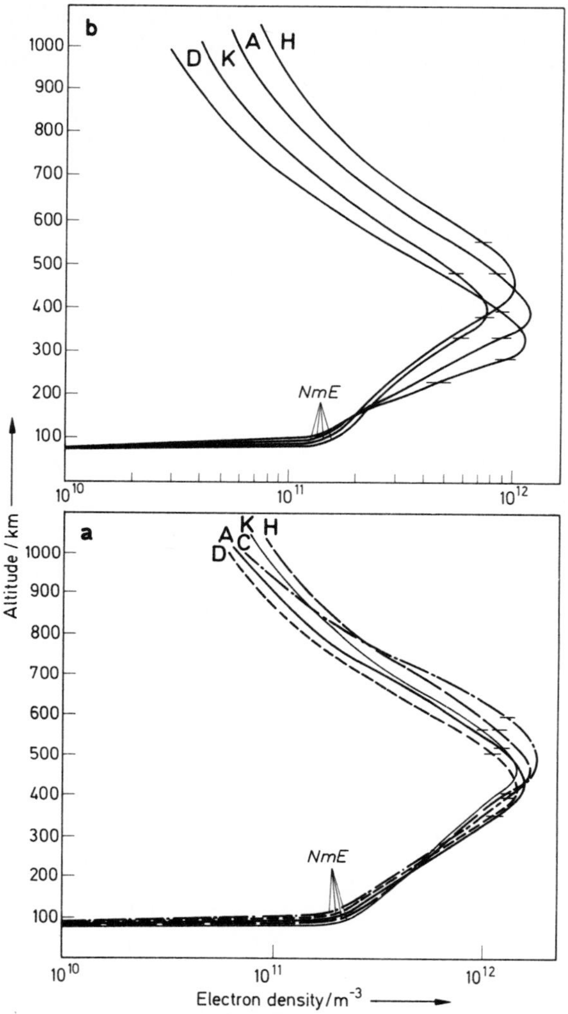

Fig. 133a, b. Average summer daytime electron density profiles at: D, Delhi (28.6° N); A, Ahmedabad (23.0°); C, Calcutta (23.0° N); H, Hyderabad (17.3° N); K, Kodaikanal (10.2° N); (**a**) high, (**b**) low solar activity[41]

(Fig. 134) undergoes important variations[41]. Also the 11 year solar cycle influences not only the magnitude but also the shape of the variations[41]. A recent survey is [42].

Unfortunately, these low-latitude features, in particular the day-time 'crests' north and south of the dip-equator are quite sharp for individual days but somewhat variable in position day-by-day[43], so that a monthly average gives a non-typical *smoothed latitude variation*; it is even more smoothed by the math-

[42] Anderson, D.N. (1981): J. Atmos. Terr. Phys. *43*, 753

[43] Vila, P. (1971): Radio Sci. *6*, 689 and 945

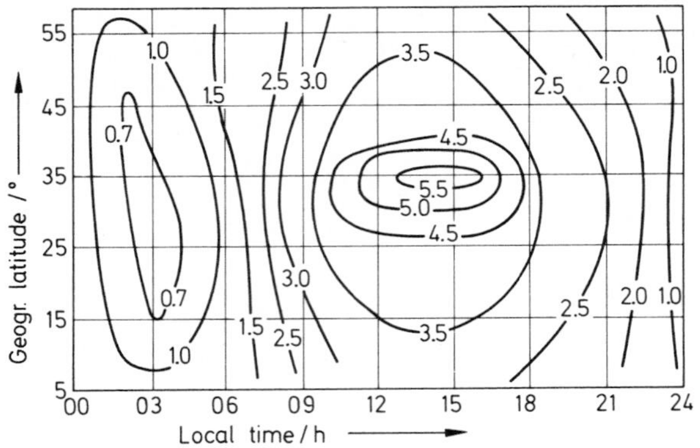

Fig. 134. Electron content/10^{17} m^{-2} (summer average, high solar activity)[41]

ematical representation methods used for mapping. In cases where horizontal gradients of electron density are important, as in certain radio wave propagation problems, it must be borne in mind that all models give much smaller gradients than are seen on any individual day.

Ground-based observation of satellite beacons [*70*] allows regional maps of electron content to be obtained averaging over a few weeks. Gradients can so much better be determined than from the data of individual sounding stations.

(vi) The *high latitudes* present a particular problem for models. The main reason is that, in the polar zones, the magnetic field lines are open and connect these zones with the tail of the magnetosphere (see Fig. 11, p. 41 of vol. 49/4 (1972) of this Encyclopedia). Thus, the polar zones have a well-defined limitation (though depending on local time). In these zones the effects of *corpuscular influx* are very important for both plasma density and heating. Furthermore, horizontal advection plays a major role, particularly in winter when large parts of the polar cap receive no solar radiation. All this has the result that for geophysical reasons descriptions of the F-region which are valid at temperate and low latitudes should not be applied inside the polar zones.

At most hours the delimitation of the polar zones is well marked by a decrease of F-region plasma density at the equatorward side and a strong poleward increase. The minimum in between is called cleft by day and trough by night. Its position may clearly be identified with narrowly spaced in-situ records of satellites[44,45]. It depends primarily on the L-coordinate[46] but, with increasing magnetic activity (Kp), is shifted equatorwards according to Fig. 135[44]

$$L_{\text{trough}} = 5.3 - 1.04\sqrt{Kp}. \tag{14.8}$$

[44] Neske, E. (1978): Space Res. *XVIII*, 237

[45] With remote sensing methods covering a large angular range (like ionospheric sounding) it is difficult to identify the phenomenon

[46] McIlwain's L is the peak radius of a magnetic field line measured in units of the Earth's radius. [See Hess, W.N. (1972): this Encyclopedia, vol. 49/4, Sect. 5, p. 124]

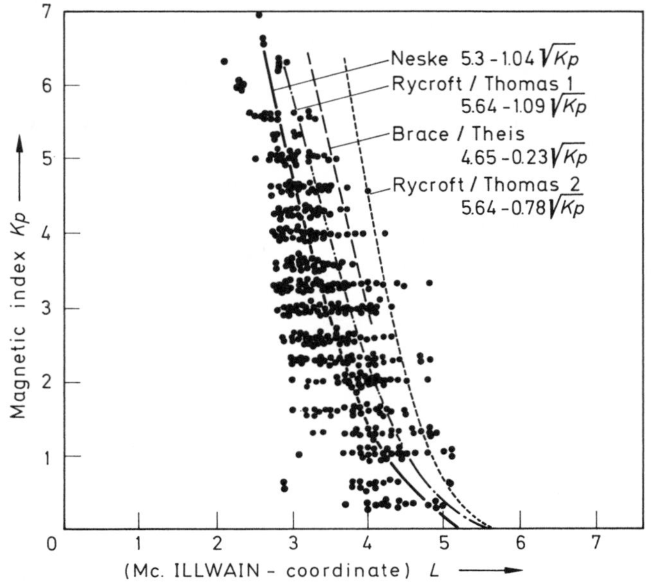

Fig. 135. Identification of the night-time trough position for 1974/1975 from in-situ measurements by the impedance-probe aboard satellite AEROS-B [*26*][44]. Abscissa, McILLWAIN's *L*-coordinate (equivalent of the invariant magnetic latitude)[46]. Ordinate, planetary magnetic figure *Kp*. Empirical relations giving *L* (trough) are indicated

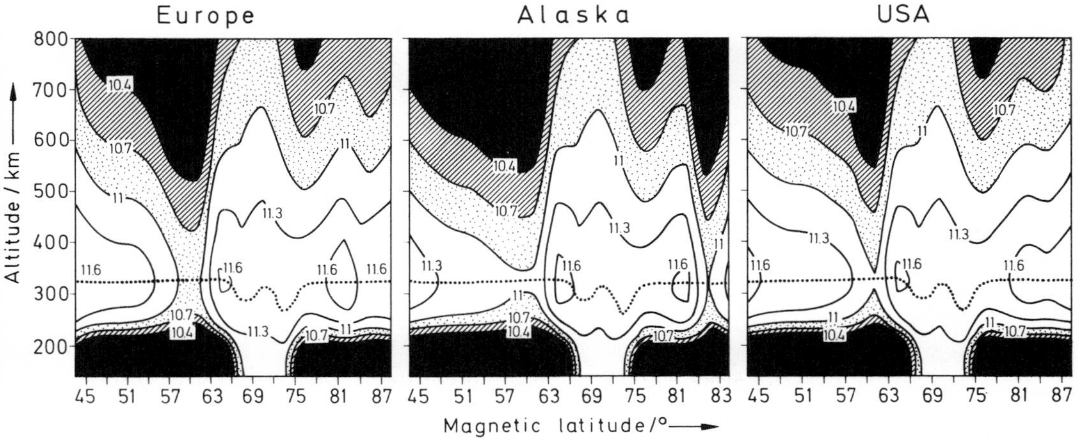

Fig. 136. Model profiles of electron density[47] as a function of the magnetic latitude (abscissa) in northern Europe (left), Alaska (centre) and USA (right). Parameter: $\log_{10}(N_e/\mathrm{m}^{-3})$

The severe effect of the phenomenon is seen in Fig. 136[47] (which was obtained by model computations taking account of strong polar convection[48]).

It is evident that the ionosphere inside the polar zones cannot be described similarly to that outside. For this reason, all present-time models cannot

[47] Sojka, J.J., Schunk, R.W. (1982): Geophys. Res. Lett. *9*, 143

[48] Sojka, J.J., Raitt, W.J., Schunk, R.W. (1981): J. Geophys. Res. *86*, 609

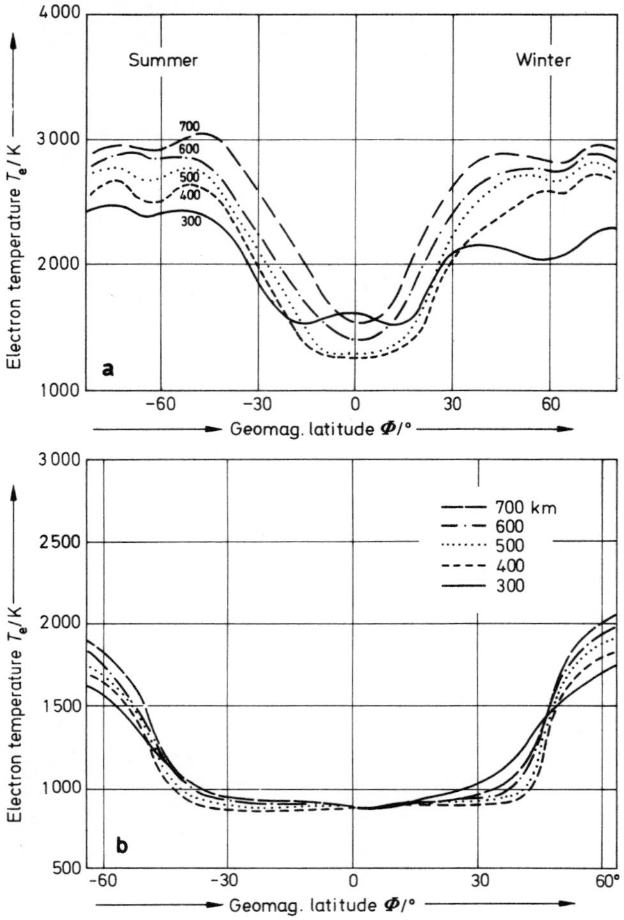

Fig. 137 a, b. Latitudinal cross sections ("geomagnetic latitude" coordinate ϕ) of electron temperature, T_e/K, at different altitudes, parameter. Average model curves[49] (see Fig. 138) derived from AEROS-A measurements [*26*], Jan–Mar 1973: (**a**) Day, 15 h LT; (**b**) night, 03 h LT

satisfactorily describe the behaviour at high latitudes but give rather crude average data at best. Separate descriptions for both areas might be appropriate.

b) Electron and ion temperature

δ) Satellite measurements are the main data source for establishing the worldwide pattern of *plasma temperatures* (see, however, Sect. 13ε). Incoherent scatter measurements (see Sect. 5) are a very useful complement to these but do not permit deduction of global variations. Latitudinal cross sections quite generally show a poleward increase (Fig. 137) at all altitudes. Only in the equatorial region do appear profiles, with a maximum somewhere below

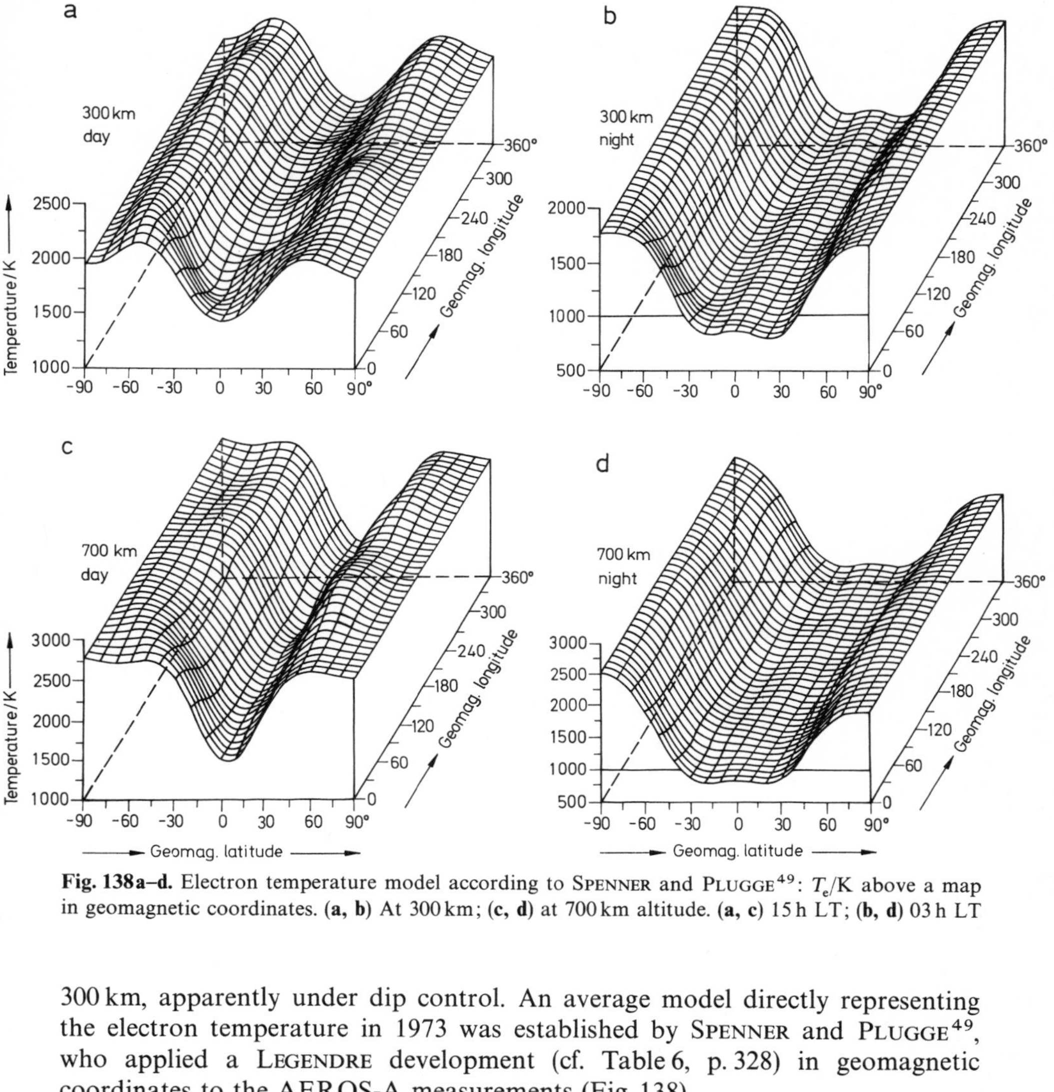

Fig. 138a–d. Electron temperature model according to SPENNER and PLUGGE[49]: T_e/K above a map in geomagnetic coordinates. (**a, b**) At 300 km; (**c, d**) at 700 km altitude. (**a, c**) 15 h LT; (**b, d**) 03 h LT

300 km, apparently under dip control. An average model directly representing the electron temperature in 1973 was established by SPENNER and PLUGGE[49], who applied a LEGENDRE development (cf. Table 6, p. 328) in geomagnetic coordinates to the AEROS-A measurements (Fig. 138).

These authors claim a small longitudinal effect which has not been confirmed since. The average behaviour as a function of height/geomagnetic latitude (Fig. 137) is described by a 3rd/2nd order development including two sets (day and night) of 36 coefficients each.

The IRI-1978 and 1979 models [*43*, *44*] use a rather simple descriptive formula based on in-situ measurements of two AEROS satellites (Fig. 139)[49,50] and on data from three incoherent scatter stations (see Sect. 5). It is composed of three terms[51], one depending on geomagnetic latitude ϕ only, another depend-

[49] Spenner, K., Plugge, R. (1979): J. Geophys. *46*, 43
[50] Spenner, K., Bilitza, D., Plugge, R. (1979): J. Geophys. *46*, 57
[51] Bilitza, D. (1981): [*44*], 11

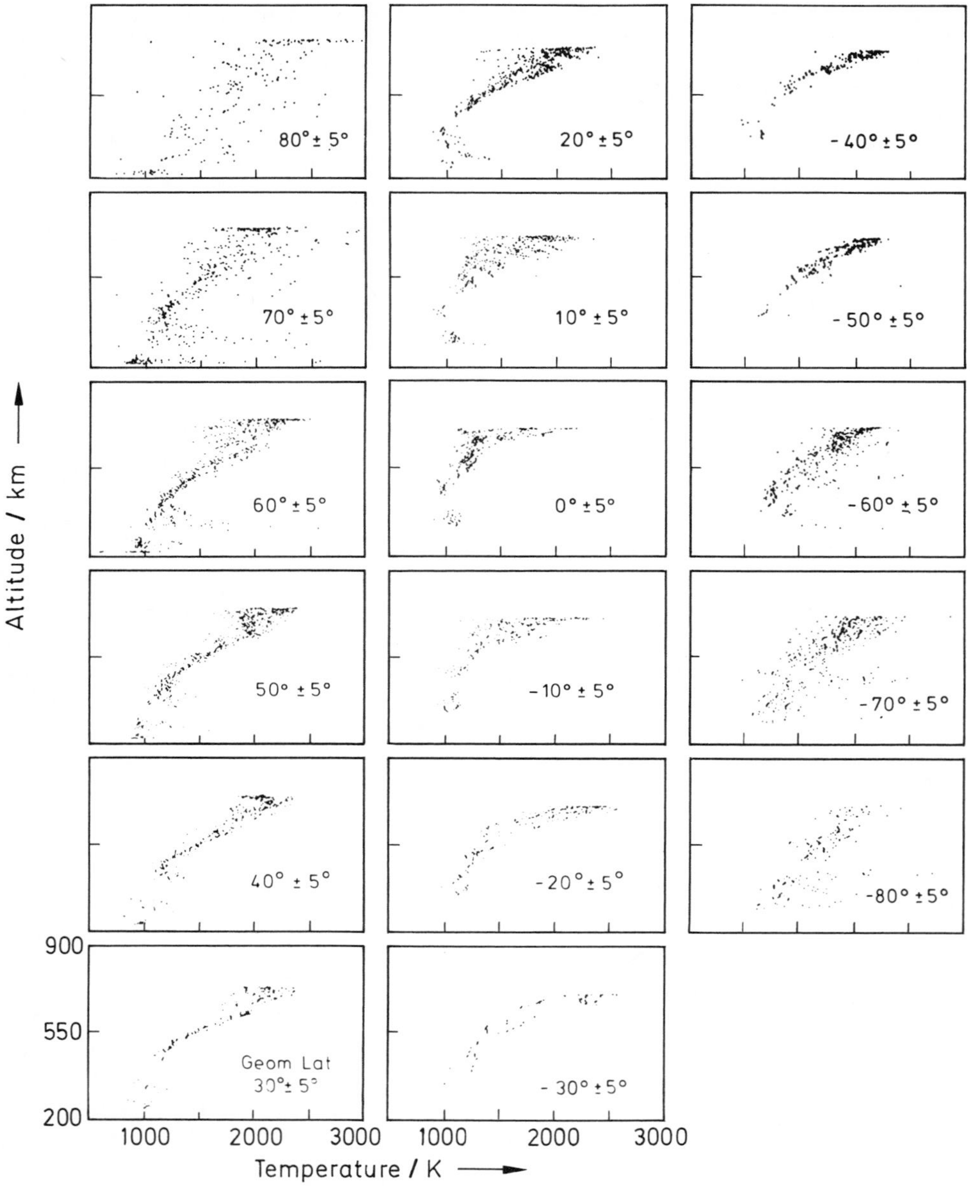

Fig. 139. Vertically displayed electron temperature measurements at 16 h 30 LT of satellite AEROS-B [26]. Each diagram covers one 10° range of geomagnetic latitude

ing on altitude z, and a mixed term which is needed for representing the low-latitude bulge (see Fig. 134) at altitude h_b $[=(70 \cdot \exp(-0.14(\phi/10°)^2)+200)\,\mathrm{km}]$

$$T_e/\mathrm{K} = A - B\frac{g(\phi/°)}{g(\phi/°)}g(\phi/°)^n + C\,\mathrm{Eps}_1(3(Z-h_b)/100\,\mathrm{km}) + D\cdot(z/\mathrm{km}-700) \qquad (14.9)$$

Table 9. Coefficients to the IRI electron temperature model [Eq. (14.9)]

	A	B	C	D	a_1	a_2	n
day	2,325	725	25,600	2	3.4	−0.014	1
night	1,600	700	0	0	0.47	+0.024	0.5

with $g(x) \equiv a_1 x + a_2 x$; $\mathrm{Eps}_1(x) \equiv \exp(x)/[1+\exp(x)]^2$; the coefficients are given in Table 9 for day and night.

In Figs. 140b, d T_e-maps for exospheric heights of 1,000 to 1,200 km are shown; these were established by applying the numerical mapping procedure described in Subsect. γ above to the data obtained aboard the japanese ISS-b satellite[35b]. The corresponding electron density maps show rather small variations with just one maximum zone, very roughly along the dip equator, see Figs. 140a, c. It is clearly visible that, in general, regions of increased electron density show decreased electron temperature and vice versa, though many features are different in both maps. Thus, these are not independent but, on the other hand, there is *no unique relation* appearing between electron density and temperature.

Since in a large height range there exists a strong *relationship* between *electron density* N_e *and temperature* T_e, independent modelling of T_e (as above) necessarily includes a rather large dispersion due to the variations of N_e. Models including such a relation should end up with smaller dispersion[52]. On the other hand, a straight-forward unique relation[53] (see Fig. 115 in Sect. 13) is not satisfactory either, see Figs. 140. From detailed studies of AEROS and AE-C data Thiemann[54] derived a formula describing the N_e versus T_e relation as an Epstein step function (applicable between 300 and 700 km altitude):

$$T_e = P_1 + (P_2 - P_1) \cdot \mathrm{Eps}_0\{(\log_{10}(N_e/\mathrm{m}^{-3}) - P_3)/P_4\}, \qquad (14.10)$$

$\mathrm{Eps}_0(x) \equiv 1/[1+\exp(-x)]$, with coefficients parabolically depending on modip μ and altitude z, however; see Table 10.

Table 10. Coefficients to Thiemann's[54] day-time model of electron temperature T_e as function of electron density N_e [Eq. (14.10)]

	abs. term	μ/rad	$(\mu/\mathrm{rad})^2$	$(z/100\,\mathrm{km})$	$(z/100\,\mathrm{km})^2$
P_1/K	220.3	−258.3	+106.5	+8.1	−0.846
P_2/K	35.7	+ 9.0	− 65.1	−6.06	+0.556
P_3	9.865	− 0.872	+ 3.304	−	−
P_4	0.446	+ 0.285	+ 0.957	−	−

[52] Mahajan, K.K., Pandey, V.K. (1978): Ind. J. Radio Space Phys. *7*, 305; (1979): J. Geophys. Res. *84*, 5885; (1980): ibidem, *85*, 213

[53] Brace, L.H., Theis, R.F. (1978): Geophys. Res. Lett. *5*, 275; (1981): J. Atmos. Terr. Phys. *43*, 1317

[54] Bilitza, D., Thiemann, H. (1982): Kleinheubacher Berichte *25*, 237

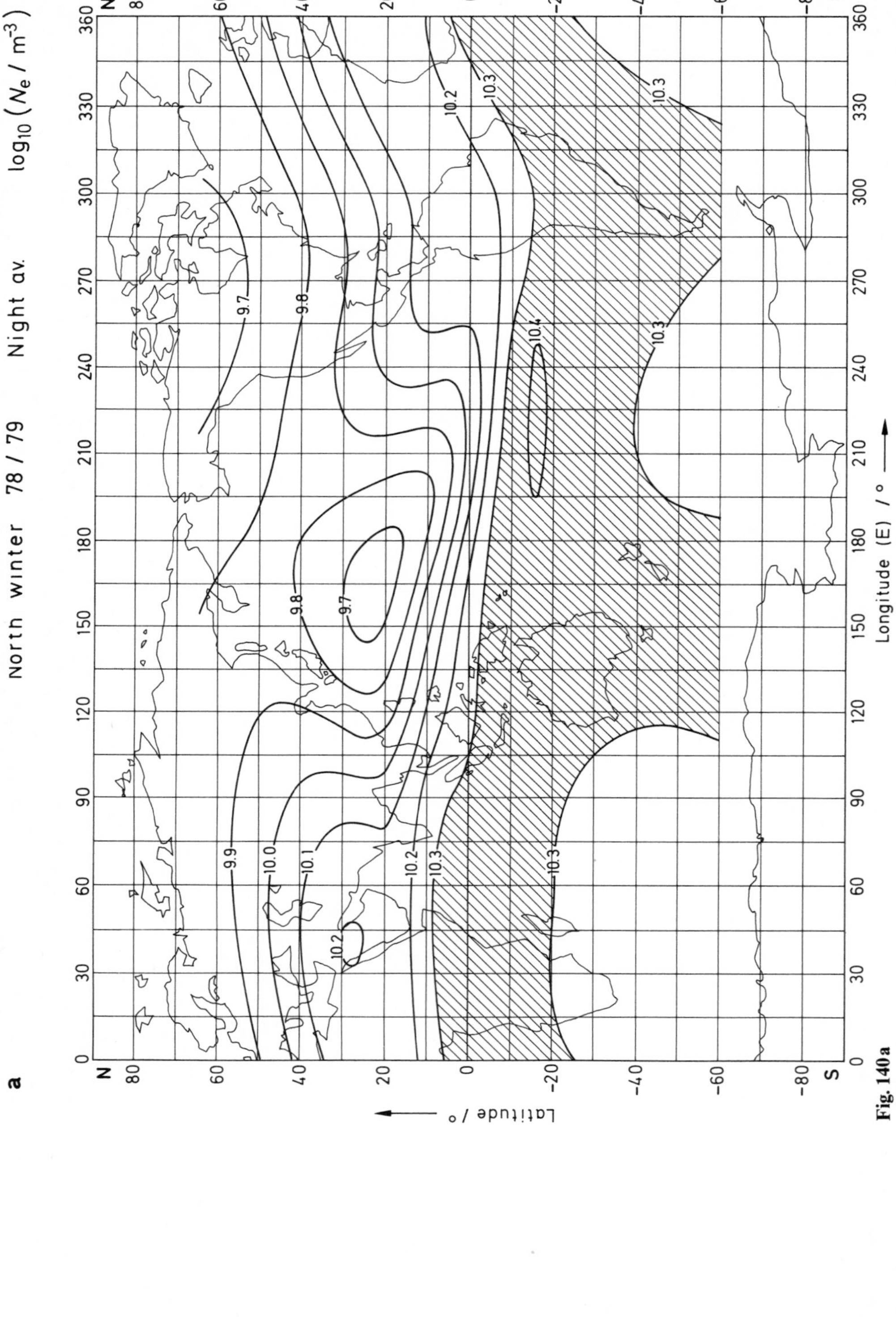
a
North winter 78 / 79 Night av. log10 (Ne / m-3)
9.7
9.8
9.9
10.0
10.1
10.2
10.3
10.4
Latitude / °
Longitude (E) / °
N
S

Fig. 140a

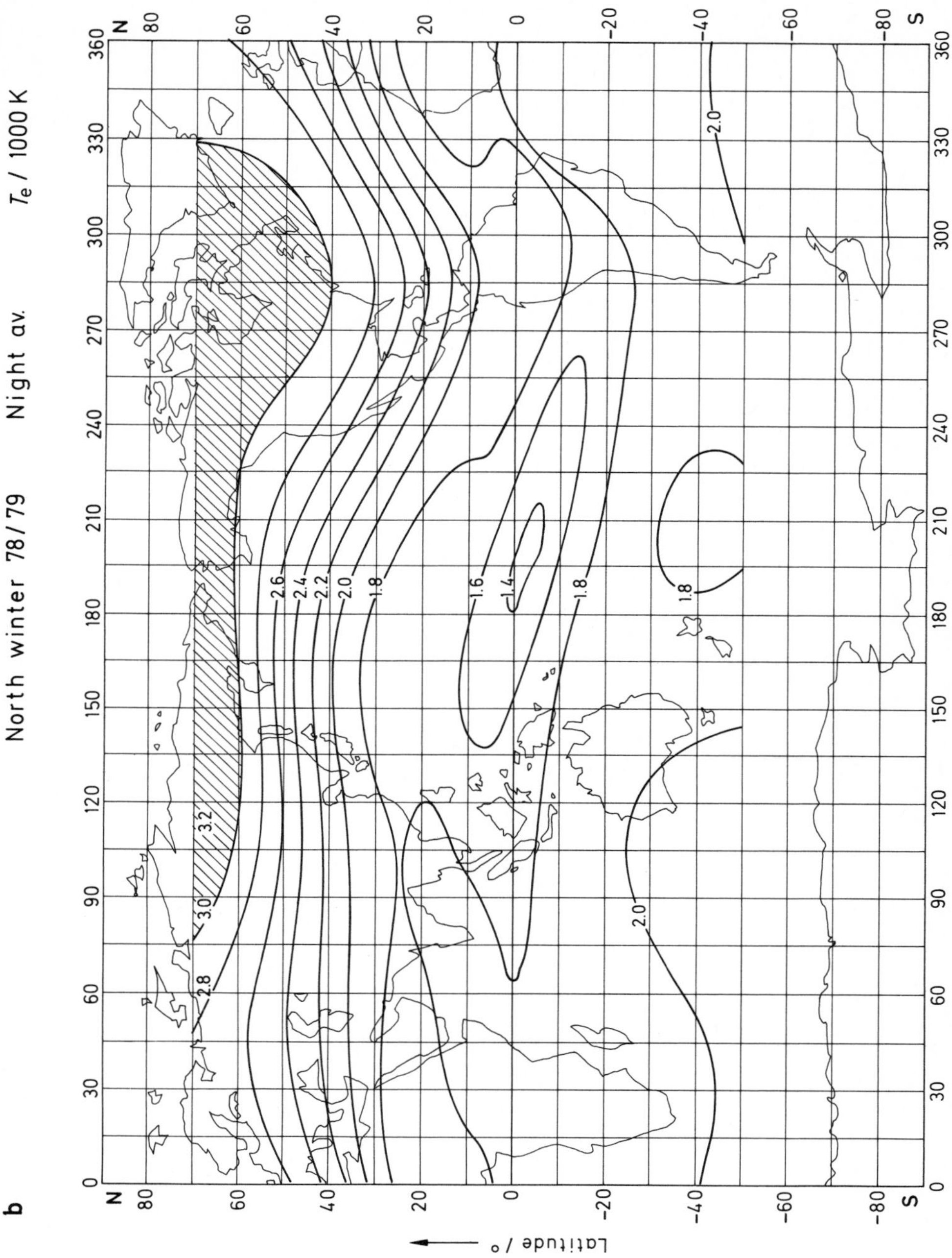

Fig. 140b (Legend see page 417)

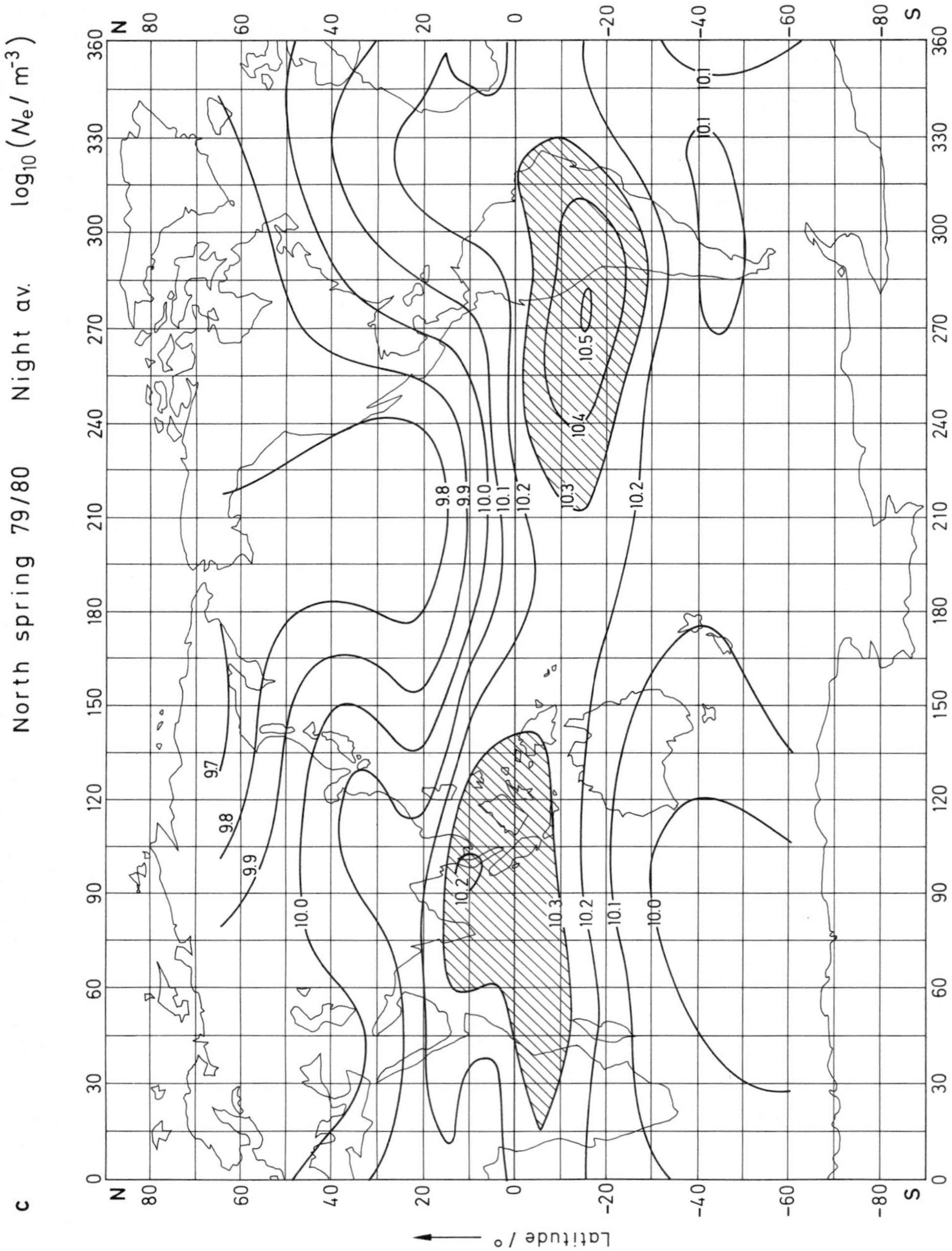
c
North spring 79/80
Night av.
$\log_{10}(N_e / m^{-3})$
Latitude / °
Longitude (E) / °
N
S
9.7
9.8
9.9
10.0
10.1
10.2
10.3
10.4
10.5

Fig. 140c

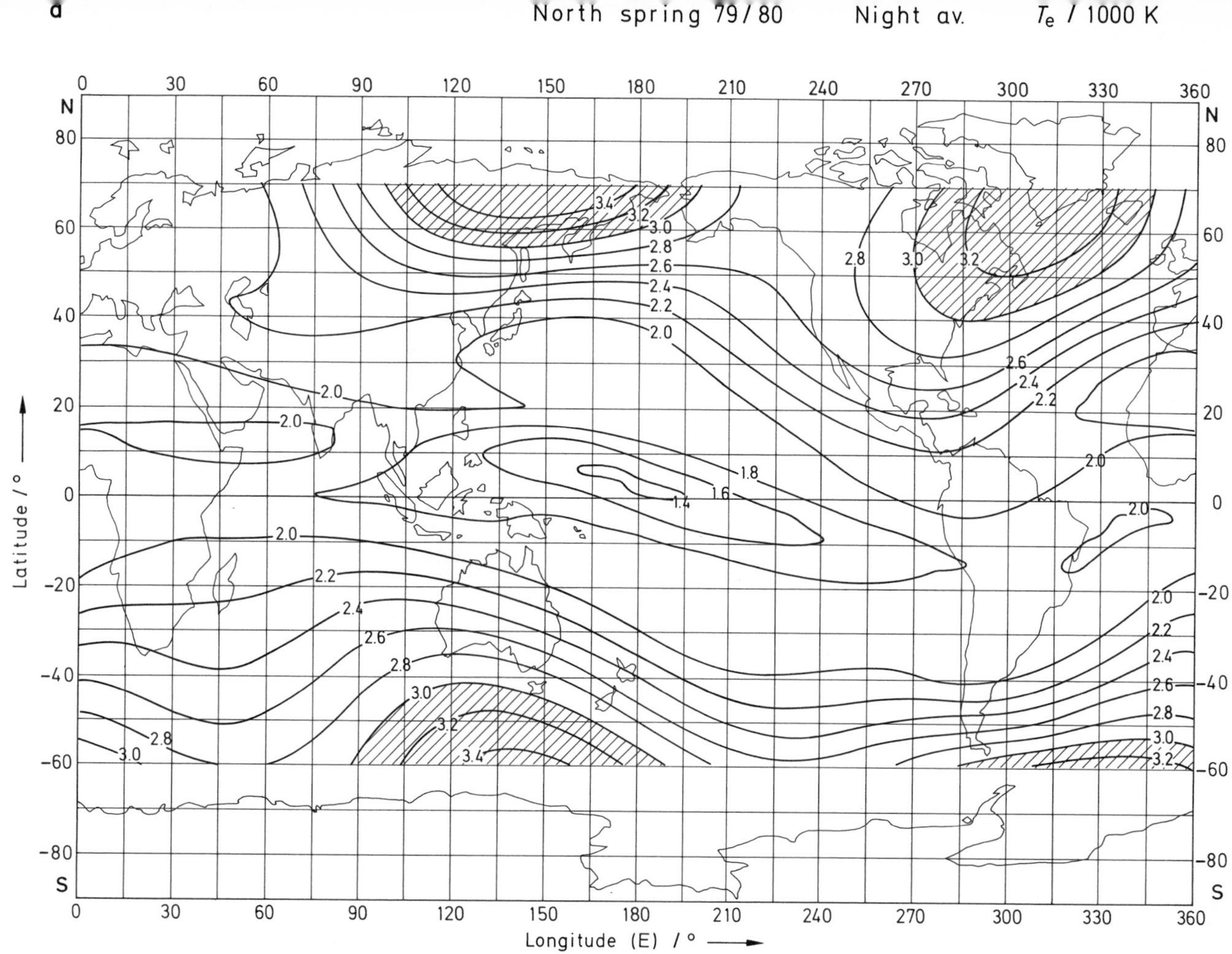

Fig. 140 a–d. World maps of electron density/temperature in the (exospheric) height range 1,000 to 1,200 km obtained from measurements of a spherical retarding potential analyzer aboard the Japanese ISS-b satellite[35b]. (**a, c**) Electron density; (**b, d**) electron temperature. (**a, b**) Winter 1978/1979; (**c, d**) spring 1980

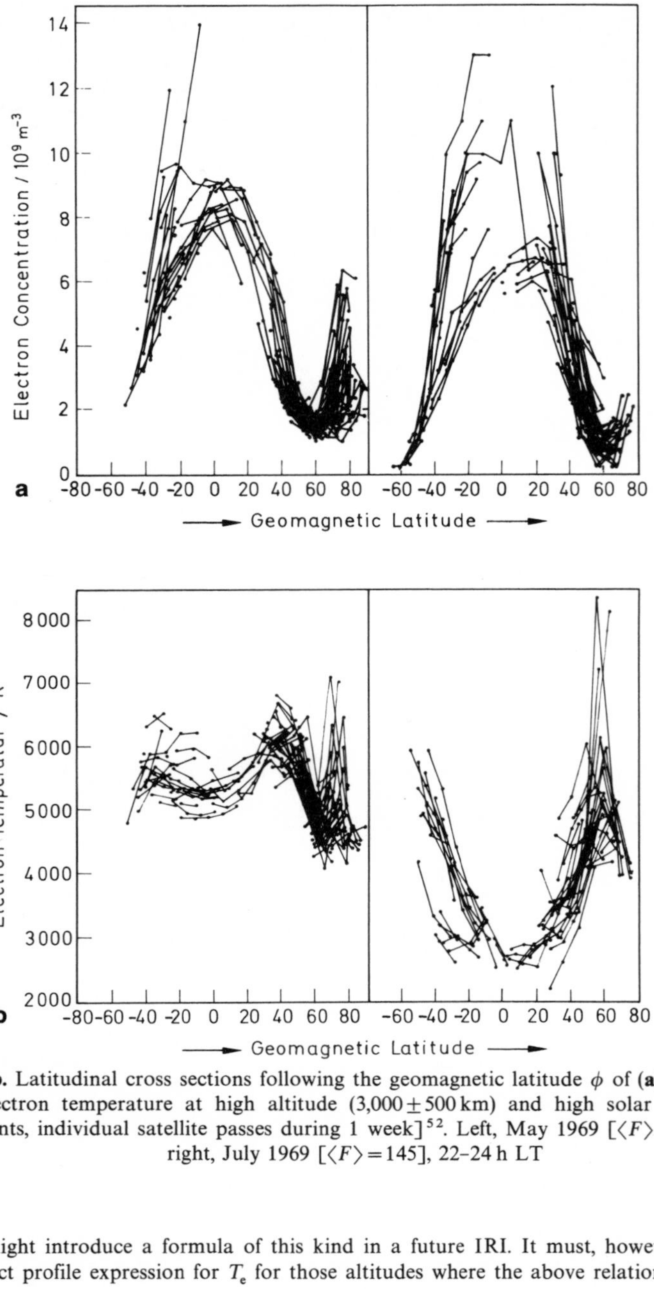

Fig. 141 a, b. Latitudinal cross sections following the geomagnetic latitude ϕ of (**a**) electron density and (**b**) electron temperature at high altitude (3,000 ± 500 km) and high solar activity. [ISIS-1 measurements, individual satellite passes during 1 week][52]. Left, May 1969 [$\langle F \rangle = 154$], 16 h, LT; right, July 1969 [$\langle F \rangle = 145$], 22–24 h LT

One might introduce a formula of this kind in a future IRI. It must, however, be combined with a direct profile expression for T_e for those altitudes where the above relation does not apply.

Conditions at very great altitudes are different since the heat balance of the plasma allows other influences there than in the ionosphere (see Sect. 16). Figure 141 shows results obtained by satellite ISIS-2 around 3,000 km[52]. There is still some relation existing between T_e and N_e but it might not be clear enough for modelling. It must be borne in mind that for the plasmasphere the effect of magnetic disturbances is so large that recovery times of many days are needed after the "pouring-out" of plasma occurring in perturbed conditions. Quite different approaches are needed for modelling the plasmasphere than may be used for the ionosphere itself. A recent global empirical model of electron temperature (with LEGENDRE polynomials in dip-latitude) uses 2 different sets of coefficients for both regions[53]. As for the *ion temperature* IRI[51] describes, first of all, the temperature gradient dT_i/dz by a sum of EPSTEIN step functions Eps_0. Integration gives

$$T_i = T_n(z_s) + M_0(z - z_s) + \sum_{j=1}^{m} (M_j - M_{j-1}) \cdot G_j\{\mathrm{Eps}_1(z; G_j, X_j) - \mathrm{Eps}_{-1}(z_s; G_j, X_j)\} \tag{14.11}$$

with $\mathrm{Eps}_{-1}(z; G, X) \equiv \log_e \dfrac{1+\exp((z-X)/G)}{1+\exp((z_0-X)/G)}$.

This model is adjusted to the marginal conditions that T_i starts with the neutral temperature T_n at $z_s = 200$ km, and that T_i does not exceed T_e at great heights. The G_j values which essentially define the slope in the different height sections are dependent on geomagnetic latitude[51].

For electron and ion temperatures the *diurnal variation* occurs mainly around sunrise and sunset, the conditions remaining fairly stable at other hours. A smooth transition between day and night conditions is achieved in IRI using [see Eq. (14.10)] EPSTEIN step functions Eps_0, with a 1 h transition time[51].

Figure 142 showing an ion temperature map at high altitude was obtained (similarly to Figs. 140) with data of the japanese ISS-b satellite[35b]. The general aspect of this map is not too different from that of the corresponding electron temperature map displaid on Figs. 140b, d but with considerably lower absolute values, particular in the low temperature zone above the lower latitudes.

c) *Ion composition*

ε) Basic data are still scarce for a well-founded description of the geographical variations in ion composition. IRI[55] actually uses two data sources: a compilation of rocket measurements by DANILOV and SEMENOV[56] and in-situ data from the AEROS satellites (Fig. 143) [*26*]. The relative percentages of the ions are given after the formalism of Eq. (14.11) (with four terms for O^+, but two only for O_2^+). NO^+ is determined by filling up to 100%. Around 300 km O^+ is largely predominant. Above that range molecular ions are neglected[57] (less than 1%) and filling up is made with light ions[58] (10% He^+, 90% H^+).

The coefficients are assumed[55], depending on the solar zenith angle χ[56]; the geographical variation is an unequivocal function of the Sun's elevation. This, certainly, is an oversimplification.

---

55 Rawer, K. (1981): [*44*], 1
56 Danilov, A.D., Semenov, V.K. (1978): J. Atmos. Terr. Phys. *40*, 1093
57 Leminov, M.G., Sitnov, Ju.S., Fatkulin, M.N. (1975): [*14*], 86
58 Michailov, Ju.M. (1976): [*15*], 124

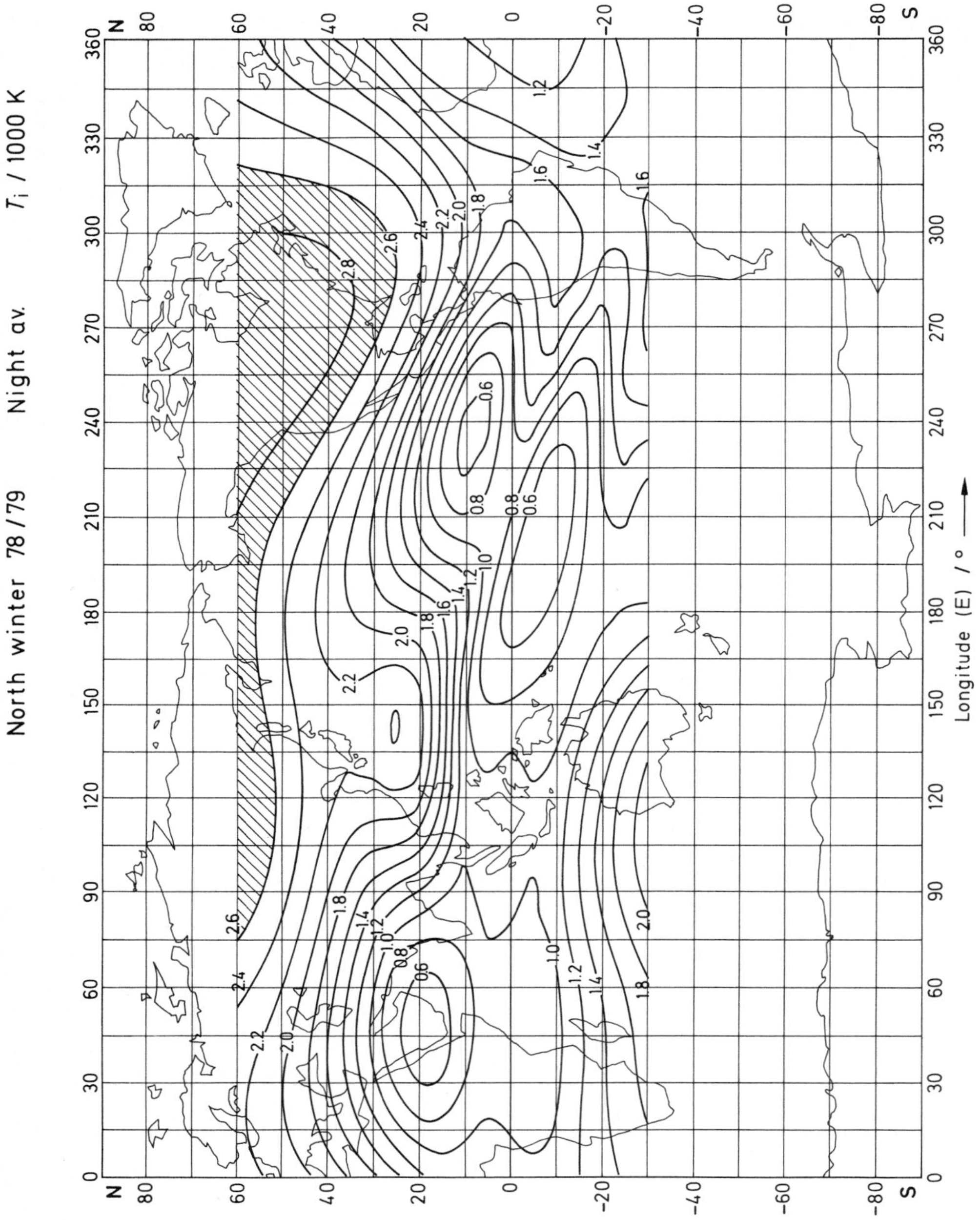

Fig. 142. World map of ion temperature (T_i/1,000 K); night average at 1,000 to 1,200 km, winter 1978/79, from measurements of a spherical retarding potential analyzer aboard the Japanese ISS-b satellite[35b]

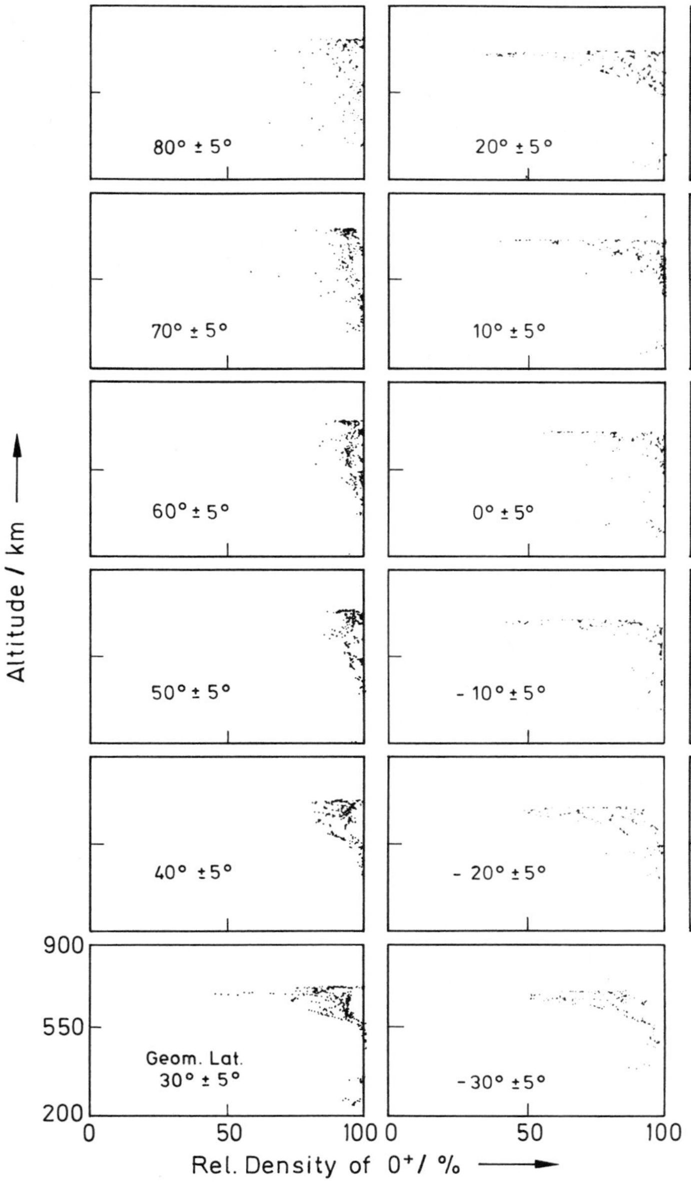

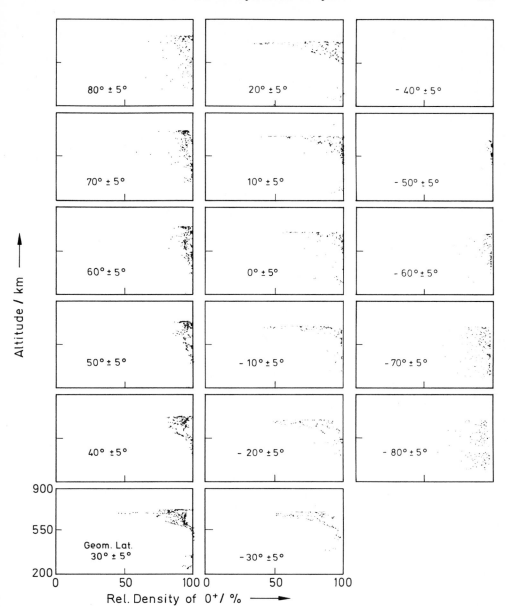

Fig. 143. Vertically displayed percentage (abscissa) of O^+ ions against total positive ions after RPA measurements aboard satellite AEROS-B [*26*], for 16 h 30 LT. Each diagram covers one 10° range of geomagnetic latitude

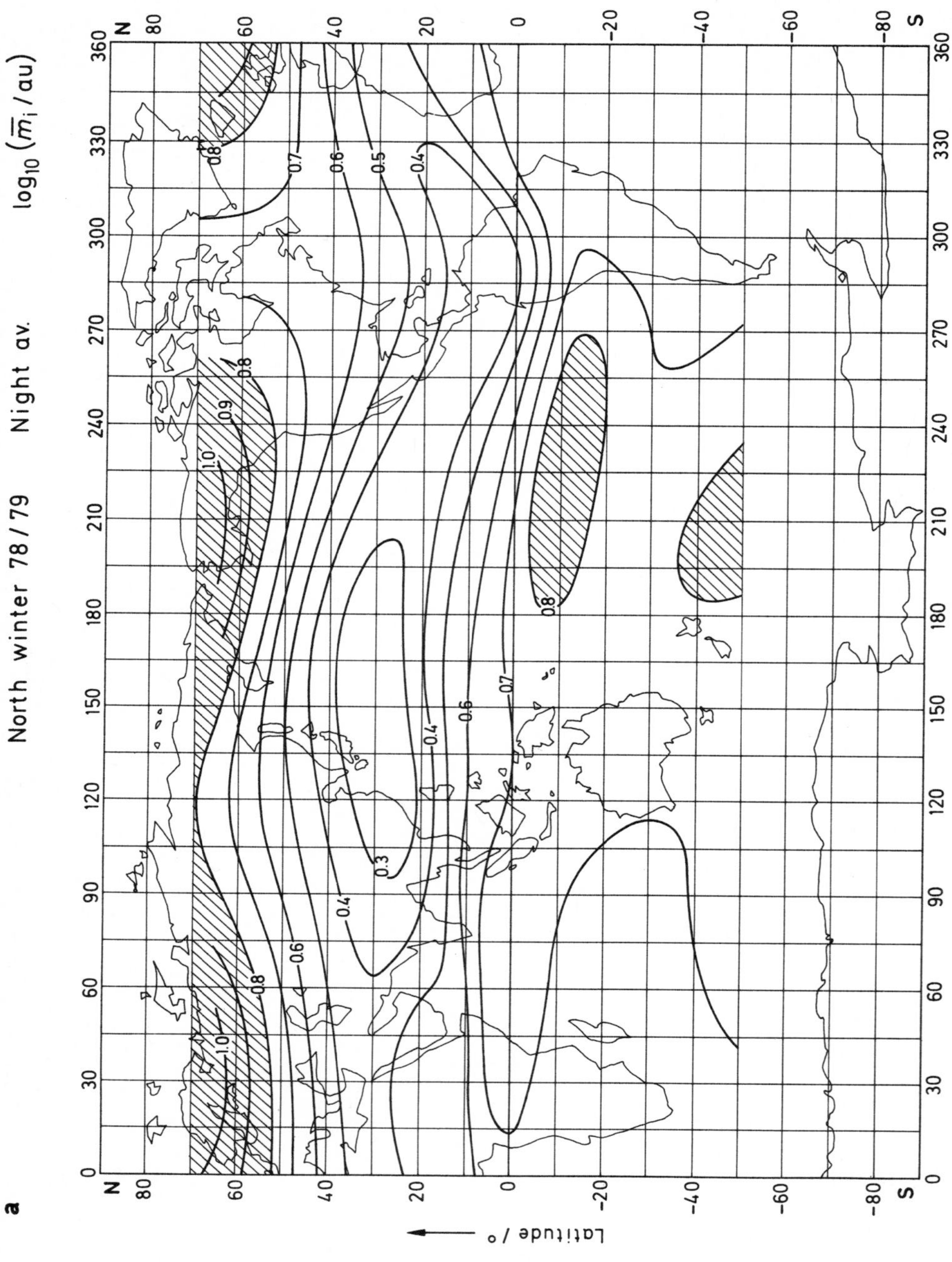
a
North winter 78/79
Night av.
$\log_{10}(\bar{m}_i/\mathrm{au})$
Latitude / °
Longitude (E) / °
N
S
80
60
40
20
0
-20
-40
-60
-80
0
30
60
90
120
150
180
210
240
270
300
330
360
0.3
0.4
0.5
0.6
0.7
0.8
0.9
1.0

Fig. 144a

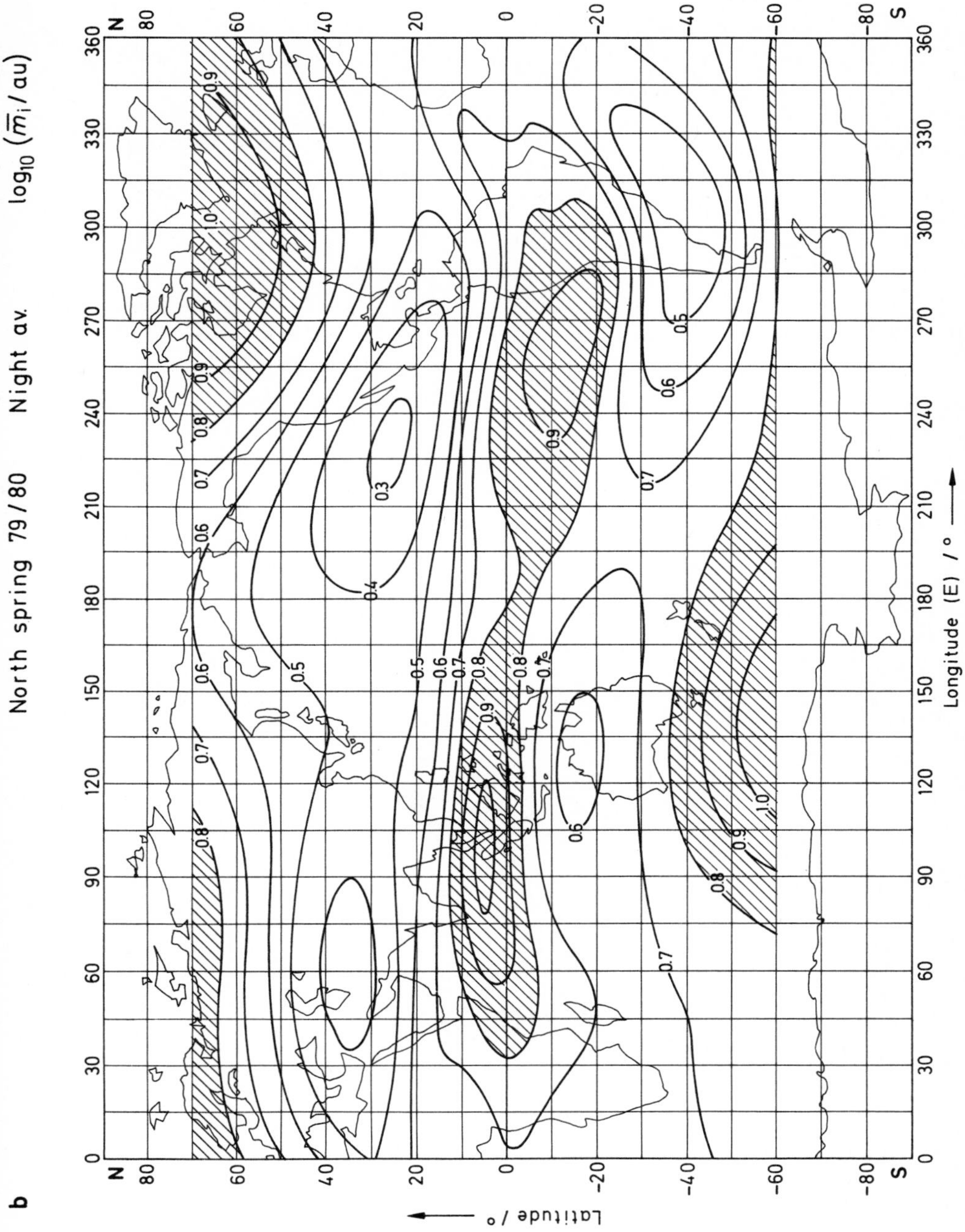

Fig. 144a, b. World maps of mean ion mass/au; night average at 1,000 to 1,200 km, from measurements of a spherical retarding potential analyzer aboard the Japanese ISS-b satellite[35b]. **(a)** Winter 1978/1979; **(b)** spring 1980

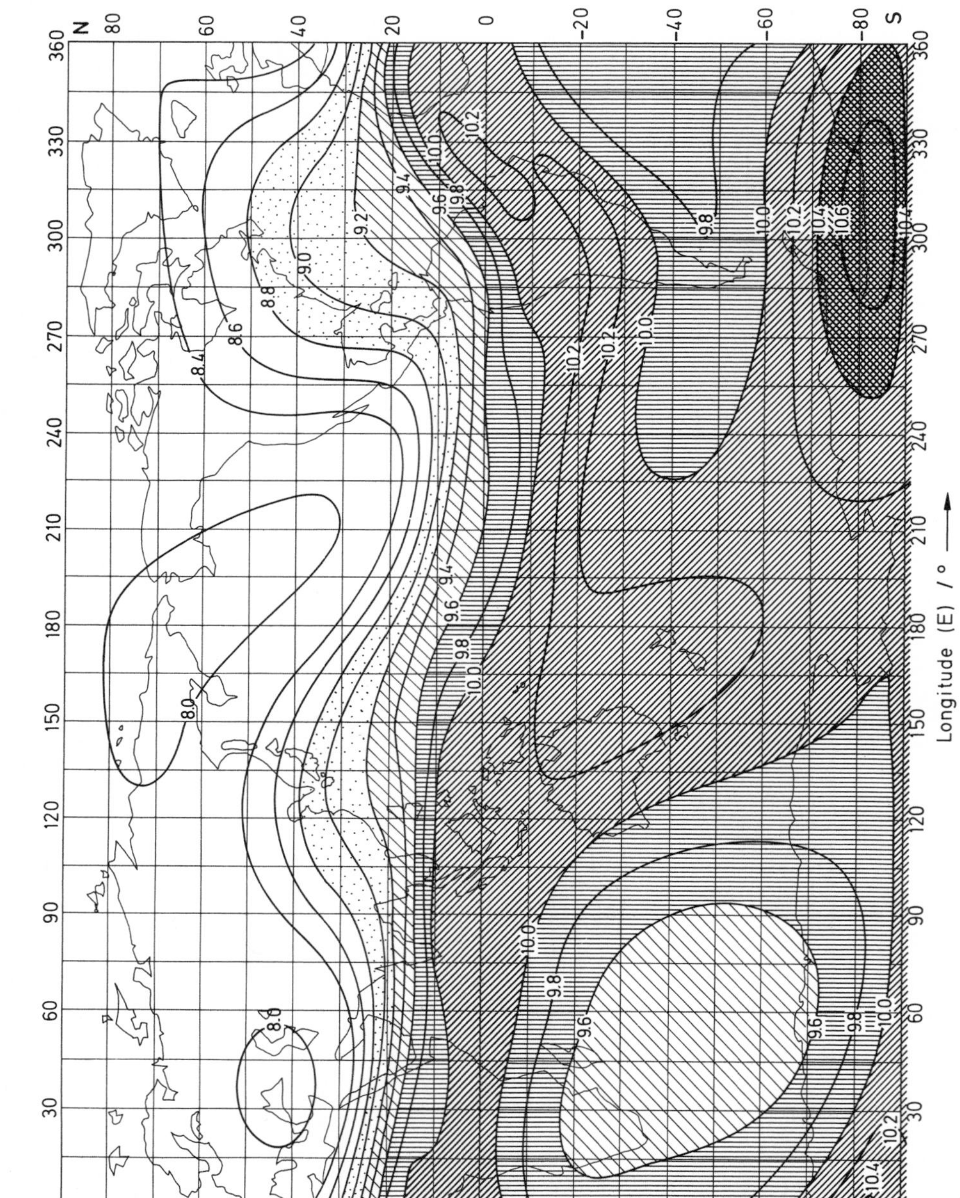

Fig. 145a

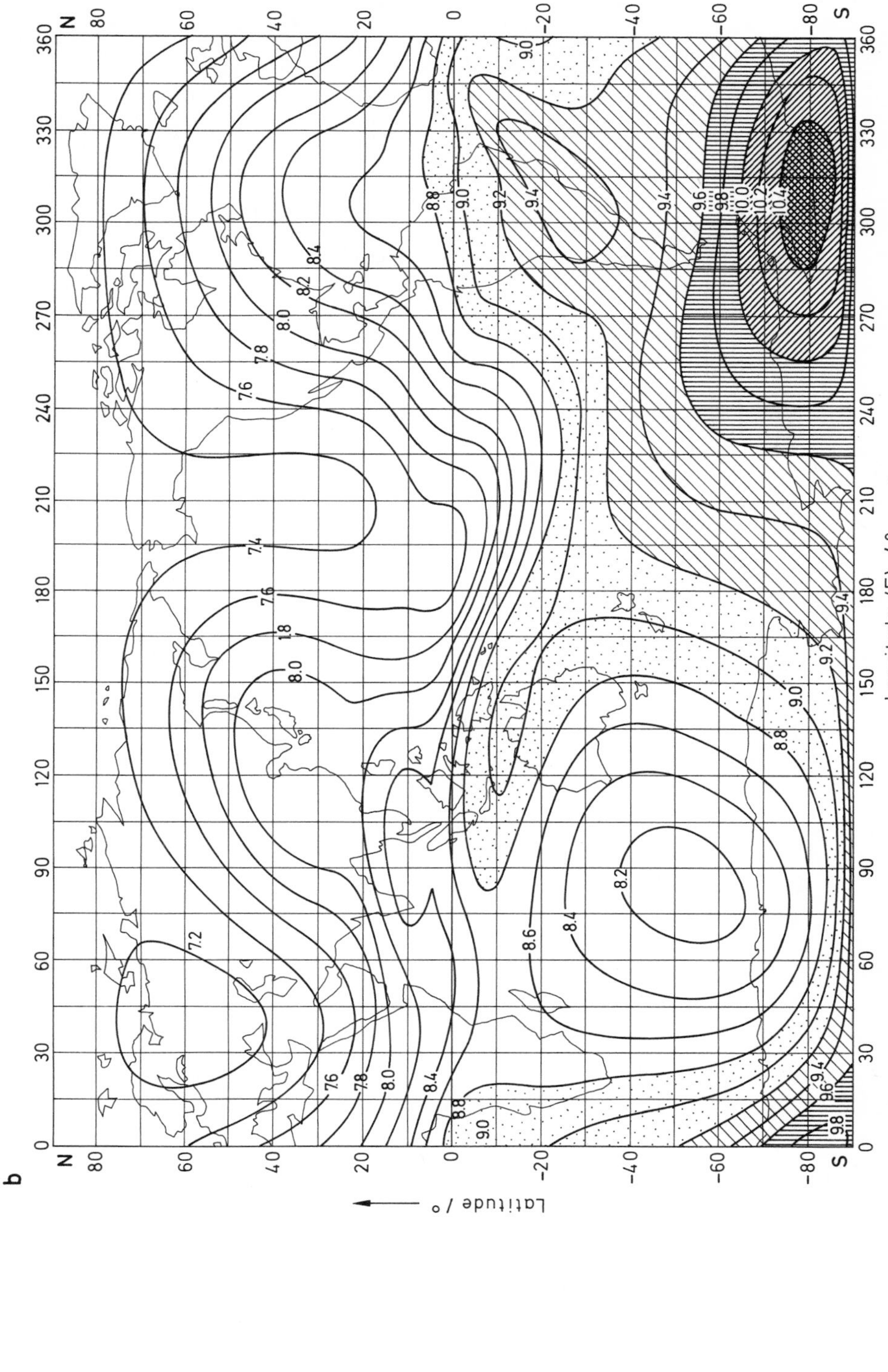

Fig. 145b (Legend see page 429)

c

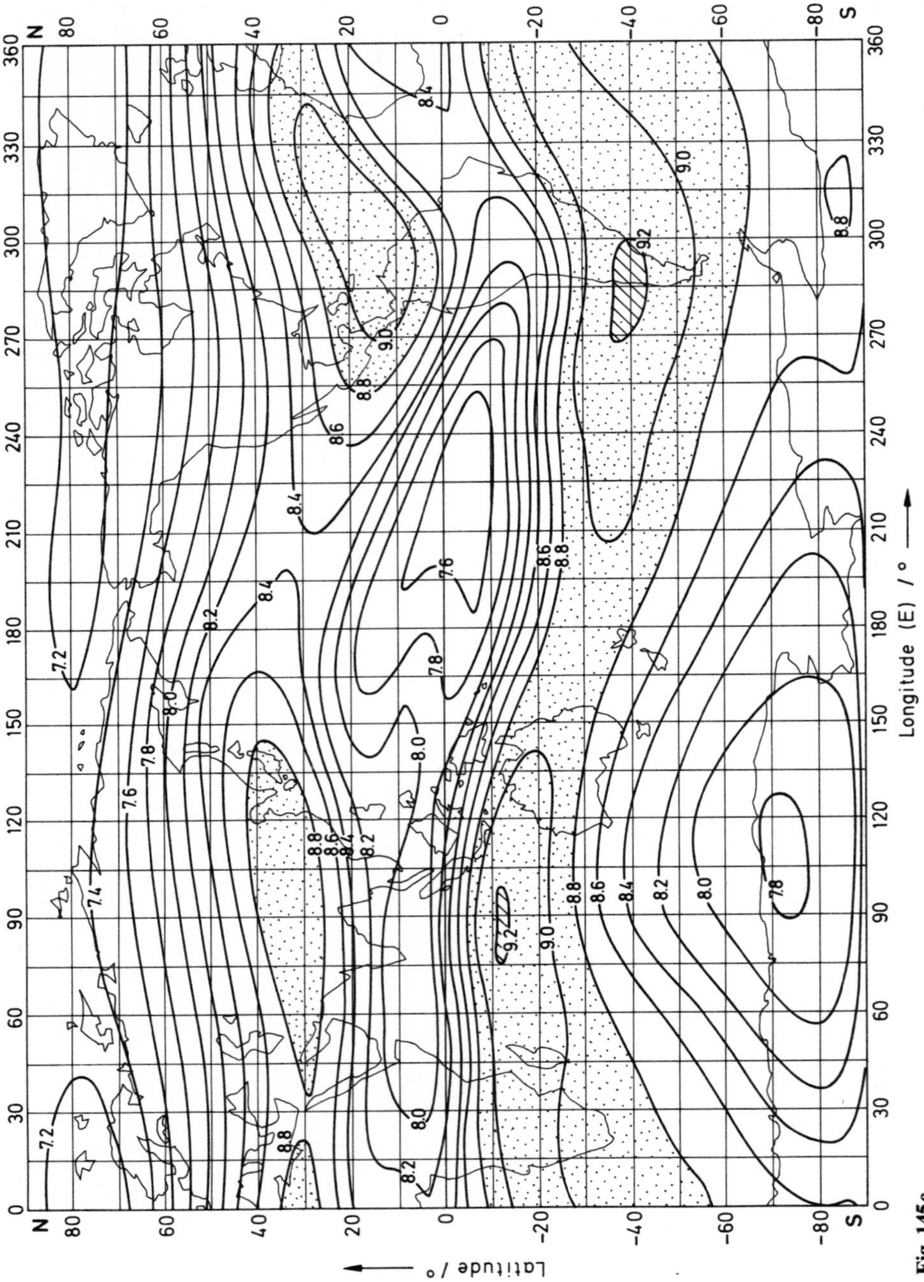
Longitude (E) / °
Latitude / °
N
S
0
30
60
90
120
150
180
210
240
270
300
330
360
80
60
40
20
0
-20
-40
-60
-80
7.2
7.4
7.6
7.8
8.0
8.2
8.4
8.6
8.8
9.0
9.2

Fig. 145c

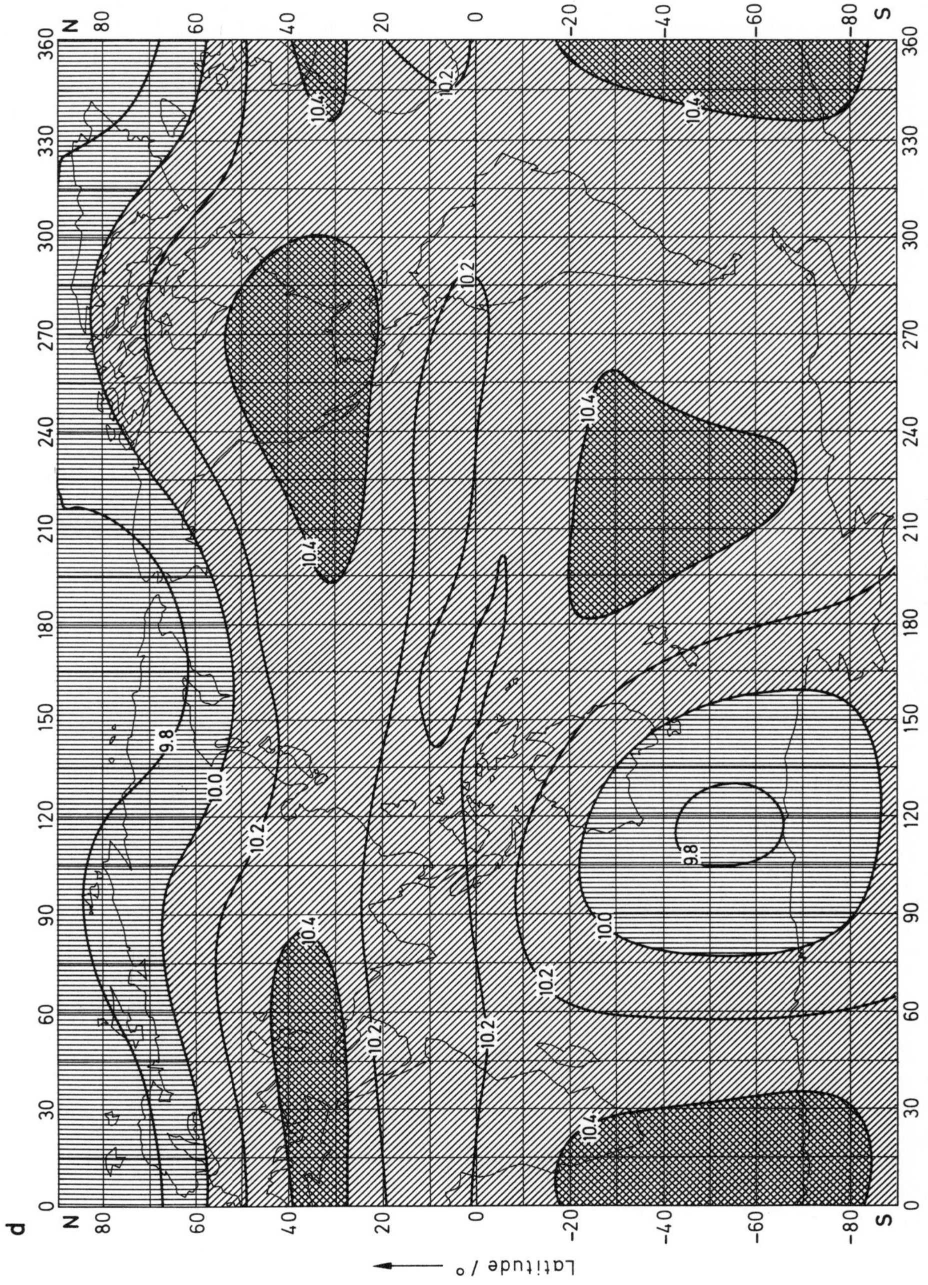

Fig. 145d (Legend see page 429)

e

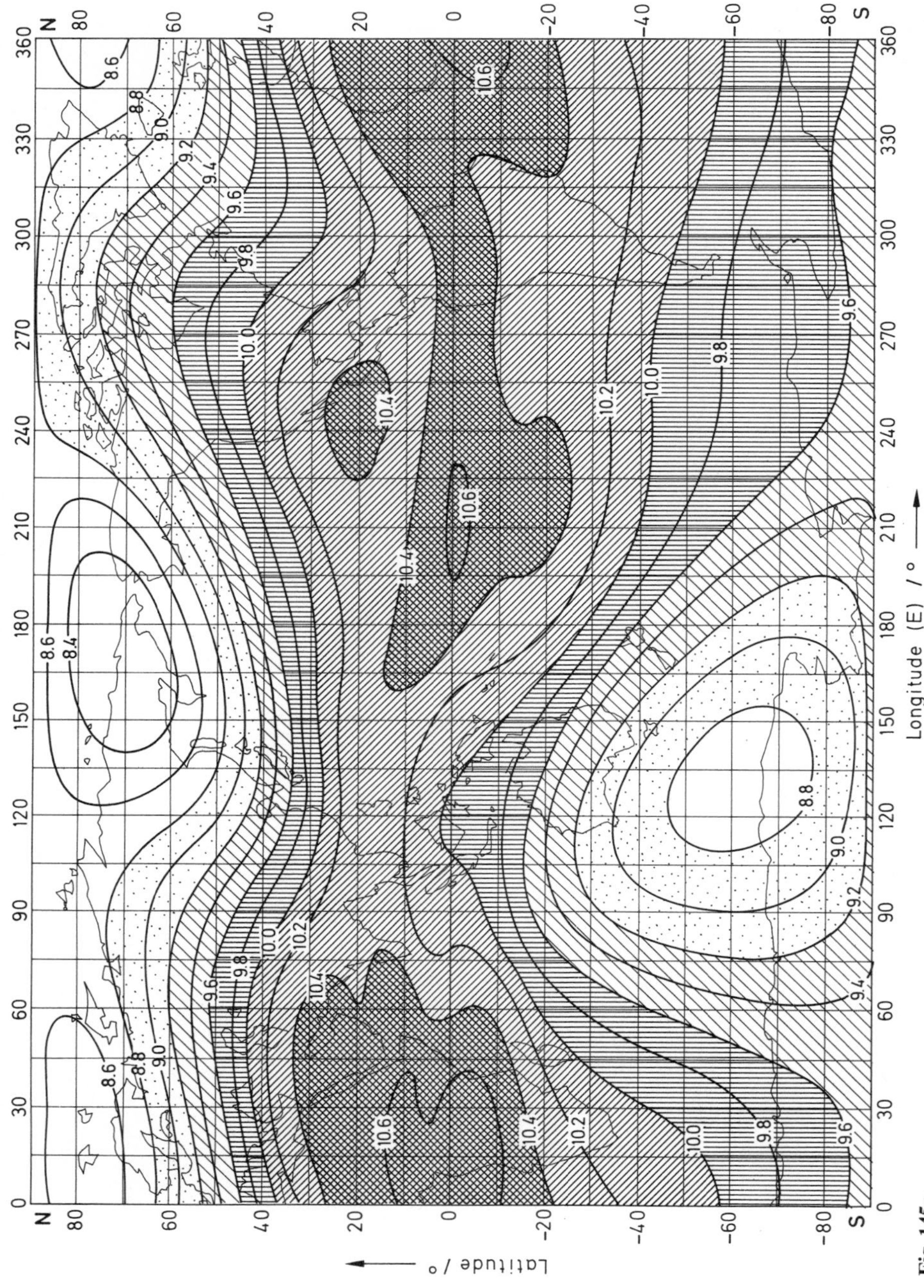

Fig. 145e

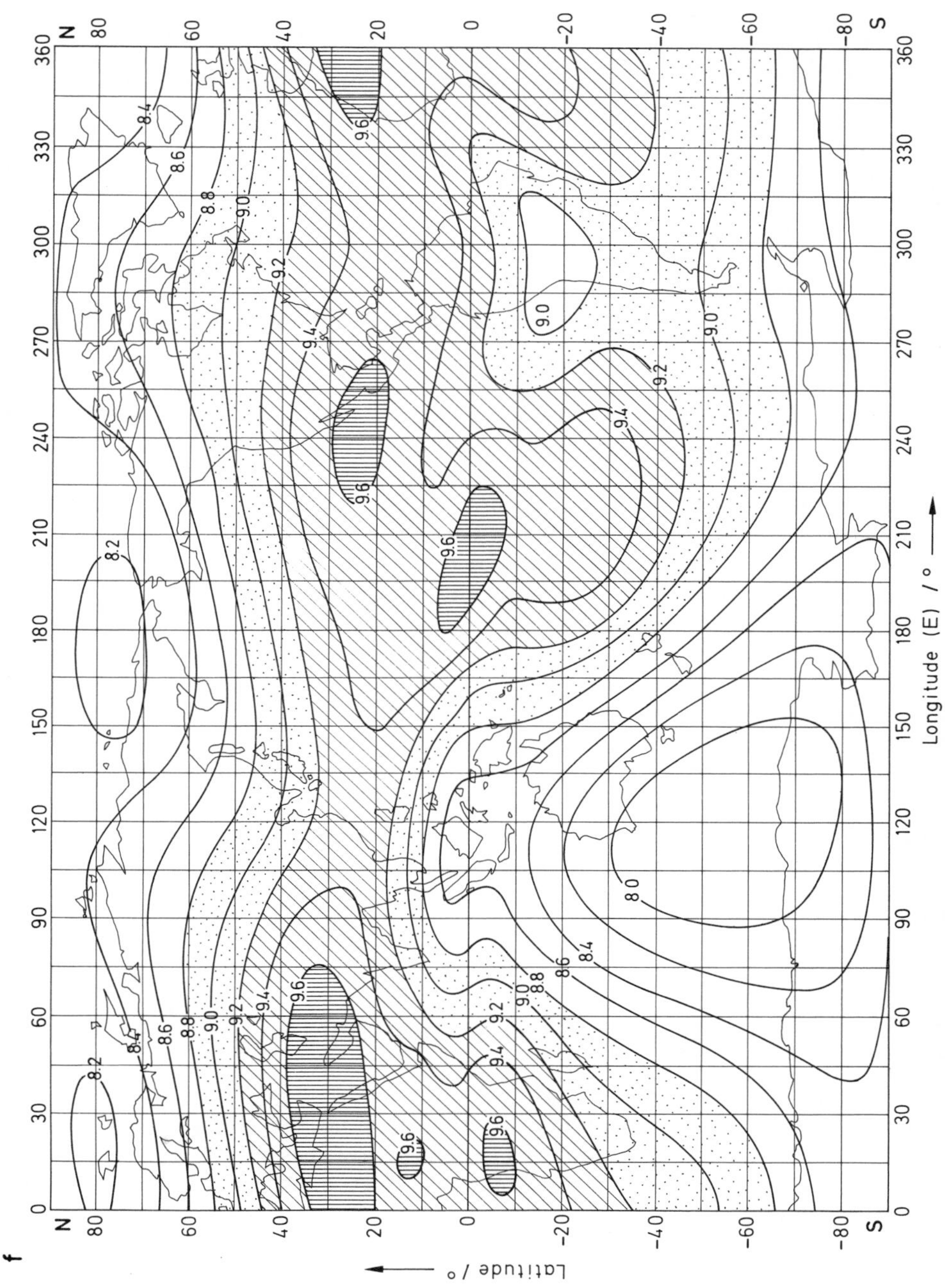

Fig. 145a–f. World maps of partial ion densities ($\log_{10} N(X^+)$) at 1,000 to 1,200 km, after mass spectrometer measurements aboard the Japanese satellite ISS-b[35c]. Fixed Local Time such that abscissa (longitude) is also Universal Time; ordinate: latitude. **(a–c)** Midmight (00 LT), **(d–f)** noon (12 LT). Each map for one ion: **(a, d)** O^+; **(b, e)** H^+; **(c, f)** He^+

Comparable satellite measurements became available during the last years only. Figure 143 [26] shows percentage values of the O^+ ion population (with the total number density of all ions as 100%). The mass plot presentation shows larger dispersion at night and at high latitudes. It also clearly exhibits a strong variation with height with almost 100% in the 250 to 500 km range. The decrease at greater heights is due to the appearance of lighter ions, in particular H^+ (see Figs. 145 below). The 'transition height' (where the light ions reach 50%) as a rule is well above the orbital height range; only at lower latitudes it may, occasionally, come down to 750 km, but normally it is much higher (cf. Fig. 117 in Sect. 13).

High altitude maps were recently established with the data of the japanese ISS-b satellite. The average ion mass is a useful relative measure since it is also independent from the absolute ion density. Figures 144a, b are maps based on data of a spherical retarding potential analyzer[35b]. Numerical values of $\bar{m}_i$ lie between 2 and 10 au. Some geomagnetical (dip-)influence seems to exist with lower values (i.e. more light ions) at both sides of the dip equator zone. The high latitude behaviour should be attributed to the particular conditions in the auroral zones where O^+ ions are lifted up (see Sect. 18), while the low latitude features should be seen in connection with the "equatorial fountain" phenomenon which also brings heavier ions to greater heights (see Sect. 15).

Specific density maps for the main ions are shown in Figs. 145a–f, also for high altitude[35c]. Since, unfortunately, absolute densities were evaluated, the diurnal variation is quite appreciable as can be seen by comparing the local midnight maps (Fig. 145a–c) with the following ones for noon. This is due to the important diurnal variation of ion density, compare Figs. 129 which are also for fixed local time. Ion density ratios (or the mean ion mass) would better characterize the composition itself and exhibit smaller variability as can be seen from Fig. 144. The relative similarity of corresponding maps for O^+ and H^+ may be taken as indicating this fact. The IRI system which gives relative densities is certainly more appropriate.

E. Aeronomical modelling

Aeronomical theory in essence aims at explaining the conditions in the upper atmosphere as resulting from cosmic influences. These are external sources of energy such that the upper atmosphere is, in fact, an *open system* which is overwhelmingly influenced from outside. This distinguishes aeronomical from, for example, classical kinetic gas theory.

The history of aeronomical theory was briefly described in vol. 49/6[1], as well as the most important phenomena which must be considered in this context.

Since it is the aim of this chapter to give an account of the rather complex situation of present-day aeronomical theories, a summary of these with ref-

[1] Rawer, K. (1982): This Encyclopedia, vol. 49/6, p. 1

erences would be useful. In Sects. 15–17 we restrict ourselves in general to the most important external energy source, namely solar extreme ultraviolet and soft X-ray irradiation[2].

The interaction of solar radiations with atmospheric gases is the fundamental phenomenon of aeronomy; it is discussed by THOMAS in vol. 49/6 of this Encyclopedia (his Chap. B, p. 16) and also by SCHMIDTKE in this volume. There are three fundamental parameters, viz:

i) The solar radiation intensities[2] in the different spectral ranges of aeronomical interest[3]; these are summarized in SCHMIDTKE's contribution in this volume.

ii) The absorption cross sections of the different photochemical reactions, in particular dissociation and ionization; these were described in much detail by THOMAS in vol. 49/6 of this Encyclopedia (his Sect. 4, p. 16).

iii) The neutral atmosphere density, composition and temperature profiles as described in Chap. C.

With these inputs the *photodissociation/ionization rate* can be established as described in Sects. 6 (p. 31) and 7 (p. 34) of THOMAS' contribution in vol. 49/6 of this Encyclopedia. This rate constitutes the primary *production term* in the balance equation of the constituent which is created by the photoreaction.

The appearance of such a new constituent is usually followed by *chemical ractions* [*13*, *47*].

These are also described in vol. 49/6 by THOMAS: at first in a more general way in his Sect. 12; then in more detail in his Chap. F for the neutrals and in Chap. G for the ion chemistry which is of major importance for the balance of electrons and of the different ionic components. Since the conditions are rather different in different height ranges he discusses separately the D-region (his Subchap. II, p. 100, with Sects. 27 on positive, 28 on negative ions), the E-region (Subchap. III, p. 110) and the F-region (Subchap. IV, p. 116)[4].

Chemical reactions provoke gain or loss of different species. For those as are directly produced by photoreactions (e.g. atomic oxygen, O) their main importance is in determining the *loss term* in the balance equation, reaction-provoked gain remaining of secondary or negligible influence. However, other species which are of importance are only (or mainly) produced by chemical reactions (e.g. atomic nitrogen, N).

Transport phenomena must also be considered in the balance of the different species: diffusion, neutral wind induced motions [*48*] and motions provoked by electric fields are to be considered[5]; see Subsect. 15b below.

---

[2] Nikol'skij, G.M. (1982): This Encyclopedia, vol. 49/6, p. 309; Schmidtke, G.: contribution in this volume

[3] Thomas, L. (1982): This Encyclopedia, vol. 49/6, his Sect. 5, p. 27 gives a detailed discussion. Some data relevant to oxygen dissociation are found in Yonezawa, T., ibidem, Sect. 16, p. 177

[4] Information about some minor neutral species of importance in this context is given by the same author in his Sect. 23 (see in particular Table 11, p. 93). Also excited species are discussed in his Chap. H, p. 120

[5] Stubbe, P. (1982): This Encyclopedia, vol. 49/6, his Sect. 9 takes account of the electric field effect in the equation of motion [his Eqs. (9.3...9)]. The equations of motion in a plasma are specified in: Rawer, K., Suchy, K. (1967): This Encyclopedia, vol. 49/2, p. 270ff. and p. 517/518

In the following we shall at first (in Sect. 15) summarize theories aimed at solving the problem by reaction and transport kinematics, restricted to local considerations (i.e. vertical profiles of the different processes). They are essentially one-dimensional in character. With such considerations the temperature is a parametrized quantity (height dependent, of course). Then in Sect. 16 we consider the heat balance and its effects, finding out that regional (or even hemispherical) considerations are needed for a direct treatment of temperature-dependent phenomena. Such advanced theories will be presented in Sect. 17. Finally, the particular phenomena due to particle influx at high latitudes, and during perturbations, will be summarized in Sect. 18.

15. Kinematic models. Under this heading we shall discuss models which consider chemistry and transport only, heating phenomena being parametrized (i.e. taken into account by a given temperature profile only). The basic equation is always that of continuity which, for each species j, holds quite generally at any level[1]:

$$\frac{\partial n_j}{\partial t} = P_j - L_j - \nabla \cdot (n_j \boldsymbol{v}_j) \tag{15.1}$$

with $\nabla \equiv \partial/\partial \boldsymbol{r}$, $\boldsymbol{r}$ being the vector coordinate.

P_j is total production [$\mathrm{m^{-3}\,s^{-1}}$], L_j total loss, the divergence term taking account of transport phenomena. As stated above, the loss term L_j at least is due to chemical reactions. In case production P_j is mainly due to a photoreaction which is described by the photoreaction rate[2]

$$Q_j(z) = \sum_i \sum_\lambda n_i(z) \cdot \sigma_j^i(\lambda) \cdot \varphi(\lambda) \tag{15.2}$$

where $\sigma_j^i(\lambda)$ is the effective cross section of the photoreaction producing species j out of i, at wavelength λ; $n_i(z)$ is the density of species i at altitude z; and $\varphi(\lambda)$ is the incoming photon flux. It increases with height due to absorption of the stream of photons at altitudes above level z:

$$\varphi(\lambda) = \varphi_0(\lambda) \cdot \exp\left[-\int_z^\infty \mathrm{d}z' \cdot \sec\chi \cdot \sum_k n_k(z') \cdot \sigma_\mathrm{T}^k(\lambda)\right] \tag{15.3}$$

φ_0 is the extraterrestrial radiation intensity, χ the solar zenith angle and σ_T^k the total absorption cross section of species k (of density n_k).

In SCHMIDTKE's contribution in this volume primary electron production rate profiles are shown (in his Fig. 29, p. 39) for six typical (observed) extreme ultraviolet emission conditions (all from 1973, a period of low solar activity). Primary production profiles for different wavelength ranges can also be found in Fig. 20, p. 36, and, for different ions, in Fig. 21, p. 37, of THOMAS' contribution in vol. 49/6 of this Encyclopedia[3]. For the lowest ionosphere the effects of the

[1] See Thomas, L. (1982): This Encyclopedia, vol. 49/6, Eq. (25.6), p. 99

[2] Ibidem. Sect. 7, Eqs. (7.1, 2), p. 34, 35, have generalized for more than one species

[3] Ibidem. Sect. 7. This author uses Heroux's solar radiation data from 1972

different sources of ionization are specified in Fig. 22, p. 38, and photodissociation profiles for oxygen and water vapour are also found in vol. 49/6 of this Encyclopedia[4].

More controversial than the production term are the two other terms on the right-hand side of Eq. (15.1). Their influence will be considered in the following.

a) Chemistry-oriented models

α) Aeronomical models of oxygen *dissociation* and ozone formation were discussed in vol. 49/6[5]. In the relevant height range (between 80 and 120 km) transport is of minor importance, (vertical) diffusion being the only transport phenomenon which might be taken into account; neglecting transport altogether one arrives at straightforward reaction equations (linking O and O_2 densities) which have proven to be quite satisfying[6]. Apparently more controversial are the neutral composition models used[7].

β) *Earlier theories* of the ionized layers of the *ionosphere* were briefly indicated in vol. 49/6 of this Encyclopedia[8]: The most important mechanism of charge removal is dissociative recombination, which is discussed there in detail: Dissociative recombination leads at lower height to a law similar to that of direct recombination (though with a much greater rate coefficient), and at greater heights to a law of attachment character for the loss term L in Eq. (15.1)[9]. Dissociative recombination can only occur with molecular ions of such species as might be dissociated when recombining with an electron:

$$XY^+ + e^- \rightarrow X' + Y', \tag{15.4}$$

meaning in the real atmosphere, for example[10]:

$$\begin{aligned} NO^+ + e^- &\rightarrow N' + O \\ O_2^+ + e^- &\rightarrow O' + O'' \\ N_2^+ + e^- &\rightarrow N' + N''. \end{aligned} \qquad (15.5\text{a–c})$$

In the lower ionosphere the primary ions are overwhelmingly molecular so that the recombination process is immediate[11]. Its probability is thus proportional to the product of ion and electron densities, $N(XY^+) \cdot N_e \approx N_e^2$. This is a classical recombination law for the loss term of Eq. (15.1):

$$L = \alpha_D \cdot N_e^2. \tag{15.5}$$

[4] Ibidem, Sect. 6, Figs. 18, 19 (pp. 32, 33)

[5] Yonezawa, T. (1982): This Encyclopedia, vol. 49/6, Chap. III, p. 177

[6] Ibidem, Eq. (19.7), p. 186 describes the combined effect of molecular (in the upper part) and eddy diffusion (in the lower part of the height range); the continuity equations without a diffusion term are given as Eqs. (19.11, 12), p. 187

[7] Ibidem, Fig. 11, p. 191

[8] Rawer, K. (1982): This Encyclopedia, vol. 49/6, p. 1

[9] Thomas, L. (1982): This Encyclopedia, vol. 49/6, Sect. 12γ, p. 46, also Sect. 25, p. 96ff. His Table 8, p. 49 gives recombination coefficients for hydrated cluster ions

[10] apostrophes indicate possible excitation; for details see ibidem Sect. 12, pp. 47, 48

[11] For recombination coefficients see ibidem, Table 8, p. 49

The situation is different where primary ions are mainly of atomic type, i.e. at altitudes above about 150 km (see Fig. 29 of SCHMIDTKE's contribution in this volume, p. 39). In that case we have to allow a reaction chain of two members:

$$\begin{aligned} &Z^+ + XY \to Z + XY^+, &&\text{(charge transfer)} \\ &XY^+ + e^- \to X' + Y' &&\text{(dissoc. recombination).} \end{aligned} \tag{15.6}$$

Where the neutral molecular density is high, the first reaction is quick, so that the probability for recombination is determined by the second one, and Eq. (15.5) still holds. At F-region heights, however, the charge transfer reaction is the slower one so that it determines the kinetic law, which is now of attachment type[12]:

$$L = \{k_{18} \cdot n(N_2) + k_{19} \cdot n(O_2)\} \cdot N_e = \beta \cdot N_e. \tag{15.7}$$

Therefore, in the upper ionosphere not the total density but that of neutral molecules only is decisive for the efficiency of recombination, in spite of the fact that at these heights they are minor constituents. Thus the effective attachment coefficient β varies with composition changes, e.g. under magnetic disturbance conditions the molecular density increases and so does β.

Computer-aided theoretical work has now largely superseded such simplifying considerations. Instead, a full system of reactions is established from which balance equations can be derived (see Sect. 15δ below for an example), one for each variable constituent. With increasing computer efficiency more and more complex systems are being considered. It must, however, been kept in mind that their reliability depends heavily on the accuracy of the different reaction rates and their dependence on temperature. Though since 1970 [*13*][13] much progress has been made in this field [*64*, *65*], some uncertainty remains.

γ) Chemistry is quite involved in the *D-region* for three reasons:

i) The solar EUV-spectrum is essentially absorbed at higher altitudes so that only longer wavelengths are available for ionization, in particular the solar hydrogen Lyman α and β emissions (and, under flare conditions, X-rays). Since their quantum energy is too weak[14] to ionize atmospheric gases other than NO the chemistry [*65*] of minor constituents is decisive for the ionospheric D-region.

ii) Collisions are frequent so that clustering of ions (mainly with water molecules) is frequent in the lower D-region.

iii) For the same reason attachment of electrons to "electronegative" molecules occurs producing negative ions (which may also form clusters)[15].

---

[12] Ibidem, Sect. 12 and Eq. (25.3), p. 98; N_2 and O_2 are the most prominent molecules but under certain conditions NO may also appear

[13] McDaniel, C.W., Cermak, V., Dalgarno, A., Ferguson, E.E. (1970): Ion-molecule reactions. New York: Wiley-Interscience; Whitten, R.C., Popoff, I.G. (1971): Fundamentals of aeronomy. New York: Wiley; Ferguson, E.E. (1974): Rev. Geophys. Space Phys. *12*, 703; (1975): Ann. Rev. Phys. Chem. *26*, 17

[14] Thomas, L. (1982): This Encyclopedia, vol. 49/6, provides quantum energies in his Tables 2 (p. 23), 3 (p. 24) and 4 (p. 26), and reaction characteristics in his Chaps. C, D, p. 39ff., in particular Tables: 9 (p. 54), 10 (p. 88), 11 (p. 93), 12 (p. 109), 13 (p. 111), 14 (p. 114)

[15] Arnold, F., Krankowsky, D. (1977): [*18*], 93

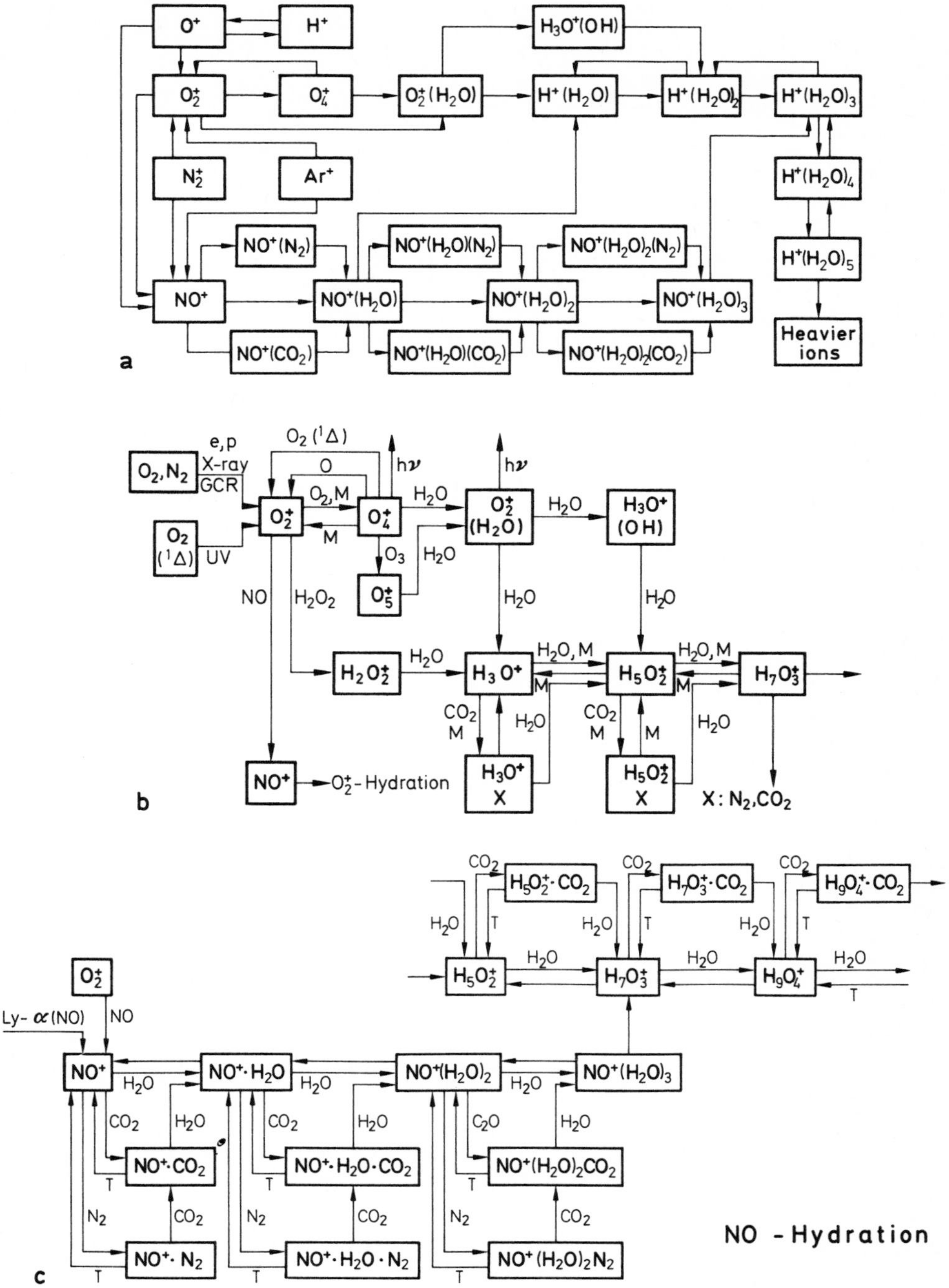

Fig. 146a–c. Chemical reaction schemes for positive ions in the D-region. (**a**) After DYMEK[16]: production of water clusters; (**b, c**) after KOPP[21]: hydration; (**b**) via O_2^+; (**c**) via NO^+ [see also Fig. 42, p. 102, and Fig. 44, p. 105, in vol. 49/6 (THOMAS)]

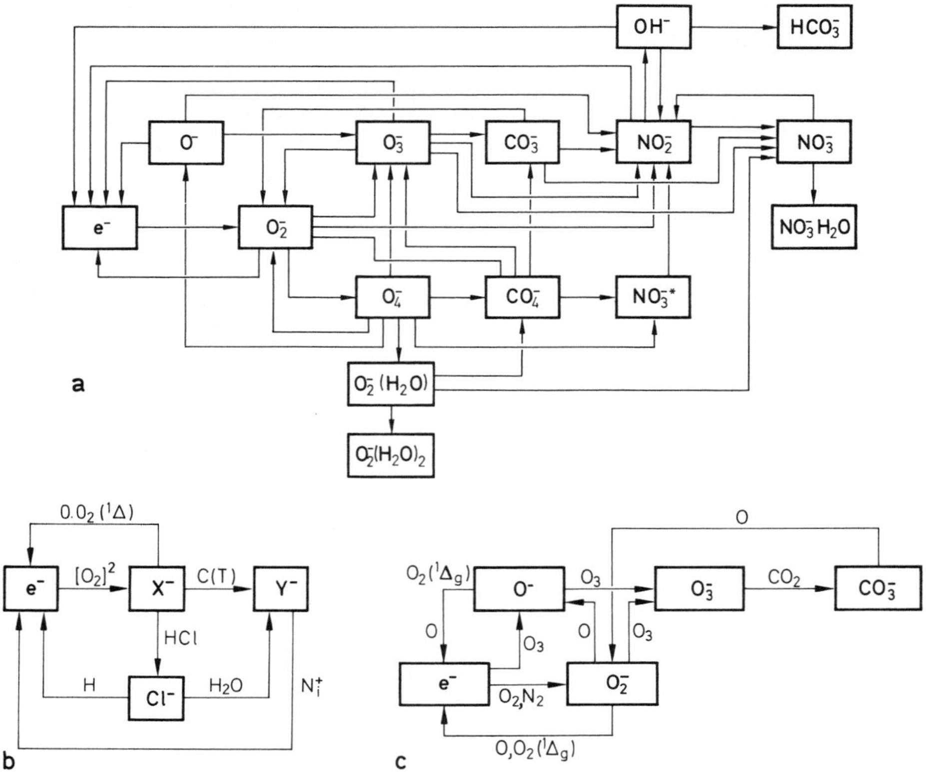

Fig. 147a–c. Chemical reaction schemes for negative ions in the D-region. (**a**) After Dymek[16]; (**b**) after Kopp[21]; (**c**) after Sears et al.[26]. [see also Fig. 46 in vol. 49/6 (Thomas), p. 108]

In vol. 49/6 of this Encyclopedia Thomas, apart from the classical reactions of Eqs. (15.4), considers not less than 12 charged reactions with nitrogen (his Table 11, p. 93), a further 24 negative ion reactions (his Table 12, p. 109) and, finally, the cluster ion chemistry as described in his Sects. 27 and 28. For modelling purposes one might aim at reducing this number to the most important reactions only, but this decision is not so easy because in-situ measurements of the different species are difficult and not yet uniform (see our Sect. 3). Thomas[14] gives three reaction schemes[16] describing the formation of positive water cluster ions from O_2^+ (his Fig. 44, p. 105) and one for negative ions (his Fig. 46, p. 108).

Trying to simplify the system and to avoid "hypothetical" reaction steps Mitra and Rowe[17] introduced a six-ion model with O_2^+, NO^+, O_4^+ as positive molecular ions, one kind of positive clusters, namely $H^+(H_2O)_n$, and two negative species, namely O_2^- and X^- (all others)[18]. It was found that the (not too well known) density of NO molecules is a critical input; it is important for the N_e ledge at about 80 km. Hydrated protons have a recombination coefficient[9], which is estimated to be ten times that of NO^+ or O_2^+. Hydrated NO^+ is lacking in this model. Also the negative ion chemistry is oversimplified.

---

[16] Dymek, M. (1980): Artificial satellites (Warszawa-Lodz: Polska Akademia Nauk) *15*, 29 provides a listing of 154 reactions including rate coefficients and their dependence on temperature

[17] Mitra, A.P., Rowe, J.N. (1972): J. Atmos. Terr. Phys. *34*, 795

[18] O_2^- is considered specifically, and all other negative ions are lumped together as X^-; O_4^+ is included because of the reaction $O_4^+ + O \rightarrow O_2^+ + O_3$

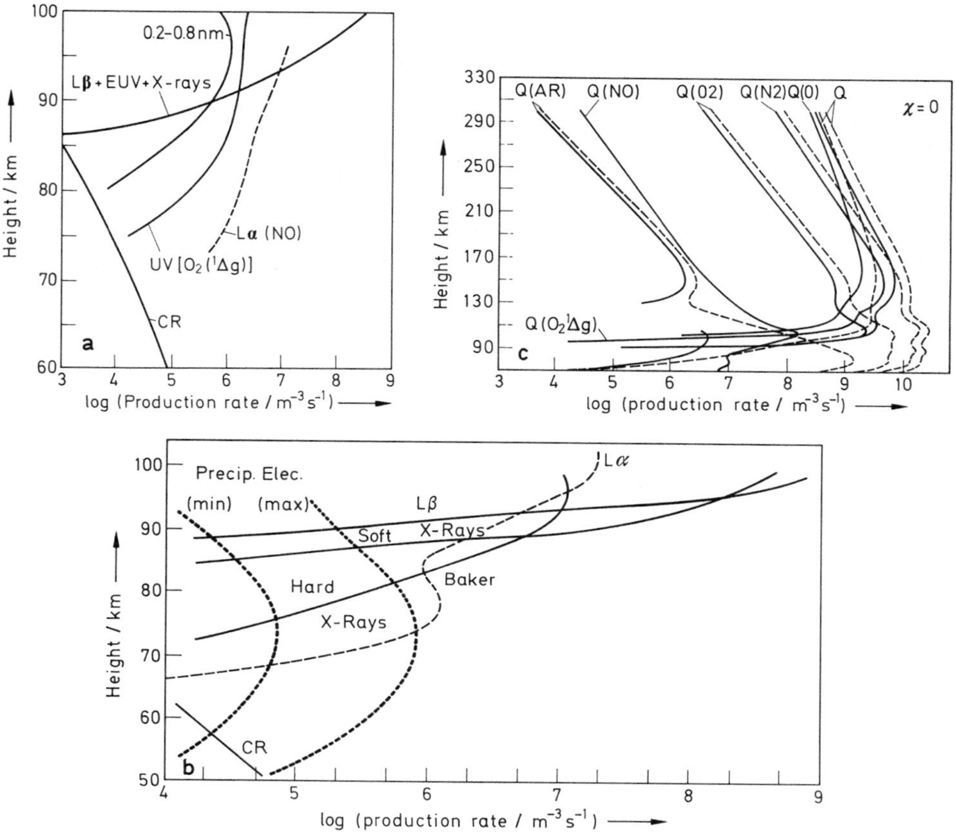

Fig. 148a–c. Ionization sources in the D-region (rates for undisturbed conditions, near noon). (**a**) After THOMAS (1971) [*61*]; (**b**) after SEARS et al. (1982)[26]. [CR, cosmic rays; dotted line, precipitating electrons]; BAKER's Lα estimate is nearer to that of THOMAS above (broken line); that of SEARS himself would be higher [by a factor of ten]; (**c**) after DYMEK (1980)[16] [full curves for quiet conditions, broken curves during strong solar flare[20]]

More evidence about positive and negative cluster ions has since been gathered experimentally. Figure 146a[16] gives a more recent scheme for reactions of any kind leading to positive water clusters (it is, essentially, a combination of THOMAS' Figs. 42 and 44, p. 102ff. in vol. 49/6). The negative reaction scheme[19] of Fig. 147a[16] differs only slightly from THOMAS' Fig. 46.

In order to investigate the effects of variable solar irradiation two groups of constituents were distinguished: one comprises constituents taking part in photo-ionization and follow-up reactions as well, viz. N_2, O_2, O, NO, Ar, He and the metastable species O_2 ($^1\Delta_g$); the other group covers chemically active minor constituents which play an important role in the chemical follow-up reactions only, viz. H_2O, CO_2, O_3, NO_2, OH, HO_2, N, H. The density profiles of the major constituents N_2, O_2, O and also of Ar, He, O_3 were parametrized and taken from the CIRA 1972

[19] First proposed by the "Pennstate team": Rowe, J.N., Mitra, A.P., Ferraro, A.J., Lee, H.S. (1974): J. Atmos. Terr. Phys. *36*, 755

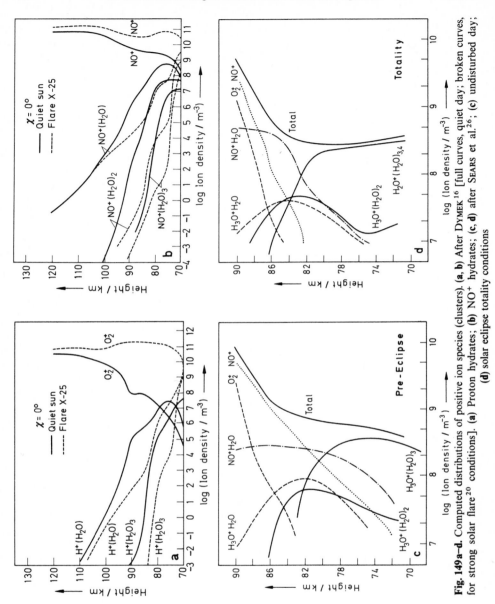

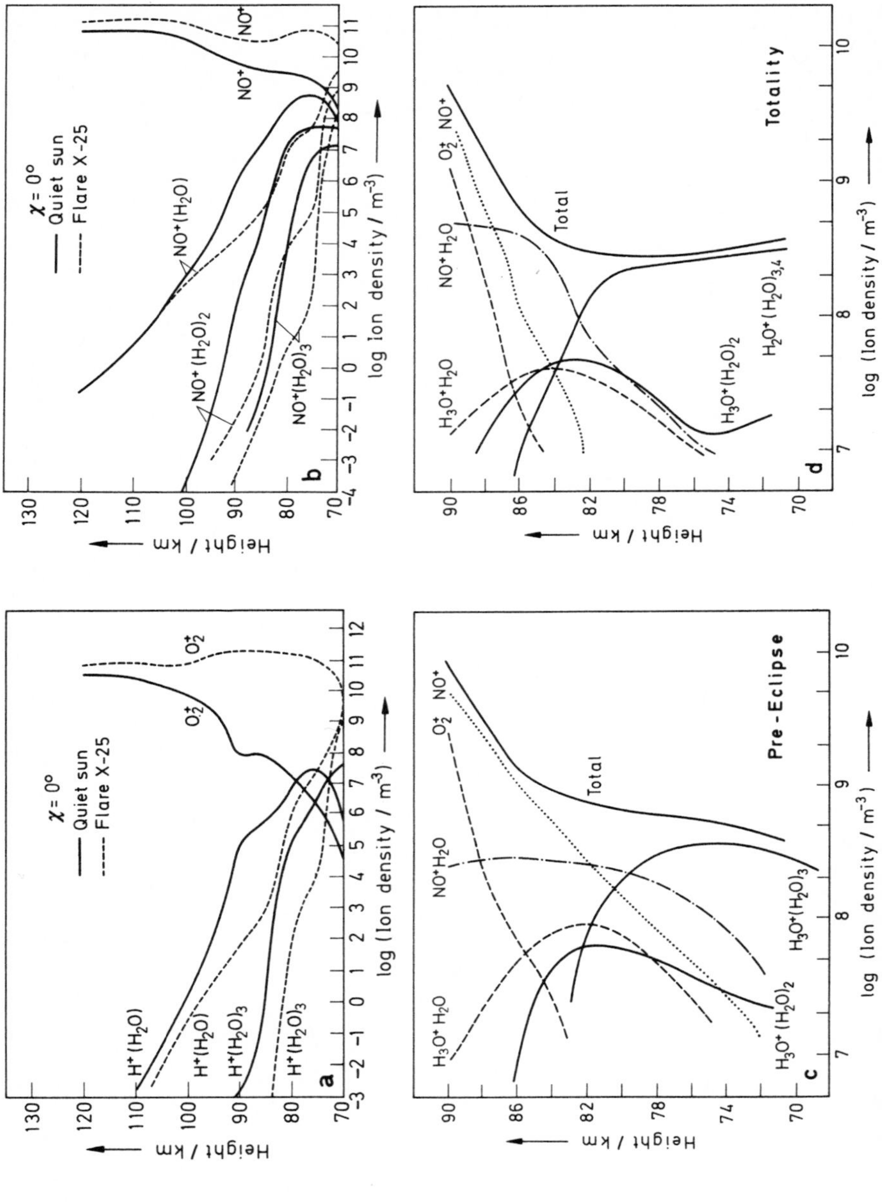

Fig. 149a–d. Computed distributions of positive ion species (clusters). **(a, b)** After DYMEK[16] [full curves, quiet day; broken curves, for strong solar flare[20] conditions]. **(a)** Proton hydrates; **(b)** NO^+ hydrates; **(c, d)** after SEARS et al.[26]; **(c)** undisturbed day; **(d)** solar eclipse totality conditions

[*12*] model. An equation system for 38 ions was resolved in the 70–300 km height range[16]. Figures 148–150 show some results of such computations, namely production rates (Fig. 148c) and ion density profiles (Figs. 149a, b, 150) under quiet day and under solar flare[20] conditions.

---

[20] The very strong flare X-25 (12 March 1969) produced a solar soft X-ray emission (0.1–0.8 nm range) of 2.5 mW m^{-2}, quiet values remaining below 0.01 mW m^{-2}; see Nikolśkij, G.M.: This Encyclopedia, vol. 49/6, Table 5, p. 349; [printing error in the caption: mW m^{-2} instead of nWm^{-2}]

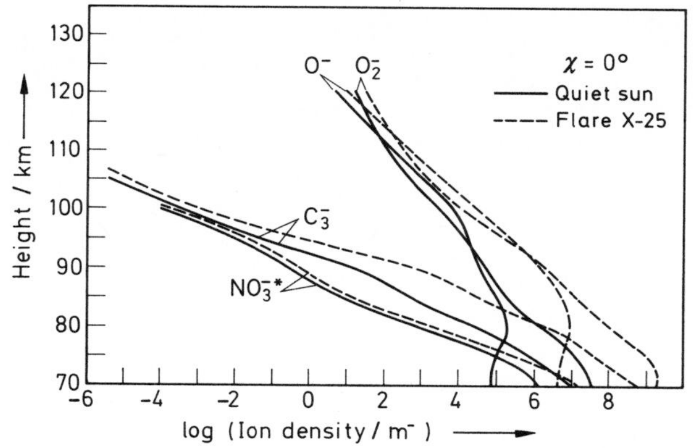

Fig. 150. Profiles of negative ion species computed after DYMEK[16] [full curves for quiet day, broken curves for strong flare[20] conditions]

Quite recently KOPP[21], comparing his own and ARNOLD's[22] measurements, has summarized these by establishing his schemes of clustering. While his reaction networks for positive clusters (Fig. 146b, c) are only slightly more involved than DYMEK's Fig. 146a the situation is different for the negative ones[22,23] where, unlike Fig. 147a[16], Cl^- seems to play an important role[22,24] in the meso- and stratosphere (it stems probably from HCl). Figure 147b shows a three negative ion model[21] as an extension of earlier models comprising only two of these[17]. It must, however, be noted that basic experimental evidence is still scarce. Apparently, negative ion modelling just starting with known laboratory data is not yet feasible. The main check point with in-situ measurements is the *negative to positive ion ratio.*

$$\lambda = \sum_j N_j^- / \sum_i N_i^+. \tag{15.8}$$

A summary of experimental evidence was given in Sect. 13(c). Figs. 119–122.

As for the positive cluster ions which, by day, prevail below an upper limit at about 82–86 km (Figs. 121a, b in Sect. 13) the main difference against earlier theories is that metastable excited molecules O_2 ($^1\Delta_g$) play an important role; they are directly ionized by Lyman alpha (122 nm) radiation acting upon NO.

Apparently, the densities of these minor constituents are rather variable and temperature variations have large effects. The big problem, however, is to find a fast reaction path for clustering with water molecules; this can be obtained via nitrogen oxides (see Fig. 146c)[21].

---

[21] Kopp, E. (1984): [*45*], 140

[22] Arnold, F., Krankowsky, D. (1971): J. Atmos. Terr. Phys. *33*, 1693

[23] Narcisi, R.S., Bailey, A.D., Della Luca, L., Sherman, C., Thomas, D.M. (1971): J. Atmos. Terr. Phys. *33*, 1147

[24] Kopp, E., André, L., Eberhardt, P., Herrmann, U. (1980): Trans. Amer. Geophys. Un. Eos *61*, 311

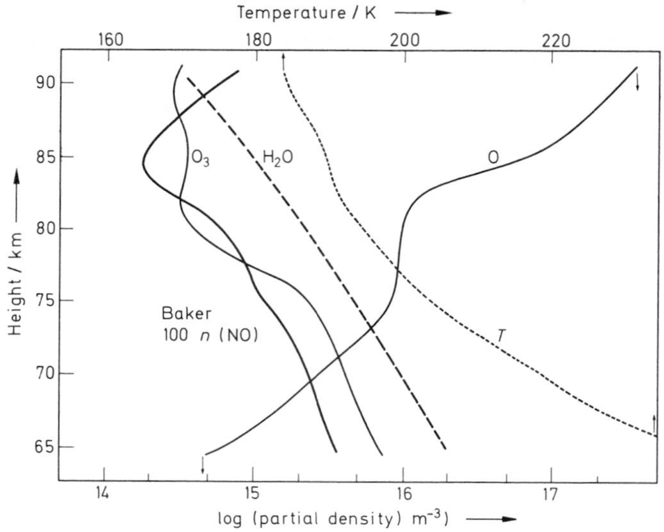

Fig. 151. Assumed height profiles of minor neutral species in the D-region, and temperature (dotted, upper abscissa scale) after SEARS et al.[26] The measured NO profile marked by BAKER[28] was found more appropriate than earlier estimates giving higher values. Atomic and ozone densities are the ones calculated by DAIRCHEM[25] during eclipse totality

A computerized D-region model called DAIRCHEM[25] was tested by SEARS et al.[26]. It takes account of corpuscular and solar X+EUV radiation inputs; see Fig. 148b. The assumed temperature and minor constituent profiles are presented in Fig. 151. The computation scheme for negative ions uses the reactions shown in Fig. 147c; it admits four ions (O^-, O_2^-, O_3^-, CO_3^-) and is, essentially, an oxygen scheme, neglecting nitrogen oxides (see Fig. 147a) and also Cl (see Fig. 147b). Results of the computations are seen in Fig. 149c, d (individual positive ion densities) and in Fig. 152 (total electron and positive ion densities). Apparently, while electron density is correctly predicted, the transition height with negative ions (where N_e and $\sum N_i^+$ begin to differ) is computed too low by more than 15 km. This should be due to unrealistic assumptions about the negative ion chemistry; see [21].

δ) The fundamental aeronomical problem of the *E-region* (90–150 km) is dissociation of oxygen. This is discussed in detail by YONEZAWA in vol. 49/6 (his Chap. III, p. 177), taking into account eddy and molecular diffusion in the lower and upper height ranges, respectively. The continuity equations [Eqs. (19.11, 19.12) in YONEZAWA] must be resolved by numerical methods since their shape is too involved due to the temperature dependence of rates.

Ion chemistry in the E-region is easier than in the D-region since the three particular conditions noted at the beginning of Sect. 15γ are not valid. THOMAS[27] gives reaction equations for six charge transfer and three dissociative recombination reactions with ions O_2^+, NO^+ and N_2^+.

[25] The Dairchem model was developed by D.W. HOOCK and M.G. HEAPS at U.S. Army Atmos. Sci. Lab., White Sands Missile Range, N.Mex., USA

[26] Sears, R.D., Heaps, M.G., Niles, F.E. (1981): J. Geophys. Res. *86*, 10073

[27] Thomas, L. (1982): This Encyclopedia, vol. 49/6, Table 13, p. 111 [rate coefficients are also given there]

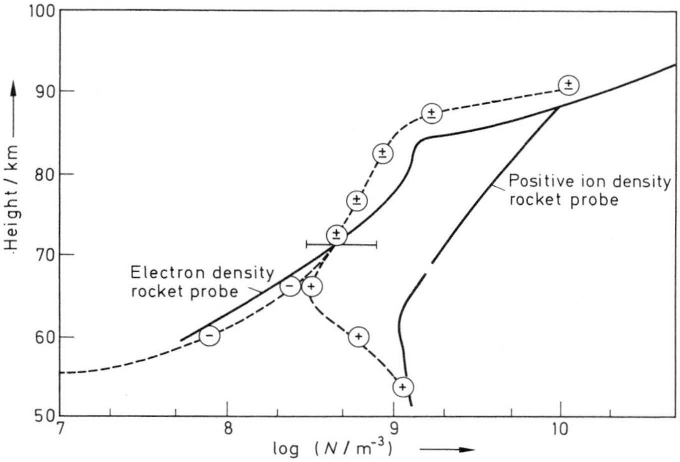

Fig. 152. Electron and positive ion density profiles in the D-region computed by the DAIRCHEM program[25] (circles and broken lines) and mean experimental data for the same conditions (full curves). [Typical experimental error range is by a factor of 3]

It is important to note that of these three NO^+ has a minimum ionization threshold. Therefore, charge transfer leading to NO^+ is almost always a final stage. In spite of negligible direct ionization NO^+ is so produced by chemical reactions and must not be overlooked [*65*]. Recent observations with a kryogenic mass spectrometer (due to KOPP and coworkers) have in fact shown that the NO^+ density outnumbers that of O_2^+ everywhere in the height range from 60 to 170 km.

The reaction scheme corresponding to [27] is shown in Fig. 153a, from which a four ion set of balance equations is directly obtained:

$$\begin{aligned}
\frac{d}{dt} N_1 &= Q_1 + [k_4 \cdot N_4 + k_9 \cdot N_3] \cdot n_1 - [k_6 \cdot n_2 + k_7 \cdot n_5 + \alpha_{D1} \cdot N_e] \cdot N_1, \\
\frac{d}{dt} N_2 &= \qquad k_5 \cdot n_3 \cdot N_4 + (k_6 \cdot n_2 + k_7 \cdot n_5) \cdot N_1 \\
&\qquad + \; k_8 \cdot n_4 \cdot N_3 - \alpha_{D2} \cdot N_e \cdot N_2, \\
\frac{d}{dt} N_3 &= Q_3 - [k_8 \cdot n_4 + k_9 \cdot n_1 + \alpha_{D3} \cdot N_e] \cdot N_3, \\
\frac{d}{dt} N_4 &= Q_4 - [k_4 \cdot n_1 + k_5 \cdot n_3] \cdot N_4,
\end{aligned} \tag{15.9a-d}$$

indices $i = 1 \dots 5$ designating species O_2, NO, N_2, O, N. Neutral densities are written as n_i charged densities as N_i. [See [27] for rate coefficients.]

A considerable simplification can be drawn from the fact that in the E-region one member (that with k_8) prevails in the N_2^+ loss term, viz. formation of NO^+ under charge transfer[28]. With this in mind THOMAS[27] obtains simpler expressions [his Eqs. (29.1–29.3)] for equilibrium which is rapidly obtained during day.

[28] Baker, K.D., Nagy, A.F., Olsen, R.V., Oran, E.S., Randhawa, J., Strobel, D.F., Tohmatsu, T. (1977): J. Geophys. Res. *82*, 3281

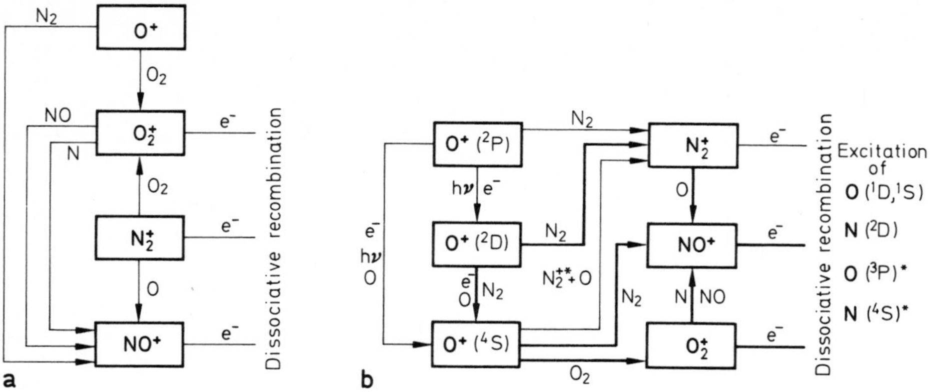

Fig. 153a, b. Reaction scheme for positive ions. (**a**) In the E-region after THOMAS[27]; (**b**) in the F-region after TORR et al.[32]

The set of Eqs. (15.9) is a coupled system of differential equations which nowadays is resolved by numerical methods. Height profiles of the neutral constituents N_2, O_2, O, N are taken as inputs from models, the production terms Q are derived from an assumed solar spectrum and the height-dependent densities of the main ions are computed. A more involved theory had to deal simultaneously with oxygen dissociation and ionization processes.

ε) Transport is a major influence in the *F-region*; see Sect. 15b) below. Nevertheless chemistry can explain a limited part of the observed phenomena. The chemical part of the balance equations is essentially the same as shown in Eqs. (15.9). Some simplification might be earned from the fact that atomic oxygen O is the most important neutral constituent so that the production term $Q_4(O^+)$ prevails. The most important removal processes are those with the terms with k_4, k_5 for charge transfer which are followed by very fast dissociative recombination (with α_{D1} and α_{D2}, respectively)[29].

Unfortunately, theory along these lines was unable to reproduce observed seasonal variations[30]. Discrepancies became even more serious at the time when ion composition measurements were made in situ by the AEROS and Atmospheric Explorer satellites: computed N_2^+ densities greatly exceeded the observed values. Thus, a stronger sink is needed for N_2^+. Re-examining the possible solutions[31], TORR et al.[32] concluded that a high rate could be obtained by charge transfer of vibrationally excited N_2^+:

$$N_2^{+*} + O \rightarrow O^+(^4S) + N_2. \tag{15.10}$$

[29] See: Yonezawa, T. (1982): This Encyclopedia, vol. 49/6, Sect. 23, p. 193

[30] See: Rawer, K., Suchy, K. (1967): This Encyclopedia, vol. 49/2, Sect. 37, p. 288

[31] Thomas, L. (1982): This Encyclopedia, vol. 49/6 on p. 117, 118 mentions an earlier theory of TORR which is superseded by the new one [32]

[32] Torr, D.G., Richards, P.G., Torr, M.R. (1981): [53], 18-1

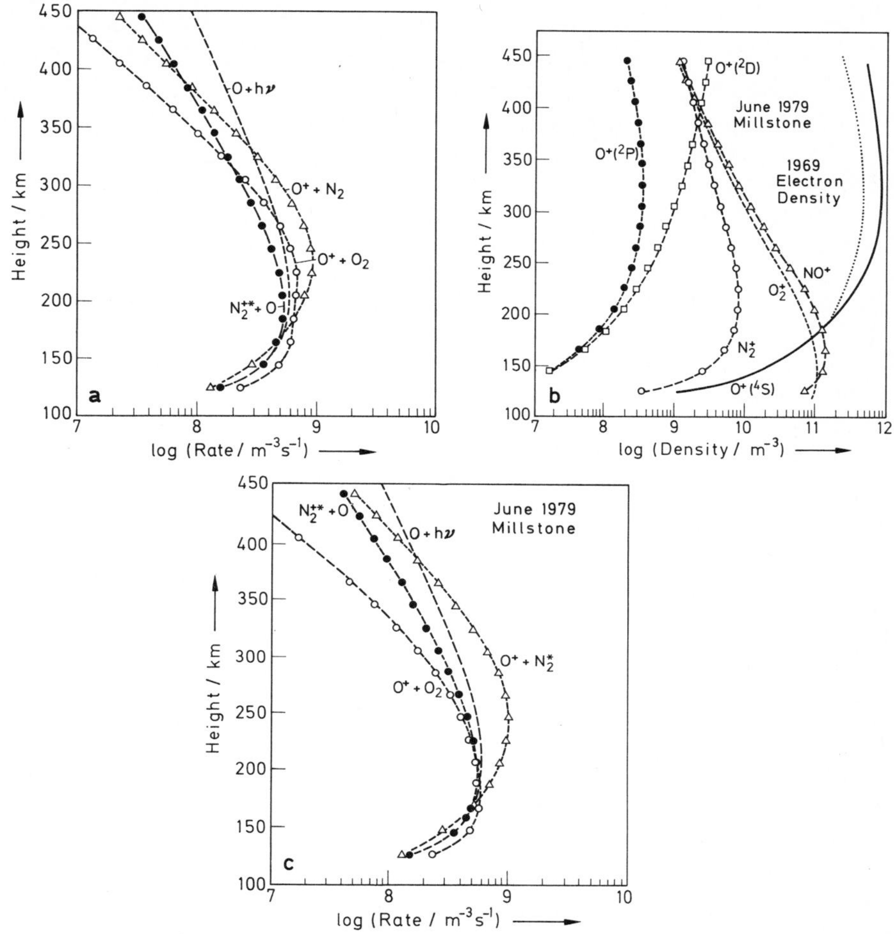

Fig. 154a–c. Ion chemistry profiles for a summer day (June 1979, solar activity maximum) [electron profile from incoherent scatter measurements at Millstone Hill, 42.6° N, 288.5° E]. The computations used the reaction scheme of Fig. 153b and included the charge transfer reaction Eq. (15.10)[32]. (**a**) $O^+(4S)$ production and loss rates; (**b**) ion concentrations [computed from (**a**)]; (**c**) same as (**a**) but including vibrationally excited N_2^+

Using a computation scheme due to Orsini[33] the partial density of N_2^{+*} could be determined[32]. It is not only sufficient to account for the observed low partial densities of N_2^+ but explains also the production of metastable O^+ ions. Torr's[32] reaction scheme is displayed in Fig. 153b. Figure 154 shows typical loss rate and density profiles for a mid-latitude summer day. The authors claim to have definitely resolved the problem of the seasonal anomaly (decreased electron densities in summer) which was earlier[34] believed to be uniquely due to changes in the neutral composition.

In fact, the ratio of partial densities of O and N_2 is larger by a factor of 2 in winter[35].

[33] Orsini, N. (1977): The vibrationally excited N_2^+ ion ... = Techn. Report 10554/13716. Ann. Arbor. Mi., USA: Univ. of Michigan

[34] Rishbeth, H., Setty, C.S.G.K. (1961): J. Atmos. Terr. Phys. *20*, 263; Rishbeth, H. (1967): Proc. Inst. Electr. Electron. Eng. *55*, 16

[35] Krankowsky, D., Kasparzak, W.T., Nier, A.O. (1968): J. Geophys. Res. *73*, 7291

b) Transport-oriented models

ζ) The transport phenomena mentioned at the beginning of this chapter are of primary importance at higher altitudes where the time to reach chemical equilibrium is of the order of hours, i.e. in the F-region. There, the ratio γ_j of the collision frequency of charged particles against their gyropulsation ω_{Bj} is quite small, such that free motion is on helical orbits with the magnetic field as axis. Thus, transport by free motion is *field aligned*. This is so for diffusion as well as for motions due to momentum transfer from the neutrals, i.e. wind-induced transport. Only motions provoked by (transverse) electric fields, i.e. by the $\boldsymbol{E} \times \boldsymbol{B}$ force are *perpendicular* to the magnetic field direction [see Sects. 16, 17, 19, 20 in YONEZAWA's contribution in vol. 49/6]. This has important consequences, depending on the local magnetic dip.

Diffusion causes regions where the concentration of some constituent is high to spread and thin out. A thorough discussion of molecular and eddy diffusion (of importance only below the turbopause) was given by YONEZAWA in vol. 49/6 of this Encyclopedia (pp. 129–245). Motions provoked by *neutral winds* onto the plasma components and vice versa [*48*] are discussed in the same volume by STUBBE (pp. 247–308). The effect of *electric fields* is summarized by the same author in his Sect. 35. The important result is that, in the F-region, their effect on the vertical distribution of electrons and ions is negligible except for a small belt along the dip equator where the magnetic field is essentially horizontal. However, important *electric currents* may be provoked and these may initiate secondary effects[36]. The importance of such currents (and the limitations of unique explanations) are illustrated by Fig. 155. One has to take account of the combined action of both, electric fields and winds, in particular in the 'dynamo region' between 90 and 160 km of altitude.

The three transport mechanisms are illustrated in Fig. 156. Transport is described in Eq. (15.1) by the last (divergence) term, which splits up into three contributions after these mechanisms.

While diffusion is a mechanism that has been known for a very long time, it is KOHL and KING[37] who have the credit for introducing plasma motions induced by neutral winds. Figure 156b shows how horizontal wind, in the presence of a magnetic field, provokes vertical plasma displacement. The field has a similar effect to the "sword" of a sailing boat; for this also cause (wind) and effect (sailing) have different directions. The theory is explained in vol. 49/6 by STUBBE, who also gives a survey on wind data. For aeronomical purposes, mean winds are mostly applied and these are deduced from the pressure/temperature maps of current atmospheric models. This is dangerous because the actual winds are almost always faster than these mean values. So the wind field is the critical point in such computations; see [*30*][38,39] (Fig. 157).

Under the conditions of the F-region the equation of motion leads to the following expressions for the vertical plasma motion induced by an electric

[36] Forbes, J.M. (1981): Rev. Geophys. Space Phys. *18*, 469

[37] Kohl, H., King, J.W. (1965): Nature *206*, 699; Rishbeth, H., Megill, L.R., Cahn, J.H. (1965): Ann. Géophys. *21*, 235

[38] Cho, H.R., Yeh, K.C. (1970): Radio Sci. *5*, 881

[39] Emery, B.A. (1978): J. Geophys. Res. *83*, 5691, 5704

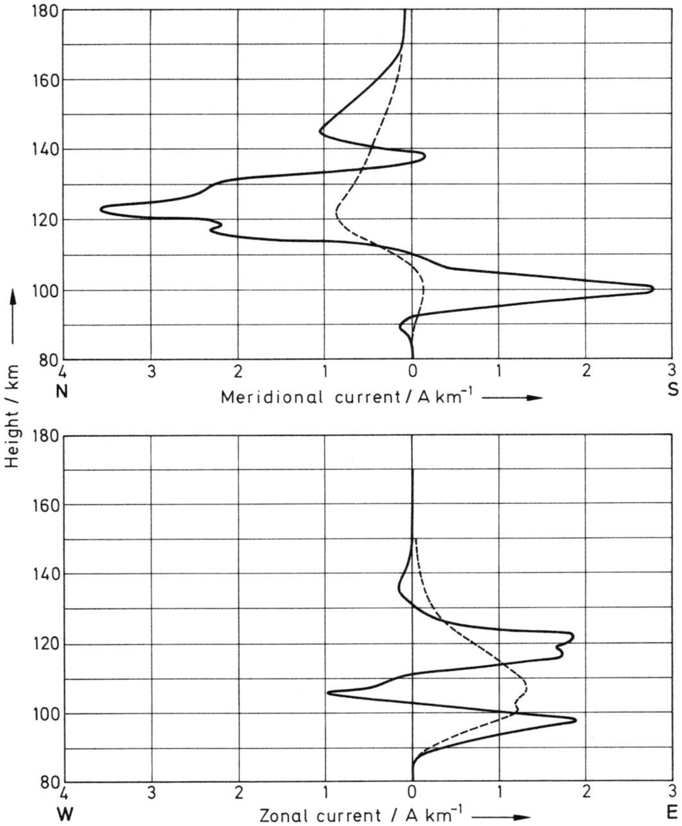

Fig. 155. Horizontal currents inferred from incoherent scatter observations at Arecibo (18.30° N, 66.75° W) [10 August 1974, 11 h LT] as full curves. The broken curves were deduced from the electric field effect only

field $\boldsymbol{E}$ and a horizontal neutral wind of components $v_{n\xi}$, $v_{n\eta}$ [ξ coordinate in the magnetic meridian, η perpendicular to it][40]:

$$v_{iz} = \cos I \cdot E_\eta / B + \sin I \cdot \cos I \cdot [v_{n\xi} + \textit{diffusion term}], \tag{15.11}$$

with I magnetic inclination (dip ψ), B magnetic field $\equiv |\boldsymbol{B}|$.

Only the (magnetically) meridional wind component $v_{n\xi}$ and the zonal component of the electric field, E_η, provoke vertical displacements of the plasma. It can be shown that the plasma drift provoked by the electric field

[40] Stubbe, P. (1982): This Encyclopedia, vol. 49/6, p. 262 [Eqs. (9.3-9)]; these equations are in geographical coordinates immediately showing the declination effect provoked by the zonal wind component; we use his Eq. (35.1), p. 303, which is in local magnetic coordinates ξ (meridional), η and z (vertical)

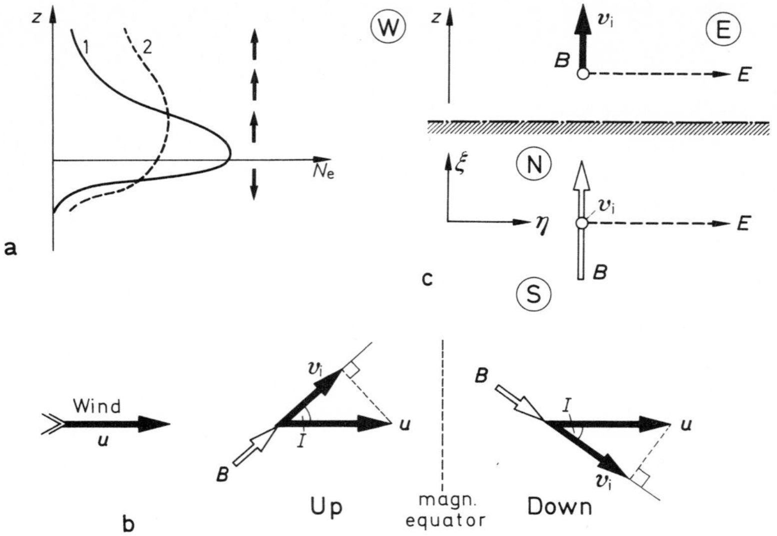

Fig. 156a–c. Transport mechanisms in the F-region by (**a**) diffusion [shape 1 goes towards 2; velocities indicated by arrows]; (**b**) momentum transfer from neutral wind $\boldsymbol{u}$ [ion transport $\boldsymbol{v}_i$ along magnetic field $\boldsymbol{B}$]; (**c**) electrical field $\boldsymbol{E}$, force: $\boldsymbol{E}\times\boldsymbol{B}$ [at magnetic equator $\boldsymbol{E}$ and $\boldsymbol{B}$ horizontal, $\boldsymbol{v}_i$ vertical]

causes an additional meridional neutral velocity, which in turn provokes another wind-driven vertical component v'_{iz} just compensating the first term in Eq. (15.11)[41]. Thus, taking into account all the effects together, wind and diffusion alone are important for vertical transport in the F-region. Separating the divergence term in Eq. (15.1) but neglecting diffusion we get[42]:

$$\frac{\partial N}{\partial t}+\boldsymbol{v}_\perp\cdot\nabla N=P-L-\nabla\cdot(N\,\boldsymbol{v}_\parallel), \tag{15.12}$$

a further term $N\cdot\nabla\cdot\boldsymbol{v}_\perp$ being negligible. The region around the dip equator must, however, be excluded from this reasoning; see Sect. 15ϑ below.

η) The *computational methods* applied in kinematic theory will now be demonstrated following the particularly clear formulation given by RÜSTER[43]. He uses a four-ion scheme (see Sect. 15δ) admitting eight reaction equations (five for charge transfer and three with dissociative recombination) leading to the following expressions for production and loss terms [indices: 1 for O_2, 2 for NO, 3 for N_2, and 4 for O; symbol n for neutral, N for ion densities; see

[41] Stubbe, P. (1982): This Encyclopedia, vol. 49/6, Sect. 35, p. 303 and Fig. 25, p. 304

[42] Anderson, D.N. (1973): Planet. Space Sci. *21*, 409; see further: Anderson, D.N., Matsushita, S. (1974): J. Atmos. Terr. Phys. *36*, 2001; Anderson, D.N. (1979) in: Low latitude aeronomical processes. p. 93. New York: Plenum Press; Anderson, D.N. (1981): J. Atmos. Terr. Phys. *43*, 753. The post-sunset enhancement of upward drift though quite common is rather variable from day to day; see Farley, D.T., Woodman, R.F., Calderon, C. (1979): J. Geophys. Res. *84*, 5792

[43] Rüster, R. (1971): J. Atmos. Terr. Phys. *33*, 137

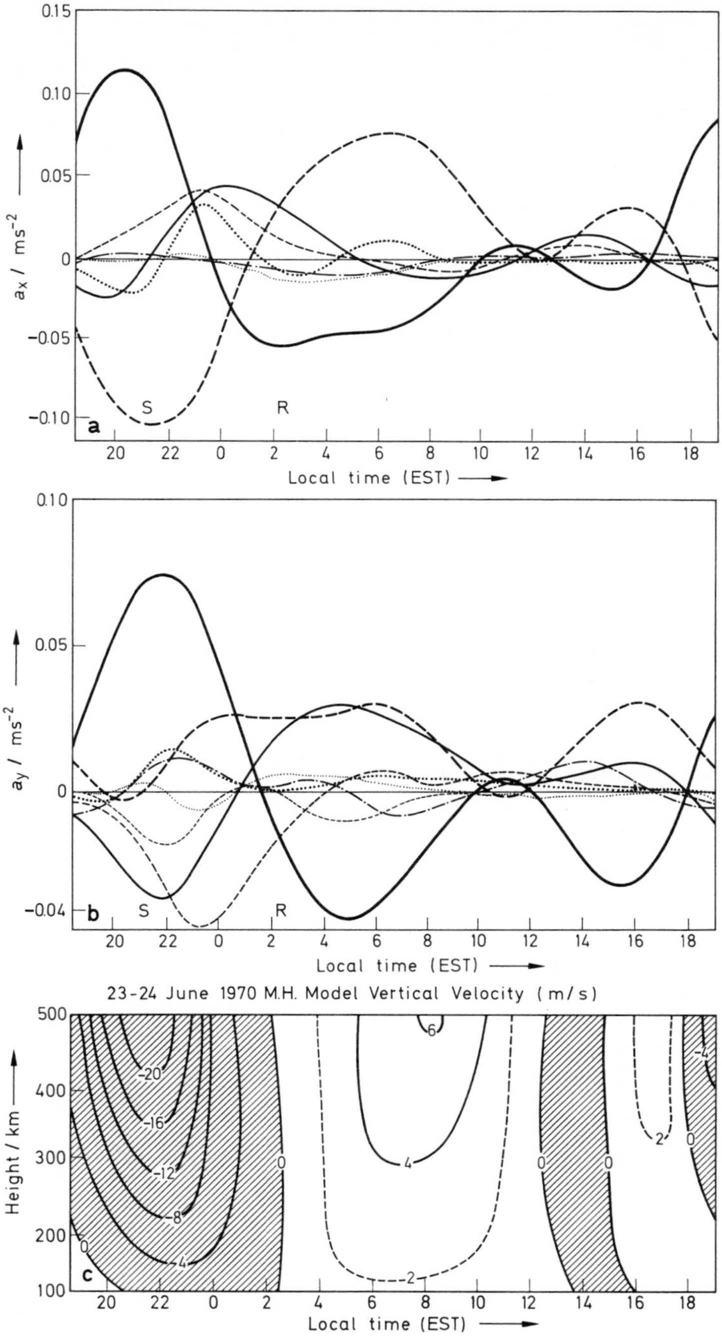

Fig. 157a–c. Effects of different influences upon wind calculations[39]. (**a**) Acceleration terms at 300 km altitude in the zonal, (**b**) in the meridional momentum equation (positive sign means towards E and N, respectively). Linear terms: —— pressure gradient; - - - ion drag; —— Coriolis; -·- acceleration; --- viscosity. Non-linear terms: **......** EW; ...... NS components. (**c**) Contour plot of vertical velocity (unit ms^{-1}, positive upwards) for Millstone Hill (Mass., USA) [computed from an aeronomical model fitted with data from incoherent scatter observations (see Sect. 5) of 23/24 June 1970]

Eqs. (15.9)]:

$$
\begin{aligned}
P_1 &= Q_1 + (k_4 N_4 + k_9 N_3) \cdot n_1; \\
P_2 &= k_5 N_4 n_3 + k_6 N_1 n_2 + k_8 N_3 n_4; \\
P_3 &= Q_3; \\
P_4 &= Q_4; \\
L_1 &= (k_6 n_2 + \alpha_{D1} N_e) \cdot N_1; \\
L_2 &= \alpha_{D2} N_e N_2; \\
L_3 &= (k_8 n_4 + k_9 n_1 + \alpha_{D3} N_e) \cdot N_3; \\
L_4 &= (k_4 n_1 + k_5 n_3) \cdot N_4.
\end{aligned}
\tag{15.13a-h}
$$

The expressions for electrons follow from these:

$$P_e = Q_1 + Q_3 + Q_4; \qquad L_e = (\alpha_{D1} \cdot N_1 + \alpha_{D2} \cdot N_2 + \alpha_{D3} \cdot N_3) \cdot N_e. \tag{15.14}$$

In the equations of motion for the ion species the acceleration term is neglected, but gravity and electrodynamic force $\boldsymbol{E} + \boldsymbol{v} \times \boldsymbol{B}$ are admitted, also a term $-\boldsymbol{k} \cdot \nabla (N_j T_i)$ representing diffusion and, finally, a friction term representing the effects of collisions between different species, $N_j F_j \cdot (\boldsymbol{v}_{ij} - \boldsymbol{v}_n)$ [F_j being a frictional coefficient].

Solving analytically and taking explicit account of the collision-to-gyro-frequency ratio γ_1, the vertical displacement velocity is found as v_{izj}:

$$
\begin{aligned}
\beta_j \cdot v_{izj} = {} & \cos I \cdot (E_\eta - \gamma_j E_\xi / \sin I)/B + (\sin^2 I + \gamma_j^2) \cdot (v_{nz} - C_j/F_j) \\
& + \cos I \cdot (\sin I \cdot v_{n\xi} + \gamma_j \cdot v_{n\eta})
\end{aligned}
\tag{15.15}
$$

where $\beta_j = 1 + \gamma_j^2$; $\gamma_j \equiv F_j / m_j \omega_{Bj}$ (vanishes at greater heights) and $\omega_{Bj} = q \cdot B / m_j$ is the gyropulsation of ion j with charge q and mass m_j.

Gravity force and diffusion quasiforce is taken account of by

$$C_j = m_j \cdot g + k \cdot \left[\frac{1}{N_j} \cdot \frac{\partial}{\partial z} (N_j T_i) + \frac{1}{N_e} \cdot \frac{\partial}{\partial z} (N_e T_e) \right]. \tag{15.16}$$

The equation of motion for the electrons includes the diffusion term and electrodynamic forces only, such that

$$\boldsymbol{k} \cdot \nabla (N_e T_e) = -q N_e \cdot (\boldsymbol{E} + \boldsymbol{v}_e \times \boldsymbol{B}) \tag{15.17}$$

from which follows the vertical (polarization) field component:

$$-E_z = \cot I \cdot E_\xi + (k/q N_e) \cdot \frac{\partial}{\partial z} (N_e T_e). \tag{15.17a}$$

An equation of motion is also established for the neutrals including pressure gradient, gravity (g) and Coriolis ($\boldsymbol{\Omega}$) forces, further friction exerted by the ions (coefficients F_j) and viscosity (ζ_n).

Thus, neutral winds are in fact computed; they depend primarily on the pressure gradients which are derived from a model of the neutral atmosphere.

$$\frac{\partial \boldsymbol{v}_n}{\partial t}+\frac{1}{\rho_n}\cdot(\boldsymbol{v}_n-\boldsymbol{v}_i)\cdot\sum_{j=1}^{4} N_j F_j=-\frac{1}{\rho_n}\nabla p_n+2(\boldsymbol{v}_n\times\boldsymbol{\Omega})+\zeta_n\cdot\frac{\partial^2 \boldsymbol{v}_n}{\partial z^2}, \tag{15.18}$$

where ρ_n, p_n are neutral mass density and pressure, respectively.

This vectorial equation together with four continuity equations for the ions [Eq. (15.1)] constitute the system of equations to be resolved. Since O^+ ions are prevalent in the F-region, the divergence term of Eq. (15.1) can be neglected for the three molecular ions[44].

The whole system of coupled non-linear partial differential equations is finally brought into the shape:

$$\begin{aligned}
&\frac{\partial}{\partial t}N_1=f_5\cdot N_1+f_6; \qquad \frac{\partial}{\partial t}N_2=f_9\cdot N_2+f_{10};\\
&\frac{\partial}{\partial t}N_3=f_7\cdot N_3+f_8;\\
&\frac{\partial}{\partial t}N_4=\left(f_1\cdot\frac{\partial^2}{\partial z^2}+f_2\cdot\frac{\partial}{\partial z}+f_3\right)N_4+f_4;\\
&\frac{\partial}{\partial t}v_{n\xi}=\left(f_{11}\cdot\frac{\partial^2}{\partial z^2}+f_{12}\cdot\frac{\partial}{\partial z}+f_{13}\right)v_{n\xi}+f_{14};\\
&\frac{\partial}{\partial t}v_{n\eta}=\left(f_{15}\cdot\frac{\partial^2}{\partial z^2}+f_{16}\cdot\frac{\partial}{\partial z}+f_{17}\right)v_{n\eta}+f_{18}.
\end{aligned} \tag{15.19}$$

The f_k are functions of t and z (independent variables), of six dependent variables ($N_1 \dots N_4, v_{n\xi}, v_{n\eta}$) and of various parameters which must be given.

The system Eqs. (15.19) is numerically resolved after transformation into a system of finite difference equations. The difficulties resulting from non-linearity and coupling are avoided by solving them iteratively one after the other instead of doing this simultaneously. See [43] for initial value and boundary problems.

ϑ) The situation is different in the *vicinity of the magnetic dip equator* where we have to allow in the continuity equation, Eq. (15.12), another term* $N_e\cdot\nabla\cdot\boldsymbol{v}_{i\perp}$ where $\boldsymbol{v}_{i\perp}$, the velocity perpendicular to the magnetic field $\boldsymbol{B}$, is due to the electric field $\boldsymbol{E}$. Near the dip equator, after ANDERSON[42], Eq. (15.12) takes the shape

$$\frac{\partial}{\partial t}N_j+\boldsymbol{v}_{j\perp}\cdot\nabla N_j=P_j-L_j-\nabla\cdot(N_j\boldsymbol{v}_{j\parallel})-N_j\nabla\cdot\boldsymbol{v}_{j\perp} \tag{15.20}$$

and

$$\boldsymbol{v}_{j\perp}=\boldsymbol{E}\times\boldsymbol{B}/B^2. \tag{15.21}$$

Horizontal electric fields are important when $\boldsymbol{B}$ is essentially horizontal such that $\boldsymbol{v}_{j\perp}$ is vertical.

---

* Note that $\nabla\equiv\partial/\partial\boldsymbol{r}$ is a vectorial operator

[44] Torr, D.G., Torr, M.R. (1970): J. Atmos. Terr. Phys. *32*, 15

Thus electric fields in the dynamo region which are quite strong[45] provoke vertical plasma transport. This causes a very characteristic phenomenon at the dip equator, namely an upward motion of the plasma like that of water in a fountain. This analogy holds also for the follow-up motion, which is downwards outside the jet, so at both sides of the equatorial zone the plasma, driven by pressure force, falls down along the magnetic lines of force (see Fig. 165 below). This causes the two peaks of electron density seen in daytime on equatorial maps, see Figs. 125 and 129 in Sect. 14, also Figs. 323 and 324 of RAWER and SUCHY in vol. 49/2 (p. 514, 515).

In the "dynamo region" around 100 km the tidal vertical motion of the ionospheric plasma in the magnetic field of Earth creates the well-known dynamo currents the magnetic effect of which is the magnetic diurnal variation. It is important that this mechanism is seriously influenced by electric fields as are observed there. Figure 158 shows examples how the combined influence may affect the current distribution[46].

Comparing the results of such computations with measured data ANDERSON[42] found two anomalies:

i) The diurnal variation of observed vertical electron drift during evening hours[47] could not be reproduced with tidal electric fields as had been inferred from the magnetic diurnal variations[48].

ii) Latitudinal sections at different longitudes, by day, though always showing two maxima, display differences in shape.

He takes care of the first problem by adapting the diurnal variation of field $\boldsymbol{E}$ to the observations; this necessitates a second maximum after sunset; see Fig. 159b, c.

As for the second question, the differences in the magnetic field strength B with larger values in the Asean sector cannot explain the observed asymmetries of the northern and southern peaks. ANDERSON feels that neutral winds in the equatorial zone should give an explanation. These should be symmetrical to the geographical equator but the magnetic field shows symmetry to the dip equator. Therefore, one expects a meridional wind to be present at the dip equator blowing the plasma to the other side. In the American sector the dip equator lies south to the geographical equator and, by day, the neutral wind is northwards, thus making the northern peak higher. The inverse holds in the Asean sector[42].

A recent analysis of vertical plasma drifts observed by incoherent scatter technique (see Sect. 5) at the magnetic equator (Jicamarca in Peru) and also at Arecibo (Puerto Rico)[49] has revealed

---

[45] Forbes, J.M., Lindzen, R.S. (1976): J. Atmos. Terr. Phys. *38*, 879, 911; (1977): ibidem *39*, 1369

[46] Anandarao, B.G., Raghavarao, R., Desai, J.N., Haerendel, G. (1978): J. Atmos. Terr. Phys. *40*, 157; the more general context is discussed in: Forbes, J.M. (1981): The equatorial electrojet. Rev. Geophys. Space Phys. *19*, 469–504

[47] Woodman, R.F. (1970): J. Geophys. Res. *75*, 6249

[48] Heelis, R.A., Kendall, P.C., Moffett, R.J., Windle, D.W., Rishbeth, H. (1974): Planet. Space Sci *22*, 743

[49] Fejer, B.G., Farley, D.T., Woodman, R.F., Calderon, C. (1979): J. Geophys. Res. *84*, 5792 [Peru]; Walker, J.C.G. (1983): Space Res. in Bulgaria *4*, 50 [Puerto Rico]

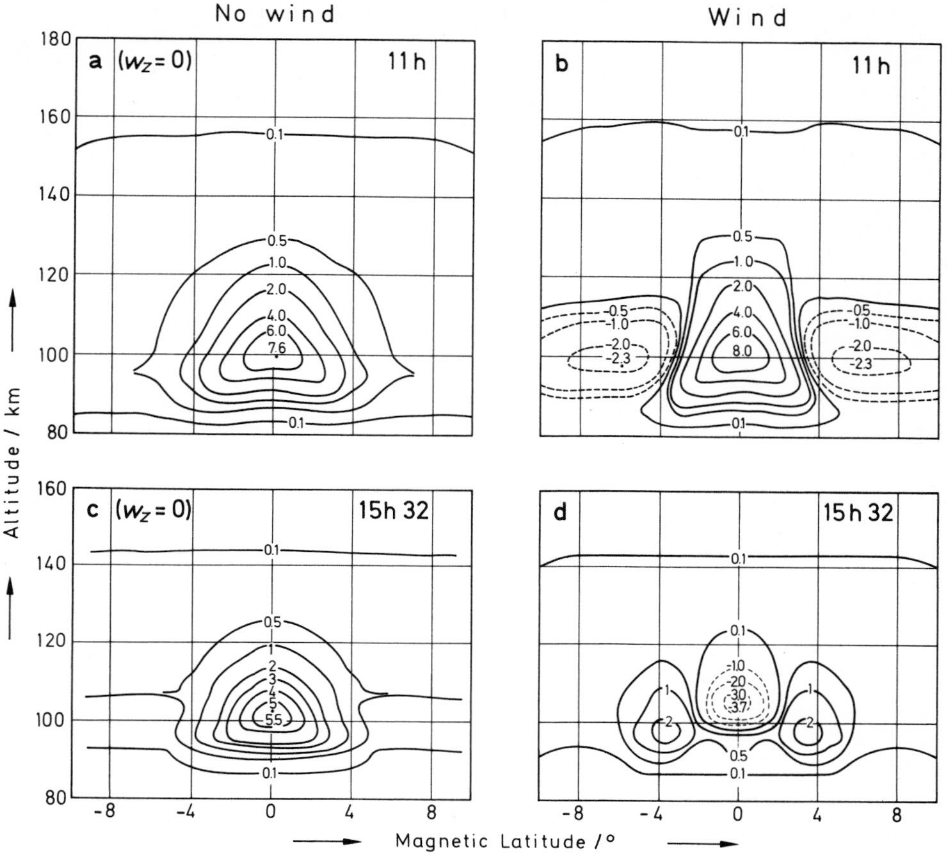

Fig. 158a–d. Eastward current density (unit A/km²) in the dynamo region computed with different assumptions[46]. (**a, b**) 11 h, (**c, d**) 15 h 30 LT; (**a**) with observed E-W wind; (**c**) with observed vertical wind; (**b, d**) no wind cases. The primary (eastward) electric field was assumed to be 0.5 mV m^{-1}

important differences between years of high and low solar activity, with larger velocities in the evening and at night around the solar minimum; see Fig. 159a. In midwinter (June) the evening reversal (see Fig. 160a) occurs earlier during the solar maximum as shown by Fig. 160b.

Transequatorial plasma drift (Fig. 161)[50] is probably particularly important during hours when one of the footpoints of the magnetic field line is sunlit while the other is in darkness[51].

c) *Ionospheres of other planets.* Ionospheres of planets were first observed with radio wave methods[52] [*3*] and in recent years also by in situ observations. Except for Venus, data are still scarce. Survey papers have been written for the terrestrial planets by SCHUNK and NAGY [*55*] and for the major planets by STROBEL [*59*]. Many conclusions are still drawn by extrapolation from terrestrial observations.

[50] Bailey, G.J., Moffett, R.J. (1979): J. Geophys. Res. *83*, 145

[51] Carlson, H.C. Jr., Walker, J.C.G. (1971): Planet, Space Sci. *20*, 141

[52] Gringauz, K.I., Breus, T.K. (1976): This Encyclopedia, vol. 49/5, p. 351

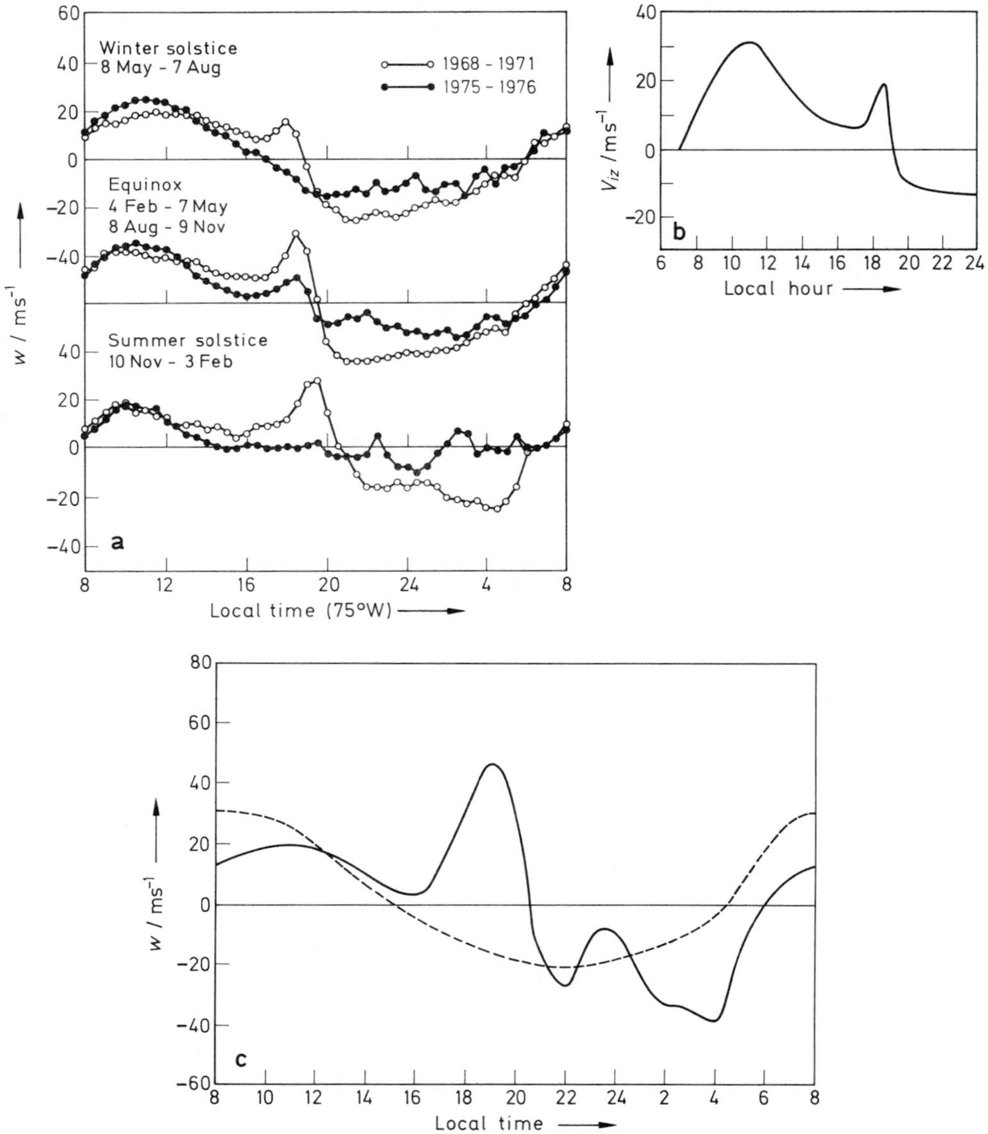

Fig. 159a–c. Diurnal variation of vertical plasma drift at the dip equator. (**a**) Observations made at Jicamarca, Peru (10,95° S, 283.13° E) during solar minimum 1968–1971 and maximum 1975/1976[49]. (**b**) Inferred diurnal variation of the electric (horizontal EW) field induced upward plasma drift[42]. (**c**) Comparison of drift predicted from tidal theory and relevant electric field (broken line) with that obtained by inclusion of an F-region polarization field (full line) [for solar activity maximum][48]

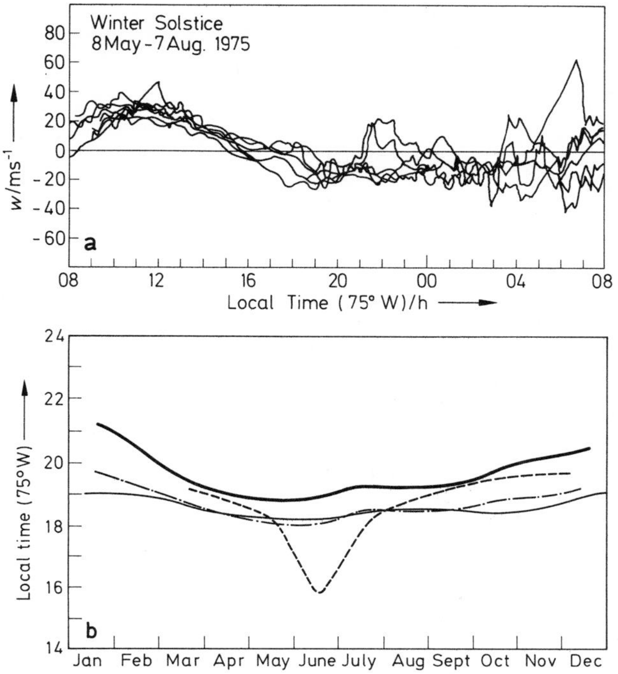

Fig. 160a, b. The evening reversal of vertical plasma drift at the dip equator [Jicamarca, Peru][49]. (**a**) Observed vertical drift velocity during winter solstice in solar minimum period (8 May to 7 Aug. 1975); (**b**) Seasonal variation of characteristic times around E-region sunset = thin line; –·– evening drift maximum; —— reversal during solar minimum 1968/1971; --- reversal during solar maximum 1975/1976

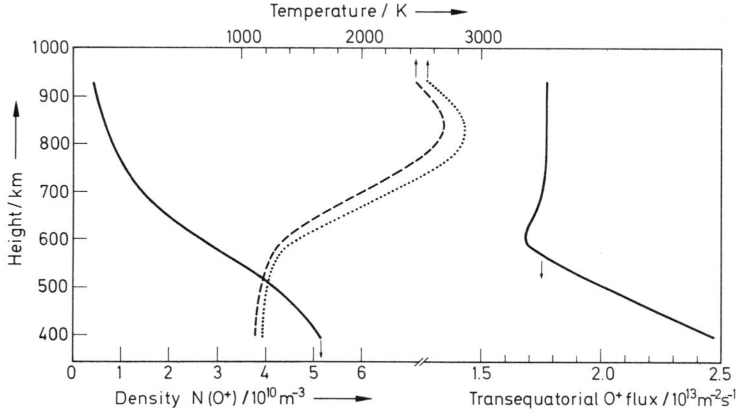

Fig. 161. Calculated profiles of O^+ ion density (full line on left) and plasma temperatures (—— T_i, ... T_e, centre, upper abscissa scale) and transequatorial O^+ flux (on right)[50]

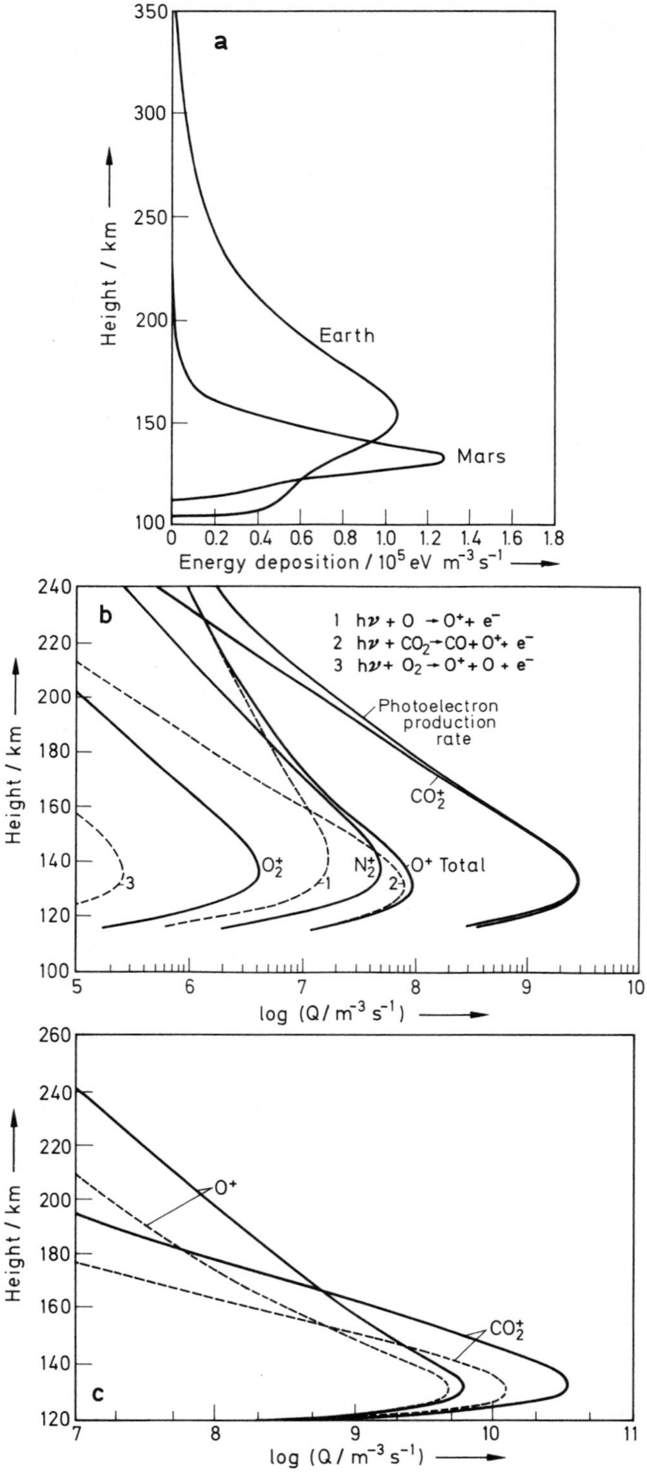
a
Height / km
Earth
Mars
Energy deposition / 10^5 eV m^-3 s^-1
b
1 hν + O → O⁺ + e⁻
2 hν + CO2 → CO + O⁺ + e⁻
3 hν + O2 → O⁺ + O + e⁻
Photoelectron production rate
CO2⁺
O2⁺
N2⁺
O⁺ Total
log (Q / m^-3 s^-1)
c
O⁺
CO2⁺
log (Q / m^-3 s^-1)

Fig. 162 a–c

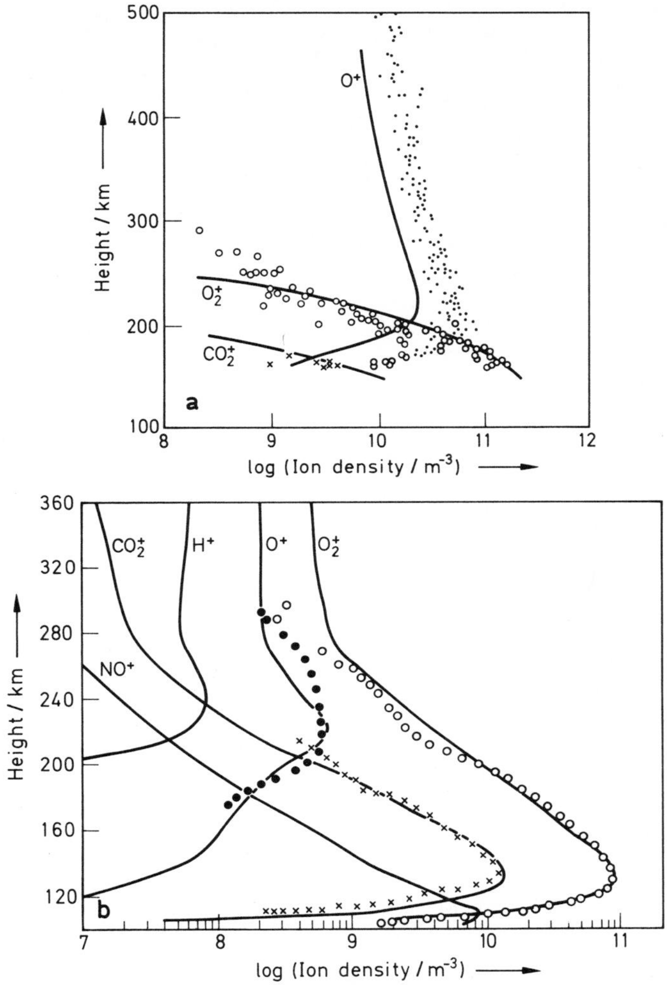

Fig. 163 a, b. Observed ion density profiles, by day. (**a**) In the Venusian upper atmosphere[55]; (**b**) in the Martian upper atmosphere[53] [symbols: ×, CO_2^+; ○, O_2^+; ●, O^+]. Curves were computed by chemistry-oriented aeronomical models

---

Fig. 162 a–c. Calculated solar energy deposition and photoionization rate profiles. (**a**) Energy deposition in the Martian atmosphere ($\chi = 45°$) compared with that for Earth ($\chi = 51°$)[53]. (**b**) Photoionization rates in the Martian ionosphere [1, from $O \to O^+ + e^-$; 2 from $CO_2 \to CO + O^+ + e^-$; 3, from $O_2 \to O^+ + e^-$][54]. (**c**) Photoionization rates ($\chi = 60°$) in the Venusian atmosphere [—— primary; ---- secondary][54] [p. 454]

---

[53] Mantas, G.P., Hanson, W.B. (1979): J. Geophys. Res. *84*, 369

[54] See (1980): J. Geophys. Res. *85*, no. 13A (Venus Orbiter Special edition), in particular: Cravens, T.E., Gombosi, T.I., Kozyra, J., Nagy, A.F., Brace, L.H., Knudsen, W.C., p. 7778 ff.

[55] Rawer, K., Spenner, K., Knudsen. W.C. (1982): Z. Flugwiss. Weltraumforsch. *6*, 147

[56] Hanson, W.B., Sanatani, S., Zuccaro, D.R. (1977): J. Geophys. Res. *82*, 4351

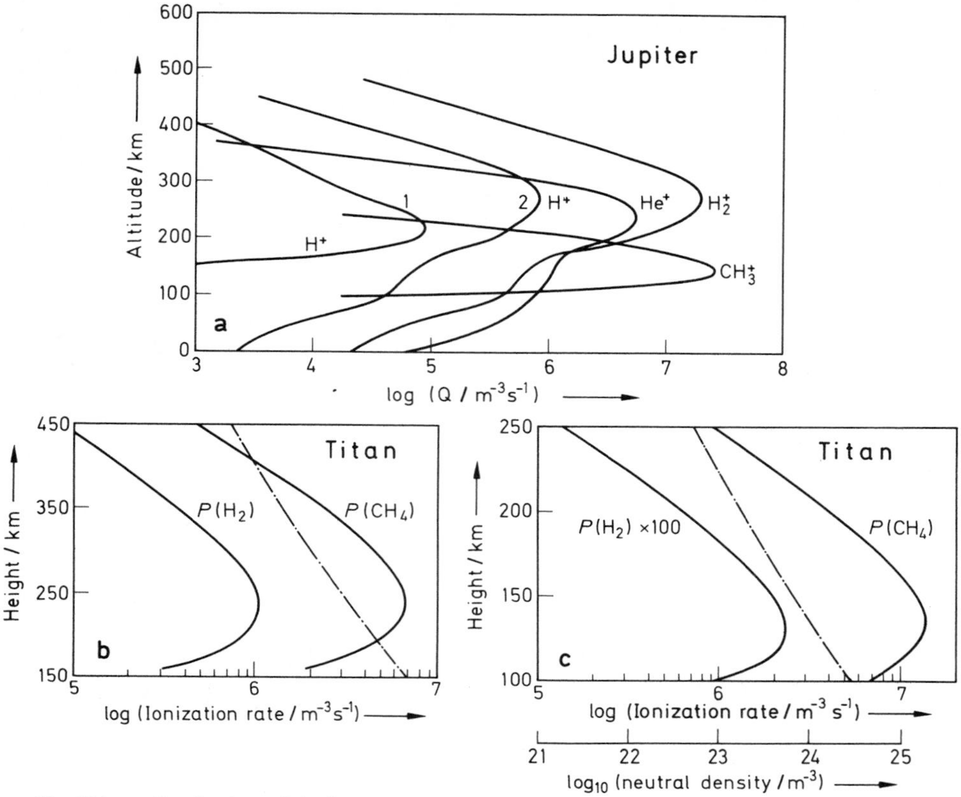

Fig. 164a–c. Production of ionization due to different causes at larger distances from the Sun. (**a**) By solar EUV in the Jovian atmosphere[57] [altitude is counted from a reference level where the numerical density reaches $10^{22}\,m^{-3}$]. Two different photoionization paths producing protons (H^+) are considered: 1, from $H \rightarrow H^+ + e^-$; 2, from $H_2 \rightarrow H^+ + H + e^-$; other ionization reactions are all direct. (**b, c**) By cosmic rays in the atmosphere of Titan (–·· neutral density with lower scale)[58]. Two different assumptions about $n(CH_4)/n(H_2)$ mixing rate r: (**b**) $r=0.5$; (**c**) $r=100$

ι) *Chemistry* is quite definitely different from that in the terrestrial atmosphere, where, due to organic life, free oxygen is present in large quantity[59]. Carbon dioxide is prevalent in the lower atmospheres of Mars and Venus. However, due to photodissociation and diffusion, atomic oxygen increases in relative importance and becomes prevalent above about 200 km. Therefore, CO_2^+ ions prevail in the lower ionosphere only (Fig. 162) but O^+ prevails at a greater height, where the ion chemical conditions are nearer to those above Earth[60] (Fig. 163; compare also SCHMIDTKE's Fig. 28 on p. 38 of this volume).

[57] Atreya, S.K., Donahue, T.M. (1976) in: Jupiter. Gehrels, T. (ed.), p. 304. Tucson, Ar., USA: University of Arizona Press

[58] Capone, L.A., Whitten, R.C., Dubach, J., Prasad, S.S., Huntress, W.T. Jr. (1976): Icarus *29*, 367

[59] Berkner, L.V., Marshall, L.C. (1965): J. Atmos. Sci. *2*, 225 (early notice). For a summary of the present ideas see: Henderson-Sellers, A. (1983): The origin and evolution of planetary atmospheres. Chapt. 4. Bristol: A. Hilger Ltd.

[60] Spenner, K., Knudsen, W.C., Whitten, R.C., Michelson, P.F., Miller, K.L., Novak, V. (1981): J. Geophys. Res. *86*, 9170

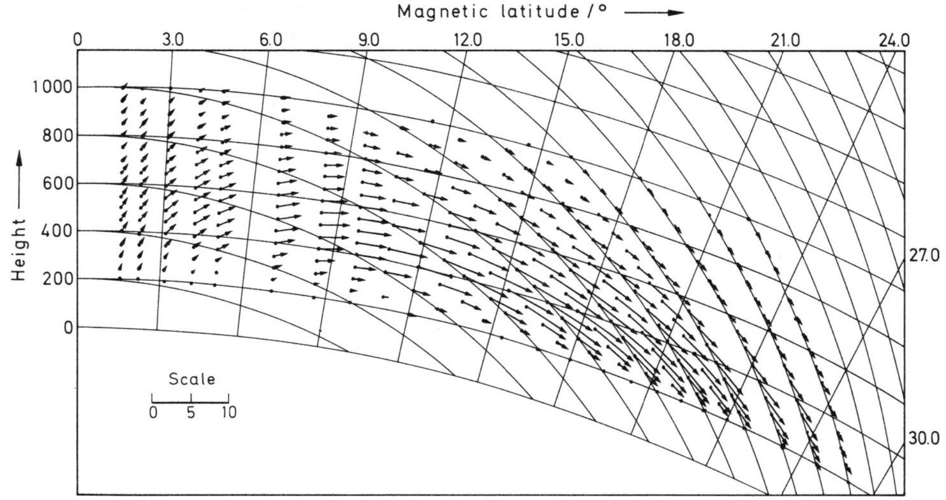

Fig. 165. Plasma drift due to combined action of electrodynamic lift and diffusion in the magnetized ionosphere of Earth[61] (see Sect. 15b for explanation) [flux in units of $10^{13}\,m^{-2}\,s^{-1}$, see scale bottom on left. Arrow length to be read after this scale]

Hydrogen and hydrocarbonic gases are very important in the atmospheres of the major planets and their natural satellites (see Fig. 164). Hydrogen chemistry plays a major role [*59*] (see Table 10, p. 88, in volume 49/6 of this Encyclopedia). At great altitudes, protons should prevail as ions but He^+ must also be present. Loss is assumed by charge transfer with hydrocarbonate or H_2 molecules [*59*].

Because precise information is lacking about the molecular ion species engaged in the dissociative recombination processes, chemical computations in planetary atmospheres are somewhat provisional.

κ) *Plasma transport* in the terrestrial upper ionosphere is largely field aligned; thus it is restricted to the meridional plane. A typical pattern is shown in Fig. 165. This should be similar for planets which have a magnetic field of their own. However, conditions are quite different with Venus, which has no noticeable field. This is of extreme importance insofar as, due to higher plasma pressure above the sunlit hemisphere, zonal plasma transport must be very important. Very high velocity flow of ions towards the terminators has in fact been observed[62].

As a consequence, the plasma above the night hemisphere stems from the day side, whence it arrives in a few minutes. Therefore, unlike Earth, only the plasma density is smaller above the night hemisphere, but composition and temperature are the same as for the day side (Fig. 166; see also Sect. 16). Very characteristic is a peak of ion (not electron) temperature above the midnight meridian (Fig. 166c); it is probably provoked by the collision of the two zonal plasma streams which cross the morning and evening terminator[62,55].

---

[61] Hanson, W.B., Moffett, R.J. (1966): J. Geophys. Res. *71*, 5559

[62] Knudsen, W.C., Spenner, K., Miller, K.L., Novak, V. (1980): J. Geophys. Res. *85*, 7803; Knudsen, W.C., Spenner, K., Miller, K.L. (1981): Geophys. Res. Lett. *8*, 241

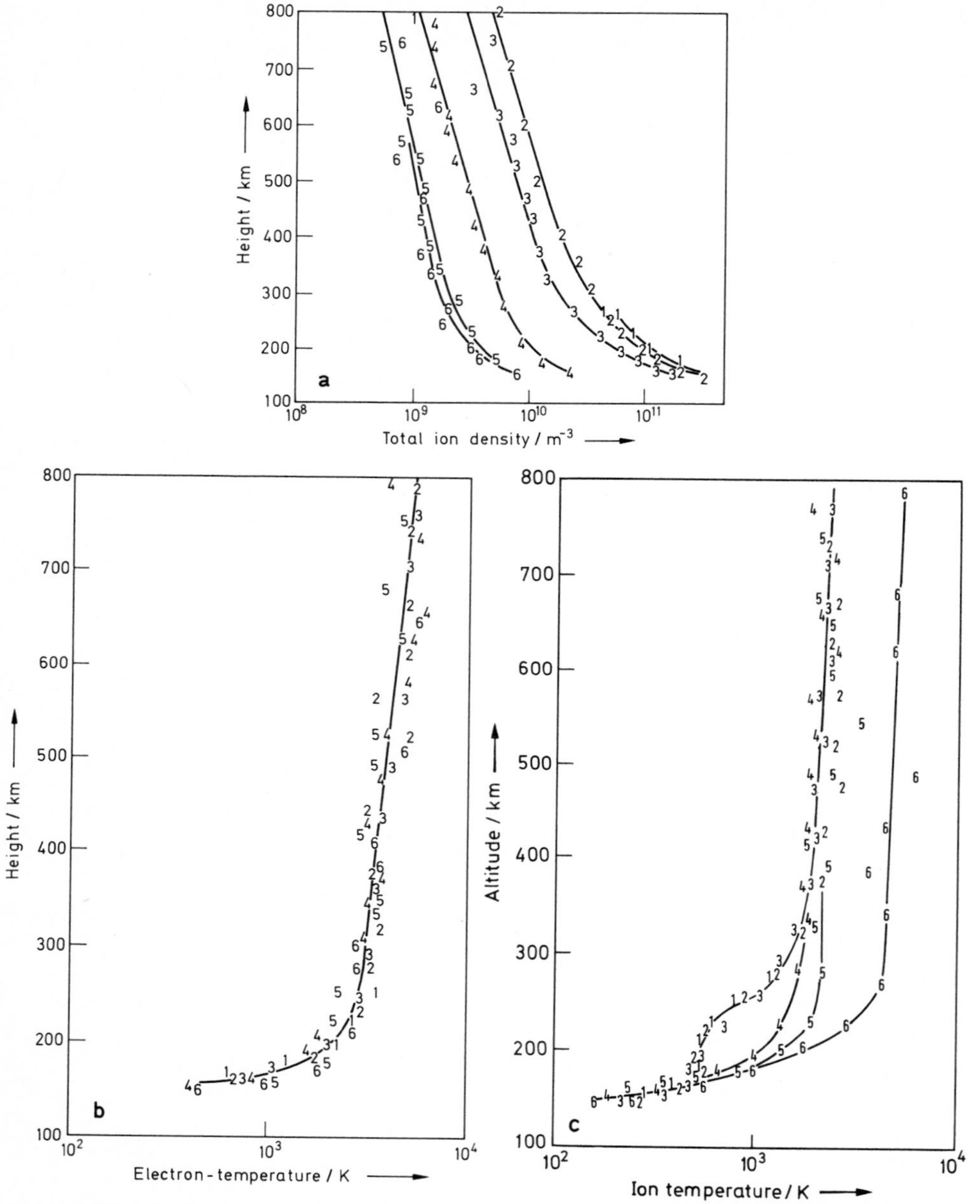

Fig. 166a–c. Venus ionosphere profiles of plasma parameters depending on solar zenith angle [numbers 1–6 designating angular ranges from 0° to 180° in 30° steps][55]. (**a**) Ion density; (**b**) electron temperature; (**c**) ion temperature

16. Thermospheric heat budget. In the early days of upper atmosphere research low temperatures were always assumed above the tropopause. In 1928 MARIS and HULBURT[1] felt that due to absorption of ionizing radiation the thermosphere should be at a temperature of around 1000 K. Under undisturbed conditions solar extreme ultraviolet (EUV) radiation is the main energy input into the thermosphere by production of ion pairs; the excess energy, i.e. the difference between quantum energy[2] $\mathscr{E}_\lambda = A/\lambda$ and ionization potential $\mathscr{E}_k$ appears as kinetic energy of the photoelectrons. These receive velocities far above the thermal velocity, thus acting as an internal heat source which mainly affects the upper thermosphere. However, the ionization energy also enters into the heat budget, because at night, when solar wave radiation is absent, the density of ionized constituents decreases so that eventually a certain part of the energy which had been stored by day becomes available (e.g. as excitation energy in the dissociative recombination processes). So, solar wave radiation is the most important source of energy for the thermosphere.

The fact that, by day, the heating goes first into a part of the electronic population has important consequences for the thermal balance between the different constituents. Since heat contact is best between electrons and electrons, less good with ions and even worse between ions and neutrals (see Sect. 13ε), and since furthermore the time it takes to reach equilibrium is long (due to low collision rates), there is *no thermal equilibrium* existing in the dayside thermosphere. This had been suspected rather early [*34*, *39*]; the first computation was made by DRUKAREV[3].

Thus, understanding the heat budget means not only in- and output of energy but also energy transfer between the different populations: hot electrons, thermal electrons, ions and neutrals. The specific temperatures are then determined by the relevant heat balance, i.e. by heating, cooling and energy flow together.

"The absorption of EUV radiation by the neutral atmosphere results in both photoionization and excitation of the neutral gases. The resulting excited atoms and molecules lose their energy in quenching collisions with electrons and other neutral particles and by radiation. Below about 250–300 km the photoelectrons lose their energy locally, but above this altitude they are able to escape from the F-region ionosphere along geomagnetic field lines. These escaping photoelectrons heat the ambient electron gas at high altitudes, and some of this energy is conducted back down into the F-region" [*54*]. The processes (locally) responsible for heating and cooling are shown in Fig. 167.

a) Energy inputs

α) The *heat deposit* by *photoelectrons* is found from their production rate Q_e [see Sect. 15, Eq. (15.2)] and the energy difference $(\mathscr{E}_\lambda - \mathscr{E}_k)$ so that one has

$$Q_h = \int_{EUV} d\lambda \cdot (\mathscr{E}_\lambda - \mathscr{E}_k) \cdot \sum n_i(z) \sigma_e^i(\lambda) \cdot \varphi(\lambda), \tag{16.1}$$

[1] Maris, H.B., Hulburt, E.O. (1928): Terr. Magn. Atmos. Elec. *33*, 229

[2] Rees, M.H., Roble, R.G. (1975): Rev. Geophys. Space Phys. *13*, 201

[3] Drukarev, G. (1946): J. Phys. USSR *10*, 81

a

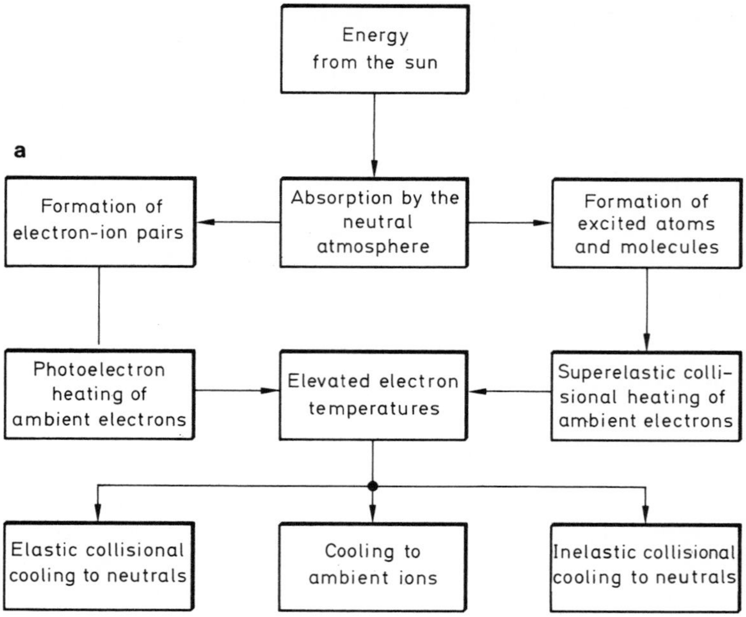
Energy from the sun
Absorption by the neutral atmosphere
Formation of electron-ion pairs
Formation of excited atoms and molecules
Photoelectron heating of ambient electrons
Elevated electron temperatures
Superelastic collisional heating of ambient electrons
Elastic collisional cooling to neutrals
Cooling to ambient ions
Inelastic collisional cooling to neutrals

c

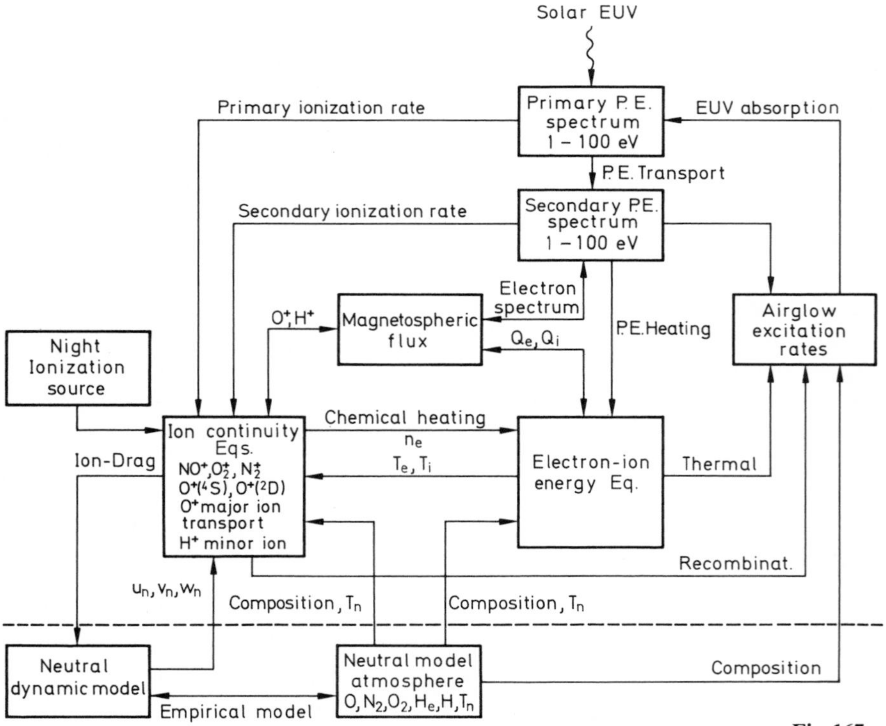
Solar EUV
Primary ionization rate
Primary P.E. spectrum 1 – 100 eV
EUV absorption
P.E. Transport
Secondary ionization rate
Secondary P.E. spectrum 1 – 100 eV
Electron spectrum
O⁺, H⁺
Magnetospheric flux
Qe, Qi
P.E. Heating
Airglow excitation rates
Night Ionization source
Chemical heating
ne
Te, Ti
Ion continuity Eqs. NO⁺, O2⁺, N2⁺ O⁺(⁴S), O⁺(²D) O⁺ major ion transport H⁺ minor ion
Ion-Drag
Electron-ion energy Eq.
Thermal
Recombinat.
un, vn, wn
Composition, Tn
Composition, Tn
Neutral dynamic model
Neutral model atmosphere O, N2, O2, He, H, Tn
Composition
Empirical model

Fig. 167a, c

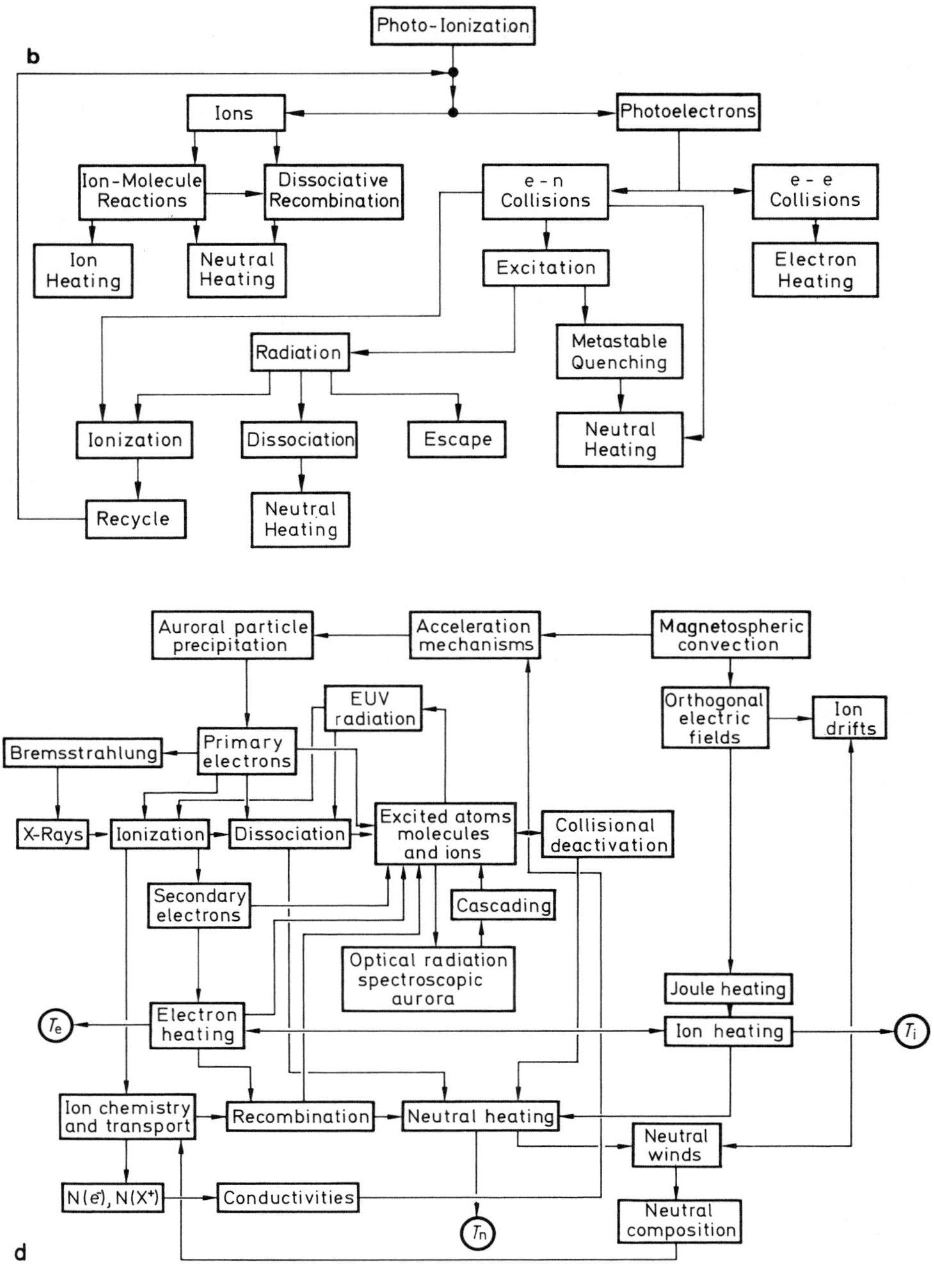

Fig. 167. (**a**) Electron energy flow in the quiet day-side ionosphere [*54*]. (**b**) Same for auroral particle precipitation and magnetospheric convection[3]. (**c**) Aeronomical computing scheme for the ionosphere of NCAR (USA)[3]. (**d**) Computing scheme for auroral disturbance conditions[3,4]

[4] Rees, M.H. (1969): Space Sci. Rev. *10*, 413; (1975): Planet. Space Sci. *23*, 1589 (see [*54*] for further references)

where i designates the different neutral constituents which are ionized and $\varphi(\lambda)$ is the photon flux given by Eq. (15.3). The subject is dealt with by SCHMIDTKE in Chap. III of his contribution in this volume; see in particular his Figs. 25 and 29, and the EUV-energy flux chart shown in his Fig. 30, p. 41.

Even by day some heating effect is also due to dissociative recombination as a consequence of the higher recombination rate for cold electrons. So dissociative recombination preferentially removes the colder electrons. However, the effect of this particular heating mechanism is only a few per cent of that by photoelectrons [5].

Deactivation of excited species by quenching [*64*], i.e. by inelastic collisions of neutral or ionized molecules/atoms in metastable states, is another heating mechanism. It was calculated for atomic oxygen ions O^+, with the result that $O^+(^2D)$ deactivation contributes only above an altitude of 300 km [6] with 5%–18% deposit described by Eq. (16.1). This problem is taken up again in Sect. 17 and in Appendix A.

Energetic electron precipitation is a very important energy source in the daytime polar cusp and nocturnal auroral oval (see Sect. 18). Energy is transferred by Coulomb collisions; see SUCHY's contribution in this volume, his Sect. 12. The primary electrons precipitate down onto the ionosphere with energies of the order of a few tens of keV, ionizing and exciting the ambient neutrals. The secondary electrons produced in these collisions have energies of tens of eV; finally, a large number of tertiary electrons reach a few eV, which is most efficient for energy transfer via electronically, vibrationally and rotationally excited molecules, atoms and ions [4,7,8]. In particular electrons of 1 and up to 2 eV transfer an important fraction of their energy in this way [9]. Flow charts are shown in Fig. 167.

Another heat source at high latitudes is electric fields provoking plasma motion perpendicular to the magnetic field. The heating rate remains small [*54*], but currents induced in the ionosphere during magnetic perturbations are an important source now known as *Joule heating*; see Sect. 18.

β) Heat transfer from more energetic electrons (e.g. photo- or precipitating electrons) to the thermal electron population depends on the flux $\Phi(\mathscr{E})$ in the energy spectrum of the "hot" electrons. A typical example of such a spectrum under rather quiet conditions is shown in Fig. 168. In earlier calculations one simply used the spectrum above [10] a certain minimum energy $\mathscr{E}_1$:

$$Q_{\mathrm{he}}(z)=\int_{\mathscr{E}_1}^{\infty} \mathrm{d}\mathscr{E}\cdot\mathscr{L}_{\mathrm{ce}}\cdot\frac{\mathrm{d}\Phi(\mathscr{E},z)}{\mathrm{d}\mathscr{E}}, \tag{16.2a}$$

where $\mathscr{L}_{\mathrm{ce}}$ is the collisional energy loss rate per unit distance of an electron of energy $\mathscr{E}$:

$$\mathscr{L}_{\mathrm{ce}}=\alpha N_{\mathrm{n}}\sum_j \alpha_j \varepsilon_j \sigma_j(\mathscr{E}),$$

[5] Hamlin, D.A., Myers, B.F. (1974): Planet. Space Sci. *22*, 1343

[6] Rohrbangh, R.P., Swartz, W.E., Simonaitis, R., Nisbet, J.S. (1973): Planet. Space Sci. *21*, 159

[7] Banks, P.M., Chappell, C.R., Nagy, A.F. (1974): J. Geophys. Res. *79*, 1459; see also Banks, P.M., Nagy, A.F. (1970): J. Geophys. Res. *75*, 1902, 6260

[8] Strickland, D.J., Book, D.L., Coffey, T.P., Fedder, J.A. (1976): J. Geophys. Res. *81*, 2755

[9] Rees, M.H., Jones, R.A. (1973): Planet. Space Sci. *21*, 1213

[10] The assumptions about the lower limit of the integral are of extreme importance, see Appendix A, p. 525

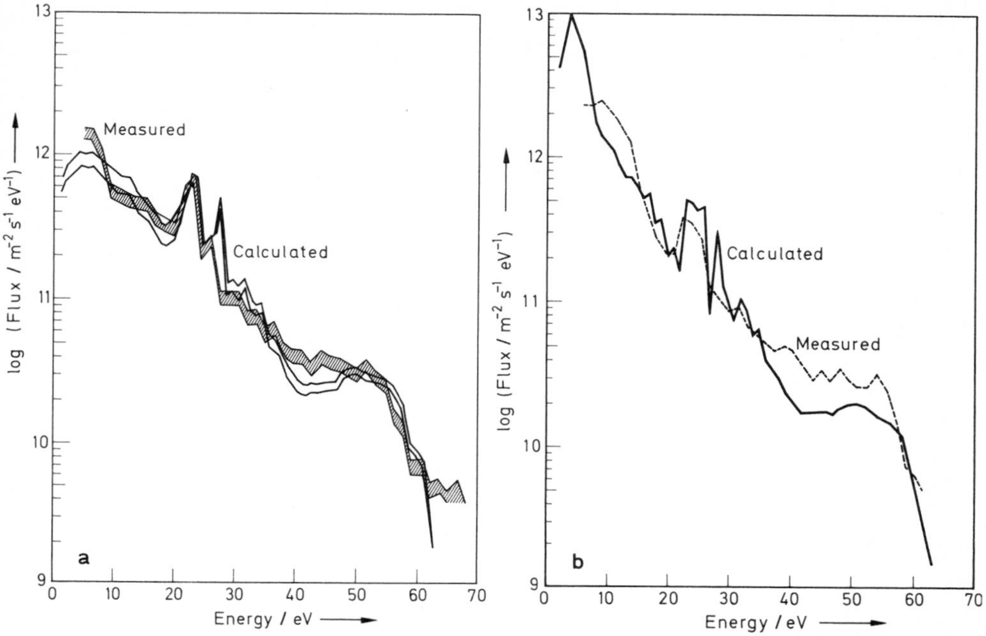

Fig. 168a, b. Measured and calculated (hemispherical upwards) photoelectron fluxes obtained in 1974 with data of the AE-C satellite. (**a**) Altitude 180 km [17 Feb.]; (**b**) altitude 360 km [10 Dec.]

α_j being an efficiency (of the order of 0.5) and ε_j the relevant excitation energy [3]. However, when applying statistical (kinetic) theory HOEGY[11] found that one more term is needed, and wrote:

$$Q_{\mathrm{he}}(z)=\int_{\mathscr{E}_T}^{\infty} \mathrm{d}\mathscr{E}\cdot\mathscr{L}_{\mathrm{ce}}\cdot\frac{\mathrm{d}\Phi}{\mathrm{d}\mathscr{E}}+(\mathscr{E}_T-\tfrac{3}{2}kT_{\mathrm{e}})\cdot\mathscr{L}_{\mathrm{ce}}\cdot\frac{\mathrm{d}\Phi(\mathscr{E}_T,z)}{\mathrm{d}\mathscr{E}}, \qquad (16.2\mathrm{b})$$

$\mathscr{E}_T$ being the energy value at which thermal and photoelectron fluxes are equal (the low energy bend in Fig. 222, Appendix A). The additional term contributes almost as much as the first one.

Furthermore, at higher energies account must be taken of quantum effects and ČERENKOV radiation. A quite involved expression was derived[12,13] from which it could be shown by numerical approximation that the (energy-dependent) heat transfer is proportional to [54][14]

$$\mathscr{E}^{-0.94}\cdot N_{\mathrm{e}}^{-0.03}[(\mathscr{E}-\mathscr{E}_{\mathrm{th}})/(\mathscr{E}-0.53\,\mathscr{E}_{\mathrm{th}})]^{2.36}, \qquad (16.2\mathrm{c})$$

[11] Hoegy, W.R. (1977): Paper presented at IAGA/IAMAP Symposium at Seattle

[12] Takayanagi, K., Itikawa, Y. (1970): Space Sci. Rev. *11*, 380. Tabulations can be found in [13]

[13] Schunk, R.W., Walker, J.C.G. (1971): J. Geophys. Res. *76*, 6159; Schunk, R.W., Hays, P.B. (1971): Planet. Space Sci. *19*, 113

[14] Swartz, W.E. (1976): J. Geophys. Res. *81*, 183

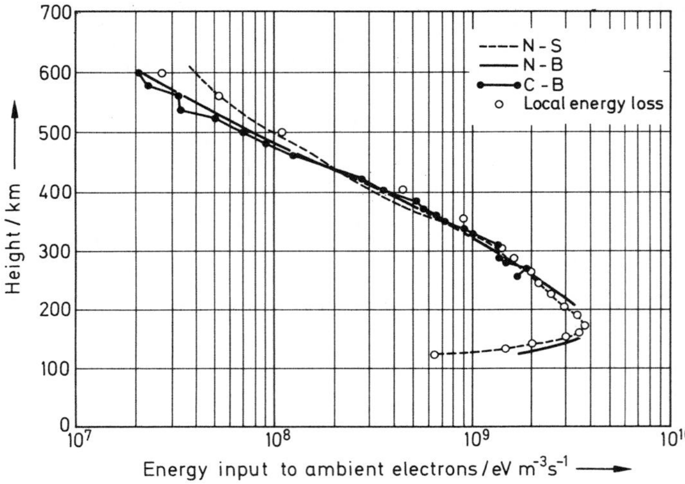

Fig. 169. Thermal electron heating profile; results of model computations by four different techniques[16] (compare Fig. 223, p. 531)

with[15] $\mathscr{E}_{\text{th}} = kT_e$, the mean energy of the thermal electrons. At higher values of $\mathscr{E}$ this makes the energy transfer decrease approximately like $1/\mathscr{E}$.

Energy loss of photoelectrons by elastic collisions with ions is negligible since the mass ratio is so great that the electron energy is almost conserved at a collision. Therefore the photoelectron degradation in a plasma is only due to electron-electron collisions.

γ) At altitudes above about 250 km the free path of photoelectrons becomes rather large so that *transport of photoelectrons* becomes important. It is, of course, mainly field aligned. There are, basically, two approaches to this problem: (a) numerical solution of some transport equation[16] (usually derived from the Boltzmann equation); (b) statistical trial methods ("Monte Carlo method")[17]. Electron heating rates obtained with different methods are shown in Fig. 169. In the so-called two-stream approach (BANKS et al.[7]) the in- and outgoing electron streams are evaluated taking account of inelastic and elastic collisions, which provokes backscatter and spectral degradation; see also[18]. Coupling between both streams is, of course, largest at the lower boundary of the transport range; see Appendix A for more details.

In other computations[19,20] concerned with lower altitudes, one distinguishes the different pitch angles (against the magnetic field); this "multistream" approach ends up with a spectrum which agrees quite well with in situ measurements[21], see Fig. 168.

---

[15] Numerically: $\mathscr{E}_e/\text{eV} = 8.617 \cdot 10^{-5}\ (T_e/\text{K})$

[16] Cicerone, R.J., Swartz, W.E., Stolarski, R.S., Nagy, A.F., Nisbet, J.S. (1973): J. Geophys. Res. *78*, 6709. More references to transport calculations can be found in [*54*]

[17] Cicerone, R.J., Bowhill, S.A. (1970): Radio Sci. *5*, 49; (1971): J. Geophys. Res. *76*, 8299

[18] Schunk, R.W. (1975): Planet Space Sci. *23*, 437; (1977): Rev. Geophys. Space Phys. *15*, 429

[19] Victor, G.A., Kirby-Docken, K., Dalgarno, A. (1976): Planet. Space Sci. *24*, 679

[20] Torr, M.R., Torr, D.G., Roble, R.G. (1981): [*53*], 22-1

[21] Doering, J.P., Bostrom, C.O., Armstrong, J.C. (1975): J. Geophys. Res. *80*, 3934

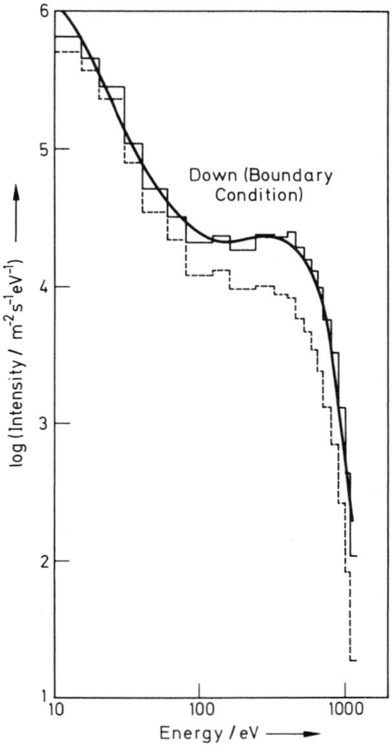

Fig. 170. 'Hemispherical' backscattered electron flux obtained from an electron beam of mean flux intensity of $2\pi \cdot 10^4 \exp[-(\frac{1}{2}(\mathscr{E}/\mathscr{E}_0-1))^2]\,\mathrm{m^{-2}\,s^{-1}\,sr^{-1}\,eV^{-1}}$ with $\mathscr{E}_0=500\,\mathrm{eV}$ incident on a $T_{ex}=1000\,\mathrm{K}$ molecular nitrogen atmosphere [*11*] with 'top' at 400 km. Smooth curve obtained by multistream, step curve by two-stream computations[22]. The dashed step curve was obtained earlier with an incorrect backscatter ratio

Another approach was made by STAMNES[22] who used an analytical solution and insists particularly on the importance of the backscatter ratio (see Fig. 170). However, with the cross sections as actually known the "two stream approach"[7] appears to be not too bad. A more detailed discussion can be found in Appendix A.

b) Transfer of energy from electrons to other constituents. Collisional interaction in the upper thermosphere is large enough to allow for energy transfer by different mechanisms. These are, however, not so important that (by day) the (thermal) electron temperature T_e would be cooled down to that of the ions, T_i, or neutrals, T_n. Inelastic collisions play a big role in the transfer to the neutrals while elastic (Coulomb) collisions are a major transfer path from thermal electrons to ions. The equilibrium value of T_e is determined by the balance of heat input and losses; this explains the rather important variations of T_e as compared with the other constituent temperatures. It must further be noted that Coulomb collisions between photoelectrons and ions, due to the large energy difference, have small effective cross sections. Thus the heat transfer to the ions goes over the thermal electrons.

[22] Stamnes, K. (1980): Planet. Space Sci. *28*, 427; (1981): J. Geophys. Res. *86*, 2405

δ) (i) *Coulomb collisions* between electrons and ions are discussed by SUCHY (his Sect. 12) in this volume. The average cooling rate of the thermal electrons (which, essentially, equals the heating rate of the ions) decreases with increasing electron temperature T_e according to:

$$L_{e,i} = \sim \frac{m_e \cdot m_i}{(m_e + m_i)^2} N_e N_i \frac{k(T_e - T_i)}{(k T_e)^{3/2}} \ln \Lambda. \tag{16.3}$$

$\ln \Lambda$ is the so-called Coulomb logarithm[23] and is of the order of 10. For a mixture of ions as in the terrestrial ionosphere one has [*54*]:

$$\begin{aligned} L_{e,i}/\text{eVm}^{-3}\text{s}^{-1} = 3.2 \cdot 10^{-14} \ln \Lambda \cdot \frac{(T_e - T_i)/\text{K}}{(T_e/\text{K})^{3/2}} \cdot \frac{N_e}{\text{m}^{-3}} \\ \cdot [N(\text{O}^+) + 4N(\text{He}^+) + 16N(\text{H}^+) \\ + \tfrac{1}{2} N(\text{O}_2^+) + 0.53 N(\text{NO}^+)]/\text{m}^{-3}. \end{aligned} \tag{16.3a}$$

Accounting only for the major ion of the upper thermosphere O^+, one may write [*17*]

$$L_{e,i}/\text{eVm}^{-3}\text{s}^{-1} = 4.8 \cdot 10^{-13} \frac{N_e}{\text{m}^{-3}} \frac{N(\text{O}^+)}{\text{m}^{-3}} (T_e/\text{K})^{-3/2} \cdot \frac{(T_e - T_i)}{\text{K}}. \tag{16.3b}$$

This is the principal heat input to the ions. It is restricted in size due to the small mass ratio m_e/m_i. On the other hand, the ion component itself is cooled by ion-neutral collisions which have a better mass ratio but a much smaller efficiency, since the induced dipole attraction as compared with the far-reaching Coulomb force is rather small (see THOMAS in vol. 49/6, pp. 41, 42, the momentum transfer cross section being given by his Eq. [10.9]).

(ii) Even smaller is the effect of *elastic collisions* between electrons and neutral atoms or molecules; this follows from the mass ratio. Taking account of the temperature-dependent collision frequencies and cross sections[23], the cooling rate is largest with hydrogen:

$$\begin{aligned} L_{e,H}/\text{eVm}^{-3}\text{s}^{-1} = 0.963 \cdot 10^{-21} \cdot \frac{N_e}{\text{m}^{-3}} \cdot \frac{n(H)}{\text{m}^{-3}} \\ \cdot [1 - 1.35 \cdot 10^{-4} T_e/\text{K}] \cdot (T_e/\text{K})^{1/2} \cdot \frac{(T_e - T_n)}{\text{K}}, \end{aligned} \tag{16.4a}$$

about equal for oxygen molecules and atoms:

$$\begin{aligned} L_{e,O}/\text{eVm}^{-3}\text{s}^{-1} = 0{,}79 \cdot 10^{-24} \cdot \frac{N_e}{\text{m}^{-3}} \frac{n(\text{O})}{\text{m}^{-3}} \\ \cdot [1 + 5.7 \cdot 10^{-4} T_e/\text{K}] \cdot (T_e/\text{K})^{1/2} \cdot \frac{(T_e - T_n)}{\text{K}}, \end{aligned} \tag{16.4b}$$

and smallest for molecular nitrogen [*54*].

---

[23] Itikawa, Y. (1973): Momentum transfer cross sections for electron collisions on atoms and molecules ... Rept. ANL-7939. Argonne, Ill. (USA): Argonne National Laboratory

ε) (i) Also small is the rate due to *collisional excitation* of *metastable electronic levels* in the atmospheric species. Most important [see curve labelled O(^{1}D) in Fig. 171 below] is the excitation of O(^{1}D) which is at the origin of the forbidden auroral and night sky red line at 630 nm [see Fig. 23, p. 48 in vol. 49/6, THOMAS]. An analytical expression is given in [24].

(ii) Larger cooling rates, in particular at lower heights, are due to *rotational excitation* of *molecules*. With BORN's well-known approximation the rotational loss rate is easily computed[25] (for a molecular constituent X) as:

$$L_{\mathrm{e,X\,rot}}/\mathrm{eV\,m^{-3}\,s^{-1}} = [1.17\cdot 10^{-16} p_Q^2 b]\cdot\frac{N_e}{\mathrm{m}^{-3}}\cdot\frac{n(\mathrm{X})}{\mathrm{m}^{-3}}(T_e/\mathrm{K})^{-\frac{1}{2}}\frac{(T_e-T_n)}{\mathrm{K}}, \quad (16.5)$$

with p_Q the quadrupole moment and b the rotational constant of the molecule. For O_2 which has a small p_Q the numerical factor becomes[26] $6.9\cdot 10^{-20}$.

Short-range interaction which is not taken account of in the BORN approximation may, however, not be negligible, particularly with O_2.

(iii) *Vibrational excitation* is of comparable importance. STUBBE and VARNUM[24] give for N_2 (which is the most important molecular constituent of the upper thermosphere):

$$L_{\mathrm{e,N_2\,vibr}}/\mathrm{eV\,m^{-3}\,s^{-1}} = 2.99\cdot 10^{-18}\frac{N_e}{\mathrm{m}^{-3}}\frac{n(\mathrm{N_2})}{\mathrm{m}^{-3}}\exp[f\cdot(5\cdot 10^{-4} - \mathrm{K}/T_e)]$$
$$\cdot[\exp(-g\cdot(T_e/T_n - 1)/(T_e/\mathrm{K}) - 1], \quad (16.6)$$

with $f = 1.06\cdot 10^4 + 7.51\cdot 10^3 \tanh[1.1\cdot 10^{-3}(T_e/\mathrm{K} - 1{,}800)]$ and $g = 3{,}300 + 1.233(T_e/\mathrm{K} - 1{,}000) - 2.056\cdot 10^{-4}(T_e/\mathrm{K} - 1{,}000)(T_e/\mathrm{K} - 4{,}000)$. The O_2 cooling rate was computed in[27].

(iv) The ground state ^{3}P of *atomic oxygen* being threefold (see Fig. 23, p. 48 in vol. 49/6, THOMAS), the relevant *fine structure* can be excited with quite low energy, so that this constitutes a particularly efficient cooling mechanism for thermal electrons[28]. Unlike older computations, which gave larger values[28], HOEGY[29] found

$$L_{\mathrm{e,O\,fine}}/\mathrm{eV\,m^{-3}\,s^{-1}} = 8.63\cdot 10^{-12}\zeta\cdot\frac{N_e}{\mathrm{m}^{-3}}\frac{n(\mathrm{O})}{\mathrm{m}^{-3}}\cdot\sum A\cdot B\cdot(T_e/\mathrm{K})^{|B-\frac{1}{2}|}$$
$$\cdot\{\varepsilon(D_x - E_x) + 5.91\cdot 10^{-9}(T_n - T_e)/\mathrm{K}$$
$$\cdot[(1+B)D_x + (F/T_e + 1 + B)E_x]\} \quad (16.7)$$

where

$$\zeta = [5 + 3\exp(-228\,\mathrm{K}/T_1) + \exp(-326\,\mathrm{K}/T_0)]^{-1}$$

is an efficiency factor. The summation goes over all three transitions: $1\rightarrow 2$, $0\rightarrow 2$ and $0\rightarrow 1$. $D_x \equiv \exp(-D)$; $E_x \equiv \exp(-E)$; temperatures T_0 and T_1 correspond to the relevant levels $J=0$ and 1. The numerical coefficients can be found in Table 11.

---

24 Stubbe, P., Varnum, W.S. (1972): Planet. Space Sci. *20*, 1121
25 Mentzoni, M.H., Rao, R.V. (1963): Phys. Rev. *130*, 2312
26 Dalgarno, A. (1968): Adv. At. Mol. Phys. *4*, 381
27 Prasad, S.S., Furman, D.R. (1973): J. Geophys. Res. *78*, 6701
28 Dalgarno, A., Degges, T.P. (1968): Planet. Space Sci. *16*, 125
29 Hoegy, W.R. (1976): Geophys. Res. Lett. *3*, 541

Table 11. Atomic oxygen fine structure cooling [numerical values to Eq. (16.7)][29]

Transition	$10^6 \cdot A$	B	$-D/\text{K}$	$-E/\text{K}$	F/K	ε
$1 \to 2$	8.58	1.008	$228/T_1$	$228/T_e$	228	0.02
$0 \to 2$	7.201	0.9617	$326/T_0$	$326/T_e$	326	0.028
$0 \to 1$	0.2463	1.1448	$326/T_0$	$98/T_e + 228/T_1$	98	0.008

In a recent experiment the low F-region was artificially heated by high-power hf radio waves emitted from a ground station. The cooling effect could be directly measured by the relaxation time after stopping the emission. The results are felt to be perhaps agreeable with the lowest cooling rates theoretically assumed, but lower rates cannot be excluded[30]; these might explain the discrepancies found when comparing satellite data with aeronomical theory[31].

As explained below under (v) recent optical data favour a much lower numerical factor in Eq. (16.7), so that the importance of the effect has become doubtful.

(v) The *relative importance* of the different cooling rates can be seen from Fig. 171. Above 200 km Coulomb collisions with ions are most important, followed by, if Eq. (16.7) is correct, $O(^3p)$ fine structure cooling. Around 150 km under the same assumption, this latter mechanism is most important followed by vibrational and rotational excitation. Elastic collisions with neutrals contribute a few per cent at best.

Excitation energy is, of course, only provisionally stored. After a more or less short lifetime it is given up either by some kind of optical emission, or in quenching collisions. Only in the latter condition, which depends on a metastable atomic of molecular state, does it help heat up the neutral gas locally. Otherwise the energy is optically emitted and disappears from the heat budget of the thermosphere.

In particular the fine structure excitation should produce a strong infrared emission at 63 μm [by J: $1 \to 2$, against which the $0 \to 1$ emission at 147 μm is of negligible intensity[32]].

Recent in-situ measurements of the 63 μm radiation[33] between 80 and 100 km show an emission peak of 8 $\mu\text{W m}^{-2}\,\text{sr}^{-1}$ which is very small compared with the older theoretical estimates given under (iv). It is now assumed that, above 120 km, infrared cooling is due to the 53 μm emission of the NO molecule rather than the fine structure of 0. Below 120 km the 15 μm emission of CO_2 is most important for the neutral heat balance[34,35].

[30] Carlson, H.C. Jr., Mantas, G.P. (1982): J. Geophys. Res. *87*, 4515

[31] Brace, L.H., Hoegy, W.R., Mayr, H.G., Victor, G.A., Hanson, W.B., Reber, C.A., Hinteregger, H.E. (1976): J. Geophys. Res. *81*, 5421

[32] Kockarts, G. (1971): [*68*], 390

[33] Grossmann, K.U., Offermann, D. (1978): Nature (London) *276*, 594

[34] Kockarts, G. (1982): paper C. 2.2.9 presented at COSPAR Gen. Ass. in Ottawa

[35] Gordiets, B.F., Kulikov, Yu.N., Markov, M.N., Marov, M.Ya. (1982): J. Geophys. Res. *87*, 4504

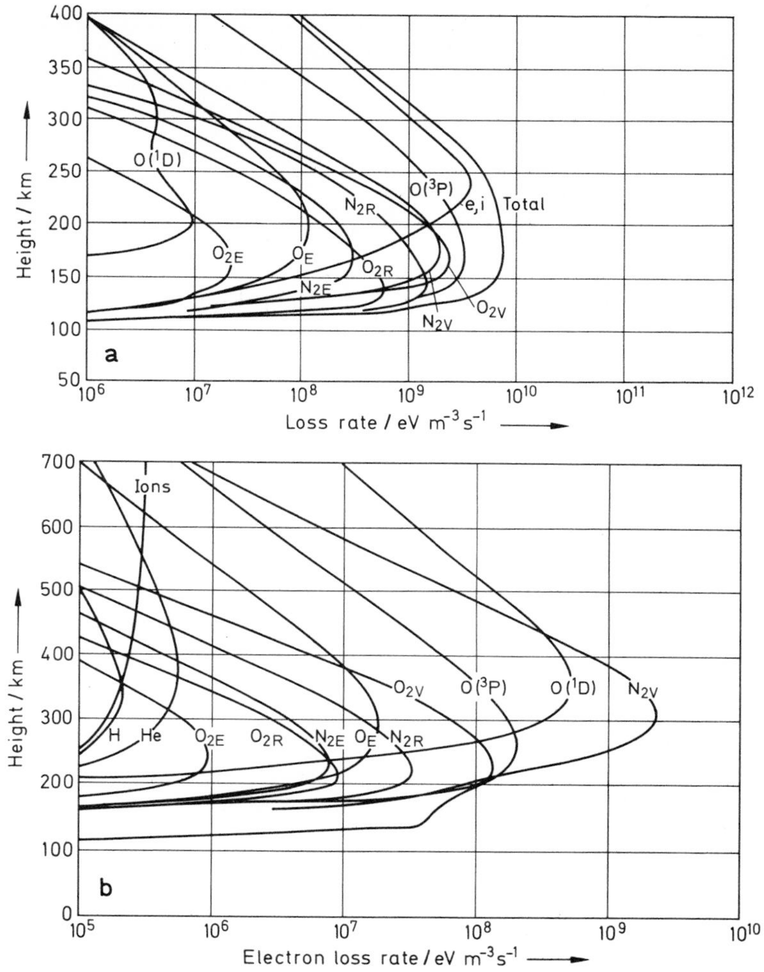

Fig. 171 a, b. Height profiles of electron energy loss rates by different mechanisms. (**a**) During (quiet) daytime[36]; (**b**) in the auroral zone during a typical SAR arc[37]. Chemical symbols used for identification; e, i, Coulomb [Sect. 16 δ(i)]; E, elastic collisions [δ(ii)]; O(^{1}D), electronic state excitation [ε(i)]; R, rotational [ε(ii)] and, V, vibrational excitation of molecules [ε(iii)]; O(^{3}P), fine structure excitation [ε(iv)]. (Note that O(^{3}P) was computed with Eq. (16.7), which probably assumes too high an efficiency of the fine structure cooling effect)

c) *Ion heat balance*

ζ) By day, the ionic population is mainly heated by the thermal electron population via Coulomb collisions. Two major heat loss mechanisms must be considered, viz. collisions with neutrals and, at higher altitudes heat conduction. Since ionic and neutral masses are comparable the heat transfer

[36] Perkins, F.W., Roble, R.G. (1978): J. Geophys. Res. *83*, 1611

[37] Rees, M.H., Roble, R.G. (1975): Rev. Geophys. Space Phys. *13*, 201

efficiency of collisions is high but the cross sections are much smaller than for Coulomb collisions [see Table 7, p. 42 in vol. 49/6, Thomas]. Also, due to the rather large mass of the ions, heat conductivity in the ion gas is much smaller than for electrons. Thus under equilibrium conditions the heating rate of Eq. (16.3) must be equal to the sum of losses of both kinds [*17*]:

$$L_{\mathrm{ei}} = L_{\mathrm{in}} - \frac{\partial}{\partial \boldsymbol{r}} \cdot \left(K_{\mathrm{i}} \frac{\partial}{\partial \boldsymbol{r}} T_{\mathrm{i}}\right), \tag{16.8}$$

and L_{in} is proportional to $(T_{\mathrm{i}} - T_{\mathrm{n}})$; see Eqs. (16.4). Equation (16.8) gives for vertical stratification

$$L_{\mathrm{ei}} = L_{\mathrm{in}} - \frac{\mathrm{d}}{\mathrm{d}z} \left(K_{\mathrm{i}} \frac{\mathrm{d}}{\mathrm{d}z} T_{\mathrm{i}}\right). \tag{16.8a}$$

The first (collisional) member on the right-hand side is quite large in the lower ionosphere, such that up to 200 km the difference $(T_{\mathrm{i}} - T_{\mathrm{n}})$ is very small. At very large heights, this term is negligible and T_{i} approaches T_{e}.

d) *Heat transport and budget.* Heat can be transported in an ionized atmosphere by conduction, by material transport and by radiation. The *temperature* change depends on the local rate of change of the total energy density, which is, according to Blum [38]:

$$\varepsilon^* = \rho\left(\tfrac{1}{2} v^2 + e_{\mathrm{i}} + u\right),$$

e_{i} being the internal energy per unit mass. ρ is mass density, $\boldsymbol{v}$ bulk velocity and u potential energy.

η) The *rate of change of energy density* (in a fixed volume element) determines the change in temperature; it is

$$\frac{\mathrm{d}\varepsilon^*}{\mathrm{d}t} \equiv \frac{\partial \varepsilon^*}{\partial t} + \frac{\partial}{\partial \boldsymbol{r}} \cdot \varepsilon^* \boldsymbol{v} \equiv \rho \frac{\partial \varepsilon}{\partial t} + \varepsilon \frac{\partial \rho}{\partial t} + \varepsilon \frac{\partial}{\partial \boldsymbol{r}} \cdot \rho \boldsymbol{v} + \rho \boldsymbol{v} \cdot \frac{\partial}{\partial \boldsymbol{r}} \varepsilon$$

$$= -\frac{\partial}{\partial \boldsymbol{r}} \cdot p \boldsymbol{v} + \frac{\partial}{\partial \boldsymbol{r}} \cdot \left(K \frac{\partial}{\partial \boldsymbol{r}} T\right) + Q, \tag{16.9}$$

with $\frac{\partial}{\partial \boldsymbol{r}} \equiv \nabla$ and $\varepsilon \equiv \varepsilon^*/\rho$. Q specifies the balance of external heat in- and output, e.g. by radiation. p is pressure and K specific heat conductivity. Using the equations of motion and of continuity one finally gets [38]:

$$\frac{\partial \varepsilon}{\partial t} = \boldsymbol{v} \cdot \frac{\partial}{\partial \boldsymbol{r}} \left(\tfrac{1}{2} v^2\right) + \frac{1}{\rho} \boldsymbol{v} \cdot \frac{\partial}{\partial \boldsymbol{r}} p - v \cdot \frac{\partial}{\partial \boldsymbol{r}} u + \frac{\partial e_{\mathrm{i}}}{\partial t} + v \cdot \frac{\partial}{\partial \boldsymbol{r}} e_{\mathrm{i}} \tag{16.10a}$$

and

$$\rho \frac{\partial \varepsilon}{\partial t} + \rho \boldsymbol{v} \cdot \frac{\partial}{\partial \boldsymbol{r}} \varepsilon = -\frac{\partial}{\partial \boldsymbol{r}} \cdot p \boldsymbol{v} + \frac{\partial}{\partial \boldsymbol{r}} \cdot K \frac{\partial}{\partial \boldsymbol{r}} T + Q, \tag{16.10b}$$

[38] Blum, P.W. (1971): [*68*], 301

which is the energy balance. Applying a few reasonable simplifications BLUM[38] finally obtains the change of temperature from the above equations (C_v is specific heat at constant volume):

$$\rho C_v \frac{dT}{dt} = -p \frac{\partial}{\partial \boldsymbol{r}} \cdot \boldsymbol{v} + \frac{\partial}{\partial \boldsymbol{r}} \cdot \left(K \frac{\partial}{\partial \boldsymbol{r}} T \right) + Q. \tag{16.10c}$$

As the first (advective) member on the right-hand side is of some importance only under particular circumstances, one has mainly a balance between heat conduction and heat in- and output Q.

When applying these considerations to the weakly ionized plasma of an ionosphere, one must distinguish the neutral and ionized components; due to strong electrostatic forces occurring with charge separation the bulk of the electron population must move with the (heavier) ions. Where plasma and neutrals are at rest in the same frame one has $\boldsymbol{v}=0$ and the first term in Eq. (16.10c) vanishes; the second (conduction) term splits up into contributions from neutrals, ions and electrons, the latter giving the largest contribution at greater heights (where the gas is not too weakly ionized). Under such conditions the heat balance is essentially determined by electronic heat conduction (as principal transport contribution), on the one side, and by energy input (absorbed radiation and follow-up processes) and output (mainly infrared radiation) on the other.

Conditions are rather different in the lower and upper thermospheres, with about 120 km as distinctive level.

ϑ) In the *lower thermosphere* the neutral density is so much larger than that of the ions that neutral heat conduction plays a very important role. At the lower boundary one has a steep temperature gradient, and heat is conducted downwards into the neutral gas. Turbulent (eddy) transport forms the largest contribution at the lower boundary, but molecular conduction still prevails around 120 km (see Table 12[35]). These transport contributions are all negative, i.e. dissipating, since the downward heat flux is only stopped at the mesopause (around 80 km). Even larger dissipative contributions stem, however, from infrared radiation[39], in particular of NO and CO_2 (in its 15 μm band).

There is (see Table 12) only one input noticeable, namely that due to dissipation of turbulent energy from lower layers which is transported by gravity waves[40] and dissipated by viscous and buoyancy forces; the relevant heat deposit peaks just below the turbopause (Fig. 172a) and may be expressed as[35]:

$$\left.\begin{matrix} Q_{\mathrm{h\,vis}} = \\ \\ Q_{\mathrm{h\,buo}} = \end{matrix}\right\} K_{\mathrm{d}} \frac{\rho g}{T} \left(\frac{\partial T}{\partial z} + \frac{g}{C_p} \right) \cdot \left\{ \begin{matrix} (R'-1) \\ \\ 1 \end{matrix} \right. \quad \text{and} \tag{16.11}$$

[39] Dickinson, R.E. (1972): J. Atmos. Sci. *29*, 1531

[40] Volland, H. (1970): J. Geophys. Res. *75*, 5618; Volland, H., Mayr, H.G. (1970): Ann. Géophys. *26*, 907; Chandra, S., Stubbe, P. (1970): Planet. Space Sci. *18*, 1021

Table 12. Vertical energy fluxes of cooling and heating[a] (daily averages)[35]

Process		Flux/W m⁻2	
Emission line $\lambda/\mu m$	Origin (excitation)	at 90 km	at 120 km
1.27	O_2 chemical	−80	0
2.8	OH chemical	−60	0
9.6	O_3 chemical	−10	0
14.4	O_3 thermal & chemical	−10	0
4.3 ... 5.3	CO_2 etc.[b] chemical	−120	−100
5.3	NO thermal	−170	−160
15	CO_2 thermal	−1,810	−150
63	O thermal	−120	−100
Molecular conduction		0	−390
Eddy conduction		−950	0
Turbulent heating		+1,250	0
Balance		−2,080	−900

[a] gain positive, loss (cooling) negative
[b] CO, NO, NO^+, N^{14}, N^{15}.

such that the sum of both terms is proportional to R' which is the ratio of the Prandtl and Richardson numbers[41]. K_d is the coefficient of heat conductivity which is identical to the eddy diffusion coefficient (see vol. 49/6, p. 144, Yonezawa); g is acceleration of gravity. Figure 172b–d shows height profiles of the different contributions after model computations[35] with the eddy diffusion coefficient at 105 km as (a very important) parameter. The turbopause appears distinctly in these profiles[42].

ι) In the *upper thermosphere* conditions are quite different; see Fig. 173. Photoionization (followed by heat transfer to the thermal electrons; see Sects. 16α, 16β above) is the most important *input* by day, O_2 dissociation becoming noticeable below 180 km and prevailing below 130 km. Conduction produces heating at those altitudes where the divergence becomes negative; these are, of course, dependent on the model assumptions, at least to some extent. While in the lower thermosphere this is the prevailing heating source at night, atmospheric adiabatic contraction is probably the main heat input in the quiet nighttime upper thermosphere.

Below 170/140 km (day/night) *cooling* is prevailingly radiative (in the 5.3 µm band of NO and 15 µm band of CO_2). Unfortunately, the degree of collisional quenching of CO_2 by O atoms is not yet well enough known, so that the

[41] See this Encyclopedia, vol. 8/2, Corrsin, p. 525, in particular p. 560. Also ibidem, Schauf, p. 591 and vol. 9, Schiffer, pp. 1–161

[42] Rosenberg, N.W. (1968): J. Atmos. Terr. Phys. *30*, 907

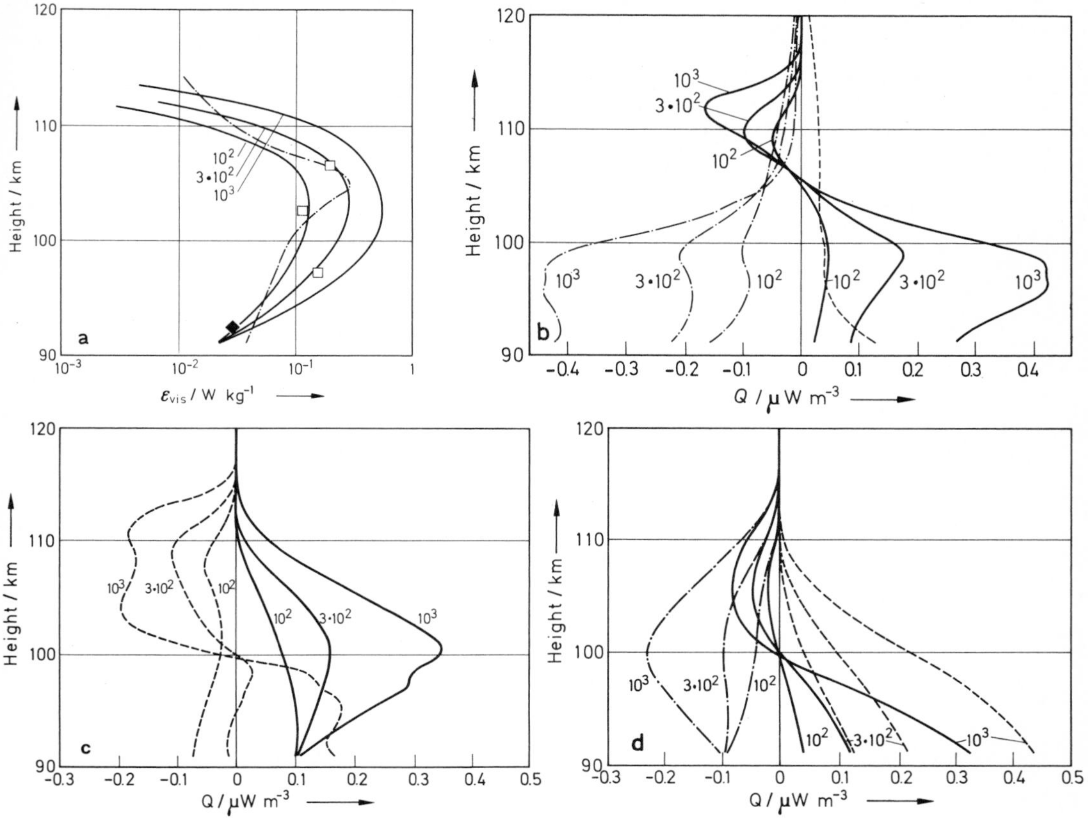

Fig. 172a–d. Heat budget in the lower thermosphere[35] calculated with different values of $k_h/\mathrm{m^2\,s^{-1}}$ (eddy diffusion coefficient at 105 km) = parameter in all figures. (**a**) Profile of viscous dissipation rate ε_{vis}. Experimental data from Roper [1966, solid square, 1974 dot-dashed curve] and of Justus [1969, open squares]. (**b**) Mean profiles of cooling and heating due to eddy turbulence (solid lines), infrared radiation (dot-dashed lines), and solar UV radiation plus chemical heating (dashed line). (**c**) Volume heating rates by dissipation of turbulence (Q_h, solid lines) and by turbulent conduction (Q_c, dotted lines). (**d**) Vertical heat fluxes due to eddy conduction (dot-dashed), to turbulent dissipation (broken) and total heat flux (solid lines)

efficiency of 15 μm cooling is numerically uncertain[35]. Above this altitude heat transport in the electron gas forms the largest contribution[43].

κ) *Electronic heat flow* is promoted by two mechanisms. First of all, as a consequence of their light mass, electrons have a high thermal conductivity K_{he} in the usual sense (i.e. at constant electron density N_e). In the real ionosphere, however, a correction term provoked by electric fields must quite often be taken into account. If μ_e is the relevant heat flow conductivity, τ_e the current flow conductivity (at constant N_e) and σ_e the electrical conductivity, then the

[43] See Annex A, p. 525

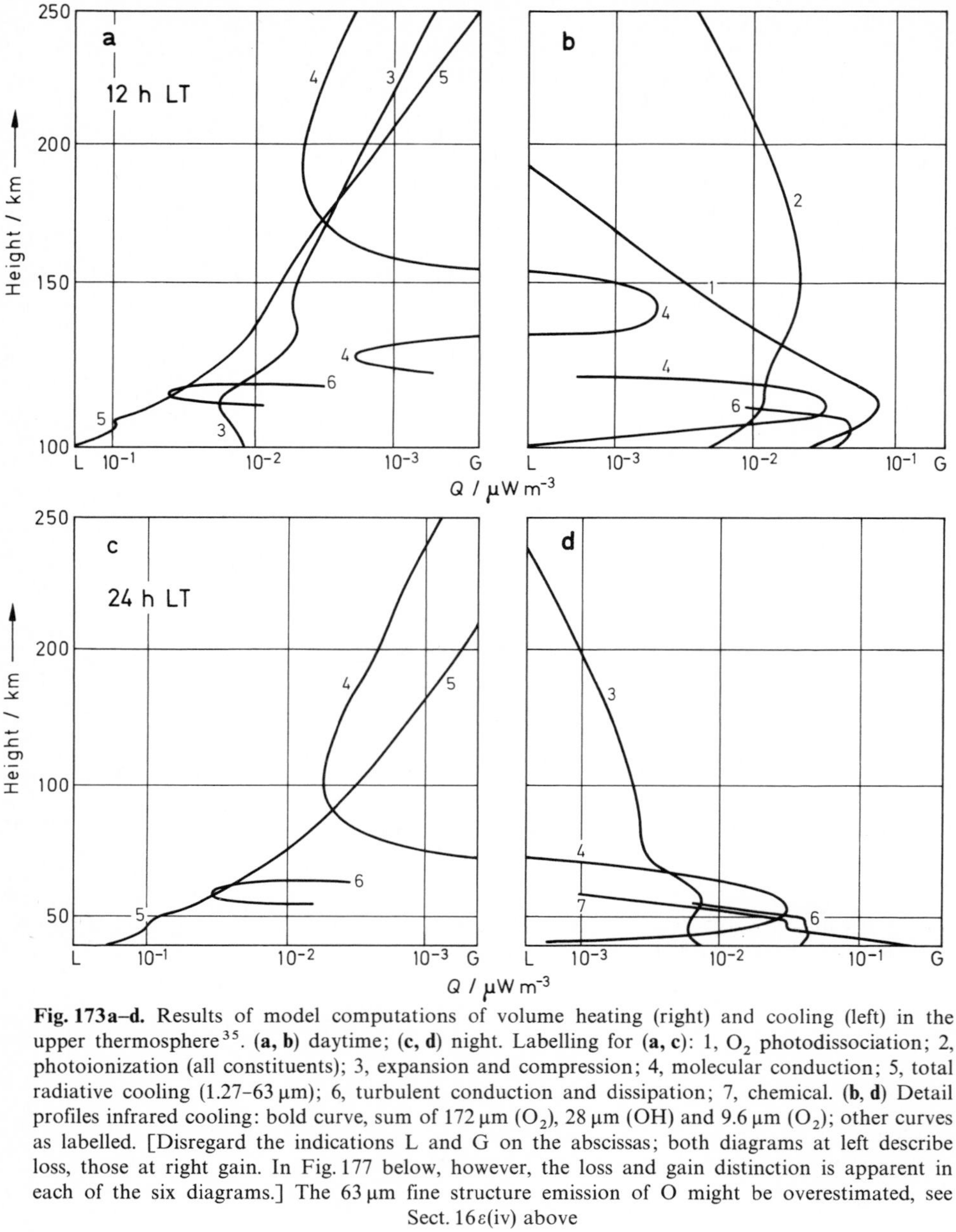

Fig. 173a–d. Results of model computations of volume heating (right) and cooling (left) in the upper thermosphere[35]. (**a, b**) daytime; (**c, d**) night. Labelling for (**a, c**): 1, O_2 photodissociation; 2, photoionization (all constituents); 3, expansion and compression; 4, molecular conduction; 5, total radiative cooling (1.27–63 μm); 6, turbulent conduction and dissipation; 7, chemical. (**b, d**) Detail profiles infrared cooling: bold curve, sum of 172 μm (O_2), 28 μm (OH) and 9.6 μm (O_2); other curves as labelled. [Disregard the indications L and G on the abscissas; both diagrams at left describe loss, those at right gain. In Fig. 177 below, however, the loss and gain distinction is apparent in each of the six diagrams.] The 63 μm fine structure emission of O might be overestimated, see Sect. 16ε(iv) above

effective thermal conductivity is [*54*]:

$$K_{\mathrm{h}}^{\mathrm{e}} = K_{\mathrm{he}} - \mu_{\mathrm{e}} \tau_{\mathrm{e}} / \sigma_{\mathrm{e}}. \tag{16.12}$$

This is the electron thermal conductivity in the limit of zero current [*58*].

The other mechanism is *thermoelectric* heat flow[44], viz. transport by an electric current of current density $\boldsymbol{J}$ which is connected with a (longitudinal) electric field. The relevant conversion factor is $\beta_e = \mu_e/\sigma_e$, so that finally the total heat flux in the electron gas is [*54*]:

$$\boldsymbol{q}_h^e = -\beta_e \boldsymbol{J} - K_h^e \cdot \frac{\partial}{\partial \boldsymbol{r}} T_e. \tag{16.13}$$

With rather involved theory the following expressions were found[45,44]

$$\beta_e = \frac{5}{2} \frac{k T_e}{q} \frac{g(\sigma_0)}{g(\mu_0)}$$
$$K_h^e = \frac{5 N_e k^2 T_e}{m_e} \frac{1}{\nu_{tot}} \left[\frac{1}{g(K_0)} - \frac{g(\sigma_0)}{2 g(\tau) \cdot g(\mu_0)} \right]. \tag{16.14a, b}$$

q and m_e are charge and mass of an electron, ν_{tot} is an average tranport collision frequency totalling collisions with ions and neutrals, whereas the different g are correction factors taking account of the velocity dependence of ν_{en} as well as of electron-ion and electron-electron effects.

Banks[46] gave an approximate expression for K_h^e deduced from simple mean free path considerations

$$K_h^e/\mathrm{eVm^{-1}s^{-1}}K^{-1} = \frac{7.7 \cdot 10^7 (T_e/\mathrm{K})^{5/2} \cdot \sin^2 \psi}{1 + 3.22 \cdot 10^8 ((T_e/\mathrm{K})^2/N_e) \cdot \sum_j n_j \cdot (\sigma_j/\mathrm{m}^2)}. \tag{16.15}$$

The presence of a magnetic field reduces the thermal conductivity by a factor of $\sin^2 \psi$ (ψ magnetic dip). The summation covers all neutral species j of density n_j and (Maxwellian) averaged momentum transfer cross section σ_j. It has been shown[44] that Eq. (16.15) is in fact accurate to within 20%, so that it is good enough in most conditions up to about 300 km.

At higher altitudes, however, in the absence of collisions one can neglect the collisions and simply have [*58*]

$$K_h^e/\mathrm{eVm^{-1}s^{-1}}K^{-1} = 7.7 \cdot 10^7 (T_e/\mathrm{K})^{5/2} \cdot \sin^2 \psi. \tag{16.15a}$$

In the upper thermosphere, all electron motions are subject to the Lorentz force exerted by the magnetic field, such that they make helical orbits with the field as axis and can therefore effectively move only along the magnetic field $\boldsymbol{B}$. Thus, only the components of $\boldsymbol{J}$ and ∇T_e parallel with $\boldsymbol{B}$ must be considered in Eq. (16.13).

Thermoelectric, i.e. current-transported heat flow depends then on field-aligned currents, the existence of which is now well proven[47]. Such a current of $10^{-5}\,\mathrm{A\,m^{-2}}$ – very usual under auroral conditions – contributes as much to the heat budget as does ordinary electron con-

[44] Schunk, R.W., Walter, J.G.C. (1970): Planet. Space Sci. *18*, 1535

[45] Shkarofsky, I.P. (1961): Can. J. Phys. *39*, 1619

[46] Banks, P. (1966): Ann. Geophys. *22*, 577

[47] Armstrong, J.C., Zmuda, A.J. (1970): J. Geophys. Res. *75*, 7122; Cole, K.D. (1971): Planet. Space Sci. *19*, 59; Zmuda, A.J., Armstrong, J.C. (1974): J. Geophys. Res. *79*, 4611; Potemra, T.A. (1977): [*18*], 337; Iijima, T., Potemra, T.A. (1978): J. Geophys. Res. *83*, 599; Banks, P. (1979): J. Geophys. Res. *84*, 6709; Brekke, A. (1981): [*53*], 13-1

ductivity[44]. At altitudes above 700 km, in the auroral zone this effect may decrease the electron temperature by as much as 1000 K[48].

λ) The *electronic heat budget* is obtained[44]:

$$\frac{3}{2}k\cdot\left(\frac{\partial}{\partial t}+\boldsymbol{v}_{\mathrm{e}}\cdot\frac{\partial}{\partial \boldsymbol{r}}\right)(N_{\mathrm{e}}T_{\mathrm{e}})=-\frac{5}{2}N_{\mathrm{e}}kT_{\mathrm{e}}\frac{\partial}{\partial \boldsymbol{r}}\cdot\boldsymbol{v}_{\mathrm{e}}-\frac{\partial}{\partial \boldsymbol{r}}\cdot\boldsymbol{q}_{\mathrm{he}}+Q_{\mathrm{he}}, \tag{16.16}$$

where q_{he} is heat flux and Q_{e} heat gain (or loss) in the electron gas. Using the continuity equation (and neglecting electron gain and loss) one gets:

$$\frac{3}{2}N_{\mathrm{e}}k(\partial T_{\mathrm{e}}/\partial t)=-N_{\mathrm{e}}kT_{\mathrm{e}}\frac{\partial}{\partial \boldsymbol{r}}\cdot\boldsymbol{v}_{\mathrm{e}}-\frac{3}{2}N_{\mathrm{e}}k\boldsymbol{v}_{\mathrm{e}}\cdot\frac{\partial}{\partial \boldsymbol{r}}T_{\mathrm{e}}-\frac{\partial}{\partial \boldsymbol{r}}\cdot\boldsymbol{q}_{\mathrm{he}}+Q_{\mathrm{he}}. \tag{16.16a}$$

In the absence of field-aligned currents (e.g. at mid-latitudes under quiet conditions) one has (for a vertical gradient of T_{e}):

$$\frac{3}{2}N_{\mathrm{e}}k\frac{\partial T_{\mathrm{e}}}{\partial t}=\sin^2\psi\frac{\partial}{\partial z}\left(K_{\mathrm{h}}^{\mathrm{e}}\frac{\partial T_{\mathrm{e}}}{\partial z}\right)+Q_{\mathrm{he}}, \tag{16.16b}$$

ψ being magnetic inclination. It follows that near the magnetic dip equator – except for horizontal gradients – the conductivity effect disappears. This has one important consequence, namely that the usual strong (inverse) relation between T_{e} and N_{e} should not hold in this particular belt (see Sect. 13b). Similar conditions are valid at low altitudes, where neutral densities and, therefore, collision frequencies are so large that $K_{\mathrm{h}}^{\mathrm{e}}$ is quite small.

Even in the ionospheric *F-region* the electron temperature T_{e} (quite different from density N_{e}) is in a thermal quasi-steady state so that the left-hand side of (16.15a, b) may be neglected. In the *absence of field-aligned currents* one then has

$$\sin^2\psi\frac{\partial}{\partial z}\left(K_{\mathrm{h}}^{\mathrm{e}}\frac{\partial T_{\mathrm{e}}}{\partial z}\right)=-Q_{\mathrm{he}}. \tag{16.16c}$$

The usefulness of this relation has been shown experimentally; see Fig. 174[49]. Fig. 175 presents rough estimates of accumulated local heating rates.

On the other hand, *at high latitudes* the thermoelectric heat flow is due to an electric current supported mainly by the electrons themselves such that $\boldsymbol{v}_{\mathrm{e}}=-\boldsymbol{J}/qN_{\mathrm{e}}$ which by insertion into Eqs. (16.12, 16.16) gives (to a good approximation [*54*]):

$$\begin{aligned}\frac{3}{2}N_{\mathrm{e}}k\frac{\partial T_{\mathrm{e}}}{\partial t}&=-J\cdot\sin\psi\cdot\left[\frac{kT_{\mathrm{e}}}{q}\left(\frac{\partial}{\partial z}\ln N_{\mathrm{e}}-\frac{3}{2}\frac{\partial}{\partial z}\ln T_{\mathrm{e}}\right)+\frac{\partial\beta_{\mathrm{e}}}{\partial z}\right]\\&\quad+\sin^2\psi\cdot\frac{\partial}{\partial z}\left(K_{\mathrm{h}}^{\mathrm{e}}\frac{\partial T_{\mathrm{e}}}{\partial z}\right)+\frac{J^2}{\sigma_{\mathrm{e}}}+Q_{\mathrm{he}}.\end{aligned} \tag{16.17}$$

---

48 Rees, M.H., Jones, R.A., Walker, J.C.G. (1971): Planet. Space Sci. *19*, 313
49 Hoegy, W.R., Brace, L.H. (1978): Geophys. Res. Lett. *5*, 269

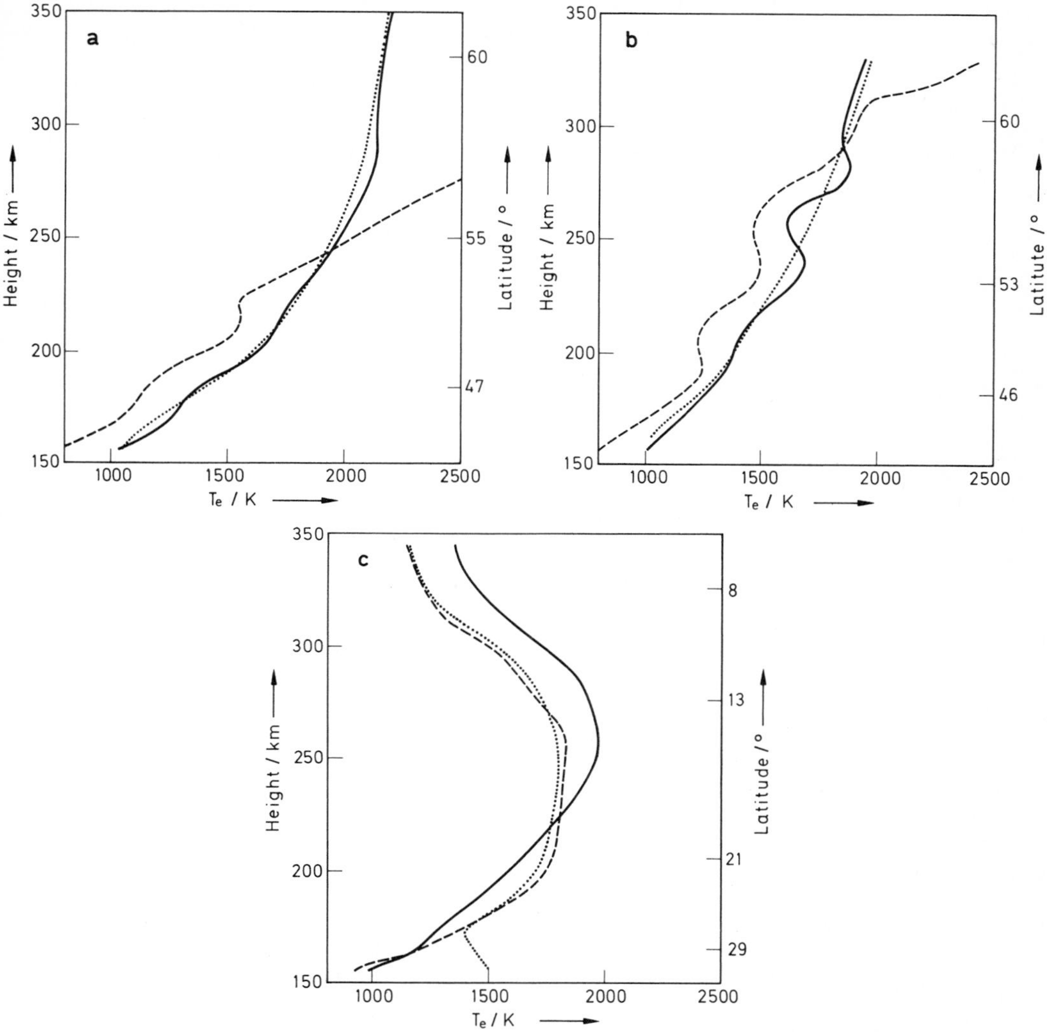

Fig. 174a–c. Electron temperature profiles measured in Jan. 1974 on USA satellite AE-C (solid lines) with profiles computed (i) from the local heating/cooling balance Q_{he} alone (broken lines); (ii) including (vertical) heat conduction in the electron gas according to Eq. (16.16) (dotted lines). (**a, b**) Mid latitude: good agreement with (ii). (**c**) low latitude: no sensible effect of conduction [in agreement with Eq. (16.16c)] [49]

Another version of this relation in which the current density J is replaced by the vertical electron flux [48] can be found in [*54*].

At *very high altitudes*, some of the above reasoning may break down because in plasmas of low density or with a high temperature gradient, there exists a limiting value for q_{he} which is independent of the gradient and is a consequence of the limited thermal velocity V_{Te} of the electrons. Unlike the

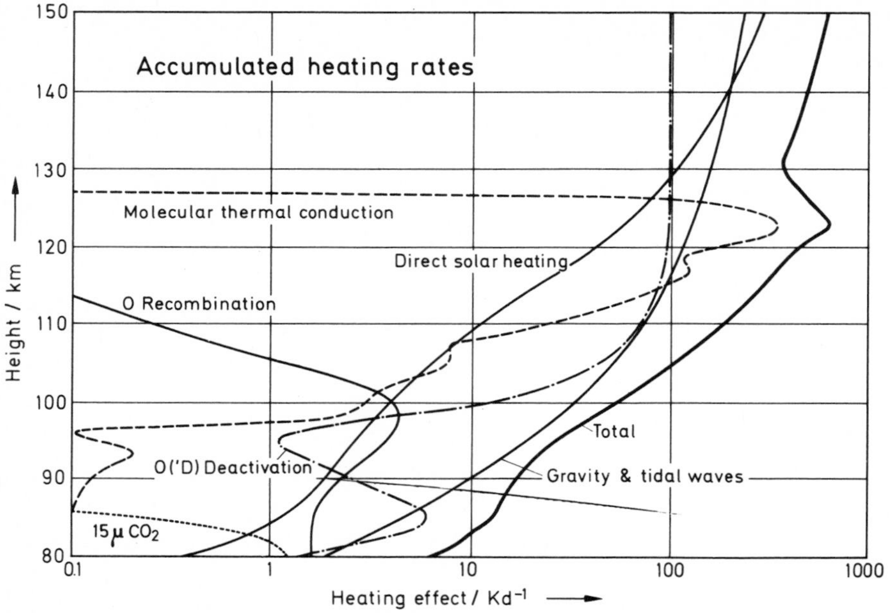

Fig. 175. Estimated accumulated (i.e. longterm) heating rates expressed as heating effect in K/d [Courtesy of NOAA, Boulder, Colorado, U.S.A.]

linear Eqs. (16.13–16.17) there must exist a saturation value q_{he}^{max} for q_{he}. An estimate[50] of the "saturation ratio" is

$$q_{he}^{max}/q_{he} = 1.05 \cdot 10^{-9} \left(\frac{N_e}{m^{-3}}\right) \cdot \left(\frac{H_{Te}}{m}\right) \Big/ (T_e/K)^2 \qquad (16.18)$$

with $H_{Te} \equiv 1/(\mathrm{d}\ln T_e/\mathrm{d}z)$ the temperature scale height.

Such conditions probably exist in the Venusian upper ionosphere (see Sect. 15), in particular at night, when the ratio may come down to 0.05 at an altitude of 700 km[51] and so better explain thermal insulation than expected theoretically. Similar reasoning might apply in the terrestrial plasmasphere.

μ) *Resuming* this section we may note two facts:

(i) The thermal balance is established quickly so that there is a *quasi-steady thermal state* even in the upper thermosphere. This implies that rather fast changes of temperature may occur, in particular with the electronic populations;

(ii) Because important contributions to the heat budget depend on the plasma density, which undergoes a *large diurnal variation*, the assumption of an invariable temperature profile is not really justified. Since many aeronomical rate coefficients are fairly strongly temperature dependent, an advanced aeronomical theory should also include the heat budget (see Sect. 17).

[50] Cowie, L.L., McKnee, C.F. (1977): Astrophys. J. *211*, 135

[51] Merrit, D., Thompson, K. (1980): J. Geophys. Res. *85*, 6778

17. Advanced aeronomical theory. Although "kinetic" aeronomical theories discussed in Sect. 15 explained quite a few features of the ionosphere, they were unable to do this quantitatively. It was then found out that several influences which were not taken account of in these theories had to be included[1].

α) *Thermal effects* were first introduced by HERMAN and CHANDRA[2], who considered steady-state solutions of the coupled differential equations for electron density N_e and the temperatures T_e, T_i and T_n. Then STUBBE [*60*] dealt with a time-dependent system of equations. Similarly to the procedure explained in Sect. 15 he used four continuity equations (for the main ions) and two equations of motion taking account of winds and of a polarization field. Additionally, however, he introduced four equations specifying the *heat balance* (for neutrals, O^+ and H^+ ions and for electrons) of general shape:

$$\frac{\partial T_j}{\partial t} = \underset{\text{(in-/output)}}{\frac{1}{n_j C_{vj}} Q_{hj}} - \underset{\text{(advection)}}{\boldsymbol{v}_j \cdot \frac{\partial}{\partial \boldsymbol{r}} T_j} + \underset{\text{(adiabatic compression)}}{(\gamma - 1) T_j \frac{\mathrm{d}}{\mathrm{d}t} \ln n_j}, \tag{17.1}$$

where c_v is specific heat (at constant volume) per particle, Q_h heat exchange (gain positive), $\gamma \equiv C_p/C_v$ ratio of specific heats and n number density. Viscosity is neglected since application is for the upper thermosphere. The heat in- and output Q_{hj} can be specified further:

$$Q_h = Q_{ph} - Q_{rad} + Q_{tr} - \frac{\partial}{\partial \boldsymbol{r}} \cdot \boldsymbol{q}_{hj}, \tag{17.2}$$

where the indices designate *ph*oton energy (absorbed), *rad*iation (emitted, mainly infrared) and *tr*ansfer between gases at different temperatures (which is a coupling term). The last term, describing the effect of conduction in a divergent heat flux, may be expressed with the thermal conductivity K_h^j as

$$\boldsymbol{q}_{nj} = -K_h^j \frac{\partial}{\partial \boldsymbol{r}} T_j. \tag{17.3}$$

Applying a few reasonable simplifications and using the connection (given by the Earth's rotation Ω) between EW-velocity and diurnal time t (at latitude φ), which is valid when geomagnetic and geographical coordinates coincide:

$$\frac{\partial}{\partial y} = \frac{1}{R_E \cos\varphi\,\Omega} \cdot \frac{\partial}{\partial t} = a \cdot \frac{\partial}{\partial t} \quad \text{with } a \cdot \cos\varphi = 2.159 \text{ s km}^{-1}$$

(with $R_E = 6370$ km). When only vertical gradients are present [*60*] one finds:

$$(1 + a \cdot v_{yj}) \cdot \frac{\partial T_j}{\partial t} = \frac{K_h^j}{M_j} \cdot \frac{\partial^2 T}{\partial z^2} + \left(\frac{1}{M_j} \frac{\partial K_h^j}{\partial z} - v_{jz}\right) \cdot \frac{\partial T_j}{\partial z}$$
$$+ (\gamma - 1) T_j \frac{\mathrm{d}}{\mathrm{d}t} \ln n_j + \frac{1}{M_j} (Q_{ph} - Q_{rad} + Q_{tr}), \tag{17.4}$$

where $M_j \equiv 1/(n_j c_{vj})$.

---

[1] Rishbeth, H. (1968): Rev. Geophys. *6*, 33, further [*47*, *48*]

[2] Herman, J.R., Chandra, S. (1969): Planet. Space Sci. *17*, 815

Specific conditions for the four constituents considered are:

(i) Neutrals (j=n). Q_{tr} usually contains contributions from all three other constituents (e^-, O^+, H^+). HERMAN and CHANDRA[2] specify these three by an expression (l designating the different neutrals)

$$Q_{tr}(\mathrm{n}, \mathrm{ion}^+) = (T_i - T_n) \cdot 3k\, N_i \sum_l \frac{\mathrm{m_i} \cdot \mathrm{m}_l}{(\mathrm{m_i} + \mathrm{m}_l)^2}\, \nu_l,$$

and they give quite an involved expression for $Q_{tr}(\mathrm{n}, \mathrm{e})$ which takes account of collisional heat transfer (as above) including the temperature dependence of the collision frequencies ν_l.

Neutral loss, apart from the adiabatic term, is assumed to occur by fine structure emission of atomic oxygen; see Sect. 16η(iv).

(ii) Electrons (j=e). In the third term on the right-hand side in Eq. (17.1) $\left[(1 + a \cdot v_{ey}) \cdot \frac{\partial}{\partial t} + v_{ez} \frac{\partial}{\partial t}\right] \ln N_e$ is written for $\frac{\mathrm{d}}{\mathrm{d}t} \ln N_e$, and then the values of $\partial N_e/\partial t$ and $\partial N_e/\partial z$ may be inserted from the simultaneously resolved continuity equations. For the thermal conductivity K_h^e of an electron gas embedded in a neutral gas, Eq. (16.15) is used with the following cross sections[2]

$$\begin{aligned} &(\text{with } O_2){:} && \sigma_{e1} = [2.2 \cdot 10^{-20} + 7.9 \cdot 10^{-22} (T_e/\mathrm{K})^{1/2}]\ \mathrm{m}^2 \\ &(\text{with } N_2){:} && \sigma_{e3} = [2.8 \cdot 10^{-21} - 3.4 \cdot 10^{-25} (T_e/\mathrm{K})^{3/2}]\ \mathrm{m}^2 \\ &(\text{with } O){:} && \sigma_{e4} = 3.4 \cdot 10^{-20}\ \mathrm{m}^2. \end{aligned}$$

Heat input is a described in Eq. (16.1), followed by collisional heat transfer to ion and neutrals. Energy input is, however, computed with standard values per photon (in two height classes) after[3]. Heat loss is due to transfer to the (positive) ions and to the neutrals. Again using an approximation given by HERMAN and CHANDRA[2]:

$$Q_{tr}(\mathrm{e}, O^+)/\mathrm{eV\, m^{-3}\, s^{-1}} = A \cdot (T_i - T_e)\, N_i\, N_e \cdot T_e^{-3/2}$$

with $A = 4.8 \cdot 10^{-13}$ and $7.7 \cdot 10^{-12}\ \mathrm{m^6\, K^{1/2}}$ for O^+ and H^+, respectively. Loss to neutrals $Q_{tr}(\mathrm{e}, \mathrm{n}) = -Q_{tr}(\mathrm{n}, \mathrm{e})$ [see (i) above] is quite small in the upper thermosphere.

(iii) Ions (j=i). A similar procedure as in (ii) is applied but with K_h^i according to [58]

$$K_h^i/\mu\mathrm{W m^{-1} K^{-1}} = 7.4 \cdot 10^{-7} (T_i/\mathrm{K})^{5/2} \cdot \sin^2 \psi / (m_i/\mathrm{au})^{1/2},$$

with m_i = ion mass.

The coefficient is $4.6 \cdot 10^6$ when $\mathrm{eV\, s^{-1}}$ is used instead of µW. Heat gain and loss occur exclusively by collisional transfer. The program admits a temperature difference $\Delta T_i = T(H^+) - T(O^+) \neq 0$ and so $Q_{tr}(O^+, H^+) \neq 0$.

[3] Dalgarno, A., McElroy, M.B., Rees, M.H., Walker, J.C.H. (1968): Planet. Space Sci. *16*, 1371

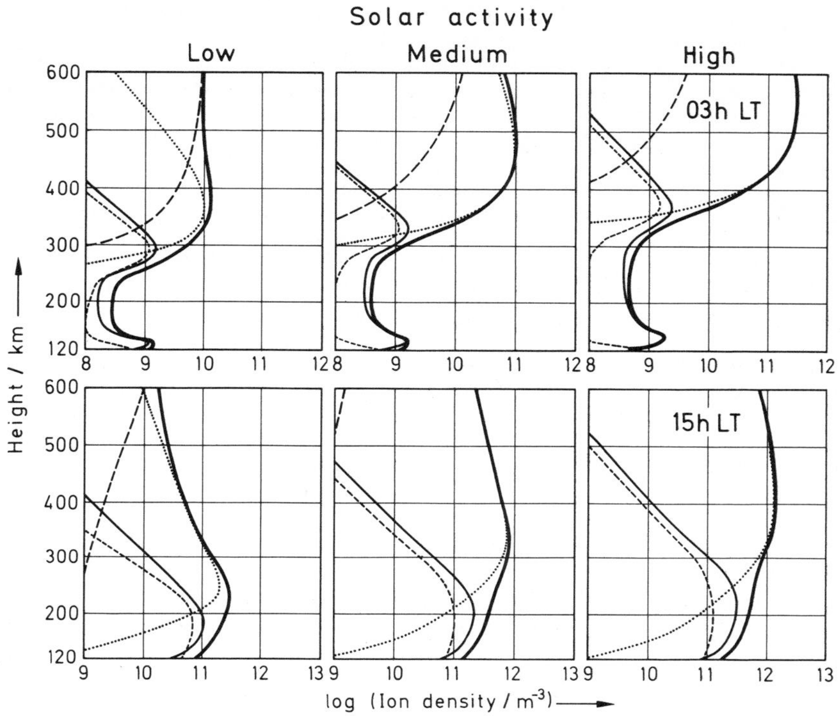

Fig. 176. Ion density profiles computed by STUBBE [*60*] for mid-latitude ($\varphi = 51.6°$ N; $\psi = 67.0°$) equinox conditions and three levels of solar activity (horizontally displayed). Top line 03 h; bottom line, 15 h LT. Long broken curve, H^+; short broken curve, O_2^+; dotted curve, O^+; thin solid curve, NO^+; bold solid curve, electron density

The final set of ten differential equations has the general shape of Eqs. [15.19] and is also resolved[4] with finite differences (of 2 km and 2 min), one after the other in the order $N(O^+)$, $N(H^+)$, $N(O_2^+)$, $N(NO^+)$, v_{nx}, v_{ny}, T_n, $T(O^+)$, $T(H^+)$. This requires a very long computing time on a large computer, so that the model is not easily accessible.

Boundary conditions are:

(a) At $z = 120$ km (lower boundary): densities from the continuity equation without the divergence term; temperatures from local production and transfer equilibrium and $T_n = 355$ K; no neutral or energy transport.

(b) At $z = 1{,}500$ km (upper boundary): vertical particle flux zero; vertical ionized temperature gradient (T_e like T_i) of 0.5 K km^{-1} by day, zero by night; neutral temperature and velocity gradients zero.

A few mid-latitude results may be discussed with Fig. 176. Ion densities below the peak are very well reproduced, only that, above their peaks, the $N(O_2^+)/N(NO^+)$ ratio, after observed data

[4] A fully implicit finite difference method is described by Laasonen, P. (1949): Acta Math. *31*, 309

(see for example Fig. 118 in Sect. 13) might be smaller. As a superposition of the NO^+ and the O^+ layers an F1-layer (at mid-latitudes, by day) appears; however, this is at an altitude which is too large by high solar activity. This feature is critically dependent on the charge transfer reaction from O^+ to NO^+, the rate for which was in fact adapted so as to reproduce the $NmF2/NmF1$ ratio. (Heat conduction parameters were also adjusted.) No valley is indicated by day but the night-time valley is probably too important, particularly with high solar activity. The topside decrease of plasma density is too slow, and the crossing-over level between O^+ and H^+ is very low for low solar activity. This shows that the upper boundary conditions are probably not realistic (see Sect. 17γ below). A few improvements were introduced in a later paper by CHANDRA and STUBBE [5].

β) *Similar theories* have since been established by different authors [6–10]. They differ in their assumptions about motions, different rate 'constants' (in particular those in the heat budget) and boundary conditions.

The plasma heat balance, in particular for the O^+ population, was reconsidered by several authors [11–15]. The difficulties with the heat balance may be seen by comparing Fig. 173 with Fig. 177 [15], which is for mid-latitude night conditions [16], assuming different values of parameters. These authors introduced an additional 'frictional heating' term F_{in} which takes account of collisions between ions and neutrals due to their relative motion:

$$F_{in}=\frac{m_n\cdot m_i}{m_n+m_i}N_i\,v_{in}\cdot\{(w_\parallel-v_{nx}\cos\psi)^2+v_{ny}^2\}. \tag{17.5}$$

With F_{in} they find that at night (under certain conditions) they can even account [15] for an ion temperature which is up to 10 K greater than that of the electrons - as has occasionally been observed [12].

While all the above theories are for mid-latitudes, other authors have considered the particular conditions in the belt around the (true) *magnetic-dip equator* [17–19]. As appears from Eq. [15.11] around $\psi=0$ wind- and diffusion-induced vertical transport disappears in the F-region so that only electric fields (by their $\boldsymbol{E}\times\boldsymbol{B}$ force) are able to initiate vertical motion of the plasma. Now, it is well known that the dynamo theory of magnetic diurnal variations requires *electric fields*, particularly in the equatorial belt.

However, the diurnal variation theoretically computed with this theory does not fit with the observed plasma motions during the evening hours. Thus, ANDERSON [19] had to introduce an empirical expression for these variations. For more information see Sect. 15ϑ.

---

[5] Chandra, S., Stubbe, P. (1972): J. Atmos. Terr. Phys. *34*, 1627
[6] Rüster, R., Dudeney, J.R. (1972): J. Atmos. Terr. Phys. *34*, 1317
[7] Deshpande, M.R. (1972): Ann. Géophys. *28*, 809
[8] Eccles, D., Burge, J.D. (1973): J. Atmos. Terr. Phys. *35*, 1927
[9] Matuura, N. (1973): J. Geophys. Res. *79*, 4679; J. Atmos. Terr. Phys. *36*, 1963
[10] Roble, G.R., Emery, B.A., Salah, J.E., Hays, P.B. (1974): J. Geophys. Res. *79*, 2868
[11] Stubbe, P. (1971): J. Sci. Ind. Res. *30*, 379
[12] Mazaudier, C., Bauer, P. (1976): J. Geophys. Res. *81*, 3447
[13] St. Maurice, J.P., Schunk, R.W. (1977): Planet. Space Sci. *25*, 907
[14] Bailey, G.J., Moffett, R.J. (1978): J. Geophys. Res. *83*, 145
[15] Bailey, G.J., Footitt, R.J., Moffett, R.J. (1979): Planet. Space Sci. *27*, 1063
[16] The heat transfer rates of [14] are numerically found in their Table 1, p. 1065
[17] Deshpande, M.R. (1973): Ann. Geophys. *29*, 43
[18] Anderson, D.N., Matsushita, S., Tarpley, J.D. (1973): J. Atmos. Terr. Phys. *35*, 753
[19] Anderson, D.N., Matsushita, S. (1974): J. Atmos. Terr. Phys. *36*, 2001; Anderson, D.N. (1981): J. Atmos. Terr. Phys. *43*, 753

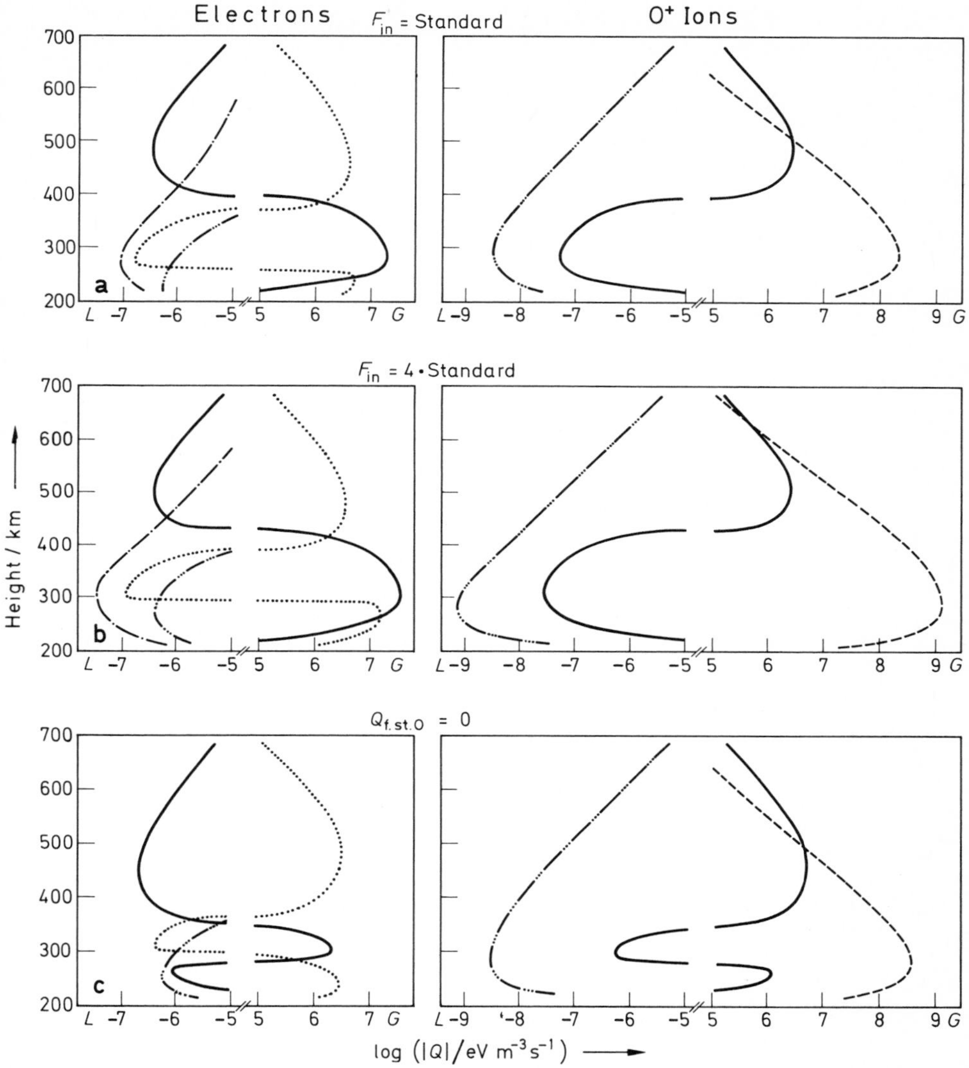

Fig. 177a–c. Contributions to the stationary heat balance of electrons (left) and O^+ ions (right) computed by [14]. Different from Fig. 173 above in *each* of the diagrams gain (G) is counted as positive, i.e. to the right, and loss (L) to the left. (**a**) With "standard" rate coefficients [15]; (**b**) with F_{in} (frictional heating) four times increased [makes at 250 km T_i up to 20 K greater than T_e]; (**c**) with "standard", but neglecting Q_{eO}, the atomic oxygen fine structure cooling [Sect. 16ε(iv)]. Solid curve, $Q_{tr}(e^-, O^+)$; broken curve, $Q_{tr}(O^+, O)$; dotted curve: heat conduction; –·– $Q_{tr}(e^-, n)$; –···– $Q_{tr}(O^+, n)$

As all these theories depend on many physical processes and the relevant rates, one might ask two questions, namely whether the established system is unique (whether the same effects could not also be provoked by another combination) and how important numerically the different influences are.

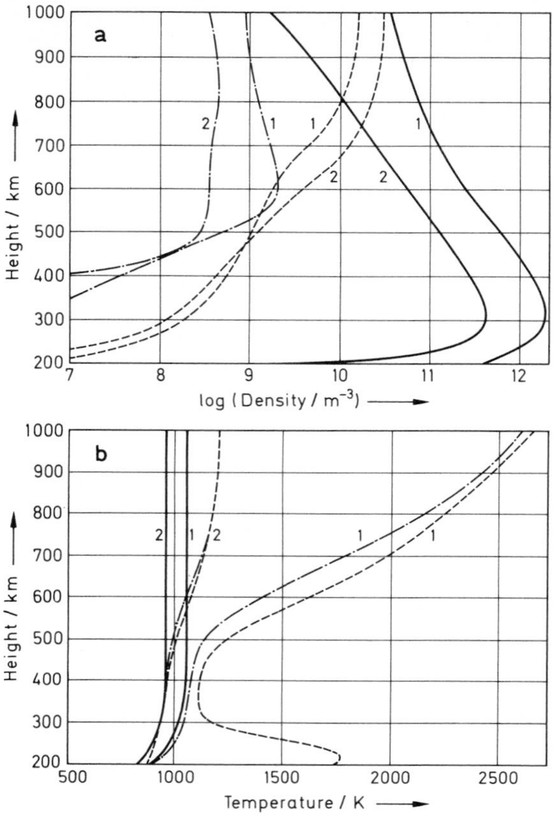

Fig. 178a, b. Model profiles computed for Delhi ($\varphi = 29°$ N, $\psi = 33°$) with adjusted parameters[21]. Curves labelled: 1, noon; 2, midnight. (**a**) Densities [solid lines, $N(O^+)$; broken line $N(H^+)$; dash-dotted lines, $N(He^+)$]; (**b**) Temperatures [solid lines, T_n; broken lines, T_e; dash-dotted lines T_i]. Note that the computed ion temperature T_i as compared with T_e is higher than found experimentally

Generalizing an earlier aeronomical model[20] (built for a low latitude site – outside the belt of low dip), a [*60*]-type model program was run with systematic variation of the different inputs[21]. Doubling the EUV flux makes an almost proportional increase in ion densities and an up to 200 K increase in T_e; increasing $n(H)$ by a factor 4 gives a proportional increase in $N(H^+)$; doubling the heating efficiency does not influence the ion densities and above 300 km has only a small effect upon T_e but increases by 400 K around 200 km. By comparing with observed data the authors conclude[21] that: EUV fluxes should be increased by 50% against Hinteregger (1970)[22], $n(H)$ should be increased by a factor of 3 against CIRA 72 [*12*] and heating efficiency should be increased by a factor of 2 against[23]. Model profiles so obtained for Delhi are shown in Fig. 178. [Apparently the bulge of T_e appears at rather a low height, and, above 300 km, the difference $(T_e - T_i)$, is too small.]

[20] Lumb, H.M., Setty, C.G.S.K. (1973): Ind. J. Radio Space Phys. *2*, 261

[21] Goel, M.K., Rao, B.C.N. (1980): Ind. J. Radio Space Phys. *9*, 164

[22] Hinteregger, H. (1970): Ann. Géophys. *26*, 547; see also Schmidtke's contribution in this volume, his Sect. 4, p. 18

[23] Swartz, W.E., Nisbet, J.S. (1972): J. Geophys. Res. *77*, 6259; (1973): *78*, 5640

While all these theories require a considerable computational effort, NISBET with his "Pennstate models" aimed at an easier, "user-friendly" computing program. His "MK 1" model[24] is essentially a static model, but a few dynamic effects are cared for by using in the program empirically obtained values of the characteristics *hmF2* and *NmF2* for the F2-peak. In particular, *NmF2* is taken from the CCIR program [*9*] (see Sect. 14γ), and the solar radiation flux is adjusted so that this value is in fact obtained. The danger with this procedure is that wind effects which are not directly admitted reappear as (large) variations in the solar radiation, which in turn may provoke considerable distortion of the profile shape. On the other hand the peak values, which are most important at applications, cannot 'run away'. The night-time profile is given as an empirical function, but during daylight hours a heat balance is determined with input via photoelectrons (similar to [*60*]) and losses as discussed in Sects. 16δ(i) and 16ε(ii–iv). The electron temperature T_e is determined from this balance[24]. Heat efficiency and solar radiation fluxes were later reexamined[23].

A more appropriate semi-theoretical model allowing for neutral winds and adjusting these (not the solar flux) has recently been worked out and is reported in Sect. 14γ(iv), p. 405. It has yet another advantage insofar as *NmF2* though in general taken over from [*9*] is improved on physical grounds in the station-free oceanic regions of Earth.

γ) The *upper boundary conditions* are most critical in such computations. In fact, the zero flux condition - assumed in most theories until the mid-seventies - cannot be maintained. Several authors[25–27] have noted that a plasma flow from the plasmasphere down into the ionosphere has a stabilizing influence on the F2-peak density. By incoherent scatter technique (see Sect. 5) downward plasma fluxes are observed by night with about $10^{11}\,\mathrm{m}^{-2}\,\mathrm{s}^{-1}$ around 700 km, but upward fluxes up to 10 times larger are observed by day[28,29]. Also there must be a downward heat flux in the electron gas in order to explain the observed T_e profiles, which increase steadily with height[30]. Recent (low-latitude) estimates[31] assume a daytime maximum of $4 \cdot 10^{13}\,\mathrm{eV\,m}^{-2}\,\mathrm{s}^{-1}$ and between 2 and $8 \cdot 10^{12}$ by night. Measured values[32] are about the same. The effects of such variation at the upper boundary on plasma temperatures are demonstrated[21] in Fig. 179a, b, while plasma flux effects are larger for the densities (see Fig. 179c, d). A schematic diagram showing the main steps of the computation schedule can be found in Fig. 167c.

---

[24] Nisbet, J.S. (1971): Radio Sci. *6*, 437

[25] Evans, J.V. (1965): J. Geophys. Res. *70*, 4331

[26] Mayr, H.G. (1968): The dynamical coupling between the F2-region and the protonosphere. Doc. X-621-68-300. Greenbelt, Md., USA: NASA GSFC

[27] Park, C.G., Banks, P.M. (1974): J. Geophys. Res. *79*, 4661

[28] Evans, J.V. (1971): Radio Sci. *6*, 843

[29] Jain, A.R., Williams, P.J.S. (1974): J. Atmos. Terr. Phys. *36*, 417

[30] Evans, J.V. (1970): J. Geophys. Res. *75*, 4803; see also: Evans, J.V., Mantas, G.P. (1968): J. Atmos. Terr. Phys. *30*, 563

[31] Titheridge, J.E. (1976): Planet. Space Sci. *24*, 247

[32] Ho, M.C., Moorcroft, D.R. (1975): Planet. Space Sci *23*, 315

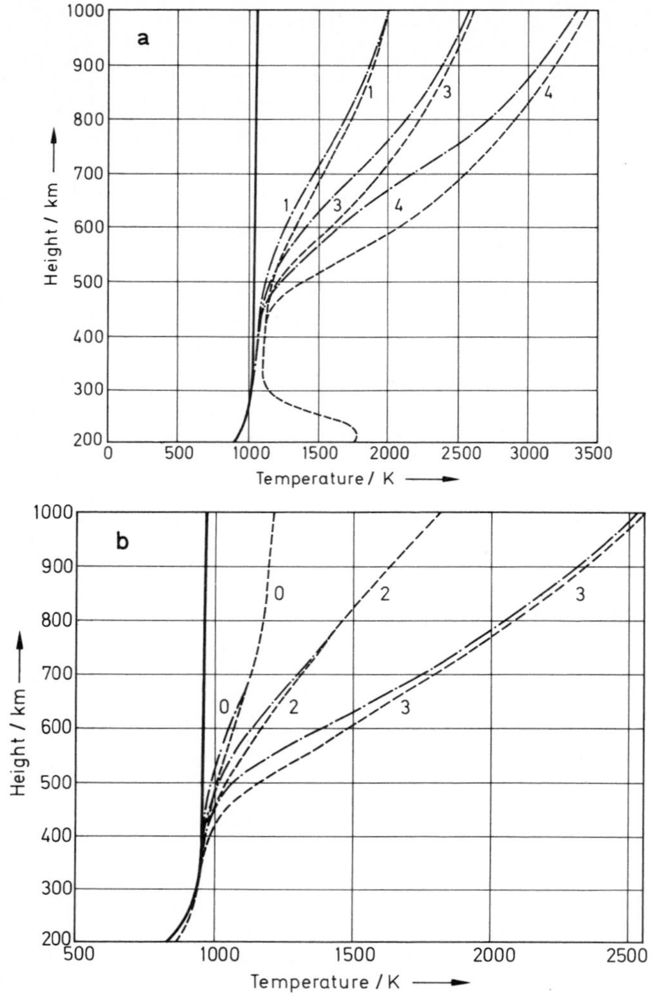

Fig. 179 a, b

The plasma flux at the upper boundary is in fact a crucial problem. The primary photoelectrons should have an isotropic distribution at their origin. However, with upward decreasing neutral and plasma density, collisions (see Sects. 16β, δ) are more frequent for downward travelling electrons, so that the medium has a filtering effect favouring upward flux. Furthermore, since the electrons move on helical orbits around the magnetic field line, the spatial collision number along this is higher for small pitch angles such that a differentiation by this angle is obtained, with preference for higher pitch angles. MANTAS' detailed computations[33], taking account of this phenomenon, finally justified the results obtained earlier with the "two-stream" model of BANKS

[33] Mantas, G.P. (1975): Planet. Space Sci. *23*, 337; Mantas, G.P., Bowhill, S.A. (1975): Planet. Space Sci. *23*, 355

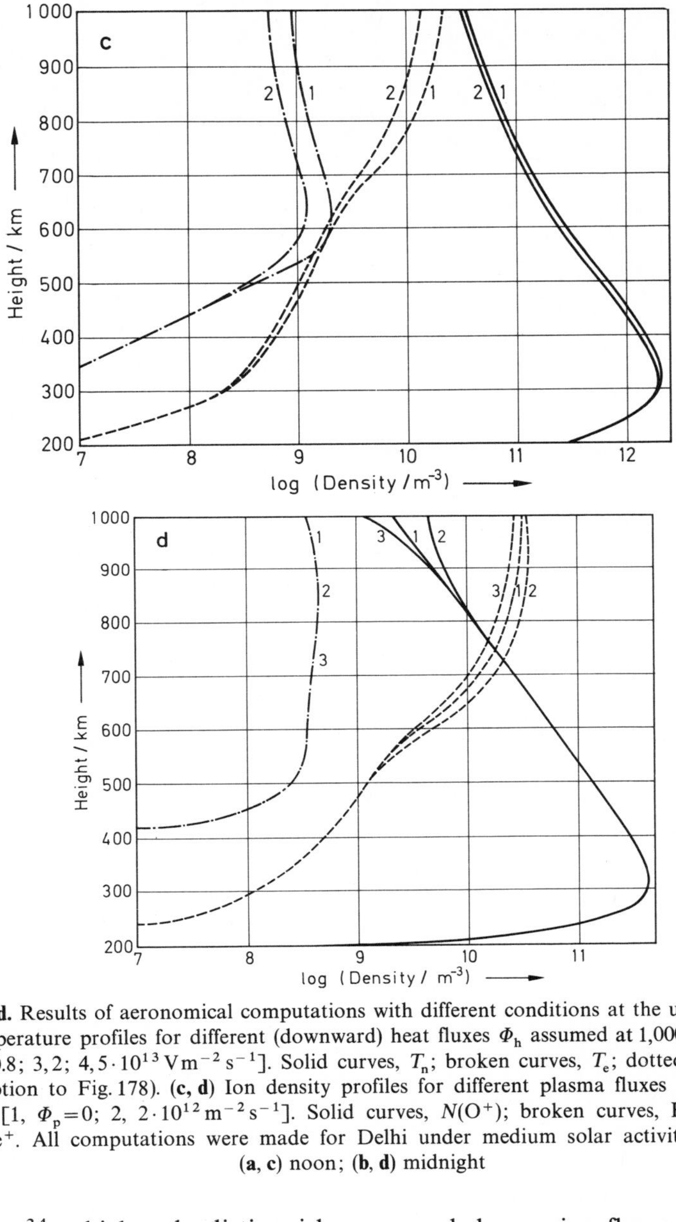

Fig. 179a–d. Results of aeronomical computations with different conditions at the upper boundary. **(a, b)** Temperature profiles for different (downward) heat fluxes Φ_h assumed at 1,000 km. [0, $\Phi_h = 0$; 1, 0.75; 2, 0.8; 3, 2; 4, $5 \cdot 10^{13}\,\mathrm{V\,m^{-2}\,s^{-1}}$]. Solid curves, T_n; broken curves, T_e; dotted-dashed curves T_i (see caption to Fig. 178). **(c, d)** Ion density profiles for different plasma fluxes Φ_p at the same boundary [1, $\Phi_p = 0$; 2, $2 \cdot 10^{12}\,\mathrm{m^{-2}\,s^{-1}}$]. Solid curves, $N(O^+)$; broken curves, H^+; dot-dashed curves, He^+. All computations were made for Delhi under medium solar activity conditions[21]; **(a, c)** noon; **(b, d)** midnight

and NAGY[34], which only distinguishes up- and downgoing fluxes. See Sect. 16 and, in particular, Appendix A, p. 525, for comparison with more involved computations[35].

[34] Banks, P.M., Nagy, A.F. (1970): J. Geophys. Res. *75*, 1902, 6260; see also: Nagy, A.F., Fontheim, E.G., Stolarski, R.S., Beutler, A.E. (1969): J. Geophys. Res. *74*, 4667

[35] Cicerone, R.J., Swartz, W.E., Stolarski, R.S., Nagy, A.F., Nisbet, J.S. (1973): J. Geophys. Res. *78*, 6709

The phenomena influencing the electron fluxes are: loss by inelastic collisions (see Sect. 16ε), elastic scattering (see Sect. 16δ) without direct energy loss but changing the pitch angle and retro-scattering, which presents a kind of coupling between up- and downgoing electrons, see Appendix A. Energy-dependent cross sections must be applied (see SUCHY's contribution in this volume, his Chap. BI, II). Input is, of course, the primary photoelectron population of a few eV; these produce, however, secondary electrons, etc. (see Sect. 16α); as a consequence of their smaller kinetic energy these contribute almost equally to both directions.

It is important to note that from the upper boundary (at about 1,000 km) the faster electrons often reach the conjugate hemisphere, acting there as a heat source. Others are reflected by Coulomb collisions so that electrons go up- and downwards as well. As an average coupling between hemispheres one has a flux (Φ) transmission ratio[36] of about 0.4, which means a boundary condition (at low latitudes):

$$\Phi^{\downarrow}(1{,}000\text{ km}) = 0.4 \cdot \Phi^{\uparrow}(1{,}000\text{ km}). \tag{17.6}$$

Estimates for higher latitudes give, of course, a smaller transmission, but the computations are rather uncertain.

δ) In view of this uncertainty is makes sense to adopt an *empirical fit* at the *upper boundary*. ROBLE [*49, 50*] has taken this step using fitting with actual incoherent scatter measurements of vertical plasma motions. While the aeronomical theories referred to in Sects. 17α and 17β intend to describe an average picture, he aims at explaining the facts for a *specific day*. By applying the upper boundary fit with measured data of an individual day the computations are avoided to 'run away' from realistic conditions; at the same time one has the advantage that the smoothing is avoided, which is inevitable with monthly or seasonally average data. Further, the well-known simplifications of the neutral wind computation from average models are at least in part avoided by establishing a specific daily wind and transport map in the computer program. Apart from this the different collisional reactions described in Sects. 17α and 17β (and in Sect. 16) are included as well as the different heat conduction and transfer effects[37] (compare Figs. 180, 181). The total number of reactions is large, and the computing time is even more than with STUBBE's [*60*] program.

A few typical results of such computations are reproduced in Figs. 27 and 28 (p. 38) of SCHMIDTKE's contribution in this volume.

ε) So ROBLE has been first to take *large scale circulation* effects into account including the actual heat input into the auroral zone (by empirical relations with the disturbance activity). This is a very important improvement, as can be seen from Fig. 180. The perturbation effect, viz. a whirl cell appearing in the winter hemisphere, makes a very large change, even inverting the meridional wind and transport direction at greater heights almost in the whole hemisphere. This cell is driven by auroral heating. The seasonal effect in the

[36] Lejeune, G., Wormser, F. (1976): J. Geophys. Res. *81*, 2900; note that twice this value was indicated by: Mantas, G.P., Carlson, H.C., Wickwar, V.B. (1978): J. Geophys. Res. *83*, 1 (the latter authors do not take account of Coulomb collisions in the plasmasphere)

[37] In SCHMIDTKE's contribution in this volume the method is used to investigate the effect of variable solar EUV input as observed for specific days

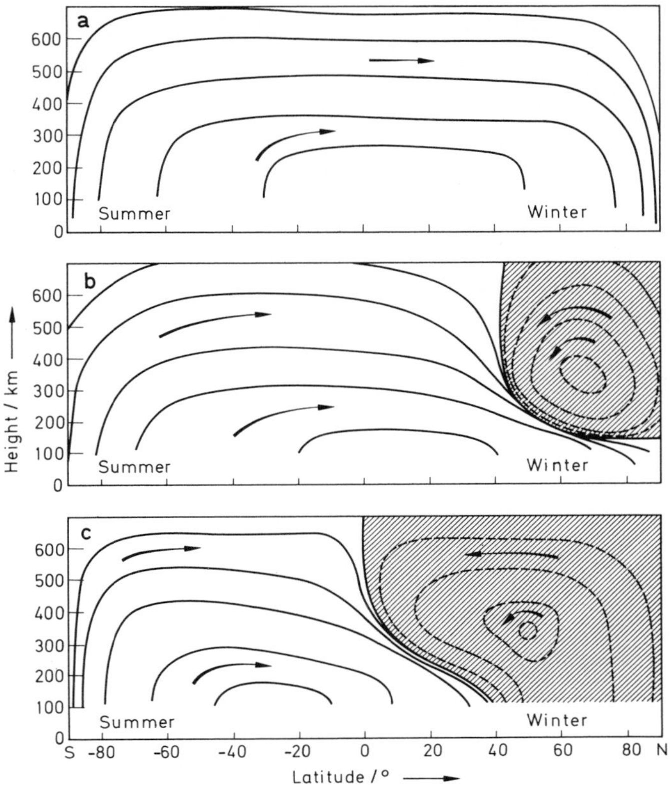

Fig. 180a–c. Mean meridional circulation in the thermosphere (northern winter solstice) for three levels of geomagnetic activity: (**a**) quiet; (**b**) moderately disturbed; (**c**) severely disturbed. [Courtesy of R.G. Roble]

circulation is quite large; in particular during equinox a practically discontinuous transition from a symmetrical equinoctial situation (see Fig. 181a) to the typical winter situation (Fig. 181b–d) occurs, – and an inverse one at the beginning of the equinox period[38]. So the thermospheric circulation has some kind of similarity to large-scale meteorological conditions, where changes also often occur suddenly. This also shows that the simple and smooth picture of heat sources leading for example to the temperature maps shown in Figs. 80 (Sect. 10), 87 and 93 (Sect. 11) is certainly oversimplified and only useful as a zero-order approximation.

These findings also show that classical dynamics (tidal theory)[39] even in its most modern form [*16*] [*28*] [*69*][40,41] (Fig. 182) encounters serious limitations

---

[38] Roble, R.G., Dickinson, R.E., Ridley, E.C. (1977): J. Geophys. Res. *82*, 5493; (1975): J. Atmos. Sci. *32*, 1737 and *34*, 178

[39] Kertz, W. (1957): This Encyclopedia, vol. 48, p. 928ff. describes the classical theory which is still useful in view of mathematical resolution methods but has been since superseded by the more recent theories

[40] Volland, H., Mayr, H.G. (1974): Radio Sci. *9*, 263

[41] Lindsen, R.S., Hong, S.S. (1974): J. Atmos. Sci. *31*, 1421

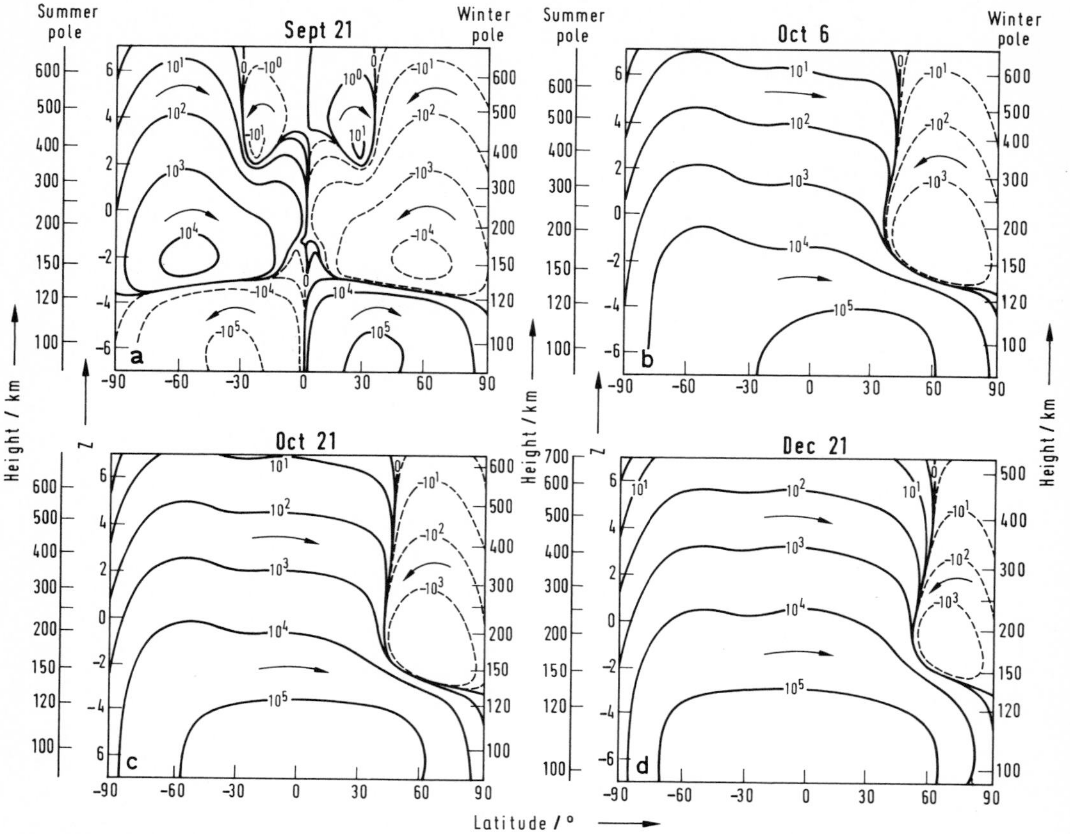

Fig. 181a–d. Meridional thermospheric circulation computed by ROBLE's method for four individual days as indicated[38]. The labels identify total meridional mass transport [units of $kg\,s^{-1}$] above the level as given by the relevant curve

as a consequence of corpuscular heating in the polar caps, which may change completely from one day to the next. This might explain why the patterns of planetary waves, although appearing, are not consistent over many days.

Zonal plasma convection at high latitudes (see Fig. 183[42]) is now a well-established phenomenon. It also has important consequences upon circulation insofar as it considerably modifies the diurnal neutral gas temperature distribution and circulation [*51*]. The displacement of the magnetic poles against the rotation axis of Earth is of some importance in this context. The convection is due to an electric field transferred from the tail of the magnetosphere[43] into the auroral zone. The effects upon temperature and circulation are quite large, particularly in the polar caps as can be seen by comparison of Figs. 184 and 185.

[42] Banks, P.M., Foster, J.C., Doupnik, J.R. (1981): J. Geophys. Res. *86*, 6869; see also: Foster, J.C., Doupnik, J.R., Stiles, G.S. (1981): J. Geophys. Res. *86*, 11357

[43] Volland, H. (1979): J. Atmos. Terr. Phys. *41*, 853

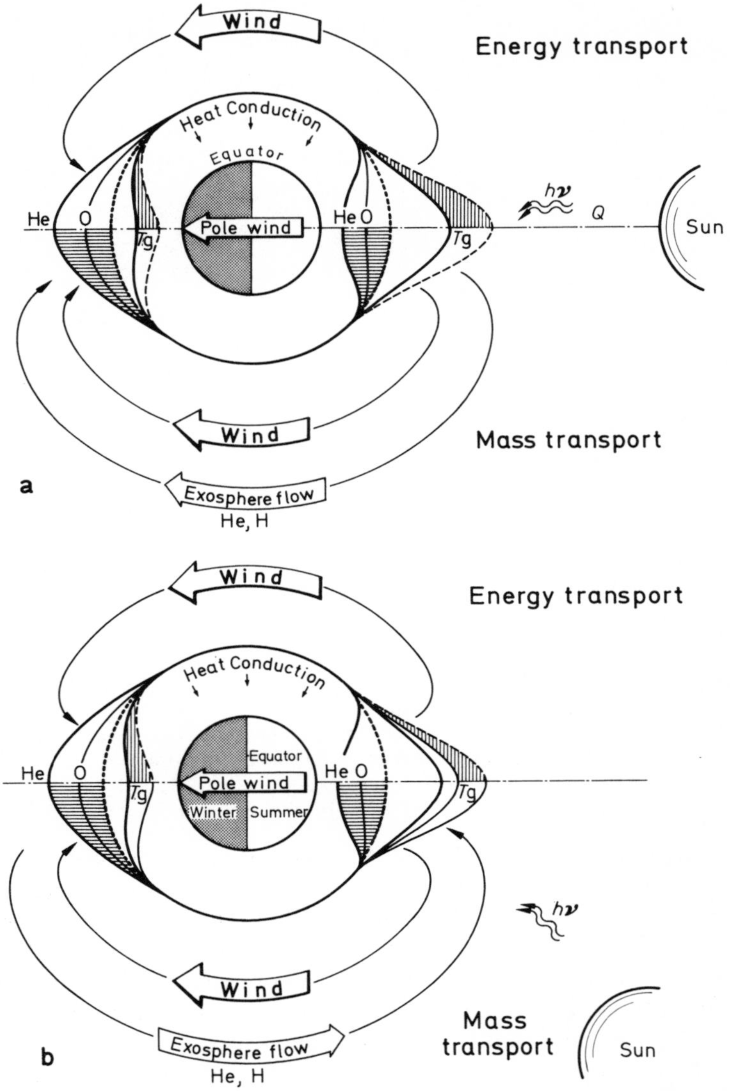

Fig. 182a, b. Schematic illustration of (**a**) diurnal, (**b**) annual atmospheric variations in the neutral atmosphere [*55*]. Broken lines illustrate temperature and density variations when horizontal transport is neglected, which is taken into account by the solid lines. Dotted lines, global average. Shaded areas represent energy (T_g) and mass redustribution (mainly O, He) as a consequence of winds and exospheric flow

ζ) The aeronomical role of excited species was discussed by THOMAS in vol. 49/6 (his Chap. H, p. 120). It has become more and more evident during the past years that reactions with such species are of much greater importance than earlier assumed [*66*]. This idea was first voiced by DALGARNO[44], who

[44] Dalgarno, A. (1970): Ann. Géophys. *26*, 601

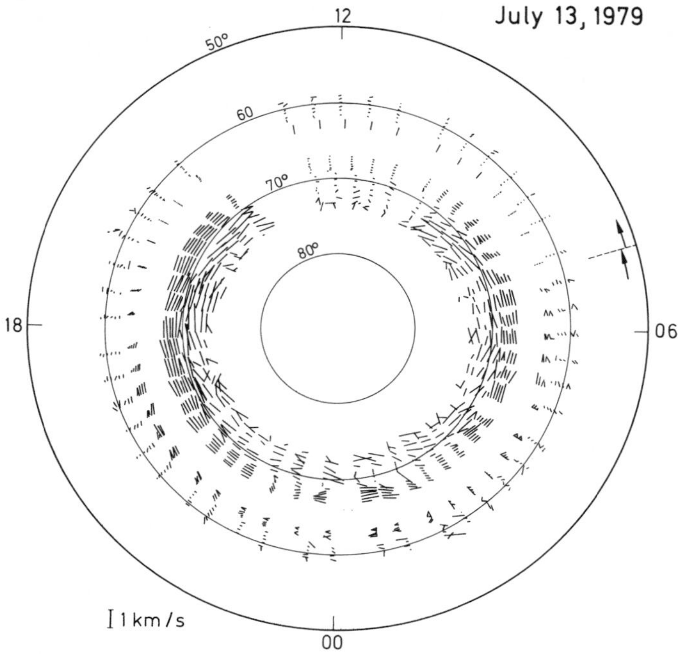

Fig. 183. Schematic high latitude plasma convection pattern[42] inferred from experimental data obtained by high latitude incoherent scatter radar measurements of plasma drifts. The observations yield a Joule-heating rate of about 100 mW m^{-2}, more than 30 times the solar EUV-input. Local time and invariant latitude as polar coordinates

found that 'the metastable $O_2(a\ ^1\Delta_g)$ molecule is an important source of O_2^+ ions in the D region, the $N(^2D)$ atom is an important source of NO and possibly of N^+ ions in the E-region and the $O^+(^2D)$ ion is an important source of N_2^+ ions in the F-region. The $N_2(A^3\Sigma_u^+)$ molecule may significantly modify the atmospheric chemistry during auroral bombardment', see Sect. 18δ.

Recently Torr and co-workers[45] established a computing program into which enter all important major and minor species including excited ones; 30 reaction equations are applied and the upper boundary is adapted to photoelectron flux between hemispheres by the 'two-stream method' (see Appendix A below). However, neutral wind effects which are so important in other computing schedules (see Sect. 15ζ) are not taken into account. The authors arrive at an estimate of the importance of the different energy transfer processes in the whole thermosphere. Heat loss of the electrons is said to be: 33% from dissociation in the Schumann-Runge bands (5.08 eV per process), 17% by airglow (O-emissions between 130 and 99 nm, and O_2 emission at 63 μm, see Sect. 16ε) and the rest only by kinetic energy transfer. As for exothermal reactions the estimate is 30% for electrons exciting metastable $O(^1D)$ and 14%

[45] Torr, D.G., Richards, P.G., Torr, M.R. (1980): J. Geophys. Res. *85*, 6819; reaction coefficient were published by: Torr, D.G., Torr, M.R. (1979): J. Atmos. Terr. Phys. *41*, 797

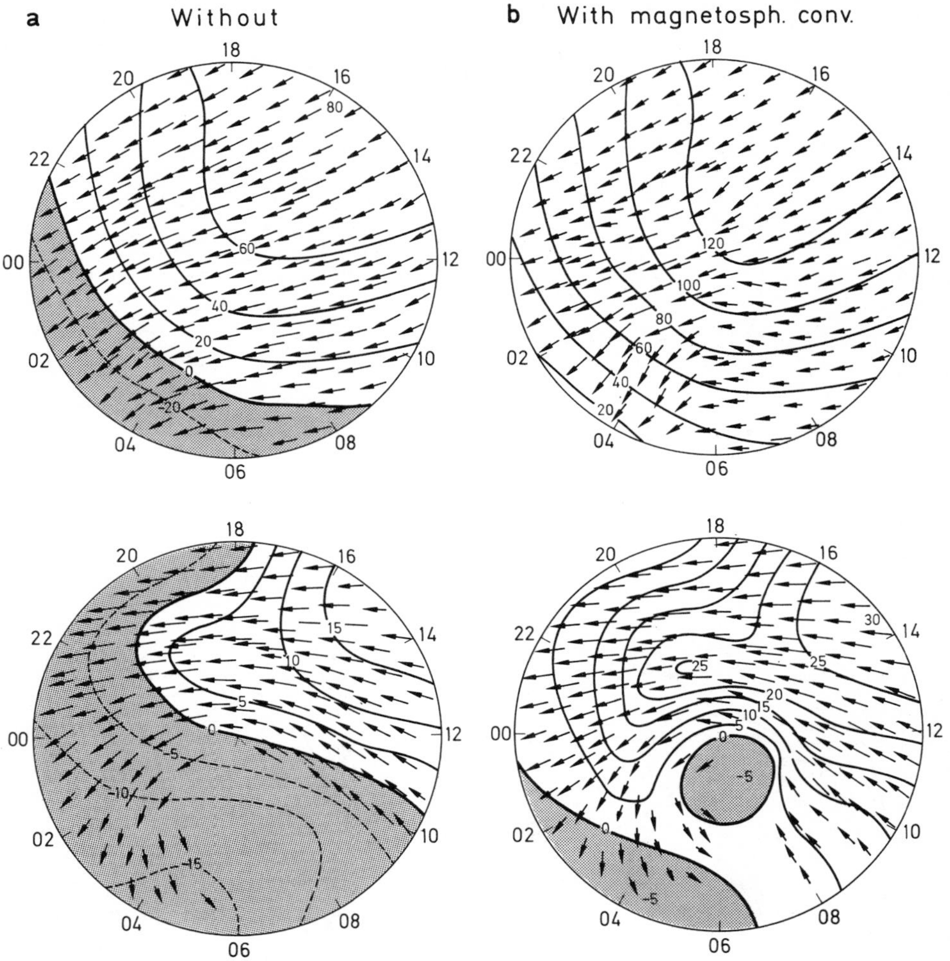

Fig. 184a, b. Contours of perturbation temperature/K (isolines) and wind vectors computed in [*51*] for the northern polar cap for altitudes of 300 km (top) and 130 km (bottom). Angular coordinate is local time. The limiting circle corresponds to 45° N latitude. Maximum wind speed 150 ms^{-1}. (**a**) Solar heating only; (**b**) including magnetospheric convection with 20 kV cross-tail potential, but assuming coincident poles. Shaded areas with negative temperature effect

only in reactions with ground states. According to TORR the great importance of excited species must now be admitted[46]. Also 'odd nitrogen' [$N(^2D)$, $N(^4s)$, NO] is quite important[47] for vertical energy transport (see THOMAS in vol. 49/6, his Sect. 33γ and δ). The effects of varying the rate coefficients are discussed in [*66*]. Figures 186 and 187 show a few results.

[46] See Torr, D.G. (1981): [*53*], 18-1

[47] See for example: Rusch, D.W. (1977): [*18*], 261 (with many references)

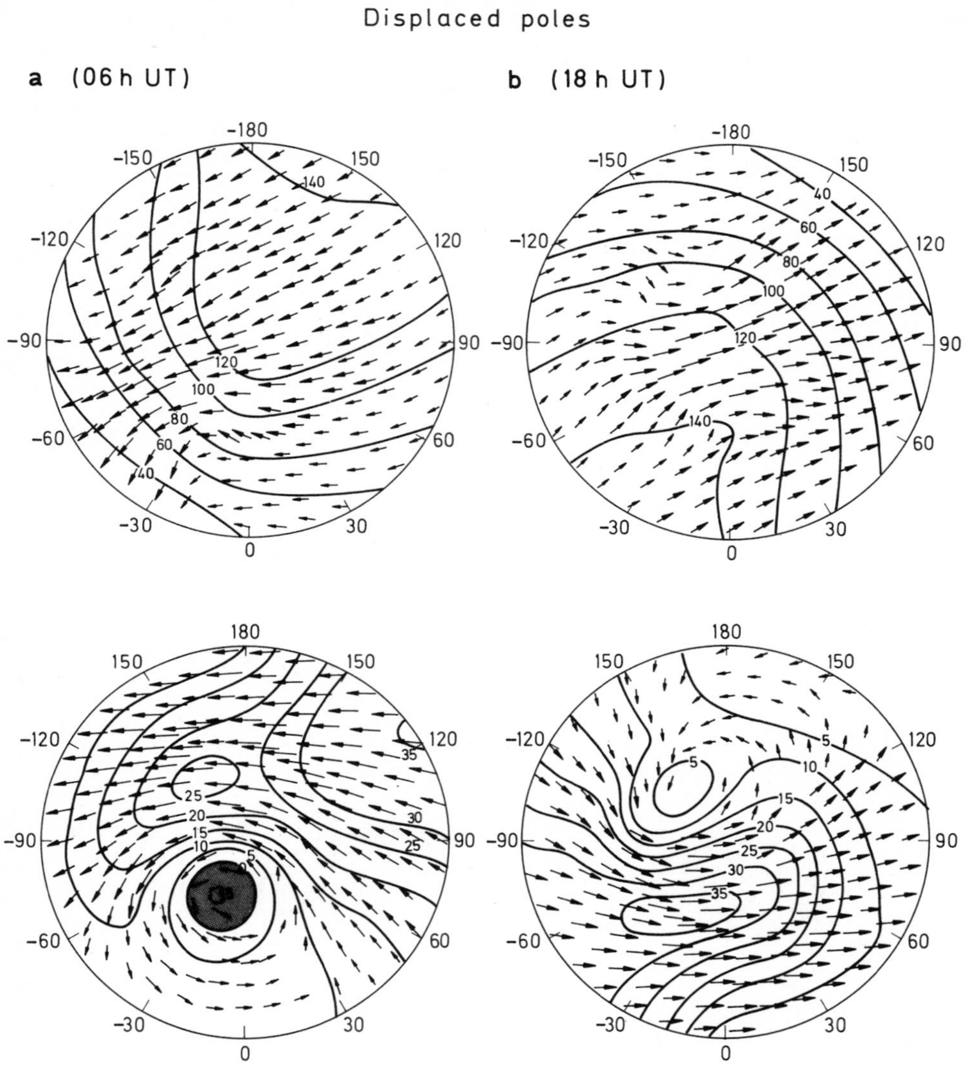

Fig. 185a–e. Contours of perturbation temperature/K (isolines) and wind vectors computed in [*51*] for the northern polar cap; time 18 and 06h UT as indicated, and altitudes of 300 km (top) and 130 km (bottom). Angular coordinate is longitude. The limiting circle corresponds to 45° N. Maximum wind speed 219 m s^{-1} in (**a**), 222 m s^{-1} in (**b**), 395 and 373 m s^{-1} (top and bottom) in (**c**), 100 and 109 m s^{-1} (top and bottom) in (**d**) and 147 and 76 m s^{-1} (top and bottom) in (**e**). (**a, b**) As in Fig. 184b but with displaced poles; (**c, d**) same but with 60 kV cross-tail potential; (**e**) same as (**c**) but taking only Joule heating into account, not ion drag

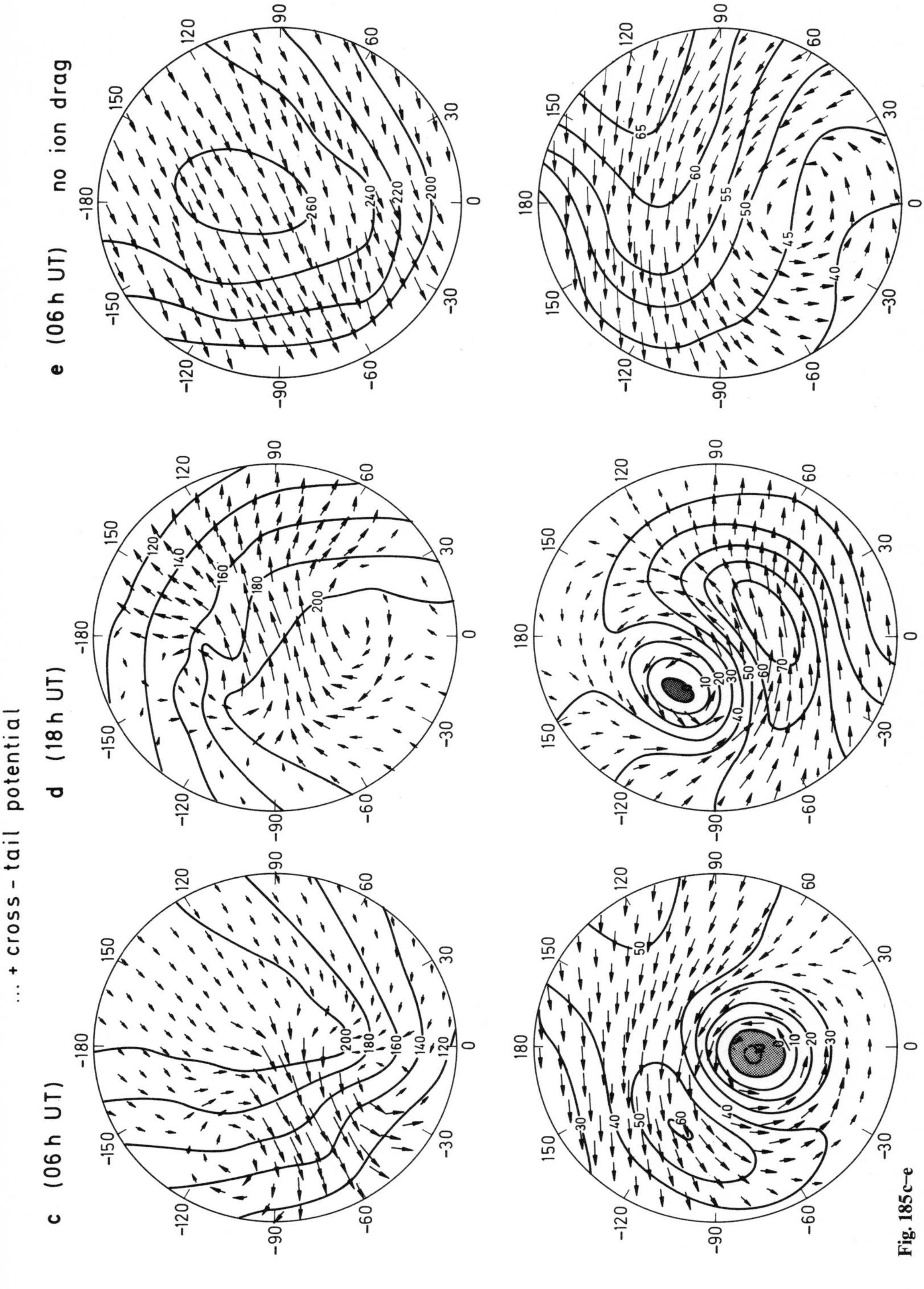

Fig. 185c–e

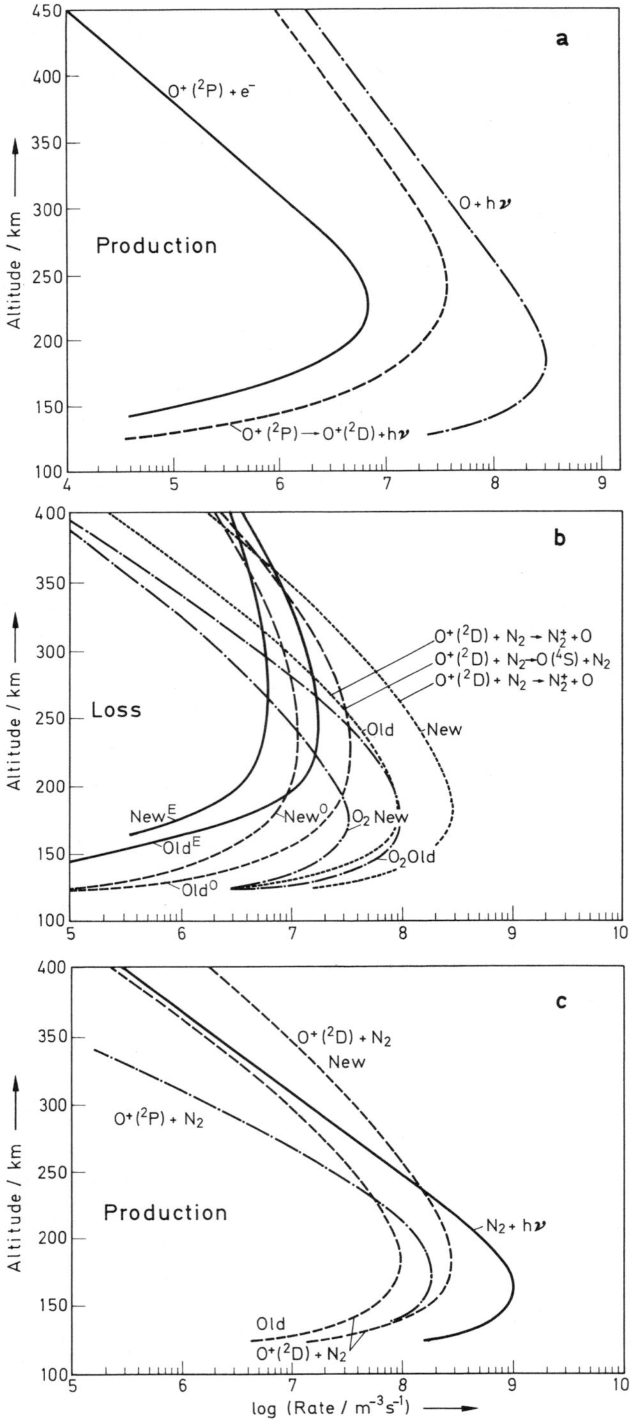
a
Production
O⁺(²P) + e⁻
O + hν
O⁺(²P) → O⁺(²D) + hν
Altitude / km
b
Loss
O⁺(²D) + N₂ → N₂⁺ + O
O⁺(²D) + N₂ → O(⁴S) + N₂
O⁺(²D) + N₂ → N₂⁺ + O
Old
New
New^E
Old^E
New^O
Old^O
O₂ New
O₂ Old
c
Production
O⁺(²D) + N₂
New
O⁺(²P) + N₂
N₂ + hν
Old
O⁺(²D) + N₂
log (Rate / m⁻³s⁻¹)

Fig. 186 a–c

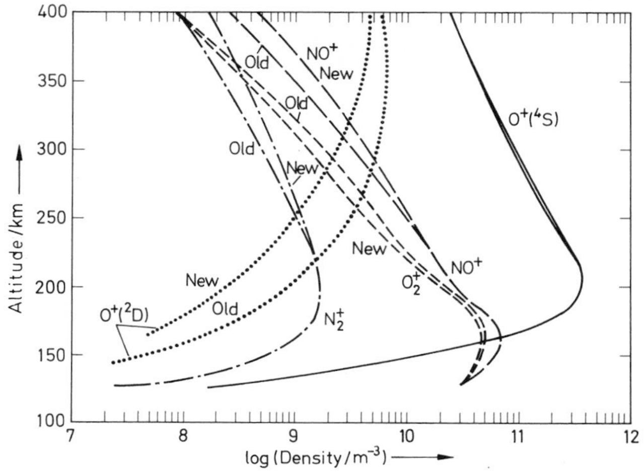

Fig. 187. Comparing densities of major species computed by a photochemical model [*66*] using low ("Old") and large ("New") rates for the charge exchange $O^+(^2D)+N_2$, and introducing the new process N_2^+ (vib) $+O \rightarrow O^+(^4S)+N_2$ (New)

18. The polar caps: coupling with the magnetosphere. The upper atmosphere above the polar caps of Earth is, fundamentally, in another geophysical environment than that at mid- and low-latitudes. This is due to the solar wind, which 'blows away' the outer part of the terrestrial magnetic field and so shapes the magnetosphere[1]. It can be seen from Fig. 188 that the fieldlines connected with the polar caps are 'brushed' backwards to form the tail of the magnetosphere. The interaction between solar wind and terrestrial auroral and disturbance phenomena occurs essentially in this tail. Strong electric fields are transferred from the tail into the auroral zones. Recent experience has shown that the most important acceleration of charged particles occurs at an altitude of about 1 Earth radius. Thus, the situation of the polar ionosphere is characterized by two influences connected with the magnetospheric tail:

(i) A corpuscular bombardment by particles travelling down along the open fieldlines.

(ii) Strong electric fields transferred from the magnetospheric tail.

α) Of the *auroral phenomena*, which are all provoked by *particle bombardment* (mainly electrons), the optical ones are most impressive[2]. However, this is only part of the whole bundle of phenomena which are initiated by this bombardment - and certainly not that which affords the highest energy sup-

[1] Whalen, J.A. (1981): in [*7*]

[2] See Akasofu, S.-I., Chapman, S., Meinel, A.B. (1966): This Encyclopedia, vol. 49/1, p. 17

Fig. 186a–c. Chemical modelling for 1979 (northern USA): production and loss rate profiles after Torr [*66*]. (**a**) Production * of $O^+(^2D)$; (**b**) loss * of $O^+(^2D)$; (**c**) production of N_2^+ showing increase of $O^+(^2D)$ channel by new, increased rates. (* see Table 90 in [*66*])

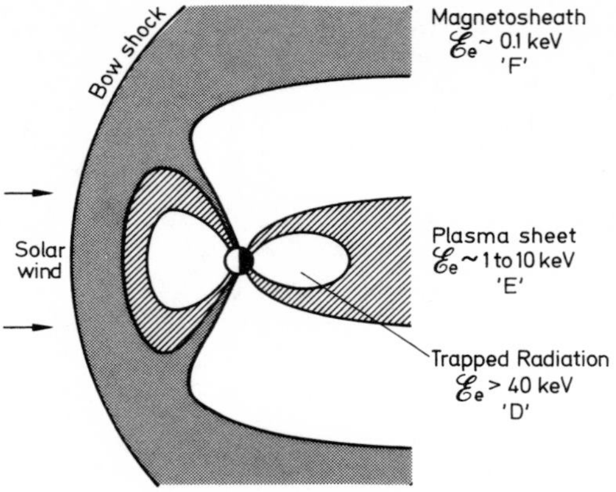

Fig. 188. Principal domains in the Earth's magnetosphere[1]. Open field lines (in the white ranges) and plasma sheet in between form the tail which extends a very great distance (see POEVERLEIN, H. (1972): Fig. 11, p. 41 in vol. 49/4 of this Encyclopedia)

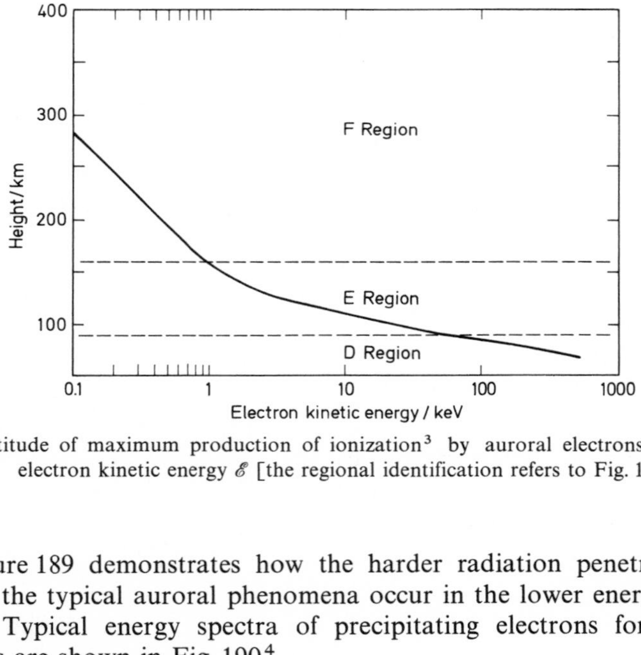

Fig. 189. Altitude of maximum production of ionization[3] by auroral electrons as a function of electron kinetic energy $\mathscr{E}$ [the regional identification refers to Fig. 192]

ply[1]. Figure 189 demonstrates how the harder radiation penetrates to lower altitudes; the typical auroral phenomena occur in the lower energy part of this diagram. Typical energy spectra of precipitating electrons for rather quiet conditions are shown in Fig. 190[4].

It is almost impossible to study the phenomenon as whole with fixed stations only, except for a statistical picture, which might, however, deteriorate by smoothing. But during the past decade many satellites have investigated the magnetosphere, so that the incoming radiation and its distribution are now

[3] Rees, H.H. (1964): Planet. Space Sci. *15*, 209

[4] Meng, CH.-I. (1981): J. Geophys. Res. *86*, 2149

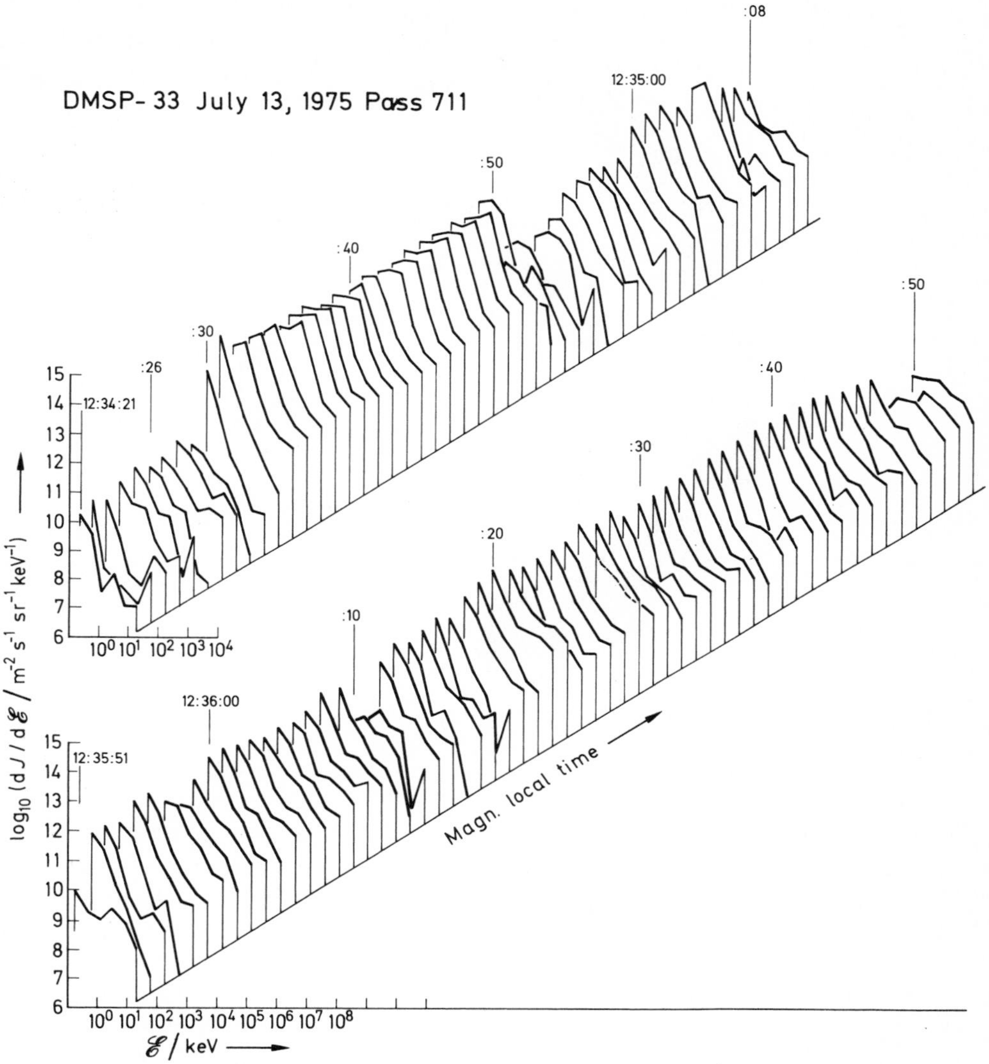

Fig. 190. Measured spectra of precipitating electrons[4]

much better known. Furthermore, on the terrestrial side, aircraft studies have made a systematic investigation of the auroral oval as a whole possible.

At the high latitude of the oval a jet aircraft can either keep up with Earth's rotation and so stay at fixed (geomagnetic) local time, or go around the auroral oval and study it as function of local time. BUCHAU, GASSMANN and co-workers[5] at AFGL have made such investigations during many years with the "flying ionospheric observatory" and have so obtained a clear picture of individual auroral events (see Fig. 191 for example).

[5] Buchau, J., Gassmann, G.J., Pike, C.P., Wagner, R.A., Whalen, J.A. (1972): Ann. Géophys. *28*, 443

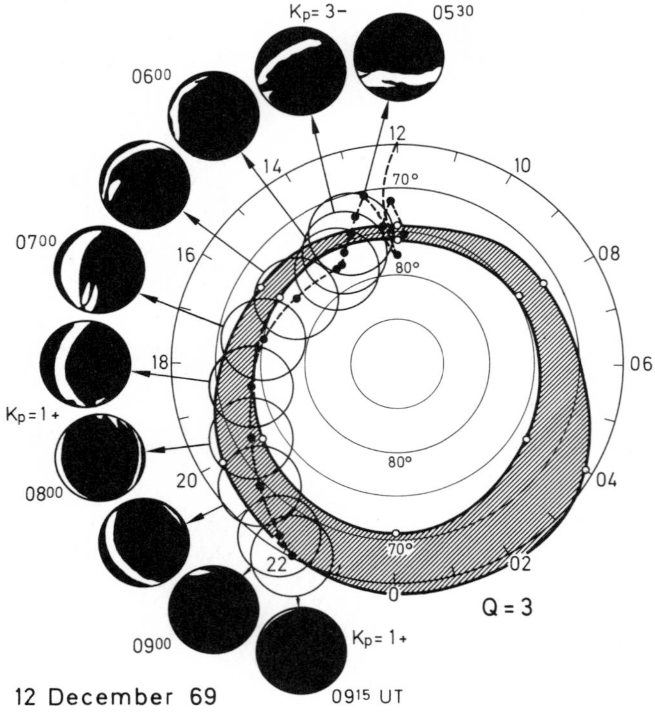

Fig. 191. The continuity of the auroral oval (shaded) documented with all-sky camera[2] auroral pictures obtained by the "flying ionospheric observatory" on the (broken) flight path[5]

The result of such observations is shown schematically in Fig. 192. As required by the Störmer theory[6] the harder radiation penetrates to a lower latitude, provoking (see Fig. 189) an intense ionization in the D-region (and absorption of hf-radio waves). The softer radiation ionizes only above 100 km and produces a thick auroral E-layer, called Ea [37]. The auroral phenomena occur mainly at the high-latitude side of the oval; their character depends on local (geomagnetic) time; see Fig. 193. The auroral ionospheric layer, Ea, accurately follows the oval; see Fig. 194. Figure 195[1] summarizes the phenomena as function of the hardness of the precipitation. The energy ranges "F, E, D" correspond to the relevant ionospheric layers of maximum particle absorption; their origin is indicated with the same letters as in Fig. 189.

Electric fields play a major role in the auroral zones. Apart from local fields connected with the spatial structure of the precipitating electrons there is a global field transferred from the tail of the magnetosphere. Since, under the very impact of the precipitation, a plasma layer is formed near the 100 km level

[6] See Akasofu, S.-I., Chapman, S., Meinel, A.B. (1966): This Encyclopedia, vol. 49/1, p. 84; also Hess, W.N. (1972): ibidem 49/4, Chap. B, p. 116. The original theories are: Störmer, C. (1907/11/13): Arch. Sci. Genève *24*, 5, 113, 221, 317; *32*, 33, 163; *35*, 483. Lemaître, G., Vallarta, M.S. (1933): Phys. Rev. *43*, 87; Lemaître, G., Bossy, L. (1946): Bull. Cl. Sci. Acad. Roy. Belg. *31*, 357

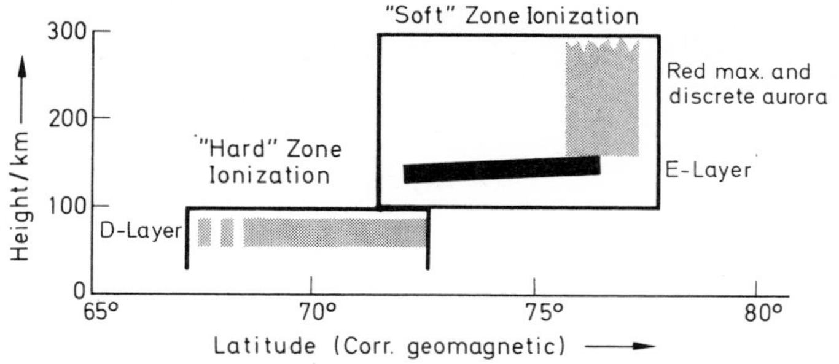

Fig. 192. Schematic of noon meridian auroral observations in an altitude versus latitude cross section. In flight observations[5] were combined with satellite observations of "hard" (above 21 keV) and "soft" (0.08–21 keV) electron precipitation[7]

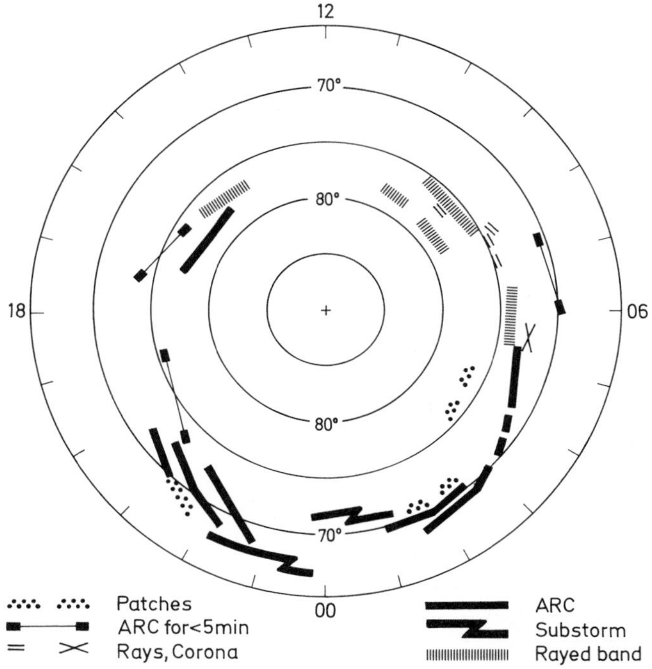

Fig. 193. Ground and aircraft observations of auroral phenomena in the whole oval[5]

(see Fig. 216 below), it has rather high conductivity so that one has an appreciable current flow in such auroral layers; see Fig. 196 for a typical example[8]. Electrical fields can (indirectly, by the relevant plasma motions) be observed with incoherent scatter technique. Some results are shown in Figs. 197–199, for subauroral[9] and auroral latitudes[10]. An idealized electrical potential pattern in

[7] Sharp, R.D., Carr, L.D., Johnson, R.G. (1969): J. Geophys. Res. *74*, 4618

[8] Schlegel, K. (1982): Phys. Blätter *38*, 81

[9] Wand, R.H., Evans, J.H. (1981): J. Geophys. Res. *86*, 103

[10] Foster, J.C., Doupnik, J.R., Stiles, G.S. (1981): J. Geophys. Res. *86*, 2143

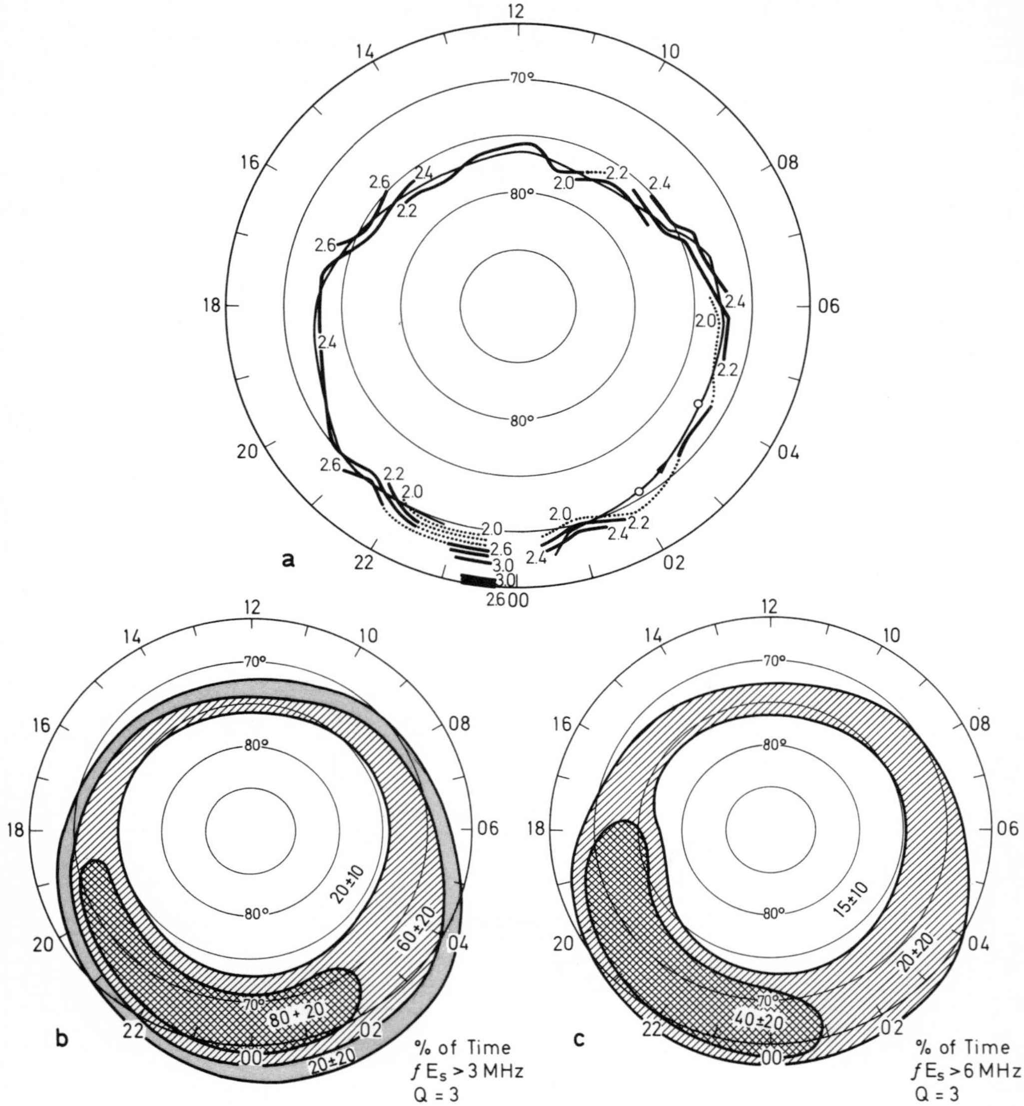

Fig. 194a–c. Oval-aligned ionospheric Ea-layer (auroral E) resulting from soft particle (1–5 keV) bombardment[5]; observation under "quiet" magnetic conditions: (**a**) Critical frequency/MHz (labels) observed during a circum-oval flight. (**b, c**) Statistical results from ground stations about the Ea-probability (labels: % probability); (**b**) for fEa > 3 MHz; (**c**) for fEa > 6 MHz

the polar caps is shown in Fig. 200; in the auroral zones the field is essentially meridional, its direction depending on that of the interplanetary magnetic force[11]. By $\boldsymbol{E} \times \boldsymbol{B}$ forces these fields drive a zonal plasma convection stream;

[11] Heppner, J.P. (1977): J. Geophys. Res. *82*, 1115; see also Volland, H. (1975): Ann. Géophys. *31*, 154

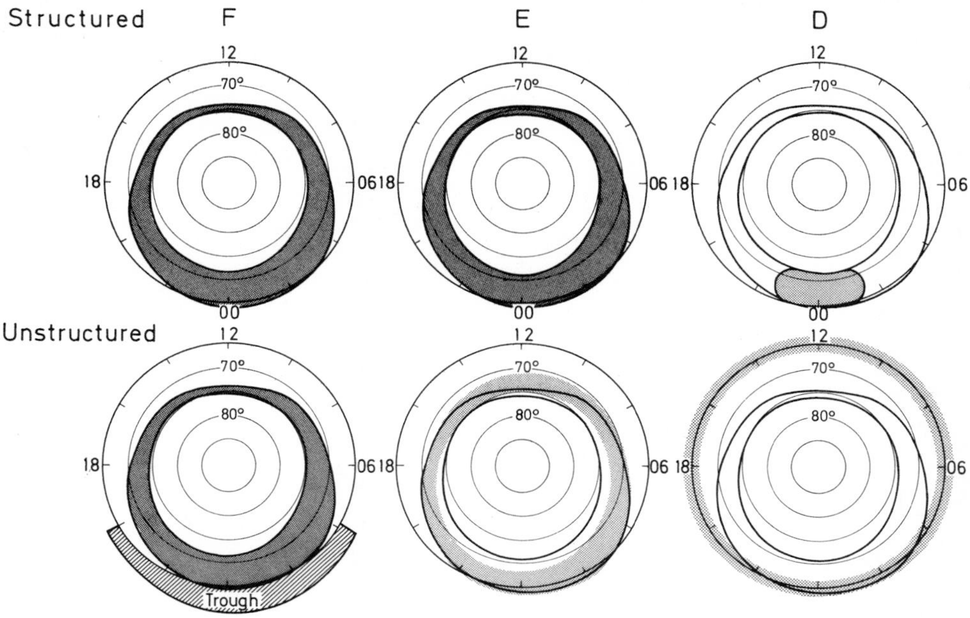

Fig. 195. Regions where auroral phenomena (shaded areas) are typically produced by precipitating electrons of different hardness (rows): F ~0.1 keV, E ~1–10 keV, D ~ >40 keV. Top, structured auroral forms; bottom, unstructured forms. Corrected geomagnetic coordinates (12 h on top). The ionospheric "trough" appears between 20 and 04 h and is due to soft radiation. The index Q=3 standard oval is indicated by the heavy solid lines[1]

see Fig. 183 in Sect. 17. At high latitudes, this is a major transport phenomenon which, together with diffusion, determines the spatial distribution of the plasma in the polar caps at (invariant magnetic) latitudes greater than 60°. It is, however, important to know fairly accurately the spatial structure of the electric field which derives from a potential as roughly shown in Fig. 200. In particular, in order to explain the very localized phenomenon of the so-called "trough" (see Sect. 14γ(v) and Fig. 135) one needs, apparently, a narrow annular zone with inversed meridional field direction at the mid-latitude border of the polar cap. This zone should cover about 12°, peaking with a field strength of about 50 V/km somewhere around 70°; the trough is then formed at a somewhat lower invariant latitude; see Eq. (14.8). Under such assumptions the particular structure of the observed electron density distribution can be reproduced with aeronomical theory. Figure 201 shows results of such computations[12] for $Kp=3$ under winter conditions when in the polar cap corpuscular ionization is the only important direct source of ionization while at lower latitude ionization by solar EUV is not negligible. The zone of low ionization someway below 70° appears clearly throughout the night.

[12] Larina, T.N., Maksimova, N.M., Mozaev, A.M., Osipov, N.K., Cernysova, S.P. (1982): Phys. Solariterr, Potsdam no. 18, 47

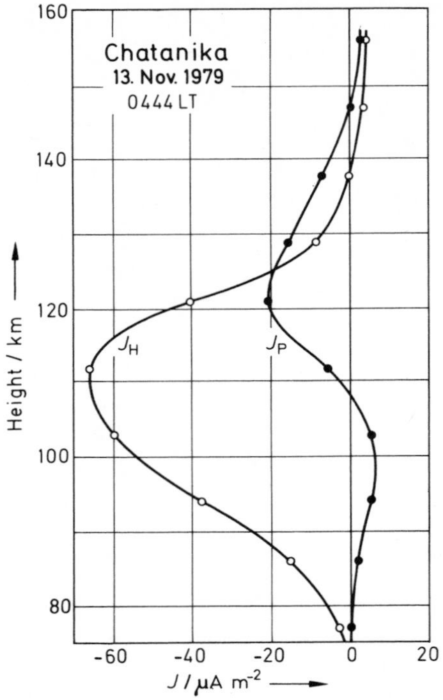

Fig. 196. Current densities inferred from observations of an incoherent scatter station in Alaska[8]. J_P, Pedersen, J_H Hall-current

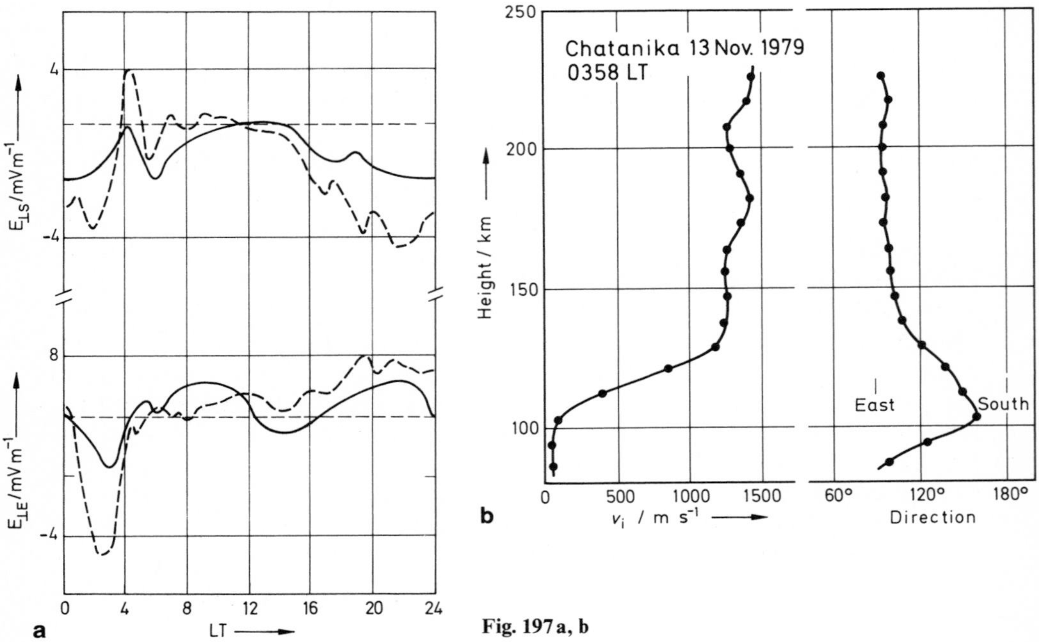

Fig. 197 a, b

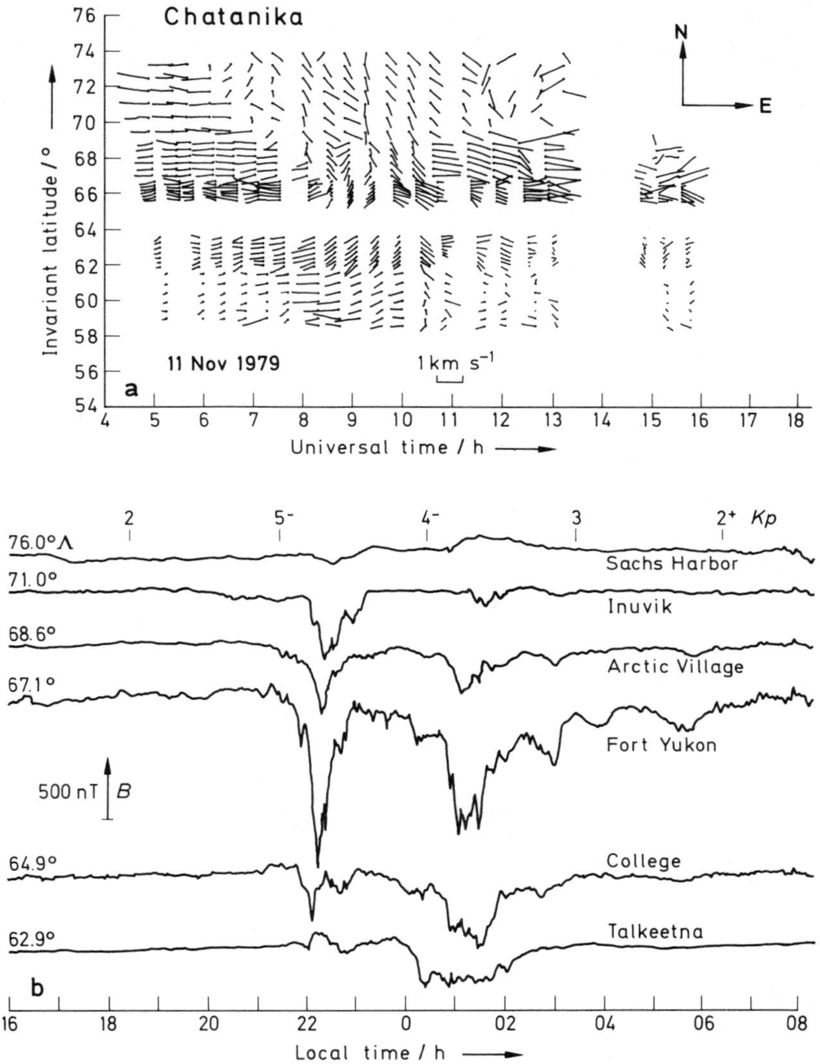

Fig. 198. (**a**) Plasma motion above Chatanika (Alaska) determined by incoherent scatter technique[10]. (**b**) Simultaneous magnetic records from neighbouring stations ordered after their invariant magnetic latitude

Fig. 197a, b. Electric fields determined from cross-field plasma motion observed by an incoherent scatter radar. (**a**) at Millstone Hill (Ma., USA)[9]. Solid curves are all-year averages for quiet days and dashed curves for disturbed days in the period May 1976–Nov 1977. An eastward field, $E_{\perp E}$ provokes a (perpendicular) northward ion drift (and a southward field an eastward drift). 2 mV/m produce a drift of about 50 m s^{-1}. (**b**) Height profile of the perpendicular ion drift observed with incoherent scatter technique in Alaska[8]. From the observed drift a field of 75 mV m^{-1} is deduced

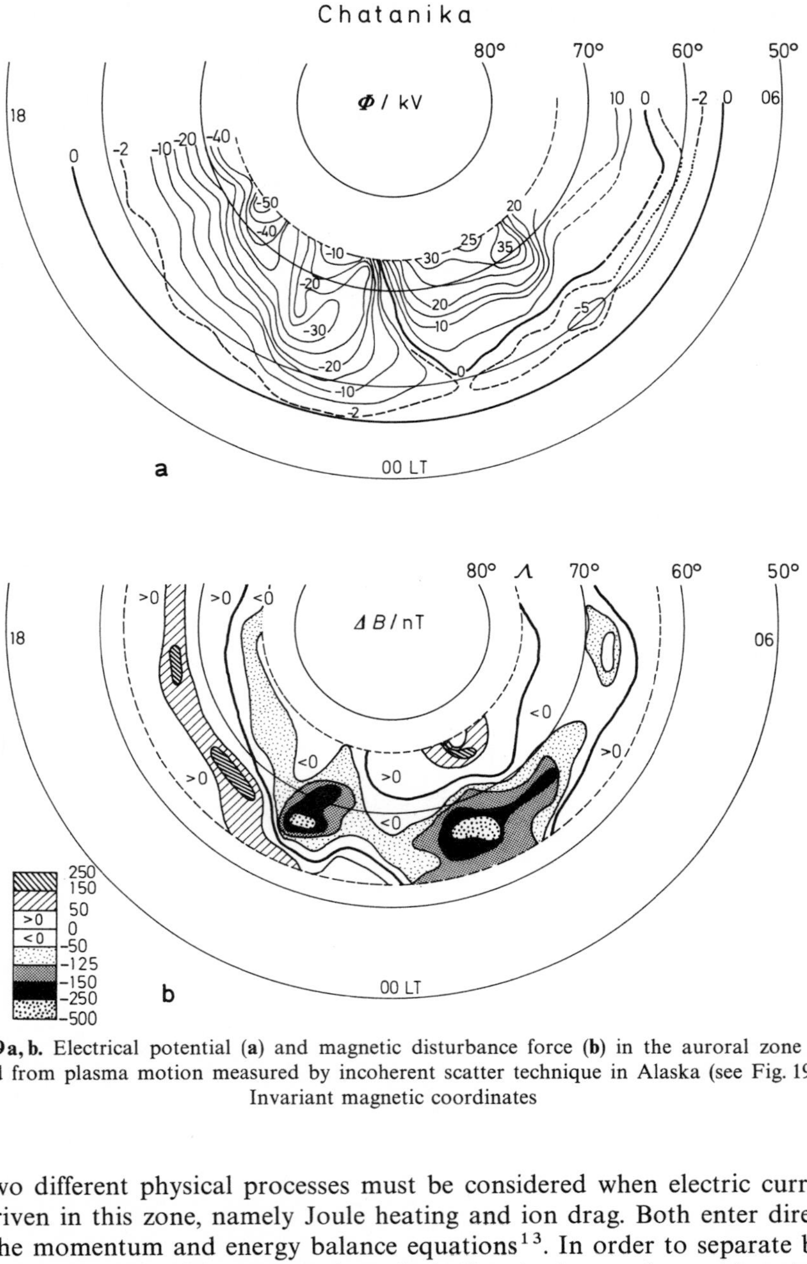

Fig. 199a, b. Electrical potential (**a**) and magnetic disturbance force (**b**) in the auroral zone both inferred from plasma motion measured by incoherent scatter technique in Alaska (see Fig. 198)[10]. Invariant magnetic coordinates

Two different physical processes must be considered when electric currents are driven in this zone, namely Joule heating and ion drag. Both enter directly into the momentum and energy balance equations[13]. In order to separate both effects one uses model computations including both together with suitable estimates of the conductivity and electric field maps. By comparison with motions and plasma data observed by incoherent scatter technique (see Sect. 5),

[13] See for example: Mayr, H.G., Harris, I. (1978): J. Geophys. Res. *83*, 3327 and their references; also Cole, K.D. (1977): [*18*], 206 and Brekke, A., ibidem 313

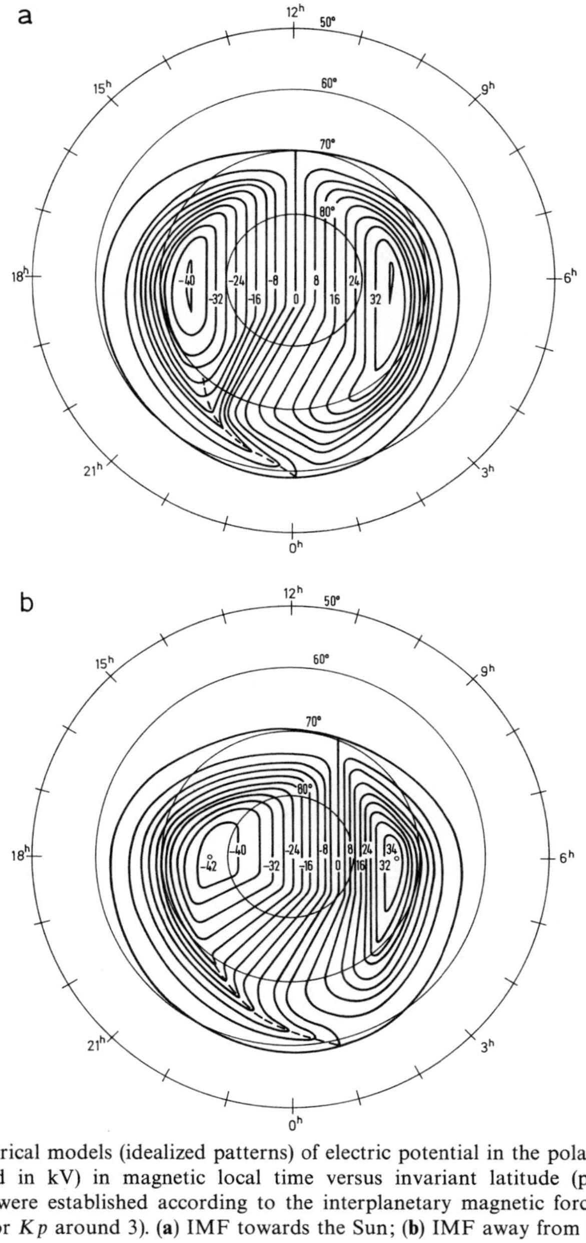

Fig. 200a, b. Empirical models (idealized patterns) of electric potential in the polar caps: equipotential lines (labelled in kV) in magnetic local time versus invariant latitude (polar coordinates). Separate models were established according to the interplanetary magnetic force (IMF) direction (both for Kp around 3). (**a**) IMF towards the Sun; (**b**) IMF away from the Sun

one may then fit theoretical and observed data and so finally infer the Joule heating. Figure 202 shows results obtained for the convection pattern shown in Fig. 183[14]. Joule heating resulting from such reasoning and data is at least as

[14] Banks, P.M., Foster, J.C., Doupnik, J.R. (1981): J. Geophys. Res. *86*, 6869

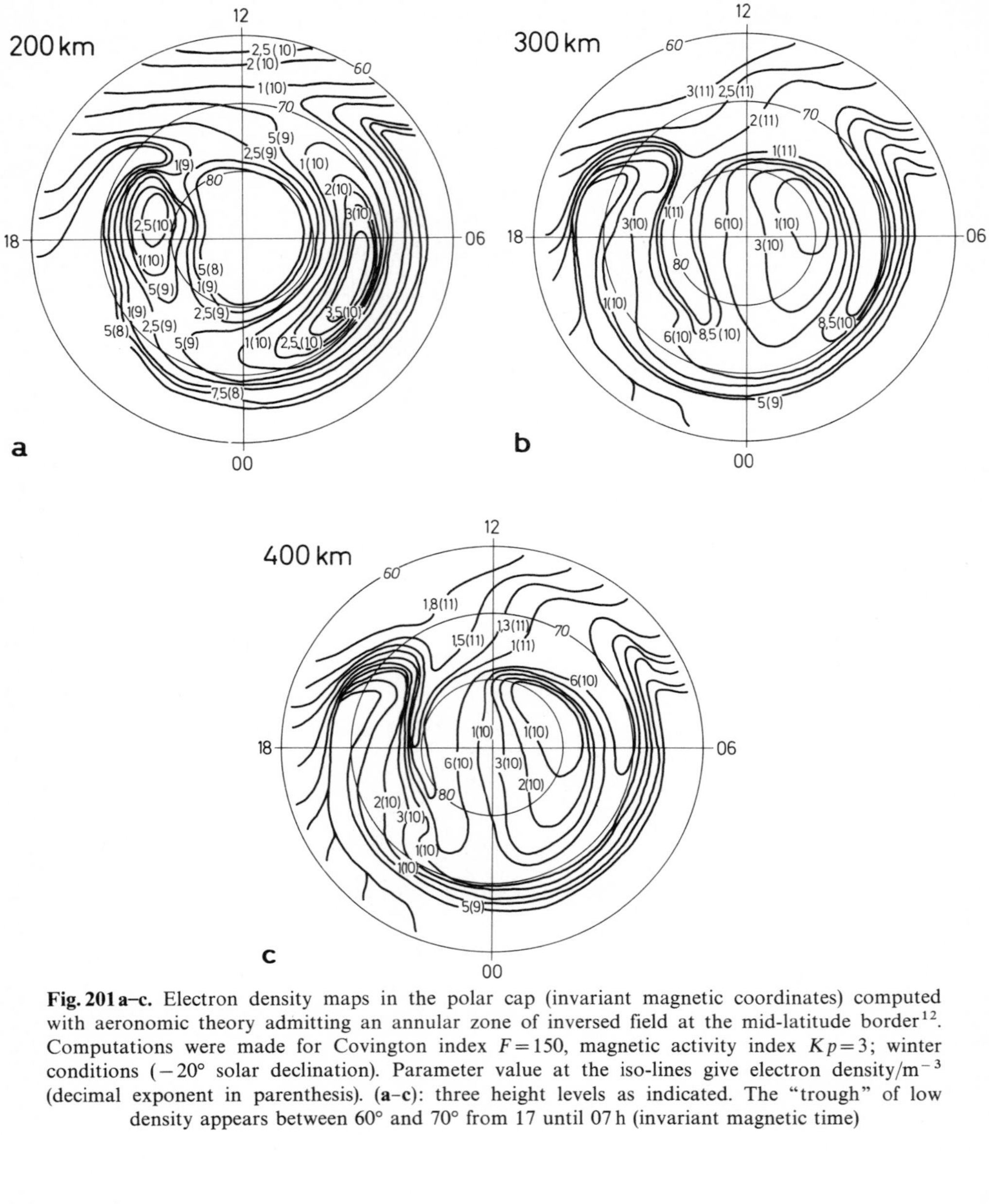

Fig. 201 a–c. Electron density maps in the polar cap (invariant magnetic coordinates) computed with aeronomic theory admitting an annular zone of inversed field at the mid-latitude border[12]. Computations were made for Covington index $F=150$, magnetic activity index $Kp=3$; winter conditions ($-20°$ solar declination). Parameter value at the iso-lines give electron density/m^{-3} (decimal exponent in parenthesis). (**a–c**): three height levels as indicated. The "trough" of low density appears between 60° and 70° from 17 until 07 h (invariant magnetic time)

important a heat source in the polar caps as direct heat transfer from the precipitating electrons by 'individual' transfer.

Energy fluxes might be estimated from all these effects via semi-empirical relations. In a latitudinal section at noon the peak is found at about 75° corrected geomagnetic latitude (i.e. in range "E" of Fig. 189) with about 100 $\mu W\,m^{-2}$ for average quiet conditions. Thus, even during quiet conditions, there is always a non-negligible total energy input of the order of 100 MW into

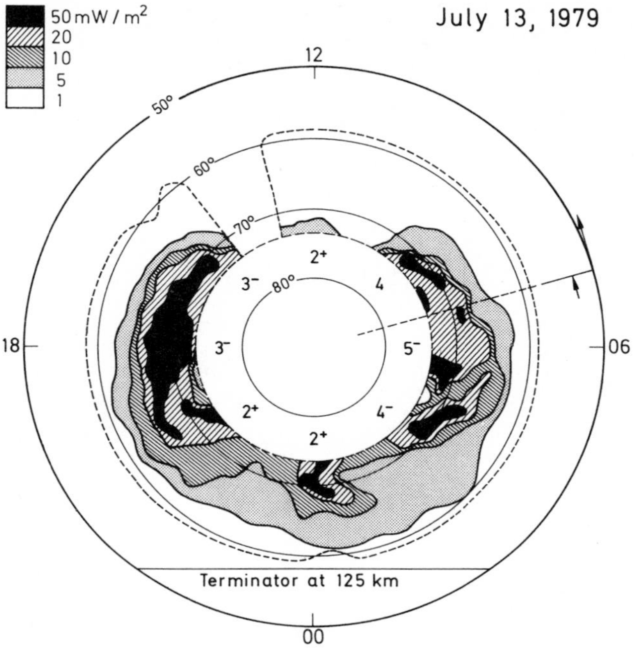

Fig. 202. Joule heating inferred from ion velocity measurements [uniform Pedersen-conductivity of 10 S was assumed]. Heating rates: from 1 to 50 mW m^{-2} (1-5-10-20-50) shown by increasing darkness of hatching[14]. [The inner, dotted limitation is due to the position of the radar]

the auroral oval which must be taken into account when establishing temperature and wind maps (see Sect. 17ε).

Since the precipitating electrons feed an appreciable amount of energy into the auroral zones, *strong vertical interchange* occurs quite regularly therein. It has important consequences and makes the upper atmosphere in these zones quite different from that at middle and low latitude. In particular, heavier ions are transported upwards and can be found at heights where otherwise only lighter ions are found; thus molecular ions may be found near 300 km and O^+ ions near 1,000 km. Figures 203–206 describe some of the phenomena[15–19]. Upgoing ion beams could recently be identified (see Fig. 204), and even double charge O^{++} ions were found (see Fig. 205). Somewhat similar phenomena occur between plasmasheet and inner plasmasphere, where H^+ and He^+ ions are travelling along the magnetic field lines; this phenomenon seems to be a characteristic of post-disturbance periods; see Fig. 206[16].

[15] Strauss, J.M. (1978): Rev. Geophys. Space Phys. *16*, 183

[16] Horwitz, L.L. (1982): Rev. Geophys. Space Phys. *20*, 929

[17] Horwitz, J.L. (1981): J. Geophys. Res. *86*, 9225

[18] Gorney, D.J., Clarke, A., Croley, D., Fennell, J., Luhmann, J., Mizera, P. (1981): J. Geophys. Res. *86*, 83

[19] Geiss, J., Young, D.T. (1981): J. Geophys. Res. *86*, 4739

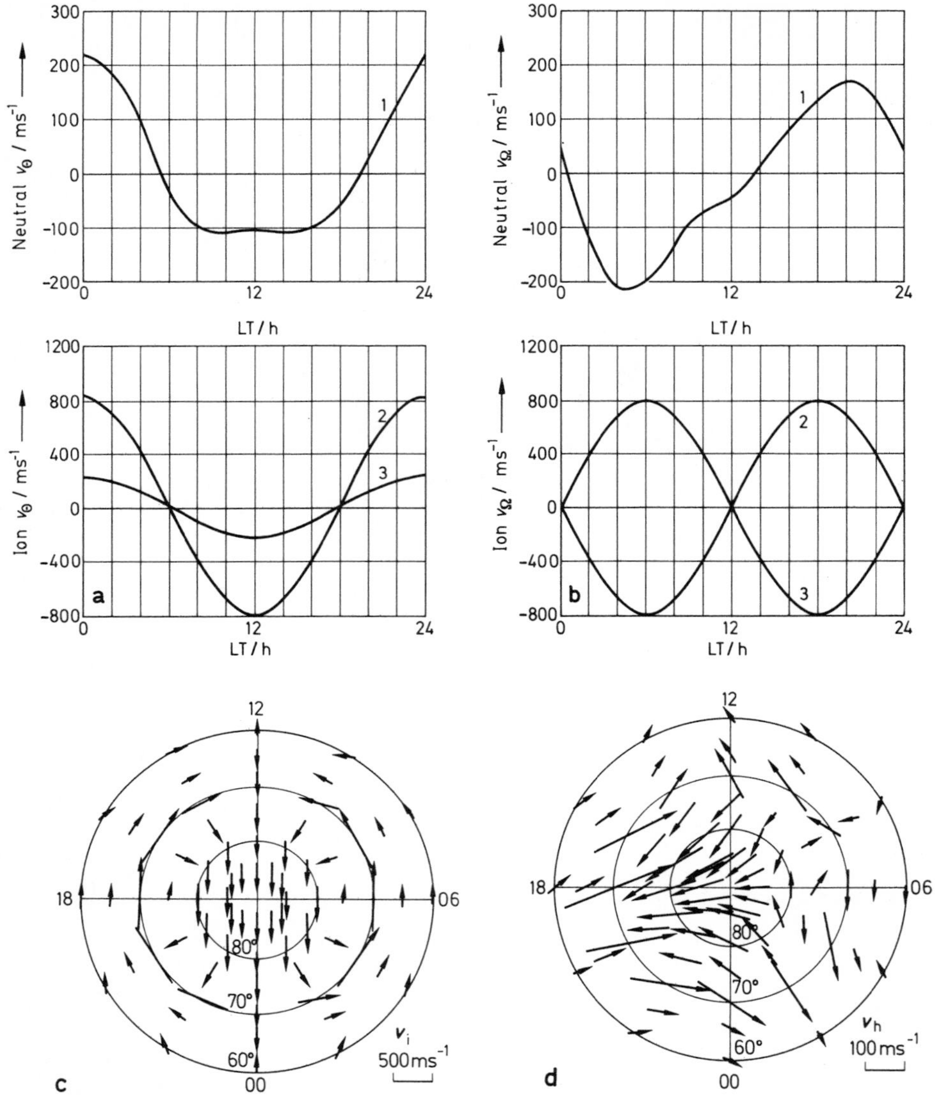

Fig. 203a–d. Neutral and plasma drifts in the polar caps[15]. (**a, b**) Typical diurnal motions at 250 km of altitude. Top: neutral wind (computed from pressure gradients driven by solar heating (1). Bottom: ion drift at 63° magnetic latitude (3) and in the polar cap (2) computed using VOLLAND's[11] electric field model (with components E_ϕ, E_θ of 50 mV/m both). (**a**) Meridional winds and drifts; (**b**) zonal winds and drifts. (**c, d**) Geographical distribution of (**c**) ion drifts and (**d**) neutral winds at 200 km calculated by H. MAEDA for diurnally varying ion density distribution. [Sources: VOLLAND, H. (1975): Ann. Géophys. *31*, 154; MAEDA, H. (1976): J. Atmos. Terr. Phys. *38*, 197]

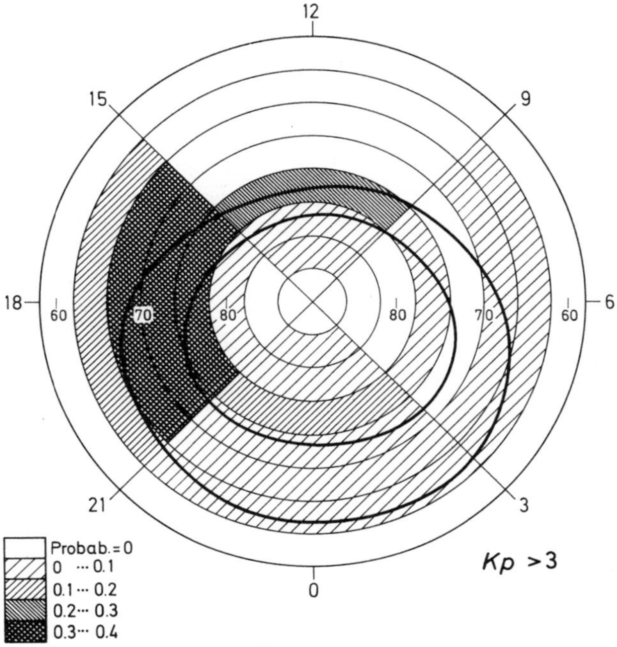

Fig. 204. Polar plot of occurrence probability (hatched surfaces) of upgoing ion beams observed at heights up to 8,000 km aboard U.S. satellite S3-3 during magnetically active periods ($Kp>3$)[18]. FELDSTEIN's (1969) statistical auroral oval is indicated for comparison (bold curves)

β) During *magnetic disturbances* the energy input is much larger; different estimates[20] arrive at about 100 GW[21] for magnetic activity[22] $Kp=6$. This energy input produces a large perturbation heating rate with peaks of almost 100 W/kg at an altitude of about 250 km[20] (see Figs. 207, 209). The neutral temperature follows with some delay and smoothing. In Fig. 209 the two heating peaks appearing in Fig. 208 are smoothed out. The computation was made with ROBLE's [*49*, *50*] computerized model (see Sect. 17ε) by inserting the above heat source into a "dynamic" model[23] essentially based on the zonally averaged pressure coordinate equations of dynamic meteorology, incorporating, however, all the physical processes discussed in Sect. 17. Ion drag, heat and momentum sources are parametrized in this model. Computations were started with a first run for quiet conditions; then the perturbation was admitted. The changes in mass transport due to magnetic perturbations were shown in Fig. 180 (Sect. 17).

---

[20] Roble, R.G., Dickinson, R.E., Ridley, E.C., Kamide, Y. (1979): J. Geophys. Res. *84*, 4207

[21] In view of the ionosphere these figures should not be compared with the *total* solar radiation flux falling onto *Earth* ($1.65 \cdot 10^{17}$ W) but only with the part of *EUV* radiation (on wavelengths below 100 nm) onto the (about 10^6 km^2) *auroral oval*, which lies between 1 and 10 MW. The UV part (below 200 nm) is much larger than this (about 100 TW) but it is absorbed at lower altitudes

[22] For definition see Siebert, M. (1971): This Encyclopedia, vol. 49/3, p. 222

[23] Dickinson, R.E., Ridley, E.C., Roble, R.G. (1975/1977): J. Atmos. Sci. *32*, 1737, *34*, 178

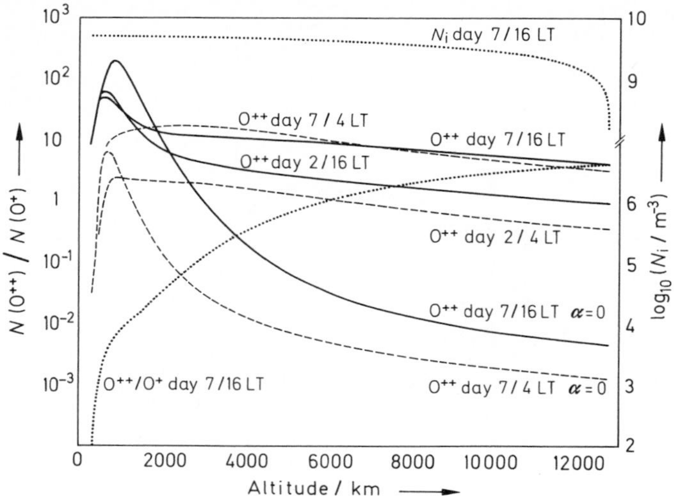

Fig. 205. Calculated plasmaspheric height profiles of O^+ and O^{++} densities (right hand ordinate, logarithmic) and of the $n(O^{++})/n(O^+)$ ratio (left hand ordinate) along an $L=3$ flux tube[19]. The different profiles refer to selected days and hours (local time). N_i = total O^{++} content in the flux tube between the given altitude and the equatorial plane

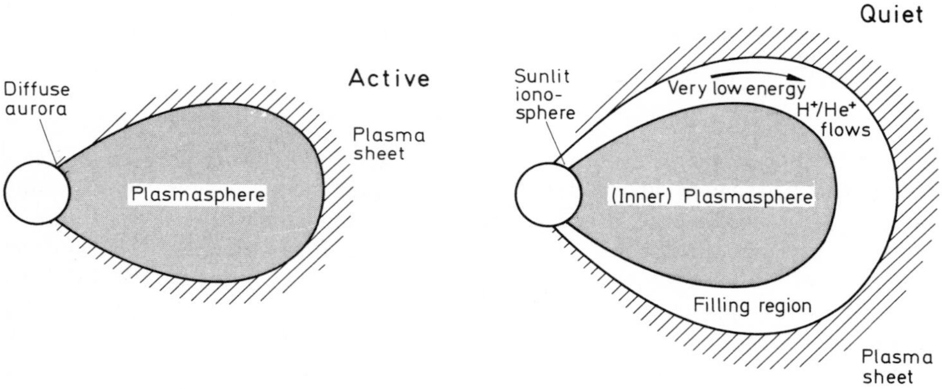

Fig. 206. Schematic presentation of the plasmapause and of the inner boundary of the plasma sheet during magnetically active periods (left) and during qiet periods (right) when refilling goes on with an outward/upward flow of H^+ and He^+ ions along that boundary[16]. [A plasma counterstreaming model was recently established: RICHARDS, P.G., SCHUNK, R.W., SOJKA, J.J. (1983): J. Geophys. Res. *88*, 7879]

These results are confirmed by electron temperature data from satellite AE-C[24]; Fig. 210 shows maximum and minimum values of T_e during a 100 d period when the satellite was in a circular 300 km orbit. The heated auroral zone appears clearly in the maximum data (and is at least indicated in the minimum data). There is a heating hole around the magnetic winter pole and this is also an ionization hole[25]; this is quite understandable by the absence of ionization sources over this pole (not, of course, at the summer pole, where solar radiation arrives).

[24] Brinton, H.C., Grebowsky, J.M., Brace, L.H. (1978): J. Geophys. Res. *83*, 4767
[25] Thomas, J.O., Andrews, M.K. (1969): Planet. Space Sci. *17*, 433

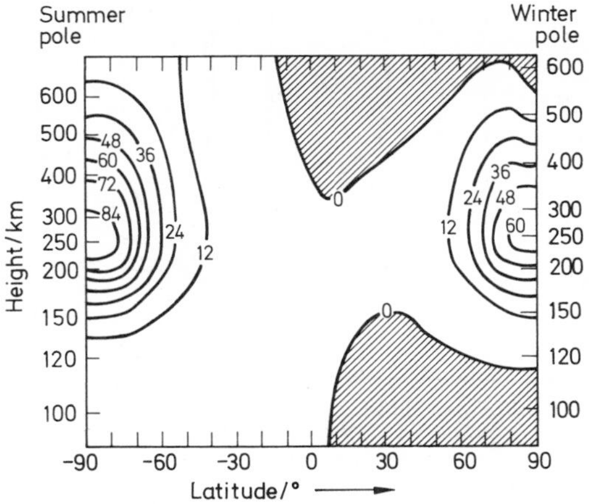

Fig. 207. Storm-generated high-latitude heating rate[20]: latitude versus height cross section for 9 Nov 1969, 14h50 UT. Curves labelled in units of W kg^{-1}

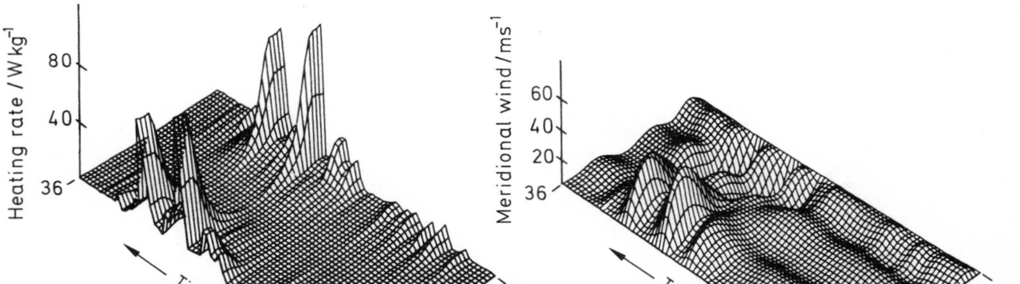

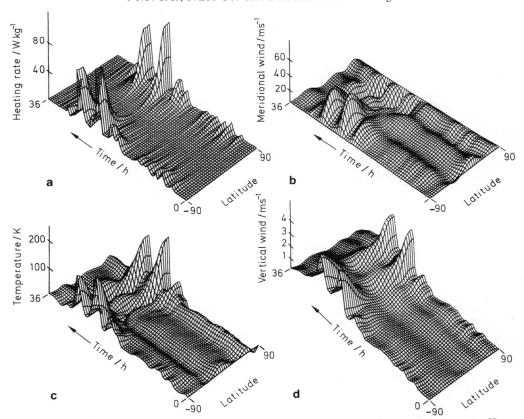

Fig. 208a–d. Storm-generated high-latitude heating and its effects at an altitude of about 250 km[20] during a 36 h period including 8 Nov (quiet day) and storm day 9 Nov 1969 [Fig. 209 is showing the absolute temperature profile during the same storm]. (**a**) Perturbation heating rate; (**b**) perturbation meridional wind; (**c**) perturbation temperature change; (**d**) perturbation vertical wind

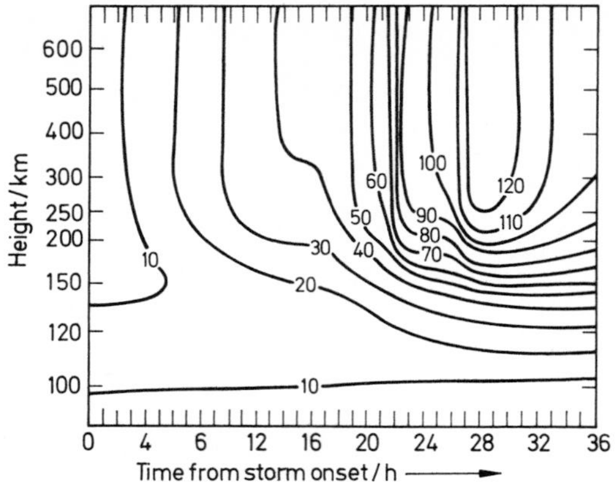

Fig. 209. Time variation of model calculated[20] neutral temperature/K (labels) during the same period as Fig. 208

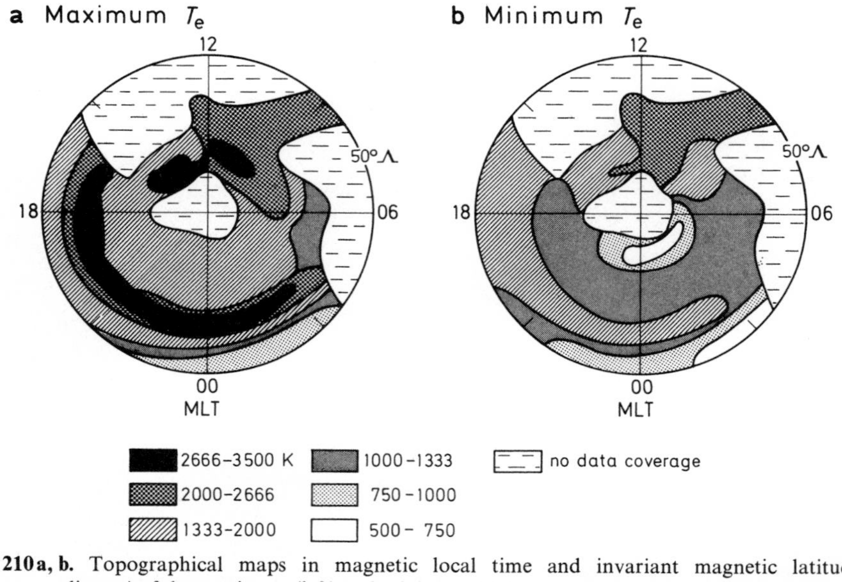

Fig. 210a, b. Topographical maps in magnetic local time and invariant magnetic latitude[26] (polar coordinates) of the maximum (left) and minimum (right) electron temperature values measured at 300 km altitude aboard satellite AE-C[24]. Maps cover a 90°–50° latitude range around the southern ("invariant") pole. Measuring period from May to Aug 1976 (100d)

The so-called "magnetic" perturbations are in fact full-scale atmospheric perturbations in the polar caps though a non-negligible corpuscular influence appears there in any case, even during quiet periods[27]. At subauroral latitudes

[26] For definition see Hess, W.N. (1972): This Encyclopedia, vol. 49/4, p. 126

[27] Really quiet periods with very low level of corpuscular effects are much rarer than perturbations. P.N. Mayaud [(1956): Ann. Géophys. *12*, 84] found about 2 days/year

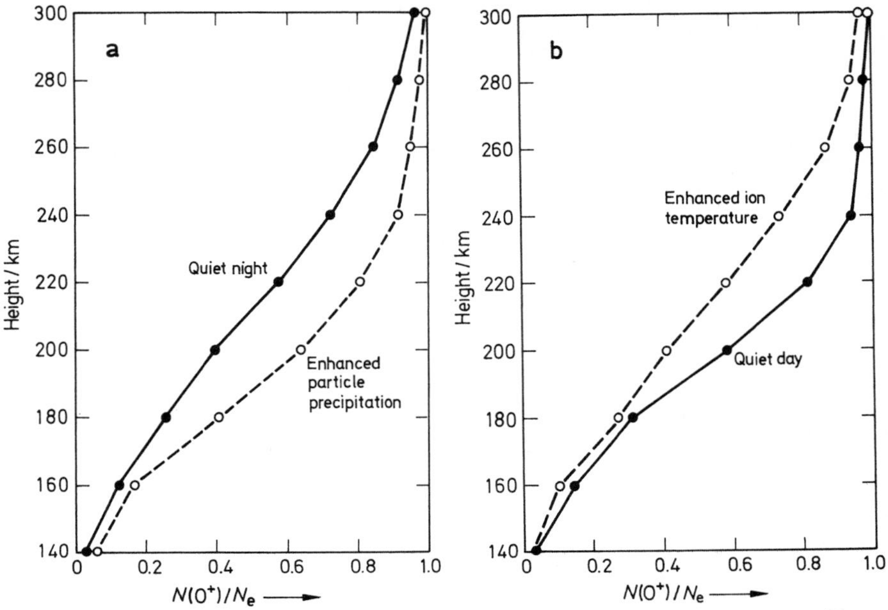

Fig. 211 a, b. Model computed[31] composition profiles giving the $N(O^+)/N_e = N(O^+)/\sum N_i$ ratio. (**a**) Quiet night (solid curve) and effect of auroral ionization due to enhanced precipitation (dashed line): the percentage of O^+ ions is increased. (**b**) Quiet day (solid curve) and effect of Joule heating increasing T_i (dashed line): the percentage of O^+ ions is decreased

equatorward neutral winds as high as $640\,\mathrm{m\,s^{-1}}$ together with westward winds of 100–$200\,\mathrm{m\,s^{-1}}$ were inferred[28] for a very severe perturbation with[22] $Kp=9$. The combined action of such neutral winds with the E to W plasma convection may explain the afternoon increase of the peak electron density on the 1st day of a storm[29]. The subsequent and more noticeable negative effects on $NmF2$ at subauroral latitudes[30] may be due to changes in the chemical composition, increasing dissociative recombination probability, in particular to the increased density of molecules as compared with atomic oxygen, which is provoked by the heating-up of the atmosphere.

Changes in ion composition were in fact observed during perturbations, with different tendencies, however[30]. It seems to be important to distinguish between the opposed effects of just particle precipitation, on the one hand, which in the range around 200 km increases the ratio of atomic to molecular ions, and Joule heating, on the other hand, which increases the ion temperature and so decreases this ratio (see Fig. 211). The first effect should only occur in the auroral zone, strictly speaking where the precipitating particles arrive. Joule heating, however, reaches further as it is provoked by electric fields and

[28] Hernandez, G., Roble, R.G. (1976): J. Geophys. Res. *81*, 5173
[29] Anderson, D.N. (1976): Planet. Space Sci. *24*, 69
[30] Blanc, M. (1981): [*53*], 21-1 (many references); Brekke, A., ibidem 14-1
[31] Kelly, J.D., Wickwar, V.B. (1981): J. Geophys. Res. *86*, 7617

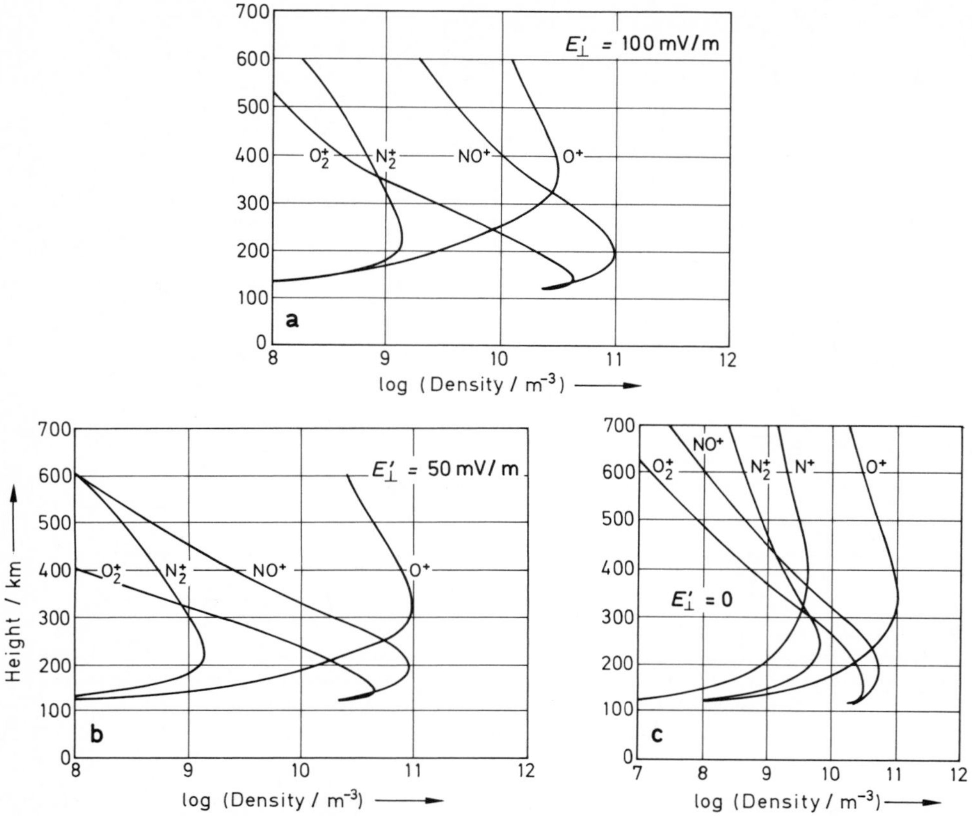

Fig. 212a–c. Ion density profiles by day, calculated for daytime high latitude (**a**) in the presence of a convection electric field of $100\,\mathrm{mV\,m^{-1}}$; (**b**) with one of $50\,\mathrm{mV\,m^{-1}}$; (**c**) without such a field [55]

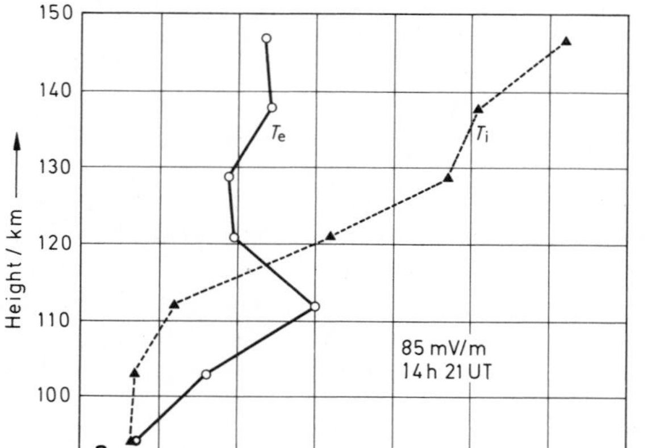

Fig. 213 a

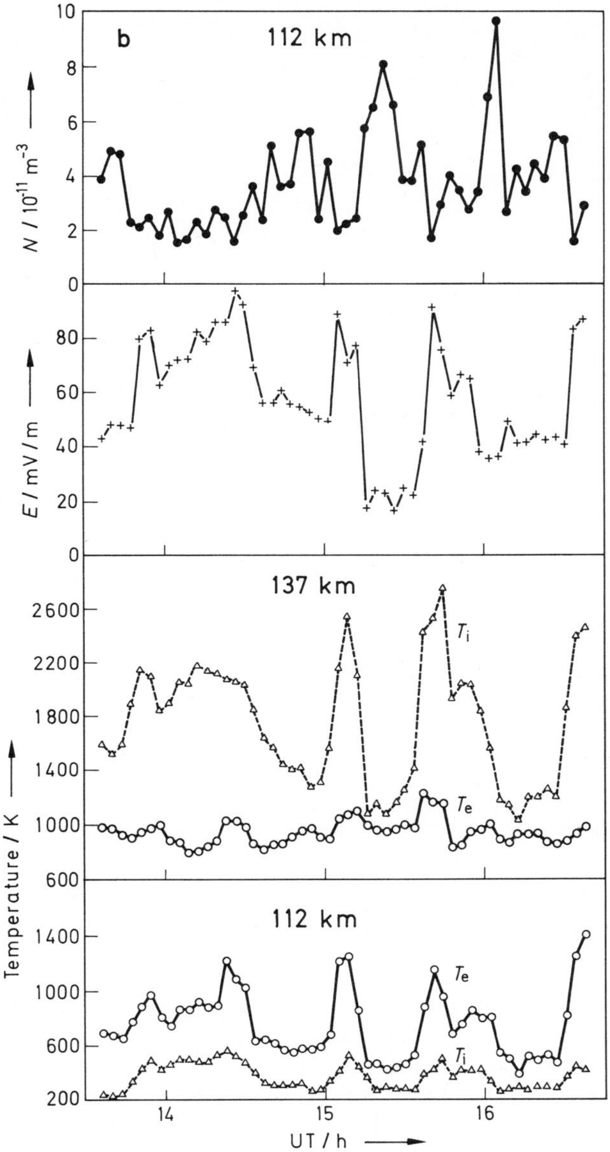

Fig. 213a, b. Electron and ion temperatures in the polar E-region in the presence of a strong electric field[32] (13 Nov 1979, afternoon). (**a**) Height profile [at about 04 h 20 LT]. (**b**) Variation in time of: electron density, electric field (southward) and temperatures at 137 and 112 km (top to bottom)

motions reaching a larger area. Thus, the prevailing effect is an increase of the relative part of molecular ions in the F-region and so increased recombination of the electrons and ions. The 'transition level' (where atomic and molecular

[32] St. Maurice, J.P., Schlegel, K., Banks, P.M. (1981): J. Geophys. Res. *86*, 1453

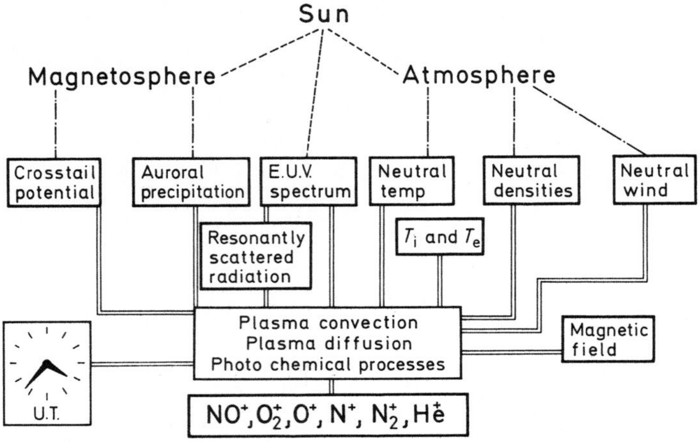

Fig. 214. Schematic representation of polar ionosphere F-region aeronomy[35]

ions have equal densities) is thereby lifted by the effect of the convection electric field (see Fig. 212).

Some perturbation temperature increase exists also in the E-region. Incoherent scatter measurements made at Chatanika, Alaska, as a result of electron bombardment show a thick Ea layer [*37*] of much higher electron density than on quiet days, but also extraordinary behavior of the plasma temperatures. While usually one has T_e greater than T_i, a typical profile through the polar E-region in the presence of a strong electric field exhibits an increase of T_i over T_e above about 120 km, which is directly related to the instantaneous electric field, thus with Joule heating (see Fig. 213)[32]. The relevant heating rates are estimated to be 1 W m^{-3}, which is extremely high. Such rates cannot be achieved by classical heat sources, not even by Joule heating. The authors[32] feel that unstable plasma waves generated by a modified two-stream instability[33] might be an explanation.

Finally, we should mention a far-reaching effect of the heat deposit in the auroral zones, namely gravity waves propagating almost meridionally toward the equator[34].

γ) The quite different mechanisms of energy transfer, viz. directly by precipitating particles on the one hand and by Joule heating on the other lead to the consequence that a detailed *theory* is difficult to establish *for perturbed conditions.* Both mechanisms have in fact quite different effects upon electron and ion temperatures (see above) and a fortiori upon ion composition. In a semi-empirical approach different computation procedures were applied according to whether Joule heating was important (rate greater than 5 mW m^{-2}) or not[30].

---

[33] Buneman, O. (1963): Phys. Rev. Lett. *10*, 285; Farley, D.T. (1963): J. Geophys. Res. *68*, 6083

[34] Klostermeier, J. (1969): J. Atmos. Terr. Phys. *31*, 25; Testud, J., Vasseur, G. (1969): Ann. Géophys. *25*, 525; Wilson, C.R. (1971): Auroral infrasonic substorms in: The radiating atmosphere. p. 374. Dordrecht (N.L.): D. Reidel. A summary of additional observations is found in: Hunsucker, R.D. (1982): Rev. Geophys. Space Phys. *20*, 293 and in: Jones, T.B. [*53*], 30-1

[35] Sojka, J.J., Raitt, W.J., Schunk, R.W. (1981): J. Geophys. Res. *86*, 2206

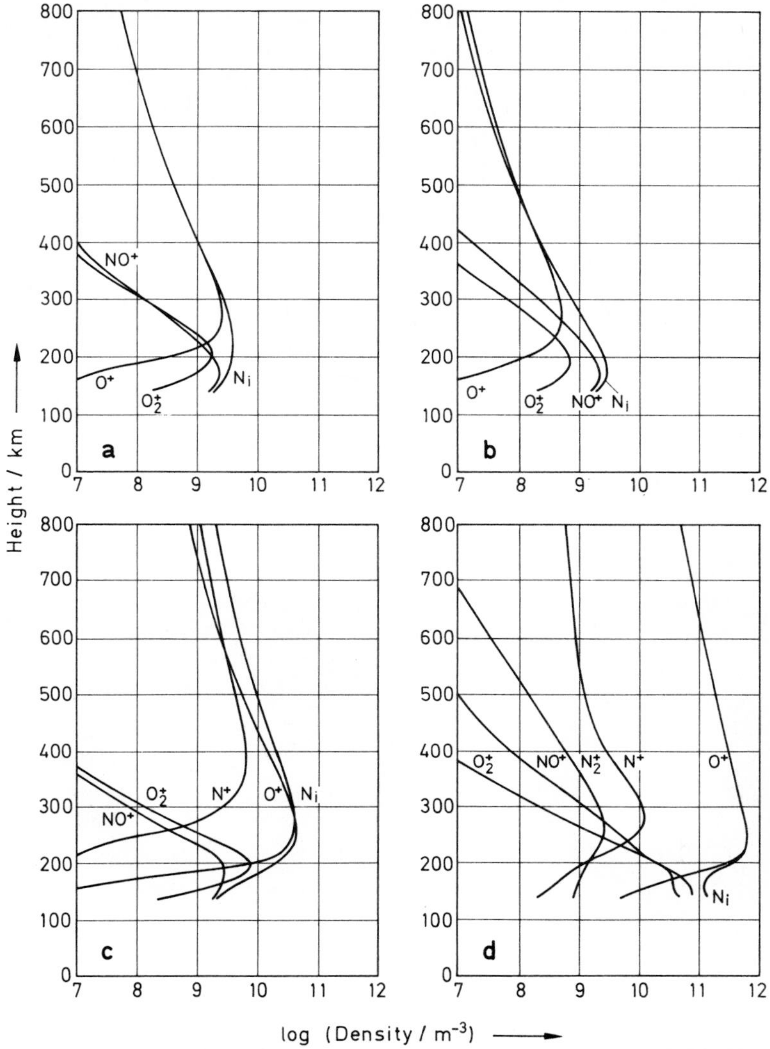

Fig. 215a–d. Typical ion density profiles (N_i = sum of all positive ions) computed by the 'high latitude dynamical model'[35] under conditions of moderate magnetic activity (magnetospheric cross-tail potential 20 kV) at different locations and universal times. (**a**) In the polar hole at 24 h UT; (**b**) same at 05 h UT; (**c**) in the polar cap adjacent to the oval at midnight local time and 24 h UT; (**d**) in the oval at midnight local time and 06 h UT[35]

Most important is certainly that the plasma convection is correctly taken account of.

First efforts simply assumed a uniform dawn-dusk electric field (of magnetospheric origin, instead of the pattern shown in Fig. 183) in a dipole magnetic field with offset between geographical and geomagnetic poles. This offset introduces marked universal time variations (or longitude

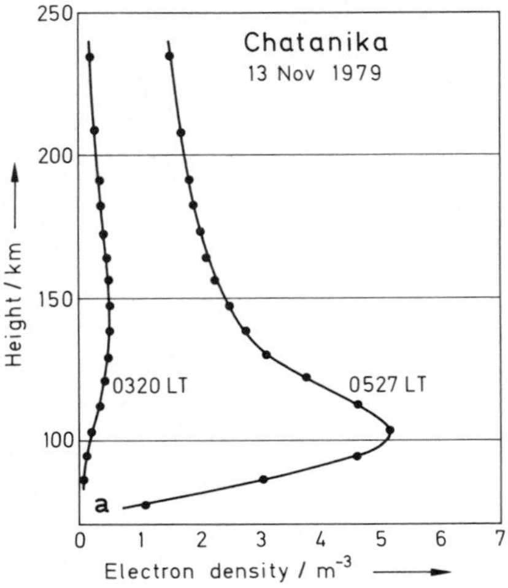

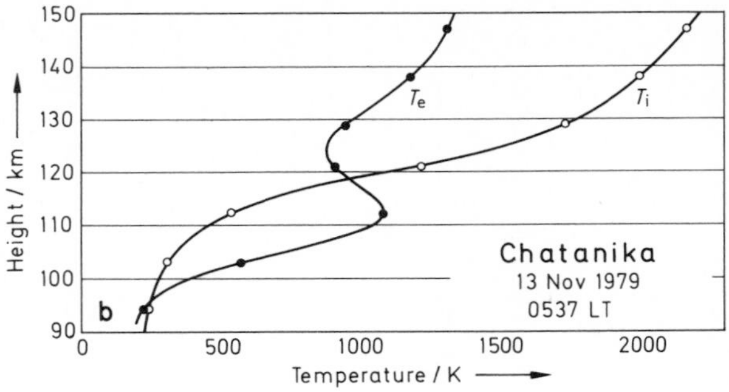

Fig. 216 a, b. Typical profiles inside the auroral zone. (**a**) Electron[35] density before and after the onset of an auroral event. (**b**) Electron and ion temperatures with "crossing over" of the latter during the event

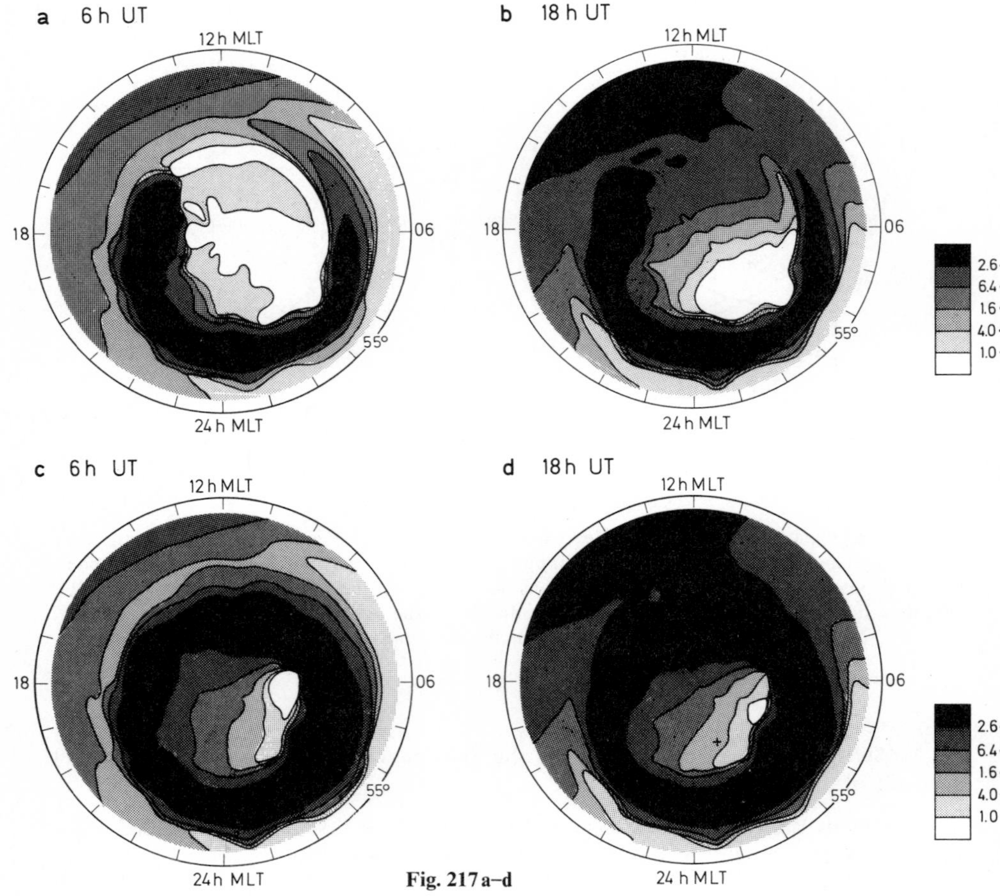

Fig. 217 a–d

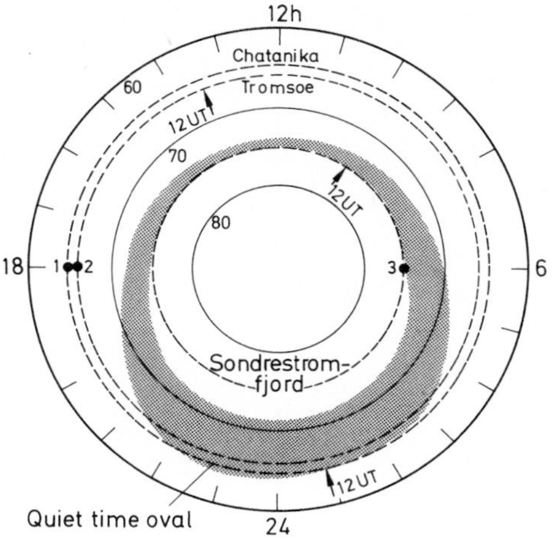

Fig. 218. Magnetic quasi-inertial frame [i.e. magnetic local time versus magnetic latitude] with quiet time auroral oval (shaded). Stations Chatanika (Alaska) and Tromsoe (Norway) have about the same latitude but differ in longitude by almost 180°; therefore Tromsoe is in the oval around midnight, but Chatanika around noon. Sondre Stromfjord (Greenland), however, is marginally in the oval around noon, but otherwise in the polar cap[35]

effects). The magnetospheric electric field boundary had to be shifted antisunward (along the magnetic meridian) in order to obtain better agreement with the observed data[36].

Then, by combining this plasma convection model with a time-dependent atmosphere-ionosphere density model[37], a 'high-latitude dynamic model' was established[38]. This latter model includes ion chemistry and transport equations (see Sect. 15) but not the energy balance of electrons.

Thus electron temperature profiles had to be parametrized according to the different geographical regions occurring in the pattern of auroral phenomena (see, for example, Fig. 195 above). The ion temperature, however, was obtained from a solution of the relevant energy balance equation, including coupling to the neutrals and frictional (Joule) heating; thermal conduction and coupling to the electrons were neglected but then taken into account by a semi-empirical reshaping procedure.

As for production, in addition to solar EUV radiation, auroral particle production and resonantly scattered radiation must be considered[15]. The latter must only be taken account of in the dark hemisphere, outside the auroral

◄

Fig. 217a–d. Computed O^+ density maps in the magnetic quasi-inertial frame for two universal times: (**a, c**) 06 h, (**b, d**) 18 h UT, and variable magnetic local time (MLT = angular coordinate). The geographical pole is marked by a cross. [(**c, d**) were computed with turning off of the auroral precipitation in the late morning sector of the oval, otherwise like (**a, b**)][38]

---

[36] Sojka, J.J., Raitt, W.J., Schunk, R.W. (1979): J. Geophys. Res. *84*, 5943; Sojka, J.J., Schunk, R.W. (1983): J. Geophys. Res. *88*, 2112

[37] Schunk, R.W., Raitt, W.J. (1980): J. Geophys. Res. *85*, 1255

[38] Sojka, J.J., Raitt, W.J., Schunk, R.W. (1981): J. Geophys. Res. *86*, 609; a recent improvement is: Schunk, R.W., Sojka, J.J. (1982/84): J. Geophys. Res. *87*, 5169 and *89*, 2348

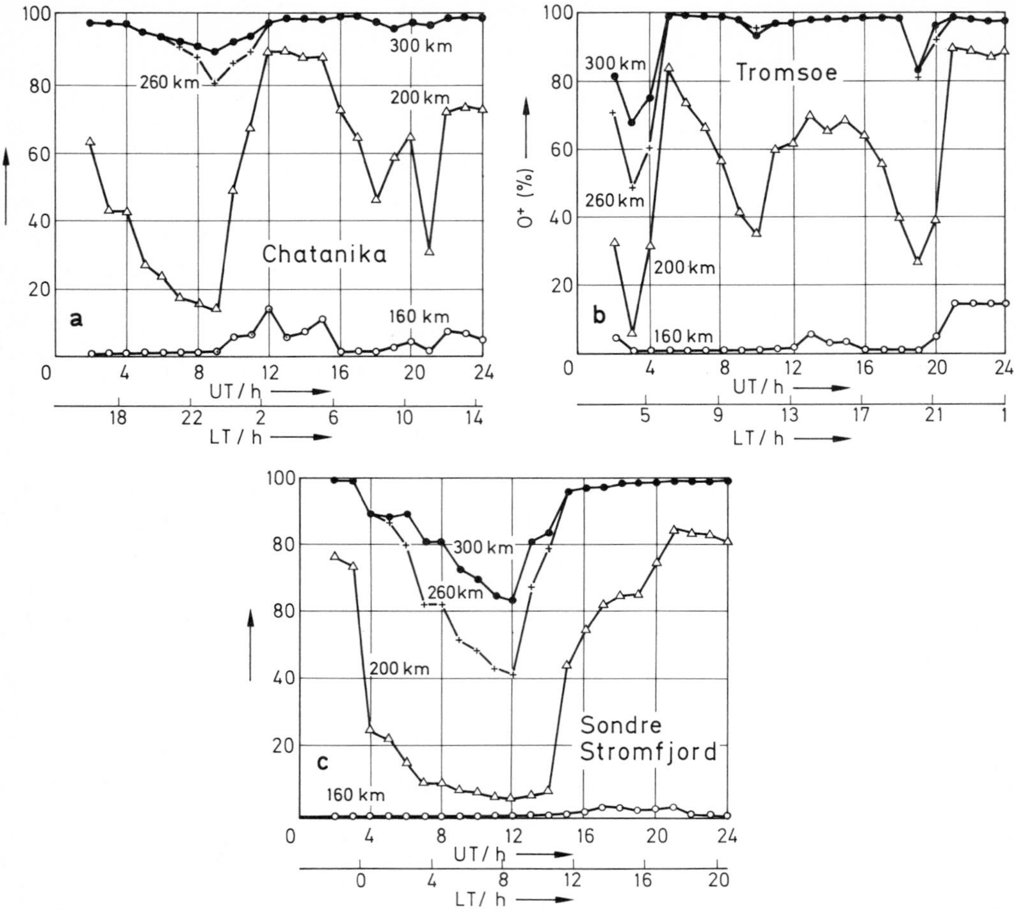

Fig. 219a–c. Diurnal "twin hour" variations for the three sites shown in Fig. 218 (UT and local solar time LT): Relative O^+ density (percentage) for four different altitudes (parameter of curves). Computations were made with the "high latitude dynamical model"[39] for quiet conditions (with the oval shown in Fig. 218). (**a**) Chatanika; (**b**) Tromsoe; (**c**) Sondre Stromfjord

zone. Auroral production rates in the oval were taken from[39]. Figure 214 shows the general scheme of these computations[35]. A few results of such model computations of ion composition are shown in Fig. 215 (vertical profiles of ion composition at different locations). A typical electron density profile inside the auroral zone is reproduced in Fig. 216a while Fig. 216b shows the ion and electron temperature profiles during the same auroral event.

The spatial distribution is exemplified in the maps represented in Fig. 217 (in geographical coordinates). It appears from these that there is a wide variability of conditions according to location and (local magnetic and uni-

[39] Knudsen, W.C., Banks, P.M., Winningham, J.D., Klumpar, D.M. (1977): J. Geophys. Res. *82*, 4784

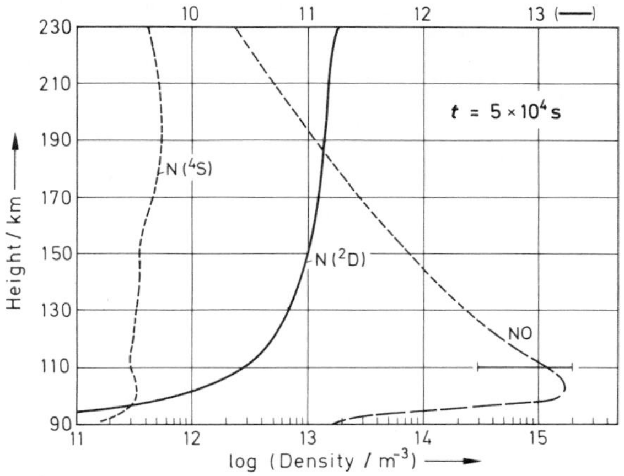

Fig. 220. Profiles of odd nitrogen species NO (broken), N(^{4}S) (dashed) and N(^{2}D) (solid curve) during a bright aurora: results of semi-empirical calculations (see text). Lower abscissa scale for NO and N(^{4}S), upper one for excited N(^{2}D)

versal) time[36]. Thus, since there is a 'twin hour dependence', the situation is quite involved, particularly when diurnal variations are considered. Due to the UT-dependence, even locations of similar latitude (Fig. 218) show appreciable differences (see Fig. 219)[35].

δ) Other models are *chemistry oriented* and more on the empirical side. The particular conditions depending on location and twin hour (see above) are taken account of by using data from in-situ measurements (obtained by a rocket launched simultaneously with the passing of a suitable satellite).

Thus, the actual energy spectrum of precipitated electrons and the neutral composition were both known from measurements. The production rates were computed by an analytical expression depending on mass density, penetration depth and primary energy[35]. The chemistry model, however, is quite involved, applying 34 reactions: six collisional ionization reactions, seven with charge transfer, two only of dissociative recombination, three with collisional excitation of metastable species, and 16 "odd nitrogen" and oxygen reactions, most of which are with metastable species[40]. The idea behind this is that on average higher energy of the precipitating electrons (as compared with solar EUV quanta) must considerably increase the populations in such elevated states, so that they should not be disregarded. Figure 220 gives calculated densities for a few odd nitrogen species under conditions of a bright aurora[41].

The "high-latitude dynamic model" (see Sect. 18γ above) has also been equipped with N^+ ions and the relevant reactions (but not with metastable

[40] Gérard, J.-C. (1970): Ann. Géophys. *26*, 777; the results of this method in computing ionization rate profiles compare well with those obtained with a more involved "Monte-Carlo" calculation by Berger, M.J., Saltzer, S.M., Maeda, K. (1970): J. Atmos. Terr. Phys. *32*, 1015

[41] Gérard, J.-C., Rusch, D.W. (1979): J. Geophys. Res. *84*, 4335 [the 34 reactions with their rates are given in Table 2, p. 4337]

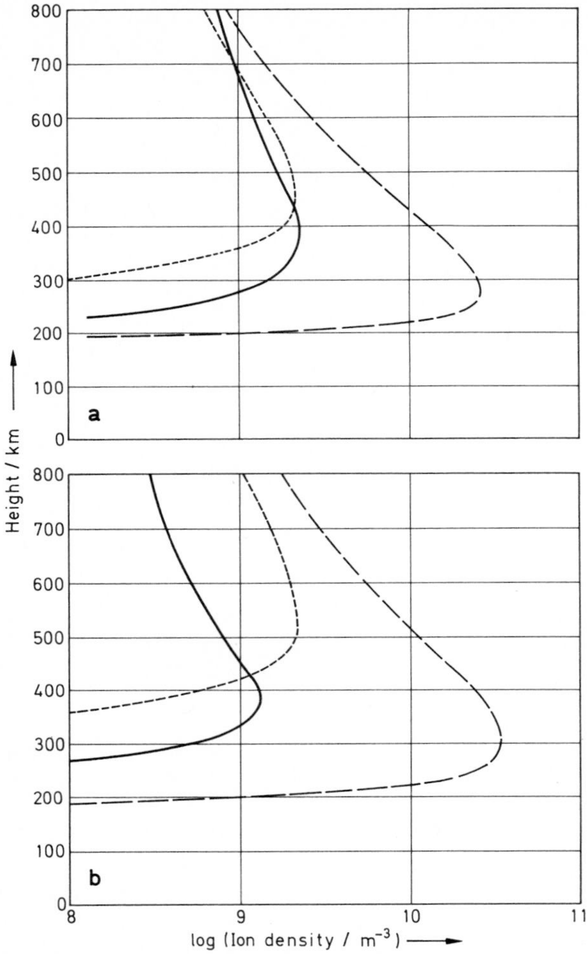

Fig. 221 a, b. Atomic ion profiles at low geomagnetic activity during winter computed with the "high latitude dynamic model"[35] but improved after[42]. N^+, solid curve; O^+, broken curve: He^+, dashed curve. (**a**) Solar minimum and eastward electric field of $15\,mV\,m^{-1}$; (**b**) solar maximum and no field[42]

species). SCHUNK et al.[42] find N^+ ions to reach 10% of O^+ at greater heights, but He^+ is more important there (Fig. 221).

The most detailed review of the role of metastable species from a viewpoint of atmospheric chemistry has been made by TORR and TORR [*66*]. These authors have also studied the effects of precipitating O^+ ions which have been identified during magnetic storms. At auroral/subauroral latitudes, during such events, energy fluxes up to $0.4\,mW\,m^{-2}\,sr^{-1}$ have been measured so that this is an additional, non-negligible source of auroral heating[43].

[42] Schunk, R.W., Raitt, W.J. (1980): J. Geophys. Res. *85*, 1255

[43] Torr, M.R., Torr, D.G., Roble, R.G. (1981): [*53*], 22-1

Appendix A: The Role of Photoelectrons in the Heat Balance of the Upper Atmosphere

By

D. BILITZA

In a large height range, by day, the ionosphere is heated up by the action of photoelectrons. The energy of primary photoelectrons is explained as the difference between quantum energy of incoming solar extreme ultraviolet light (EUV) and the ionization potential of the different atmospheric atoms and molecules. By elastic and inelastic collisions these electrons (with energies of between a few and about 100 eV) lose more and more energy, so that there is a population of "suprathermal" electrons with an energy spectrum reaching down to thermal energy (of the order of 0.1 eV). Collision partners are the ambient neutrals, ions and thermal electrons. The direct heating effect results from elastic collisions. Energy transfer via elastic collisions increases with decreasing mass difference and velocity difference of the interacting particles. Therefore the most effective heating occurs at low suprathermal energies and is due to collisions with the thermal electrons.

Direct heating of ions and neutrals by suprathermal electrons is only of minor importance. They are, however, heated up by elastic collisions with the more numerous thermal electrons, the temperature of which increases by direct heating from the suprathermal population.

In fact, two electron populations can be distinguished:

1. Thermal electrons. Due to the slowness of the recombination processes and since energy interchange between the low-energy electrons is particularly large, this population is essentially in an internal thermal equilibrium, and a Maxwellian velocity distribution can be assumed.

2. Suprathermal electrons. These have no Maxwellian distribution; there is a continuous degradation process taking place, starting with photoelectrons, which are eventually slowed down to thermal energies. The energy spectrum of ionospheric electrons comprises both populations, the thermal part of the spectrum merging at transition energies with the suprathermal 'tail'.

Theoretical description may start from a Boltzmann equation for all ionospheric electrons. Heat-balance equations are derived from its second velocity moment. However, only few authors[1-3] have applied this rather cumbersome approach. In order to make the calculations easier it has become customary to consider the thermal and suprathermal electrons separately. Most authors distinguish between the two populations just by cutting the electron spectrum into a thermal and a suprathermal part at an energy, $\mathscr{E}_t$, which is not known a

---

[1] Ashihara, O., Takayanagi, K. (1974): Planet. Space Sci. *22*, 1201

[2] Jasperse, J.R. (1976): Planet. Space Sci. *24*, 33

[3] Jasperse, J.R. (1977): Planet. Space Sci. *24*, 743

priori and must be determined while calculating the two distribution functions[4]. Unfortunately, the boundary $\mathscr{E}_t$ affects all Boltzmann collision integrals. A rigorous treatment is, however, beyond the scope of theoretical calculations. Usually a constant value (e.g. $\mathscr{E}_t = 0.1$ eV, $\mathscr{E}_t = 1$ eV) is assumed and is introduced into the collision integral describing the interaction of thermal and suprathermal electrons. In some calculations $\mathscr{E}_t$ is chosen as that energy where the two iteratively determined distribution functions become equal[5,6]. A different approach has been proposed quite recently[7].

Assuming Maxwellian velocity distribution of the thermal electrons and taking the second velocity moment for this population, the well-known 'heat-conduction equation' is obtained (see the contribution by YONEZAWA in volume 49/6 of this Encyclopedia, his Eq. (7.1), p. 259), which determines the electron temperature. This equation is coupled with the actual distribution function of suprathermal electrons by the so-called 'electron heating rate', describing the energy interchange between the two electron populations.

In Boltzmann's equation for the suprathermal electrons it is convenient to replace the distribution function by electron fluxes. In the lower thermosphere, due to the high neutral density, the photoelectrons lose their energy locally. At these heights the mean free path of electrons is small compared with the atmospheric scale height. Therefore, the flux equation can be further simplified by ignoring transport effects. Quite unlike this, photoelectrons travel over large distances in the upper thermosphere, but, at least in the magnetized terrestrial atmosphere, only along the magnetic field lines, thus anisotropically. Electrons with a high enough energy are even able to escape from the local ionosphere and travel through the plasmasphere to the conjugate ionosphere (at the other end of the field line). Looking at the electron path in more detail, its helical character is apparent (see the contribution by HESS in volume 49/4 of this Encyclopedia, his Chap. B, p. 116). The smaller the 'pitch angle' of an electron (between the helical orbit and the field line), the greater is its 'effective free path' and 'effective velocity' along the field line. Therefore, one expects different suprathermal electron fluxes at different pitch angles. In the flux equation the pitch angle has to be considered as an additional variable. Few numerical studies have made an effort to reproduce the detailed pitch angle structure of the electron fluxes[8,9]. In both studies the description of directional changes during collisions was simplified by assuming either isotropic or forward scattering of the colliding electron.

Analytical approaches to photoelectron transport[10,11] apply the 'discrete ordinate method' of CHANDRASEKHAR[12]. These studies provide useful insight

---

[4] Krinberg, I.A. (1973): Planet. Space Sci. *21*, 523

[5] McCormick, P.T., Michelson, P.F., Pettibone, D.W., Whitten, R.C. (1976): J. Geophys. Res. *81*, 5196

[6] Stamnes, K., Rees, M.H. (1983): Geophys. Res. Lett. *10*, 30

[7] Bilitza, D. (1983): Doctorate Thesis, University Freiburg (F.R.G.)

[8] Mantas, G.P. (1975): Planet. Space Sci. *23*, 337; Mantas, G.P., Bowhill, S.A. (1975): Planet. Space Sci. *23*, 35

[9] Oran, E.S., Strickland, D.J. (1978): Planet. Space Sci. *26*, 1161

[10] Stolarski, R.S. (1972): J. Geophys. Res. *77*, 2862

[11] Stamnes, K. (1977): J. Geophys. Res. *82*, 2391

[12] Chandrasekhar, S. (1960): Radiative Transfer. New York: Dover

into some aspects of the transport problem, but they are inadequate for representing the actual conditions.

Numerical and analytical methods have also been used for computing both transport and energy degradation of auroral electrons[13,14].

In most heat-balance computations, however, a distinction is made only between integrated upward and downward electron fluxes. This so-called *two-stream model* was introduced by BANKS and NAGY in 1970[15,16]. The 'two-stream equations' are obtained by integrating the electron flux equation over the upper- and lower-pitch-angle hemisphere.

Comparisons between different approaches revealed limitations of the simple two-stream approach in reproducing the suprathermal electron fluxes[17-19], though the differences are less than 10%. No major discrepancies were reported between the electron-heating rates obtained with the different approaches[17]. Nevertheless, suggestions have been made for improving the two-stream representation[7,18].

Looking at the interaction between suprathermal electrons and ambient particles in more detail, account has to be taken of the following processes (in brackets the assumptions which are usually made are given):

1. Elastic collision with

a) the neutrals O, O_2 and N_2 [no energy degradation, only directional changes of the electron],

b) the ions O^+, NO^+ and O_2^+ [negligible in two-stream calculations, due to the small directional changes caused by Coulomb collisions; must be considered if a more detailed pitch-angle distribution is to be taken into account[7,8]],

c) the thermal electrons. [Usually the "continuous-slowing-down approximation[6,20] is applied, wherein the collisions with the thermal electrons are introduced as a frictional force on the left side of Boltzmann's equation[8]. The FOKKER-PLANCK formalism is more involved and rarely used[1-4,21]. As for the computed upward and downward fluxes, both methods yield almost identical results[7]].

2. Inelastic collisions [no directional changes, only energy degradation in the order of increasing electron energy loss]:

d) Excitation of fine-structure states of O [up to now usually ignored[20]].
e) Excitation of rotational and vibrational states of O_2 and N_2.
f) Excitation of electronic states of O, O_2 and N_2.
g) Ionization [and/or dissociation] of O, O_2 and N_2.

[13] Strickland, D.J., Book, D.L., Colfey, T.P., Fedder, J.A. (1976): J. Geophys. Res. *81*, 2755
[14] Stamnes, K. (1980): Planet. Space Sci. *28*, 427
[15] Banks, P.M., Nagy, A.F. (1970): J. Geophys. Res. *75*, 190
[16] Nagy, A.F., Banks, P.M. (1970): J. Geophys. Res. *75*, 6260
[17] Cicerone, R.J., Swartz, W.E., Stolarski, R.S., Nagy, A.F., Nisbet, J.S. (1973): J. Geophys. Res. *78*, 6709
[18] Swartz, W.E. (1976): J. Geophys. Res. *81*, 183
[19] Swartz, W.E., Stamnes, K. (1977): J. Geophys. Res. *82*, 2401
[20] Stamnes, K., Rees, M.H. (1983): J. Geophys. Res. *88*, 6301
[21] Krinberg, I.A., Garifullina, L.A., Atkatova, L.A. (1974): J. Atmos, Terr. Phys. *36*, 1727

h) Excitation of rotational, vibrational and electronic states of O^+, O_2^+ and N_2^+ (can be ignored[7]).

For numerical simulation the *cross-sections* of all the above processes over the energy range up to 100 eV are needed. For inelastic collisions, most authors have used an analytical description of the effective cross-section as function of energy which was established by GREEN and co-workers[3,22–24]. A large amount of valuable data has been obtained during the past decade by laboratory experiments applying specific techniques (see the contribution by THOMAS in volume 49/6 of this Encyclopedia, his Chap. D, p. 55). For the highly reactive atomic oxygen, which is the major constituent in the whole upper thermosphere, theoretical computations have to be relied on. Apart from GREEN's compilation, two more recent sets of cross-sections were used in electron-degradation models. A comparison of the three sets was made by STAMNES and REES[20], who computed the relevant photoelectron spectra and electron-heating rates. Below 20 eV they found some differences in the energy spectrum but only minor differences in the heating rates.

The strength of the coupling between electron fluxes in different pitch-angle directions is determined by the differential (elastic) cross-sections used. In the two-stream approach the coupling factor β between the upward and downward fluxes is closely related to these cross-sections. For β, which is the probability of an electron leaving its pitch-angle hemisphere upon collision, different analytical energy dependences have been proposed[7,16,25]. But they are rather similar and only minor differences (in electron fluxes) result from this ambiguity[7].

In numerical computation two *boundary conditions* are needed. In the lower thermosphere transport effects can be neglected, and isotropic pitch-angle distribution may be assumed. Therefore, the fluxes at the lower boundary are easily calculated if the neutral densities and the solar EUV intensities are known. If the computation is made along the magnetic field line (from the local thermosphere into the magnetically conjugate thermosphere), the second boundary lies in the conjugate thermosphere and can easily be determined. The computational problems encountered in such estimates have only recently been overcome[26,27,28].

In order to set a realistic upper boundary condition in the local ionosphere, the flux of electrons escaping from the local ionosphere has to be assessed. One way of defining the boundary condition is by specifying the downward/upward flux ratio at the boundary, which might be interpreted as a kind of reflection coefficient (or transmission coefficient of the plasmaspheric field tube between the local and conjugate ionosphere if both can be assumed to be not too different).

For low-latitude conditions (but off the magnetic equator such as at Arecibo, Puerto Rico), good agreement between observed and computed electron fluxes was reached when assuming a flux ratio of 0.8 at 1000 km[7].

Theoretical computations earlier predicted about half that value[29]. More recently, MANTAS, CARLSON, and WICKWAR[30] came out with a ratio of 0.8

[22] Green, A.E.S., Swada, T. (1972): J. Atmos. Terr. Phys. *34*, 1719

[23] Green, A.E.S., Stolarski, R.S. (1972): J. Atmos. Terr. Phys. *34*, 1703

[24] Banks, P.M., Chappell, C.R., Nagy, A.F. (1974): J. Geophys. Res. *79*, 1459

[25] Stamnes, K. (1981): J. Geophys. Res. *86*, 2405

[26] Young, E.R., Richards, P.G., Torr, D.G. (1980): J. Computational Phys. *38*, 141

[27] Young, E.R., Torr, D.G., Richards, P., Nagy, A.F. (1982): Planet. Space Sci. *28*, 881

[28] Chandler, M.O., Behnke, R.A., Nagy, A.F., Fontheim, E.G., Richards, P.G., Torr, D.G. (1983): J. Geophys. Res. *88*, 9187

[29] Lejeune, G., Wormser, F. (1976): J. Geophys. Res. *81*, 2900

[30] Mantas, G.P., Carlson, H.C., Wickwar, N.B. (1978): J. Geophys. Res. *83*, 1

when taking account of a type of 'quasi-trapping' of suprathermal electrons in the plasmasphere. They feel that multiple backscattering between both thermospheres is rather frequent and that Coulomb collisions in the plasmasphere end up with a pitch angle redistribution which is favourable for trapping.

In Fig. 222 the downward electron fluxes measured aboard the satellite AE-E[31] are compared with theoretical results obtained by the two-stream approach. These latter used solar EUV photon fluxes measured aboard AE-E[32] and AEROS-A[33].

First comparisons between measured and calculated electron fluxes were restricted to the lower (transportless) thermosphere. In general the shape of the measured spectrum was excellently reproduced, but the absolute values of the computed fluxes were below the measured values by a factor of approximately 1.5–2.0[34–36]. Other computations ended up with somewhat better absolute agreement[37, 38].

In both flux measurements (electrons as well as photons) an uncertainty of several tens of percent must be admitted (mainly due to calibration problems). At altitudes below 200 km, the calculated electron fluxes depend critically on the neutral density profiles.

Once the suprathermal electron fluxes have been computed, the heating rate of the thermal electrons due to the suprathermals may be calculated. To this end, use can be made of a well-developed formalism describing the energy loss per unit length of a charged particle in a plasma*[39–42]. In integrating over all suprathermal energies, different values $\mathscr{E}_0$ for the lower boundary have been used in ionospheric heat-balance studies. As an obvious choice for $\mathscr{E}_0$ one might take $\mathscr{E}_t$, i.e. the energy level separating thermal and suprathermal electrons. But this level is difficult to determine, such that usually a constant lower boundary value is assumed. HOEGY has shown [*54*] that a more thorough evaluation of the second velocity moment of the Boltzmann collision operator for electron-electron collision leads to an additional term [see p. 463, Eq.

* The continuous slowing down approximation (mentioned above) is based upon this formalism

[31] Lee, J.S., Doering, J.P., Potemra, T.A., Brace, L.H. (1980): Planet. Space Sci. *28*, 973 and 947

[32] Reference EUV-spectrum for low solar activity (F74113): Heroux, L., Hinteregger, H.E. (1978): J. Geophys. Ress. *83*, 5305

[33] Typical spectra for weak$^{(0)}$ and strong$^{(+)}$ solar irradian during the AEROS-A mission: Schmidtke, G. (1978): Scientific rep. W.B.3, Institut für physikalische Weltraumforschung, Freiburg, F.R.G. (see also [7] for tabulated data)

[34] Victor, G.A., Kirby-Docken, K., Dalgarno, A. (1976): Planet. Space Sci. *24*, 221

[35] Nagy, A.F., Doering, J.P., Peterson, W.K., Torr, M.R., Banks, P.M. (1977): J. Geophys. Res. *82*, 5099

[36] Hernandez, S.P., Doering, J.P., Abreu, V.J., Victor, G.A. (1983): Planet. Space Sci. *31*, 221

[37] Richards, P.G., Torr, P.G. (1981): Geophys. Res. Lett. *8*, 99

[38] Jasperse, J.R., Smith, E.R. (1978): Geophys. Res. Lett. *5*, 84

[39] Larkin, A.I. (1960): Soviet Phys. JETP *10*, 186

[40] Itikawa, Y., Aono, O. (1966): Physics Fluids *9*, 1259

[41] Schunk, R.W., Hays, P.B. (1971): Planet. Space Sci. *19*, 113

[42] Swartz, W.E., Nisbet, J.S., Green, A.E.S. (1971): J. Geophys. Res. *76*, 8425

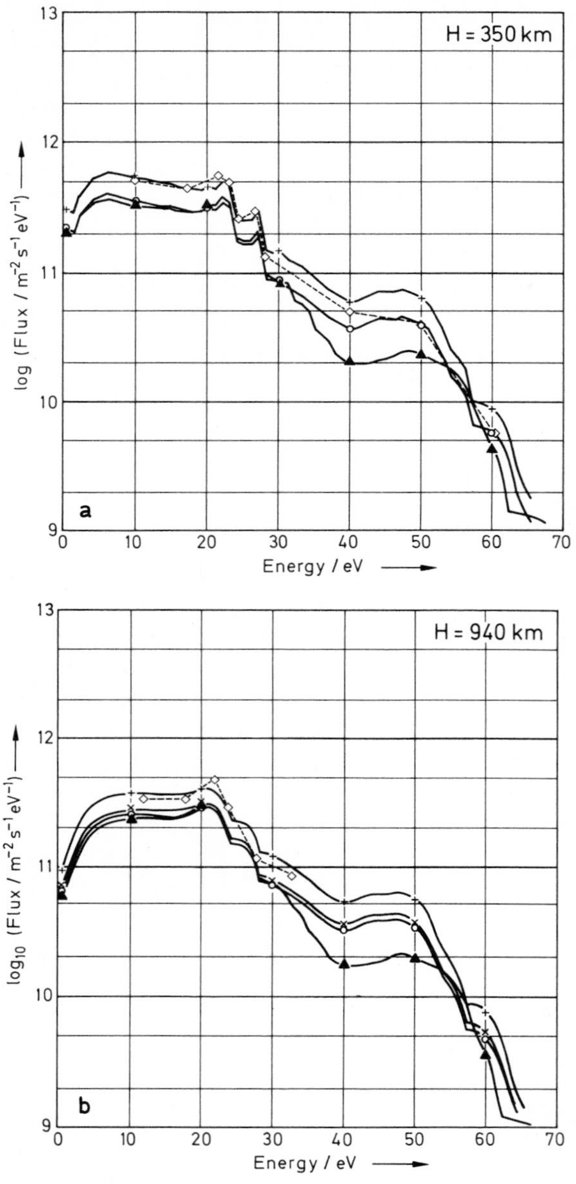

Fig. 222a, b. Energy spectra of downward electron flux at 350 (*top*) and 940 km (*bottom*) above Arecibo (Puerto Rico). Comparison of measured data from satellite AE-E (-◇-)[31] with computations by the two-stream approximation for different (observed) solar EUV radiation conditions (▲; ○; +)[32,33] for conditions around the minimum of the solar cycle[7]

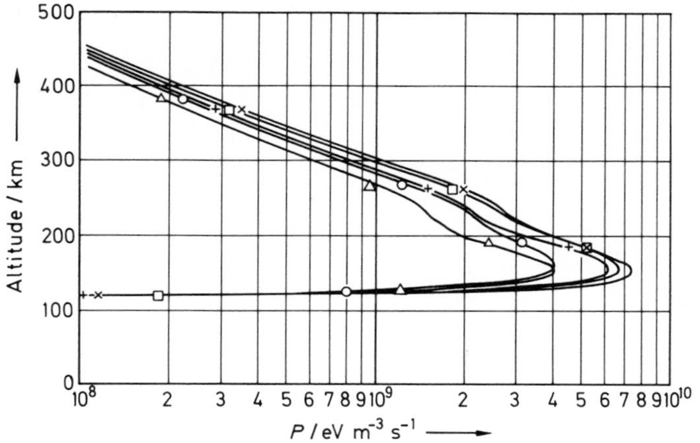

Fig. 223. Profiles of electron-heating rates calculated[7] with different assumptions: $\mathscr{E}_0=0.5$ eV ($\times$), 1 eV ($\circ$), $\mathscr{E}_t$ ($\triangle$) and 1.2 kT_e ($\square$)[7], and with HOEGY's formula ($+$) (Compare Fig. 109, p. 464)

(16.2b)] in the electron-heating rate; its contribution may be as high as about 40%. Figure 223 shows electron-heating rates calculated for different lower boundary values and formulas.

When applying a lower boundary of, say, 1 eV or when putting $\mathscr{E}_0=\mathscr{E}_t$ without HOEGY's additional term, the heating rate is underestimated by up to 50%.

Such misinterpretation of the electron-heating rate leads to a severe underestimation of the electron temperature when it is calculated from the 'heat conduction equation', In earlier calculations authors compensated for this error by artificially increasing the solar EUV photon fluxes[43,44]. Figure 224 shows that with the revised electron-heating rate the mean values of plasma temperatures measured aboard the satellites Atmospheric Explorer C and AEROS-B are in good agreement with the calculated temperatures.

Summarizing, it may be stated that the following processes need to be considered in the heat-conduction equations for electrons and ions[45]:

Electrons:

Heating by elastic collisions with the suprathermal electrons

Cooling by
1. Elastic collisions with the neutrals O, H, He, O_2 and N_2
2. Elastic collisions with the ions O^+, H^+, He^+, O_2^+ and NO^+
3. Rotational and vibrational excitation of the molecules O_2 and N_2
4. Fine-structure excitation of atomic oxygen O

[43] Roble, R.G. (1975): Planet. Space Sci. *23*, 1017
[44] Stolarski, R.S., Hays, P.B., Roble, R.G. (1975): J. Geophys. Res. *80*, 2266
[45] Schunk, R.W., Walker, J.C.G. (1973) in: High temperature physics and chemistry. Rouse, C.A. (ed.), p. 2–62. New York: Pergamon Press

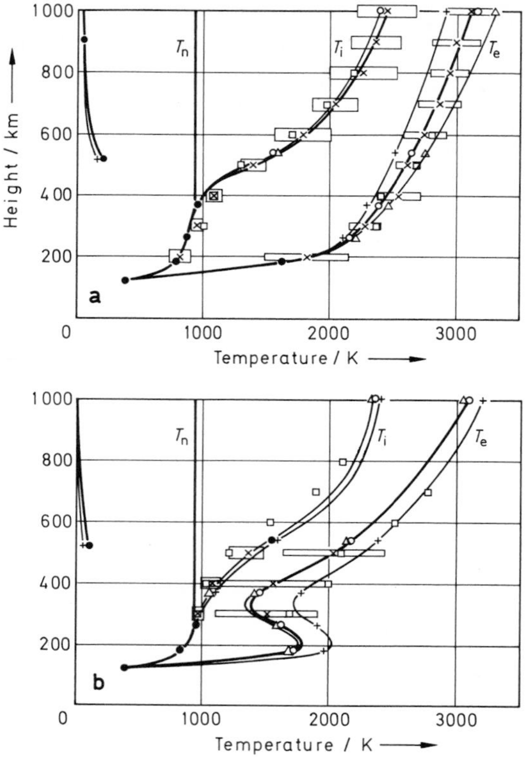

Fig. 224a, b. Theoretically computed electron and ion (O^+) temperature profiles (curves) compared with values from in situ measurements aboard satellites AE-C (×) and AEROS-B (□). Rectangles give range of satellite observations. Conditions: low solar activity, summer daytime. Lefthand curve shows calculated temperature *difference* between H^+ and O^+ ions[7]. (**a**) Northern middle latitude (Millstone Hill, Mass., USA). Different heat fluxes in the electron gas at 1000 km assumed in the computations, viz. 2.3 (+), 3.4 (○) and 4.4 (△)$\cdot 10^{13}$ eV $m^{-3} s^{-1}$. (**b**) Northern low latitude (Arecibo, Puerto Rico). Different solar EUV radiation spectra assumed in the computations: AE reference spectrum[46] for low activity (○); AEROS typical (low-activity) spectrum[47] for low (△) and high (+) solar irradiance[7]

Ions:

Heating by elastic collisions with the thermal electrons

Cooling by

1. Elastic collisions with neutrals (induced dipole interaction) and with other ion species
2. Resonant-charge exchange with the neutral parent gas
3. Quasi-resonant charge exchange of O^+ in hydrogen and of H^+ in oxygen gas

---

[46] Heroux, L., Hinteregger, H.E. (1978): J. Geophys. Res. *83*, 53

[47] Schmidtke, G. (1978): Scientific Report W.B.3, Institut für physikalische Weltraumforschung, Freiburg, F.R.G.; (1979): Ann. Géophys. *35*, 141

At higher thermospheric altitudes the calculated profiles depend strongly on the heat flux which is conducted down from the plasmasphere (see Fig. 224a). Suprathermal electrons again play a critical role in the determination of this flux insofar as they may escape from the sunlit local or magnetic conjugate ionosphere and then heat the plasmaspheric thermal electrons.

General references

[*1*] Banks, P.M., Kockarts, G. (1973): Aeronomy, parts A, B. New York: Academic Press

[*2*] Barlier, F., Berger, C., Falin, J.L., Kockarts, G., Thuillier, G. (1978): A thermospheric model based on satellite drag data. Ann. Géophys. *34*, 9

[*3*] Bauer, S. (1973): Physics of planetary ionospheres. Berlin & Heidelberg: Springer-Verlag

[*4*] Bauer, S.J. (1974): Rocket and satellite measurements in the magnetosphere. Greenbelt (Md., USA): NASA-GSFC

[*5*] Belrose, J.S. (ed.) (1981): Medium, long and very long wave propagation (at frequencies less than 3000 kHz) (=AGARD Conf. Proc. No. 305). Paris: AGARD

[*6*] Booker, H.G., Smith, E.K. (1970): A comparative study of ionospheric measurement techniques. J. Atmos. Terr. Phys. *32*, 467–497 [in the issue on 'Electromagnetic probing of the upper atmosphere', p. 457–736]

[*7*] Chang, T.S., Copp, B., Jasperse, J.R. (1981): Physics of space plasmas. (=SPI Conf. Proc. and Reprint Series *4*). Cambridge (Ma., USA): Scientific Publishers

[*8*] Chapman, S., Lindzen, R.S. (1970): Atmospheric tides. Dordrecht: D. Reidel

[*9*] Comité Consultatif International des Radiocommunications [CCIR] (1967): CCIR Atlas of ionospheric characteristics (=Rep. 340) [and later revisions, last edition 1983]. Geneva: Union Internationale des Télécommunications

[*10*] Committee on Space Research (COSPAR), Working Group II (1964): Sounding rocket research techniques (=IQSY Instruction Manual No. 9). London: IQSY Secretariat; (meanwhile: Paris: COSPAR Secretariat)

[*11*] Committee on Space Research (1965): CIRA 1965, COSPAR International Reference Atmosphere 1965. Amsterdam: North-Holland Publ. Comp.

[*12*] Committee on Space Research (1972): CIRA 1972, COSPAR International Reference Atmosphere 1972. Berlin: Akademie-Verlag

[*13*] Danilov, A.D. (1970): Chemistry of the ionosphere. New York: Plenum Press

[*14*] Fatkulina, M.N., Deminova, M.G. (eds.) (1975): Fizika i modelirovanie ionosfery (Physics and modelling of the ionosphere). Moscow: Izdatel'stvo "Nauka"

[*15*] Fatkulina, M.N., Deminova, M.G. (eds.) (1976): Fizika i empiričeskoe modelirovanie ionosfery (Physics and empirical modelling of the ionosphere). Moscow: Izdatel'stvo "Nauka"

[*16*] Forbes, J.M., Garret, H.B. (1979): Theoretical studies of atmospheric tides. Rev. Geophys. Space Phys. *17*, 1951; Forbes, J.M. (1982): Atmospheric tides. J. Geophys. Res. *87*, 5222–5252

[*17*] Giraud, A., Petit, M. (1975): Physique de l'ionosphère. Paris: Presses Universitaires de France

[*18*] Grandal, B., Holtet, J.A. (eds.) (1977): Dynamical and chemical coupling between the neutral and ionized atmosphere. Dordrecht (N.L.): de Reidel

[*19*] Hedin, A.E., Mayr, H.G., Reber, C.A., Spencer, N.W., Carignan, G.R. (1974): Empirical model of global thermospheric temperature and composition based on data from the OGO 6 quadrupole mass spectrometer. J. Geophys. Res. *79*, 215

[*20*] Hedin, A.E., Salah, J.E., Evans, J.V., Reber, C.A., Newton, G.P., Spencer, N.W., Kayser, D.C., Alcaydé, D., Bauer, P., Cogger, L., McClure, J.P. (1977): A global thermospheric model based on mass spectrometer and incoherent scatter data MSIS. (1) J. Geophys. Res. *82*, 2139; Hedin, A.E., Reber, C.A., Newton, G.P., Spencer, N.W., Brinton, H.C., Mayr, F.G., Potter, W.E. (2) Ibidem 2148

[*21*] Hvostikov, I.A. (1963): Fizika ozonosfery i ionosfery (=Ionosfernye Issled. No. 11). Moscow: Izdatel'stvo "Nauka"

[22] Ivanov-Holodnyi, G.S., Nikol'skij, G.M. (1969): Solnce i ionosfera. Moscow: Izdatel'stvo "Nauka" [translation (1972): The Sun and the ionosphere. Jerusalem: Israel Program for Scientific Translations]
[23] Jacchia, L.G. (1971): Revised static models of the thermosphere and exosphere with empirical temperature profiles (=Astrophys. Obs. Spec. Rep. No. 332). Washington (D.C., USA): Smithsonian Institution
[24] Jacchia, L.G. (1977): Thermospheric temperature, density and composition: new models (=Astrophys. Obs. Spec. Rep. No. 375). Washington (D.C., USA): Smithsonian Institution
[25] Kallmann-Bijl, H. (ed.) (1961): CIRA 1961, COSPAR International References Atmosphere 1961. Amsterdam: North-Holland Publ. Comp.
[26] Lämmerzahl, P., Rawer, K., Roemer, M. (1979): Ergebnisse des AEROS-Satellitenprogramms (=Rep. MPI H-1980-V3). Heidelberg: Max-Planck-Institut für Kernphysik
[27] Landmark, B. (ed.) (1963): Meteorological and astronomical influences on radio wave propagation. New York: Academic Press
[28] Lindzen, R.S. (1979): Atmospheric tides. Ann. Rev. Earth Planet. Sci. *7*, 199
[29] Mayr, H.G., Harris, I., Spencer, N.W. (1978): Some properties of upper atmosphere dynamics. Rev. Geophys. Space Phys. *16*, 539
[30] Mayr, H.G., Harris, I. (1979): F-region dynamics. Rev. Geophys. Space Phys. *17*, 492 [many references]
[31] McConnel, J.C. (1976): The ionospheres of Mars and Venus Ann. Rev. Earth Planet. Sci. *4*, 319
[32] McKinley, D.W.R. (1961): Meteor science and engineering. New York: McGraw-Hill
[33] Morfitt, D.G., Ferguson, J.A., Snyder, F.P. (1981): Numerical modelling of the propagation medium at ELF/VLF/LF. [5], 32-1
[34] Mitra, S.K. (1952): The upper atmosphere. 2nd ed. Calcutta: The Royal Asiatic Soc. of Bengal
[35] Nicolet, M. (1978): Etude des réactions chimiques de l'ozone dans la stratosphère. Bruxelles: Institut Royal Météorologique de Belgique
[36] Paetzold, H.-K., Regener, E. (1957): Ozon in der Erdatmosphäre. This Encyclopedia, vol. 48, 370–426
[37] Piggott, W.R., Rawer, K. (1961, 1972): URSI Handbook of ionogram interpretation and reduction. [1] (1961): Amsterdam: Elsevier; [2] (1972): (=Rep. UAG-23): Boulder (Co., USA): NOAA, WDC-A (STP) [(1978): (=UAG-23A) revision of Chaps. 1–4]
[38] Ratcliffe, J.A. (ed.) (1960): The physics of the upper atmosphere. London: Academic Press
[39] Rawer, K. [1] (1953): Die Ionosphäre. Groningen (N.L.): P. Noordhoff N.V.; [2] (1956): The ionosphere. New York: F. Ungar
[40] Rawer, K. (ed.) (1965): Results of ionospheric drift observations. Ann. Intern. Geophys. Y. (vol.) *33*
[41] Rawer, K. (ed.) (1968): Winds and turbulence in stratosphere, mesosphere and ionosphere. Amsterdam: North-Holland Publ. Comp. New York: Interscience Publishers
[42] Rawer, K. (ed.) (1974): Methods of measurements and results of lower ionosphere structure. Berlin: Akademie-Verlag
[43] Rawer, K., Bilitza, D., Ramakrishnan, S. (1978): International Reference Ionosphere 1978. Bruxelles: Union Radioscientifique Internationale
[44] Rawer, K. (Chmn.) (1981): International Reference Ionosphere – IRI 79 (=Rep. UAG-82, eds. Lincoln, J.V., Conkright, R.O.). Boulder (Co., USA): NOAA, WDC-A (S.T.P.)
[45] Rawer, K., Minnis, C.M. (1983): Experience with, and proposed improvements of the International Reference Ionosphere (=Rep. UAG-88). Boulder (Co., USA): NOAA, WDC-A (S.T.P.). The proceedings of two more recent workshops on the same subject can be found in (1982/84): Adv. Space Res. *2* (10) and *4* (1)
[46] Revah, I., Spizzichino, A. (1963): Etude des cisaillements de vents dans la basse ionosphère par l'observation des traînées météoriques. Ann. Géophys. *19*, 43 and *20*, 248
[47] Rishbeth, H., Garriott, O.K. (1969): Introduction to ionospheric physics. New York: Academic Press
[48] Rishbeth, H. (1972): Thermospheric winds and the F-region. A review. J. Atmos. Terr. Phys. *34*, 1
[49] Roble, R.G. (1975): The calculated and observed diurnal variation of the ionosphere over Millstone Hill on 23–24 March 1970. Planet. Space Sci. *23*, 1017; (1976): Solar EUV flux variation during a solar cycle as derived from ionospheric modelling considerations. J. Geophys. Res. *81*, 265

[*50*] Roble, R.G. (1977): Variations of the mean meridional circulation in the thermosphere. [*18*], 217; Roble, R.G., Salah, J.E., Emery, B.A. (1977): The seasonal variation of the diurnal thermospheric winds over Millstone Hill during solar cycle maximum. J. Atmos. Terr. Phys. *39*, 503

[*51*] Roble, R.G., Dickinson, R.E., Ridley, E.C. (1982): Global circulation and temperature structure of the thermosphere with high-latitude plasma convection. J. Geophys. Res. *87*, 1599

[*52*] Roemer, M. (1971): Structure of the thermosphere and its variations, deduced from satellite decay. [*68*], 229

[*53*] Schmerling, E. (ed.) (1981): The physical basis of the ionosphere in the solar-terrestrial system (=AGARD Conf. Proc. No. 295). Paris: AGARD

[*54*] Schunk, R.W., Nagy, A.F. (1978): Electron temperatures in the F-region of the ionosphere: theory and observations. Rev. Geophys. Space Phys. *16*, 355

[*55*] Schunk, R.W., Nagy, A.F. (1980): Ionospheres of terrestrial planets. Rev. Geophys. Space Phys. *18*, 813 [with long listing of potential reactions and the relevant rate coefficients]

[*56*] Smith, E.K., Matsushita, S. (eds.) (1962): Ionospheric sporadic E. London: Pergamon Press. Compare also later reports by same authors as special issues of Radio Sci. (1966, 1971)

[*57*] Soicher, H. (1978): Operational modelling of the Aerospace propagation environment (=AGARD Conf. Proc. No. 238). Paris: AGARD [vol. 1 and 2]

[*58*] Spitzer, L. (1962): Physics of fully ionized gases. New York: Interscience

[*59*] Strobel, D.F. (1979): The ionospheres of the major planets. Rev. Geophys. Space Phys. *17*, 1913

[*60*] Stubbe, P. (1970): Simultaneous solution of the time dependent, coupled continuity equations, heat conduction equations, and eqations of motion for a system consisting of a neutral gas, an electron gas and a four component ion gas. J. Atmos. Terr. Phys. *32*, 865

[*61*] Thomas, L. (1971): The lower ionosphere. J. Atmos. Terr. Phys. *33*, 157

[*62*] Thullier, G., Falin, J.L., Wachtel, C. (1977): Experimental global model of the exospheric temperature based on measurements from the Fabry-Perot interferometer on board the OGO-6 satellite. J. Atmos. Terr. Phys. *39*, 399 [model M1]

[*63*] Thullier, G., Falin, J.L., Barlier, F. (1977): Global experimental model of the exospheric temperature using optical and incoherent scatter measurements. J. Atmos. Terr. Phys. *39*, 1195 [model M2]

[*64*] Torr, D.G., Torr, M.R. (1978): Review of rate coefficients of ionic reactions determined from measurements made by Atmospheric Explorer satellites. Rev. Geophys. Space Phys. *16*, 327

[*65*] Torr, D.G. (1979): Ion chemistry. Rev. Geophys. Space Phys. *17*, 510 [143 references]

[*66*] Torr, M.R., Torr, D.G. (1982): The role of metastable species in the thermosphere. Rev. Geophys. Space Phys. *20*, 91 [385 references, many helpful tables]

[*67*] Vassy, E. (1956/59/66): Physique de l'atmosphère [Tome I/II/III]. Paris: Gauthier-Villars

[*68*] Verniani, F. (ed.) (1971): Physics of the upper atmosphere. Bologna: Editrice compositori

[*69*] Volland, H., Mayr, H.G. (1977): Theoretical aspects of tidal and planetary wave propagation. Rev. Geophys. Space Phys. *15*, 203

[*70*] Wernik, A.W. (ed.) (1981): Scientific and engineering uses of satellite radio beacons. Warszawa, Lodz: Panstwowe Wydawnictwo Naukowe

[*71*] Whitten, R.C., Colin, L. (1974): The ionospheres of Mars and Venus. Rev. Geophys. Space Phys. *12*, 155

Subject Index

Errata (to volumes XLIX/3–6)

Handbuch der Physik, Band XLIX/2 and 3

See volume XLIX/4, errata on pages 581 and 582, for corrections;

further to

Handbuch der Physik, Band XLIX/3

Correction to "Worldwide Magnetic Storm Phenomena"

p. 156, line 37	Add minus sign to one side of the Eq. (Maxwell's second eq.)

Handbuch der Physik, Band XLIX/4

Correction to "The Earth's Magnetosphere"

p. 9, 20th and 19th lines from bottom	After "highly conductive" insert "in the sense of magnetodynamic theory".
p. 19, footnote 8	Instead of "See contribution" read "An additional viscosity tensor is discussed"
p. 29, footnote 2	Instead of "*the* Reynolds number" read "*the* magnetic Reynolds number".
p. 56, footnote 6	Replace "p. 216" by "p. 218".
p. 65, 24th line	Replace "Van Allen" by "Owens and Frank".
p. 76, 19th and 20th lines	Instead of "through a non-axially-symmetric field" read "in connection with an electric field".
p. 83, 2nd line	Delete the parentheses in (1 mV/m).

Corrections to "Waves and Resonances in Magneto-active Plasma"

p. 405, line after Eq. (4.2)	Instead of "als" read "also".
p. 406, Eq. (4.7 a)	Instead of $1/(2\pi)^4$ read $1/(2\pi)^2$.
p. 449, Eq. (18.15a), line 8 of Sect. 19	First denominator: read $\bar{v}$ *instead of* v. Instead of "anisotropic" read "isotropic".
p. 450, Eq. [18.6], middle member	Invert, so that it reads $V_{T\mathrm{e}}/\omega_{N\mathrm{e}}$.
p. 466, Eq. (22.1), second line	Add minus sign to second member, so that it reads $-\varepsilon_{yx}^{(0)}/\varepsilon_0$.
p. 474, Fig. 4	Note that the values of the zeroes of n^2 written on the abscissa are taken from approximations [see Eqs. (25.15), (25.16)]. ($\vartheta \equiv \Theta$)
p. 490, Table 6, first column, No. 4	Read "whistler" instead of "whistlr".

Handbuch der Physik, Band XLIX/5

Correction to "Linear Internal Gravity Waves in the Atmosphere"

p. 181, Eq. (2.16)	Instead of i/ρ_0 read $1/\rho_0$.

Handbuch der Physik, Band XLIX/6

Corrections to "The Neutral and Ion Chemistry of the Upper Atmosphere"

p. 84/85, Eqs. (22.1), (22.2), (22.4)	Last term should read $\frac{1}{T}\frac{\partial T}{\partial h}$ instead of $\frac{1}{T}\frac{\partial T}{\partial z}$.
p. 97, Eq. (25.1)	In last term read $\frac{\partial \lambda}{\partial t}$ instead of $\frac{\mathrm{d}\lambda}{\mathrm{d}t}$.
p. 111, Table 13, line R2	Replace second $\rightarrow$ by $+$ $[N+0+2.76\,\mathrm{eV}]$.

Correction to "Extreme Ultraviolet Observational Data on the Solar Spectrum"

p. 349, Table 5, heading	Instead of $\mathrm{nW\,m^{-2}}$ read $\mathrm{mW\,m^{-2}}$

Physics and Chemistry in Space

Editors: **L. J. Lanzerotti, J. T. Wasson**

Volume 11
H. Volland
Atmospheric Electrodynamics
1984. 120 figures. IX, 205 pages. ISBN 3-540-13510-3

Contents: Introduction. - Plasma Component of the Air. - Thunderstorms and Related Phenomena: Global Electric Circuit. Thunderstorms. Thunderstorm Electrification. Lightning. Sferics. - Tidal Wind Interaction with Ionospheric Plasma: Electric Sq und L Currents. Ionospheric Dyamo. - Solar Wind Interaction with Magnetosphere: Coupling Between Solar Wind and Geomagnetic Field. Global Magnetospheric Electric Fields and Currents. Theory of Wave Propagation Within the Magnetosphere. Waves in the Magnetosphere. Appendix. - References. - Subject Index.

Volume 10
A. v. Gurevich
Nonlinear Phenomena in the Ionosphere
Translated from the Russian by J. G. Adashko
1978. 76 figures, 17 tables. X, 370 pages. ISBN 3-540-08605-6

Contents: Plasma Kinetics in an Alternating Electric Field. - Self-Action of Plane Radio Waves. - Interaction of Plane Radio Waves. - Self-Action and Interaction of Radio Waves in an Inhomogeneous Plasma. - Excitation of Ionosphere Instability.

Volume 9
A. Nishida
Geomagnetic Diagnosis of the Magnetosphere
1978. 119 figures, 1 table. VIII, 256 pages. ISBN 3-540-08297-2

Contents: Geomagnetic Field Under the Solar Wind. - Power Supply Through the Interplanetary Field Effect. - Implosion in the Magnetotail. - Dynamic Structure of the Inner Magnetosphere. - Magnetosphere as a Resonator.

Volume 8
A. Hasegawa
Plasma Instabilities and Nonlinear Effects
1975. 48 figures. XI, 217 pages. ISBN 3-540-06947-X

Contents: Introduction to Plasma Instabilities. - Microinstabilities - Instabilities Due to Velocity Space Nonequilibrium. - Macroinstabilities - Instabilities Due to Coordinate Space Nonequilibrium. - Nonlinear Effects Associated with Plasma Instabilities.

Volume 7
M. Schulz, L. J. Lanzerotti
Particle Diffusion in the Radiation Belts
1974. 83 figures. IX, 215 pages. ISBN 3-540-06398-6

Contents: Adiabatic Invariants and Magnetospheric Models. - Pitch-Angle Diffusion. - Radial Diffusion. - Prototype Observations. - Methods of Empirical Analysis.

Springer-Verlag
Berlin
Heidelberg
New York
Tokyo

Physics and Chemistry in Space

Editors: **L. J. Lanzerotti, J. T. Wasson**

Volume 6
S. J. Bauer
Physics of Planetary Ionospheres
1973. 89 figures. VIII, 230 pages. ISBN 3-540-06173-8

Contents: Neutral Atmospheres. - Sources of Ionization. - Thermal Structure of Planetary Ionospheres. - Chemical Processes. - Plasma Transport Processes. - Models of Planetary Ionospheres. - The Ionosphere as a Plasma. - Experimental Techniques. - Observed Properties of Planetary Ionospheres. - Appendix: Physical Data for the Planets and their Atmospheres.

Volume 5
A. J. Hundhausen
Coronal Expansion and Solar Wind
1972. 101 figures. XII, 238 pages. ISBN 3-540-05875-3

Contents: History and Background. The Identification and Classification of Some Important Solar Wind Phenomena. - The Dynamics of a Structureless Coronal Expansion Chemical Composition of the Expanding Coronal and Interplanetary Plasma. High-Speed Plasma Streams and Magnetic Sectors. Flare-Produced Interplanetary Shock Waves. Concluding Remarks.

Volume 4
A. Omholt
The Optical Aurora
1971. 54 figures. XIII, 198 pages. ISBN 3-540-05486-3

Volume 3
I. Adler, J. I. Trombka
Geochemical Exploration of the Moon and Planets
1970. 129 figures. X, 243 pages. ISBN 3-540-05228-3

Volume 2
J. G. Roederer
Dynamics of Geomagnetically Trapped Radiation
1970. 94 figures. XIV, 166 pages. ISBN 3-540-04987-8

Volume 1
J. A. Jacobs
Geomagnetic Micropulsations
1970. 81 figures. VIII, 179 pages. ISBN 3-540-04986-X

Forthcoming Titles

S. K. Atreya
Atmospheres and Ionospheres of the Outer Planets

A. Gurevich, E. E. Tsedelina
Very Long Propagation of High Frequency Radio Waves

V. A. Krasnopolsky
Photochemistry of Mars' and Venus' Atmospheres

L. L. Lazutin
X-Ray Emissions of Auroral Electrons and Magnetospheric Dynamics

Springer-Verlag
Berlin
Heidelberg
New York
Tokyo